MUSHROOMS

of Northeastern North America

MUSHROOMS

of Northeastern North America

Alan E. Bessette

Arleen R. Bessette

David W. Fischer

Syracuse University Press

Title page (page ii): *Pete Griffith collecting in the Adirondacks.*

LIBRARY OF CONGRESS CATALOGING-IN-PUBLICATION DATA

Bessette, Alan.
    Mushrooms of northeastern North America / Alan E. Bessette, Arleen
R. Bessette, David W. Fischer.
        p.   cm.
    Includes bibliographical references (p.   ) and index.
        ISBN 0-8156-2707-6 (cloth: alk. paper). — ISBN 0-8156-0388-6
(pbk. : alk. paper)
        1. Mushrooms—Northeastern States—Identifications. I. Bessette,
Arleen Rainis, 1951–   . I. Fischer, David W. (David William),
1959–   . III. Title.
    QK617.B483   1996
    589.2'22'097485—dc20   96–5729

Book design by Christopher Kuntze.
Printed in Hong Kong through PrintNet.

To mushroom hunters everywhere,
with appreciation for your enthusiasm,
curiosity, and joy in discovery

Although *Mushrooms of Northeastern North America* includes information regarding the edibility of various mushrooms, this book is *not* intended to function as an adequate manual for the accurate identification and safe consumption of wild mushrooms. Readers interested in consuming wild fungi should consult other sources of information, including experienced mycologists or other literary sources or both, before eating any wild mushrooms. Neither the authors nor the publisher are responsible for any undesirable outcomes that may occur for those who fail to read or heed this warning.

Contents

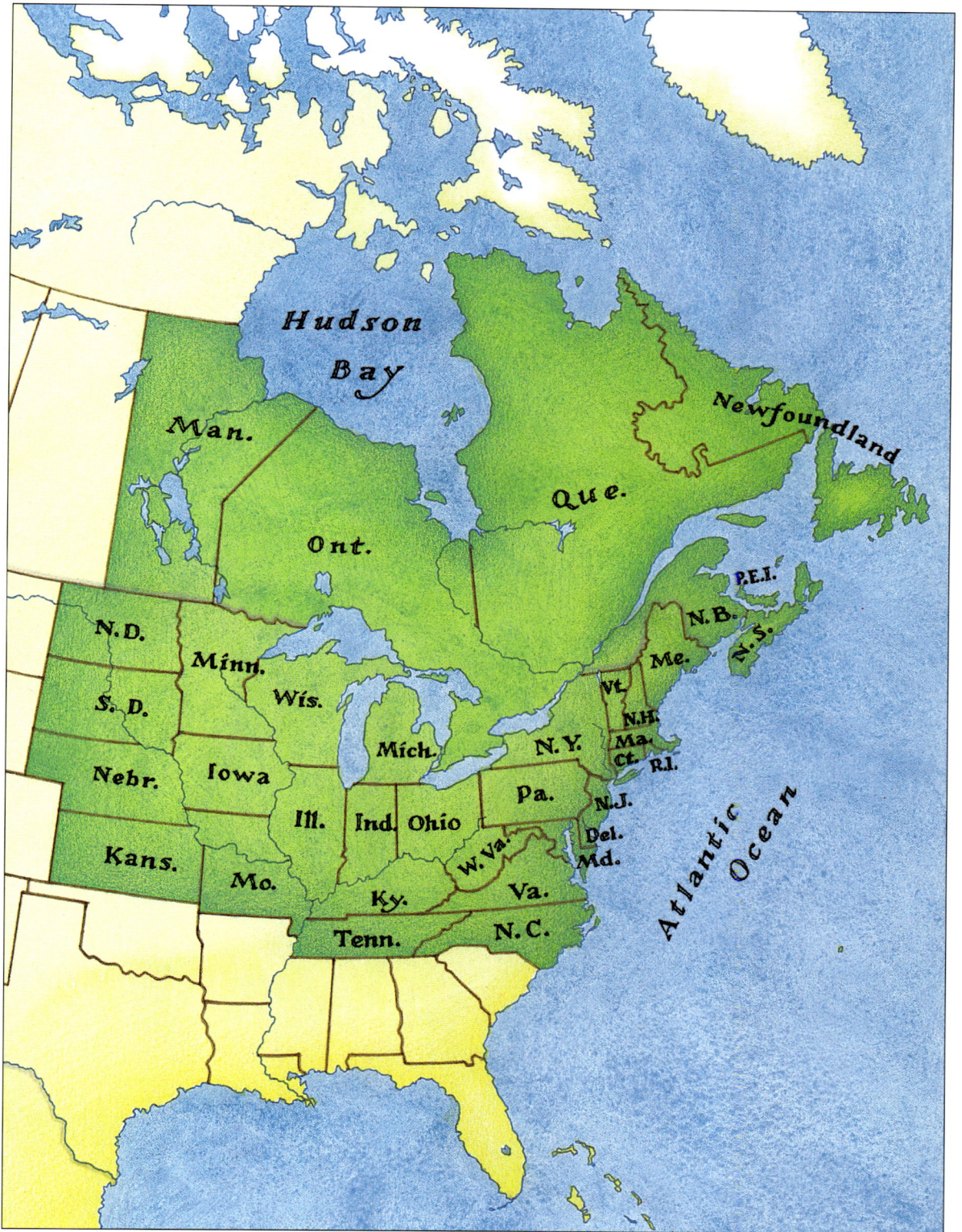

Distribution Range for Mushrooms of Northeastern North America

Volvariella bombycina

Preface

In 1868, botanist Charles Horton Peck began studying the fungi of northeastern North America. Before his pioneering efforts, few of this continent's endemic mushrooms had been described and named. Over the next forty years, Peck described and named more than three thousand species of fungi—the lion's share of this continent's native mushrooms. For this reason he has been called "the father of modern American mycology." Some of these were specimens sent to Peck by other collectors, but the bulk were collected by Peck himself. When Peck retired in 1915 after suffering a stroke, his work was nowhere near completion. Even today the region's diverse ecosystems yield mushrooms that are as yet unnamed.

Northeastern North America boasts a wealth of extraordinarily diverse habitats for collecting and studying mushrooms. The botanical, geological, and climatic variations between one area of this region and another are, in many cases, remarkable. For example, some mushroom species that are common to the White Mountains of New Hampshire or the Adirondack Mountains of New York are rare or unknown in the pine barrens of New Jersey or the coastal pine-oak forests of Massachusetts. Even between areas that are relatively close to each other, there can be surprising variation between their mycofloras. From bogs to sand plains, mountaintops to coastal lowlands, the Northeast is virtually unrivaled in its ecological diversity.

Countless professional and amateur mycologists have continued Peck's work. Although vast numbers of specimens have been deposited in various herbaria throughout the region, no single work devoted solely to the diversity of its fungi has yet been published. This book introduces this mycoflora. Beautiful color photographs, combined with nontechnical descriptions and easy-to-follow keys, are provided to assist both experienced and beginning mushroom hunters with accurate identification of species.

It is our hope that whatever the reason for your interest in mushrooms—whether it be for scientific study, the search for edible species, or sheer appreciation of their beauty—this book will serve as a trustworthy and inspiring guide to mushrooms of northeastern North America.

Acknowledgments

Many people have assisted us with this work. We are grateful for their myriad contributions. We thank the following persons for mycological notes, technical information, and assistance with species identification: Timothy J. Baroni, Harold H. Burdsall, Jr., Edward Bosman, Ernst E. Both, William R. Burk, Raymond M. Fatto, Robert Gilbertson, John H. Haines, Richard L. Homola, Bruce Horn, Richard P. Korf, Currie D. Marr, Orson K. Miller, Jr., Gregory Mueller, Clark Ovrebo, Ronald Petersen, Donald Pfister, Scott Redhead, Samuel S. Ristich, Clark Rogerson, William C. Roody, Walter E. Sturgeon, Rodham E. Tulloss, Eugene Varney, and James J. Worrall. Thanks also to Sheldon Cushing, Raymond M. Fatto, Emily Johnson, Peter Katsaros, Richard Kay, Samuel S. Ristich, William C. Roody, and Walter E. Sturgeon for contributing slides that greatly enhance this book. We thank the following individuals who made valuable mycological contributions of specimens for photography and study: William K. Chapman, David Harris, Nancy Hinman, Alma and Robert Ingalls, Peter Molesky, Sally Reymers, Jessica Scialdo, and Helen and Ralph Wagner. We are grateful to Sam Norris for the beautiful mushroom illustration included in the introduction. We thank the members of the mushroom clubs who have invited us to share their fungi and their knowledge of them. We greatly appreciate the efforts and contributions of Ernst E. Both who reviewed the bolete section of the manuscript, Bettie McDavid Mason who copyedited the manuscript, and Christopher Kuntze who designed the book, all of whom made valuable comments and suggestions for its improvement. We are especially grateful to Robert Mandel and his staff at Syracuse University Press, who made this book possible.

MUSHROOMS

of Northeastern North America

Introduction to Mycology

BIOLOGICAL BASICS: MUSHROOM FACTS AND FALLACIES

Mushrooms, being neither plant nor animal, belong to their own kingdom, the kingdom Fungi. Fungi lack chlorophyll and cannot produce food for themselves. They obtain nutrients through a process of external digestion and absorption. Some, as decomposers, extract what they need from dead and decaying materials and are called **saprobes.** Those that attack live plants, animals, or other fungi are called **parasites.** The third group exists in a mutually beneficial relationship with living trees or other plants that is called a **mycorrhizal relationship,** one in which each partner obtains part of what it needs from the other. Learning which food source or **substrate** a particular kind of mushroom requires greatly improves the likelihood of finding it. Now you know where you should search: on the fallen hardwood or conifer logs, beneath the ancient hemlocks, in open fields and meadows, or elsewhere.

Still, what exactly is a mushroom? If you go by names, you will surely be confused. Many mushrooms have at least two names: a common name and a scientific name. The common name may be widely used or regional. It may describe the appearance of the mushroom or honor the person after whom the mushroom was named. For example, *Boletus frostii* has two common names: Frost's Bolete (in honor of Charles Christopher Frost) and the Apple Bolete (describing its appearance). *Boletus edulis* has no fewer than six common names! Some species lack common names altogether, but each mushroom has a scientific name. Scientific names always have two parts. The first part is the genus, or "generic name," the first letter of which is always capitalized. The second part is the species, or "specific name," with all letters lowercased. Due to disagreement among taxonomists, there are discrepancies as to some mushrooms' correct scientific names. Therefore more than one scientific name may be assigned to a single mushroom.

Imagine a vast underground network of fine filaments that are interconnected and interwoven. When conditions are correct (temperature, moisture, nutrients, pH, daylight length), this living mat, called a **mycelium,** produces an aboveground "fruit": a mushroom. Like any other fruit, mushrooms have "seeds," microscopic reproductive structures known as **spores.** If carefully done, picking a mushroom has no more ecological impact than picking a piece of fruit from a tree.

FUNGAL ANATOMY: PARTS OF A MUSHROOM

When you are first starting out as a mushroom collector, you would be wise to become acquainted with mushroom anatomy because it differs from that of other organisms and will likely be unfamiliar to you. Refer to the accompanying illustrations as you read about the basic macroscopic features described.

As previously stated, a mushroom is the fruiting body that arises from the larger fungal organism, the **mycelium,** which is typically underground or within decaying wood. A mushroom begins as an immature form called a **button.** Depending on the species, the button may initially be entirely surrounded by a membranous structure known as the **universal veil.** As the mushroom expands, it stretches and tears the universal veil, often leaving remnants on the cap. These remnants are referred to as **patches** or **warts.** There may be a cup-like remnant of the universal veil called a **volva** around the base of the mushroom stalk. Most mushrooms, however, lack a universal veil and therefore have neither patches/warts nor a volva.

The typical mature mushroom has a **cap** and a **stalk.** The stalk may be attached to the cap centrally, off-center, or at the side of the cap. On the underside of the cap there may be **gills,** which are blade-like structures upon which spores are produced. In place of gills, there may be **teeth/spines** or **tubes** (the open end of each tube is called a **pore**). The tubes are packed closely together; their collective pores are known as the **pore surface.** Both teeth/spines and tubes serve the same basic reproductive

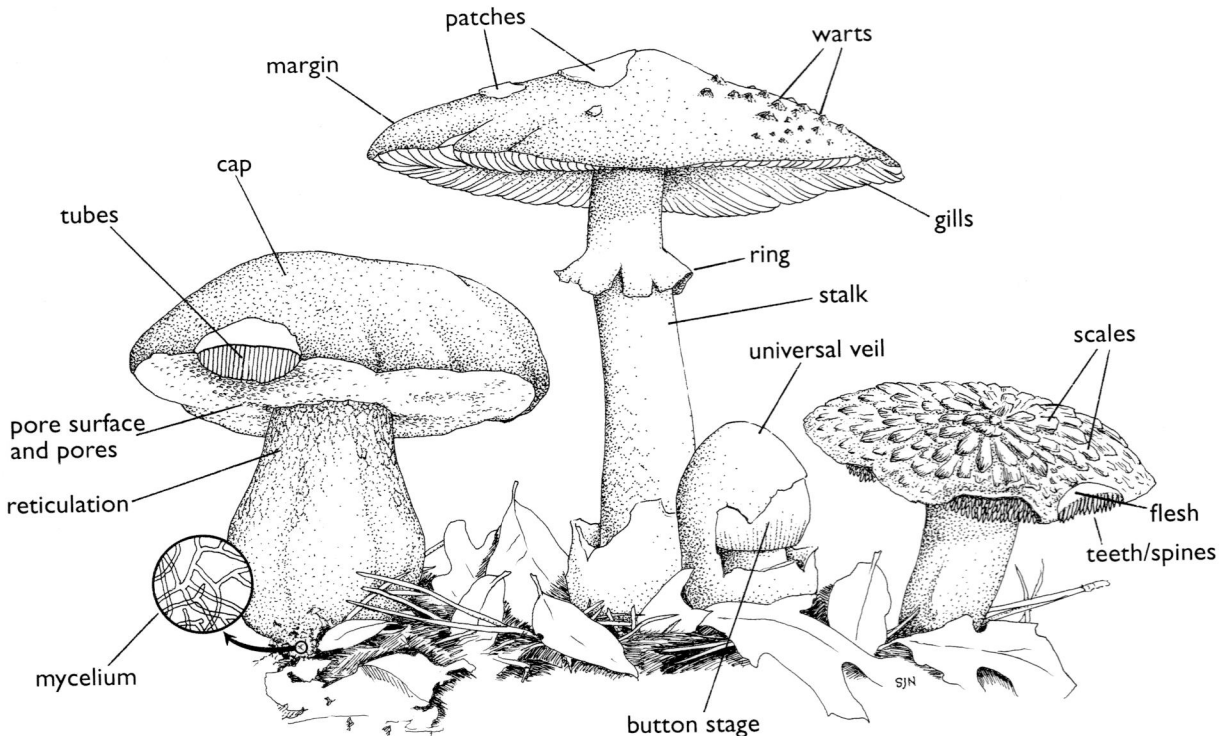

function as gills. In some species the underside of the immature cap is covered by a piece of tissue stretching from the cap's edge, or **margin,** to the stalk. This tissue, the **partial veil** (not shown), covers and protects the developing gills or tubes. As the mushroom cap expands, the partial veil tears, often leaving remnants on the cap margin or adhering to the stalk, where it forms a **ring.**

Refer to the glossary for more precise definitions of these structures and for other mycological terms that you encounter in this book.

TOOLS & TIPS FOR COLLECTING: CLOTHING & EQUIPMENT

One of the advantages of collecting mushrooms is that the required equipment is both minimal and inexpensive. Basics include a basket or box in which to store your finds, a knife, and comfortable clothing appropriate to weather and terrain. The following list includes items that we have found to be essential or useful:

1. A basket or sturdy-sided container (plastic buckets or bags are not recommended because specimens tend to become crushed)
2. A sturdy knife for digging, cutting and cleaning
3. Waxed paper, brown paper bags, or waxed paper sandwich bags for wrapping specimens and keeping them separate
4. Comfortable clothing appropriate to the season and terrain, including raingear
5. Insect repellant
6. Compass, map, and whistle
7. Pencil and paper for field notes and spore prints
8. Camera equipment and film
9. Walking stick
10. Food and beverage

CHANGING SEASONS: WHEN TO COLLECT MUSHROOMS

Mushroom fruiting patterns are affected by numerous conditions including humidity, temperature, daylight length, and precipitation. While it is impossible to predict exactly when mushrooms will fruit, there are some basic guidelines that, if followed, will help ensure a successful foray.

Some species of mushrooms have only one fruiting season, while others have split or multiple fruiting periods. These must be discovered for the particular geographic area in which you are collecting. Generally, the collecting season in northeastern North America begins in April and extends through October for most species. Some, like *Flammulina velutipes* and certain polypores, can be collected later in the fall and winter, or even year-round.

Typically, the best time to collect is from two to five days after a significant rainfall (a "soaker") or sooner if rains have been falling at frequent intervals. Sunny, windy, dry days may assist with spore dispersal but they reduce fruiting by minimizing the moisture essential to it. Some mushrooms fruit optimally during hot, humid weather, while others prefer cooler temperatures. Late summer and early fall are usually the most productive seasons for collecting. Of course, there are always exceptions. For example, the Yellow Morel fruits from April through early June. Once summer's warm days have arrived, you have missed your chance to pick morels until the next spring.

Take the time to learn weather and fruiting patterns for your own area. Keep notes of when and where you collect species to refer to next year. In this way you have the best chance of keeping your basket filled.

MUSHROOM HABITATS: WHERE TO COLLECT MUSHROOMS

Now that you know when to collect mushrooms, the next question is where. Again, the answer is dependent on several factors: local weather conditions, the type of mushroom being sought, time of year, and geographic location.

Since most mushrooms require moisture to fruit, during times of extended dry weather the best locations to search for them are in naturally moist areas: in bogs; along the shorelines of ponds and lakes; along the banks of streams, creeks and rivers; and in cool ravines. Fallen trees and stumps, which often retain moisture longer than the surrounding soil, are good places to explore. However, after several days of rain, these same locations may be too wet, and you might do better to search in drier locations: hillsides, meadows, and sandy areas.

Because mushrooms require rather specific substrates, and because many exist in mycorrhizal relationships with specific trees or other plants, the kind of mushroom you are hunting for will affect where you should look: on the ground or on trees; beneath conifers or hardwoods; in meadows; or in bogs. It is extremely helpful to learn what the mushroom requires in order to know where to look for it.

Since fruiting patterns are usually seasonal, some mushrooms, such as morels, are found only in the spring. Others, such as boletes, are most abundant during summer and fall. Still others—*Tricholoma* species, for example—are most abundant during fall. Fruiting patterns vary from one geographic location to another throughout North America. In the Northeast, spring and early summer collecting is usually best in hardwood and mixed forests, while late summer and fall collecting is often best in mixed and coniferous forests.

FILLING THE BASKET: HOW TO COLLECT MUSHROOMS

While you are collecting, your primary concern is to keep the specimens in as good condition as possible so that key identifying features are not lost or damaged by storage and transport. It is also important to keep species from becoming mixed together. Therefore we recommend wrapping collections of a single species together in either waxed paper or in brown paper sandwich bags. These materials allow the mushrooms to "breathe," unlike plastic bags and wraps, which cause moisture to build up, hastening decay.

For identification purposes, it is also important to collect whole specimens, including beneath-the-ground features. Carefully dig specimens up rather than pull them up or cut them off at the base. Gather specimens in various stages of development whenever possible to help ensure accurate identification. Note the substrate, location, and nearby tree types. Wrap your notes with your collections. You might add an extra sheet of paper in order to make an "in-the-field" spore print as described in the next section.

SPORE PRINTS: THE ESSENTIAL HIDDEN FEATURE

Our good friend Bill Roody often speaks about "Coyote Mushrooms"—mushrooms that trick you into thinking that they are something they are not. Sometimes key field features alone will not guarantee accurate identification. Spore prints, like a "missing link," often bridge the gap between macroscopic and microscopic identification, facilitating more definitive differentiation between otherwise confusing fungi.

Spore prints are formed when the mushroom spores are allowed to drop undisturbed onto a surface. They are simple to make. Cut the cap from the stalk, place it gill- or pore-side down on a piece of clean white paper and cover it with a cup or dish to prevent disturbance by drafts. Allow eight hours or so for a good, thick spore deposit. You might also wrap one or two mature caps, with fertile surface down, on a piece of white paper while you are collecting in the field. In this way, you might have an adequate spore print ready for your next step in identification once you arrive home.

A spore print is also useful if you intend to do microscopic work, as it is the best source of mature spores.

KEYS AND DESCRIPTIONS: HOW TO USE THIS BOOK

The mushroom species illustrated in this book are arranged in major groups based on similarities in their appearance. The sequence in which those species occur in this book corresponds to the sequence in which they appear in the keys. They are not intentionally arranged by order, family, or genus. Representatives of each of the twenty-two major groups are illustrated in part 2, "Color Key to the Major Groups of Mushrooms." The color key and accompanying brief descriptions constitute the foundation upon which this entire work is based.

If you know the identification of a species and wish to read about it, consult the index. If, however, you wish to identify an unknown mushroom, follow the steps presented below. Before attempting the identification procedure, be sure that you have collected as many different stages of growth as possible (in as *fresh* condition as possible), made notes about the habitat and substrate, and obtained a spore print if one is obtainable. Identifying mushrooms can sometimes be a very difficult task, and every bit of information is useful.

Mushroom Identification Procedure

1. Always start at the beginning of the color key and determine which major group best describes the mushroom you are attempting to identify.
2. Turn to the page indicated for the major group and read the introductory information presented.
3. Start at the beginning of that major group's key. Read all the descriptions preceded by the same number and determine which choice most closely matches your specimen. Proceed through the key until a species, genus, or subgroup is identified.
4. In some instances, the unknown specimen will key out to a species without reference to additional information and without a color illustration. However, in many cases you are referred to a full description and an accompanying color illustration. If so, read the description, examine the color illustration, and compare it with your unknown. Color illustrations are arranged alphabetically at the end of the descriptions of each major group.
5. In some instances, the unknown specimen will key out to a genus. If so, you are referred to pages where species within the genus are described and illustrated or where additional keys to species are located. Read the descriptions, examine the color illustrations, and compare them with your unknown.
6. In some instances, the unknown specimen will key out to a subgroup. If so, you are referred to additional keys or to pages where species are described and illustrated. Read the descriptions, examine the color illustrations, and compare them with your unknown. Be sure to read the "Comments" section when it is provided.

When examining keys, read each entry carefully and pay particular attention to qualifiers ("usually," "sometimes," "frequently," "typically," "when young," and "at maturity," among others). "Frequently" does not mean "always," and "typically" means "usually, but not always." We have endeavored to avoid such qualifiers, but in many cases they are unavoidable.

Information for some species in the keys is fairly lengthy, especially for species that are not illustrated. We have attempted to provide sufficient information to allow you to differentiate between very similar species and to provide useful information such as the fruiting season, habitat, substrate, and edibility.

The language used in this book is nontechnical, except in instances where simpler definitions become cumbersome or less descriptive, and in the "Microscopic features" sections, as needed. The technical terms used in this work are defined in the glossary. References to exemplary color or black-and-white illustrations are made whenever possible.

Notes on Descriptions of Illustrated Species

SCIENTIFIC NAME: A Latin scientific name is provided for each species. The name used may not be the same as is commonly found in other field guides; it may reflect a recent taxonomic change. Often, an alternative name is listed in the "Comments" section. Scientific names always have two parts. The first part is the genus or generic name, the first letter of which is always capitalized. The second part is the species or specific name, with all letters in lower case. Due to disagreement among taxonomists, there are discrepancies as to some mushrooms' correct scientific names. Therefore, more than one scientific name may be assigned to a single mushroom.

We have also provided unabbreviated author citations for the convenience of advanced mycologists who may find this information useful. In some instances, the author citation is simple — for example, *Pholiota subcaerulea* Smith and Hesler was described and named by Alexander H. Smith and Lexemuel R. Hesler. More often, however, the names of the original authors are enclosed by parentheses and followed by the names of those who later reclassified those fungi. In many instances, a colon is used to separate names within parentheses. *Lentinus torulosus* (Persoon : Fries) Lloyd was first described as *Agaricus torulosus* by Christiaan Persoon; Elias Fries later validated Persoon's concept of the species. Still later, Curtis Lloyd concluded that the species should be placed in the genus *Lentinus*.

COMMON NAME: One or more common names are provided wherever they have been previously published or reported. No new "common" names have been coined for this book.

MACROSCOPIC FEATURES: The appearance of the fruiting body is described, including size, shape, color, staining reactions, odor, and taste. The morphological features of some mushrooms, such as puffballs and jelly fungi, are described under the single heading of "Fruiting body." Others have morphological features described under separate headings such as "Cap," "Gills," "Pore surface" and "Stalk." Many mushrooms have distinctive odors; this information is noted if useful. The flesh of some species has a distinctive taste and is indicated if known. *If you choose to taste the tissues of a mushroom, be advised that some mushrooms taste hot and peppery and may irritate, burn, or numb your mouth if chewed for an extended period.*

Note also that there is no significant risk in properly tasting mushrooms ***unless you swallow the tissue!*** To safely taste mushrooms, place a small piece in your mouth, chew it for only a few seconds and spit it out. If the taste is mild (not bitter or peppery), wait a minute and then chew a second small piece for 15–30 seconds and again spit it out, as some mushrooms' bitter or acrid tastes are subtle.

SPORE PRINT: Spore print color is a valuable characteristic for mushroom identification, especially for gilled mushrooms and boletes. It is reliable, usually easy to obtain, and is used as a differentiating character in many of the keys.

MICROSCOPIC FEATURES: Information about spore size, shape, surface features, and microscopic color is presented here. Additional information, including length of **asci,** shape of **paraphyses,** presence of **setae,** and other useful microscopic characters, is included where appropriate.

FRUITING: The habit of mushroom growth (solitary, scattered, in groups or clusters), the substrate, habitat, fruiting period, and frequency are described here. The fruiting period is stated as a month-to-month range and describes the time during which the mushroom is likely to occur. On occasion, mushrooms will appear outside of their expected fruiting period due to unusual weather conditions. Frequency is estimated for the entire region; species listed as "occasional" may be locally abundant in some areas or rare to absent in others.

EDIBILITY: Species known to be so are listed as "edible." The term "inedible" is used for mushrooms that are too "hot" or bitter, or too fibrous or woody, for consumption. Any species described as "poisonous," "not recommended," or "edibility unknown" ***should not be eaten!***

COMMENTS: This section includes brief descriptions of similar species, alternate names, cautions and warnings, explanations, and other useful or interesting information.

Color Key to the Major Groups of Mushrooms

Split Gill and Ally (see p. 27)

Small, stalkless fruiting bodies; undersurfaces gill-like but longitudinally split or distinctly crimped, often forked and vein-like; on wood.

Plicaturopsis crispa (see p. 27)

Schizophyllum commune (see p. 27)

Chanterelles and Allies (see p. 29)

Fleshy fruiting bodies with caps and stalks or funnel-like shape; undersurfaces either with blunt, gill- to vein-like ridges that are often forked and crossveined, or nearly smooth; usually on the ground.

Cantharellus ignicolor (see p. 33)

Craterellus foetidus (see p. 35)

Gilled Mushrooms (Agarics) (see p. 39)

Undersurfaces with knifeblade-like gills radiating from a stalk or, on stalkless species, from point of attachment to substrate; growing on a variety of substrates.

Paxillus atrotomentosus (see p. 220)

Phyllotopsis nidulans (see p. 230)

Boletes (see p. 315)

Fleshy fruiting bodies with caps and typically central stalks; cap undersurface with a sponge-like layer of vertically arranged tubes, each terminating in a pore; the sponge-like layer usually separating easily from the cap tissue; on the ground or sometimes on wood.

Gyrodon merulioides (see p. 349)

Strobilomyces floccopus (see p. 354)

Polypores (see p. 373)

Woody "conks" and fleshy to leathery mushrooms with pores on their undersurfaces (the pores are sometimes minute; use a hand lens); the pore layer typically not separating easily from the cap tissue; shape varying from cap-and-stalk to stalkless and shelf-like, or rather complex; usually on wood.

Bondarzewia berkeleyi (see p. 385)

Fomitopsis pinicola (see p. 388)

Lenzites betulina (see p. 380)

Polyporus squamosus (see p. 392)

Tooth Fungi (see p. 401)

Fleshy, corky, or leathery mushrooms with downward-oriented spines or "teeth"; shape varying from cap-and-stalk to branched and icicle-like, fan-shaped or shelf-like; on the ground, on wood, or on fallen pine cones.

Auriscalpium vulgare (see p. 406)

Hericium coralloides (see p. 407)

Hydnum umbilicatum (see p. 409)

Steccherinum pulcherrimum (see p. 411)

Cauliflower Mushrooms (see p. 415)

Rather large, round, cauliflower- or lettuce-like clusters; branches leaf-like, lacking pores on the undersurface (use a hand lens); usually on the ground at the base of trees.

Sparassis crispa (see p. 415)

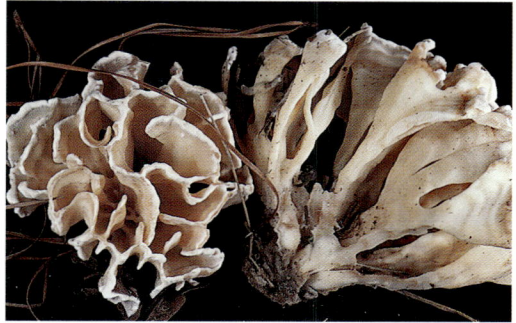

Sparassis herbstii (see p. 415)

Branched and Clustered Corals (see p. 417)

Fruiting bodies of two types: bundles of erect, worm-like, typically unbranched appendages, often fused at their bases; or erect, coral-like, repeatedly branched appendages; on the ground or on wood.

Clavulinopsis fusiformis (see p. 421)

Clavaria zollingeri (see p. 420)

Fiber Fans and Vases (see p. 425)

Leathery or fibrous-tough fan- to vase-shaped or coral-like fruiting bodies, often with split or torn margins; typically some shade of brown at maturity, with or without whitish tips or margins; fertile surfaces smooth, wrinkled or warty but lacking pores (use a hand lens); on the ground or enveloping roots, branches, seedlings or mosses.

Thelephora regularis var. *multipartita* (see p. 425)

Thelephora terrestris (see p. 426)

Jelly Fungi (see p. 427)

Distinctly gelatinous fruiting bodies, usually rubbery but sometimes soft, with considerable variation in shape and color; (one tiny species has a rounded, dry, finely cracked head, a thin stalk, and resembles a matchstick); on the ground or on wood.

Dacrymyces palmatus (see p. 431)

Phlogiotis helvelloides (see p. 432)

Crust and Parchment Fungi and Allies (see p. 437)

Fruiting bodies typically hard, thin, spreading, crust-like to leathery or papery; some nearly flat, others with small, projecting shelf-like caps; fertile surfaces rough, warted, wrinkled, cracked, toothed or smooth, but lacking pores (use a hand lens); usually on decaying wood.

Cystostereum murraii (see p. 441)

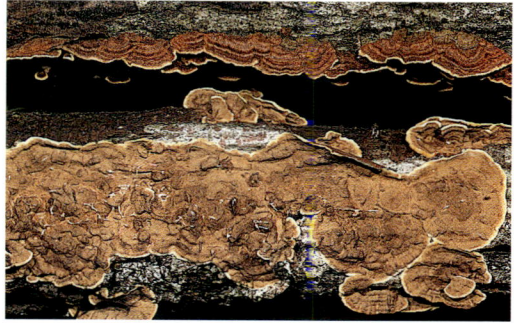

Hymenochaete tabacina (see p. 441)

Laxitextum roseo-carneum (see p. 442)

Serpula lacrimans (see p. 442)

Puffballs, Earthballs, Earthstars, and Allies (see p. 445)

Round, oval, pear- to turban-shaped, irregularly rounded or star-shaped fruiting bodies, usually with powdery interiors at maturity; usually stalkless but occasionally stalked; on the ground or decaying wood, sometimes partially or completely buried.

Geastrum triplex (see p. 453)

Lycoperdon perlatum (see p. 454)

Scleroderma citrinum (see p. 456)

Tulostoma brumale (see p. 457)

Stinkhorns (see p. 461)

Fruiting body erect and phallus-like with a stalk and head, or pear-shaped to nearly round and stalkless, or squid-like with arched, tapered arms; usually coated with a foul-smelling slime; on the ground, mulch, woodchips, or decaying wood.

Phallogaster saccatus (see p. 463)

Dictyophora duplicata (see p. 462)

Phallus ravenelii (see p. 463)

Pseudocolus fusiformis (see p. 463)

Bird's-Nest Fungi (see p. 465)

Small, cylindric to vase-shaped fruiting bodies containing numerous egg-like peridioles (tiny spore sacs); on wood chips, mulch, branches, dung, and other organic debris.

Crucibulum laeve (see p. 465)

Cyathus striatus (see p. 466)

Blueberry Galls and Azalea Apples (see p. 467)

Gall-like swellings on flowers, leaves and shoots; surface white and powdery to mold-like, red and powdery or smooth, or green and shiny to white in age; parasitic on species of the heath family: blueberry, huckleberry, bilberry, wild azalea, and others.

Exobasidium rhododendri (see p. 467)

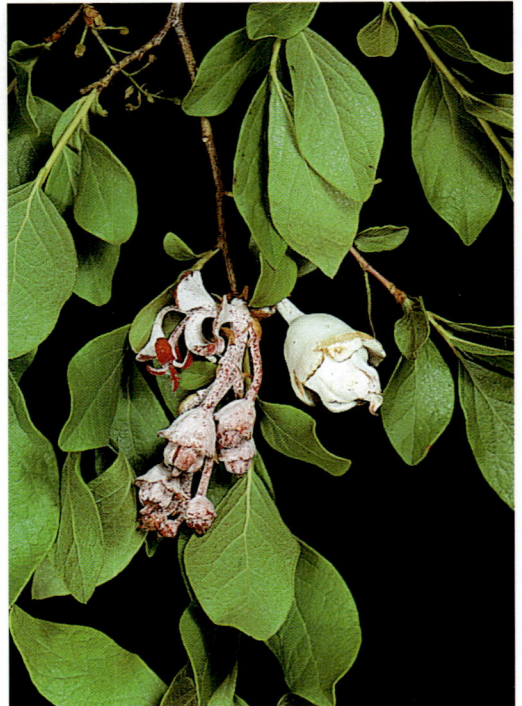

Exobasidium vaccinii (see p. 467)

Rusts and Smuts (see p. 469)

Fruiting bodies variable: tiny, powdery, rusty orange to yellow patches on cinquefoils and Jack-in-the-Pulpits; or tumor-like, whitish to silvery gray with black tints and a dark brown powdery interior at maturity, on ears of corn; or gall-like, with orange, jelly-like "horns" on cedar.

Gymnosporangium juniperi-virginiana (see p. 469)

Pucciniastrum potentillae (see p. 470)

Uromyces ari-triphylli (see p. 470)

Ustilago maydis (see p. 470)

Morels, False Morels, and Allies (see p. 471)

Fruiting bodies with conic to bell-shaped caps with pits and ridges; or caps brain-like, saddle-shaped or irregularly lobed; stalks typically hollow or multichambered, indistinct to massive; on the ground or on decaying wood.

Helvella crispa (see p. 475)

Morchella elata complex (see p. 476)

Cup and Saucer Fungi (see p. 481)

Fruiting bodies resembling small cups or saucers; flesh thin and brittle; with or without a stalk; on the ground or on decaying wood.

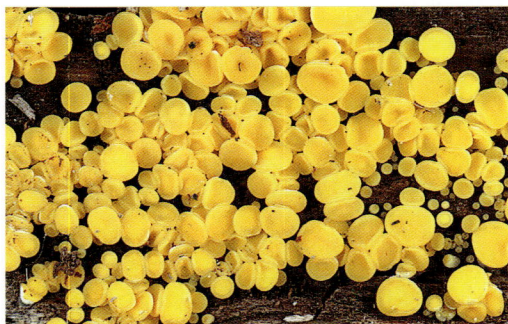

Bisporella citrina (see p. 490)

Sarcoscypha coccinea (see p. 496)

Earth Tongues, Earth Clubs, and Allies (see p. 503)

Erect fruiting bodies resembling tongues or clubs; includes species with only a cylindric to irregular stalk and species with a stalk and "head"; fertile surfaces not roughened like sandpaper, and lacking gills, veins, spines, or pores; on the ground or on decaying wood.

Clavaria argillacea (see p. 508)

Geoglossum difforme (see p. 509)

Leotia atrovirens (see p. 510)

Spathularia velutipes (see p. 512)

Cordyceps, Claviceps, and Allies (see p. 517)

Fruiting body curved and spindle-shaped to cylindric, hard, purplish to brownish black, growing on the inflorescences of grasses; or cylindric to oval or club- to spindle-shaped, yellow to reddish orange or brown, attached to buried insects or buried false truffles; or cylindric to club-shaped, whitish to yellowish or brownish orange, fertile surfaces roughened like sandpaper, on the ground or on decaying wood.

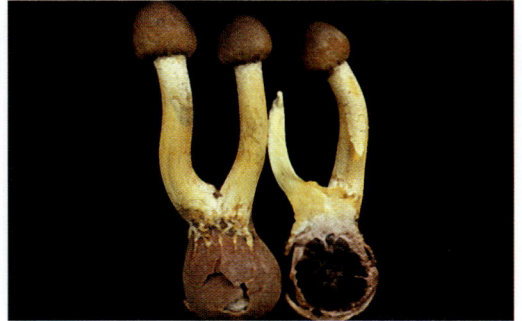

Cordyceps capitata (see p. 518)

Cordyceps militaris (see p. 518)

Claviceps purpurea (see p. 518)

Cordyceps ophioglossoides (see p. 518)

Carbon and Cushion Fungi (see p. 521)

Fruiting bodies extremely variable, mostly cushion-shaped to round, sometimes crust-like and spreading; or erect and cylindric to club-shaped or antler-like; if erect, fruiting body also hard and black, at least on the lower half; fertile surfaces roughened like sandpaper, wrinkled or furrowed; fibrous-tough to woody or hard and carbonaceous, or sometimes gelatinous; creamy white, yellow, green, brick-red, rusty brown, or black; usually on decaying wood or leaves.

Creopus gelatinosus (see p. 524)

Daldinia concentrica (see p. 524)

Ustulina deusta (see p. 525)

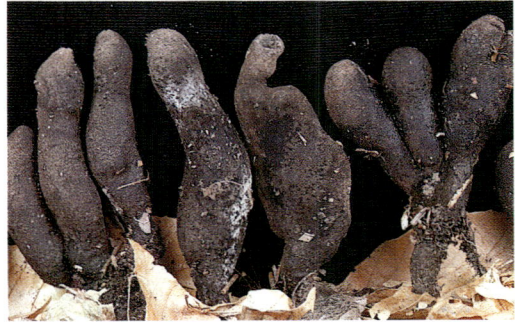
Xylaria polymorpha (see p. 526)

Hypomyces and Allies (see p. 529)

Parasitic fungi that cover and usually disfigure gilled mushrooms, boletes, polypores, and some Ascomycetes; roughened, sandpaper-like, moldy, feathery, or powdery appearance and texture.

Hypomyces chrysospermus (see p. 530)

Hypomyces hyalinus (see p. 531)

Hypomyces lactifluorum (see p. 531)

Hypomyces luteovirens (see p. 531)

FAMILY, GENUS,

AND SPECIES DESCRIPTIONS WITH

KEYS AND ILLUSTRATIONS

Split Gill Family

The Split Gill family includes less than a dozen genera, two of which are commonly known. Fruiting bodies grow on wood and are stalkless or have only a short lateral stalk. Their fertile surfaces are gill-like and split longitudinally or are distinctly crimped, often forked and vein-like.

Key to the Split Gill and Ally

1. Fruiting body white to grayish white, fan-shaped to shell-shaped, with gill-like folds that are split lengthwise along their free edge — *Schizophyllum commune* Fries (see p. 27).
1. Fruiting body yellow-orange to reddish brown or yellow-brown, fan-shaped to shell-shaped, with gill-like folds that are crimped, often forked and vein-like — *Plicaturopsis crispa* (Fries) Reid (see p. 27).

Plicaturopsis crispa (Fries) Reid (see photo, p. 9)
  COMMON NAME: Crimped Gill.
  CAP: ⅜–⅞" (1–2.3 cm) wide, fan-shaped to shell-shaped, concentrically zoned; surface yellow-orange to reddish brown or yellow-brown, finely tomentose, dry; margin undulating, lobed, scalloped, decurved to inrolled, whitish to pale yellow.
  FLESH: membranous, thin, flexible when moist, hard and brittle when dry; odor and taste not distinctive.
  GILL-LIKE FOLDS: crimped, often forked and vein-like, frequently anastomosing, moderately distant, narrow, whitish to grayish.
  STALK: very short, a narrow central extension of the cap, sometimes absent.
  SPORE PRINT: white.
  MICROSCOPIC FEATURES: spores 3–4 × 1–1.5 μm, sausage-shaped to elliptic, smooth, hyaline, often containing two oil drops.
  FRUITING: in overlapping clusters or groups on branches and trunks of hardwoods, especially beech and birch; year-round; infrequent.
  EDIBILITY: inedible.
  COMMENTS: also known as *Trogia crispa*. *Stereum* species lack gills and have a smooth undersurface. Although not as common, widely distributed or well-known as the Common Split Gill, the Crimped Gill has distinctive markings which make it easy to identify.

Schizophyllum commune Fries (see photo, p. 9)
  COMMON NAME: Common Split Gill.
  CAP: ⅜–1¾" (1–4.5 cm) wide, fan-shaped to shell-shaped; surface white to grayish white,

covered by a dense layer of fine hairs, dry; margin incurved to inrolled, becoming nearly flat when moist, wavy and usually torn in age.

FLESH: thin, leathery, flexible, whitish or grayish, often with brownish tones in age.

GILL-LIKE FOLDS: split lengthwise along their free edge, often serrated or torn, subdistant, narrow, white to gray or pinkish gray.

STALK: absent or a short, narrow extension of the cap.

SPORE PRINT: white.

MICROSCOPIC FEATURES: spores 5–7.5 × 2–3 μm, cylindric, smooth, hyaline.

FRUITING: solitary, scattered, or in overlapping clusters on decaying hardwoods; year-round; common.

EDIBILITY: inedible.

COMMENTS: easily recognized by its densely hairy, fan-shaped cap and split gills. The Common Split Gill is among the most widely distributed mushrooms in the world.

Chanterelles and Allies

Members of this small group produce fruiting bodies that are often funnel- to vase-shaped at maturity. Many resemble gilled mushrooms but their spores are not produced on true gills. Their fertile surfaces are typically blunt, gill- to vein-like ridges that are often forked or joined together by crossveins, and a few have nearly smooth fertile surfaces. Chanterelles usually grow on the ground and several are believed to be mycorrhizal with various trees. Many are popular edibles. Although a few species produce gastrointestinal upset, none is known to be fatally poisonous.

Key to Species of Chanterelles and Allies

1. Fruiting body a large cluster of fan- to spoon-shaped caps up to 24" (61 cm) or more wide; cap ¾–4" (2–10 cm) wide, funnel-shaped, deeply depressed; surface dark blue to dull purple or purplish gray; margin wavy, often lobed; fertile surface decurrent, with blunt, vein-like ridges, colored like the cap or grayer; spores 6–9 × 5–8 μm; on the ground under conifers; June–October; edible—*Polyozellus multiplex* (Underwood) Murrill.
1. Fruiting body solitary or in clusters but not dark blue to dull purple; cap ⅛–4" (3–10 cm) wide, cylindric and truncate with a depressed center when young, becoming funnel-shaped to somewhat flattened in age, often perforated at the center; violet when young, becoming tan to brownish yellow in age, smooth, lacking prominent scales; margin uplifted, wavy, often lobed at maturity; fertile surface decurrent, with blunt, vein-like ridges and crossveins, violet to grayish violet, becoming dull ochre to tan in age; scattered, in groups or clusters on the ground under conifers—*Gomphus clavatus* (Fries) S. F. Gray (see p. 35).
1. Fruiting body solitary or in groups but not in large clusters; cap ⅜–1¾" (1–4.5 cm) wide, reddish orange; fertile surface decurrent, with distant, forking, blunt, gill-like ridges with crossveins, pale reddish orange to orange-salmon; stalk smooth, colored like the cap, sometimes whitish at the base; solitary, scattered, or in groups on the ground in mixed woods or hardwoods, especially under oak—*Cantharellus cinnabarinus* Schweinitz (see p. 32).
1. Fruiting body not with the above combinations of characters → 2.
    2. Cap surface with coarse scales; vase-like to funnel-shaped; fertile surface strongly decurrent, with blunt, vein-like ridges → 3.
    2. Cap surface lacking coarse scales, smooth or with tiny scales or fibrils, some shade of brown or gray, lacking distinct yellow or orange-yellow tones; convex or vase- to funnel-shaped; fertile surface with blunt, vein- to gill-like ridges, slightly to strongly decurrent → 4.

2. Cap surface lacking coarse scales, smooth or with tiny scales or fibrils; cap **or** stalk some shade of yellow or orange; fertile surface slightly to strongly decurrent, with blunt, gill- to vein-like ridges, or nearly smooth → 5.

3. Upper surface bright orange, yellow-orange or reddish orange to orange-brown— *Gomphus floccosus* (Schweinitz) Singer (see p. 36).

3. Upper surface creamy white to tan or pale pinkish cinnamon with brown to dark brown coarse scales—*Gomphus kauffmanii* (Smith) Corner (see p. 36).

4. Cap ⅜–2" (1–5 cm) wide, convex, becoming nearly plane, with a depressed center; surface pale to dark yellowish brown, darkest at the center, paler outward, becoming dingy yellow to yellowish orange in age; fertile surface decurrent, with blunt gill-like ridges, pale yellow to pale orange-yellow; stalk nearly equal or tapering downward, pale yellowish brown, developing yellow-orange tones in age; on the ground in hardwoods or mixed woods—*Cantharellus appalachiensis* Petersen (see p. 32).

4. Cap ⅜–3⅛" (1–8 cm) wide, funnel-shaped and deeply depressed; upper surface grayish brown to dark brown or blackish, with darker radiating fibers or small fibrous scales; fertile surface decurrent, smooth or with shallow, blunt, vein-like ridges, gray to brown or blackish, often with ochre-orange tints; stalk indistinct, a short extension below the fertile surface, hollow; flesh thin, brittle to fibrous; odor pleasant, often somewhat fruity; on the ground in hardwoods or mixed woods—*Craterellus fallax* Smith (see p. 34).

4. Cap 1⅜–4" (3.5–10 cm) wide, funnel-shaped with a deep central depression; upper surface grayish brown with darker, minute fibrous scales; fertile surface decurrent, with very shallow, blunt, forked, vein-like ridges and crossveins, gray to grayish brown, often with ochre-orange tints; stalk 1⅛–3" (3–7.5 cm) long below the fertile surface, ⅜–1⅛" (1–3 cm) thick at the apex, tapering toward the base, solid or sometimes hollow, grayish buff to grayish brown; flesh thin, brittle to fibrous; odor strongly fragrant; on the ground under oak—*Craterellus foetidus* Smith (see p. 35).

4. Cap ⅝–3½" (1.5–9 cm) wide, vase- to funnel-shaped with a deep central depression; upper surface smooth to wrinkled, dark brown, fading to grayish brown; fertile surface decurrent, with shallow, forked, blunt, vein-like ridges and crossveins, pale to dark gray fading to pale grayish brown; stalk 1–2" (2.5–5 cm) long, ¼–⅝" (6–16 mm) thick at the apex, tapering downward, hollow, grayish brown; flesh thin, brittle to fibrous; odor not distinctive; spore print white; spores 7–9 × 5–6.5 μm, elliptic, smooth, hyaline; on the ground among mosses, usually under conifers; July–October; edibility unknown—*Craterellus caeruleofuscus* Smith.

4. Cap ⅝–2¾" (1.6–7 cm) wide, vase- to funnel-shaped with a deep central depression; upper surface smooth except for small, elevated fibrous scales on the margin, blackish brown fading to gray-brown; fertile surface decurrent, with raised, prominently forked, blunt, vein-like ridges and distinct crossveins, brownish gray; stalk 1½–3½" (4–9 cm) long, up to ⅝" (1.6 cm) thick at the apex, tapering downward, sometimes compressed, hollow, dark brownish gray; flesh thin, fibrous to brittle; odor not distinctive; taste disagreeable; spore print white; spores 8–11 × 5–6.5 μm, elliptic, smooth, hyaline; scattered or in clusters on the ground in mixed woods; July–October; edibility unknown—*Craterellus cinereus* (Fries) Quélet.

5. Cap ⅜–3⅛" (1–8 cm) wide, convex with a shallow central depression, soon perforated; surface scurfy with tiny erect fibrous scales, especially along the margin, sometimes nearly smooth at maturity, orange to yellow-orange, becoming pale yellow-orange to

brownish orange or pale yellow-brown in age; fertile surface slightly decurrent, with gill- to vein-like blunt ridges, often forking, sometimes with crossveins, orange-yellow, developing a pinkish tinge in age; stalk orange to yellow-orange or yellow—*Cantharellus ignicolor* Petersen (see p. 33).

5. Cap ⅜–2¾" (1–7 cm) wide, convex to nearly plane or funnel-shaped, with a depressed center that often becomes perforated; surface roughened with dark brown radiating fibers and fibrous scales, or sometimes smooth, pale orange-yellow to orange buff beneath the scales or yellow if the surface is smooth; fertile surface decurrent, smooth or with very shallow, blunt vein-like ridges, pale yellow to pale orange, often with a pinkish tint; stalk pale orange to pale orange-yellow, usually with a white base—*Cantharellus xanthopus* (Persoon) Duby and *Cantharellus luteocomus* Bigelow (see p. 34).

5. Not with the above combinations of characters; cap yellowish brown to dark yellow-brown or blackish brown; stalk yellow to yellow-orange or brownish orange → 6.

5. Not with the above combinations of characters; cap *and* stalk some shade of yellow or orange → 7.

    6. Cap blackish brown to dark olive-brown with a brownish yellow margin, fading to pale brown or brownish yellow, smooth, ⅜–2¾" (1–7 cm) wide, convex to nearly plane or funnel-shaped, with a depressed center; fertile surface decurrent, with forked, blunt, gill-like ridges and crossveins, grayish or pale violaceous gray, appearing waxy; stalk smooth, brownish orange on the upper portion, yellow below, sometimes with a whitish basal mycelium—*Cantharellus infundibuliformis* Fries, *sensu* Smith and Morse (see p. 33).

    6. Cap reddish brown, fading to pale yellow-brown, smooth or roughened with small fibrous scales, especially along the margin, ⅜–3" (1–7.5 cm) wide, convex to nearly plane or funnel-shaped, with a depressed center, sometimes perforated; fertile surface decurrent, with blunt, often forked, gill-like ridges and crossveins, yellowish when young, becoming violaceous or violaceous gray; stalk smooth, brownish orange on the upper portion, yellow below, sometimes whitish at the base, with or without mycelium; spores 8–12 × 6–10 μm, granular; among sphagnum mosses in bogs and on soil in wet areas under conifers; July–October; edible—*Cantharellus tubaeformis* Fries.

7. Cap ⅝–5½" (1.6–14 cm) wide, convex to nearly plane, sometimes with a depressed center; surface dry, nearly smooth, orange-yellow to yellow; fertile surface decurrent, with blunt, forked, gill-like ridges, usually lacking crossveins, pale yellow to yellow or pale orange; stalk pale yellow to orange-yellow—*Cantharellus cibarius* Fries (see p. 32).

7. Fruiting body nearly identical to the previous choice except that the fertile surface is smooth or has very shallow forked, blunt, vein-like ridges and crossveins, especially near the cap margin—*Cantharellus lateritius* (Berkeley) Singer (see p. 34).

7. Cap ¼–1⅛" (6–30 mm) wide, convex to nearly plane and slightly depressed at the center, sometimes funnel-shaped in age; surface smooth, yellow-orange to orange, fading to pale yellowish orange then pale yellow; fertile surface decurrent, with forked, blunt, gill-like ridges and crossveins, orange to yellow-orange; stalk ⅝–2⅜" (1.6–6 cm) long, up to ⅜" (1 cm) thick, orange to yellow-orange or yellow; spores 6–12 × 4–6.5 μm; scattered or in groups on the ground or among mosses in hardwoods, especially beech and maple; July–September; edible—*Cantharellus minor* Peck.

Cantharellus appalachiensis Petersen

CAP: ⅜–2" (1–5 cm) wide, convex, becoming nearly plane with a depressed center, not perforated; surface with some matted fibers at first, becoming smooth, moist, pale to dark yellowish brown, darkest at the center, paler outward, becoming dingy yellow to yellowish orange in age; margin incurved when young, becoming elevated and wavy at maturity, sometimes crimped.

FLESH: thin, brittle at first, becoming softer, pale yellowish brown to buff; odor and taste not distinctive.

FERTILE SURFACE: decurrent, with forked, blunt, gill-like ridges, often with crossveins in age, pale yellow to pale orange-yellow.

STALK: ⅝–2" (1.6–5 cm) long, up to ½" (1.3 cm) thick at the apex, nearly equal or tapering downward, solid at first, becoming hollow in age, nearly smooth, pale yellowish brown, developing yellow-orange tones in age.

SPORE PRINT: pale ochraceous salmon.

MICROSCOPIC FEATURES: spores 6–10 × 4–6 μm, elliptic to oblong, smooth, hyaline.

FRUITING: solitary, scattered, in groups or sometimes clustered on the ground in mixed woods or hardwoods; July–September; uncommon.

EDIBILITY: unknown.

Cantharellus cibarius Fries

COMMON NAME: Chanterelle, Golden Chanterelle.

CAP: ⅝–5½" (1.6–14 cm) wide, convex to nearly plane, sometimes with a depressed center; surface dry, nearly smooth, orange-yellow to yellow; margin thin, incurved to inrolled when young, often remaining so for a long time, becoming uplifted and wavy in age, sometimes crimped or lobed.

FLESH: thick, firm, white; odor fragrant like apricots or not distinctive; taste peppery or not distinctive.

FERTILE SURFACE: decurrent, with forked, blunt, gill-like ridges, with or without crossveins, pale yellow to yellow or pale orange.

STALK: ⅝–2¾" (1.6–7 cm) long, up to 1" (2.5 cm) thick, equal or enlarged at either end, smooth, pale yellow to orange-yellow.

SPORE PRINT: pinkish cream to pale buff.

MICROSCOPIC FEATURES: spores 8–11 × 4.5–6 μm, elliptic, smooth, hyaline.

FRUITING: solitary, scattered, in groups or sometimes clustered on the ground in woods; July–September; fairly common.

EDIBILITY: edible, choice.

COMMENTS: compare with the Jack O'Lantern, *Omphalotus olearius* (poisonous), which has true gills with sharp edges and grows on wood or buried wood, typically in large overlapping clusters.

Cantharellus cinnabarinus Schweinitz

COMMON NAME: Cinnabar-red Chanterelle.

CAP: ⅜–1¾" (1–4.5 cm) wide, convex, becoming broadly convex and depressed, then funnel-shaped; surface smooth or nearly so, reddish orange, fading to pale pinkish orange in age; margin incurved when young, becoming uplifted and wavy in age.

FLESH: whitish, tinged reddish orange near the cap surface; odor not distinctive, taste not distinctive or slightly acrid.

FERTILE SURFACE: decurrent, with distant, forking, blunt, gill-like ridges with crossveins; pale reddish orange to orange-salmon.

STALK: ¾–2" (2–5 cm) long, ⅛–⅜" (3–10 mm) wide, nearly equal or tapered downward, often curved, reddish orange, fading to pale pinkish orange.

SPORE PRINT: pinkish cream.

MICROSCOPIC FEATURES: spores 6–10.5 × 4–6 μm, elliptic to oblong, smooth, hyaline.

FRUITING: scattered or in groups on soil or among mosses; June–October; fairly common.

EDIBILITY: edible.

Cantharellus ignicolor Petersen (see photo, p. 9)

COMMON NAME: Flame-colored Chanterelle.

CAP: ⅜–3⅛" (1–8 cm) wide, convex with a shallow central depression, soon perforated; surface scurfy with tiny erect fibrous scales, especially along the margin, sometimes nearly smooth at maturity, orange to yellow-orange, becoming pale yellow-orange to brownish orange or pale yellow-brown in age; margin incurved when young, becoming uplifted, wavy and crimped to lobed at maturity.

FLESH: thin, flexible, orange to yellow-orange, with brownish tints in age; odor of citronella or not distinctive; taste not distinctive.

FERTILE SURFACE: slightly decurrent, with gill- to vein-like blunt ridges, often forking, sometimes with crossveins, orange-yellow, developing a pinkish tinge in age.

STALK: ¾–2⅜" (2–6 cm) long, ⅛–⅜" (3–10 mm) thick, nearly equal, smooth, orange to yellow-orange or yellow, sometimes with white to yellowish matted fibers at the base.

SPORE PRINT: pale ochraceous salmon.

MICROSCOPIC FEATURES: spores 9–13 × 6–9 μm, elliptic to broadly elliptic, smooth, hyaline.

FRUITING: scattered, in groups or clusters on the ground in mixed woods or among sphagnum mosses in bogs; July–October; fairly common.

EDIBILITY: unknown.

COMMENTS: commonly misidentified; use a hand lens and look for the tiny erect fibrous scales, especially along the margin.

Cantharellus infundibuliformis Fries, *sensu* Smith and Morse

CAP: ⅜–2¾" (1–7 cm) wide, convex to nearly plane or funnel-shaped with a depressed center, sometimes perforated; surface smooth, blackish brown to dark olive-brown, paler toward the margin, fading to pale brown or brownish yellow; margin wavy, often irregular, sometimes grooved or crimped in age.

FLESH: thin, watery, yellow-brown; odor faintly fragrant or not distinctive; taste not distinctive.

FERTILE SURFACE: decurrent, with forked, blunt, gill-like ridges and crossveins, grayish or pale violaceous gray, appearing waxy.

STALK: ¾–3" (2–7.5 cm) long, up to ⅜" (1 cm) thick at the apex, tapering downward, smooth, brownish orange on the upper portion, yellow below, sometimes with a whitish basal mycelium.

SPORES PRINT: pale pinkish buff.

MICROSCOPIC FEATURES: 9–12 × 6.5–8 μm, broadly elliptic to nearly round, smooth, hyaline.

FRUITING: scattered or in groups on the ground in woods or among sphagnum mosses; July–October; occasional to frequent.

EDIBILITY: not recommended, according to A. H. Smith.

Cantharellus lateritius (Berkeley) Singer

COMMON NAME: Smooth Chanterelle.

CAP: ⅝–4¾" (1.6–12 cm) wide, convex to nearly plane, sometimes with a low, broad umbo or slightly depressed; surface dry, nearly smooth, yellow-orange to yellow; margin thin, incurved when young, becoming uplifted and wavy in age, often crimped or lobed.

FLESH: thick, firm, white; odor fragrant like apricots or not distinctive; taste not distinctive.

FERTILE SURFACE: decurrent, with forked, blunt, gill-like ridges and crossveins, pale yellow to yellow or orange.

STALK: ⅝–3⅛" (1.6–8 cm) long, up to 1" (2.5 cm) thick at the apex, tapering downward, smooth, pale yellow to orange-yellow.

SPORE PRINT: pinkish yellow to pale yellow-orange.

MICROSCOPIC FEATURES: spores 7.5–12 × 4.5–6.5 μm, elliptic, smooth, hyaline.

FRUITING: scattered, in groups or clusters on the ground in hardwoods, especially under oaks; July–September; fairly common.

EDIBILITY: edible, choice.

COMMENTS: compare with the Jack O'Lantern, *Omphalotus olearius* (poisonous), which has true gills with sharp edges and grows on wood or buried wood, typically in large overlapping clusters.

Cantharellus xanthopus (Persoon) Duby

COMMON NAME: Yellow-footed Chanterelle.

CAP: ⅜–2¾" (1–7 cm) wide, convex to nearly plane or funnel-shaped, with a depressed center that often becomes perforated; surface roughened with dark brown radiating fibers and fibrous scales, sometimes nearly smooth, pale orange-yellow to orange-buff beneath the scales; margin thin, incurved when young, becoming uplifted, wavy and crimped to lobed in age.

FLESH: thin, flexible or brittle, pale buff to yellow-orange; odor and taste not distinctive.

FERTILE SURFACE: decurrent, smooth or with very shallow, blunt, vein-like ridges; pale yellow to pale orange, often with a pinkish tint.

STALK: ¾–2" (2–5 cm) long, up to ½" (1.3 cm) thick at the apex, tapering downward, becoming hollow in age, pale orange to pale orange-yellow, usually with a white base.

SPORE PRINT: pale orange-buff.

MICROSCOPIC FEATURES: spores 9–11 × 6–7.5 μm, elliptic to broadly elliptic, smooth, hyaline.

FRUITING: scattered, in groups or clusters, typically among mosses in wet woods; July–October; frequent to common.

EDIBILITY: edible.

COMMENTS: also known as *Cantharellus lutescens*. Howard Bigelow proposed the name *Cantharellus luteocomus* (edible) for specimens with a smooth yellow cap and slightly larger spores, 8.5–13 × 6–8.5 μm.

Craterellus fallax Smith

COMMON NAME: Black Trumpet.

CAP: ⅜–3⅛" (1–8 cm) wide, funnel-shaped and deeply depressed; upper surface grayish brown to dark brown or blackish, with darker radiating fibers or tiny fibrous scales; margin inrolled at first, becoming arched, wavy, and irregular.

FLESH: thin, brittle to fibrous, colored like the surface; odor pleasant, often somewhat fruity; taste not distinctive.

FERTILE SURFACE: decurrent, smooth or with shallow, blunt, vein-like ridges, gray to brown or blackish, often with ochre-orange tints, bruising blackish.

STALK: indistinct, a short extension below the fertile surface, often hollow, dark brown to blackish.

SPORE PRINT: ochraceous orange to ochraceous buff.

MICROSCOPIC FEATURES: spores 11–18 × 7–11 μm, broadly elliptic, smooth, hyaline.

FRUITING: scattered, in groups or clusters on the ground in hardwoods or mixed woods; July–September; occasional to fairly common.

EDIBILITY: edible and very popular.

COMMENTS: *Craterellus cornucopioides* (edible) is nearly identical but has a whitish spore print.

Craterellus foetidus Smith (see photo, p. 9)

COMMON NAME: Fragrant Black Trumpet.

CAP: 1⅜–4" (3.5–10 cm) wide, funnel-shaped, with a deep central depression; upper surface grayish brown with darker, minute, fibrous scales; margin incurved at first, becoming arched, wavy and irregular.

FLESH: thin, brittle to fibrous, colored like the surface; odor strongly fragrant; taste not distinctive.

FERTILE SURFACE: decurrent, with very shallow, blunt, forked, vein-like ridges and crossveins, gray to grayish brown, often with ochre-orange tints.

STALK: 1⅛–3" (3–7.5 cm) long below the fertile surface, ⅜–1⅛" (1–3 cm) thick at the apex, tapering toward the base, solid or sometimes hollow, somewhat fibrous to nearly smooth, grayish buff to grayish brown.

SPORE PRINT: ochraceous orange to ochraceous buff.

MICROSCOPIC FEATURES: spores 8.5–12 × 5–7 μm, elliptic to oblong, smooth, hyaline.

FRUITING: scattered, in groups or clusters on the ground under oaks; July–September; infrequent to occasional.

EDIBILITY: edible and very popular.

Gomphus clavatus (Fries) S. F. Gray

COMMON NAME: Pig's Ear Gomphus.

CAP: 1⅛–4" (3–10 cm) wide, cylindric and truncate with a depressed center when young, becoming funnel-shaped to somewhat flattened in age, often perforated at the center; surface smooth, lacking prominent scales, violet when young, becoming tan to brownish yellow in age; margin uplifted, wavy, often lobed at maturity.

FLESH: thick, soft, brittle to fibrous, whitish to pale buff; odor and taste not distinctive.

FERTILE SURFACE: decurrent, with blunt, vein-like ridges and crossveins, violet to grayish violet, becoming dull ochre to tan in age.

STALK: up to 2" (5 cm) long below the fertile surface, up to ¾" (2 cm) thick at the apex, tapering downward, often fused with adjacent stalks, solid, colored like the fertile surface.

SPORE PRINT: ochraceous.

MICROSCOPIC FEATURES: spores 10–13 × 4–6.5 μm, narrowly elliptic to spindle-shaped, minutely warted, hyaline.

FRUITING: scattered, in groups or clusters on the ground under conifers; August–October; infrequent to rare.

EDIBILITY: edible, choice.

Gomphus floccosus (Schweinitz) Singer

COMMON NAME: Scaly Vase Chanterelle.

CAP: 1½–6¼" (4–16 cm) wide, funnel- to vase-shaped and deeply depressed; upper surface moist, with numerous coarse orange to reddish orange scales over a paler orange ground color, becoming orange-brown in age; margin thin, wavy, typically lobed at maturity.

FLESH: moderately thick, fibrous, whitish; odor and taste not distinctive.

FERTILE SURFACE: strongly decurrent, with forked, blunt, vein-like ridges and crossveins; pale yellow to creamy white, becoming ochre tinged in age.

STALK: up to 4" (10 cm) long below the fertile surface, up to 2" (5 cm) thick at the apex, tapering downward, solid, becoming hollow in age, nearly smooth, pale yellow to creamy white, browning in age.

SPORE PRINT: ochraceous.

MICROSCOPIC FEATURES: spores 11.5–14.5 × 7–8 μm, elliptic, minutely warted, hyaline.

FRUITING: solitary, scattered or in groups on the ground in mixed woods, especially under hemlock; June–September; fairly common.

EDIBILITY: not recommended; although enjoyed by some, it is a common cause of gastrointestinal upset.

Gomphus kauffmanii (Smith) Corner

CAP: 1–8" (2.5–20.5 cm) wide, funnel- to vase-shaped and deeply depressed; upper surface dry, with numerous coarse brown to dark brown erect or recurved scales over a creamy white to tan or pale pinkish cinnamon ground color; margin thin, wavy, often lobed or split at maturity.

FLESH: thin, fibrous, whitish; odor and taste not distinctive.

FERTILE SURFACE: strongly decurrent, with forked, blunt, vein-like ridges and crossveins, creamy white to pale brown, darkening in age, staining pinkish purple when bruised.

STALK: up to 4¾" (12 cm) long below the fertile surface, up to 1⅜" (3.5 cm) thick at the apex, tapering downward, solid, becoming hollow in age, nearly smooth, creamy white at first, becoming pale to dark brown in age, staining pinkish purple when bruised.

SPORE PRINT: ochre-yellow.

MICROSCOPIC FEATURES: spores 12.5–18.5 × 6–7.5 μm, elliptic to spindle-shaped, minutely warted, hyaline.

FRUITING: solitary, scattered or in groups on the ground under conifers, especially hemlock; July–September; rare to infrequent.

EDIBILITY: not recommended; although enjoyed by some, it is a common cause of gastrointestinal upset.

Cantharellus appalachiensis

Cantharellus cibarius

Cantharellus cinnabarinus

Cantharellus infundibuliformis

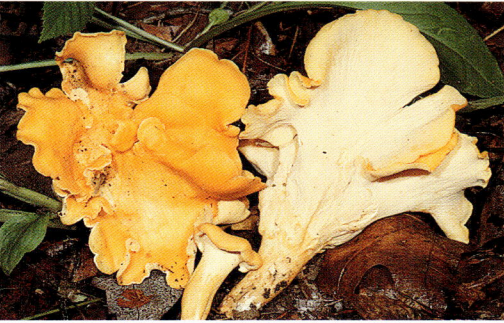

Cantharellus lateritius

Cantharellus xanthopus

Craterellus fallax

Gomphus clavatus

Gomphus floccosus

Gomphus kauffmanii

Gilled Mushrooms

Gilled mushrooms, also known as agarics, belong to a very large group of fungi that have caps with knifeblade-like gills on the undersurface. Many have a central to eccentric stalk; others are laterally-stalked or stalkless. The cap diameter of some species rarely exceeds ⅛" (3 mm) at maturity, while others attain a diameter of 10" (25.5 cm) or more. A universal veil surrounds the button stage of some gilled mushrooms, often leaving warts and patches on the cap and a volva surrounding the stalk base. Many species produce a partial veil, which often leaves a ring on the stalk at maturity.

Gilled mushrooms occur in a seemingly endless array of colors and sometimes change color as they mature. They grow on a wide varity of substrates, including soil, humus, wood, sawdust, straw, cones, fruits, manure, and other mushrooms. Many are excellent edibles, many are poisonous (some are deadly!), and the edibility of the vast majority is unknown.

Many species are easy to identify using only macroscopic features, while others are more difficult and require the use of a microscope. The principle key to gilled mushrooms and the additional keys to species emphasize macroscopic features. We provide some microscopic information to assist readers with access to a microscope. Additional keys to species are not provided when all the species in a genus included in this work are identified using the principle key to gilled mushrooms, or when a particular genus is too complicated for a work such as this book, or when the genus has been insufficiently studied.

For the convenience of the reader, we have narrowed the concept of some genera to exclude characteristics of certain species that do not occur in this region. Obtaining a spore print and determining its color are critical steps in the identification process and should be performed before attempting to use the following key.

KEY TO THE GENERA OF GILLED MUSHROOMS

1. Stalk central to eccentric → 2.
1. Stalk absent to lateral → 26.
    2. Gills attached to decurrent; gills, cap flesh, or stalk exuding latex when cut; universal veil, partial veil and ring absent; spore print white, cream, or yellow to ochre; spores with various amyloid ornamentations—Genus *Lactarius* (see p. 156).
    2. As above, except latex absent; gills white to pale orange; lamellulae few or absent in many species; stalk lacking vertical fibers, snapping somewhat like a piece of

chalk; flesh brittle and crumbly; cap cuticle membranous, detachable (at least near cap margin), sometimes white but often colorfully pigmented (pink, orange, red, purple, green); spore print color and spores as above—Genus *Russula* (see p. 244).

2. Not as in either of the above choices, but spore print white to cream → 3.
2. Spore print pink, tan, yellow, or darker → 4.

3. Universal veil slimy to glutinous, cap and lower stalk likewise; gills free or nearly so, white; partial veil present or absent; spores smooth, inamyloid, typically globose, 6 μm long at most—Genus *Limacella* (see p. 193).

3. Universal veil present, usually leaving remnants (warts on cap or stalk, or volva); partial veil present in young specimens or margin striate or both; gills free or nearly so; terrestrial; never clustered; spores globose to elliptic, smooth, amyloid or inamyloid—Genus *Amanita* (see p. 57).

3. Entire mushroom usually very moist; most species semitranslucent and colorful (yellow, orange, red, purple) with colors fading conspicuously as specimens dry out; gills appearing waxy, thickened, attached, often distant and crossveined; when rubbed, gills typically leaving a waxy residue on your fingers; partial veil rarely present; most species terrestrial; not usually clustered; spores smooth, inamyloid—Genus *Hygrophorus* (see p. 126).

3. Cap coated with loose granules; stalk sheathed halfway or farther up from below, the sheath sometimes flaring at the top; gills variously attached but never free; spores smooth, thin-walled, amyloid or inamyloid—Genus *Cystoderma* (see p. 115).

3. Cap white, tan, brownish, or reddish, usually distinctly scaly in age; gills free, white, close; partial veil present, usually leaving a ring on stalk; terrestrial, usually growing on dead plant debris (e.g., leaves, needles, wood chips); spores smooth, dextrinoid, amyloid, or inamyloid—Genus *Lepiota* and Allies (see p. 185).

3. Spore print white to cream, but mushroom not otherwise as in any of the above choices; gills attached; other characters exceedingly variable → 32.

4. Spore print buff to pink to salmon or pinkish brown → 5.
4. Spore print pale yellowish cream to orangish yellow → 8.
4. Spore print lilac or lilac-tinted, lilac-gray or violet-gray; cap often pinkish, usually finely scaly when dry; gills attached to decurrent, pinkish or flesh-colored to purplish, usually appearing thick and/or waxy; stalk fibrous, tough; spores inamyloid, minutely spiny except smooth in one species—Genus *Laccaria* (see p. 152).
4. Spore print greenish brown to yellowish brown; gills attached to decurrent, crossveined to almost poroid, yellowish at first; gill layer easily separable from the cap flesh; cap surface blueing with ammonia; spores smooth, asymmetric, inamyloid; cystidia typically abundant, clamp connections absent—Genus *Phylloporus* (see p. 332).
4. Spore print greenish, lacking brown tones—Genus *Lepiota* and Allies (see p. 185).
4. Spore print with an orange to red tint when fresh, ranging from bright orange to rust or reddish brown → 9.
4. Spore print yellowish brown to brown, lacking an orange to red tint → 15.
4. Spore print dark purplish brown → 20.
4. Spore print gray to black → 24.

5. Gills distinctly free; saccate volva present; partial veil absent; growing on wood, sawdust, and compost or on other mushrooms; spores smooth, thick-walled, inamyloid—Genus *Volvariella* (see p. 268).

5. Gills distinctly free; volva and partial veil both absent; growing on wood, sawdust, or other woody substrate; spores smooth, inamyloid—Genus *Pluteus* (see p. 232).

5.   Gills free; partial veil present, usually leaving a ring on the stalk; terrestrial → 6.

5.   Gills attached but sometimes appearing free; partial veil absent; cap conic to broadly conic when young, becoming bell-shaped to nearly flat with an umbo in age, dark brown, hairy; stalk dark brown, hairy; base of stalk with bristle-like hairs; spores 13–16 × 7–9 μm, angular in all views; solitary, scattered or in groups on leaf litter or decaying hardwood; edibility unknown—*Pouzarella nodospora* (Atkinson) Mazzer.

5.   Gills attached but often appearing free; partial veil absent; cap usually conic, thin-fleshed; stem slender, often twisted, fragile, usually not white, base typically coated with white mycelium; spores angular, with a pointed apex—Genus *Nolanea* (see p. 213).

5.   Gills attached, sinuate or decurrent; partial veil absent → 7.

     6.   Cap smooth, white, not scaly; spore print white to pale pink; growing on lawns or grassy areas; spores with an apical pore—*Lepiota naucinoides* Peck (see p. 189).

     6.   Cap less than 3" (7.5 cm) wide, slightly scaly when mature; spores smooth, amyloid, inamyloid, or dextrinoid, without an apical pore—Genus *Lepiota* (see p. 185).

7.   Cap less than 3" (7.5 cm) wide and more or less flat at maturity with a sunken center and tiny scales; overall colors and/or staining reactions sometimes striking (e.g., teal, pink, blue to violet or black) but often more or less brown; gills more or less decurrent, sometimes with colored edges; stalk slender, fragile, less than ¼" (7 mm) thick; lower stalk usually white-coated; odor often pronounced and/or odd (e.g., like burnt rubber, mice, bathroom cleanser, bubble gum, or farinaceous); spores angular, with a pointed apex—Genus *Leptonia* (see p. 189).

7.   Cap 1–3" (2.5–7.5 cm) wide, flesh-colored to apricot to reddish pink, surface wrinkled, veined, or netted; gills attached; mushroom growing on wood; spores globose or nearly so, minutely warty or spiny, inamyloid—*Rhodotus palmatus* (Bulliard : Fries) Maire (see p. 243).

7.   Cap thin-fleshed, less than 2" (5 cm) wide, center depressed to sunken; gills decurrent; stalk ⅛" (3 mm) thick at most; spores smooth, inamyloid—Genus *Chrysomphalina* and Allies (see p. 78).

7.   Spore print brownish pink, brownish salmon or pinkish; gills often sinuate; spores angular in all views—Genus *Entoloma* and Allies (see p. 117).

7.   Spore print pinkish cream or pinkish buff, lacking a brownish tint; cap usually white, gray, tan, brown, typically not colorful, often sunken to funnel-shaped; gills thin, usually sinuate or decurrent; spores smooth to finely warty, typically inamyloid (amyloid in only a few species)—Genus *Clitocybe* and Allies (see p. 80).

     Note: Some species of other genera in the Tricholoma family also have spores that appear somewhat pinkish in mass. If a specimen does not key out in *Clitocybe* and Allies, try keying it out from → 32.

     8.   Mushroom tough, corky to fibrous or leathery, not readily decaying; cap becoming sunken at the center; gills descending the stalk; stalk solid, tough, usually densely hairy; found on decaying wood, which may be buried; spores cylindric, smooth, inamyloid—Genus *Lentinus* (see p. 183).

     8.   Cap smooth, convex to flat, often with an umbo, texture like leather, white to yellowish to dark brown, often hygrophanous; gills crowded, attached, never decurrent, white; stalk usually tall, slender, longitudinally striate; often found on humus, sometimes on lawns, never on decaying wood; spores warty, with a plage, and amyloid—Genus *Melanoleuca* (see p. 203).

     8.   Cap usually less than 3" (7.5 cm) wide, sunken at the center in age; gills sometimes forked, always descending stalk; stalk narrow, brittle; usually found among mosses, lichens, or liverworts, but sometimes on soil or wood; spores smooth, inamyloid—Genus *Chrysomphalina* and Allies (see p. 78).

9. Cap 2" (5 cm) wide at most, surface dry, coated with short, erect, brown scales over a grayish brown to yellowish ground color; scales fragile, soon powdery and easily removed; gills free, bright to dark red, becoming brown; partial veil membranous, leaving remnants on upper stalk and on the cap margin; stalk scurfy to nearly smooth; spore print dull red when fresh, drying purplish brown; spores 5–7 × 2–3 μm—*Melanophyllum echinatum* (Roth : Fries) Singer (see p. 204).

9. Gills free, close to crowded, yellowish at first; cap viscid, glabrous, becoming striate; partial veil absent; spores smooth, with an apical pore—Genus *Bolbitius* (see p. 72).

9. Gills free to deeply notched, close; cap viscid, glabrous, conic to campanulate, more or less brown; partial veil absent; stalk with a long, tapering root; exclusively under conifers; spores roughened to finely wrinkled, lacking a pore, often with a snout-like projection—Genus *Phaeocollybia* (see p. 221).

9. Not as in any of the above choices; growing on wood → 10.

9. Not as in any of the above choices; growing on the ground → 12.

9. Not as in any of the above choices; growing on decaying remains of another mushroom, the Shaggy Mane *(Coprinus comatus)*—*Psathyrella epimyces* (Peck) Smith.

  10. Cap margin distinctly inrolled when young; gills decurrent, forked, distinctly crossveined to almost pore-like at the stalk, gill layer easily separable from cap flesh; stalk eccentric to almost lateral, distinctly velvety; spores smooth, lacking a pore—*Paxillus atrotomentosus* (Bataille : Fries) (see p. 220).

  10. Gills often mottled; stalk slender and decidedly brittle, easily snapping in half; partial veil sometimes evident; spores smooth to roughened, with an apical pore—Genus *Psathyrella* (see p. 236).

  10. Cap convex, less than 4" (10 cm) wide; cap and stalk scaly to powdery or granular; partial veil more fibrous than membranous, leaving at most a zone of fibers near the top of the stalk; spores smooth, with or without an apical pore—*Phaeomarasmius erinaceellus* (Peck) Singer (see p. 222).

  10. Gills becoming bright orange, spore print bright orange; flesh bitter; cap blackish with KOH; spores roughened to warty, lacking an apical pore and lacking a plage—Genus *Gymnopilus* (see p. 120).

  10. Cap typically convex, 2½" (6.5 cm) wide at most, hygrophanous, usually with tiny white veil patches, especially near the margin; fibrous or membranous partial veil present when young; spore print pale yellowish to cinnamon-brown; spores smooth, lacking a pore—Genus *Tubaria* (see p. 267).

  10. Not as in any of the above choices → 11.

11. Gill edges whitish, finely serrate; partial veil absent; cap minutely powdery or velvety; spores smooth, lacking an apical pore—*Simocybe centunculus* (Fries) Karsten (see p. 247).

11. Cap typically glabrous; gills usually notched or slightly decurrent, often white-fringed; membranous partial veil present when young; stalk ⅛" (3mm) thick at most; spores warty or at least roughened, but with a plage—Genus *Galerina* (see p. 118)

11. Cap usually scaly, often viscid; fibrous to membranous partial veil present, usually leaving a ring on the stalk or remnants on the cap margin; lower stalk scaly; mushrooms often robust and in large clusters on decaying wood; spores smooth, usually with an apiculus and/or an apical pore that, in some species, causes the spore to appear truncate—Genus *Pholiota* (see p. 223).

  12. Cap margin distinctly inrolled when young; gills decurrent, forked, distinctly crossveined to almost pore-like at stalk, gill layer easily separable from cap flesh; spores smooth, lacking a pore—*Paxillus involutus* (Bataille : Fries) Fries (see p. 220).

12. Cap usually brown but sometimes white to yellowish or lilac; cap radially fibrous, often splitting at the margin, often umbonate, usually less than 2½" (6.5 cm) wide; gills with a pale, fringed edge; partial veil a cortina, rarely leaving a ring on the stalk; odor often spermatic, sometimes fruity; spores smooth to bumpy, sometimes angular, lacking an apical pore—Genus *Inocybe* (see p. 150).

12. Gills becoming distinctly rust-colored, spore print distinctly rust-colored; young specimens with an obvious cortina, usually leaving at most a fibrous annular zone on the stalk; stalk often with a bulbous base; spores warty to finely wrinkled—Genus *Cortinarius* (see p. 105).

12. Cap brownish yellow to yellowish brown with a white bloom, especially at the center; gills becoming distinctly rust-colored, spore print distinctly rust-colored; membranous partial veil present, leaving a membranous ring on the stalk; spores warty to wrinkled, dextrinoid—*Rozites caperata* (Fries) Karsten (see p. 243).

12. Cap viscid; gills typically sinuate to notched, with a white margin; odor often radish-like; spores smooth, thick-walled, dextrinoid—Genus *Hebeloma* (see p. 123).

12. Not as in any of the above choices; stalk slender and fragile or brittle → 13.

12. Not as in any of the above choices; stalk neither fragile nor brittle → 14.

13. Gills often mottled; stalk slender and decidedly brittle, easily snapping in half; partial veil sometimes evident; spores smooth to roughened, with an apical pore—Genus *Psathyrella* (see p. 236).

13. Stalk quite slender and fragile but not brittle as described above; spores smooth, with an apical pore, the apex often flattened—Genus *Conocybe* (see p. 97).

14. Cap slimy, brown, with dry fibrous scales; partial veil whitish, leaving remnants on the cap margin and sometimes leaving a ring on the stalk; stalk dark brown; growing in clusters on the ground; spores 4.5–7 × 3.5–4.5 μm, smooth, with a distinct apiculus and a minute but distinct apical pore—*Pholiota terrestris* Overholts.

14. Cap typically glabrous; gills usually notched or slightly decurrent, often white-fringed; membranous partial veil present when young; stalk ⅛" (3 mm) thick at most; spores warty or at least roughened, with a plage—Genus *Galerina* (see p. 118).

14. Cap typically convex, 2½" (6.5 cm) wide at most, hygrophanous, usually with tiny white veil patches, especially near the margin; fibrous or membranous partial veil present when young; spore print pale yellowish to cinnamon-brown; spores smooth, lacking an apical pore—Genus *Tubaria* (see p. 267).

15. Partial veil membranous (check young specimens) → 16.

15. Partial veil fibrous to cortinate (check young specimens) → 17.

15. Partial veil absent even in very young specimens → 18.

16. Cap usually scaly, often viscid; gills attached; fibrous to membranous partial veil present, usually leaving a ring on the stalk or remnants on the cap margin; lower stalk scaly; often robust and in large clusters on decaying wood; spores smooth, usually with an apiculus and/or an apical pore that, in some species, causes the spore to appear truncate—Genus *Pholiota* (see p. 223).

16. Cap usually thick-fleshed and robust; gills close to crowded, free or nearly so, white or pale gray at first often becoming pink and always turning dark brown to black, with or without a purple tint when mature; stalk cleanly separable from the cap; spores smooth, lacking an apical pore or with only an obscure apical pore—Genus *Agaricus* (see p. 51).

16. Cap moderately thick-fleshed, sometimes robust, often wrinkled or cracked over the disc at maturity; gills attached; stalk usually with conspicuous white rhizo-

morphs at the base; often found in troops or clusters in woody soil, on humus, dung, or lawns, or especially on wood chips, but rarely on logs or stumps; spores smooth, typically with a wide pore—Genus *Agrocybe* (see p. 55).

16. Cap glabrous and hygrophanous, often appearing zoned; usually in clusters on wood; spores smooth, usually with an apiculus and/or an apical pore that, in some species, causes the spore to appear truncate—Genus *Pholiota* (see p. 223).

17. Cap usually scaly, often viscid; fibrous to membranous partial veil present, usually leaving a ring on the stalk or remnants on the cap margin; lower stalk scaly; often robust and in large clusters on decaying wood; spores smooth, usually with an apiculus and/or an apical pore that, in some species, causes the spore to appear truncate—Genus *Pholiota* (see p. 223).

17. Cap brown, sometimes white to yellowish or lilac, radially fibrous, often splitting at the margin, often umbonate, usually less than 2½" (6.5 cm) wide; gills with a pale, fringed edge; partial veil a cortina, rarely leaving a ring on the stalk; odor often spermatic, sometimes fruity; spores smooth to bumpy, sometimes angular, lacking an apical pore—Genus *Inocybe* (see p. 150).

17. Not as in either of the above choices → 19.

18. Cap margin distinctly inrolled when young; gills decurrent, forked, distinctly crossveined to almost pore-like at the stalk, gill layer easily separable from the cap flesh; spores smooth, lacking a pore—Genus *Paxillus* (see p. 219).

18. Cap less than 2" (5 cm) wide, typically almost fleshless, distinctly striate, often splitting radially at maturity, usually with fine, clear hairs (use a hand lens); gills typically well spaced; spores smooth, with an apical pore—Genus *Coprinus* (see p. 99).

18. Gill edges whitish, finely serrate; partial veil absent; cap minutely powdery or velvety; spores smooth, lacking an apical pore—*Simocybe centunculus* (Fries) Karsten (see p. 247).

18. Not as in any of the above choices → 19.

19. Gills often mottled; stalk slender and decidedly brittle, easily snapping in half; partial veil sometimes evident; spores smooth to roughened, with an apical pore—Genus *Psathyrella* (see p. 236).

19. Cap glabrous, usually yellowish or with a yellow tint; gills pallid to greenish at first, becoming smoky gray at maturity; usually growing on wood or humus or in moss; spores usually smooth with an apical pore—Genus *Hypholoma* (see p. 146).

19. Cap viscid; gills typically sinuate to notched, with a white margin; odor often radish-like; spores smooth, thick-walled, dextrinoid—Genus *Hebeloma* (see p. 123).

19. Cap usually thick-fleshed, sometimes robust, often cracked at maturity; gills attached; stalk usually sturdy; often found in troops or clusters in woody soil, on humus, dung, lawns, or especially on wood chips, but rarely on logs or stumps; spores smooth, typically with a wide pore—Genus *Agrocybe* (see p. 55).

19. Cap margin adorned with long, coarse hairs; cap usually becoming sunken at the center in age; gills often crossveined, white to dull pinkish when young; stalk fragile, whitish, hollow; spores small (6 μm maximum), round or nearly so, with minute spines or bumps—*Ripartites tricholoma* (Albertini and Schweinitz : Fries) Karsten.

20. Partial veil more or less membranous (check young specimens) → 21.

20. Partial veil more or less fibrous or cortinate (check young specimens) → 22.

20. Partial veil absent even in young specimens → 23.

21. Cap 2" (5 cm) wide at most, surface dry, coated with short, erect, brown scales over a grayish brown to yellowish ground color; scales fragile, soon powdery and easily

removed; gills free, bright to dark red, becoming brown; partial veil membranous, leaving remnants on the upper stalk and cap margin; stalk scurfy to nearly smooth; spore print dull red when fresh, drying purplish brown; spores 5–7 × 2–3 μm—*Melanophyllum echinatum* (Roth : Fries) Singer (see p. 204).

21. Gills close, attached, often notched, edges often whitish, finely serrate; partial veil present, usually leaving a ring on the stalk and sometimes remnants on the cap margin; rhizomorphs often attached to base of stalk; spores smooth, with a truncate apical pore—Genus *Stropharia* (see p. 248).

21. Not as in either of the above choices → 23.

22. Gill edges whitish, finely serrate; partial veil absent; cap minutely powdery or velvety; spores smooth, lacking an apical pore—*Simocybe centunculus* (Fries) Karsten (see p. 247).

22. Cap smooth, usually viscid; gill edges smooth, often remaining whitish at maturity; partial veil sparse, fibrous, usually evident only on young specimens, not leaving a ring; stalk often staining blue to greenish blue when bruised; spores smooth, with a truncate apical pore—Genus *Psilocybe* (see p. 240).

22. Not as in either of the above choices → 23.

23. Cap usually thick-fleshed and robust; gills close to crowded, free or nearly so, white or pale gray at first, often becoming pink and always turning dark brown to black, with or without a purple tint when mature; stalk cleanly separable from the cap; spores smooth, without an apical pore or with only an obscure apical pore—Genus *Agaricus* (see p. 51).

23. Gills often mottled; stalk slender and decidedly brittle, easily snapping in half; partial veil sometimes evident; spores smooth to roughened, with an apical pore—Genus *Psathyrella* (see p. 236).

23. Cap glabrous, usually yellowish or with a yellow tint; gills pallid to greenish at first, becoming smoky gray at maturity; partial veil evident or not; usually growing on wood, on humus, or in moss; spores usually smooth with an apical pore—Genus *Hypholoma* (see p. 146).

24. Gills thick, widely spaced and distinctly decurrent, yellowish to orange or salmon at first; flesh of lower stalk colored buff to orange; spores smooth, long, and narrow; flesh amyloid—Genus *Chroogomphus* (see p. 76).

24. Cap viscid or slimy; gills thick, widely spaced, and distinctly descending the stalk, white or whitish at first; flesh white; spores smooth, cylindric; flesh inamyloid—Genus *Gomphidius* (see p. 119).

24. Not as in either of the above choices → 25.

25. Gills extremely crowded; gills and sometimes cap dissolving into a black, ink-like fluid at maturity; spores smooth, with an apical pore—Genus *Coprinus* (see p. 99).

25. Gills often mottled; stalk slender and decidedly brittle, easily snapping in half; partial veil sometimes evident; spores smooth to roughened, with an apical pore—Genus *Psathyrella* (see p. 236).

25. Cap smooth, dry to viscid, usually gray to brown or black; faces of gills becoming black-dotted in age, edges often whitish; partial veil absent; typically found on dung or in manured areas such as pastures, but sometimes on soil or in moss; spores smooth, with a flattened end and an apical pore—Genus *Panaeolus* (see p. 216).

25. Cap less than 2" (5 cm) wide, typically almost fleshless, distinctly striate, often splitting radially at maturity, usually with fine, clear hairs (use a hand lens); gills typically well spaced; spores smooth, with an apical pore—Genus *Coprinus* (see p. 99).

26. Spore print white to cream → 28.

26. Spore print yellowish; cap smooth to finely velvety in age, up to 4" (10 cm) wide,

variously yellow to green or purple in color; gills yellow, neither forked nor cross-veined; mushroom tough, not decaying readily; found only in autumn after frosts, on decaying wood; spores smooth, sausage-shaped, amyloid—*Panellus serotinus* (Fries) Kühner (see p. 218).

26. Spore print yellowish olive to olive-yellow when fresh, drying yellowish cinnamon; cap smooth to finely velvety, 3" (7.5 cm) wide at most, yellow overall; gills orangish yellow, forked, crossveined and distinctly corrugated, wrinkled or wavy; gill layer easily separable from the cap flesh; odor unpleasant; spores 3–3.5 × 1.5–2 μm, ellipsoid, smooth, inamyloid—*Paxillus corrugatus* Atkinson (see p. 220).

26. Spore print yellow or yellowish; cap greenish yellow to brownish; gills yellow, forked and crossveined and corrugated or wrinkled; gill layer easily separable from cap flesh; spores 4–6 × 3–4 μm, elliptic, smooth, inamyloid or dextrinoid—*Paxillus panuoides* (Fries : Fries) Fries (see p. 221).

26. Spore print pale yellowish cream to orangish yellow; otherwise not as in the previous choice; spores smooth, cylindric, inamyloid—Genus *Lentinus* (see p. 183).

26. Spore print buff to pink to salmon or pinkish brown → 27.

26. Spore print light grayish lilac; spores smooth, cylindric or nearly so, inamyloid—Genus *Pleurotus* (see p. 231).

26. Spore print dull brown to yellowish brown or pinkish brown; spores smooth to roughened or appearing dotted, globose to elliptic or almond-shaped, inamyloid—Genus *Crepidotus* (see p. 113).

27. Cap about 1–3" (2.5–7.5 cm) wide, flesh-colored to apricot to reddish pink, surface wrinkled, veined, or netted; gills attached; mushroom growing on wood; spores globose or nearly so, minutely warty or spiny, inamyloid—*Rhodotus palmatus* (Bulliard : Fries) Maire (see p. 243).

27. Cap distinctly fuzzy, yellow to orange; odor and taste disagreeable; spores smooth, cylindric, inamyloid—*Phyllotopsis nidulans* (Persoon : Fries) Singer (see p. 230).

27. Growing on other mushrooms; spores smooth, angular in all views—*Claudopus parasiticus* (Quélet) Ricken (see p. 79).

27. Not as in either of the above choices; odor often farinaceous; spores more or less elliptic, with longitudinal ridges, appearing angular only in end view—Genus *Clitopilus* (see p. 88).

27. Odor not farinaceous; not growing on other mushrooms; spores distinctly angular in all views—Genus *Claudopus* (see p. 79).

27. Macroscopically not as in any of the above choices; spores smooth to roughened or appearing dotted, globose to elliptic or almond-shaped, inamyloid—Genus *Crepidotus* (see p. 113).

28. Cap and gills orange overall; gills somewhat decurrent, repeatedly and regularly forked but not crossveined; growing on or about decaying conifer wood or needle litter; spores elliptic to cylindric, smooth, mostly dextrinoid—*Hygrophoropsis aurantiaca* (Wulfen : Fries) Maire (see p. 126).

28. Gill edges appearing distinctly white-fringed (use a hand lens); spores smooth, inamyloid, usually elliptic; gills with prominent cheilocystidia—Genus *Tricholomopsis* (see p. 265).

28. All parts staining or bruising blackish; spores smooth to finely warty or spiny, round to elliptic or cylindric but sometimes appearing triangular, inamyloid—Genus *Lyophyllum* (see p. 194).

28. Gills strongly decurrent; entire mushroom orange overall, normally luminescing green when fresh (view in complete darkness for 5–10 minutes); spores smooth,

globose to subglobose, inamyloid—*Omphalotus olearius* (De Candolle : Fries) Singer (see p. 216).

28.   Not as in any of the above choices → 29.

29.   Cap white to gray or brownish, smooth to minutely velvety or scaly; flesh typically gelatinized or rubbery; gill edges neither serrate nor fringed (use a hand lens); spores smooth or appearing finely pitted or dotted, inamyloid—Genus *Hohenbuehelia* (see p. 125).

29.   Cap hairy to scaly, tan to pale brown, less than 1½" (4 cm) wide; mushroom tough, not decaying readily; taste quite acrid; normally luminescing green when fresh (view in complete darkness for 5–10 minutes); spores smooth, sausage-shaped, amyloid—*Panellus stipticus* (Bulliard : Fries) Karsten (see p. 219).

29.   Cap dry, finely hairy, bluish black, typically less than ½" (1.3 cm) wide; flesh rubbery-gelatinous; gills gray to nearly black; found on the undersurface of decaying logs; spores round, smooth, inamyloid—*Resupinatus applicatus* (Bataille : Fries) S. F. Gray (see p. 242).

29.   Not as in any of the above choices; gills serrate and/or cap leathery to corky → 30.

29.   Not as in any of the above choices; gills not serrate; cap not leathery to corky → 31.

30.   Gills purplish, not serrate; stalk, if present, very tough and usually hairy; spores smooth, elliptic, inamyloid—Genus *Lentinus* (see p. 183).

30.   Gills decurrent, serrate; taste bitter or acrid; spores finely warted or spiny, amyloid—Genus *Lentinellus* (see p. 181).

31.   Cap brown, less than 1" (2.5 cm) wide, becoming minutely velvety to hairy in age; thin, membranous partial veil present in very young specimens; spores smooth, cylindric, weakly amyloid—*Tectella patellaris* (Fries) Murrill (see p. 252).

31.   Cap more or less white, fairly robust, up to 6" (15 cm) wide, typically cracked or with visible water spots in age; stalk present; usually growing on living hardwoods; spore print cream; spores smooth, globose to elliptic, inamyloid—Genus *Hypsizygus* (see p. 149).

31.   Cap usually smooth, white to brown, up to 6" (15 cm) or more wide, thick-fleshed; gills decurrent, broad, white to cream; spore print white to cream or grayish lilac; spores more or less cylindric, smooth, inamyloid—*Pleurotus ostreatus* complex (see p. 232).

31.   Cap 4" (10 cm) wide at most, white, thin-fleshed, pliant; spore print white; gills narrow, crowded, white to yellowish; stalk virtually absent; typically found in groups or almost clustered on dead conifer logs, especially hemlock; spores globose or nearly so, smooth, inamyloid—*Pleurocybella porrigens* (Persoon : Fries) Singer (see p. 231).

31.   Cap white, less than 1" (2.5 cm) wide, smooth to minutely hairy, soft-fleshed; gills finely fringed (use a hand lens); spores round to rounded-angular, inamyloid—*Cheimonophyllum candidissimus* (Berkeley and Curtis) Singer (see p. 75).

31.   Cap 1" (2.5 cm) wide at most, usually white to brown or purplish; stalk, if present, rudimentary, typically minutely velvety—Genus *Panellus* (see p. 218).

32.   Gill edges serrate (use a hand lens) → 33.

32.   Gills repeatedly and regularly forked → 34.

32.   Not as in either of the above choices; partial veil present → 35.

32.   Not as in any of the above choices; partial veil absent → 36.

33.   Flesh bitter or acrid; spores finely warted or spiny, amyloid—Genus *Lentinellus* (see p. 181).

33.   Flesh mild to bitter; spores smooth, inamyloid—Genus *Lentinus* (see p. 183).

34.   Cap gray overall, 3" (7.5 cm) wide at most; gills staining reddish; growing in

haircap moss; spores smooth, somewhat spindle-shaped, amyloid—*Cantharellula umbonata* (Gmelin : Fries) Singer (see p. 74).

34. Cap and gills orange overall; gills somewhat decurrent, repeatedly and regularly forked but not crossveined; growing on or about decaying conifer wood or needle litter; spores elliptic to cylindric, smooth, mostly dextrinoid—*Hygrophoropsis aurantiaca* (Wulfen : Fries) Maire (see p. 126).

34. Cap pinkish at first, fading to buff; gills decurrent, white to pinkish, some distinctly forked, typically crossveined; growing on or about decaying conifer wood or needle litter; odor strongly fragrant, reminiscent of bubble gum; spores 3–5 × 2–3 μm, elliptic, smooth, dextrinoid; edibility unknown—*Hygrophoropsis olida* (Quélet) Métrod.

35. Solitary to clustered on deciduous wood; gills decurrent, white discoloring yellowish, covered at first by a white membranous veil; cap 2–5" (5–12.5 cm) wide, coated with tiny, matted, grayish fibrils on a whitish ground color, becoming slightly scurfy and whitish to dull yellowish tan overall in age; flesh white; odor fragrant to slightly pungent; taste not distinctive; stalk eccentric to central, whitish, sometimes with a sparse, membranous, white superior ring; edible—*Pleurotus dryinus* (Persoon : Fries) Kummer (see p. 231).

35. Cap and lower stalk densely coated with rusty brown, pointed, recurved scales, dry; margin incurved and often remaining so at maturity, coated with rusty brown fibers; gills notched, close, white, edges finely scalloped; spores 5–6 × 3.5–4 μm, elliptic, smooth, hyaline, amyloid; scattered, in groups or clusters on decaying wood; edibility unknown—*Leucopholiota decorosa* (Peck) O. K. Miller, Jr., Volk and Bessette.

35. Cap yellow to tan or brown, with erect hairs, at least over the center; gills attached, usually slightly decurrent; ring usually prominent, often yellow- to brown-edged; typically found in large clusters on or about dead or decaying trees; spores smooth to very finely wrinkled, inamyloid—Genus *Armillaria* (see p. 68).

35. Lower stalk markedly swollen, cylindric to club-shaped, mostly buried; spores smooth, elliptic, inamyloid—*Squamanita umbonata* (Sumstine) Bas (see p. 248).

35. Partial veil distinctly two-layered, essentially composed of two separate partial veils—*Catathelasma ventricosa* (Peck) Singer (see p. 75).

35. Not as in any of the above choices; found on the ground, usually under conifers, aspen, or oak trees—Genus *Tricholoma* (see p. 252).

36. Found growing on other mushrooms or on decaying remains of other mushrooms → 37.

36. Found growing on cones or nut hulls → 38.

36. Not as in either of the above choices → 39.

37. Gills close; stalk attached to a reddish brown, appleseed-like tuber; spores smooth, elliptic, inamyloid—*Collybia tuberosa* (Bulliard : Fries) Kummer (see p. 97).

37. As in the previous choice except tuber yellowish orange, more or less round; spores smooth, elliptic to oval or lacrymoid, inamyloid—*Collybia cookei* (Bresadola) Arnold.

37. Gills widely spaced and poorly formed or absent; cap covered with brown powder when mature; spores smooth, oval, inamyloid—*Asterophora lycoperdoides* (Bulliard : Mérat) Ditmar (see p. 71).

37. Gills well formed; cap silky, not powdery, white to grayish or pale tan; spores smooth, elliptic, inamyloid—*Asterophora parasitica* (Bulliard : Fries) Singer (see p. 71).

38. Found on pine cones or other conifer cones; gills white, crowded, and narrow; base of stalk with long, coarse hairs; spores smooth, elliptic, less than 5 μm long, amyloid—*Baeospora myosura* (Fries) Singer (see p. 71).

38. Found on walnut hulls; spores smooth to minutely roughened, elliptic, amyloid— *Mycena luteopallens* (Peck) Saccardo (see p. 212).

38. Found on magnolia cones or sweetgum fruit; spores smooth, elliptic, inamyloid— *Strobilurus conigenoides* (Ellis) Singer (see p. 248).

39. Cap pinkish at first, fading to buff; gills decurrent, white to pinkish, some distinctly forked, typically crossveined; growing on or about dead conifer wood, needles, etc.; odor strongly fragrant, reminiscent of bubble gum; spores 3–5 × 2–3 μm, elliptic, smooth, dextrinoid; edibility unknown—*Hygrophoropsis olida* (Quélet) Métrod.

39. Not as in the previous choice; growing on stumps, logs, or twigs, etc. → 40.

39. Not as in either of the previous choices; growing on the ground, twigs, needles, leaves, humus → 43.

    40. Cap 3" (7.5 cm) wide at most; stalk ³⁄₁₆" (5 mm) wide at most → 41.

    40. Usually growing in clusters of 10 or more specimens; caps yellowish to pinkish brown, with minute, erect hairs at the center; gills slightly decurrent; stalk base usually tapered; spores smooth, inamyloid—*Armillaria tabescens* (Scopoli) Emel (see p. 70).

    40. Entire mushroom very tough, fibrous to leathery or corky, purplish when young, becoming tan to brown in age; cap smooth; stalk finely hairy when young; spores smooth, inamyloid—*Lentinus torulosus* (Persoon : Fries) Lloyd (see p. 185).

    40. Entire mushroom very tough, fibrous to leathery or corky, usually found growing on living hardwoods; cap surface smooth at first, becoming cracked and/or water-spotted at maturity; spores smooth, globose to elliptic, inamyloid—Genus *Hypsizygus* (see p. 149).

    40. Not as in any of the above choices → 42.

41. Cap and stalk bright yellow, gills cream to yellow; cap scurfy to granular-mealy; growing on decaying deciduous logs or sticks; spores smooth, oval elliptic, inamyloid— *Cyptotrama asprata* (Berkeley) Redhead and Ginns (see p. 114).

41. Gills extremely crowded, lavender; cap also lavender or lavender-tinted, at least when young; spores smooth, amyloid—*Baeospora myriadophylla* (Peck) Singer (see p. 72).

41. Cap ½–1½" (1.2–4 cm) wide, zoned with long radially arranged hairs; gills close, narrow, nearly free from the stalk; stalk hairy, hollow; spores 4–6 × 3–5 μm; on decaying hardwood—*Crinipellis zonata* (Peck) Patouillard (see p. 114).

41. Cap ⁵⁄₁₆–⁵⁄₈" (8–15 mm) wide, entire fruiting body very similar to the previous choice, cap depressed over the disc with a tiny nipple-like projection at maturity; flesh whitish, odor spicy or not distinctive, taste not distinctive; spores 6–9 × 4–6 μm; scattered or in groups on decaying stems and leaves of grasses and other plants, sometimes on twigs; edibility unknown—*Crinipellis scabella* (Albertini and Schweinitz : Fries) Murrill = *C. stipitaria* (Fries) Patouillard.

41. Not as in any of the above choices → 42.

    42. Usually growing in clusters of 10 or more specimens; cap viscid, yellowish brown to reddish brown; stalk dark brown and velvety at the base; spores smooth, elliptic, inamyloid—*Flammulina velutipes* (Fries) Karsten (see p. 117).

    42. Cap fibrous to finely scaly, usually yellow to reddish orange; flesh typically distinctly yellowish; gills often yellowish or orangish, gill edges often appearing ragged or fringed; spores smooth, inamyloid—Genus *Tricholomopsis* (see p. 265).

    42. Not as in either of the above choices → 43.

43. Cap cuticle like a thick, rubbery membrane; gills white, sometimes with darker edges; stalk with a long, tapering tap root; spores smooth to finely roughened, oval to elliptic

to lemon- or almond-shaped, sometimes with a prominent apiculus, inamyloid—Genus *Xerula* (see p. 270).

43. Mushroom typically white overall; cap dry, smooth, thick-fleshed; gill layer readily separable from flesh of the cap; base of stalk attached to copious white mycelium, which binds together a substantial mass of dead leaves, needles, etc.; odor often disagreeable or farinaceous; taste bitter or farinaceous; spores amyloid-warted to variously amyloid-ornamented, plage absent—Genus *Leucopaxillus* (see p. 191).

43. Cap often pinkish, usually finely scaly when dry; gills attached to decurrent, pinkish or flesh-colored to purplish, usually appearing thick and/or waxy; stalk fibrous, tough; spores inamyloid, minutely spiny except smooth in one species—Genus *Laccaria* (see p. 152).

43. Cap variously colored, often scaly or viscid but sometimes smooth and/or dry; gills sinuate with few exceptions, usually white, yellow, or grayish; spores smooth, fusoid to subglobose, inamyloid (if amyloid, see *Porpoloma umbrosum,* p. 236)—Genus *Tricholoma* (see p. 252).

43. Not as in any of the above choices → 44.

    44. Cap gray to grayish brown, with darker radial fibers; gills white, very broad; stalk white, with thick, white cords attached to the base; found on or about well-decayed logs and stumps; spores oval, smooth, inamyloid—*Megacollybia platyphylla* (Persoon : Fries) Kotlaba and Pouzar (see p. 203).

    44. Cap usually white, gray, tan, brown, not typically colorful, often sunken to funnel-like; gills thin, usually distinctly decurrent; spores smooth to finely warty, typically inamyloid (amyloid in only a few species)—Genus *Clitocybe* and Allies (see p. 80).

    44. Cap variously colored, usually flat at maturity, margin typically incurved to inrolled at first; gills variously attached but never decurrent, typically white, narrow, and close; stalk slender but not hair-like; spores smooth, inamyloid or dextrinoid, usually elliptic to lacrymoid—Genus *Collybia* and Allies (see p. 88).

    44. Cap variously colored, typically 2" (5 cm) wide at most, often conic or bell-shaped, cap margin usually striate when fresh; gills variously attached; stalk typically slender, ⅛" (3 mm) thick, and fragile; spores smooth, amyloid or inamyloid—Genus *Mycena* and Allies (see p. 204).

    44. Not as in any of the above choices → 45.

45. Dried mushrooms reviving when moistened; cap convex to umbilicate to radially grooved, like an umbrella, smooth to finely velvety, white, gray, or brown to orangish or reddish; flesh typically so thin as to be virtually nonexistent; gills variously attached to the stalk or to a collar; stalk typically bristle-like, always thin, often less than 1⁄16" (2 mm) thick; usually growing on dead plant matter (wood, leaves, needles, etc.); spores smooth, cylindric to oval, inamyloid—Genus *Marasmius* and Allies (see p. 196).

45. Cap smooth, convex to flat, often with an umbo, texture like leather, white to yellowish to dark brown, often hygrophanous; gills crowded, attached, never decurrent, white; stalk usually tall, slender, longitudinally striate; often found on humus, sometimes on lawns, never on decaying wood; spores warty, with a plage, amyloid—Genus *Melanoleuca* (see p. 203).

45. Cap flesh-pink to pale vinaceous pink, becoming pale pinkish brown to pinkish tan or yellowish tan at the center, less than 2" (5 cm) wide; margin usually inrolled at first; flesh thin, white; gills white to cream, close to crowded, attached at first, becoming decurrent in age, finely scalloped, becoming eroded in age; stalk less than 2" (5 cm) long, no more than ¼" (7 mm) thick, pink overall at first, becoming dingy yellow to

yellowish tan, with a narrow white zone at the apex, typically coated with long, white hairs near or at the base; often growing in clusters; spores smooth, elliptic to oval, inamyloid—*Calocybe persicolor* (see p. 74).

45. As above except cap bright pale pink to flesh-pink becoming yellowish tan; stalk base sometimes coated with shorter matted whitish fibrils; not growing in clusters—*Calocybe carnea* (Bulliard : Fries) Donk (see comments under *Calocybe persicolor,* p. 74).

45. Cap fleshy, white to grayish to brownish, often bruising blackish; sometimes abundant in a small area, often clustered; gills variously attached, often staining and/or bruising blackish; usually growing in woody dirt or on dirty wood; spores variously shaped, smooth or ornamented, inamyloid—Genus *Lyophyllum* (see p. 194).

Genus Agaricus

Agaricus, a very large genus, includes more than one hundred species in North America. They are small to very large terrestrial mushrooms with a central stalk that separates easily from the cap. Their spore print colors range from chocolate-brown to purplish brown or blackish brown, and their spores are smooth, with or without an apical pore. Members of this genus have fibrillose to scaly caps, and their gills are free or nearly so. Many are excellent edibles, some are poisonous, and the edibility of the majority is unknown. At present, no comprehensive work exists for northeastern North America. Some species are easy to recognize, but many are very difficult to identify. Because the genus is complex and includes a large number of species, a comprehensive key to species is beyond the scope of this book. We have included a sampling of some of the more common and distinctive species that occur in this region. Identification based on the use of this key should be regarded as tentative at best.

Key to Species of *Agaricus*

1. Cap predominantly white, whitish or grayish white, sometimes bruising yellowish → 2.
1. Cap more richly colored, especially on the disc → 5.
    2. Growing in lawns, fields and other grassy areas → 3.
    2. Growing in woods → 4.
3. Cap 1–4" (2.5–10 cm) wide, not bruising yellowish; flesh not bruising yellowish; odor and taste not distinctive—*A. campestris* Linnaeus : Fries (see p. 53).
3. Cap 3–7" (7.5–18 cm) wide, bruising yellowish; flesh bruising yellow; odor fragrant, like anise or almonds; taste slightly of almonds or not distinctive—*A. arvensis* Schaeffer : Secretan (see p. 52).
    4. Stalk base abruptly bulbous; cap 2–6" (5–15.5 cm) wide, convex to nearly flat, smooth, silky, dry, white when young, becoming pale yellow in age, typically staining reddish in age when wet; flesh white, odor fragrant, like anise or almonds; stalk white, bruising yellowish; partial veil with cottony patches on the undersurface; ring superior, membranous, white; spores 6–8 × 4–5 μm; edible—*A. abruptibulbus* Peck.
    4. Stalk base slightly enlarged to bulbous but not abruptly bulbous; nearly identical to the previous choice but with smaller spores, 5–6.5 × 3.5–4.5 μm—*A. silvicola* (Vittadini) Peck (see p. 54).
5. Cap grayish brown to blackish brown on the disc, with tiny grayish brown to blackish brown scales on a whitish ground color; flesh white, typically staining yellow in the stalk base when cut and rubbed; odor like phenol or not distinctive; taste not distinctive or disagreeable; partial veil white, membranous, usually with cottony patches, with or

without brown droplets or brown stains on the unbroken undersurface; ring superior, single- or double-layered—*A. placomyces* Peck complex (see p. 53).

5.   Not as above → 6.

  6.   Underside of partial veil or ring with conspicuous cottony patches; ring not double-edged → 7.

  6.   Underside of partial veil or ring lacking conspicuous cottony patches; ring single- or double-edged → 8.

7.   Cap and stalk flesh white, staining reddish brown when cut and rubbed; odor and taste not distinctive; cap covered with tiny, flattened fibrils that aggregate into flattened, scale-like tufts, dull pinkish brown to dark brown, darkest on the flattened disc, 1½–5⅛" (4–13 cm) wide; stalk nearly equal down to a somewhat bulbous base, smooth, white, with a white superior ring; spores 5–6 × 3–4 μm; in woods; edible—*A. silvaticus* Schaeffer complex.

7.   Cap and stalk flesh white, not staining reddish brown when cut and rubbed; odor fragrant, like almonds; taste like almonds or not distinctive; cap tawny to ochre-tawny—*A. subrufescens* Peck (see p. 54).

  8.   Ring very large and distinctive, typically median, double-edged; upper edge flaring or free; lower edge free or projecting slightly at the top of a white basal sheath; flesh odor and taste not distinctive; cap covered with tiny, matted fibrils, sometimes cracked, white to dingy pale yellowish, typically coated with dirt, 1½–5½" (4–14 cm) wide; stalk very firm, smooth, white to whitish; spores 5–7 × 4–5 μm; fruiting above or below the ground, in groups or scattered on hard-packed soil, sometimes in gardens and on compost piles; edible—*A. bitorquis* (Quélet) Saccardo.

  8.   Ring small, fragile, single-edged, often disappearing; flesh odor and taste of almonds or anise-like; cap dry, coated with silky matted fibrils; fibrils yellowish brown to grayish brown on a whitish ground color, staining yellow to rusty orange when bruised; ¾–2¾" (2–7 cm) wide; stalk white, staining yellow to rusty orange when bruised; spores 4.5–6 × 3–4 μm; in lawns, fields, and other grassy areas; edible—*A. micromegethus* Peck.

  8.   Ring fragile, sheathed upward, sometimes torn; flesh white, quickly staining crimson rosy to vinaceous; odor and taste not distinctive; cap coated with a dense layer of radiating, dark brown fibrils—*A. fuscofibrillosus* (Møeller) Pilát (see p. 53).

  8.   Ring sparse, persistent or evanescent; flesh odor and taste not distinctive; cap white to grayish or grayish brown—*A. campestris* Linnaeus : Fries (see p. 53).

Agaricus arvensis Schaeffer : Secretan

COMMON NAME: Horse Mushroom.

CAP: 3–7" (7.5–18 cm) wide, somewhat oval when young, becoming convex to broadly convex in age; surface nearly smooth at first, becoming slightly scaly at maturity, white, bruising yellowish, staining yellow when KOH is applied to the surface; margin sometimes rimmed with veil remnants; flesh white, bruising yellow; odor like anise or almonds; taste slightly of almonds or not distinctive.

GILLS: free, close, white to grayish white when young, becoming grayish brown to dark brown.

STALK: 2–5" (5–12.5 cm) long, ⅜–1" (1–2.5 cm) thick, nearly equal overall or with an enlarged base; smooth, white, staining yellowish when bruised; partial veil membranous, white, typically leaving a persistent superior ring.

SPORE PRINT: dark brown.

MICROSCOPIC FEATURES: spores 7–9 × 4–6 μm, elliptic, smooth, pale brown.

FRUITING: scattered, in groups or sometimes in arcs and fairy rings on the ground in lawns, fields, and other grassy areas; June–October; fairly common.

EDIBILITY: edible and often rated choice.

Agaricus campestris Linnaeus : Fries

COMMON NAME: Meadow Mushroom, Pink Bottom.

CAP: 1–4" (2.5–10 cm) wide, convex to nearly flat in age; surface fibrillose or nearly smooth, sometimes fibrillose-scaly, dry, white to grayish or grayish brown; flesh white; odor and taste not distinctive.

GILLS: free, crowded, pink when young, becoming dark brown in age.

STALK: 1–2⅜" (2.5–6 cm) long, ⅜–⅝" (1–1.6 cm) thick, nearly equal or enlarged or tapered at the base, smooth, white; partial veil membranous, white, leaving a sparse, persistent or evanescent superior ring.

SPORE PRINT: dark brown.

MICROSCOPIC FEATURES: spores 6–9 × 4–6 μm, elliptic, smooth, pale brown.

FRUITING: scattered, in groups or sometimes in clusters, arcs, or fairy rings on lawns, pastures, golf courses, and other grassy areas; June–September; common.

EDIBILITY: edible and often rated choice.

Agaricus fuscofibrillosus (Moeller) Pilát

CAP: 1½–2¾" (4–7 cm) wide, convex, becoming nearly plane, often somewhat umbonate and flattened on the disc; surface dry, coated with a dense layer of radiating dark brown fibrils on a hazelnut ground color, often darker brown on the disc, lacking conspicuous scales; flesh thin, white, quickly staining crimson rosy to vinaceous when cut, especially at the stalk apex and center of the cap; odor and taste not distinctive.

GILLS: free or nearly so, narrow, crowded, pale pinkish vinaceous at first, soon vinaceous, and finally dark brown at maturity; edges paler than the faces.

STALK: 1½–3⅛" (4–8 cm) long, 5⁄16–½" (8–12 mm) thick, nearly equal or slightly enlarged at the base, not bulbous, hollow, whitish when fresh, becoming pale brown in age, fibrillose; partial veil membranous, white, leaving a fragile superior ring that is sheathed upward and sometimes torn.

SPORE PRINT: dark brown.

MICROSCOPIC FEATURES: spores 5–7 × 4–4.5 μm, ovoid, smooth, brown.

FRUITING: scattered or in groups on the ground in hardwoods and mixed woods; August–October; infrequent.

EDIBILITY: edible.

COMMENTS: *Agaricus haemorrhoidarius* (edible) and *A. silvaticus* (edible) have wider scaly caps, up to 4" (10 cm) and thicker stalks, up to ¾" (2 cm), usually with a bulbous base. The flesh of *A. haemorrhoidarius* immediately stains pinkish vinaceous when cut. The flesh of *A. silvaticus* quickly stains dark crimson when cut. There is much conflicting information in the literature about these species, primarily because of differences in interpretation and the lack of a critical study of eastern North American material.

Agaricus placomyces Peck complex

CAP: 1–3½" (2.5–9 cm) wide, convex, becoming broadly convex with a low broad umbo, sometimes flattened on the disc; surface covered with tiny, grayish to grayish brown scales on a white ground color; disc grayish brown; flesh white; odor and taste not distinctive.

GILLS: free, crowded, white at first, becoming pink and finally dark brown in age.

STALK: 1⅜–4" (3.5–10 cm) long, ¼–½" (5–13 mm) thick, nearly equal down to a bulbous or sometimes abruptly bulbous base, smooth, white; flesh staining yellow in the base when cut and rubbed; partial veil membranous, white, with cottony patches, with or without brown droplets or brown stains on the unbroken undersurface, leaving a superior, single-layered ring with cottony patches on the undersurface.

SPORE PRINT: dark brown.

MICROSCOPIC FEATURES: spores 4.5–6 × 3.5–4.5 μm, oval to elliptic, smooth, pale brown.

FRUITING: scattered or in groups on the ground in woods, especially hardwoods, in grassy areas with trees, or on sawdust piles; June–September; fairly common.

EDIBILITY: unknown.

COMMENTS: this is a complex of highly variable characters that makes identification of species extremely difficult. *Agaricus pocillator* (edibility unknown) differs by having a double-layered ring and lacking brown liquid or brown stains on the unbroken undersurface of the partial veil. However, we have found collections with a double-layered ring with brown droplets and brown stains on the unbroken undersurface of the partial veil. *Agaricus meleagris* (poisonous) differs by having flesh with a strong odor of phenol and lacking the brown droplets and brown stains. Clearly, this group needs additional study.

Agaricus silvicola (Vittadini) Peck

COMMON NAME: Woodland Agaricus.

CAP: 2–6" (5–15.5 cm) wide, convex, becoming broadly convex to nearly flat in age; surface smooth, silky, dry, white, becoming dingy white to pale yellowish in age; staining yellow when KOH is applied to the surface; margin sometimes with veil remnants; flesh white; odor fragrant like anise or almonds; taste somewhat of almonds or not distinctive.

GILLS: free, crowded, whitish at first, becoming pinkish and finally chocolate-brown.

STALK: 3–6 " (7.5–15.5 cm) long, ⅜–¾" (1–2 cm) thick, nearly equal down to a slightly enlarged or bulbous base, dry, smooth, white; partial veil membranous, white, with cottony patches on the undersurface, leaving a superior, white ring.

SPORE PRINT: chocolate-brown.

MICROSCOPIC FEATURES: spores 5–6.5 × 3.5–4.5 μm, elliptic, smooth, pale brown.

FRUITING: solitary, scattered, or in groups on the ground in woods and in grassy areas with trees; July–October; fairly common.

EDIBILITY: edible.

Agaricus subrufescens Peck

CAP: 2¾–7" (7–18 cm) wide, convex, often with a flattened disc when young, becoming nearly plane in age; surface dry, covered with smooth, matted fibrils when young, soon breaking up into tiny, fibrillose scales; scales tawny to ochre-tawny on a whitish to yellowish ground color, usually darker on the disc; flesh white; odor fragrant, like almonds; taste like almonds or not distinctive.

GILLS: free, close, whitish when young, becoming pinkish brown and finally dark vinaceous brown.

STALK: 2–6" (5–15.5 cm) long, ⅜–¾" (1–2 cm) thick, nearly equal down to an enlarged base which is often buried in the substrate, silky and white to pinkish above the ring, covered with silky, white fibrils and scattered, coarse, cottony patches and zones from the base up; partial veil membranous, white, with cottony patches on the undersurface, leaving a persistent, superior, skirt-like ring.

SPORE PRINT: dark brown.

MICROSCOPIC FEATURES: spores 6–8 × 4–5 μm, elliptic, smooth, pale brown.

FRUITING: scattered or in groups in rich humus, often under conifers or on compost piles.

EDIBILITY: edible.

Genus *Agrocybe*

Agrocybe is a fairly small genus of small to large mushrooms that grow on various substrates, including lawns and other grassy areas; humus in woods; gardens; wood chips; and dung. Their spore prints are rusty brown to dark brown and their spores are smooth, often with an apical pore. They have smooth caps that often become cracked in age. A few species are edible, but the edibility of the majority of species is unknown.

Key to Species of *Agrocybe*

1. Partial veil absent, lacking veil remnants on the cap margin or a ring on the stalk; cap ¾–3⅛" (2–8 cm) wide at maturity, blackish brown to grayish brown when fresh; growing on decaying wood—*A. firma* (Peck) Singer (see p. 57).

1. Partial veil present on young specimens, leaving veil remnants on the cap margin (sometimes sparse) or a ring on the stalk → 2.

    2. Cap dry, white at first, becoming buff to pale tan at least on the disc, soft, smooth and suede-like, becoming cracked to deeply fissured in age; gills close, whitish when young, becoming grayish, then dark purple-brown; edges typically whitish in age; stalk ¼–½" (5–13 mm) thick, white, nearly smooth, usually with white rhizomorphs at the base; partial veil thin, submembranous, leaving remnants on the cap margin or an evanescent superior ring; spores 10–14 × 6–8 μm, truncate with an apical pore; on lawns, fields, and other grassy areas; edible—*A. dura* (Fries) Singer.

    2. Cap viscid when fresh, dark reddish brown overall when young, fading to ochre to brownish orange, often with a brown margin; stalk dull brown at maturity; in lawns or gardens and on the ground in woods—*A. erebia* (Fries) Kühner (see p. 56).

    2. Cap viscid to slightly glutinous and suede-like when fresh, becoming dry and shiny, often cracked on the disc in age; ochre to rusty ochre at first, becoming ochre-yellow to yellow or buff in age, ⅜–2" (1–5 cm) wide, hemispheric when young, becoming broadly convex; gills pale brownish at first, becoming orange-brown to pale rusty brown at maturity; stalk ⅛–½" (2–12 mm) thick, nearly equal, pale yellow to ochre, fibrillose, typically slightly bulbous at the base, often with white rhizomorphs; flesh odor farinaceous; taste farinaceous or somewhat bitter to disagreeable; spores 10–14 × 7–10 μm; on lawns and fields, on wood mulch, or sawdust mixed with manure; edible but of poor quality—*A. semiorbicularis* (Bulliard) Fayod = *A. pediades* (Fries) Fayod.

    2. Cap moist or dry but not viscid, whitish, tinged yellow, tan or rusty brown to brown on the disc and overall in age; ring small, thin, superior to median, sometimes falling away at maturity; on lawns, fields, and other grassy areas, sometimes

on the ground in woods or on decaying wood—*A. praecox* (Fries) Fayod complex (see p. 57).

2. Cap moist or dry but not viscid; color variable, ochre-yellow to yellowish cinnamon or yellow-brown; ring large, persistent, pendant, superior; on decaying hardwood logs, stumps, and wood chips, especially common in urban areas where hardwood chips are used for landscaping—*A. acericola* (Peck) Singer (see p. 56).

Agrocybe acericola (Peck) Singer

COMMON NAME: Maple Agrocybe.

CAP: 1–4" (2.5–10 cm) wide, convex, becoming nearly flat in age, sometimes shallowly depressed on the disc, moist and smooth or wrinkled when fresh, becoming dry and cracked to deeply fissured in age; surface color variable, ochre-yellow to yellowish cinnamon or yellow-brown when young, fading to yellowish tan to tan in age; flesh white; odor farinaceous; taste farinaceous to bitter.

GILLS: attached, close, whitish to buff when young, becoming grayish brown to smoky brown at maturity; edges entire, not finely scalloped at maturity.

STALK: 1½–5" (4–12.5 cm) long, ¼–¾" (6–20 mm) thick, nearly equal or enlarged downward, smooth, white at first, becoming dark grayish to grayish brown from the base up in age, often with white rhizomorphs at the base; partial veil white, membranous, leaving a large, pendant, persistent, superior ring that is radially striate on the upper surface and stained cinnamon by the spores.

SPORE PRINT: cinnamon-brown.

MICROSCOPIC FEATURES: spores 8–11 × 5–6.5 μm, elliptic, with an apical pore, smooth, pale brown.

FRUITING: scattered, in groups or clusters on decaying hardwood logs, stumps, and wood chips; April–September; common, especially in urban areas where hardwood chips are used for landscaping.

EDIBILITY: inedible, typically bitter and unpleasant.

Agrocybe erebia (Fries) Kühner

CAP: ¾–2⅜" (2–6 cm) wide, convex, becoming nearly plane in age, often with a low, broad umbo; surface viscid when fresh, becoming dry, smooth to wrinkled, dark reddish brown overall when young, fading toward the margin in age, often remaining darker brown on the disc and margin, and ochre to brownish orange in between; flesh pale brown; odor and taste not distinctive.

GILLS: attached to slightly decurrent, close to subdistant, often crossveined, pale brown at first, becoming dull brown at maturity.

STALK: 1½–3⅛" (4–8 cm) long, ⅛–⅜" (4–10 mm) thick, enlarged downward or nearly equal, whitish to pale brown when young, becoming darker dull brown in age, whitish to pale brown and pruinose above the ring, fibrillose below; partial veil white, membranous, leaving a superior to median ring with striations on the upper surface.

SPORE PRINT: dull brown.

MICROSCOPIC FEATURES: spores 11–15 × 5–7 μm, elliptic, lacking an apical pore, smooth, pale brown.

FRUITING: scattered or in groups on lawns and fields and on the ground in woods; June–October; fairly common.

EDIBILITY: unknown.

Agrocybe firma (Peck) Singer

CAP: ¾–3⅛" (2–8 cm) wide, convex, becoming nearly flat in age, sometimes with a low, broad umbo; surface smooth, blackish brown when young, becoming grayish brown, fading to tawny to ochre-yellow with a darker disc in age; flesh pale yellow; odor and taste not distinctive or slightly farinaceous.

GILLS: attached, close, brownish, becoming yellowish brown to grayish brown.

STALK: 1½–3⅛" (4–8 cm) long, ¼–⅜" (6–10 mm) thick, nearly equal, whitish at first, becoming yellowish brown, white-pruinose overall or at least near the apex, with white rhizomorphs at the base; partial veil and ring absent.

SPORE PRINT: cinnamon-brown.

MICROSCOPIC FEATURES: spores 6–9 × 4–5 μm, elliptic, smooth, pale brown.

FRUITING: scattered or in groups on decaying hardwood; June–September; occasional.

EDIBILITY: unknown.

Agrocybe praecox (Persoon : Fries) Fayod complex

COMMON NAME: Spring Agrocybe.

CAP: 1⅛–2¾" (2–7 cm) wide, convex, becoming nearly plane; surface smooth, moist, whitish, tinged yellow, tan, or rusty brown to brown on the disc and overall in age; flesh white; odor farinaceous; taste not distinctive or somewhat farinaceous.

GILLS: notched or attached with short decurrent lines, close, whitish at first, becoming tinged grayish and finally brownish to rusty brown; edges whitish and finely scalloped at maturity.

STALK: 1⅛–4" (3–10 cm) long, ⅛–½" (3–12 mm) thick, equal or slightly enlarged downward, smooth to longitudinally striate, white, tinged yellowish to brownish in age; partial veil whitish, thin, frail, leaving a small, thin superior ring that often falls away by maturity; ring usually stained cinnamon by the spores.

SPORE PRINT: rusty brown.

MICROSCOPIC FEATURES: spores 8–13 × 5–7 μm, elliptic, with or without an apical pore, smooth, pale brown.

FRUITING: scattered or in groups on lawns, fields, and other grassy areas, sometimes on the ground in woods or on decaying wood; April–June; fairly common.

EDIBILITY: edible.

COMMENTS: this is a complex of several forms and possibly several species. Specimens collected in lawns (see photo) typically have smaller, paler caps and more slender stalks than those collected on the ground in woods or on decaying wood. This complex needs further study.

Genus *Amanita*

The Amanitas are among our handsomest agarics but are perhaps more famous for their deadly species, such as the Death Cap *(A. phalloides)* and the Destroying Angels *(A. virosa* complex). They are fairly easy to identify to genus. The white spore print, free or nearly free gills, and usually conspicuous remnants of a universal veil in the form of warts or volva make them easy to recognize. They are beautiful, sometimes poisonous, and important as mycorrhizal partners with numerous tree species.

Seasoned collectors often discover specimens that cannot be identified, and there is still far more taxonomic work to be done in this genus. Rodham Tulloss, one of the leading authorities on Amanitas of northeastern North America, has records of more than sixty species or varieties (including about twenty as yet unnamed) collected from New York state alone. He estimates that there are more than one hundred Amanitas in the Northeast. Several of the European species names used here may be misapplied in North America. Members of Section Lepidella are particularly difficult to identify accurately. Recognizing these limitations and hoping to inspire more collectors to help study this important genus, we have strived to present sufficient information in this key to enable the reader to identify specimens at least to section.

Key to Sections of Amanita

1. Membranous saccate volva around base of stalk; cap margin striate; lamellulae truncate; spores inamyloid → 4 *(Section Vaginatae).*
1. Membranous saccate volva around base of stalk; cap margin usually nonstriate; lamellulae attenuate in most species, spores amyloid; warts or patches (if present) also membranous → 6 *(Section Phalloidae).*
1. Membranous saccate volva around base of stalk; spores amyloid; warts usually more powdery/mealy than membranous; cap margin usually appendiculate; ring usually absent → 9 *(Section Amidella).*
1. No membranous saccate volva around base of stalk → 2.
    2. Cap margin striate; lamellulae usually truncate; spores inamyloid; ring and basal bulb both absent → 4 *(Section Vaginatae).*
    2. Cap margin striate; lamellulae truncate; spores inamyloid; ring or partial veil usually but not always present; basal bulb present → 10 *(Section Amanita).*
    2. Cap margin nonstriate; lamellulae usually attenuate; spores amyloid; basal bulb present → 3.
3. Marginate or submarginate bulb present → 6 *(Section Phalloidae).*
3. Warts or powdery patches usually visible on cap and/or on stalk base; cap usually white, cream, or gray; basal bulb usually large → 15 *(Section Lepidella).*
3. Small warts or powdery patches usually visible on cap or on stalk base; cap usually distinctly colored; basal bulb usually small → 14 *(Section Validae).*

Section Vaginatae

    4. Ring pale yellow to pale orange; volva white, saccate; cap reddish orange to yellowish orange; stalk and gills pale yellow—*A. caesarea* complex (see p. 62).
    4. Ring white; cap grayish brown; edibility unknown—*A. spreta* (Peck) Saccardo.
    4. Ring absent; volva not saccate; cap tannish brown, darker at the center; warts powdery, grayish brown; edibility unknown—*A. ceciliae* (Berkeley and Broome) Bas.
    4. Ring absent; volva saccate → 5.
5. Cap pale olive-tan to brownish olive, with a usually darker umbo; volva grayish, especially inside—*A. sinicoflava* Tulloss (see p. 67).
5. Cap grayish, with a usually darker umbo; volva and stalk white; edible—*A. vaginata* (Bulliard : Fries) Vittadini.
5. Cap tannish orange to deep orange; stalk pallid but usually covered with orange fibers; edibility unknown—*A. crocea* (Quélet) Singer.
5. Cap reddish brown to orangish brown; stalk white; edible—*A. fulva* (Schaeffer) : Persoon.

Section Phalloidae

6.   Cap white to pallid overall; volva saccate → 7.
6.   Cap white to pallid overall, sometimes slightly yellowish at the center; basal bulb marginate, with several vertical clefts; odor usually of raw potatoes—*A. brunnescens* var. *pallida* Krieger (see comments under *A. brunnescens* var. *brunnescens,* p. 62).
6.   Cap typically yellowish green, darker at the center and paler at the margin; ring white—*A. phalloides* Link : Fries (see p. 66).
6.   Cap distinctly colored; volva not saccate → 8.
7.   Entire mushroom white to off-white—*A. virosa* complex (see p. 67 or go on to 7a).
7a. Cap not staining yellow with KOH—*A. verna* (Bulliard : Fries) LaMarck (see comments under *A. virosa* LaMarck, p. 67).
7a. Cap staining yellow with KOH; basidia mostly four-spored—*A. virosa* LaMarck (see p. 67).
7a. Cap staining yellow with KOH; basidia mostly two-spored—*A. bisporigera* LaMarck (see comments under *A. virosa* LaMarck, p. 67).
8.   Cap brownish gray; warts pale gray; ring grayish; odor of raw potatoes—*A. porphyria* (Albertini and Schweinitz : Fries) Secretan (see p. 66).
8.   Cap grayish brown; warts white to pallid; ring white or off-white; basal bulb marginate, with one to several vertical clefts—*A. brunnescens* var. *brunnescens* Atkinson (see p. 62).
8.   Cap greenish yellow; warts grayish lavender; ring greenish yellow; bulb marginate; odor of raw potatoes—*A. citrina* f. *lavendula* Coker (see p. 62).
8.   Cap greenish yellow; warts pallid to pale yellow; ring yellowish; bulb marginate; odor of raw potatoes—*A. citrina* f. *citrina* (Schaeffer) : Roques (see comments under *A. citrina* f. *lavendula,* p. 63).

Section Amidella

9.   Cap white; warts pinkish; ring absent; volva large, saccate, pinkish; edibility unknown —*A. peckiana* Kauffman.
9.   Cap whitish, sometimes brownish at the center; warts whitish; volva white; edibility unknown—*A. volvata* (Peck) Lloyd.

Section Amanita

10. Partial veil absent (not evident even in very young specimens); cap deep red or scarlet to reddish orange at center, paler toward margin; warts mostly at center of cap, powdery, dingy yellowish; gills pale yellow, edges often very powdery, lamellulae truncate; stalk usually pale yellow and powdery; flesh white, but orange immediately beneath cap cuticle; basal bulb rounded, sometimes rooting slightly; edibility unknown—*A. parcivolvata* (Peck) Gilbert.
10. Not as above; base of stalk with an upper margin or warts or rings (remnants of universal veil); cap brown, orange, red, or yellow → 12.
10. Not as above; base of stalk as in the previous choice, but cap not brown, orange, red, or yellow → 13.
10. Not as above; cap whitish to pale gray, often tinged yellowish; warts randomly

distributed, thin, whitish; bulb round to oval; ring usually missing at maturity; edibility unknown—*A. crenulata* Peck.

10.   Not as in any of the above choices → 11.

11.   Cap less than 3" (7.5 cm) wide, whitish gray, usually coated with fine, powdery/mealy, brownish gray to dark gray warts; ring absent; bulb fairly small, nearly round, whitish; edibility unknown—*A. farinosa* Schweinitz.

11.   Cap up to 4¾" (12 cm) wide, yellowish orange to pinkish orange, with slightly paler warts; stalk pale yellow; partial veil yellow; ring yellow (but often missing or present only as remnants adhering to cap margin); stalk pale yellow; bulb slight; edibility unknown—*A. wellsii* (Murrill) Saccardo.

12.   Cap large, up to 10" (25 cm) wide, pale brownish yellowish overall, more brownish at the center; warts white to cream, often concentrically arranged; ring white, thin, membranous; rim or collar of white tissue above bulb; bulb usually oval, sometimes slightly rooting; edibility unknown—*A. pantherina* var. *velatipes* (Atkinson) Jenkins (see p. 65).

12.   Cap medium, less than 4" (10 cm) wide, bright orange; loose, powdery/mealy warts on cap and around lower stalk usually yellowish; ring yellowish; gills white, sometimes edged with yellow; stalk white to yellowish; bulb small; edibility unknown—*A. frostiana* (Peck) Saccardo.

12.   Cap large, up to 7" (18 cm) wide, yellow to orangish yellow, paler at margin; warts tan to yellowish, usually somewhat concentrically arranged; lower stalk with several ascending rings of universal veil tissue—*A. muscaria* var. *formosa* (Persoon : Fries) Bertillon (see p. 64).

12.   Cap medium, up to 4" (10 cm) wide, pale yellow to pale yellowish brown; warts and stalk white to pale cream; ring white but often missing; edibility unknown—*A. gemmata* (Fries) Bertillon complex.

13.   Cap white to pale yellow; ring absent; warts white or pallid; stalk whitish; bulb moderate, usually with a ring of universal veil tissue at the top; edibility unknown—*A. albocreata* (Atkinson) Gilbert.

13.   Cap whitish; warts tannish; stalk white, slowly bruising yellowish; lower stalk usually decorated with several ascending rings of universal veil tissue—*A. muscaria* var. *alba* Peck (see comments under *A. muscaria* var. *formosa*, p. 65).

13.   Cap white overall, more tan at the center; warts white; ring white, sometimes tinted yellowish; bulb rounded, usually with a thin ring of universal veil tissue at the top; edibility unknown—*A. pantherina* var. *multisquamosa* (Peck) Jenkins.

Section Validae

14.   Flesh slowly staining red where bruised (base of stalk almost always stained reddish); cap bronze to reddish brown; warts tannish; ring white to pale tan; bulb slight—*A. rubescens* var. *rubescens* (Persoon : Fries) S. F. Gray (see p. 66).

14.   Flesh slowly staining reddish where bruised; cap golden yellow to brownish yellow with yellowish warts and patches; ring yellowish; bulb slight—*A. flavorubescens* Atkinson (see p. 63).

14.   Flesh not staining reddish where bruised; cap orangish yellow; warts orangish yellow to yellowish orange; gills often yellow-edged; ring yellowish orange; stalk white—*A. flavoconia* Atkinson (see p. 63).

14. Flesh not staining red where bruised; cap brownish overall, reddish brown at the center, paler (even whitish) at the edge; warts whitish to grayish, sometimes almost pyramidal, often powdery near the cap edge; ring white to cream; stalk whitish to pale cream, often enlarged downward, smooth to fibrous or scaly; bulb slight; edibility unknown—*A. excelsa* (Fries) Bertillon = *A. spissa* (Fries) Kummer.

Section Lepidella

15. Cap, stalk, and warts white to cream or tannish, though pinkish or reddish stains may be present → 16.

15. Cap and stalk pale pinkish orange; warts tiny, spiny to conic, colored like the cap; odor sweetish but also reported as resembling old ham or nauseating; bulb very large, up to 5" (12.5 cm) or wider, often vertically split; cap large, reportedly up to 12" (30 cm) wide; edibility unknown—*A. daucipes* (Montagne) Lloyd.

15. Cap grayish olive to brownish gray but sometimes nearly white; warts pale grayish olive to pale brownish gray; gills grayish olive to brownish gray; ring thick but delicate, usually disintegrating; stem grayish; bulb sometimes rooting; edibility unknown— *A. pelioma* Bas.

15. Not as in any of the above choices → 17.

   16. Cap and stalk whitish to tannish; gills white to pale cream, lamellulae relatively few, truncate; gills, stalk and flesh all quickly staining pinkish when bruised; odor often of anise; ring white to pale yellowish; bulb moderate in size, marginate; edibility unknown—*A. mutabilis* Beardslee.

   16. Cap white; warts small, white, conic/pointed, slender; gills white; ring white, membranous; bulb rounded, abrupt (flattened at the top); edibility unknown— *A. abrupta* Peck.

   16. Mushroom white overall; cap 3–8" (7.5–21 cm) wide; warts whitish, conic at cap center, more of a powdery layer near the cap edge; partial veil very thick but very delicate, typically leaving no ring but often some bits adhering to the cap margin; odor bleach-like; bulb almost round to oval or elliptic; edibility unknown— *A. polypyramis* (Berkeley and Curtis) Saccardo.

   16. Cap white, 3–6" (7.5–15.5 cm)wide; warts pyramidal, white to tannish, rather large; ring white, membranous, often two-layered; bulb usually somewhat rooting; edibility unknown—*A. cokeri* (Gilbert and Kühner) Gilbert.

   16. Not as in any of the above choices → 17.

17. Cap dingy white to pale gray but typically appearing gray to gray-brown due to a conspicuous layer of dark gray to brownish gray, pyramidal/conic, slender, pointed warts; stalk gray to brownish gray near base; bulb somewhat elongated, usually deeply rooting but always at least slightly rooting—*A. onusta* (Howe) Saccardo (see p. 65).

17. Mushroom white overall, but cap gray to grayish brown at the center, finely powdery, 1–3" (2.5–8 cm) wide; gills barely to distinctly attached, whitish, edges powdery; lamellulae abundant, attenuate to almost truncate; partial veil powdery-fibrous, quickly disintegrating; stalk radicating somewhat, typically flattened and/or doglegged, sometimes with reddish stains or spots—*A. longipes* Bas (see p. 64).

17. Cap white to creamy white; warts large (up to 6 mm wide), coarse, brownish to tannish cream, flattened, rather scale-like at cap margin; gills pale cream to yellowish cream; ring thick, delicate; bulb large, rooting; edibility unknown—*A. ravenelii* (Berkeley and Curtis) Saccardo.

Amanita brunnescens var. **brunnescens** Atkinson

COMMON NAME: Cleft-foot Amanita.

CAP: 1½–6" (3.5–15 cm) wide, convex to flat with a broad umbo; surface grayish brown with whitish radial fibers, often with darker brownish stains, paler at the margin, typically smooth, dry to sticky, margin nonstriate; with a small number of randomly arranged, floccose to felt-like, pallid warts; flesh pallid, slowly staining brown; odor occasionally of raw potatoes.

GILLS: free, crowded, white, with generally attenuate lamellulae; partial veil pallid to dingy white.

STALK: 2¼–6" (6.5–15 cm) long, ⅜–¾" (8–21 mm) thick, tapering slightly toward the top, white with brownish stains, smooth to slightly flocculent or fibrous, typically stuffed; ring superior, pendant, readily adhering to stalk, white to dingy white, membranous; bulb large, subglobose, marginate, typically with several (usually four) vertical clefts.

SPORE PRINT: white.

MICROSCOPIC FEATURES: spores 7.8–9.4 × 7.5–8.6 μm, globose to subglobose, smooth, thin-walled, hyaline, amyloid.

FRUITING: scattered or grouped, terrestrial in mixed woods; July–October; frequent to common.

EDIBILITY: allegedly edible but not recommended because of possible confusion with poisonous species and because relatively few people have eaten it.

COMMENTS: *A. brunnescens* var. *pallida* (not recommended) is identical except that the cap is white to pale yellow. In the authors' experience, it also has a very consistent odor of raw potatoes.

Amanita caesarea (Scopoli : Fries) Greville complex

COMMON NAME: Caesar's Mushroom.

CAP: 2½–9" (6–22 cm) wide, convex to flat, usually with a distinct umbo; surface reddish orange at the center to yellowish orange at the margin, usually rather smooth, dry to sticky, with a distinctly striate margin; sometimes with one to three white patches of universal veil tissue; flesh white; odor not distinctive.

GILLS: free, crowded, fairly broad, pale yellow, with truncate lamellulae; partial veil thin, membranous, pale yellow to pale orange.

STALK: 3½–9" (9–23 cm) long, ⅜–1¼" (1–3 cm) thick, tapering somewhat toward the top, pale yellow, sometimes streaked with orange, smooth to scaly/roughened, stuffed to hollow; ring superior, pale yellow to pale orange, thin, membranous; volva large, white, thick, membranous, lobed; bulb absent.

SPORE PRINT: white.

MICROSCOPIC FEATURES: spores 7.8–9.4 × 5.5–6.7 μm, elliptic, smooth, thin-walled, hyaline, inamyloid.

FRUITING: solitary to loosely grouped, terrestrial in mixed woods; July–October; infrequent to locally frequent.

EDIBILITY: edible.

COMMENTS: North American collections of *A. caesarea* are often called *A. umbonata*, *A. jacksonii*, or *A. hemibapha*; the group needs more study.

Amanita citrina f. **lavendula** Coker

COMMON NAME: Citron Amanita.

CAP: 2–4½" (5–12 cm) wide, convex to nearly flat; surface pale greenish yellow, often with tannish stains, smooth, dry to sticky, margin smooth or slightly striate in age; usually coated with a moderate number of randomly arranged, floccose to fibrous, grayish lavender warts; flesh moderately thick, pallid, sometimes staining tan or gray to lavender; odor strongly of raw potatoes.

GILLS: free or very finely attached, crowded, moderately broad, white to whitish, with truncate lamellulae; partial veil greenish yellow, sometimes with lavender tints.

STALK: 2¼–5" (6–13 cm) long, ¼–¾" (6–20 mm) thick, tapering slightly toward the top, white with tannish stains, typically smooth and silky, stuffed to hollow; ring superior, pendant, readily adhering to stalk, pale greenish yellow, membranous, thin, upper surface finely striate; with a conspicuous marginate basal bulb; lower stalk often decorated with floccose lavender warts.

SPORE PRINT: white.

MICROSCOPIC FEATURES: spores 5.5–7 × 5.5–7 μm, globose to subglobose, smooth, thin-walled, hyaline, amyloid.

FRUITING: solitary to scattered or grouped, terrestrial in mixed woods; August–October; occasional.

EDIBILITY: extremely suspect.

COMMENTS: this variety is often not distinguished from *A. citrina* f. *citrina* (edibility suspect), which is identical in most respects. The latter lacks the lavender tints (but those are sometimes difficult to discern without bright light); it is a rather common fall mushroom in northeastern woods. Both varieties are often mistakenly identified as *A. phalloides* by novices who misinterpret the edge of the marginate basal bulb as the edge of a saccate volva.

Amanita flavoconia Atkinson

COMMON NAME: Yellow Patches.

CAP: 1¼–4" (3–9 cm) wide, convex to flat; surface yellow tinted orange, especially near the center, smooth, dry to sticky, margin nonstriate unless eroded; with numerous, randomly arranged, bright yellow to orangish, floccose warts; flesh white; odor not distinctive.

GILLS: free or nearly so, moderately crowded, moderately broad, white, with rounded to attenuate lamellulae; partial veil yellow to orangish yellow.

STALK: 2–4½" (5–12 cm) long, ¼–½" (7–14 mm) thick, tapering slightly toward the top, white, coated with yellow flocculence, especially near the base, stuffed to fairly solid; ring superior, pendant, white to yellow, membranous, thin; bulb small, ovoid, often with bright yellow warts at the base.

SPORE PRINT: white.

MICROSCOPIC FEATURES: spores 7.8–8.6 × 5.4–8.6 μm, elliptic, smooth, thin-walled, hyaline, amyloid.

FRUITING: scattered to grouped, terrestrial in hardwoods or mixed woods; July–October; rather common.

EDIBILITY: toxicity suspected.

COMMENTS: easily misidentified as the less frequent *A. frostiana*, but that species has a striate margin, truncate lamellulae, and inamyloid spores; it also tends to be daintier. Both species are among the loveliest of our Amanitas.

Amanita flavorubescens Atkinson

COMMON NAME: Yellow Blusher.

CAP: 1½–4⅜" (4–11 cm) wide, convex, becoming nearly plane; surface smooth, somewhat viscid when moist, golden yellow to brownish yellow, typically with yellowish warts and patches; margin lacking striations; flesh whitish, slowly staining reddish when bruised; odor and taste not distinctive.

GILLS: free or nearly so, close to crowded, creamy white; partial veil whitish to yellowish.

STALK: 2–5½" (5–14 cm) long, ⅜–¾" (1–2 cm) thick, enlarged slightly downward to a club-shaped bulb and often tapered below the bulb, whitish to yellowish, usually with reddish stains near the base, nearly smooth; flesh yellowish, slowly staining reddish when exposed; ring superior, thin, membranous, pendant, yellowish; bulb slight, ovoid, sometimes coated with yellowish patches.

SPORE PRINT: white.

MICROSCOPIC FEATURES: spores 8–10 × 5.5–7 μm, elliptic, smooth, hyaline, amyloid.

FRUITING: solitary, scattered, or in groups on the ground in mixed woods, especially with oak; June–October; occasional.

EDIBILITY: unknown, toxicity suspected.

COMMENTS: *A. rubescens* (see p. 66) is similar but lacks the brighter yellow coloration.

Amanita longipes Bas

CAP: 1–3" (2.5–8 cm) wide, hemispheric when young, becoming boadly convex; surface dry, coated with a dense layer of fine powder, white, pale grayish buff to pale grayish brown on the disc; margin nonstriate, incurved when young, typically rimmed with flaps of torn partial veil; flesh white; odor and taste usually not distinctive.

GILLS: barely to distinctly attached, close, whitish; edges powdery; lamellulae abundant; partial veil fibrous-cottony, white, not forming a ring.

STALK: 2–5½" (5–14 cm) long, ¼–¾" (6–20 mm) thick, enlarged downward and club-shaped, then tapered abruptly and radicating somewhat, typically flattened and/or doglegged, dry, powdery to scurfy, white, sometimes with reddish stains or spots.

SPORE PRINT: white.

MICROSCOPIC FEATURES: spores 8–18 × 4–7 μm, elliptic, smooth, hyaline, amyloid.

FRUITING: solitary, scattered, or in groups on sandy soil in mixed woods, especially oak-pine; July–October; fairly common.

EDIBILITY: unknown.

COMMENTS: the stalk is often deeply buried and coated with sand.

Amanita muscaria var. formosa (Persoon : Fries) Bertillon

COMMON NAME: Yellow-orange Fly Agaric.

CAP: 1¾–7" (4.5–18 cm) wide, convex to flat or slightly sunken at the center; surface pale yellow to orangish yellow, typically deeper orange at the center and fading toward the margin, smooth, dry to sticky, margin distinctly striate; coated with numerous, concentrically to irregularly arranged, floccose, pallid or tannish to yellowish warts; flesh thick, white; odor not distinctive.

GILLS: free or very finely attached, crowded, rather broad, white to cream, sometimes slightly yellowish on the edges, with abundant truncate lamellulae; partial veil white to yellowish, thin.

STALK: 1¾–6½" (4–15 cm) long, ¼–1¼" (7–30 mm) thick, tapering slightly toward the top, white to pale yellowish, staining yellowish when handled, decorated with fine, delicate fibers or small scales, sometimes rather roughened, stuffed; ring superior, pendant, white to pale yellow, membranous, thin but with a thickened edge, often evanescent; lower portion of stalk decorated with several pallid to pale yellowish ascending rings of universal veil tissue; bulb subglobose, sometimes with a very slight radicating point.

SPORE PRINT: white.

MICROSCOPIC FEATURES: spores 9–13 × 6–8 μm, typically elliptic, smooth, thin-walled, hyaline, inamyloid.

FRUITING: scattered to grouped, usually gregarious, terrestrial in woods or under trees, in grassy areas; June–October; very common.

EDIBILITY: poisonous (contains ibotenic acid and muscimol).

COMMENTS: an easily identified variety of an easily identified species; this is one of our most common and conspicuous mushrooms, often fruiting in huge quantities under aspen or Norway spruce. Experienced mushroom hunters note that its fruitings under the latter often coincide with equally prodigious fruitings of *Boletus edulis*. Another variety, *A. muscaria* var. *alba* (poisonous), has a whitish cap with tannish warts; the flesh of its stalk sometimes stains yellow where bruised.

Amanita onusta (Howe) Saccardo

CAP: 1–4" (2.5–10 cm) wide, convex to flat to slightly sunken at the center, sometimes with a low umbo; surface dry, dingy white to pale gray, almost completely covered near the center by conic to pyramidal gray warts and at the margin by fine, gray flocculence; margin conspicuously appendiculate, nonstriate unless eroded; flesh whitish to pale gray; odor not distinctive or bleach-like.

GILLS: free or very finely attached, crowded, moderately broad, white to creamy yellowish, with attenuate lamellulae; partial veil white to pale gray, extremely delicate, floccose, usually leaving appendiculate remnants on cap margin.

STALK: 1½–6" (3.5–15 cm) long, ¼–¾" (6–15 mm) thick, tapering slightly toward the top, whitish near the top, grayer below, the lower portion coated with gray flocculence or delicate fibers, nearer ground level covered with flocculent gray warts or scales; ring delicate, floccose, white to gray, usually disintegrating; bulb usually somewhat spindle-shaped, with a slightly to deeply radicating root.

SPORE PRINT: white.

MICROSCOPIC FEATURES: spores 8–12 × 5–8.5 μm, typically broadly elliptic, smooth, thin-walled, hyaline, amyloid.

FRUITING: solitary or scattered to grouped, not abundant, terrestrial in mixed woods; July–August; infrequent to rare.

EDIBILITY: unknown.

COMMENTS: a very distinctive species more frequently collected in the southern part of the region. The stalk can be very deeply radicating in loose, rich soil.

Amanita pantherina var. **velatipes** (Atkinson) Jenkins

CAP: 3–7½" (7–18 cm) wide (but see comment below), convex to flat; surface dry, pale brown or pale reddish brown at center, fading to creamy yellow at margin, smooth, dry to sticky, margin decidedly striate; with numerous floccose, pallid warts that are often concentrically arranged; flesh thin, pallid; odor not distinctive.

GILLS: free, crowded, white, with numerous truncate lamellulae; partial veil pallid to dingy white.

STALK: 3–8" (8–20 cm) long, ⅜–⅞" (8–20 mm) thick, tapering slightly toward the top, white, somewhat smooth but fibrous or roughened near the base; ring superior to median to inferior, flaring to pendant to collapsed, white, thin, membranous, persistent; bulb subglobose to ovoid, with a ring of tissue that gives the stalk a rolled-up-sock appearance.

SPORE PRINT: white.

MICROSCOPIC FEATURES: spores 8–13 × 6–8 μm, broadly elliptic to elongate, smooth, thin-walled, hyaline, inamyloid.

FRUITING: usually in groups, terrestrial in mixed woods; July–September; occasional to frequent.

EDIBILITY: toxic (contains ibotenic acid and muscimol).

COMMENTS: one of the Northeast's largest Amanitas; we usually find this species near oaks and have collected specimens with caps nearly 10" (25 cm) in diameter.

Amanita phalloides Link : Fries

COMMON NAME: Death Cap.

CAP: 2¼–6" (6–16 cm) wide, convex to nearly flat; surface yellowish green to olivaceous, often browner near the center and with a paler margin, with radial fibers, smooth, dry to sticky, margin nonstriate; with or more often without one or two white membranous patches; flesh thick at center, white; odor not distinctive to unpleasant.

GILLS: almost free, crowded, white; partial veil white.

STALK: 3–6" (7–15 cm) long, ⅜–⅞" (9–20 mm) thick, tapering toward the top, white, smooth to minutely scaly; ring superior, pendant, white, thin, membranous, fairly persistent; bulb large, globose to subglobose; volva white, saccate, fairly thin, membranous, torn into several lobes.

SPORE PRINT: white.

MICROSCOPIC FEATURES: spores 8.5–10 × 7–8 μm, subglobose to elliptic, smooth, thin-walled, hyaline, amyloid.

FRUITING: usually in groups, terrestrial in mixed woods or under trees in grassy areas; September–October; rare but locally common.

EDIBILITY: deadly poisonous (contains amatoxins and phallotoxins).

COMMENTS: compare *A. citrina* forms *lavendula* and *citrina*, which lack true saccate volvas.

Amanita porphyria (Albertini and Schweinitz : Fries) Secretan

CAP: 1¾–4½" (4–12 cm) wide, convex to nearly flat; surface dark brownish gray, darkest toward the center, sometimes with a slight umbo, smooth, dry to sticky, margin smooth or vaguely striate; with a small number of randomly arranged floccose gray warts; flesh whitish; odor not distinctive.

GILLS: almost free or clearly attached, crowded, nearly white to pale gray, with subattenuate lamellulae; partial veil gray.

STALK: 2–4½" (5–12 cm) long, ¼–¾" (6–18 mm) thick, tapering slightly toward the top, whitish with gray fibers, nearly smooth, stuffed to hollow; ring subsuperior to median, usually adhering to the stalk, pale gray, membranous, thin, upper surface striate; with a subglobose, marginate bulb; lower stalk often decorated with floccose gray warts.

SPORE PRINT: white.

MICROSCOPIC FEATURES: spores 7–10 × 6.5–10 μm, typically globose, smooth, thin-walled, hyaline, amyloid.

FRUITING: solitary to scattered or grouped, terrestrial in mixed woods; July–October; occasional to infrequent.

EDIBILITY: toxicity suspected.

Amanita rubescens (Persoon : Fries) S. F. Gray

COMMON NAME: The Blusher.

CAP: 1½–8" (4–20 cm) wide, convex to flat; surface bronze with brown or red tints, typically smooth and dry, margin nonstriate to faintly striate, with numerous randomly arranged tan to reddish tan floccose warts; flesh thick, white, rather slowly bruising reddish; odor not distinctive.

GILLS: free or very finely attached, crowded, broad, white, often with slight reddish discolorations, with abundant truncate to subattenuate lamellulae; partial veil white to tan.

STALK: 3–8" (7–2.5 cm) long, ¼–1½" (7–40 mm) thick, tapering slightly toward the top, white to pale tan, slowly bruising reddish, smooth to slightly floccose or fibrous, solid to stuffed; ring superior, pendant, white to pale tan, membranous, thin, often torn; lower portion of stalk infrequently with indistinct rings of floccose universal veil tissue; bulb elongate, normally with distinct reddish stains.

SPORE PRINT: white.

MICROSCOPIC FEATURES: spores 7–9 × 5–7 μm, elliptic, smooth, thin-walled, hyaline, amyloid.

FRUITING: scattered to grouped, terrestrial in mixed woods or under trees in grassy areas; July–October; very common.

EDIBILITY: edible but not recommended.

COMMENTS: an easily identified species by virtue of its reddish stains and warts that are not yellowish. This mushroom is frequently parasitized by *Hypomyces hyalinus.*

Amanita sinicoflava Tulloss

CAP: 1–2½" (2.5–6.5 cm) wide, bell-shaped to broadly convex, with a pronounced umbo; surface tan to brown with an olive tinge, the umbo often darker, usually smooth, dry to sticky, with a distinctly striate margin; lacking warts; flesh moderately thin, white; odor not distinctive.

GILLS: free or very finely attached, close, fairly slender, white to cream, often tinged pale orangish, with truncate lamellulae; partial veil absent.

STALK: 2 ¾–5½" (7–14 cm) long, ¼–½" (7–13 mm) thick, conspicuously tapering toward the top, white to gray, bruising darker, finely fibrous, hollow; ring absent; volva membranous but fragile, very pale gray outside, pale gray inside, sometimes with speckled, brownish red stains, often collapsed; bulb absent.

SPORE PRINT: white.

MICROSCOPIC FEATURES: spores 9–12 × 8–11.5 μm, typically subglobose, smooth, thin-walled, hyaline, inamyloid.

FRUITING: widely scattered to loosely grouped, usually gregarious, terrestrial in mixed woods; June–October; infrequent to locally frequent.

EDIBILITY: toxicity suspected.

COMMENTS: this is an effective look-alike for edible Amanitae such as *A. fulva* and *A. vaginata*, though it may be more closely related to *A. ceciliae* (suspect).

Amanita virosa LaMarck

COMMON NAME: Destroying Angel.

CAP: 1⅛–5⅛" (3–13 cm) wide, convex to nearly flat; surface white, sometimes with a discolored center, smooth, dry to sticky, margin nonstriate; no warts present; cap staining yellow with KOH; flesh white; odor and taste not distinctive.

GILLS: free or very finely attached, crowded, white, with attenuate lamellulae; partial veil white, thin, membranous.

STALK: 2⅜–8" (6–20 cm) long, ¼–¾" (7–20 mm) thick, tapering slightly toward the top, white, smooth to roughened, solid; ring superior, pendant, readily adhering to stalk, white, membranous, thin and delicate, often torn; bulb small, usually fairly round; volva large, white, saccate, fairly thin, membranous, sometimes multilobed.

SPORE PRINT: white.

MICROSCOPIC FEATURES: spores 9–11 × 7–9 μm, subglobose to globose, smooth, thin-walled, hyaline, amyloid; basidia mostly four-spored.

FRUITING: usually solitary, scattered or in groups on the ground in mixed woods; June–November; common.

EDIBILITY: deadly poisonous (contains amatoxins and phallotoxins).

COMMENTS: it, as well as *A. bisporigera* and *A. verna*, are equally deadly. *Amanita bisporig-era* has mostly two-spored basidia and also stains yellow in KOH; in our experience, it is also usually less robust, has a thinner, smoother stalk, and is more frequent in early summer. *Amanita verna* does not stain yellow with KOH.

Genus *Armillaria*

This genus has recently undergone some major trimming by systematic mycologists. Many of the species formerly placed here have been reclassified in other genera, especially *Tricholoma*. *Armillaria* is now limited to *A. tabescens* and the *A. mellea* (Honey Mushroom) complex. They are nonetheless ubiquitous. The mycophile can quickly find the telltale "shoestring" rhizo-morphs (which are sometimes bioluminescent) of these fungi in most any forest, regardless of whether any Honey Mushrooms are fruiting at the time. Although *A. tabescens* is easily identi-fied to species, the *A. mellea* complex is another matter. This group of species has been inten-sively studied, but considerable confusion remains regarding the use of field characters for defining species. The following key is based on several scientific publications; any identifica-tions based on its use should be deemed tentative because the diagnostic characters of the species remain somewhat confused.

Honey Mushrooms are generally considered fine edibles. Some people get gastric upsets after eating some collections, but such reports are rare in this region. It is important that these mushrooms be thoroughly cooked.

Key to Species of *Armillaria*

1. Partial veil and ring absent—*A. tabescens* (Scopoli) Emel (see p. 70).
1. Partial veil present; cap predominantly yellow, viscid, with a few darkish hairs, especially near the center of the cap, which may be either erect or prone; gills attached, typically with fine decurrent lines descending to the ring, white to pale buff; partial veil white on the upper surface, yellowish on the lower surface; stalk fibrous, often becoming scaly just below the ring, white at first, becoming yellowish to brown or olive in age, staining brownish where bruised; in clusters on the ground, especially at the base of trees; rhizo-morphs flattened; spores 8.5–12 × 6–7 μm, broadly elliptic to oval, with an apiculus, smooth, hyaline, inamyloid; basidial clamp connections entirely absent—*A. mellea* (Vahl : Fries) Kummer (see p. 69).
1. Partial veil present; cap not predominantly yellow; clamp connections present on at least some basidia → 2.
   2. Partial veil cortinate, evanescent, colorless; cap tan to pinkish brown, dry, with blackish hairs or scales; gills whitish to pinkish, attached, typically with fine, decur-rent lines descending to the ring; stalk clavate at first, becoming more cylindric in age, whitish to pinkish near the top, staining yellow where bruised; typically not clustered, appearing terrestrial; rhizomorphs cylindric, not flattened; spores 8.5–12 × 6–7 μm, broadly elliptic to oval, with an apiculus, smooth, hyaline, inamyloid; edible if thoroughly cooked—*A. lutea* Gillet = *A. bulbosa* (Barla) Kile and Watling.
   2. Partial veil membranous to fibrous, thin, delicate, yellowish overall, whitish, or whitish with a yellow lower surface; growing in clusters; rhizomorphs cylindric, not flattened → 3.
   2. Partial veil membranous to fibrous, thick, white, with a fluffy brown margin; growing in clusters; rhizomorphs cylindric or flattened → 4.
3. Cap finely fibrous, not glabrous, orangish brown to brown, decorated with dark brown

to blackish hairs and mustard-yellow warts from a universal veil; gills subdecurrent to decurrent, whitish at first, becoming cream to pinkish tan in age; partial veil fibrous to membranous, thin, delicate, usually collapsed against the stalk in age, whitish to yellow on the upper surface, yellow on the lower surface; stalk fibrous, clavate, grayish brown to brownish red over a white ground color, sometimes staining brown, typically with mustard-yellow universal veil remnants; growing solitary or in small clusters (up to five mushrooms) on or about stumps or trees; spores 8–10 × 6–8 μm, broadly elliptic to oval, with an apiculus, smooth, hyaline, inamloid; edibility unknown—*A. sinapina* Bérubé and Dessureault.

3. Cap very finely fibrous or almost glabrous, variably brown, only moderately decorated with golden yellow hairs or fine scales; gills attached, subdecurrent to strongly decurrent, whitish at first, becoming light brownish in age; partial veil or ring fibrous to almost membranous, thin, quite delicate, usually collapsed against the stalk in age, whitish to grayish, sometimes brownish with golden yellow tints; stalk clavate to bulbous, fibrous, whitish to brownish or yellowish, sometimes staining brownish where bruised; growing solitary or in small clusters (up to eight mushrooms) on or about stumps or trees, especially maple; spores 8.5–10 × 5–7 μm, broadly elliptic to oval, with an apiculus, smooth, hyaline, inamyloid; edibility unknown—*A. calvescens* Bérubé and Dessureault.

4. Cap tan to yellowish brown or reddish brown but typically rather dark brown, densely hairy, the hairs dark reddish brown to blackish; partial veil or ring thick, membranous, whitish with a fluffy brown margin; stalk cylindric, fibrous, the fibers orangish to reddish brown, entire stalk sometimes stained those colors overall, with a yellowish apex and yellow mycelial growth at the extreme base; typically growing in large clusters but sometimes solitary on or about stumps or trees; rhizomorphs flattened; spores 8–11 × 5.5–7 μm, broadly elliptic to oval, with an apiculus, smooth, hyaline, inamyloid—*A. ostoyae* (Romagnesi) Herink (see p. 70).

4. Cap variably brown, with blackish hairs; partial veil or ring thick, membranous, whitish with a fluffy, brown edge; stalk enlarged downward at first but soon becoming equal, fibrous, staining dark brown to blackish; typically growing in large clusters but sometimes solitary on or about stumps or trees; rhizomorphs cylindric, not flattened; spores 8–10 × 5–7 μm, broadly elliptic to oval, with an apiculus, smooth, hyaline, inamyloid; edible if thoroughly cooked—*A. gemina* Bérubé and Dessureault.

Armillaria mellea (Vahl : Fries) Kummer

COMMON NAME: Honey Mushroom.

CAP: 1½–4" (4–10 cm) wide, convex to nearly flat; surface yellowish brown at first but soon becoming predominantly yellow with a darker center, sometimes dry but typically viscid, usually with a few darkish hairs, especially near the center of the cap, which may or may not be erect; flesh white, moderately thick at the center; odor and taste not distinctive.

GILLS: attached to subdecurrent, typically with fine, decurrent lines descending to the ring, white to pale buff; partial veil thick, membranous, whitish on the upper surface, yellowish on the lower surface, leaving a neat disc-like ring.

STALK: 2–6" (5–15 cm) long, about ⅜" (8–10 mm) thick, fibrous, often becoming scaly just below the ring, white at first, becoming yellowish to brown or olive in age, staining brownish where bruised; rhizomorphs flattened.

SPORE PRINT: pale cream.

MICROSCOPIC FEATURES: spores 8.5–12 × 6–7 μm, broadly elliptic to oval, smooth, hyaline, inamyloid; basidial clamp connections absent.

FRUITING: in clusters on the ground, especially at the base of trees or stumps, sometimes directly on wood; June–September; occasional to common.

EDIBILITY: edible if thoroughly cooked.

COMMENTS: this, the most easily identified species in the *A. mellea* complex, is a virulent parasite of mixed hardwood forests.

Armillaria ostoyae (Romagnesi) Herink

CAP: 2–4" (5–10 cm) wide, variously convex to nearly flat; surface dry, tan to yellowish brown to more typically dark reddish brown, densely covered with dark reddish brown to blackish hairs; flesh firm, white, rather thick at the center; odor and taste not distinctive.

GILLS: attached to subdecurrent, close, white to cream at first, becoming grayish orange to cinnamon; partial veil thick, membranous, leaving a whitish ring with a fluffy brown margin.

STALK: 2–8" (5–20 cm) long, about ⅝" (1.5 cm) thick at the apex, typically quite thickened downward at first, becoming cylindric in age, often quite tapered at the very base, fibrous, the fibers generally orangish to reddish brown; entire stalk staining mahogany to blackish; often with adhering bits of the partial veil; with yellow mycelial growth at the extreme base; rhizomorphs flattened.

SPORE PRINT: pale cream.

MICROSCOPIC FEATURES: spores 8–11 × 5.5–7 μm, broadly elliptic to oval, smooth, hyaline, inamyloid; clamp connections present at the bases of some basidia.

FRUITING: typically growing in large clusters but sometimes solitary on or about stumps or trees; July–November; occasional to common, more frequent northward.

EDIBILITY: a fine and often abundant edible if thoroughly cooked.

COMMENTS: this species can be quite difficult to distinguish from *A. gemina* (edible); examining the rhizomorphs and/or spores is essential.

Armillaria tabescens (Scopoli) Emel

COMMON NAME: Ringless Honey Mushroom.

CAP: 1¼–4" (3–10 cm) wide, convex to broadly convex with a very broad umbo; surface dry, orangish brown to brown, with darker, cottony hairs or tufts of fibers; flesh white to brownish; odor strong; taste astringent.

GILLS: subdecurrent, whitish at first, becoming pinkish brown.

STALK: 3–8" (7.5–20 cm) long, ¼–⅝" (5–15 mm) thick, typically tapered toward the base; vertically scurfy-fibrous; whitish near the top, yellowish to brownish below.

SPORE PRINT: cream.

MICROSCOPIC FEATURES: spores 6.5–8 × 4.5–5.5 μm, elliptic, smooth, hyaline, inamyloid.

FRUITING: in clusters on or about trees or stumps, especially oak; August–October; occasional to common, generally more frequent southward.

EDIBILITY: edible if thoroughly cooked.

COMMENTS: some people get upset stomachs from eating this tasty mushroom; it should be cooked thoroughly.

Genus *Asterophora*

Asterophora lycoperdoides (Bulliard) Ditmar

COMMON NAME: Powder Cap.

CAP: ⅜–¾" (1–2 cm) wide, rather round; surface dry, cottony, becoming covered with dense brown powder; odor and taste farinaceous.

GILLS: typically malformed, attached, distant, narrow, thick, often forked, whitish.

STALK: ¾–1¼" (2–3 cm) long, ⅛–⅜" (3–10 mm) thick; white, becoming brownish; minutely hairy or silky; stuffed, becoming hollow.

SPORE PRINT: white.

MICROSCOPIC FEATURES: basidiospores $3-6 \times 2-4$ μm, elliptic to oval, smooth, hyaline, inamyloid; cap surface also produces chlamydospores that are strongly warted or with blunt spines, oval to subglobose, $13-20 \times 10-20$ μm (excluding ornamentation), thick-walled, pale brown, inamyloid.

FRUITING: in clusters on decaying mushrooms, especially *Russula* and *Lactarius* species; July–November; occasional to common but often overlooked.

EDIBILITY: unknown.

Asterophora parasitica (Bulliard : Fries) Singer

CAP: ¼–¾" (8–20 mm) wide, round to convex, becoming nearly flat in age; smooth when young, coated with fine white fibrils; surface white to pale gray, becoming grayish brown in age; margin incurved at first; flesh thin, whitish to brownish; odor unpleasant; taste typically farinaceous.

GILLS: whitish to grayish brown, often poorly developed, occasionally forked, attached to subdecurrent, edges typically finely granular.

STALK: ⅜–1¼" (1–3 cm) long, up to ⅛" (2–3 mm) thick; with fine, white fibers on a grayish brown ground color; base sometimes minutely velvety and white; solid at first, becoming hollow.

SPORE PRINT: white.

MICROSCOPIC FEATURES: basidiospores $5-6 \times 3-4$ μm, elliptic, smooth, hyaline; gills also produce chlamydospores that are spindle-shaped to oval, $12-17 \times 9-10$ μm, typically thick-walled, hyaline, inamyloid.

FRUITING: typically gregarious and grouped on decomposing *Russula* and *Lactarius* species, sometimes on others; infrequent to occasional.

EDIBILITY: unknown.

Genus *Baeospora*

Baeospora myosura (Fries) Singer

COMMON NAME: Conifer-cone Baeospora.

CAP: ¼–¾" (5–20 mm) wide, convex; surface cinnamon to pale tan, moist, smooth; flesh very thin; odor and taste not distinctive.

GILLS: attached, quite crowded and narrow, white.

STALK: ⅜–2" (1–5 cm) long, 1⁄32–1⁄16" (1–2 mm) thick, whitish to brownish, minutely hairy above, with longer hairs at the base, hollow.

SPORE PRINT: white.

MICROSCOPIC FEATURES: spores $3-4.5 \times 1-2.5$ μm, elliptic, smooth, hyaline, amyloid; cheilocystidia clavate to bowling-pin shaped.

FRUITING: in groups on conifer cones, especially white pine and Norway spruce, also on magnolia cones; September–October; infrequent to occasional.

EDIBILITY: unknown.

COMMENTS: similar gilled mushrooms on cones lack the distinctly crowded gills and very small amyloid spores of this genus.

Baeospora myriadophylla (Peck) Singer

COMMON NAME: Lavender Baeospora.

CAP: ⅜–1½" (1–4 cm) wide, convex to nearly plane, sometimes depressed on the disc; surface moist, smooth, lavender to dingy lavender, fading to ochre-brown with a lavender margin, then ochre-buff overall in age; flesh grayish brown; odor and taste not distinctive.

GILLS: attached, becoming nearly free in age, close to crowded, lavender to dingy lavender.

STALK: ¾–2" (2–5 cm) long, ¹⁄₁₆–⅛" (1–3 mm) thick, dry, smooth, lavender to brownish lavender, with long hairs on the base.

SPORE PRINT: white.

MICROSCOPIC FEATURES: spores 3.5–4.5 × 2–3 μm, elliptic, smooth, hyaline, amyloid.

FRUITING: scattered or in groups on decaying conifer and hardwoods, especially hemlock; June–October; occasional.

EDIBILITY: unknown.

Genus *Bolbitius*

Bolbitius is a small genus of fewer than ten species in North America. They are fragile, small to medium mushrooms that grow on fertilized grass, manure, manured soil, or decaying wood. Their spore prints range from bright or dull rusty brown to ochre. They have a viscid cap that is usually conspicuously striate at maturity, nearly free gills that are close to crowded, and smooth spores that are often truncate with an apical pore. Their stalks are fragile, hollow, and lack a ring. None is commonly collected for the table.

Key to Species of *Bolbitius*

1. Cap up to 1" (2.5 cm) wide at maturity, viscid, bell-shaped to conic when young, becoming nearly plane, often with an umbo in age, bright yellow-orange, fading to vinaceous buff with a yelow-orange disc and cracked in age; stalk yellow to whitish; spores 10–14 × 6–8 μm; edibility unknown—*B. titubans* (Bulliard : Fries) Fries.
1. Cap more than 1" (2.5 cm) wide at maturity → 2.
   2. Cap white to whitish, viscid when fresh, conical to bell-shaped, becoming nearly plane at maturity, glabrous, sulcate-striate on the margin in age; stalk hollow, fragile, conspicuously floccose with snow-white fibrils, veil absent; spores 8–11 × 5–6 μm; on wood, sawdust, wood mulch, and woody debris; edibility unknown—*B. sordidus* Lloyd.
   2. Cap vinaceous gray to grayish pink or grayish salmon → 3.
   2. Cap yellow to orange-yellow or dark brownish olive to pale grayish olive, sometimes with yellow-orange tints → 4.
3. Cap viscid, vinaceous gray to lilac with a violaceous black disc, reticulate from the margin to the disc (use a hand lens), ⅜–1½" (1–4 cm) wide; stalk white; spores 8–12 × 5–6 μm; edibility unknown—*B. reticulatus* (Persoon : Fries) Ricken.

3. Cap viscid, grayish pink to grayish salmon when young, fading to pale grayish brown in age; not reticulate, sulcate from the margin to the disc in age, often with a low, broad umbo, 1–2⅜" (2.5–6 cm) wide; stalk white; spores 12–15 × 6–8 μm; edibility unknown— *B. coprophilus* (Peck) Hongo.

    4. Cap color variable, dark brownish olive to pale grayish olive, sometimes with orange-yellow tints, viscid, smooth or wrinkled, striate when young, becoming sulcate at maturity; conic to convex when young, becoming bell-shaped in age, ¾–2⅜" (2–6 cm) wide; gills whitish to pale yellow when young, becoming rusty tawny at maturity; stalk pale yellow or sometimes whitish— *B. variicolor* Atkinson (see p. 73).

    4. Cap orange-yellow to yellow, viscid, conic to convex when young, becoming bell–shaped in age, faintly striate on the margin when young, becoming sulcate in age and sometimes splitting, ⅜–2¾" (1–7 cm) wide; gills pale yellow when young, becoming dark rusty tawny at maturity; stalk whitish to pale yellow, often darker yellow in age; spores 12–15 × 6–7 μm; edible— *B. vitellinus* (Persoon : Fries) Fries.

Bolbitius variicolor Atkinson

CAP: ¾–2⅜" (2–6 cm) wide, conic to convex when young, becoming bell-shaped in age; surface viscid, smooth or wrinkled, striate on the margin when young, becoming sulcate at maturity; color variable, dark brownish olive to pale grayish olive or greenish olive, sometimes with orange-yellow tints; flesh whitish; odor and taste not distinctive.

GILLS: attached, becoming free in age, close, narrow, whitish to pale yellow when young, becoming rusty tawny at maturity.

STALK: 1½–4" (4–10 cm) long, ⅛–⁵⁄₁₆" (3–8 mm) thick, nearly equal overall, hollow, finely scurfy overall, pale yellow or sometimes whitish.

SPORE PRINT: rusty tawny to rusty ochre.

MICROSCOPIC FEATURES: spores 11–14 × 6–7 μm, elliptic with a conspicuous apical pore, smooth, pale brown.

FRUITING: scattered, in groups or clusters on manured areas, in gardens, fertilized grass, compost piles and along horse trails; May–October; common.

EDIBILITY: unknown.

COMMENTS: also known as *B. vitellinus* var. *olivaceus*.

Genus *Callistosporium*

Callistosporium purpureomarginatum Fatto and Bessette

CAP: ⅝–1½" (1.5–4 cm) wide, convex to nearly flat in age, often with a slightly depressed center; surface smooth, purplish red to brownish violet when fresh and moist, drying to a pinkish tan from the center toward the edge but retaining a bluish to brownish purple band at the margin; band usually sharply delineated, about ⅛" (2–3 mm) wide; flesh thin, yellowish; odor not distinctive; taste slowly bitter with a metallic aftertaste.

GILLS: attached, close, pinkish to yellow with purple overtones; edges reddish purple, with several tiers of lamellulae.

STALK: ¾–1½" (2–4 cm) long, ¹⁄₁₆–⅛" (1.5–4 mm) thick, equal, pruinose at the apex, smooth below, usually curved, reddish purple with yellow toward the base or apex, aging brownish, hollow.

SPORE PRINT: white.

MICROSCOPIC FEATURES: spores 4–6 × 3–4 μm, elliptic, smooth, hyaline, inamyloid, many staining vinaceous in KOH; dried pieces of gill tissue impart a conspicuous vinaceous tinge when mounted in KOH.

FRUITING: solitary, scattered or in groups on well-decayed oak logs; July–September; occasional.

EDIBILITY: unknown.

COMMENTS: easily distinguished by the reddish purple gill edges, the thin violet band at the cap margin, a usually curved stalk because of growing out of the lower sides of decaying oak logs, and its collybioid stature. *Callistosporium luteo-olivaceum* (edibility unknown) is typically yellowish brown overall, with gill edges that are only occasionally reddish brown in age and grows on decaying conifer wood. *Baeospora myriadophylla* (edibility unknown) has crowded, thin, lavender gills.

Genus *Calocybe*

Calocybe persicolor (Fries) Singer

CAP: 5/8–13/4" (1.6–4.5 cm) wide; surface flesh-pink to pale vinaceous pink, becoming pale pinkish brown to pinkish tan or yellowish tan on the disc at maturity and nearly overall in age; margin typically inrolled when young, often remaining so at maturity, becoming uplifted and wavy in age; flesh thin, white; odor and taste not distinctive.

GILLS: attached when young, becoming decurrent in age, close to crowded, finely scalloped and often eroded in age, white to cream.

STALK: 3/4–13/4" (2–4.5 cm) long, 1/8–1/4" (2–7 mm) thick, flesh-pink to vinaceous pink, becoming dingy yellow to yellowish tan in age, with a narrow white zone at the apex, typically coated with long, white hairs on the lower one-third or on the base.

SPORE PRINT: white.

MICROSCOPIC FEATURES: spores 4–6 × 2–3 μm, elliptic to oval, smooth, hyaline, inamyloid; basidia with siderophilous granules.

FRUITING: typically in clusters, solitary, or in groups on decaying humus or soil in grassy areas or woods; August–October; occasional.

EDIBILITY: unknown.

COMMENTS: *Calocybe carnea* (edibility unknown) is nearly identical and may actually be a variety of *C. persicolor*. It differs by growing solitary or in groups rather than in clusters, having brighter pink cap color and shorter, more matted, whitish fibrils on the stalk base.

Genus *Cantharellula*

Cantharellula umbonata (Fries) Singer

COMMON NAME: Grayling.

CAP: typically 3/4–2" (2–5 cm) wide, occasionally becoming almost 3" (7.5 cm) wide; convex at first, becoming flat to sunken; most specimens with a small, pointed umbo; margin incurved at first, becoming upturned and wavy in age; surface gray to grayish brown overall, often appearing to have whitish blotches; dry to moist; smooth to minutely hairy; flesh white; odor and taste not distinctive.

GILLS: close to crowded, decurrent, repeatedly and regularly forked; whitish, developing spot-like reddish or sometimes yellow stains in age.

STALK: 1–5" (2.5–12.5 cm) long, 1/8–1/4" (3–7 mm) thick often with swollen portions; somewhat flexible, often bent, curved, and/or twisted; white to gray; silky above, stuffed, usually with whitish mycelium binding the lower stalk to moss; often water-saturated near the base.

SPORE PRINT: white.

MICROSCOPIC FEATURES: spores 8–11 × 3–4.5 μm, somewhat spindle-shaped, smooth, hyaline, amyloid.

FRUITING: Scattered to gregarious, sometimes in fairy rings, quite exclusively in haircap (*Polytrichum*) moss; August–November; common.

EDIBILITY: a fine edible if young and fresh.

COMMENTS: formerly classified as *Cantharellus umbonatus* because of the forked gills, but not even closely related to that genus.

Genus *Catathelasma*

Catathelasma ventricosa (Peck) Singer

CAP: 3–6" (7.5–15 cm) wide, convex; surface dry, smooth at first, becoming patchy-scaly, whitish to gray, the margin usually adorned with bits of tissue from the partial veil; flesh thick, very firm, white; odor not distinctive; taste unpleasant.

GILLS: decurrent, close to subdistant, whitish to buff.

STALK: 2–4" (5–10 cm) long, 1–2" (2.5–5 cm) thick, tapering toward the base and usually mostly buried, dry, white above the double ring, brownish yellow below it; ring two-layered, the upper one fibrous, the lower one membranous.

SPORE PRINT: white.

MICROSCOPIC FEATURES: spores 9–11 × 4–6 μm, elliptic, smooth, hyaline, amyloid.

FRUITING: solitary to grouped or scattered on the ground under conifers; August–October; infrequent.

EDIBILITY: edible, but mediocre at best.

COMMENTS: sometimes spelled *C. ventricosum*.

Genus *Cheimonophyllum*

Cheimonophyllum candidissimus (Berkeley and Curtis) Singer

COMMON NAME: White Oysterette.

CAP: ⅜–¾" (1–2 cm) wide, hemispheric to kidney-shaped; surface with a white layer of minute hairs on a yellowish background; margin incurved at first, becoming slightly striate in age; flesh thin, soft, whitish to yellowish; odor and taste not distinctive.

GILLS: white to yellowish, subdistant to fairly close, rather broad.

STALK: whitish, lateral, stubby; sometimes absent.

SPORE PRINT: white.

MICROSCOPIC FEATURES: spores 5–6 × 4.5–5.5 μm, globose to subglobose, smooth, hyaline, inamyloid; with branched, thread-like cheilocystidia.

FRUITING: solitary or in groups, on dead deciduous wood; July–October; occasional and easily overlooked.

EDIBILITY: unknown.

COMMENTS: previously classified as *Pleurotus candidissimus* and easily mistaken at first glance for a *Crepidotus*.

Genus *Chlorophyllum*

Chlorophyllum molybdites Massee

COMMON NAME: Green-spored Lepiota, Green-spored Parasol.

CAP: 2¾–12" (7–30 cm) wide, round, becoming convex to nearly flat; surface dry, white, covered when young with large, pinkish brown to cinnamon patches that become scale-like in age, usually clustered toward the disc; flesh white, thick, not staining when cut or bruised, or occasionally slowly staining reddish; odor and taste not distinctive.

GILLS: free from the stalk, close, broad; white, becoming greenish to grayish green in age, staining yellow to brownish when cut or bruised.

STALK: 4–10" (10–25 cm) long, ⅜–1" (1–2.5 cm) thick, nearly equal or enlarged downward, smooth, white, staining brownish when bruised; partial veil white, membranous, forming a superior ring with a fringed or double edge; ring becoming brownish on the underside in age, often movable.

SPORE PRINT: green.

MICROSCOPIC FEATURES: spores 8–13 × 6.5–8 μm, elliptic, smooth, hyaline, dextrinoid.

FRUITING: solitary, scattered, in groups or fairy rings in lawns, meadows, and gardens; August–October; occasional to fairly common.

EDIBILITY: poisonous (see comments below).

COMMENTS: this mushroom is one of the most frequent causes of serious mushroom poisoning in eastern North America; its toxins can cause such severe vomiting that hospitalization is required to prevent acute, life-threatening dehydration.

Genus *Chromosera*

Chromosera cyanophylla (Fries) Redhead, Ammirati, and Norvell

CAP: ¼–1" (6–25.5 mm) wide, broadly convex with a flattened disc that often becomes depressed in age; surface smooth, viscid when fresh, translucent-striate, grayish lavender when very young, soon becoming dingy yellow with a paler margin; flesh pale yellow; odor and taste not distinctive.

GILLS: subdecurrent, subdistant to close, bright lilac when young, fading to pale dull lilac in age; edges even, with two tiers of lamellulae.

STALK: ⅜–1⅜" (1–3.5 cm) long, about 1/16" (1–1.5 mm) thick, nearly equal, smooth, slimy when fresh, fragile, lilac at first, fading to yellowish with a lilac tint in age, sometimes coated with a lilac basal mycelium.

SPORE PRINT: white.

MICROSCOPIC FEATURES: spores 6–7 × 3–4 μm, elliptic, smooth, hyaline, inamyloid.

FRUITING: scattered or in groups on decaying conifer wood, especially balsam fir and hemlock; May–October; uncommon.

EDIBILITY: unknown.

COMMENTS: formerly known as *Mycena lilacifolia*.

Genus *Chroogomphus*

Chroogomphus species or "Pine Spikes," as many call them, are terrestrial mushrooms associated with various conifers, especially pines. It is a small genus with fewer than a dozen known species in all of North America. They typically have smoky gray to blackish spore prints, long and narrow spores, decurrent yellowish to orange or salmon gills and a central stalk that tapers downward or is nearly equal. Their cap surfaces are smooth and often sticky, but not slimy, and their flesh is buff, ochraceous, pale salmon, pinkish salmon, or reddish.

Key to Species of *Chroogomphus*

1. Cap medium to large, ¾–3⅛" (2–8 cm) wide; stalk ¼–¾" (6–20 mm) thick → **2**.
1. Cap small to medium, ⅜–1½" (1–4 cm) wide; stalk ⅛–⁵⁄₁₆" (3–8 mm) thick → **3**.
   2. Cap dull orange-brown to ochraceous brown or pale reddish brown, sticky when fresh; gills pale ochre when young, becoming pale cinnamon-brown in age; stalk ochraceous buff to tawny-orange; partial veil pale ochraceous, fibrillose, sometimes leaving a thin layer of fibrils on the upper portion of the stalk—*C. rutilus* (Schaeffer : Fries) O. K. Miller, Jr. (see p. 77).
   2. Cap color variable, vinaceous red, orange-red or yellow-brown, often dark vinaceous red in age, sticky when fresh; gills buff to pale orange when young, becoming dull ochraceous, then smoky brown in age; stalk pale ochraceous, becoming orange-buff to vinaceous red in age; partial veil pale ochraceous, fibrillose, sometimes leaving a thin layer of fibrils on the upper portion of the stalk; spores 17–23 × 4.5–7.5 μm; cystidia with thickened walls in the midportion; cuticular hyphae 6–7 μm wide; on the ground under conifers; August–November—*C. vinicolor* (Peck) O. K. Miller, Jr. (see p. 78).
   2. Nearly identical to the previous choice except the spores measure 17–20 × 4.5–6 μm, the cystidia are more uniformly thickened, up to 5 μm thick, and the cuticular hyphae are 2–5 μm wide; edibility unknown—*C. jamaicensis* (Murrill) O. K. Miller, Jr.
3. Stalk whitish and nearly smooth on the upper one-third, lemon-yellow and fibrillose on the lower two-thirds, nearly equal; cap surface moist but not sticky when fresh, vinaceous to pinkish or orange-vinaceous; gills pale salmon to buff when young, developing grayish tints in age; spores 18–29 × 6–8.5 μm; on the ground under conifers in bogs; August–September; edibility unknown—*C. flavipes* (Peck) O. K. Miller, Jr.
3. Stalk bright yellowish orange to bright ochraceous, becoming reddish near the base, nearly equal; cap surface sticky when fresh, bright yellowish orange to bright ochraceous; gills salmon to ochraceous salmon, often becoming reddish near the stalk at maturity; spores 14–20 × 4.5–7 μm; on soil or humus under conifers; August–November; edibility unknown—*C. ochraceus* (Kauffman) O. K. Miller, Jr.

Chroogomphus rutilus (Schaeffer : Fries) O. K. Miller, Jr.
   COMMON NAME: Brownish Chroogomphus.
   CAP: ¾–3⅛" (2–8 cm) wide, obtuse to convex, sometimes with a small, pointed umbo; surface sticky when fresh, dull orange-brown to ochraceous brown or pale reddish brown; margin incurved when young; flesh pale salmon to pale ochraceous or pinkish tinged; odor and taste not distinctive.
   GILLS: thick, decurrent, close to subdistant, broad, pale ochre when young, becoming pale cinnamon-brown in age.
   STALK: 2–6" (5–15.5 cm) long, ¼–1" (6–25 mm) thick, solid, tapered toward the base or nearly equal, ochraceous buff to tawny-orange, often with vinaceous tints in age; partial veil pale ochraceous, fibrillose, sometimes leaving a thin layer of fibrils on the upper portion of the stalk.
   SPORE PRINT: smoky gray to blackish.
   MICROSCOPIC FEATURES: spores 14–22 × 6–7.5 μm, elliptic, smooth, pale gray-brown; cystidia thin-walled.
   FRUITING: scattered or in groups on the ground under conifers; August–October; occasional.
   EDIBILITY: edible.

Chroogomphus vinicolor (Peck) O. K. Miller, Jr.

COMMON NAME: Wine-cap Chroogomphus.

CAP: ¾–3⅛" (2–8 cm) wide, convex to obtuse or sometimes with a small umbo; surface smooth, sticky when fresh, shiny when dry, color variable, vinaceous red, orange-red or yellow-brown, often dark vinaceous red in age; flesh orange in young specimens, fading to ochraceous buff or pale salmon in age; odor and taste not distinctive.

GILLS: subdistant to distant, decurrent, buff to pale orange when young, becoming dull ochraceous then smoky brown in age.

STALK: 2–5" (5–10 cm) long, ¼–¾" (6–20 mm) thick at the apex, tapered downward, dry, nearly smooth, pale ochraceous, becoming orange-buff to vinaceous red in age; partial veil pale ochraceous, fibrillose, sometimes leaving a thin layer of fibrils on the upper portion of the stalk.

SPORE PRINT: smoky gray to dull olive-gray.

MICROSCOPIC FEATURES: spores 17–23 × 4.5–7.5 μm; cystidia with thickened walls, up to 7.5 μm thick in the midportion; cuticular hyphae 6–7 μm thick.

FRUITING: scattered or in groups on the ground or among mosses under conifers, especially pines; August–November; occasional.

EDIBILITY: edible.

COMMENTS: *Chroogomphus jamaicensis* (edibility unknown) is nearly identical but has slightly smaller spores, 17–20 × 4.5–6 μm, cystidia with more uniformly thickened walls, up to 5 μm thick, and cuticular hyphae that measure 2–5 μm wide.

Genus *Chrysomphalina* and Allies

This group includes several white- or pale-spored genera that share obvious field characters: habitat usually either on wood or among mosses, lichens, or liverworts; rather small size; caps that are depressed at the center at maturity; decurrent gills; and no partial veils. The key includes a few fairly common species.

Key to Species of *Chrysomphalina* and Allies

1. Cap no more than 1½" (4 cm) wide and stalk no more than ⅛" (3 mm) thick at maturity → 2.
1. Cap wider and/or stalk thicker at maturity than in the previous choice → 3.
    2. Growing with a lichen (*Botrydina vulgaris* Brébeck), though the lichen may be obscured by moss; cap brownish or reddish at first, fading in age to yellowish white, margin distinctly and regularly scalloped in age; stalk about 1⁄16" (1–3 mm) thick, brownish at first, fading below to pale yellow; spores 7–9 × 4–6 μm, broadly elliptic, smooth, inamyloid, clamp connections absent; edibility unknown—*Phytoconis ericetorum* (Fries) Redhead and Kuyper = *Omphalina ericetorum* (Fries) M. Lange.
    2. Growing on conifer wood; gills pale orangish yellow; cap minutely fibrous, scaly or scurfy, brownish orange at the center, yellowish brown at the edge, fading slightly in age, the margin not scalloped as in the previous choice; stalk yellow at first, darker in age; base typically with a tuft of white mycelium; spores 8–15 × 5–6 μm—*Chrysomphalina chrysophylla* (Fries) Clémençon (see p. 79).
    2. Growing on decaying wood in conifer or mixed woods; gills pale grayish brown; cap convex when young, becoming depressed on the disc and finally broadly funnel-shaped, margin not scalloped; surface smooth, hygrophanous, grayish brown, translucent-striate; flesh grayish, odor and taste not distinctive; stalk smooth, dark

brown to grayish brown; spores 6–9 × 3.5–5 μm—*Omphalina epichysium* (Persoon) Quélet (see p. 215).

3.   Growing on rotting conifer wood, especially hemlock; cap 1–2⅜" (2.5–6 cm) wide, brownish yellow to yellow-brown, covered with minute blackish brown to reddish brown matted fibers and scales that often disappear in age; gills pale yellow; stalk yellowish, staining brownish from handling, spores 6.5–8 × 3.5–5 μm—*Omphalina ectypoides* (Peck) Bigelow (see p. 215)..

3.   Growing on wood; cap grayish brown to yellowish brown, typically at least 1½" (4 cm) wide at maturity; gills pale yellowish; stalk ⅛–¼" (3–7 mm) thick, pale yellow to pale gray or white, finely hairy; spores elliptic, 7–9 × 3–5 μm, smooth, inamyloid, clamp connections present; edibility unknown—*Gerronema strombodes* (Berkeley and Montagne) Singer.

Chrysomphalina chrysophylla (Fries) Clémençon

CAP: ⅜–1½" (1–4 cm) wide, convex, depressed at the center; surface minutely fibrous, scaly or scurfy, moist, brownish orange at the center, yellowish brown at the edge, fading slightly in age; margin incurved; flesh thin, brownish; odor and taste not distinctive.

GILLS: decurrent, subdistant to distant, fairly broad, sometimes slightly corrugated, pale orangish yellow.

STALK: ¾–1½" (2–4 cm) long, about 1/16" (1.5–3 mm) thick, moist, typically smooth, yellow at first, darker in age; base sometimes with a tuft of white mycelium.

SPORE PRINT: pale yellowish.

MICROSCOPIC FEATURES: spores 8–15 × 5–6 μm, elliptic to almost cylindric, smooth, hyaline, inamyloid.

FRUITING: in groups or small clusters on conifer wood; July–September; occasional to frequent.

EDIBILITY: unknown.

COMMENTS: previously classified as *Gerronema chrysophylla* (Fries) Singer.

Genus *Claudopus*

Claudopus parasiticus (Quélet) Ricken

CAP: ⅛–5/16" (3–8 mm) wide, hemispheric to convex when young, becoming broadly convex or somewhat shell-shaped in age; surface coated with tiny fibrils, white when young, becoming buff to pale tan, often tinted pinkish brown from falling spores; margin often irregular and somewhat lobed; flesh very thin, whitish; odor not distinctive; taste unknown.

GILLS: attached to slightly decurrent, subdistant, pinkish buff when young, becoming flesh-pink to pale salmon.

STALK: about ⅛" (2–3 mm) long, about 1/16" (1–2 mm) thick, nearly central when very young, soon eccentric and nearly lateral at maturity, tapered downward or nearly equal, translucent to whitish when young, soon becoming flesh-pink, typically coated at the base with a dense layer of white mycelium.

SPORE PRINT: pink to salmon.

MICROSCOPIC FEATURES: spores 10–13 × 8–10 μm, angular in all views, smooth, hyaline.

FRUITING: parasitic on other mushrooms (e.g., *Cantharellus cibarius,* polypores) or growing on well-decayed wood; July–September; uncommon.

EDIBILITY: unknown.

COMMENTS: this genus needs further study.

Genus *Clitocybe*

Clitocybe is a very large genus, with more than two hundred species known from North America. They are small to very large mushrooms with a white, cream, or pinkish spore print and spores that may be amyloid or inamyloid (*sensu* Bigelow). Their gills are thin, not thick and waxy, and are broadly attached when young and often decurrent at maturity. They typically have a central stalk and lack a partial veil and ring. Although they grow on a wide variety of substrates, most occur on soil and humus, leaf and needle litter, and wood debris.

A few species are edible, some are poisonous and the edibility of the majority is unknown. Because the genus is complex and includes a large number of species, a comprehensive key to species is beyond the scope of this book. The following is a key to some of the most distinctive and common species found in this region. For additional information, consult the two volumes of *North American Species of Clitocybe* by Howard E. Bigelow. If your specimen cannot be identified using this key, and the cap diameter of mature specimens is less than 3" (7.5 cm) wide, see the key to *Chrysomphalina* and Allies, p. 78.

Key to Species of *Clitocybe* and Allies

1. Stature resembling a Collybia; cap ⅜–1½" (1–4 cm) wide; radially streaked and often conspicuously so; typically growing in very dense clusters on decaying wood; gills attached to subdecurrent, white to pale grayish; stalk white, whitish, or grayish; cap dark brownish buff, pale to dark grayish brown or grayish to smoky gray, at least on the disc; flesh whitish, odor and taste not distinctive; spores amyloid and smooth—*Clitocybula* species (see key to species of *Collybia* and Allies, p. 88).

1. Not as above; cap bright orange to yellow-orange, 2¾–7" (7–18 cm) wide; gills decurrent, close, narrow, thin, yellow-orange; stalk yellow-orange; growing in clusters at the base of hardwood trees and stumps, especially oak, or on the ground attached to buried wood—*Omphalotus olearius* (De Candolle : Fries) Singer (see p. 216).

1. Not with the above combinations of characters → 2.
   2. Cap predominantly white, whitish, grayish white, buff, tan, yellowish tan, or pale butterscotch when fresh and mature, sometimes darker on the disc → 3.
   2. Cap with darker, richer colors when fresh and mature → 12.

3. Cap large to very large at maturity, 3½–15" (9–39 cm) wide → 4.
3. Cap small to medium at maturity, ¾–3½" (2–9 cm) wide → 5.
   4. Odor pungent and unpleasant; taste rancid to unpleasant; cap up to 7" (18 cm) wide at maturity—*C. robusta* Peck (see p. 85).
   4. Odor fragrant; taste not distinctive; cap up to 5⅛" (13 cm) at maturity, whitish when young, becoming buff to pinkish buff; gills whitish when young, becoming pale pinkish to pinkish buff—*C. irina* (Fries) Bigelow and Smith (see p. 84).
   4. Odor weakly farinaceous or not distinctive; taste not distinctive; cap up to 8" (20 cm) wide at maturity, white to whitish; gills white to buff, short to moderately decurrent; stalk ⅝–1⅜" (1.5–3.5 cm) thick at the apex; spores 6–8.5 × 3–4.5 μm; edibility unknown—*C. candida* Bresadola.
   4. Odor weakly to distinctly farinaceous; taste not distinctive or disagreeable; cap up to 15" (39 cm) wide at maturity, white to buff; gills whitish to buff, moderately to long decurrent; stalk ¾–2⅜" (2–6.5 cm) thick at the apex; spores 6–8 × 3–4.5 μm; edibility unknown—*C. gigantea* (Fries) Quélet.

5. Odor fragrant or anise- to licorice-like, or pungent and unpleasant like coal tar, to strongly fishy → 6.
5. Odor mildly farinaceous, fragrant but not like anise or licorice, or not distinctive → 7.

6. Odor fragrant; taste bitter or mild; cap 1⅛–3⅛" (3–8 cm) wide, white, often with a satiny luster; gills pale buff; stalk enlarged downward, fused at the base, often under the soil surface, forming clusters in conifer and hardwoods; spores 4.5–6 × 3–4 μm; edibility unknown—*C. subconnexa* Murrill.

6. Odor anise- to licorice-like; taste mild or anise-like; cap ¾–3⅜" (2–8.5 cm) wide, white with a buff disc—*C. odora* (Fries) Kummer (see p. 85).

6. Odor pungent and unpleasant like coal tar to strongly fishy; taste unpleasant; cap buff with a cinnamon disc, cinnamon overall when very young, distinctly funnel-shaped, ¾–3" (2–7.5 cm) wide; spores 4.5–8 × 3–4.5 μm; edibility unknown—*C. phaeophthalma* (Persoon) Kuyper = *C. hydrogramma* (Fries) Kummer.

7. Cap 2⅜" (6 cm) wide or more at maturity → 8.

7. Cap less than 2⅜" (6 cm) wide at maturity → 9.

8. Growing on the ground under conifers and hardwoods; stalk grayish white to buff, paler toward the base, enlarged downward and distinctly bulbous to club-shaped, ¼–⅝" (7–15 mm) thick at the apex; cap grayish white to brownish tan, often darker on the disc—*C. subclavipes* Murrill (see p. 86).

8. Growing on the ground under conifers and hardwoods; stalk longitudinally streaked with olive-gray fibrils over a whitish ground color, enlarged downward and distinctly bulbous to club-shaped, ⅛–½" (4–12 mm) thick at the apex; faded caps tan to buff in age, usually accompanied by some fresh caps that are gray-brown to olive-brown—*C. clavipes* (Fries) Kummer (see p. 83).

9. Growing on decaying wood → 10.

9. Growing among grasses in lawns and pastures or on leaves or needles in woods → 11.

10. Gills decurrent at maturity, not forming a collar, occasionally forked, typically not crossveined, white, crowded to close; stalk nearly equal, typically ⅛" (2–4 mm) thick at the apex, creamy white to white, smooth and shiny; base covered with conspicuous white hairs; cap ¾–2" (2–5 cm) wide, creamy white, fading to white, funnel-shaped or sometimes nearly plane; gills attached at first, becoming decurrent, crowded; spores 5–8 × 2.5–4 μm; edibility unknown—*C. leptoloma* (Peck) Peck.

10. Gills decurrent at maturity, ends typically equal on the stalk and forming a collar, usually forked and crossveined, pale buff, yellowish or pinkish buff; stalk nearly equal or enlarged in either direction, typically less than ¼" (5 mm) thick at the apex, pale pinkish buff; base often curved, usually coated with matted buff hairs and rhizomorphs; cap ⅜–1¾" (1–4.5 cm) wide, pinkish buff to pinkish cinnamon when young, soon fading to buff or whitish, convex, sometimes with a depressed disc; in groups or clusters, sometimes scattered; spores 4–6 × 2.5–4 μm; edibility unknown—*C. americana* Bigelow.

10. Gills attached to subdecurrent, not forming a collar, not forked or crossveined, whitish; stalk nearly equal or slightly enlarged downward, typically less than ¼" (5 mm) thick at the apex, white to pale buff; base often curved, with few or no rhizomorphs; cap ⅜–1¾" (1–4.5 cm) wide, white, convex, sometimes with a depressed disc; usually scattered or in groups; spores 3.5–5 × 2.5–4 μm; edibility unknown—*C. truncicola* (Peck) Saccardo.

11. Growing among grasses in lawns and pastures, often in arcs; stalk nearly equal, typically less than ¼" (5 mm) thick at the apex, often curved, usually eccentric, white to buff; base coated with white mycelium, sometimes with white rhizomorphs; cap ⅝–2" (1.5–5 cm) wide, white to grayish white; gills white to pale buff; spores 3–5.5 × 2–3.5 μm; poisonous—*C. dealbata* (Fries) Kummer ssp. *sudorifica* (Peck) Bigelow.

11. Growing on needle litter under pine or spruce; stalk slightly enlarged downward, not distinctly bulbous, typically less than ¼" (6 mm) thick at the apex, whitish or buff to pale butterscotch; cap ¾–1½" (2–4 cm) wide, yellowish buff to pale butterscotch; spores 5–8 × 2.5–4.5 μm; edibility unknown—*C. coniferophila* Bigelow.

11. Growing on decaying leaves in hardwoods or sometimes on needle litter under conifers; stalk nearly equal or slightly enlarged downward, not distinctly bulbous, about ¹⁄₁₆–⅛" (1–3 mm) thick at the apex, watery whitish to yellowish; cap ⅜–1⅛" (1–3 cm) wide, with a white canescence over a watery white to yellowish ground color when fresh, becoming whitish and opaque on drying, convex at first, becoming nearly plane with a slightly depressed disc at maturity; gills attached at first, becoming decurrent, ends typically equal on the stalk and forming a collar, white to whitish; spores 4.5–6 × 2.5–4 μm—*C. candicans* (Fries) Kummer (see p. 83).
    12. Growing on decaying wood → 13.
    12. Growing on the ground, among mosses, on conifer needle litter, or in gardens, lawns, and fertilized areas → 14.

13. Cap distinctly funnel-shaped, brownish ochraceous, streaked with blackish brown to reddish brown matted fibers and minute scales that often disappear in age—*Omphalina ectypoides* (Peck) Bigelow (see p. 215) = *C. ectypoides* (Peck) Saccardo.

13. Cap broadly funnel-shaped, dark brown, moist, hygrophanous and fading to pale grayish brown; gills brown, becoming grayish brown to pale brown, usually forming a collar on the stalk, usually crossveined; stalk enlarged downward or nearly equal, brown, occasionally with white rhizomorphs at the base; spores 7–12 × 5–7 μm; edibility unknown—*Pseudoclitocybe cyathiformis* (Fries) Singer = *C. cyathiformis* (Fries) Kummer.
    14. Odor moderately to strongly farinaceous, anise- to licorice-like or resembling green corn (check buttons and mature specimens) → 15.
    14. Odor fragrant or not distinctive, rarely weakly farinaceous → 16.

15. Odor anise- to licorice-like; cap bluish to bluish green to greenish—*C. odora* (Fries) Kummer (see p. 85).

15. Odor like green corn, especially in the button stage, somewhat farinaceous in older specimens; taste mild to disagreeable; cap pale pinkish cinnamon when young, soon darker pinkish brown to orange-brown or walnut-brown; gills pale pinkish brown; stalk dingy white, becoming pale pinkish brown; spores 4–6 × 2–3.5 μm; on conifer needle litter; spore print pinkish; edibility unknown—*C. martiorum* Favre.

15. Odor farinaceous; cap small, ⅜–1½" (1.5–4 cm) wide, convex, becoming funnel-shaped in age, grayish, grayish brown, or olive-brown to blackish—*C. trullaeformis* (Fries) Quélet (see p. 87).

15. Odor farinaceous; cap medium, 1–2⅛" (2.5–5.5 cm) wide, convex when young, becoming nearly flat with a shallowly depressed center at maturity, grayish brown, becoming brownish tan in age, translucent-striate when moist; on conifer needle litter—*C. subditopoda* Peck (see p. 86).

15. Odor and taste strongly farinaceous; cap medium, ¾–2¾" (2–6.5 cm) wide, convex when young, becoming nearly flat, often with a shallowly depressed disc; surface orange-brown, darker on the disc; stalk nearly equal or tapered downward; base coated with white matted hairs; spores 7–10 × 4–6 μm; edibility unknown—*C. sinopica* (Fries) Kummer.
    16. Stalk distinctly enlarged downward, coated with olive-gray fibrils on a whitish to dull yellowish ground color; base club-shaped or bulbous; gills strongly decurrent, white to pale yellow; cap gray-brown to olive-brown; on the ground, often among mosses—*C. clavipes* (Fries) Kummer (see p. 83).

16. Stalk equal or often bulbous to abruptly bulbous at the base, whitish, pale violet or lavender, bruising dark lavender, ⅜–1⅛" (1–3 cm) thick at the apex, fairly short; gills notched to sinuate, not decurrent, pale lavender to violet or lilac; cap violet to lilac-gray to pinkish buff; in woods, on lawns and compost piles—*C. nuda* (Fries) Bigelow and Smith (see p. 84).

16. Stalk equal or tapered downward, streaked with white fibrils over a watery brown ground color, ⅛–⅜" (2–10 mm) thick at the apex; gills attached when young, becoming decurrent in age, pinkish buff; cap vinaceous brown to violaceous brown, fading to pale cinnamon to fawn, often darker on the disc; on the ground in cultivated and fertilized areas, including gardens, lawns, and fields—*C. tarda* Peck (see p. 86).

16. Stalk equal or slightly enlarged at the base, orange-cinnamon to brownish orange and colored like the cap or darker—*C. squamulosa* (Fries) Kummer (see p. 85).

16. Stalk equal or slightly enlarged at the base, whitish and not colored like the cap—*C. gibba* (Fries) Kummer (see p. 84).

Clitocybe candicans (Fries) Kummer

CAP: ⅜–1⅛" (1–3 cm) wide, convex when young, becoming nearly plane with a slightly depressed disc at maturity; surface with a white canescence over a watery white to yellowish ground color when fresh, becoming whitish and opaque on drying; margin incurved to inrolled when young; flesh thin, watery, whitish; odor and taste not distinctive.

GILLS: attached at first, becoming decurrent, close, narrow, ends typically equal on the stalk and forming a collar, white to whitish.

STALK: ⅜–1⅛" (1–3 cm) long, about 1⁄16–⅛" (1–3 mm) thick at the apex, nearly equal or slightly enlarged downward, hollow in age, nearly smooth, watery whitish to yellowish, often coated with whitish mycelium at the base.

SPORE PRINT: white.

MICROSCOPIC FEATURES: spores 4.5–6 × 2.5–4 μm, ellipsoid, smooth, hyaline, inamyloid.

FRUITING: scattered, in groups, or in clusters on decaying leaves in hardwoods or sometimes on needle litter under conifers; August–October; occasional to fairly common.

EDIBILITY: unknown.

Clitocybe clavipes (Fries) Kummer

COMMON NAME: Fat-footed Clitocybe.

CAP: ¾–3½" (2–9 cm) wide, convex when young, becoming flat with a shallowly depressed center in age, with or without a low, broad umbo; surface moist, smooth, sometimes with minute, matted fibrils over the disc, gray-brown to olive-brown, paler toward the margin, fading to buff in age; margin becoming elevated and wavy in age; flesh whitish; odor fragrant; taste not distinctive.

GILLS: strongly decurrent, close to subdistant, whitish when young, becoming pale yellowish to cream in age, often forked and crossveined.

STALK: 1⅜–2⅜" (1.5–6 cm) long, ⅛–½" (4–12 mm) thick, enlarged downward, stuffed, moist, longitudinally streaked with olive-gray fibrils over a whitish ground color; base bulbous or club-shaped, often covered with white mycelium.

SPORE PRINT: white.

MICROSCOPIC FEATURES: spores 6–10 × 3.5–5 μm, oval, smooth, hyaline, inamyloid.

FRUITING: solitary, scattered or in groups on the ground under conifers or hardwoods or in mixed woods; July–November; fairly common.

EDIBILITY: edible.

Clitocybe gibba (Fries) Kummer

COMMON NAME: Funnel Clitocybe.

CAP: 1¼–3½" (3–9 cm) wide, flat to slightly depressed on the disc when young, becoming funnel-shaped in age; surface dry to moist, smooth, pinkish cinnamon to ochraceous salmon, fading to pale pinkish cinnamon to pinkish tan in age; margin uplifted; flesh white; odor not distinctive; taste not distinctive or slightly farinaceous.

GILLS: decurrent, close, narrow, forked, whitish.

STALK: 1⅛–3⅛" (3–8 cm) long, ⅛–½" (3–12 mm) thick, equal, sometimes enlarged at the base, hollow in age, dry, smooth, whitish, covered with white mycelium at the base.

SPORE PRINT: white.

MICROSCOPIC FEATURES: spores 5–10 × 3.5–6 μm, elliptic, smooth, hyaline, inamyloid.

FRUITING: solitary, scattered, or in groups on the ground in hardwoods and mixed woods; July–October; fairly common.

EDIBILITY: edible.

Clitocybe irina (Fries) Bigelow and Smith

CAP: 1⅝–5⅛" (4–13 cm) wide, convex when young, becoming broadly convex to flat with a low, broad umbo in age; surface smooth, sticky, soon becoming dry, sometimes with water spots or pitted, whitish when young, becoming buff to pinkish buff in age; margin inrolled and cottony when young, becoming expanded and smooth in age, often lobed or wavy; flesh whitish to pinkish; odor fragrant; taste not distinctive.

GILLS: attached to subdecurrent, crowded, whitish when young, becoming pale pinkish buff in age.

STALK: 1⅝–3⅛" (4–8 cm) long, ⅜–1" (1–2.5 cm) thick, nearly equal or enlarged downward, solid, dry, whitish to pale tan, bruising or aging brownish, smooth to scurfy, often longitudinally striate in age; base sometimes bulbous or club-shaped.

SPORE PRINT: pale pinkish buff.

MICROSCOPIC FEATURES: spores 7–10 × 4–5 μm, elliptic, finely roughened or smooth, hyaline, inamyloid.

FRUITING: scattered or in groups on the ground under conifers, especially spruce, in mixed hardwoods or borders of swamps; August–October; fairly common.

EDIBILITY: edible.

Clitocybe nuda (Fries) Bigelow and Smith

COMMON NAME: Blewit.

CAP: 1⅝–5⅞" (4–15 cm) wide, convex when young, becoming broadly convex to nearly flat in age; surface smooth, hygrophanous, slightly viscid and shiny when moist, dull when dry, sometimes finely cracked over the disc, with or without a low, broad umbo, violet to lilac-gray to pinkish buff, fading to pinkish tan or buff in age; margin inrolled when young, becoming expanded and occasionally uplifted in age, wavy, faintly striate when moist; flesh pale violet or lilac; odor fragrant; taste not distinctive.

GILLS: notched to sinuate, close to crowded, pale lavender to violet or lilac, becoming lilac-buff to brownish in age.

STALK: 1⅛–3" (3–7.5 cm) long, ⅜–1⅛" (1–3 cm) thick, equal, often abruptly bulbous or club-shaped at the base, solid, dry, fibrillose to scurfy, whitish, pale violet or lavender, bruising dark lavender, becoming brownish in age.

SPORE PRINT: pinkish buff.

MICROSCOPIC FEATURES: spores 5.5–8 × 3.5–5 μm, elliptic, roughened or sometimes smooth, hyaline, inamyloid.

FRUITING: solitary, in groups, or in clusters on the ground under hardwoods and conifers, in meadows and lawns, on decaying vegetable matter, and near compost piles; August–November; fairly common.

EDIBILITY: edible.

COMMENTS: when collecting for the table, avoid confusion with possibly poisonous look-alikes, especially *Cortinarius* species, which have a rusty brown spore print.

Clitocybe odora (Fries) Kummer

COMMON NAME: Anise-scented Clitocybe.

CAP: ¾–3⅜" (2–8.5 cm) wide, convex when young, becoming broadly convex to nearly flat in age, often shallowly depressed over the disc; surface moist, radially streaked with tiny fibers, bluish to bluish green to greenish to white, with a buff disc, sometimes with a whitish bloom when young; flesh whitish to buff; odor and taste like anise or licorice.

GILLS: attached to subdecurrent, close, broad, whitish to buff.

STALK: ⅝–3½" (1.5–9 cm) long, ⅛–⅝" (4–16 mm) thick, equal or sometimes enlarged at the apex or base, solid, becoming hollow in age, moist, covered with matted fibrils, whitish to buff; base covered with white mycelium.

SPORE PRINT: pinkish to pinkish cream.

MICROSCOPIC FEATURES: spores 5–9 × 3.5–5 μm, elliptic to oval, smooth, hyaline, inamyloid.

FRUITING: solitary, scattered, or in groups on the ground under hardwoods and mixed woods; July–September; occasional.

EDIBILITY: edible.

Clitocybe robusta Peck

CAP: 2–7" (5–18 cm) wide, convex when young, becoming broadly convex to nearly flat in age, sometimes shallowly depressed at the center; surface smooth, satiny, dry, viscid when wet, white, becoming pale buff to pale tan in age; margin incurved to inrolled when young, becoming expanded, wavy, and lobed in age; flesh white; odor pungent and unpleasant; taste rancid or unpleasant.

GILLS: attached to subdecurrent, close to crowded, sometimes forked or crossveined, white, becoming pale buff to pale pinkish buff in age.

STALK: 1⅝–4" (4–10 cm) long, ⅜–1⅜" (1–3.5 cm) thick, nearly equal down to a bulbous or club-shaped base, solid, becoming hollow in age, whitish, covered with matted fibrils, sometimes striate.

SPORE PRINT: pale yellow.

MICROSCOPIC FEATURES: spores 5–8 × 2.5–5 μm, elliptic, smooth, hyaline, inamyloid.

FRUITING: scattered, in groups, or sometimes clustered on the ground under hardwoods and conifers; August–October; occasional to fairly common.

EDIBILITY: unknown.

Clitocybe squamulosa (Fries) Kummer

CAP: ¾–3⅛" (2–8 cm) wide, broadly convex to flat when young, becoming deeply depressed and funnel-shaped in age; surface dry, smooth, sometimes minutely scaly over the disc or with tiny fibrils overall, orange-cinnamon to brownish orange, sometimes darker over the disc; flesh whitish to pale buff; odor and taste not distinctive or weakly farinaceous.

GILLS: decurrent, close to subdistant, forked, sometimes crossveined, cream to pale buff.

STALK: 1⅛–2¾" (3–7 cm) long, ⅛–⅜" (3–10 mm) thick, equal or slightly enlarged at the base, hollow in age, dry, smooth or longitudinally striate, colored like the cap or somewhat darker; base often covered with white mycelium.

SPORE PRINT: white.

MICROSCOPIC FEATURES: spores 5–7.5 × 3–4.5 μm, elliptic, smooth, hyaline, inamyloid.

FRUITING: solitary, scattered, or in groups on the ground or among mosses under conifers, sometimes in lawns; July–October; occasional.

EDIBILITY: unknown.

Clitocybe subclavipes Murrill

CAP: 1⅛–3⅛" (3–8 cm) wide, convex when young, becoming flat with a shallowly depressed center in age; surface dry, dull, with tiny, matted fibrils overall, occasionally with water spots when wet, grayish white to brownish tan, often darker over the disc; margin usually uplifted and broadly lobed in age; flesh whitish; odor and taste not distinctive or slightly farinaceous.

GILLS: decurrent, close to subdistant, narrow, sometimes forked and crossveined, whitish to cream.

STALK: 1⅛–2⅜" (3–6 cm) long, ¼–⅝" (7–15 mm) thick, enlarged downward to a bulbous or club-shaped base, often curved, stuffed, fibrillose, grayish white to buff; base paler and sometimes coated with white mycelium.

SPORE PRINT: white.

MICROSCOPIC FEATURES: spores 4.5–7 × 3–4.5 μm, elliptic, smooth, hyaline, inamyloid, rarely dextrinoid.

FRUITING: solitary, scattered, or in groups on the ground under hardwoods or conifers; July–October; occasional.

EDIBILITY: unknown.

Clitocybe subditopoda Peck

CAP: 1–2⅛" (2.5–5.5 cm) wide, convex when young, becoming nearly flat with a shallowly depressed center in age; surface smooth, hygrophanous, shiny and silky when moist, becoming dull and opaque, grayish brown with a faint purplish tint when young, becoming yellowish to brownish tan or olive-brown in age, remaining darkest over the disc; margin inrolled when young, becoming expanded in age, translucent-striate when moist; flesh brownish; odor and taste farinaceous.

GILLS: decurrent, close, sometimes crossveined, grayish white, becoming grayish brown to dull brownish tan.

STALK: ¾–2⅜" (2–6 cm) long, ⅛–⅜" (3–10 mm) thick, nearly equal or enlarged at the base or apex, hollow in age, often curved near the base, colored like the cap, coated with minute, whitish fibrils when young.

SPORE PRINT: white.

MICROSCOPIC FEATURES: spores 3.5–6 × 2.5–4 μm, elliptic, smooth, hyaline, inamyloid.

FRUITING: in groups or clustered, often in arcs or fairy rings, on needle litter under conifers, especially pine and spruce; September–November; fairly common.

EDIBILITY: unknown.

Clitocybe tarda Peck

CAP: ¾–3" (2–8 cm) wide, convex when young, becoming broadly convex to flat in age, sometimes shallowly depressed on the disc, with or without a low, broad umbo; surface smooth, moist, hygrophanous, vinaceous brown to violaceous brown, fading to pale cinnamon to fawn, often darker on the disc; margin translucent-striate when fresh and moist; flesh thin, watery brownish; odor and taste not distinctive.

GILLS: attached when young, subdecurrent in age, close to subdistant, pinkish to pale violet-tan or pinkish buff.

STALK: ¾–3⅛" (2–8 cm) long, ⅛–⅜" (2–10 mm) thick, equal or tapered downward, solid, flexible, streaked with white fibrils over a watery brown ground color.

SPORE PRINT: pinkish buff.

MICROSCOPIC FEATURES: spores 6–8 × 3.5–5 μm, elliptic, with minute warts, hyaline, inamyloid.

FRUITING: in groups or clusters on the ground in cultivated or fertilized areas such as gardens, lawns, fields, compost piles, and occasionally sawdust piles, rarely in woods; August–November; occasional.

EDIBILITY: unknown.

COMMENTS: *Clitocybe tarda* var. *alcalina* (edibility unknown) has an alkaline odor; it has been reported from Tennessee.

Clitocybe trullaeformis (Fries) Quélet

CAP: ⅜–1½" (1–5 cm) wide, convex when young, becoming flat to funnel-shaped in age; surface dry, velvety over the disc, sometimes minutely scaly, becoming cracked overall in age, grayish, grayish brown, olive-brown to blackish, darkest over the disc; flesh whitish; odor and taste farinaceous.

GILLS: decurrent, close to subdistant, narrow, often forked, sometimes crossveined, whitish to ivory-yellow.

STALK: ⅜–1¾" (1–4.5 cm) long, 1/16–⅜" (1.5–5 mm) thick, nearly equal or tapering downward, solid, smooth, sometimes striate, pale tan to pale olive-buff, usually with white mycelium at the base.

SPORE PRINT: white.

MICROSCOPIC FEATURES: spores 4–7 × 2.5–4 μm, elliptic, smooth, hyaline, inamyloid.

FRUITING: scattered or in groups on the ground under conifers and hardwoods; June–October; occasional.

EDIBILITY: unknown.

Genus *Clitocybula*

Clitocybula lacerata (Scopoli) Métrod

CAP: ¾–2⅜" (2–6 cm) wide, convex, becoming broadly convex, typically with a depressed disc; margin becoming uplifted, wavy, and torn in age; surface conspicuously radially streaked, at least on the disc, moist, color variable, pale to dark grayish brown or brownish gray when young, fading to pale dull brown to tan with a whitish to buff margin; flesh whitish; odor and taste not distinctive.

GILLS: attached to subdecurrent, subdistant, often crossveined, white with a grayish tint in age.

STALK: ¾–2" (2–5 cm) long, ⅛–3/16" (2–5 mm) thick, nearly equal or tapered downward, often curved, smooth, hollow, white, becoming pale gray in age.

SPORE PRINT: white.

MICROSCOPIC FEATURES: spores 6–8 × 4.5–6 μm, elliptic to broadly elliptic, smooth, hyaline, amyloid.

FRUITING: usually in very dense clusters on decaying logs or stumps, sometimes attached to buried wood or on soil in conifer or hardwoods; July–September; occasional.

EDIBILITY: unknown.

Genus *Clitopilus*

Clitopilus prunulus (Scopoli : Fries) Kummer

COMMON NAME: Sweetbread Mushroom.

CAP: 1¼–4" (3–10 cm) wide, convex to flat or even slightly sunken in age; surface dry, felty, whitish to grayish, with an inrolled margin that becomes wavy and irregular in age; flesh white, firm; odor and taste farinaceous or similar to bread dough.

GILLS: decurrent, close, white, becoming pinkish in age.

STALK: 1–3" (2.5–7.5 cm) long, ⅛–⅝" (3–15 mm) thick, dry, dull white to pale grayish, often somewhat eccentric.

SPORE PRINT: salmon-pink.

MICROSCOPIC FEATURES: spores 9–12 × 4–7 μm, elliptic, with longitudinal ridges in profile, appearing angular in end view, inamyloid.

FRUITING: scattered or in groups on the ground in open woods; June–September; occasional to frequent.

EDIBILITY: edible with caution.

COMMENTS: this is an edible species, but it must be accurately identified to rule out poisonous *Entoloma* species. The pink, elliptic, longitudinally striate spores are the diagnostic hallmark of the genus *Clitopilus*.

Genus *Collybia*

Members of the genus *Collybia* are small to medium mushrooms with a central cartilaginous stalk, thin flesh, and a white, cream, or pinkish buff spore print. They have gills that are variously attached but never decurrent, and they lack a partial veil and ring on the stalk. Various odors are produced when the flesh is crushed, but none has a farinaceous odor. Spores are mostly inamyloid, but a few are dextrinoid.

Although some are edible, a few are inedible, and the edibility of the majority is unknown. Large species of the genus *Marasmius,* as well as species of *Clitocybula* and *Callistosporium* that macroscopically resemble *Collybia* species, are identified at the beginning of the key.

Key to Species of *Collybia* and Allies

1.   Growing on the ground, on decaying leaves or in grass → 2.
1.   Growing on decaying wood; solitary, or in groups but not in dense clusters → 3.
1.   Growing on decaying wood in dense clusters → 4.
1.   Not with the above combination of characters → 5.
      2.   Cap light brown to dull yellowish brown, fading to pale tan or pale yellowish white, with a brownish umbo that does not fade substantially in age, 1⅛–4⅜" (3–11 cm) wide; scattered or in groups on the ground in mixed woods; odor and taste not distinctive; gills white to dull yellow, lamellulae numerous; stalk ⅛–½" (3–12 mm) thick, dry, chalky white, typically vertically striate; spores 6–7 × 3–4.5 μm; edibility unknown—*Marasmius nigrodiscus* (Peck) Halling.
      2.   Cap light yellow, sometimes tinted orangish in spots (especially at the edge), ¾–2½" (2–6.5 cm) wide; growing on fallen deciduous leaves; odor and taste either not distinctive or somewhat radishy; gills yellowish white to pale orangish yellow, lamellulae numerous; stalk up to ⅜" (9 mm) thick, coated with minute hairs, whitish to pale yellow, hollow; rhizomorphs absent, but with an abundant pale yellow basal mycelium that usually forms a mat of the leaves on which it grows; spores 6–8 × 3–5 μm—*Marasmius strictipes* (Peck) Singer (see p. 202).

2. Cap tan, brownish or yellowish white, ⅜–1¾" (1–4.5 cm) wide; growing in grass on lawns, often in fairy rings or arcs; flesh odor sweet or fragrant, like almonds or not distinctive; taste not distinctive; gills pale tannish to yellowish white, lamellulae numerous; stalk smooth or longitudinally twisted-striate, ¹⁄₁₆–³⁄₁₆" (1.5–4.5 mm) thick, white to yellowish or faintly orangish near the apex, more brownish below; rhizomorphs absent; spores 6.5–10 × 3.5–6 μm—*Marasmius oreades* (Bolton : Fries) Fries (see p. 200).

2. Not with the above combination of characters → 5.

3. Cap olive-brown when young, becoming yellowish brown in age, ¾–2½ (2–6.5 cm) wide; growing on decaying conifer wood; flesh odor not distinctive; taste not distinctive or unpleasant to bitter; gills yellow to golden yellow, becoming dark red to reddish brown when dried; stalk often somewhat flattened, colored like the cap or darker, becoming dark reddish brown from the base up as the surface dries; spores 4.5–6.5 × 3–4.5 μm, many staining vinaceous in KOH; dried pieces of gill tissue impart a conspicuous vinaceous tinge when mounted in KOH—*Callistosporium luteo-olivaceum* (Berkeley and Curtis) Singer.

3. Cap purplish red to brownish violet when fresh and moist, drying to pinkish tan from the center toward the edge but retaining a bluish to brownish purple band at the margin; growing on decaying oak logs; flesh odor not distinctive; taste slowly bitter with a metallic aftertaste; gills pinkish to yellow with purple overtones; edges reddish purple; stalk usually curved, reddish purple with yellow toward the base; spores 4–6 × 3–4 μm, many staining vinaceous in KOH; dried pieces of gill tissue impart a conspicuous vinaceous tinge when mounted in KOH—*Callistosporium purpureomarginatum* Fatto and Bessette (see p. 73).

3. Not with the above combination of characters → 5.

4. Cap pale to dark gray-brown on the disc, progressively paler toward the margin, with radial streaks that are sometimes indistinct, convex at first, becoming broadly convex and often shallowly depressed; scaly or fibrillose-scaly on the disc; flesh whitish, odor and taste not distinctive; gills subdistant, crossveined; edges uneven to serrate and torn in age; stalk fibrillose-scaly to punctate (use a hand lens); spores 5–6.5 × 4–5.5 μm, subglobose; on decaying hardwood logs and stumps; edibility unknown—*Clitocybula oculus* (Peck) Singer.

4. Cap color variable, pale to dark grayish brown or brownish gray when young, fading to pale dull brown to tan with a whitish to buff margin, conspicuously radially streaked, at least on the disc; flesh whitish, odor and taste not distinctive; gills subdistant—*Clitocybula lacerata* (Scopoli) Métrod (see p. 87).

4. Cap dark brownish buff to dark grayish buff, with radial streaks that are sometimes indistinct, convex, sometimes with a small umbo; flesh whitish, odor and taste not distinctive; gills close to crowded; stalk scurfy, base coated with white hairs; spores 3.5–5 × 3.5–5 μm, globose; on decaying conifer logs—*Clitocybula familia* (Peck) Singer.

4. Cap whitish with a grayish to brownish gray or fuscous disc, convex at first, becoming somewhat depressed, with radial streaks that are sometimes indistinct; flesh whitish, odor and taste not distinctive; gills close, sometimes crossveined; stalk smooth to slightly scurfy; spores 4.5–6.5 × 3.5–5.5 μm, subglobose; on decaying conifer or hardwood—*Clitocybula abundans* (Peck) Singer.

4. Not with the above combinations of characters → 5.

5. Stalk covered over most of its length by conspicuous short or long hairs that may be erect or matted, apex sometimes smooth → 6.

5. Stalk smooth or nearly so to the unaided eye, sometimes longitudinally striate, pruinose, or furrowed, occasionally coated with short or long hairs only near the base or with inconspicuous hairs overall → 9.
    6. Cap surface conspicuously wrinkled and sulcate from the margin nearly to the disc at maturity → 7.
    6. Cap surface smooth to slightly wrinkled or obscurely wrinkled to sulcate only on the margin at maturity → 8.
7. Gills reddish purple to reddish lilac; stalk pale pinkish lilac, tinged brownish on the lower portion, coated with whitish hairs; cap reddish lilac to vinaceous purple, ⅜–1⅛" (1–3 cm) wide; flesh odor pungent and unpleasant, taste unpleasant; KOH stains cap, gills, and stalk bright blue; spores 6.5–8.5 × 3–4 μm; on humus and leaf litter; edibility unknown—*Collybia iocephala* (Berkeley and Curtis) Singer.
7. Gills whitish to pinkish buff, typically subdistant to distant; stalk buff to pinkish buff on the upper portion when young, becoming brown to dark brown from the base up and blackish brown on the base in age, nearly smooth at the apex, with whitish to grayish hairs below; cap cinnamon-brown, fading to cinnamon in age, ⅜–1½" (1–4 cm) wide; flesh odor not distinctive, taste bitter to somewhat acrid, rarely mild; spores 8.5–12 × 3–4.5 μm; on leaf litter, humus, and decaying twigs—*Collybia subnuda* (Ellis : Peck) Gilliam (see p. 96).
7. Gills whitish to pale pinkish buff, close; stalk whitish near the apex, orange-cinnamon to reddish brown below, coated with whitish hairs nearly overall, or whitish hairs on the upper portion and dull brownish orange hairs below; hairs often dull brownish orange overall in age; cap reddish brown, fading to cinnamon-brown then pale pinkish brown in age, ⅜–1⅛" (1–3 cm) wide; flesh odor and taste not distinctive; spores 6–9 × 3–4 μm; on the ground or sometimes on needle or leaf litter; edibility unknown—*Collybia biformis* (Peck) Singer.
7. Gills white, becoming pinkish tan in age, crowded; stalk whitish at the apex and pinkish tan below when young, becoming dingy pinkish tan overall in age, coated with matted to erect hairs from the base to the apex, with white rhizomorphs at the base; cap dull dingy tan, fading to pinkish buff, with a reddish brown disc, ¾–2" (2–5 cm) wide; flesh odor and taste not distinctive; spores 6–8 × 3–4 μm; on leaf litter in hardwoods; edibility unknown—*Collybia kauffmanii* Halling.
    8. Gills crowded to close, pinkish buff then cream; stalk pale pinkish buff near the apex, becoming pale cinnamon toward the base, coated with a dense layer of whitish hairs, covered with white mycelium at the base; cap reddish brown, fading to pale pinkish cinnamon, ⅜–2" (1–5 cm) wide; spores 7–11 × 3.5–5 μm; in dense clusters, groups, or scattered on leaf litter and humus—*Collybia confluens* (Persoon : Fries) Kummer (see p. 94).
    8. Gills subdistant, whitish; stalk brown to dark reddish brown, lower one-third covered by reddish brown matted hairs; cap orange-brown, fading to cinnamon, then pale dingy brown in age, ¼–1" (5–25 mm) wide; scattered or in groups on humus, among fallen deciduous leaves; edibility unknown—*Collybia semihirtipes* (Peck) Halling (see p. 96).
    8. Gills close to crowded, white to creamy white; stalk whitish and nearly smooth at the apex, reddish brown below, coated from the base to near the apex with dull reddish orange to orange-tawny, matted to erect hairs; base often thickened and spongy; cap reddish brown, fading to grayish orange with a whitish margin, ⅜–1½" (1–4 cm) wide; KOH and NH₄OH stain stalk surface green; spores 6–8.5 × 3–4 μm; scattered, in groups or clusters on leaf and needle litter; edibility unknown—*Collybia spongiosa* (Berkeley and Curtis) Singer.

9. Cap small, ⅛–⅝" (3–16 mm) wide at maturity; stalk narrow, ¹⁄₁₆" or less (1–1.5 mm) thick; often on the blackened remains of decaying mushrooms but also on humus or decaying wood → 10.

9. Cap medium to large, typically greater than 1" (2.5 cm) at maturity; growing on various substrates but not on the blackened remains of decaying mushrooms → 11.

   10. Stalk attached to a dark reddish brown, shiny, oval, appleseed-like sclerotium— *Collybia tuberosa* (Bulliard : Fries) Kummer (see p. 97).

   10. Stalk attached to a small yellowish to orange-ochraceous, rounded to irregular, often wrinkled sclerotium measuring ⅛–⅜" (3–8 mm) in diameter; stalk surface cinnamon-buff, covered with a thin coating of hairs near the base; cap pinkish buff to orange-buff, fading to dingy whitish and darker on the disc in age, ⅛–⅜" (2–10 mm) wide; margin sometimes striate; flesh odor and taste not distinctive; gills white to pinkish buff; spores 4.5–6 × 3–3.5 μm; edibility unknown—*Collybia cookei* (Bresadola) Arnold.

   10. Stalk not attached to a sclerotium, whitish to pale dingy orange, coated with a sparse layer of tiny hairs, covered at the base with white mycelium, often forming whitish rhizomorphs; cap white with a slight pinkish to orange tinge on the disc, becoming dingy pale orange in age, ⅛–⅜" (3–10 mm) wide; margin translucent-striate when moist; flesh odor and taste not distinctive; gills white to pinkish buff; spores 5–6 × 2–3 μm; edibility unknown—*Collybia cirrhata* (Persoon) Quélet.

11. Odor of crushed flesh like garlic, onion, or rotting cabbage → 12.

11. Odor fragrant or not distinctive → 13.

   12. Growing on woodchips, in groups or clusters; gills pale brown when young, becoming pale pinkish buff in age; cap sulcate nearly to the disc, dark reddish brown to pale pinkish cinnamon and darker in the grooves, ⅜–2⅜" (1–6 cm) wide; flesh odor like old garlic or onions, taste onion-like to slightly unpleasant; stalk dark reddish brown when young, fading to pale reddish brown in age; spores 8–9 × 3–4 μm; edibility unknown—*Collybia dysodes* Halling.

   12. Growing on hardwood leaf litter; gills white; cap smooth, dark reddish brown when young, fading in age to tan on the disc and pinkish gray to pinkish buff on the margin, finally whitish with pale yellowish tan on the disc—*Collybia polyphylla* (Peck) Singer (see p. 96).

   12. Growing on conifer needles; gills buff to cinnamon-buff; cap smooth except wrinkled to sulcate on the margin, dull brown, fading to orange-buff on the margin in age, ¼–1" (6–25 mm) wide; flesh odor like onion, garlic, or rotting cabbage, taste strongly of garlic; stalk ochraceous buff at the apex, pale reddish brown toward the base; spores 7.5–10 × 3–5 μm; edibility unknown—*Collybia pinastris* (Kauffman) Mitchel and Smith.

13. Flesh taste bitter, slightly acrid, metallic, or distinctly unpleasant → 14.

13. Flesh taste not distinctive or slightly unpleasant → 15.

   14. Taste slightly acrid or unpleasant; growing on woodchips, decaying wood, or buried wood in lawns; stalk conspicuously twisted-striate—*Collybia luxurians* Peck (see p. 95).

   14. Taste metallic and unpleasant or bitter; odor not distinctive or slightly unpleasant; growing on moist soil; gills pale orange-yellow when young, becoming pale ochraceous salmon in age, close to crowded; cap moist, smooth, dark brown when young, fading to dull orange-buff in age, ⅜–1¾" (1–4.5 cm) wide; KOH stains cuticle slightly olivaceous; stalk dull orange-buff when young, becoming yellowish orange at the apex and orange-brown toward the base, stalk base attached to a ball

of soil; spores 5.5–7 × 3–3.5 μm; April–June; edibility unknown—*Collybia earleae* (Murrill) Murrill.

14. Taste bitter; growing on the ground on well-decayed wood under conifers and hardwoods; gills yellow, sometimes orange-yellow, fading to cream in age, crowded; cap honey-yellow when fresh, fading to pale yellow in age, ⅝–1⅝" (1.5–4 cm) wide; stalk pale yellow, becoming ochraceous to tawny ochraceous in age, with pinkish buff to pale orange rhizomorphs at the base; spores 5–6.5 × 2.5–3.5 μm; edible—*Collybia subsulphurea* Peck.

14. Taste bitter; growing in humus or on buried wood under conifers; cap large, 2–6" (5–15.5 cm) wide, pinkish buff, becoming whitish in age, developing reddish brown to rusty streaks and spots in age—*Collybia maculata* (Albertini and Schweinitz : Fries) Kummer (see p. 95).

14. Nearly identical to the previous choice except the gills are distinctly yellow and the stalk may be white or yellow—*Collybia maculata* var. *scorzonerea* (Fries) Gillet.

15. Gills distinctly yellow, cinnamon-brown or brown → 16.

15. Gills white, whitish, pinkish buff, or orange-buff → 17.

16. Gills yellow, sometimes orange-yellow, fading to cream in age, crowded; cap honey-yellow when fresh, fading to pale yellow in age, ⅝–1⅝" (1.5–4 cm) wide; flesh odor fragrant or not distinctive; stalk pale yellow, becoming ochraceous to tawny-ochraceous in age, with pinkish buff to pale orange rhizomorphs at the base; spores 5–6.5 × 2.5–3.5 μm; edible—*Collybia subsulphurea* Peck.

16. Gills reddish brown to purple-brown or brown, blackish when dry; cap and stalk reddish brown to dark purple-brown or blackish; growing under hardwoods—*Collybia alkalivirens* Singer (see p. 93).

16. Gills cinnamon-brown, close; cap smooth, moist to slippery, becoming striate in age, dark reddish brown, fading to cinnamon-brown in age; growing on needle litter and humus under pines; flesh odor not distinctive; stalk cinnamon-brown on the upper portion, grayish white below; spores 9–10 × 3–4 μm; edibility unknown—*Collybia putilla* (Fries) Singer.

17. Growing among sphagnum mosses in dense clusters in bogs; cap reddish brown to chestnut-brown with a paler margin, fading in age to pale reddish brown with a light tan or buff margin; stalk brown—*Collybia acervata* (Fries) Kummer (see p. 93).

17. Growing on decaying wood or woodchip mulch and the surrounding soil → 18.

17. Growing on humus or needle litter under conifers, hardwoods, or mixed woods → 19.

18. Stalk longitudinally striate, nearly equal, not bulbous or flattened, whitish on the upper portion, dingy white to brownish toward the base, partially buried in the wood, with thick rhizomorphs at the base; cap dark reddish brown to purplish brown when young, fading to pale brown on the disc and darker toward the margin, ¾–2" (2–5 cm) wide; spores 5.5–7 × 3–4.5 μm; edibility unknown—*Collybia egregia* Halling.

18. Stalk longitudinally striate to twisted-striate or smooth, buff near the apex, becoming ochraceous buff toward the base, ⅛–¼" (2–7 mm) thick, base somewhat bulbous, with white mycelium and rhizomorphs; cap ⅜–1½" (1–4 cm) wide, surface lubricous, hygrophanous, medium brown when young and fresh, fading to cream-buff as it dries; KOH stains cuticle grayish; flesh odor and taste not distinctive; gills white to ivory, edges even at first, becoming eroded or finely toothed in age; spore print white; spores 6–7 × 2.5–3.5 μm; growing in groups or clusters on woodchip mulch and the surrounding soil; edibility unknown—*Collybia brunneola* Vilgalys and Miller.

18. Stalk smooth, not longitudinally striate, bulbous at the base, typically enlarged near the apex and base, and narrowed in the middle, sometimes flattened—*Collybia dichrous* (Berkeley and Curtis) Gilliam (see p. 94).

18. Stalk smooth, not longitudinally striate, equal, brown; growing in dense clusters—*Collybia acervata* (Fries) Kummer (see p. 93).

18. Stalk smooth, not longitudinally striate, nearly equal or enlarged downward, sometimes bulbous, whitish near the apex, becoming pale yellow to orange-yellow or darker toward the base—*Collybia dryophila* (Bulliard : Fries) Kummer (see p. 95).

19. Gill edges serrate in buttons, mature and aged caps; gills close, whitish; stalk whitish, somewhat translucent, enlarged toward the base or nearly equal; cap brownish pink to pale cinnamon when young, fading to pale pinkish cinnamon, then pinkish buff, ¾–2" (2–5 cm) wide; flesh odor not distinctive; spores 5–7.5 × 3–4 μm, dextrinoid; on humus under conifers; edibility unknown—*Collybia lentinoides* (Peck) Saccardo.

19. Gill edges not serrate, straight and even when young, typically becoming uneven, somewhat scalloped to eroded or cut in age; gills close to crowded, white, sometimes pinkish in age; stalk enlarged downward, longitudinally striate to sulcate, pale cinnamon-brown, whitish at the base; cap reddish brown to bay-brown when young, fading to reddish cinnamon in age, smooth and slippery when moist, ¾–2⅜" (2–6 cm) wide; flesh odor not distinctive; spore print pale pink; spores 7–10 × 3.5–5 μm; edible—*Collybia butyracea* (Bulliard : Fries) Kummer.

19. Gill edges smooth and even, not serrate; growing in dense clusters; stalk brown—*Collybia acervata* (Fries) Kummer (see p. 93).

19. Gill edges smooth and even, not serrate; growing scattered, in groups or clusters; stalk whitish near the apex, becoming pale yellow to orange-yellow or darker toward the base—*Collybia dryophila* (Bulliard : Fries) Kummer (see p. 95).

Collybia acervata (Fries) Kummer

COMMON NAME: Clustered Collybia.

CAP: 1⅝–2" (1.5–5 cm) wide, convex when young, becoming broadly convex to flat in age; surface smooth, dry to moist, hygrophanous, reddish brown to chestnut brown when young, paler along the margin, fading in age to pale reddish brown with a light tan or buff margin, opaque, sometimes translucent-striate; flesh whitish; odor and taste not distinctive.

GILLS: attached or notched, close to crowded, whitish with a pinkish tinge.

STALK: 1½–4" (4–10 cm) long, ⅛–¼" (2–6 mm) thick, equal, hollow, brittle, dry, smooth, shiny, pale yellow-brown to purple-brown, reddish brown or brown, with white hairs toward the base.

SPORE PRINT: white to cream.

MICROSCOPIC FEATURES: spores 5.5–7 × 2.5–3 μm, lacrymoid to elliptic, smooth, hyaline, inamyloid.

FRUITING: in dense clusters or groups among sphagnum mosses in bogs, on decaying wood, or on rich humus in conifer or mixed woods; July–October; occasional.

EDIBILITY: unknown.

Collybia alkalivirens Singer

COMMON NAME: Little Brown Collybia.

CAP: ⅜–1⅜" (1–3.5 cm) wide, hemispheric when young, becoming convex to nearly flat with a low broad umbo in age; surface smooth, hygrophanous, dry to moist, dull, dark purple-brown to reddish brown when moist, fading to red-brown to yellow-brown when dry; margin striate; flesh whitish to reddish brown; odor and taste not distinctive.

GILLS: attached to nearly free, close, sometimes crossveined, reddish brown to purple-brown to brown, blackish when dry.

STALK: 1⅛–3⅛" (3–8 cm) long, 1/16–⅛" (1–3 mm) thick, nearly equal or enlarged slightly at the base, hollow, smooth, dull, dark purple-brown to brown to blackish, sometimes paler at the apex, covered with matted fibers at the base.

SPORE PRINT: white.

MICROSCOPIC FEATURES: spores 5–9 × 2–5 μm, lacrymoid, smooth, hyaline, inamyloid.

FRUITING: solitary, scattered, in groups or caespitose on the ground, on decaying wood, in leaf litter, or among mosses under hardwoods; May–July; uncommon.

EDIBILITY: unknown.

COMMENTS: all parts stain green in KOH or NH_4OH.

Collybia confluens (Persoon : Fries) Kummer

COMMON NAME: Tufted Collybia.

CAP: ⅜–2" (1–5 cm) wide, convex, becoming nearly plane; surface moist then soon dry, smooth, reddish brown, fading to pale pinkish cinnamon or whitish in age; flesh thin, whitish; odor and taste not distinctive or sometimes slightly garlic-like.

GILLS: free or attached, crowded to close, pinkish buff, then cream, fading to whitish in age.

STALK: 1–4" (2.5–10 cm) long, 1/16–¼" (1.5–6 mm) thick, typically flared at the apex and base, otherwise nearly equal, pale pinkish buff near the apex, becoming pale cinnamon toward the base, coated with a dense layer of whitish to pale grayish hairs, covered with white mycelium at the base.

SPORE PRINT: white.

MICROSCOPIC FEATURES: spores 7–11 × 3–5 μm, narrowly elliptic, smooth, hyaline, inamyloid.

FRUITING: in dense clusters, in groups, or scattered on leaf litter and humus; July–November; common.

EDIBILITY: edible.

Collybia dichrous (Berkeley and Curtis) Gilliam

CAP: ⅜–1½" (1–4 cm) wide, convex when young, becoming broadly convex to flat with an uplifted margin, sometimes shallowly depressed over the disc or with a low umbo; surface dry, smooth, radially striate and finely wrinkled from the margin to the disc or nearly so, translucent-striate when wet, dark brown to reddish brown over the disc, paler toward the margin, fading to light brown to grayish brown overall in age; flesh white; odor and taste not distinctive.

GILLS: attached at first, becoming nearly free in age, close to subdistant, crossveined, white to pale pinkish buff, sometimes with rusty brown spots.

STALK: ⅜–2" (1–5 cm) long, 1/16–¼" (1–5 mm) thick, typically enlarged near the apex and base and narrowed in the middle, sometimes flattened, bulbous at the base, hollow, flexible, dry, whitish overall at first, soon reddish brown from the base up, with tiny fibrils toward the apex, often with hairs at the base.

SPORE PRINT: white to cream.

MICROSCOPIC FEATURES: spores 10–12 × 3–4.5 μm, lacrymoid to elliptic, smooth, hyaline, inamyloid.

FRUITING: in groups or caespitose on hardwood logs and stumps; July–September; occasional.

EDIBILITY: unknown.

Collybia dryophila (Bulliard : Fries) Kummer

COMMON NAME: Oak-loving Collybia.

CAP: ⅜–2¾" (1–7 cm) wide, convex when young, becoming broadly convex to flat in age; surface smooth, moist, silky, hygrophanous, dark reddish brown when young, soon fading to orange-brown, tan, or yellowish tan, usually remaining darker over the disc; margin often uplifted in age; flesh whitish to yellowish; odor and taste not distinctive.

GILLS: attached to nearly free, crowded to close, whitish to pinkish buff or sometimes yellowish in age.

STALK: 1⅛–3½" (3–9 cm) long, ⅛–¼" (2–8 mm) thick, nearly equal or enlarged downward, base sometimes bulbous and often covered with white mycelium, flexible, hollow, dry, smooth when young, finely striate in age, whitish near the apex, becoming pale yellow to orange-yellow or darker toward the base.

SPORE PRINT: white to cream.

MICROSCOPIC FEATURES: spores 5–7 × 3–3.5 μm, lacrymoid to elliptic, smooth, hyaline, inamyloid.

FRUITING: scattered, in groups, or caespitose on humus or decaying wood in oak-pine forests; June–November; common.

EDIBILITY: edible.

COMMENTS: often confused with *C. butyracea* (edible), which has a pale pink spore print. This mushroom is often attacked by the Collybia Jelly, *Syzygospora mycetophila* (see p. 433). A variety with yellow gills has been called *C. dryophila* var. *luteifolia* and has been renamed by Roy Halling as *C. dryophila* var. *funicularis* (edible).

Collybia luxurians Peck

CAP: ⅜–3⅛" (2–8 cm) wide, convex when young, becoming broadly convex to nearly flat in age; surface smooth, dry to moist, dark reddish brown over the disc when fresh, fading to light brown or buff toward the margin, becoming light brown overall in age; margin typically uplifted and wavy in age; flesh whitish to pinkish buff; odor not distinctive; taste slightly acrid or unpleasant.

GILLS: attached, close to crowded, whitish to pinkish buff.

STALK: 1–4¾" (4–12 cm) long, ⅛–½" (2–13 mm) thick, nearly equal or tapered downward, dry, conspicuously twisted-striate, covered overall with fine hairs that brush off easily, whitish to buff toward the apex, pale brown below, base covered with white mycelium, rhizomorphs typically present.

SPORE PRINT: cream.

MICROSCOPIC FEATURES: spores 6–10 × 3–6 μm, lacrymoid, smooth, hyaline, inamyloid.

FRUITING: in dense clusters or groups on decaying wood, woodchips, or in lawns on buried wood; June–September; occasional.

EDIBILITY: unknown.

Collybia maculata (Albertini and Schweinitz : Fries) Kummer

COMMON NAME: Spotted Collybia.

CAP: 1⅜–4" (3.5–10 cm) wide, convex when young, becoming broadly convex to nearly flat in age, sometimes with a low, broad umbo; surface smooth, dry to moist, pinkish buff when young, becoming whitish in age, developing reddish brown to rusty streaks and spots overall but especially toward the center; flesh white; odor not distinctive or mildly unpleasant; taste bitter.

GILLS: attached, close to crowded, whitish to buff, developing brown to rusty spots in age.

STALK: 2–4¾" (5–12 cm) long, ⅜–½" (1–1.3 cm) thick, nearly equal, slightly rooting at the base, hollow, dry, longitudinally striate, whitish, developing rusty brown spots in age.

SPORE PRINT: pinkish buff to yellowish buff.

MICROSCOPIC FEATURES: spores 5.5–7 × 5–6 μm, globose to subglobose, smooth, hyaline, often dextrinoid.

FRUITING: solitary, scattered, in groups or fairy rings on humus or buried wood under conifers and mixed woods; July–October; fairly common.

EDIBILITY: inedible due to the bitter taste.

Collybia polyphylla (Peck) Singer

CAP: ¾–2¾" (2–7 cm) wide, convex when young, becoming broadly convex to flat in age; surface smooth, dry, dark reddish brown when young, fading in age to tan on the disc and pinkish buff on the margin, finally whitish with pale yellowish tan on the disc; flesh whitish; odor and taste like garlic.

GILLS: attached or notched, becoming nearly free in age, crowded, white.

STALK: 1¼–2½" (3–6.5 cm) long, ⅛–¼" (2–5 mm) thick, nearly equal, hollow, flexible, often pruinose at the apex, nearly smooth below, pale pinkish or yellowish to reddish brown, covered overall with inconspicuous white to pale gray hairs.

SPORE PRINT: white to cream.

MICROSCOPIC FEATURES: spores 5.5–7.5 × 3–4 μm, nearly elliptic to lacrymoid, smooth, hyaline, inamyloid.

FRUITING: scattered, in groups, or caespitose on leaf litter under hardwoods; June–September; occasional.

EDIBILITY: unknown.

Collybia semihirtipes (Peck) Halling

CAP: ¼–1" (5–25 mm) thick, convex to nearly flat in age, smooth, hygrophanous, orange-brown, fading to cinnamon, then pale, dingy brown in age; margin obscurely wrinkled to sulcate in age; flesh whitish; odor and taste not distinctive.

GILLS: attached to sinuate, subdistant, sometimes crossveined and forked, whitish; edges delicately toothed.

STALK: ¾–3½" (2–9 cm) long, ¹⁄₁₆–¼" (1.5–6 mm) thick, nearly equal, typically curved near the base, brown to dark reddish brown, lower one-third covered by reddish brown matted hairs.

SPORE PRINT: cream.

MICROSCOPIC FEATURES: spores 7.5–10 × 3–5 μm, lacrymoid, smooth, hyaline, inamyloid.

FRUITING: scattered or in groups on humus, among fallen deciduous leaves in hardwoods and mixed woods; May–June and September–October; occasional.

EDIBILITY: unknown.

Collybia subnuda (Ellis : Peck) Gilliam

CAP: ⅜–1½" (1–4 cm) wide, bluntly convex to nearly plane, sometimes with a low, broad umbo; surface dry, smooth, coated with flattened fibrils, becoming wrinkled to somewhat shallowly sulcate, especially along the margin, cinnamon-brown, fading to cinnamon in age; flesh thin, whitish; odor not distinctive; taste bitter to somewhat acrid, rarely not distinctive.

GILLS: attached, notched or nearly free, subdistant to distant, whitish to pinkish buff.

STALK: ¾–2¾" (2–7 cm) long, ¹⁄₃₂–⅛" (1–3 mm) thick, typically flared at the apex, otherwise nearly equal, dry, buff to pinkish buff on the upper portion when young, becoming brown to dark brown from the base up and blackish brown on the base in age, nearly smooth at the apex, with whitish to grayish hairs below.

SPORE PRINT: ivory-white.

MICROSCOPIC FEATURES: spores 8–11 × 3–5 μm, lacrymoid to elliptic, smooth, hyaline, inamyloid.

FRUITING: scattered, in groups, or in clusters on leaf litter, humus, and decaying twigs; July–October; fairly common.

EDIBILITY: unknown.

Collybia tuberosa (Bulliard : Fries) Kummer

COMMON NAME: Tuberous Collybia.

CAP: ⅛–⅝" (2–15 mm) wide, convex when young, becoming flat in age, often with a shallow depression over the disc or with a small broad umbo; surface dry, dull, smooth to finely hairy, pinkish buff to buff on the disc, whitish toward the margin, fading in age to whitish overall; margin occasionally striate; flesh whitish; odor and taste not distinctive.

GILLS: attached, close to subdistant, whitish to pinkish buff.

STALK: ⅜–2" (1–5 cm) long, up to ¹⁄₁₆" (1.5 mm) thick, nearly equal, hollow in age, flexible, dry, scurfy or pruinose at the apex, covered with tiny hairs at the base, whitish to pinkish buff or tannish, brusing darker when handled; base attached to a dark reddish brown, shiny, oval, appleseed-like sclerotium that measures ⅛–½" × ⅛–¼" (3–13 × 2–6 mm).

SPORE PRINT: white.

MICROSCOPIC FEATURES: spores 4–6 × 3–3.5 μm, elliptic to lacrymoid, smooth, hyaline, inamyloid.

FRUITING: in groups on the blackened remains of decaying mushrooms, especially species of *Russula* and *Lactarius,* or on humus or decaying wood; August–November; occasional.

EDIBILITY: unknown.

Genus *Conocybe*

The genus *Conocybe* is estimated to have fewer than fifty species in North America. They are small, fragile mushrooms that grow on a variety of substrates, including grass, humus, decaying wood, and dung. Their spore prints are rusty brown to reddish cinnamon, and their spores are smooth and often truncate, with an apical pore. They have a long, slender, central stalk with or without a ring.

Species in this genus are not commonly collected for the table. The edibility of most species is unknown, and one species is deadly poisonous. Many species are very difficult to identify, even with the aid of a microscope. Because of the complexity of the genus, a comprehensive key to species is beyond the scope of this book. We have included a limited key to some of the more common and distinctive species that occur in this region.

Key to Species of *Conocybe*

1.   Cap white to creamy white, becoming tinged ochraceous on the disc → 2.
1.   Cap red-brown, rusty brown, cinnamon, or dark ochre to fulvous → 3.
      2.   Gills crimped or wavy, strongly intervenose; cap surface wrinkled, not striate on the margin, ¼–1" (6–25 mm) wide; stalk 2–4" (5–10 cm) long, about ⅛" (2–4 mm) thick, white to pale ochraceous, white-pruinose, with a slightly bulbous base; spores 12–16 × 8–11 μm; in grassy areas; edibility unknown—*C. crispa* (Longyear) Singer.
      2.   Gills not crimped or wavy, not strongly intervenose; cap surface smooth, striate from the margin to the disc—*C. lactea* (Lange) Métrod (see p. 98).
3.   Ring present on the stalk, sometimes detached and falling away → 4.
3.   Ring absent → 5.
      4.   Cap surface wrinkled to strongly wrinkled, conic, becoming convex, striate on the margin, dark rusty brown, becoming fulvous or paler in age, ⅜–1" (1–2.5 cm) wide; stalk white to whitish, ¾–1½" (2–4 cm) long, ¹⁄₁₆–⅛" (1.5–4 mm) thick; ring conspicuous, membranous, whitish to pale brown, striate on the upper surface; gills yellowish, becoming tawny; edges white, finely scalloped; spores 8–11 × 4.5–6 μm; on rich humus and hardwood debris; ediblility unknown— *C. rugosa* (Peck) Watling.
      4.   Cap surface smooth, rarely wrinkled, conic, becoming convex, striate nearly to the disc, pale cinnamon to tawny-brown, to orange-brown, fading in age, ¼–1" (6–26 mm) wide; stalk orange-brown to dark brown, scurfy, ¾–2" (2–5 cm) long, about ¹⁄₁₆" (1–2 mm) thick; ring conspicuous, membranous, ochre to yellow-brown, striate on the upper surface; gills whitish, becoming fulvous to rusty brown; edges white, finely fringed; spores 7–12 × 4–6.5 μm; in grassy areas, among mosses, and on wood chips; deadly poisonous—*C. filaris* (Fries) Kühner.
5.   Stalk 2¾–5½" (7–14 cm) long, about ⅛" (1.5–3.5 mm) thick, base distinctly swollen to bulbous, sometimes abruptly bulbous, whitish when very young, soon becoming ochre, then reddish brown, darkening in age; ring absent; cap ¼–1⅜" (6–35 mm) wide, bluntly conic to bell-shaped, striate or weakly so when fresh, smooth, shiny; gills cinnamon to rusty brown; edges white, finely scalloped to fringed in age; spores 11–16 × 7–9 μm; on manured fields, in gardens, and on straw and sawdust mixed with dung; edibility unknown—*C. subovalis* (Kühner) Kühner and Watling.
5.   Stalk 1⅛–3⅛" (3–8 cm) long, ¹⁄₁₆–⅛" (1.5–4 mm) thick, base not enlarged or slightly enlarged; spores 10–14 × 5–6 μm; otherwise very similar to the previous choice— *C. tenera* (Schaeffer : Fries) Kühner complex (see p. 99).

Conocybe lactea (Lange) Métrod

COMMON NAME: White Dunce Cap.

CAP: ⅜–1⅛" (1–3 cm) wide, conic to bluntly conic, becoming bell-shaped; surface dry, smooth, striate from the margin to the disc, whitish to creamy white, tinged ochraceous on the disc; flesh whitish; odor and taste not distinctive.

GILLS: nearly free, close, very narrow, whitish when young, becoming tawny to reddish cinnamon.

STALK: 1½–4" (4–10 cm) long, ¹⁄₁₆–⅛" (1–3 mm) thick, nearly equal, usually with a basal bulb, white-pruinose or nearly smooth, white to translucent-white; partial veil and ring absent.

SPORE PRINT: reddish cinnamon.

MICROSCOPIC FEATURES: spores 10–14 × 6–9 μm, elliptic with an apical pore, smooth, pale brown.

FRUITING: scattered or in groups in lawns, fields, and other grassy areas; May–September; common.

EDIBILITY: edible.

Conocybe tenera (Schaeffer : Fries) Kühner complex

COMMON NAME: Brown Dunce Cap.

CAP: ⅜–¾" (1–2 cm) wide, conic to bell-shaped; surface shiny when fresh, striate from the margin to the disc, red-brown to rusty brown or fulvous, fading to pale tawny to pale ochraceous when dry; flesh pale reddish brown; odor and taste not distinctive.

GILLS: nearly free, close to subdistant, whitish when young, becoming pale brown and finally cinnamon to rusty brown at maturity.

STALK: 1⅛–3⅛" (3–8 cm) long, 1/16–⅛" (1.5–4 mm) thick, nearly equal or with a slightly swollen base; minutely pruinose or nearly smooth; whitish at first, darkening to brown or dark brown in age; partial veil and ring absent.

SPORE PRINT: rusty brown.

MICROSCOPIC FEATURES: spores 10–14 × 5–6 μm, elliptic, with an apical pore, smooth, pale brown.

FRUITING: scattered or in groups on rich humus in woods, in gardens, and on lawns, fields, and fertilized areas; May–September; fairly common.

EDIBILITY: unknown.

Genus *Coprinus*

Members of the genus *Coprinus* are small to medium mushrooms that grow in various habitats, including decaying wood, compost piles, dung, soil, lawns, plant debris, and gardens. More than fifty species occur in North America. Their spore print colors range from dark vinaceous brown to blackish brown to black. They have smooth or roughened spores with an apical pore, and they discolor in concentrated sulphuric acid. The gills, and usually the cap, of most species deliquesce into a black inky fluid. A few species deliquesce slowly or not at all.

Some are popular edibles, a few are poisonous to some individuals when consumed with alcohol (see comments for *C. atramentarius*), and the edibility of the majority is unknown.

Key to Species of *Coprinus*

1. Growing on dung or a mixture of dung and wood shavings, sawdust, or straw, or sometimes on dung in gardens → 2.

1. Growing on wood, in fields and lawns, on plant debris, soil, and compost piles, or in gardens → 3.

    2. Cap 1/16–⅛" (1.5–3 mm) wide and somewhat egg-shaped when young, becoming conic to convex and ¼–½" (5–12 mm) wide at maturity, whitish to gray, striate to sulcate; stalk 1/32–1/16" (0.5–1.5 mm) thick; gills and cap strongly deliquescent; odor and taste not distinctive; spores 11–15 × 6–8 μm; edibility unknown—*C. ephemerus* (Bulliard : Fries) Fries.

    2. Cap ¼–⅜" (5–10 mm) wide and somewhat egg-shaped when young, expanding to nearly plane and ¾–1" (2–2.5 cm) wide at maturity; surface granular to scurfy, often more scaly on the disc at maturity, pale to dark gray; margin striate; flesh odor pungent and unpleasant, like coal tar or very dirty socks; spores 11–14 × 5–7 μm; inedible—*C. narcoticus* (Batsch : Fries) Fries.

2.   Cap ½–1⅛" (6–30 mm) wide and somewhat conic when young, expanding to nearly plane and ¾–2" (2–5 cm) wide at maturity; surface covered with small, white scales at first, becoming smooth with a striate margin, white when covered with scales, soon gray to smoky gray, often with a brownish disc; margin plicate-striate in age; flesh odor not distinctive; spores 9–12 × 6–7 µm—*C. cinereus* (Schaeffer : Fries) S. F. Gray (see p. 102).

2.   Cap ¾–1⅛" (2–3 cm) wide, elliptic to egg-shaped when young, expanding to nearly plane and 2–2⅜" (5–6 cm) wide at maturity; surface coated with tiny scales at first, becoming more coarsely scaly in age, white to whitish; margin striate in age; stalk thickened at the base, white, with a persistent white movable ring; spores 16–22 × 10–13 µm; edibility unknown—*C. sterquilinus* (Fries) Fries.

2.   Cap ⅜–⅝" (1–1.6 cm) wide, elliptic to egg-shaped when young, expanding to nearly plane and ¾–1½" (2–4 cm) wide at maturity; surface covered with a conspicuous, snow-white, powdery covering when young, becoming nearly smooth and white at maturity; margin striate in age; stalk equal or with a slightly enlarged base, coated with a powdery covering like the cap, lacking a ring; spores 15–18 × 9–13 µm; edibility unknown—*C. niveus* (Persoon : Fries) Fries.

3.   Cap typically greater than 2⅛" (5.5 cm) wide at maturity; growing on lawns or decaying wood → 4.

3.   Cap typically less than 2⅛" (5.5 cm) wide at maturity; growing in fields and lawns, on decaying wood, on decaying vegetable matter, on compost piles, or in gardens → 5.

4.   Flesh odor and taste not distinctive, grayish white; cap gray to gray-brown— *C. atramentarius* (Bulliard : Fries) Fries (see p. 102).

4.   Flesh odor and taste not distinctive, white; cap white with a brownish disc, coated with coarse, reddish brown scales; 1⅛–2" (3–5 cm) wide, oval to cylindric when young, expanding to nearly plane and 2–3⅛" (5–8 cm) wide at maturity; stalk enlarged downward and often bulbous at the base; partial veil white, submembranous, leaving a ring on the lower portion of the stalk; spores 10–14 × 6–8.5 µm— *C. comatus* (Müller : Fries) S. F. Gray (see p. 103).

4.   Flesh odor strong and unpleasant, taste disagreeable or sometimes not distinctive, pale tan; cap 1½–3" (4–7.5 cm) wide, with conspicuous white to dingy yellow scales—*C. variegatus* Peck (see p. 105).

4.   Not with the above combinations of characters → 5.

5.   Growing on decaying wood or woody debris → 6.

5.   Growing in fields and lawns, on decaying vegetable matter, on soil, humus, and compost piles, or in gardens → 9.

6.   Stalk usually arising from a spreading mat of yellow-orange, orange, or cinnamon-brown mycelium → 7.

6.   Stalk not arising from a spreading mat of mycelium, or if so, the mycelium is white to pale yellow → 8.

7.   Veil reddish brown to dark tawny, soon breaking up into large, thick patches and scales; mature cap shallowly sulcate from the margin to the disc, up to 1" (2.5 cm) wide, pale tawny to grayish ochraceous; stalk up to 1" (2.5 cm) long, enlarged at the base, white to yellowish; spores 7–10 × 4–4.5 µm; edibility unknown—*C. laniger* Peck.

7.   Veil brown, soon breaking up into thin, tiny granules; mature cap striate from the margin to the disc, up to 2" (5 cm) wide, tawny-brown; stalk up to 3" (7.5 cm) long, equal or slightly enlarged at the base, white to pale yellow; spores 8–11 × 4–6 µm; edibility unknown—*C. radians* (Desmazieres) Fries.

7. Not with the above combinations of characters → 8.
 8. Cap minutely fibrillose and scurfy when young, becoming nearly smooth in age, bell-shaped to convex, striate to sulcate from the margin to the disc, buff or creamy buff when young, becoming violaceous gray with a buff to fulvous disc, not deliquescent; gills white at first, becoming grayish, then brown to fuscous; spores 7–10 × 4–5 μm; edibility unknown—*C. disseminatus* (Persoon : Fries) S. F. Gray.
 8. Cap and stalk covered with a dense coating of white to grayish white cottony fibers; cap white at first, becoming grayish and radially sulcate at maturity; typically persistent and not deliquescing, ⅜–2" (1–5 cm) wide; margin often split and sometimes revolute in age; spores 10–13 × 6–7 μm—*C. lagopus* (Fries) Fries (see p. 103).
 8. Mushroom nearly identical to the previous choice except for smaller spores, 7–9 × 5–7 μm; edibility unknown—*C. lagopides* Karsten.
 8. Cap covered with a sparse coating of tiny whitish granular scales on the button stage, soon becoming smooth, weakly to strongly sulcate, tawny to orange-brown on the disc, yellowish tan toward the margin; stalk smooth, white—*C. micaceus* (Bulliard : Fries) Fries (see p. 104).
 8. Cap covered with white to buff fibrils or tiny scales and soon becoming smooth, striate to conspicuously sulcate from the margin to the disc, buff to pale ochraceous buff, with a darker tawny to fulvous disc, 1–2½" (2.5–6.5 cm) wide; growing on compost piles or plant debris in gardens, sometimes on wood chips; stalk smooth, white, sometimes arising from an orange to rusty mat of mycelium; spores 7–10 × 3–5 μm, smooth; edible—*C. domesticus* (Bolton : Fries) S. F. Gray.
 8. Cap with a sparse layer of tiny fibrils, soon becoming smooth, weakly to strongly sulcate, grayish brown on the disc, paler toward the margin, ¾–2¾" (2–7 cm) wide; growing on decaying hardwood, especially sugar maple; stalk smooth, white; spores 10–14 × 6–8 μm, roughened; poisonous to some people—*C. insignis* Peck.
9. Growing on decaying plant debris including blades of grass, on compost piles or sometimes on wood chips → 10.
9. Growing on dry or wet soil or among decaying leaves in mud → 11.
9. Growing among grasses in fields or lawns → 12.
 10. Growing on decaying plant debris; cap very small, ⅛–⅜" (3–10 mm) wide, white when young, becoming grayish brown, striate to shallowly sulcate; spores 7–8 × 4.5–5.5 μm; edibility unknown—*C. brassicae* Peck.
 10. Growing on compost piles or plant debris in gardens, sometimes on wood chips; cap covered with white to buff fibrils or tiny scales and soon becoming smooth, conspicuously striate to sulcate from the margin to the disc, buff to pale ochraceous buff, with a darker tawny to fulvous disc, 1–2½" (2.5–6.5 cm) wide; stalk smooth, white, sometimes arising from an orange to rusty mat of mycelium; spores 7–10 × 3–5 μm; edible—*C. domesticus* (Bolton : Fries) S. F. Gray.
11. Growing on muddy soil or among decaying leaves in mud; cap ⅜–¾" (1–2 cm) wide, rusty tawny to orange-brown, paler toward the margin, becoming grayish rust with a somewhat persistent rusty tawny disc, finely to deeply sulcate from the margin to the disc; gills not deliquescent or very slowly so—*C. dilectus* Fries (see p. 103).
11. Growing on soil; cap ⅜–1" (1–2.5 cm) wide, rusty tawny to orange-brown or reddish brown, distinctly sulcate from the margin to the disc; spores 7–12 × 6–10 μm, angular, shaped like a keystone; edibility unknown—*C. angulatus* Peck = *C. boudieri* Quélet.

11. Not with the above combinations of characters → 12.
    12. Gills distant; cap deeply sulcate, brown to rusty tawny on the disc, gray to pale grayish brown toward the margin—*C. plicatilis* (W. Curtis : Fries) Fries (see p. 104).
    12. Gills very crowded; cap weakly to strongly sulcate, tawny to orange-brown on the disc, yellowish tan toward the margin—*C. micaceus* (Bulliard : Fries) Fries (see p. 104).
    12. Cap minutely fibrillose and scurfy when young, becoming nearly smooth in age, bell-shaped to convex, striate to sulcate from the margin to the disc, buff or creamy buff when young, becoming violaceous gray with a buff to fulvous disc, not deliquescent; gills white at first, becoming grayish then brown to fuscous; spores 7–10 × 4–5 μm; edibility unknown—*C. disseminatus* (Persoon : Fries) S. F. Gray.

Coprinus atramentarius (Bulliard : Fries) Fries

COMMON NAME: Alcohol Inky.

CAP: 1½–3" (4–7.5 cm) wide, oval to egg-shaped when young, becoming convex in age; surface smooth to slightly scurfy, sometimes forming tiny scales on the disc, dry, gray to gray-brown, often with shallow grooves on the margin; flesh grayish white; odor and taste not distinctive.

GILLS: free, very crowded, white when very young, soon gray, then black, and deliquescing to a black, inky fluid in age.

STALK: 1½–6" (4–15.5 cm) long, ⅜–¾" (1–2 cm) thick, nearly equal or tapered downward, silky, hollow in age, white, with a white annular zone toward the base.

SPORE PRINT: black.

MICROSCOPIC FEATURES: spores 8–12 × 4.5–6 μm, elliptic, smooth, with an apical pore, pale grayish black.

FRUITING: in clusters on grass, wood chips, and tree bases, May–November; very common.

EDIBILITY: edible, with caution; see comments.

COMMENTS: enjoyed without adverse reactions by most who consume it. However, some individuals experience coprine poisoning if alcoholic beverages are consumed while, or up to 48 hours before or after, eating this mushroom. Coprine poisoning signs and symptoms may include nausea, vomiting, marked flushing, rapid breathing, and severe headache.

Coprinus cinereus (Schaeffer : Fries) S. F. Gray

COMMON NAME: Gray Shag.

CAP: ¼–1⅛" (6–30 mm) wide and oval to conic at first, becoming broadly conic to nearly plane and ¾–2" (2–5 cm) wide in age; surface moist, covered with small white scales at first, becoming smooth with a striate margin, white when covered with scales, soon gray to smoky gray or blackish, often with a brownish disc; margin plicate-striate in age; flesh grayish at first, becoming black as the mushroom deliquesces in age; odor and taste not distinctive.

GILLS: attached at first then free from the stalk, crowded, grayish then black as the mushroom deliquesces.

STALK: 2–4¾" (5–12 cm) long, ⅛–⅜" (3–10 mm) thick, enlarged downward or nearly equal, sometimes with a long, tapering, rooting base (see photo), fragile, hollow except on the lower portion, slightly scaly to smooth, white.

SPORE PRINT: black.

MICROSCOPIC FEATURES: spores 9–12 × 6–7 μm, elliptic with an apical pore, smooth, brown.

FRUITING: scattered, in groups or clusters on dung, manured soil or among wood chips mixed with rich soil; May–November; fairly common.

EDIBILITY: edible.

COMMENTS: the variety shown in the photograph is sometimes called *C. macrorhizus.*

Coprinus comatus (Müller : Fries) S. F. Gray

COMMON NAME: Shaggy Mane, Lawyer's Wig.

CAP: 1⅛–2" (3–5 cm) wide and oval to cylindric at first, becoming broadly conic to nearly plane and 2–3⅛" (5–8 cm) wide in age, fragile; surface dry, white with a brownish disc, coated with coarse scales that are white to pale reddish brown and usually darkest at the tips; flesh white at first, becoming black as the mushroom deliquesces in age; odor and taste not distinctive.

GILLS: attached at first then free from the stalk, crowded, white then pinkish, and finally black as the mushroom deliquesces.

STALK: 3–12" (7.5–30 cm) long, ⅜–1" (1–2.5 cm) thick, enlarged downward to a bulbous base, sometimes rooting, hollow, glabrous to silky-fibrillose, white, fragile; partial veil white, submembranous, leaving a thin, inferior ring.

SPORE PRINT: black.

MICROSCOPIC FEATURES: spores 10–14 × 6–8.5 μm, ellipsoid, truncate, with an apical pore, smooth, purple-brown.

FRUITING: scattered, in groups or clusters in grassy areas, on soil, or in wood chips; May–November; common.

EDIBILITY: edible.

COMMENTS: *Coprinus sterquilinus* is a smaller, white to whitish mushroom that grows on dung or on manured soil, and has much larger spores that measure 16–22 × 10–13 μm.

Coprinus dilectus Fries

CAP: ⅜–¾" (1–2 cm) wide; elliptic to conic, becoming broadly conic in age; surface minutely scurfy at first, becoming nearly smooth, finely to deeply sulcate from the margin to the disc, rusty tawny to orange-brown, paler toward the margin, becoming grayish rust with a somewhat persistent rusty tawny disc; flesh orange-brown; odor and taste not distinctive.

GILLS: attached, crowded, buff to orange-buff when young, becoming gray, then dark violaceous brown, not deliquescent or very slowly so.

STALK: ⅝–1⅜" (1.6–3.5 cm) long, 1/16–⅛" (1–4 mm) thick, equal or slightly enlarged at the base, whitish pruinose at first, becoming nearly smooth, grayish hyaline to ochraceous buff.

SPORE PRINT: dark violaceous brown.

MICROSCOPIC FEATURES: spores 11–14 × 6–7 μm, elliptic with an apical pore, smooth, pale brown.

FRUITING: solitary, scattered, or in small groups on muddy soil or among decaying leaves in mud; May–September; uncommon.

EDIBILITY: unknown.

Coprinus lagopus (Fries) Fries

CAP: ⅜–2" (1–5 cm) wide, oval to elliptic at first, becoming convex to nearly plane at maturity, margin often split and sometimes revolute in age; surface covered with a dense coating of erect to recurved, white to grayish white, cottony fibers or scales, which gradually disappear in age, white at first, becoming grayish then finally black, usually with a brown disc, often persisting, becoming shallowly fissured then deeply sulcate at maturity; flesh grayish to black; odor and taste not distinctive.

GILLS: attached at first, soon free, close, white at first, becoming gray then black as the mushroom deliquesces.

STALK: 1½–4" (4–10 cm) long, ¹⁄₁₆–³⁄₁₆" (1.5–4 mm) thick, nearly equal or slightly enlarged downward, dry, hollow at maturity, floccose at first, becoming smooth in age, white.

SPORE PRINT: violaceous black.

MICROSCOPIC FEATURES: spores 10–13 × 6–7 μm, ellipsoid to almond-shaped, smooth, violaceous brown.

FRUITING: solitary, scattered or in groups on wood chips, sawdust, or other woody debris; July–October; occasional.

EDIBILITY: unknown.

Coprinus micaceus (Bulliard : Fries) Fries

COMMON NAME: Mica Cap.

CAP: ¾–2" (2–5 cm) wide, oval to egg-shaped when young, becoming bell-shaped to convex in age; surface covered with a sparse coating of tiny whitish granular scales on the button stage and soon becoming smooth, weakly to distinctly sulcate from the margin nearly to the disc, tawny to orange-brown on the disc, yellowish tan toward the margin, becoming grayish to dark grayish brown in age; flesh whitish; odor and taste not distinctive.

GILLS: attached to nearly free, very crowded, white when young, soon gray, then black, and deliquescing to a black, inky fluid in age.

STALK: 1–3⅛" (2.5–8 cm) long, ⅛–¼" (2–5 mm) thick, equal, smooth, hollow in age, white, lacking an annular zone.

SPORE PRINT: black.

MICROSCOPIC FEATURES: spores 7–10 × 4–5 μm, elliptic with an apical pore, smooth, pale grayish black.

FRUITING: in dense clusters on lawns or decaying wood; April–October; very common.

EDIBILITY: edible.

Coprinus plicatilis (W. Curtis : Fries) Fries

COMMON NAME: Japanese Umbrella Inky.

CAP: ⅜–1" (1–2.5 cm) wide, oval to egg-shaped when young, becoming convex to nearly plane or with a depressed disc; surface deeply sulcate, brown to rusty tawny on the disc, gray to pale grayish brown toward the margin; flesh grayish; odor and taste not distinctive.

GILLS: free and attached to a collar-like ring of tissue at the stalk apex, distant, grayish buff when young, becoming gray, then black, not deliquescing.

STALK: 1⅛–2¾" (3–7 cm) long, ¹⁄₃₂–⅛" (1–3 mm) thick, nearly equal or with an enlarged base, smooth, dry, hollow in age, white to somewhat hyaline.

SPORE PRINT: blackish.

MICROSCOPIC FEATURES: spores 10–13 × 7–10 μm, broadly oval with an apical pore, smooth, pale grayish black.

FRUITING: scattered or in groups in fields and lawns; May–September; fairly common.

EDIBILITY: edible.

COMMENTS: also known as *Pseudocoprinus plicatilis.*

Coprinus variegatus Peck

COMMON NAME: Scaly Inky Cap.

CAP: 1½–3" (4–7.5 cm) wide, oval to egg-shaped when young, becoming bell-shaped in age; surface grayish brown to gray, with conspicuous white to dingy yellow scales that wash off easily; flesh pale tan; odor strong and unpleasant; taste disagreeable or sometimes not distinctive.

GILLS: free, very crowded, white when young, soon grayish, then purplish brown to black and deliquescing to a black, inky fluid.

STALK: 2–4¾" (5–12 cm) long, ¼–⅜" (5–10 mm) thick, nearly equal or tapered downward, scurfy to nearly smooth on the upper portion, fibrillose from the base up to an obscure annular zone, hollow in age, whitish.

SPORE PRINT: black.

MICROSCOPIC FEATURES: spores 7.5–10 × 4–5.5 μm, elliptic with an apical pore, smooth, pale grayish black.

FRUITING: in dense clusters on decaying hardwood, often on stumps and buried wood in lawns; June–August; common.

EDIBILITY: edible, with caution; see comments.

COMMENTS: with most collections, the unpleasant odor and disagreeable taste do not disappear when this mushroom is cooked. Many who consume it state that it is unpleasant at best. It has also caused coprine poisoning in some individuals when consumed with alcohol (see comments for *C. atramentarius*, p. 102).

Genus *Cortinarius*

The genus *Cortinarius* is huge, with more than five hundred species believed to occur in North America. They are small to large, mostly terrestrial mushrooms with a partial veil that is typically cortinate and leaves a thin, fibrous annular zone. Their spore prints are rusty brown to ochraceous tawny, and their spores are roughened, wrinkled, or warty and lack an apical pore and plage. Their gill edges are usually not white and finely fringed, and cheilocystidia are typically absent. These are useful characters for separating *Cortinarius* species from *Hebeloma* species, which usually have abundant cheilocystidia and finely fringed, white gill edges.

A few species are listed as edible, some are poisonous, a few have caused fatalities, and the edibility of the vast majority is unknown. We do not recommend eating any species of *Cortinarius* because of possible poisoning due to toxins that may be cumulative over time. Because the genus is complex and includes a large number of species, a comprehensive key to species is beyond the scope of this book. We have included a sampling of some of the more common and distinctive species that occur in this region.

Cortinarius alboviolaceus (Fries) Fries

COMMON NAME: Silvery-violet Cort.

CAP: 1¼–3¾" (3–9.5 cm) wide, bell-shaped when young, becoming convex to nearly flat in age, with a low, broad umbo; surface dry, shiny, silky, coated with flattened, silvery fibrils, pale violet to lilac-buff; margin incurved when young and remaining so long into maturity; flesh pale violet; odor and taste not distinctive.

GILLS: attached, close, broad, pale violet to grayish lilac when young, becoming cinnamon-brown in age.

STALK: 1⅝–3⅛" (4–8 cm) long, ¼–¾" (5–20 mm) thick, enlarged downward to a bulbous or club-shaped base, dry, stuffed, colored like the cap; partial veil cortinate, white, leaving a thin, fibrous annular zone, sheathed from the base up to the annular zone by a silky, white veil.

SPORE PRINT: rusty brown.

MICROSCOPIC FEATURES: spores 7–10 × 4–6 μm, elliptic, roughened, pale brown.

FRUITING: solitary or in groups on the ground under conifers and hardwoods, especially beech and oak; August–October; fairly common.

EDIBILITY: edible, but not recommended.

COMMENTS: *Cortinarius obliquus* (edibility unknown) lacks the sheathing universal veil and has an abruptly bulbous base.

Cortinarius armillatus (Fries) Kummer

COMMON NAME: Bracelet Cort.

CAP: 2–4¾" (5–12 cm) wide, bell-shaped when young, becoming broadly convex to nearly flat with a low, broad umbo in age; surface moist to dry, covered with tiny fibrils, nearly smooth, reddish brown to brownish orange; margin incurved when young, expanded in age, sometimes with veil remnants attached; flesh yellowish buff to pale tawny; odor not distinctive or like radishes; taste not distinctive or slightly bitter.

GILLS: attached, distant, broad, tawny, becoming rusty brown in age.

STALK: 2¾–5⅞" (7–15 cm) long, ⅜–1" (1–2.5 cm) thick, enlarged downward to a bulbous or club-shaped base, solid, fibrous, whitish to pale brown, with multiple cinnabar-red rings or zones; partial veil cortinate, whitish, leaving a thin, fibrous annular zone on the upper portion of the stalk.

SPORE PRINT: rusty brown.

MICROSCOPIC FEATURES: spores 7–12 × 5–7 μm, elliptic to almond-shaped, with warts, pale brown.

FRUITING: solitary or in groups on the ground or among mosses under conifers and hardwoods, especially birch and pine; August–October; fairly common.

EDIBILITY: edible, but of poor quality and often riddled with larvae, and not recommended.

COMMENTS: *Cortinarius trivialis* = *C. collinitis* (edibility unknown) has a slimy to sticky, ochraceous tawny to red-brown cap and a slimy stalk, which is white on the upper portion and brown from the base up, with irregular bands of thick gluten.

Cortinarius bolaris Fries

CAP: 1⅛–2⅜" (3–6 cm) wide, convex when young, becoming broadly convex in age; surface dry, whitish, covered overall with reddish to orange-red hairs and scales; margin incurved when young, remaining so well into maturity, surpassing the gills; flesh white with tints of pale yellow; odor and taste not distinctive.

GILLS: attached, close, moderately broad, pale buff, becoming pale cinnamon-brown.

STALK: 1½–2⅜" (4–6 cm) long, ¼–⅜" (5–10 mm) thick, enlarged downward to a bulbous base, hollow in age, whitish, covered by reddish to orange-red fibers and flattened scales, staining yellow to reddish when bruised; partial veil cortinate, white, leaving a thin, fibrous annular zone on the upper portion of the stalk.

SPORE PRINT: rusty brown.

MICROSCOPIC FEATURES: spores 6–7 × 5–5.5 μm, oval to subglobose, slightly roughened, pale brown.

FRUITING: in groups or clusters on the ground under conifers and in mixed woods; August–October; infrequent.

EDIBILITY: unknown.

Cortinarius camphoratus (Fries) Fries

CAP: 1½–4" (4–10 cm) wide, convex when young, becoming broadly convex to nearly flat in age, with or without a low, broad umbo; surface dry, with a silvery whitish bloom at first, coated with tiny, flattened fibrils, buff with pale lilac tints, developing golden tones in age; margin incurved when young, becoming expanded in age, fibrillose; flesh pale lilac to purplish; odor strongly of raw potatoes; taste unpleasant.

GILLS: attached, subdistant, broad, lilac when young, becoming dull brown to cinnamon-brown in age.

STALK: 2–4" (5–10 cm) long, ⅜–¾" (1–2 cm) thick, enlarged downward to a club-shaped base, solid, sheathed from the base to the annular zone with silky, white, matted fibrils; partial veil cortinate, white, leaving a thin, fibrous annular zone.

SPORE PRINT: rusty brown.

MICROSCOPIC FEATURES: spores 9–11 × 5–6 μm, elliptic, roughened, pale brown.

FRUITING: solitary, scattered, or in groups on the ground under conifers and hardwoods; September–October; fairly common.

EDIBILITY: edible, but not recommended.

COMMENTS: also known as *C. caesiocyaneus.*

Cortinarius claricolor (Fries) Fries

CAP: 2–5⅞" (5–15 cm) wide, hemispheric when young, becoming broadly convex to nearly flat in age; surface smooth, shiny, viscid when fresh, often finely wrinkled or shallowly sulcate toward the margin in age, reddish orange overall, fading to yellowish toward the margin; margin strongly incurved when young, becoming expanded in age, often covered with patches of white veil remnants; flesh whitish; odor slightly disagreeable or not distinctive; taste not distinctive.

GILLS: attached, crowded, narrow, buff to orangish.

STALK: 2⅜–5⅞" (6–15 cm) long, ⅜–¾" (1–2 cm) thick, nearly equal or tapering downward, dry, solid, white to very pale buff or cream, white-pruinose at the apex, covered from the base to the annular zone with a white, cottony-membranous sheath; partial veil cortinate, white, leaving a thin, fibrous annular zone.

SPORE PRINT: rusty brown.

MICROSCOPIC FEATURES: spores 7–10 × 3.5–4.5 μm, almond-shaped to elliptic, nearly smooth, pale brown.

FRUITING: solitary, scattered, or in groups on the ground under conifers, especially pine; August–October; uncommon.

EDIBILITY: unknown.

COMMENTS: *Rozites caperata* (edible) is very similar but has a white, membranous partial veil that leaves a membranous superior to median ring on the stalk.

Cortinarius corrugatus Peck

CAP: 1⅜–4" (3–10 cm) wide, bluntly conic to bell-shaped, becoming broadly convex in age; surface smooth, shiny, viscid, with distinct radial corrugations, rusty ochraceous to ochre-tawny; flesh white; odor and taste not distinctive.

GILLS: attached, close, broad, purplish at first, becoming rusty cinnamon at maturity; edges often eroded in age.

STALK: 2–4⅜" (5–11 cm) long, ¼–⅝" (6–16 mm) thick, nearly equal down to an enlarged and sometimes bulbous base, typically scurfy overall, moist or dry, tawny to ochraceous; partial veil cortinate, leaving a sparse annular zone or evanescent.

SPORE PRINT: rusty brown.

MICROSCOPIC FEATURES: spores 10–14 × 7–9 μm, elliptic, roughened, pale brown.

FRUITING: solitary, scattered, or in groups on the ground under hardwoods, especially oak and beech, or in mixed woods; July–October; occasional.

EDIBILITY: unknown.

Cortinarius croceofolius Peck

COMMON NAME: Saffron-colored Cort.

CAP: 1–2" (2.5–5 cm) wide, bell-shaped when young, becoming broadly convex to nearly flat with a low, broad umbo in age; surface dry, covered with minute hairs and scales, dark orange to saffron-yellow when young, becoming dark brownish to cinnamon-orange in age, remaining darkest over the disc, fading to orange toward the margin; margin incurved when young, remaining so well into maturity; flesh yellowish; odor and taste not distinctive.

GILLS: attached, close, narrow, orange when young, becoming brownish orange to cinnamon in age.

STALK: 1–2" (2.5–5 cm) long, ¼–½" (6–12 mm) thick, enlarged downward, hollow, yellow to yellowish brown, often reddish at the base; partial veil cortinate, yellow, leaving a thin, fibrous annular zone on the upper portion of the stalk.

SPORE PRINT: rusty brown.

MICROSCOPIC FEATURES: spores 6–7.5 × 4–5 μm, elliptic, roughened, pale brown.

FRUITING: solitary, scattered, or in groups on the ground under conifers; September–November; occasional.

EDIBILITY: unknown.

COMMENTS: also known as *C. malicorius.*

Cortinarius distans Peck

CAP: ¾–2" (2–5 cm) wide, bell-shaped when young, becoming broadly convex to nearly flat with a prominent pointed to rounded umbo in age; surface minutely scurfy or with tiny scales, hygrophanous, cinnamon to dark purplish brown when moist, fading to pale golden brown when dry or in age; margin incurved when young, remaining so well into maturity, often becoming radially split in age; flesh brownish, becoming yellowish; odor not distinctive or slightly of radish or peanut butter; taste not distinctive.

GILLS: attached, distant, broad, tawny-yellow, becoming dark cinnamon-brown.

STALK: 1⅝–3⅛" (4–8 cm) long, ¼–½" (5–12 mm) thick, nearly equal, often enlarged toward the base, stuffed, buff, streaked and mottled with tawny-yellow, brown, and brownish purple, fibrillose, with whitish remnants of the universal veil; partial veil cortinate, white, leaving a thin, fibrous annular zone on the upper portion of the stalk.

SPORE PRINT: rusty brown.

MICROSCOPIC FEATURES: spores 6–8 × 5–6 μm, oval, roughened, pale brown.

FRUITING: in groups or clusters on the ground under hardwoods; July–October; infrequent.

EDIBILITY: unknown.

Cortinarius evernius Fries

CAP: 1⅛–4½" (3–11.5 cm) wide, conic to bell-shaped when young, becoming broadly convex to nearly flat with a low, broad umbo in age; surface silky, radially streaked with tiny fibrils, slippery when wet, hygrophanous, purplish brown to reddish brown when fresh, fading to ochre-brown, paler ochre over the disc; margin incurved when young, becoming expanded in age, wavy, with silky veil remnants; flesh purplish brown; odor slightly of radish or not distinctive; taste not distinctive.

GILLS: adnexed, distant, broad, purplish, becoming cinnamon-brown.

STALK: 4–8" (10–20 cm) long, ¼–¾" (6–20 mm) thick, nearly equal, solid, purplish to lilac, covered from the base to the apex by whitish streaks and patches of universal veil; partial veil cortinate, whitish, leaving a thin, fibrous annular zone on the upper portion of the stalk.

SPORE PRINT: rusty brown.

MICROSCOPIC FEATURES: spores 8–10 × 5–6 μm, elliptic, roughened, pale brown.

FRUITING: in groups or clusters on the ground or among mosses in bogs and conifer woods; July–October; fairly common.

EDIBILITY: unknown.

Cortinarius iodes Berkeley and Curtis

COMMON NAME: Viscid Violet Cort, Spotted Cort.

CAP: ¾–2⅜" (2–6 cm) wide, bell-shaped when young, becoming broadly convex in age; surface smooth, slimy, dark lilac or purplish, fading in age, developing yellowish spots or becoming yellowish over the disc; flesh lilac to violet, fading in age; odor and taste not distinctive.

GILLS: attached, close, lilac to violet when young, becoming grayish cinnamon in age.

STALK: 1½–2¾" (4–7 cm) long, ¼–⅝" (5–15 mm) thick, nearly equal or enlarged downward, solid, slimy, smooth, violet or purplish, sometimes whitish toward the base; partial veil cortinate, pale violet, leaving a thin, fibrous annular zone on the upper portion of the stalk.

SPORE PRINT: rusty brown.

MICROSCOPIC FEATURES: spores 8–10 × 5–6.5 μm, elliptic, finely roughened, pale brown.

FRUITING: scattered, in groups or clusters on the ground under hardwoods; July–September; fairly common.

EDIBILITY: edible, but not recommended.

COMMENTS: *Cortinarius iodiodes* (inedible) is nearly identical but has a bitter-tasting cap surface. *Cortinarius vibratilis* (inedible) has a small, slimy to sticky yellowish brown to orange-brown cap, white gills that become brown in age, and a white, sticky stalk that is enlarged downward.

Cortinarius luteus Peck

CAP: ¾–2" (2–5 cm) wide, conic when young, becoming convex to broadly convex to nearly flat with a low, broad umbo in age; surface dry, dull, yellow, often dark orange over the disc, covered with matted fibrils and minute scales; flesh yellow; odor and taste not distinctive.

GILLS: attached, subdistant, broad, pale yellowish, becoming yellowish buff in age.

STALK: 2–4" (5–10 cm) long, ⅜–¾" (1–2 cm) thick, nearly equal, solid, dry, silky-fibrillose with yellow fibrils over a whitish ground color, whitish in age; partial veil cortinate, yellow, leaving a thin, fibrous annular zone on the upper portion of the stalk.

SPORE PRINT: rusty brown.

MICROSCOPIC FEATURES: spores 7–8 × 6–7 μm, subglobose to broadly elliptic, roughened, pale brown.

FRUITING: solitary, scattered, or in groups on the ground or among mosses in mixed woods; July–September; occasional.

EDIBILITY: unknown.

Cortinarius marylandensis Ammirati and Smith

CAP: 1–2⅜" (2.5–6 cm) wide, bell-shaped when young, becoming broadly convex to flat with a low, broad umbo in age; surface dry, smooth, shiny or dull, somewhat felty, bright reddish orange to reddish brown, changing to rose, then purple with KOH; margin somewhat incurved when young, becoming expanded and often split in age; flesh yellowish; odor and taste not distinctive.

GILLS: attached, subdistant, broad, cinnabar-red when young, fading to cinnamon-brown in age.

STALK: ¾–2¾" (2–7 cm) long, ⅛–⅜" (3–10 mm) thick, equal or enlarged slightly downward toward the base, dry, hollow, streaked with reddish orange to reddish brown fibrils over a reddish yellow ground color, often yellowish to whitish at the base; partial veil cortinate, reddish, leaving a thin, fibrous annular zone on the upper portion of the stalk.

SPORE PRINT: rusty brown.

MICROSCOPIC FEATURES: spores 7–9 × 4–5 μm, elliptic, slightly roughened, pale brown.

FRUITING: scattered or in groups on the ground under conifers and hardwoods, especially oak and beech; August–October; occasional.

EDIBILITY: unknown.

COMMENTS: *Cortinarius cinnabarinus* (edibility unknown) is nearly identical, but its cap cuticle does not stain rose, then purple, when KOH is applied.

Cortinarius paleiferus Svrček

CAP: ⅜–1⅛" (1–3 cm) wide, conic when young, becoming broadly convex with a small, sharp umbo in age; surface dry, brown, covered overall with minute white hairs and scales; margin incurved when young, expanding in age, with white veil remnants; flesh brownish; odor fruity, similar to apples, with a strong hint of geranium leaves; taste not distinctive.

GILLS: attached, subdistant, moderately broad, violet to violet-brown when very young, soon becoming pale brown to milky coffee-brown.

STALK: 1⅛–2¾" (3–7 cm) long, ⅛–¼" (3–6 mm) thick, equal, violet overall or at least near the apex when young, becoming dark brown at maturity, solid, becoming hollow in age, sometimes with pale violet mycelium surrounding the base, covered from the base to the annular zone with a cottony-membranous sheath that breaks up, leaving partial rings and zones; partial veil cortinate, white, leaving a thin, fibrous annular zone.

SPORE PRINT: rusty brown.

MICROSCOPIC FEATURES: spores 7.5–10 × 5–6 μm, elliptic, finely roughened, pale brown.

FRUITING: solitary, scattered, or in groups on the ground and among mosses in conifer and mixed woods, also under alder; August–October; fairly common.

EDIBILITY: unknown.

COMMENTS: *Cortinarius paleaceus* (edibility unknown) is nearly identical but has crowded gills that are grayish when young and brown at maturity, a cap that may be smooth or coated with white scales, a pale ochre to brown stalk that lacks violet color when young, and spores that measure 6–9 × 4–6 μm. *Cortinarius hemitrichus* (edibility unknown) is similar but has a larger cap, ¾–2" (2–5 cm) wide, and lacks the fruity odor.

Cortinarius pholideus (Fries : Fries) Fries

CAP: 1½–4" (4–10 cm) wide, bell-shaped to hemispheric when young, becoming broadly convex to nearly flat with a low, broad umbo in age; surface dry, pale buff, becoming yellow-buff, densely covered overall with dark brown fibrils and scales; margin incurved when young, remaining so well into maturity, often with cortinate veil remnants; flesh pale violet when young, becoming buff; odor and taste not distinctive.

GILLS: adnexed, close, moderately broad, pale violet when young, becoming grayish violet, then yellowish brown in age.

STALK: 1⅝–4¾" (4–12 cm) long, ¼–½" (5–12 mm) thick, nearly equal, often enlarged toward the base, stuffed, dry, buff with pale violet tones, especially at the apex when young, buff overall in age; covered from the base to the annular zone with a dark brown cottony-membranous sheath that breaks up, leaving irregular partial rings, zones, and scales; partial veil cortinate, whitish, leaving a thin, fibrous annular zone.

SPORE PRINT: rusty brown.

MICROSCOPIC FEATURES: spores 6–8 × 5–5.5 μm, oval to broadly elliptic, roughened, pale brown.

FRUITING: scattered, in groups, or caespitose on the ground or on very decayed logs in conifer or mixed woods; August–October; infrequent.

EDIBILITY: unknown.

Cortinarius pyriodorus Kauffman

CAP: 1½–4¾" (4–12 cm) wide, hemispheric when young, becoming broadly convex to nearly flat, sometimes with a low, broad umbo in age; surface dry, coated with tiny radial fibers, shiny, deep lilac when young, fading to pale lilac with buff tones in age; margin incurved when young, becoming expanded in age, often coated with veil remnants; flesh buff with cinnamon-brown streaks throughout; odor of ripe pears; taste not distinctive to slightly bitter.

GILLS: attached, close to subdistant, moderately broad, tawny when young, darkening to dull pale brown in age.

STALK: 2–4½" (5–11.5 cm) long, ½–1" (1.2–2.5 cm) thick, enlarged downward to a bulbous or abruptly bulbous base, solid, dry, colored like the cap and densely fibrillose when young, becoming pale lilac to grayish lilac, staining brownish when handled or bruised; base often coated with white mycelium; partial veil cortinate, white, leaving a thin, fibrous annular zone on the upper portion of the stalk.

SPORE PRINT: rusty brown.

MICROSCOPIC FEATURES: spores 7–10 × 4–6 μm, elliptic, roughened, pale brown.

FRUITING: solitary, scattered, or in groups on the ground under conifers; August–October; occasional.

EDIBILITY: unknown.

COMMENTS: *Cortinarius obliquus* (edibility unknown) is very similar; it has an abruptly bulbous base but lacks the odor of ripe pears.

Cortinarius semisanguineus (Fries) Gillet

COMMON NAME: Red-gilled Cort.

CAP: ¾–2⅜" (2–6 cm) wide, bell-shaped to convex when young, becoming broadly convex to nearly flat, with a low, broad umbo in age; surface dry, covered with tiny matted fibrils, yellow-brown to cinnamon-buff, darker over the disc in age; margin incurved when young, becoming expanded; flesh whitish to yellowish; odor not distinctive; taste not distinctive or slightly bitter.

GILLS: attached, crowded, narrow, dark blood-red to cinnabar.

STALK: ¾–3" (2–7.5 cm) long, ⅛–¼" (3–6 mm) thick, equal, solid, fibrous, dull yellow, often reddish at the base and whitish at the apex, fibrillose; partial veil cortinate, yellowish, leaving a thin, fibrous annular zone on the upper portion of the stalk.

SPORE PRINT: rusty brown.

MICROSCOPIC FEATURES: spores 5–8 × 3–5 μm, elliptic, roughened, pale brown.

FRUITING: scattered or in groups on the ground or among mosses under hardwoods and conifers; July–November; fairly common.

EDIBILITY: unknown.

Cortinarius torvus (Fries) Fries

CAP: 1⅝–3⅛" (4–8 cm) wide, convex when young, becoming broadly convex to nearly flat in age, often with a low, broad umbo; surface moist, not sticky, covered with a fine, silvery bloom when young, occasionally scurfy to scaly, becoming smooth and radially wrinkled in age, often with spots, purplish brown to copper-brown at first, becoming brownish buff over the disc and buff toward the margin in age; flesh purplish gray at first, becoming brownish in age; odor sweet, aromatic; taste not distinctive to bitter.

GILLS: attached, subdistant, broad, dark purplish at first, becoming dark cinnamon-brown.

STALK: 1⅝–2¾" (4–7 cm) long, ¼–⅜" (6–8 mm) thick, enlarged downward to a bulbous or club-shaped base, solid, dry; sheathed from the base to the ring with a white, membranous universal veil that stains purplish brown in age or when handled; ring superior to median, membranous, white, sometimes flaring; purplish brown above the ring.

SPORE PRINT: rusty brown.

MICROSCOPIC FEATURES: spores 8–12 × 5–6 μm, elliptic, roughened, pale brown.

FRUITING: in groups or clusters on the ground or in decaying debris under hardwoods or in mixed woods; August–October; occasional.

EDIBILITY: unknown.

COMMENTS: an unusual species of *Cortinarius* because of the persistent membranous ring. Compare with *Rozites caperata* (edible), which has a brownish orange to brownish yellow or ochre-orange cap with a hoary sheen when young (see p. 243).

Cortinarius traganus (Weinmann : Fries) Fries

COMMON NAME: Pungent Cort.

CAP: 1½–4¾" (4–12 cm) wide, convex when young, becoming broadly convex to nearly flat in age, with or without a low, broad umbo; surface dry, covered with whitish, tiny, matted fibrils over a lilac ground color, becoming lilac overall, fading in age; margin incurved when young, expanded in age; flesh yellowish with reddish brown streaks; odor pungent and pleasant or disagreeable; taste not distinctive or slightly bitter.

GILLS: attached, subdistant, broad, yellow-brown to rusty cinnamon.

STALK: 1½–4¾" (4–12 cm) long, ⅜–1⅛" (1–3 cm) thick, enlarged downward to a bulbous or club-shaped base, solid, dry, lilac, covered with matted fibrils; partial veil cortinate, whitish lilac to grayish lilac, leaving a thin, fibrous annular zone on the upper portion of the stalk.

SPORE PRINT: rusty-brown.

MICROSCOPIC FEATURES: spores 7.5–10 × 4.5–6 μm, elliptic to almond-shaped, roughened, pale brown.

FRUITING: scattered or in groups on the ground or among mosses under conifers; August–November; occasional to fairly common.

EDIBILITY: unknown.

Cortinarius violaceus (Fries) S. F. Gray

COMMON NAME: Violet Cort.

CAP: 2–4¾" (5–12 cm) wide, convex when young, becoming broadly convex to nearly flat in age, with or without a small umbo; surface dry, covered with fibers that form minute, erect tufts or scales, especially over the disc, dark violet to purple; margin incurved when young, expanded in age; flesh dark violet to grayish; odor and taste not distinctive.

GILLS: attached, subdistant, broad, dark violet, becoming grayish cinnamon-brown.

STALK: 2¾–5½" (7–14 cm) long, ⅜–1" (1–2.5 cm) thick, enlarged downward to a club-shaped base, solid, dry, covered with tiny, matted fibrils, dark violet to purplish; partial veil cortinate, violet, leaving a thin, fibrous annular zone.

SPORE PRINT: rusty brown to rusty cinnamon.

MICROSCOPIC FEATURES: spores 12–17 × 8–10 µm, elliptic, roughened or with warts, pale brown.

FRUITING: solitary, scattered, or in groups on the ground or among mosses under conifers; September–October; fairly common.

EDIBILITY: edible, but not recommended.

Genus *Crepidotus*

Crepidotus is a large genus of more than one hundred species in North America. They are small to medium mushrooms that typically grow on decaying wood. Their spore prints are dull brown to fulvous, and their spores are smooth or finely roughened. They are stalkless or sometimes have a rudimentary stalk, and their caps are fan- to shell-shaped or kidney-shaped and nearly plane at maturity.

Little is known about the edibility of this group, and none is recommended for the table. The identification of most species is often very difficult and requires the use of a microscope. Because the genus is complex and includes a large number of species, a comprehensive key to species is beyond the scope of this book. We have included three of the more common or distinctive species that occur in northeastern North America. For additional information, we recommend *North American Species of Crepidotus* by Hesler and Smith.

Crepidotus applanatus var. **applanatus** (Persoon : Fries) Kummer
COMMON NAME: Flat Crep.
CAP: ⅜–1½" (1–4 cm) wide, shell- to fan-shaped; surface smooth, moist, occasionally pubescent, stalkless, fibrillose at the point of attachment, hygrophanous, white when young, becoming pale pinkish tan to cinnamon-brown in age, often paler at the point of attachment; margin striate when fresh; flesh white, thin; odor and taste not distinctive.
GILLS: attached, crowded to close, narrow, radiating from the point of attachment, white when young, becoming pinkish brown to brownish.
SPORE PRINT: brown to cinnamon-brown.
MICROSCOPIC FEATURES: spores 4–5.5 µm, round, punctate, pale brown.
FRUITING: in clusters on dead and decaying hardwood, rarely on conifers; July–September; common.
EDIBILITY: unknown.

Crepidotus cinnabarinus Peck
CAP: ¼–⅝" (5–15 mm) wide, shell- to fan-shaped or kidney-shaped and convex, becoming nearly flat in age; surface dry, fibrillose, bright cinnabar-red overall; flesh reddish; odor and taste not distinctive; stalkless or with a rudimentary stalk.
GILLS: attached, subdistant, moderately broad, cinnabar-red.
SPORE PRINT: dull brown.
MICROSCOPIC FEATURES: spores 6–10 × 5–6 µm, oval, punctate, pale brown.
FRUITING: in clusters on hardwoods, especially beech, basswood, and poplar; July–September; rare.
EDIBILITY: unknown.

Crepidotus mollis (Fries) Staude
CAP: ⅜–3⅛" (1–8 cm) wide, kidney- to fan-shaped or shell-shaped, somewhat convex to nearly plane; surface hygrophanous, color variable, olive-brown to ochraceous or ochraceous whitish, coated with short brownish fibers; flesh thin, whitish; odor not distinctive; taste not distinctive or bitter; stalkless or with a rudimentary stalk.
GILLS: decurrent, close to crowded, whitish at first, becoming cinnamon to grayish cinnamon; edges entire or fimbriate.
SPORE PRINT: brown.
MICROSCOPIC FEATURES: spores 7–10 × 4.5–6 μm, ellipsoid, smooth, with a double wall, pale brown.
FRUITING: in groups on decaying hardwood bark or, rarely, on conifers; July–September; occasional.
EDIBILITY: unknown.
COMMENTS: formerly known as *Agaricus (Crepidotus) fulvotomentosus*.

Genus *Crinipellis*

Crinipellis zonata (Peck) Patouillard
COMMON NAME: Zoned Crinipellis.
CAP: ½–1½" (1.2–4 cm) wide, convex to nearly flat with a small depression at the center; surface dry orangish to orangish brown from a coating of long, radially arranged, stiff reddish brown hairs on a creamy background, the hairs dextrinoid, typically appearing zoned; flesh thin, somewhat tough; odor and taste not distinctive.
GILLS: free or nearly so, close, narrow, white.
STALK: 1–2" (2.5–5 cm) long, ⅟32–⅟16" (1–2 mm) thick, decorated like the cap, tough.
SPORE PRINT: white.
MICROSCOPIC FEATURES: spores 4–6 × 3–5 μm, elliptic, smooth, hyaline, inamyloid; cheilocystidia present.
FRUITING: scattered or in groups on dead deciduous wood, especially twigs; occasional to frequent; typically August–September.
EDIBILITY: unknown.
COMMENTS: other species in this genus share the distinctive dextrinoid hairs and spores that are usually inamyloid.

Genus *Cyptotrama*

Cyptotrama asprata (Berkeley) Redhead and Ginns
COMMON NAME: Golden Scruffy Collybia.
CAP: ¼–1" (5–25 mm) wide, convex to broadly convex; surface golden at first, becoming bright yellow, finely granular-coated, often becoming wrinkled or furrowed; margin at first pressed against the upper stalk; flesh thin, pale yellowish; odor and taste not distinctive.
GILLS: attached, usually becoming slightly decurrent, distant, broad, white.
STALK: 1–2" (1–5 cm) long, ⅟16–⅛" (1.5–3 mm) thick, solid, fairly tough, scurfy, golden on a pale yellow background.
SPORE PRINT: white.
MICROSCOPIC FEATURES: spores 8–12 × 6–7.5 μm, lemon-shaped to broadly oval, smooth, hyaline, inamyloid.

FRUITING: usually several scattered on hardwood sticks or logs, especially oak;
May–October; occasional to fairly common.

EDIBILITY: unknown.

COMMENTS: also often identified as *Cyptotrama chrysopeplum.*

Genus *Cystoderma*

Cystoderma is a small genus of approximately twenty known species in North America. They are small to medium mushrooms with white spore prints and grow on the ground or on decaying wood. They have a granular cap and a stalk that is sheathed halfway or more up to a ring or fibrous annular zone with granules or tiny scales. Their gills are attached, not free, and they have smooth spores. The edibility of most species is unknown.

Key to Species of *Cystoderma*

1. Ring not flared; cap margin often rimmed with small flaps of veil tissue; cap ⅝–1¾" (1.6–4.5 cm) wide, obtuse to broadly convex, sometimes with a broad umbo, dry, granular to nearly smooth when young, becoming slightly wrinkled and breaking up to form tiny scales in age, ochre-tawny to dingy orange over the disc, orange-yellow toward the margin; flesh white, odor and taste usually not distinctive; gills whitish, becoming pale orange-yellow at maturity; stalk sheathed up to a superior to median ring with ochre-tawny to dingy tan or buff granules and tiny scales; ring usually incomplete; growing among mosses, on humus, or on conifer needles—*C. amianthinum* (Scopoli : Fries) Fayod (see p. 116).

1. Fruiting body nearly identical to the previous choice except that the cap is conspicuously radially wrinkled to reticulate and the flesh has an odor of freshly husked corn; June–October; edibility unknown—*C. amianthinum* var. *rugosoreticulatum* (Lorinser) Smith and Singer.

1. Not with the above combinations of characters → 2.
    2. Growing among mosses, on conifer needles, or on very rotten conifer logs, usually in groups or clusters, especially in the Great Lakes area; ring flared; cap margin lacking veil remnants; cap 1–2" (2.5–5 cm) wide, dry, granular; rusty brown, cinnamon-brown, rusty orange to ochre-tawny; flesh whitish, odor and taste not distinctive; stalk typically enlarged downward, whitish to pale tawny and smooth above the ring, sheathed up to a conspicuous, membranous ring with rusty brown to ochre-tawny granules and tiny scales; spores 3.5–5.5 × 3–4 μm, elliptic, smooth, amyloid, hyaline; September–October; edibility unknown—*C. fallax* Smith and Singer.
    2. Growing on decaying hardwood, scattered or in groups, widespread in the region; ring flared; cap margin lacking veil remnants; cap 1–3½" (2.5–9 cm) wide, dry, granular to scaly or sometimes wrinkled, bright to dull orange or ochre-tawny; flesh whitish, odor and taste not distinctive; stalk typically enlarged downward, whitish to tawny and often pruinose above the ring, sheathed up to a conspicuous, membranous ring with orange to ochre-tawny granules and tiny scales; spores 4–5 × 2.5–3 μm, elliptic, smooth, amyloid, hyaline; August–October; edibility unknown—*C. granosum* (Morgan) Smith and Singer.
    2. Not with the above combination of characters; ring not flaring; margin typically rimmed with small flaps of veil tissue → 3.
3. Stalk sheathed up to the ring with orange-cinnamon to rusty orange or ochre-tawny

granules and tiny scales over a dingy yellow ground color, often hollow in age; ring superior, membranous, sometimes well formed but often incomplete; cap ⅝–1½" (1.6–4 cm) wide, dry, granular to scaly or sometimes wrinkled; color variable, orange-cinnamon, rusty orange to ochre-tawny or dull orange-yellow; flesh whitish, odor and taste not distinctive; spores 3–5 × 2.5–3 μm, elliptic, smooth, hyaline, inamyloid; cheilocystidia absent—*C. granulosum* (Batsch : Fries) Fayod (see p. 116).

3. Nearly identical to the previous choice except stalk sheathed up to the ring with cinnabar-red to orange-brown granules and tiny scales over a dingy yellow ground color; cap color variable, cinnabar-red, rusty red, rusty brown to orange-brown; cheilocystidia 32–46 × 4–10 μm, fusoid-pointed with a long narrow neck; edibility unknown—*C. cinnabarinum* (Albertini and Schweinitz : Secretan) Fayod.

Cystoderma amianthinum (Scopoli : Fries) Fayod

CAP: ⅝–1¾" (1.6–4.5 cm) wide, obtuse to broadly convex, sometimes with a broad umbo; surface dry, granular to nearly smooth when young, becoming slightly wrinkled and breaking up to form tiny scales in age, ochre-tawny to dingy orange over the disc, becoming pale orange-yellow toward the margin, staining rusty brown with KOH or NH₄OH; margin often rimmed with small flaps of veil tissue; flesh white; odor and taste usually not distinctive.

GILLS: attached, close, whitish, becoming pale orange-yellow at maturity, with 2–3 tiers of lamellulae.

STALK: 1–2⅜" (2.5–6 cm) long, ⅛–¼" (2–7 mm) thick, nearly equal or enlarged downward, sheathed up to the ring with ochre-tawny to dingy tan or buff granules and tiny scales; veil fibrous-membranous, leaving an incomplete or sometimes well-formed superior to median ring; smooth to pruinose and pale yellow to pale brown above the ring.

SPORE PRINT: white.

MICROSCOPIC FEATURES: spores 4.5–7 × 2.5–3.5 μm, elliptic, smooth, hyaline, amyloid.

FRUITING: solitary, scattered or in groups among mosses, on humus, or on conifer needles; June–October; fairly common.

EDIBILITY: unknown.

Cystoderma granulosum (Batsch : Fries) Fayod

CAP: ⅝–1½" (1.6–4 cm) wide, obtuse to convex when young, becoming nearly flat in age; surface dry, granular to scaly or sometimes wrinkled; color variable, orange-cinnamon, rusty orange to ochre-tawny or dull orange-yellow; margin typically rimmed with small flaps of veil tissue; flesh whitish; odor and taste not distinctive.

GILLS: notched or attached, close, white to pale yellow, with several tiers of lamellulae.

STALK: ¾–2⅜" (2–6 cm) long, ⅛–¼" (3–6 mm) thick, nearly equal or enlarged downward, sometimes bulbous, often hollow or stuffed in age, sheathed up to the ring with orange-cinnamon to rusty orange or ochre-tawny granules and tiny scales over a dingy yellow ground color; veil fibrous-membranous, leaving an incomplete or sometimes well-formed superior ring; smooth and whitish to pale orange above the ring.

SPORE PRINT: white.

MICROSCOPIC FEATURES: spores 3.5–5 × 2.5–3 μm, elliptic, smooth, hyaline, inamyloid; cheilocystidia absent.

FRUITING: scattered or in groups among mosses, on humus, or on conifer needles in woods; August–October; fairly common.

EDIBILITY: unknown.

COMMENTS: the more orange cap colors and lack of cheilocystidia distinguish this species from the very similar *C. cinnabarinum* (edibility unknown).

Genus *Entoloma* and Allies

The Genus *Entoloma* and Allies is a highly variable group of mushrooms with pinkish (flesh-color, salmon-pink, or brownish pink) spores, attached gills, and no partial veils. Microscopic examination of the spores is often required to determine that the spores are angular. *Rhodocybe* and *Rhodocybella* spores are similar but typically appear more warty, and in *Clitopilus* the spores are longitudinally ridged. Most are very difficult to identify to species, and the entire group needs a great deal more study. In addition, there is little agreement on which of the various genera should be recognized as distinct from *Entoloma*. For these reasons, a workable and comprehensive key to the species found in northeastern North America is beyond the scope of this book.

In addition to the single species of *Entoloma* described below, see also descriptions and photographs of a few of the more common and/or distinctive species in this group: *Clitopilus prunulus* (p. 88) and *Rhodocybe mundula* (p. 242), as well as four species each of *Leptonia* (p. 189) and *Nolanea* (p. 213).

Entoloma abortivum (Berkeley and Curtis) Donk
COMMON NAME: Abortive Entoloma.
CAP: 2–4" (5–10 cm) wide, typically convex, somewhat fibrous to almost smooth, the edge inrolled and often becoming irregular in age, sometimes with a low umbo or becoming slightly depressed at the center; surface pale gray to grayish brown; flesh white, thick; odor and taste farinaceous.
GILLS: typically decurrent, close, grayish at first, soon becoming pinkish.
STALK: 1–4" (2.5–10 cm) long, ¼–⅝" (5–15 mm) thick, usually enlarged at the base, dry, solid, scurfy, whitish to grayish, the base usally coated with white mycelium.
SPORE PRINT: salmon-pink.
MICROSCOPIC FEATURES: spores 8–10 × 4.5–6 μm, elliptic and angular, typically 6-sided, hyaline, inamyloid.
FRUITING: scattered or in groups, often gregarious, on the ground, especially in humus or near rotting stumps, in woods; August–November; frequent to common.
EDIBILITY: edible with caution (see comments below).
COMMENTS: this mushroom can be best identified by, and should be eaten only if found in immediate proximity to, the aborted fruiting bodies. They are 1–2" (2.5–5 cm) high and 1–4" (2.5–10 cm) wide, chalky white masses with pinkish marbling in the spongy white inner flesh. Both the normal gilled mushrooms and the aborted fruiting bodies are good edibles. There is good evidence to suggest that the aborted form is caused by some interaction with at least one or two species of "Honey Mushrooms" (*Armillaria* spp.; see p. 68), which are usually found in the vicinity.

Genus *Flammulina*

Flammulina velutipes (Fries) Karsten
COMMON NAME: Velvet Foot, Winter Mushroom.
CAP: ¾–2¾" (2–7 cm) wide, convex becoming nearly flat; surface orangish brown to reddish yellow, darker at the center; smooth, slimy, very sticky when dry; margin incurved at first, becoming striate; flesh thin, yellowish, usually watery; odor and taste not distinctive.

GILLS: attached, becoming subdecurrent, yellowish, typically close and narrow.

STALK: usually 1–3" (2.5–7.5 cm) long but sometimes longer, ⅛–¼" (3–7 mm) thick; smooth and yellowish at the top, becoming dark reddish brown and extremely velvety starting at the base; very tough, not easily broken.

SPORE PRINT: white.

MICROSCOPIC FEATURES: spores 7–9 × 3–6 µm, elliptic, smooth, hyaline, inamyloid.

FRUITING: in clusters on deciduous trees, logs, and stumps, especially willow, aspen, poplar, and elm; usually fruiting during early spring or late fall but also during winter thaws and summer cold spells; fairly common.

EDIBILITY: edible, though the sticky caps can be difficult to clean.

COMMENTS: previously classified as *Collybia velutipes*. It has been cultivated in the Orient as Enotake or Enoki-take.

Genus *Galerina*

Galerina is a very large genus, with more than two hundred species in North America. They are small to medium mushrooms that grow on humus, among mosses, and on decaying wood. Their spore prints are ochraceous to rusty brown, and their spores are mostly roughened, with a plage, lack an apical pore and are often dextrinoid. They have slender stalks, up to ⅛" (4 mm) thick at the apex, and their gill edges are often white and finely fringed.

Some species are deadly poisonous, and the edibility of most is unknown. None is recommended for the table. Identification to species is very difficult, even with the use of a microscope. Because the genus is complex and includes a large number of species, a comprehensive key to species is beyond the scope of this book. We have included two of the more common and distinctive species that occur in this region. For additional information, we recommend *A Monograph on the Genus Galerina Earle* by Smith and Singer.

Galerina autumnalis var. **autumnalis** (Peck) Smith and Singer

COMMON NAME: Deadly Galerina.

CAP: 1–2½" (2.5–6.5 cm) wide, convex when young, becoming nearly flat in age, with or without a low, broad umbo; surface viscid, moist, smooth, hygrophanous, dark brown to dark amber, fading to yellowish orange or buff, often remaining darker over the disc; margin translucent-striate when moist; flesh brownish to buff, thin; odor not distinctive to slightly farinaceous; taste not distinctive.

GILLS: attached to very slightly decurrent, close, broad, brown becoming rusty brown.

STALK: 1⅛–3½" (3–9 cm) long, ⅛–⅜" (3–10 mm) thick, equal or slightly enlarged at the base, hollow, dry, smooth to slightly pruinose at the apex, covered with grayish white flattened fibrils over a brown to blackish brown ground color; base usually coated with white mycelium; partial veil white, membranous, leaving a persistant, membranous superior ring.

SPORE PRINT: rusty brown.

MICROSCOPIC FEATURES: spores 8.5–10.5 × 5–6.5 µm, elliptic, roughened, pale brown.

FRUITING: scattered, in groups or in clusters on decaying hardwood and conifer stumps and logs; May–November; fairly common.

EDIBILITY: deadly poisonous.

COMMENTS: this mushroom, which contains the same kinds of toxins that are found in deadly *Amanita* species, causes liver and kidney damage, coma, and death.

Galerina tibiicystis (Atkinson) Kühner

COMMON NAME: Sphagnum-bog Galerina.

CAP: ⅜–1⅜" (1–3.5 cm) wide, conic when young, becoming broadly conic with a distinct umbo in age; surface smooth, moist, hygrophanous, becoming opaque in age, translucent-striate from the margin nearly to the disc, dark amber to brownish orange, fading to yellowish buff, often remaining darker over the disc; flesh pale yellowish to buff, thin, watery; odor and taste not distinctive.

GILLS: attached, close to subdistant, narrow to moderately broad, tawny to pale cinnamon-brown.

STALK: 2⅛–8¾" (5.5–22 cm) long, ¹⁄₁₆–⅛" (2–4 mm) thick, equal, hollow, fragile, amber to pale buff, whitish on the lower portion that is embedded in moss, white-pruinose or covered with tiny fibrils when young, soon becoming naked; partial veil and ring absent.

SPORE PRINT: rusty brown.

MICROSCOPIC FEATURES: spores 8.5–14 × 5–7 μm, elliptic to oval, roughened with a minute apical callus, pale brown.

FRUITING: solitary, scattered, or in groups among sphagnum mosses in bogs and wet areas in woods; May–November; fairly common.

EDIBILITY: unknown.

Genus *Gomphidius*

Gomphidius species are small to medium terrestrial mushrooms associated with various conifers, especially pines. It is a small genus with fewer than a dozen known species in all of North America. They typically have smoky gray to blackish spore prints, long and narrow spores and decurrent, thick, white gills. Most species have a white central stalk that tapers down to a yellow base. Their cap surfaces are slimy to sticky, and their flesh is white. Several species are edible but often rated mediocre.

Key to Species of *Gomphidius*

1. Glutinous partial veil present on immature specimens; a partial veil of white fibrils may also be present → 2.
1. Glutinous partial veil absent on immature specimens; a partial veil of white fibrils may be present → 3.
    2. Cap pink to rose-red, smooth, slimy; gills decurrent, white, becoming smoky gray; stalk solid, tapered downward, white with a yellow base, with a thin, slimy superior ring that blackens as the spores are trapped; spores 14–20 × 4.5–7 μm; on the ground under conifers from Michigan to Maine, north into Canada; July–October; edible—*G. subroseus* Kauffman.
    2. Cap color variable, grayish brown to purplish gray, pale cinnamon-gray, vinaceous brown or pale salmon, slimy—*G. glutinosus* (Schaeffer : Fries) Fries (see p. 120).
3. Cap pale pinkish cinnamon to orange-cinnamon, pale reddish brown or salmon when young, typically spotted black or blackening in age, smooth, slimy; gills decurrent, white, becoming smoky gray at maturity; stalk tapered downward, white with a yellow base, coated on the lower two-thirds with dark brown to purplish black fibrils; partial veil absent; spores 14–22 × 6–8 μm; on the ground under conifers; August–October; edible—*G. maculatus* (Scopoli : Fries) Fries.
3. Cap white to pale yellow when young, soon becoming pale vinaceous to pinkish brown and finally blackening in age, sticky to slimy; gills decurrent, white, becoming smoky

gray to blackish at maturity; stalk nearly equal or tapered downward, white with a yellow or pinkish base that blackens in age; a partial veil of sparse, whitish fibrils is usually present on button stages, sometimes leaving a superior annular zone; spores 13–22 × 4.5–7 μm; on the ground under conifers; August–October; edibility unknown—
G. nigricans Peck.

Gomphidius glutinosus (Schaeffer : Fries) Fries

COMMON NAME: Slimy Gomphidius.

CAP: ¾–4" (2–10 cm) wide, convex, becoming broadly convex to nearly flat; surface smooth, slimy; color variable, grayish brown to purplish gray, pale cinnamon-gray, vinaceous brown or pale salmon, typically spotted black in age; flesh thick, whitish, often tinted pink; odor not distinctive; taste acidic.

GILLS: decurrent, thick, close, broad, pale grayish white, becoming smoky gray at maturity.

STALK: 1⅜–4" (3.5–10 cm) long, ¼–¾" (7–20 mm) thick, tapered downward, solid, white with a yellow base; partial veil slimy, with white fibrils beneath the glutinous layer, leaving a thin, slimy superior ring on the stalk, blackening as spores are trapped.

SPORE PRINT: smoky gray to blackish.

MICROSCOPIC FEATURES: spores 15–21 × 4–7 μm, elliptic, smooth, grayish brown.

FRUITING: solitary, scattered, or in groups on the ground in conifer woods; June–November; occasional.

EDIBILITY: edible.

Genus *Gymnopilus*

Gymnopilus species are small to large mushrooms that grow on decaying wood. More than seventy species have been described from North America. Their spore prints vary from orange to rusty orange to bright rusty brown. Most species have bitter flesh, and their spores are finely roughened and lack an apical pore. A drop of KOH or NH_4OH will stain the cap surface blackish. A few species are hallucinogenic, but the edibility of the majority is unknown.

Key to Species of *Gymnopilus*

1. Cap large to very large, 3–7" (7.5–18 cm) wide at maturity → 2.
1. Cap small to medium, ⅜–3⅛" (1–8 cm) wide at maturity → 3.
    2. Flesh taste not distinctive; cap up to 6" (15.5 cm) wide, covered overall with tiny, dark ochraceous scales over a pale ochraceous to yellowish ground color; stalk up to 2" (5 cm) thick; partial veil cortinate, leaving a sparse annular zone; spores 7.5–10 × 4.5–5.5 μm; scattered or in clusters on decaying hardwood; hallucinogenic—
        G. validipes (Peck) Hesler.
    2. Flesh bitter; odor anise- to licorice-like; gills attached; cap up to 7" (18 cm) wide, pale orange-yellow to ochre-orange, smooth when young, becoming scaly in age; stalk up to 1⅛" (3 cm) thick; partial veil submembranous to fibrous, typically leaving a superior ring—*G. spectabilis* (Fries) Smith (see p. 122).
    2. Flesh bitter; odor sweet, fragrant, often anise-like; gills notched to sinuate; cap up to 4" (10 cm) wide, yellow to orange-yellow to ochre-orange, smooth, lacking scales, or slightly scaly on the disc in age; stalk up to ⅝" (1.6 cm) thick; partial veil submembranous to fibrous, typically leaving a superior ring; spores 6–9 × 4–5.5 μm, scattered or in clusters on decaying wood; inedible because of the bitter taste, not hallucinogenic—*G. luteus* (Peck) Hesler.

3.   Cap dry, bright yellow, yellow, golden yellow to orange-yellow to tawny-orange or dark reddish purple to purplish red fading to rusty cinnamon or dingy yellow in age; with a whitish to yellowish partial veil present on the button stage; flesh bitter → 4.

3.   Cap moist or dry; ochraceous-orange to orange-brown or reddish brown to bay-brown; veil absent on the button stage; flesh taste not distinctive or bitter → 5.

   4.   Cap surface glabrous or nearly so; veil whitish, cortinate, sparse, evanescent; flesh white to whitish; gills yellowish white to pale yellow when young, becoming rust-spotted in age; edges typically even; cap orange-yellow to golden yellow, ¾–2" (2–5 cm) wide; spores 7–9 × 4–5.5 μm; edibility unknown—*G. penetrens* (Fries : Fries) Murrill.

   4.   Cap surface fibrillose to scaly (use a hand lens); veil yellowish, cortinate, sparse, evanescent; flesh yellow; gills yellow to orange-yellow when young, becoming brownish cinnamon to rusty yellow, edges minutely fringed; cap golden yellow to tawny, paler on the margin, ¾–2¾" (2–7 cm) wide—*G. sapineus* (Fries) Maire (see p. 122).

   4.   Cap covered with conspicuous dark reddish purple to purplish red scales on a tawny ground color when young, fading to rusty cinnamon to fulvous or dingy yellow at maturity; veil yellowish, fibrillose, sparse; flesh pale purplish vinaceous to pale pinkish salmon on young specimens, yellow at maturity; gills yellow to rusty orange—*G. luteofolius* (Peck) Singer (see p. 121).

5.   Cap reddish brown to bay-brown, ⅜–2" (1–5 cm) wide; flesh reddish brown to bay-brown, taste not distinctive or bitterish; gills attached to decurrent, nearly free in age, bright yellow when young, becoming ochraceous and sometimes staining rusty, close—*G. picreus* (Fries) Karsten (see p. 122).

5.   Cap orange-brown to ochraceous orange, ¾–2¾" (2–7 cm) wide; flesh pale orange to tawny-yellow, distinctly bitter; odor not distinctive or resembling raw potatoes; gills attached to notched, becoming nearly free in age, pale orange-yellow when young, becoming brownish orange; edges minutely fringed; spores 7–10 × 4–5.5 μm; edibility unknown—*G. liquiritiae* (Persoon : Fries) Karsten.

Gymnopilus luteofolius (Peck) Singer

CAP: ¾–3⅛" (2–8 cm) wide, obtuse to convex, becoming broadly convex in age, margin fibrillose to fibrillose-scaly, incurved and remaining so well into maturity, even; surface dry, covered with conspicuous erect to slightly recurved scales; scales dark reddish purple to purplish red on a tawny ground color when young, fading to rusty cinnamon to fulvous or dingy yellow at maturity; flesh pale purplish vinaceous to pale pinkish salmon on young specimens, yellow at maturity; odor not distinctive; taste bitter.

GILLS: attached to decurrent, close, dull yellow at first, becoming dingy orange-yellow then rusty orange at maturity.

STALK: 1⅛–3¾" (3–9.5 cm) long, ⅛–½" (3–12 mm) thick, nearly equal or enlarged downward, dry, hollow in age, longitudinally striate, fibrillose, apex pale to dull yellow, instantly staining pinkish red when bruised, dark reddish purple below when young, becoming dull reddish brown in age; partial veil fibrillose, yellowish, leaving a thin superior fibrillose annular zone; mycelium whitish to pale yellow.

SPORE PRINT: bright rusty orange.

MICROSCOPIC FEATURES: spores 5.5–8.5 × 3.5–4.5 μm, ellipsoid to subovoid, smooth, ochraceous, dextrinoid.

FRUITING: in dense clusters on decaying wood chips, logs, stumps and sawdust; June–November; occasional.

EDIBILITY: hallucinogenic.

COMMENTS: this mushroom is sometimes mistaken for *Tricholomopsis rutilans* (edible but of poor quality) which is similarly colored but has a white spore print.

Gymnopilus picreus (Fries) Karsten

CAP: ⅜–2" (1–5 cm) wide, convex to bell-shaped, becoming broadly convex in age; surface smooth, then scurfy, moist, hygrophanous, dark reddish brown to bay-brown, fading somewhat in age; margin inrolled when young; flesh reddish brown to bay-brown; odor not distinctive; taste not distinctive or slightly bitter.

GILLS: attached to decurrent, becoming nearly free in age, close, narrow, bright yellow when young, becoming ochraceous and sometimes staining rusty.

STALK: 1⅛–3⅛" (3–8 cm) long, ⅛–¼" (2–5 mm) thick, enlarged slightly at the base, smooth, brown to yellow-brown and typically darkening from the base up; apex white-pruinose; partial veil and ring absent.

SPORE PRINT: ochraceous orange.

MICROSCOPIC FEATURES: spores 7–10 × 4.5–6 μm, elliptic to subovoid, slightly roughened, lacking an apical pore, pale brownish.

FRUITING: scattered or in groups on decaying logs and stumps in mixed woods; July–October; infrequent.

EDIBILITY: unknown.

Gymnopilus sapineus (Fries) Maire

CAP: ¾–2¾" (2–7 cm) wide, hemispheric when young, becoming broadly convex; surface coated with minute scales, golden yellow to tawny, paler on the margin; flesh yellow; odor pungent or not distinctive; taste bitter or not distinctive.

GILLS: attached, close, broad, yellow to orange-yellow when young, becoming brownish cinnamon to rusty yellow; edges minutely fringed.

STALK: 1–2¾" (2.5–7 cm) long, ⅛–⅜" (3–10 mm) thick, nearly equal or tapered toward the base, yellowish, becoming brownish near the base, often hollow in age; partial veil present on the button stage, cortinate, sparse, yellowish, sometimes leaving a thin annular zone.

SPORE PRINT: rusty orange to bright rusty brown.

MICROSCOPIC FEATURES: spores 7–10 × 4–5.5 μm, elliptic, slightly roughened, lacking an apical pore, pale brown.

FRUITING: scattered or in groups on decaying wood and sawdust; July–October; fairly common.

EDIBILITY: unknown.

Gymnopilus spectabilis (Fries) Smith

COMMON NAME: Big Laughing Gym.

CAP: 3–7" (7.5–18 cm) wide, convex, becoming broadly convex to nearly plane; surface typically dry, smooth when young, becoming scaly in age, pale orange-yellow to ochre-orange; flesh pale yellow when young, becoming darker yellow in age; odor anise- to licorice-like, sometimes not distinctive; taste bitter.

GILLS: attached to decurrent, close to crowded, ochraceous buff at first, becoming yellow to rusty; edges minutely fringed.

STALK: 2–7" (5–18 cm) long, ⅜–1⅛" (1–3 cm) thick, typically enlarged downward to the base, then abruptly tapered, streaked with tiny fibrils; colored like the cap and often brownish near the base; partial veil submembranous to fibrous, typically leaving a superior ring.

SPORE PRINT: rusty orange to rusty brown.

MICROSCOPIC FEATURES: spores 7–10 × 4.5–6 μm, elliptic to oval, roughened, lacking an apical pore, pale brownish.

FRUITING: in clusters or scattered on decaying wood or on the ground attached to buried wood; July–October; occasional to fairly common.

EDIBILITY: hallucinogenic.

Genus *Hebeloma*

Hebeloma is a genus of small to medium mushrooms that are terrestrial under conifers and hardwoods. Their spore prints are shades of rusty brown to cinnamon-brown, and their spores are roughened. They have a viscid cap, a dry, central stalk with a white-pruinose to scurfy apex and flesh that often has a radish-like odor or taste. Their gill edges are usually white and finely fringed, and cheilocystidia are typically present and abundant. These are useful characters for separating *Hebeloma* species from *Cortinarius* species, which usually lack cheilocystidia and finely fringed, white gill edges.

Some species are poisonous, and the edibility of the majority is unknown. Identification of species is difficult, even with the use of a microscope. We have included four of the more common or distinctive species that occur in northeastern North America.

Hebeloma crustuliniforme (Bulliard) Quélet

COMMON NAME: Poison Pie.

CAP: 1⅛–3⅛" (3–8 cm) wide, convex, becoming nearly flat at maturity; margin inrolled when young, becoming uplifted and wavy in age; surface smooth, viscid, creamy white with dull yellow-brown to reddish brown tones at the center; flesh white, odor radish-like; taste bitter to disagreeable.

GILLS: notched, close to crowded, whitish at first, becoming grayish, then finally grayish brown at maturity; edges white, finely fringed, with watery to brownish droplets when fresh, brown-dotted when dry.

STALK: 1⅛–3⅛" (3–8 cm) long, ¼–¾" (7–20 mm) thick, nearly equal or bulbous at the base, white-pruinose to scurfy, at least on the upper portion, white to buff, becoming brownish from the base upward in age or when handled, hollow, often with white mycelium at the base; partial veil and ring absent.

SPORE PRINT: yellow-brown to brown.

MICROSCOPIC FEATURES: spores 9–13 × 5–7.5 μm, elliptic, slightly roughened, pale brown; dextrinoid.

FRUITING: scattered, in groups, or sometimes in arcs or fairy rings on the ground under hardwoods or conifers, often found under bushes and hedges or in lawns in residential areas; August–November; fairly common.

EDIBILITY: poisonous, causing gastric distress.

COMMENTS: *H. sinapizans* (poisonous) is very similar with a larger cap, up to 6" (15.5 cm) wide, a radish-like odor and taste, a stouter stalk, up to 1" (2.5 cm) thick, which becomes scaly on the upper portion, and spores that measure 10–14 × 6–8 μm.

Hebeloma mesophaeum (Persoon : Fries) Quélet

CAP: ¾–2⅜" (2–6 cm) wide, convex, becoming broadly convex, often depressed at the center, sometimes with an umbo; margin inrolled at first, becoming decurved well into maturity, sometimes uplifted and wavy in age, typically rimmed with cortinate veil remnants; surface viscid, fibrillose, especially near the margin, reddish brown to orange-brown over the center, brownish yellow to yellowish tan toward the margin; flesh pale watery brown; odor not distinctive; taste radish-like, then bitter.

GILLS: sinuate to notched, close, whitish at first, becoming grayish brown at maturity; edges whitish and finely fringed in age.

STALK: 1⅛–3½" (3–9 cm) long, ⅛–⅜" (3–10 mm) thick, nearly equal, dry, fibrillose-striate, white at first, becoming dull brown from the base upward in age; partial veil cortinate, pale yellow, soon stained brownish by falling spores, leaving veil remnants on the cap margin or a fibrillose, superior annular zone.

SPORE PRINT: rusty brown.

MICROSCOPIC FEATURES: spores 8–11 × 4–7 μm, broadly elliptic, slightly roughened, pale brown.

FRUITING: scattered or in groups on soil or grassy areas, often found under bushes and hedges in residential areas; April–November; fairly common.

EDIBILITY: poisonous, causing gastric distress.

Hebeloma radicosum (Bulliard : Fries) Ricken

CAP: 2–4" (5–10 cm) wide, convex, becoming broadly convex to nearly plane in age; surface viscid when fresh, fibrillose to fibrillose-scaly, various shades of yellowish tan, golden brown, ochre, or pale cinnamon, paler toward the margin; margin incurved when young, often rimmed with partial veil fibers; flesh white; odor of marzipan or bitter almonds; taste bitter.

GILLS: sinuate to nearly free, close, whitish when young, soon becoming ochre and finally pale golden brown to reddish brown; edges often finely scalloped or fringed in age.

STALK: 3–7" (7.5–18 cm) long, ½–1" (1.3–2.5 cm) thick, typically swollen near the middle and tapered in either direction, usually rooted, sheathed from the base up to the ring with pale ochre-brown to pale cinnamon fibers and cottony scales over a whitish to cream ground color, solid, firm, white near the apex; partial veil whitish, fibrous, leaving a superior ring.

SPORE PRINT: rusty brown to cinnamon-brown.

MICROSCOPIC FEATURES: spores 8–10 × 5–6 μm, almond-shaped, warted, pale brown.

FRUITING: scattered or in groups on the ground or well-decayed wood in hardwood forests, especially beech, sometimes under conifers; July–November; uncommon.

EDIBILITY: poisonous.

Hebeloma velatum Peck

CAP: ¾–2¾" (2–7 cm) wide, obtuse to convex, becoming nearly plane, sometimes slightly depressed or with a low, broad umbo; surface viscid when moist, smooth, becoming silky on drying and often coarsely cracked in age, grayish brown to chestnut with a purplish red disc, fading to dull ochraceous in age; margin decorated with conspicuous whitish to yellowish veil remnants; flesh white; odor of radish; taste not distinctive.

GILLS: notched, close, whitish when young, becoming yellow-brown at maturity; edges minutely white-fimbriate.

STALK: 1⅛–3⅛" (3–8 cm) long, ⅛–⅜" (3–10 mm) thick, nearly equal, dry, hollow in age, densely fibrillose from the base up to a conspicuous annular zone, pruinose-floccose at the apex, whitish when young, becoming dingy ochraceous from the base up to a paler apex at maturity; partial veil cottony-membranous, whitish to yellowish, leaving remnants on the cap margin and a fibrillose annular zone.

SPORE PRINT: dull yellowish brown.

MICROSCOPIC FEATURES: spores 9–12 × 5–8 μm, narrowly elliptic, slightly roughened, pale brown.

FRUITING: scattered or in groups on the ground, typically in sandy soil, under conifers, especially pines, or in mixed woods; September–November; infrequent.

EDIBILITY: unknown.

COMMENTS: the presence of conspicuous whitish to yellowish veil remnants (see photo, p. 286) on the cap margin is diagnostic.

Genus *Hohenbuehelia*

Members of this genus are small to medium-sized mushrooms that are stalkless or have rudimentary lateral stalks. They are fan- to shell-shaped or somewhat funnel-shaped agarics that grow on decaying wood, sawdust, or woody debris. They have white spore prints and very thick-walled cystidia that are often incrusted. Accurate identification to species often requires the use of a microscope. They have an unusual adaptation: recent research has shown that, like *Pleurotus,* these otherwise innocuous agarics have developed a way to trap nematodes and can therefore be considered to be among the few known carnivorous fungi.

Key to Species of *Hohenbuehelia*

1. Cap pale pink to pinkish buff or tan to pale brown or grayish brown → 2.
1. Cap usually mouse-gray to bluish gray or bluish black, sometimes fading to whitish or olivaceous to brownish in age → 3.
    2. Cap pale pink to pinkish buff, surface rubbery and slippery; stalk rudimentary, lateral; flesh odor and taste not distinctive; gills crowded; spores 3–4 μm, globose; on woody debris and sawdust piles; edibility unknown—*H. angustatus* (Berkeley) Singer.
    2. Cap tan to pale brown or grayish brown, sometimes with a whitish bloom, surface smooth to minutely hairy; stalk lateral, often well differentiated or sometimes rudimentary; flesh odor and taste not distinctive; gills crowded; spores 7–9 × 4–5 μm, elliptic; on wood or woody debris; edible—*H. petaloides* (Bulliard : Fries) Schulzer (see p. 125).
3. Cap mouse-gray, fading to pale gray or whitish in age, surface covered with a gelatinous layer of ridges that form a reticulum; stalk absent; flesh odor strongly fungoid to farinaceous, taste unpleasant to farinaceous; gills subdistant; spores 6–9 × 5–6 μm, ellipsoid; on wood and woody debris; edibility unknown—*H. mastrucatus* (Fries) Singer.
3. Cap bluish gray to bluish black, sometimes olivaceous to brownish, surface velvety-tomentose; stalk absent; flesh odor and taste not distinctive; gills crowded; spores 7–9 × 3.5–5.5 μm, ellipsoid; on decaying hardwoods; edibility unknown—*H. atrocaerulea* (Fries : Fries) Singer.

Hohenbuehelia petaloides (Bulliard : Fries) Schulzer

COMMON NAME: Leaf-like Oyster.

CAP: 1–3″ (2.5–7.5 cm) wide, roughly fan-shaped or somewhat funnel-shaped, minutely hairy (use a hand lens) to smooth, tan to pale brown or grayish brown, sometimes with a whitish bloom; flesh thin, whitish, gelatinous; odor and taste not distinctive.

GILLS: decurrent to base of stalk, crowded, narrow, whitish to grayish, with minutely fringed edges (use hand lens).

STALK: short, stubby, lateral, hairy, whitish.

SPORE PRINT: white.

MICROSCOPIC FEATURES: spores 7–9 × 4–5 μm, elliptic, smooth, hyaline, inamyloid; cystidia very large, thick-walled, with rounded, incrusted ends.

FRUITING: solitary, in groups or clusters on wood or woody debris; June–September; occasional.

EDIBILITY: edible.

COMMENTS: *H. geogenia* (edible) has a pale yellow-brown to hazel-brown cap, a farinaceous odor and taste, and smaller spores that measure 5–6 × 4–5 μm.

Genus *Hygrophoropsis*

Hygrophoropsis aurantiaca (Wulfen : Fries) Maire

COMMON NAME: False Chanterelle.

CAP: ¾–3" (2–7.5 cm) wide, convex, becoming flat to slightly sunken at the center; surface usually orange overall but sometimes yellow or orangish brown, darker at the center than at the margin, dry to slightly moist, typically minutely velvety but sometimes appearing rather smooth; margin incurved, frequently remaining so; flesh orangish; odor and taste not distinctive.

GILLS: close to almost crowded, decurrent, narrow but typically with thickened, blunt edges; repeatedly and regularly forked; pale orangish yellow to deep orange.

STALK: 1–4" (2.5–10 cm) long, ¼–¾" (6–20 mm) thick; typically curved, central to eccentric to almost lateral; dry, orange overall; usually finely hairy; fairly tough.

SPORE PRINT: white.

MICROSCOPIC FEATURES: spores 5.5–8 × 2.5–4.5 μm, elliptic, smooth, hyaline, dextrinoid.

FRUITING: single to scattered or gregarious on wood, needles, or other debris of various conifers, especially spruce; August–November; common.

EDIBILITY: edible, with caution.

COMMENTS: considered a look-alike of *Cantharellus cibarius* because of its blunt, forked gills and similar color. Reported hallucinations resulting from this mushroom suggest confusion with *Gymnopilus* species; caution is nonetheless advised. At any rate, the flavor of *Hygrophoropsis aurantiaca* is mediocre at best.

Genus *Hygrophorus*

The genus *Hygrophorus,* also known as Waxy Caps, is a large genus with more than two hundred known species. They are small to medium-sized mushrooms that often have brightly colored caps and usually grow on the ground. Species of *Hygrophorus* have a central stalk, a white spore print, and thick gills that feel waxy when crushed. A partial veil is absent or evanescent and rarely leaves a ring on the stalk.

Many species are edible, a few are inedible or poisonous, and the edibility the remainder is unknown. None has been reported to cause fatalities. Although many are edible, few are rated as good or choice, and most are considered rather bland. Some authors divide the genus *Hygrophorus* into several genera based primarily on microscopic characters. Genera sometimes recognized include *Hygrocybe, Camarophyllus, Hygrotrama,* and *Neohygrophorus,* among others. This work emphasizes macroscopic features, and therefore only the genus *Hygrophorus* is recognized.

Key to Species of *Hygrophorus*

1. Cap ⅜–1⅛" (1–3 cm) wide; surface slimy to sticky, color variable, green to olive-green at first, becoming amber-orange, orange, pinkish orange or pinkish, translucent-striate initially, becoming opaque in age; gills pale yellow-orange, yellow, greenish or reddish; stalk slimy to sticky—*H. psittacinus* Fries (see p. 143).

1. Cap ¼–1⅜" (5–35 mm) wide; surface slimy to sticky, color variable, shell-pink to peach,

orange-red to orange to olive-orange or shades of violet-gray, yellow, or cream; translucent-striate; gills colored like the cap; stalk slimy to sticky; flesh odor often skunk-like to fishy—*H. laetus* (Fries) Fries (see p. 141).

1.   Cap 1½–4⅜" (4–11 cm) wide; surface smooth, sticky when moist, pinkish buff to pale flesh-colored, pale pinkish salmon or pale tan, often cracked, especially or the disc; margin incurved when young; flesh white or tinged pinkish red; odor not distinctive or fragrant; taste not distinctive or turpentine-like; gills attached to slightly decurrent, subdistant, white to pale salmon, crossveined; stalk ⅜–¾" (1–2 cm) thick white to buff, sometimes tinged pinkish red, with tiny white scales or points near the apex that quickly stain yellow in KOH; on the ground in conifer woods or bogs—*H. pudorinus* (Fries) Fries (see p. 143).

1.   Cap 1¼–4¾" (3–12 cm) wide; surface dry, covered with minute, flattened fibers, appearing streaked to somewhat scaly, dark purplish brown, red-brown or cinnamon on the disc, pale pink, pinkish tan to whitish toward the margin; stalk silky and white at the apex, scurfy and colored like the cap below, tapered at the base, with a white partial veil on young specimens, leaving a faint superior ring that becomes purplish red in age—*H. purpurascens* (Fries) Fries (see p. 143).

1.   Not with the above combinations of characters → 2.

    2.   Cap white, whitish, buff, cream, ivory, or grayish white → 3.

    2.   Cap pale pink to pink, shell-pink, vinaceous pink, coral-pink to coral-red, grayish pink to brownish pink, salmon-pink, or vinaceous → 6.

    2.   Cap pale to bright yellow, lemon-yellow, golden yellow, honey-yellow, orange-yellow to apricot-yellow, or amber-yellow → 9.

    2.   Cap yellow-orange to salmon-orange, amber-orange to orange, reddish orange to orange-red, or red → 12.

    2.   Cap tan, olive, pinkish cinnamon, pinkish brown to reddish brown, orange-brown, yellow-brown to olive-brown, rusty, gray-brown, dark brown, ashy gray to gray, fuscous, or blackish → 20.

3.   Cap margin and stalk apex covered with yellow-orange to golden yellow granules that often form golden droplets in wet weather—*H. chrysodon* (Fries) Fries (see p. 137).

3.   Cap margin and stalk apex not covered with yellow-orange to golden yellow granules; stalk apex usually beaded with pale yellow drops when fresh, stalk viscid or glutinous; cap viscid or glutinous, white bcoming pinkish buff; gills white at first, becoming yellow and finally brownish in age, drying dark reddish brown; spores 7–9 × 3.5–4.5 µm; on the ground under hardwoods, especially beech; August–October; edibility unknown—*H. chrysaspis* Métrod.

3.   Cap margin and stalk apex not covered with yellow-orange to golden yellow granules or drops; cap, gills, and stalk pure white when young and fresh, sometimes becoming sordid in age → 4.

3.   Cap margin and stalk apex not covered with yellow-orange to golden yellow granules or drops; cap, gills, and stalk not pure white when young and fresh, often tinted yellowish, pale tan, grayish, or brownish, especially on the disc → 5.

    4.   Cap 2½–7" (6.5–18 cm) wide, surface sticky to dry; stalk ⅝–1" (1.5–2.5 cm) thick, dry, often tapered downward and abruptly tapered at the base; spores 6–8 × 4–5.5 µm; on the ground in hardwoods; edible—*H. sordidus* Peck.

    4.   Cap ¾–2¾" (2–7 cm) wide; stalk ⅛–⅝" (3–15 mm) thick; cap and stalk surface slimy to sticky; cap convex when young; on the ground under conifers or hardwoods—*H. eburneus* (Fries) Fries (see p. 139).

    4.   Cap ¾–2" (2–5 cm) wide; stalk ¼–¾" (6–20 mm) thick; cap and stalk surface dry

to moist but not slimy or sticky; cap typically with a broad, distinct umbo when young; spores 4.5–6 × 3–4.5 μm; on the ground in mixed woods; edible—*H. angustifolius* (Murrill) Hesler and Smith.

5. Stalk moist or dry but not slimy; cap ⅜–1¾" (1–4.5 cm) wide, moist and lubricous, pale yellow to pale tan over the disc, remainder whitish; margin translucent-striate; gills white, crossveined; stalk ⅛–5⁄16" (3–8 mm) thick, whitish, moist to dry; spores 7–11 × 4.5–7 μm; on the ground in conifer or mixed woods; edible—*H. borealis* Peck (*Hygrophorus niveus* Fries and *H. virgineus* Fries are likely synonyms of *H. borealis* Peck).

5. Stalk slimy; cap ¾–3½" (2–9 cm) wide, slimy, creamy buff to pale yellow on the disc, whitish on the margin; gills and stalk colored like the cap; growing on the ground in conifer woods—*H. gliocyclus* Fries (see p. 140).

5. Stalk slimy; cap ¾–3½" (2–9 cm) wide, slimy, color on the disc variable, brownish, smoky gray, or yellowish, becoming whitish to grayish white on the margin; gills white at first, becoming creamy in age; stalk colored like the cap, apex whitish pruinose; spores 6–8 × 3.5–5 μm; on the ground in hardwoods and mixed woods; edibility unknown—*H. occidentalis* Smith and Hesler.

6. Cap sharply conic when young, becoming broadly convex with a sharply pointed umbo in age—*H. calyptraeformis* Berkeley and Broome (see p. 136).

6. Cap convex to nearly plane, lacking an umbo or with a low, broad umbo → 7.

7. Cap ⅝–1¾" (1.5–4.5 cm) broad, convex to nearly plane; surface dry, scurfy, and streaked with tiny fibrils and scales, especially on the margin; gills distant, decurrent, whitish; stalk ⅛–⅜" (4–10 mm) thick, equal or tapered downward, white when young, with pinkish streaks in age; spores 5.5–8 × 4–5 μm; on the ground in mixed woods; edibility unknown—*H. subrufescens* Peck.

7. Cap greater than 2" (5 cm) wide at maturity → 8.

8. Gills close to subdistant; cap surface slimy to sticky; stalk ⅜–¾" (1–2 cm) thick, tapered at the base, dry, scurfy overall or at least on the upper portion, white and beaded with moisture drops at the apex when fresh, white below with pinkish red fibrils and dots; gills whitish to pinkish buff when young, becoming spotted pinkish red; cap color variable, spotted and streaked vinaceous pink to purplish red over a pale pink to white ground color—*H. erubescens* (Fries) Fries (see p. 139).

8. Gills distant to subdistant; cap surface sticky when fresh; gills whitish to pale pink at first, becoming pinkish red overall and darkening in age; cap 1⅛–2¾" (3–7 cm) wide, pinkish red to vinaceous overall, with darker pinkish brown streaks and spots, somewhat scurfy with tiny fibrils and scales; margin incurved and white when young; stalk ¼–⅝" (6–15 mm) thick, tapered at the base, coated with tiny fibrils, dry, apex whitish, colored like the cap below; spores 6–8 × 4–5.5 μm; on the ground under spruce in woods and in bogs; edibility unknown—*H. capreolarius* (Kalchbrenner) Saccardo.

8. Gills close to crowded; cap surface sticky at first but soon dry; gills white when young, soon developing purplish red spots and flushed purplish red in age, cap 2–4¾" (5–12 cm) wide, with tiny fibrils and scales, color variable, spotted and streaked vinaceous pink to purplish red over a pale pink to white ground color, typically white along the margin and darkest on the disc; margin incurved and white when young; stalk ⅝–1⅜" (1.5–3.5 cm) thick, usually not tapered at the base, dry, pruinose at the apex, smooth below, white at first, soon colored like the cap; spores 6–8 × 3–5 μm; on the ground in hardwoods, especially oak, or in oak-pine forests—*H. russula* (Fries) Quélet (see p. 145).

9. Cap ⅜–1⅛" (1–3 cm) wide; surface sticky, orange-yellow to amber-yellow or pale yellow; margin translucent-striate; gills subdistant to distant, decurrent, whitish; stalk ¹⁄₁₆–⅛" (1.5–3 mm) thick, colored like the cap—*H. parvulus* Peck (see p. 142).

9. Cap ¾–2¾" (2–7 cm) wide; surface sticky, covered with thick gluten, honey-yellow to reddish yellow or pale yellow, at least on the disc, becoming pale yellow to whitish on the margin—*H. flavodiscus* Frost (see p. 140).

9. Not as above; cap becoming depressed at the center; gills decurrent → 10.

9. Not as above; cap hemispheric to convex, conic to bell-shaped, often becoming plane, sometimes with a low broad umbo; gills notched → 11.

10. Cap ⅜–1½" (1–4 cm) wide, smooth, sticky, bright yellow to apricot-yellow, fading to pale yellow or whitish; flesh yellowish to whitish, odor and taste not distinctive; gills subdistant to distant, yellow to pale yellow, often with darker yellow edges, crossveined; stalk 1⅛–3⅛" (3–8 cm) long, ⅛–¼" (2–5 mm) thick, smooth, sticky, colored and fading like the cap, hollow; spores 6.5–9 × 4–6 μm; on humus, in wet soil, and among mosses in conifer or hardwood forests and in bogs; edibility unknown—*H. nitidus* Berkeley and Curtis.

10. Nearly identical to the previous choice but cap and stalk not sticky and not fading to pale yellow or whitish and flesh yellow with a somewhat fragrant odor; edibility unknown—*H. vitellinus* Fries.

11. Stalk moist or dry, not sticky, smooth, dull, 1½–2¾" (4–7 cm) long, ¼–½" (6–12 mm) thick, yellow to orange-yellow, often with a whitish base; cap 1–2½" (2.5–6.5 cm) wide; surface smooth, sticky, yellow to orange-yellow, darkest on the disc; margin often split in age; gills pale yellow with whitish edges; spores 6–8 × 4–6.5 μm; on soil and humus in conifer or hardwoods; edible—*H. flavescens* (Kauffman) Smith and Hesler.

11. Stalk moist and sticky when fresh, cap bright lemon-yellow, otherwise nearly identical to the previous choice; edible—*H. chlorophanus* (Fries) Fries.

11. Stalk moist or dry, not sticky; flesh yellow, odor and taste not distinctive; cap orange-yellow to bright golden yellow; gills orange-yellow; stalk pale yellow to orange-yellow—*H. marginatus* var. *concolor* Peck (see p. 141).

11. Stalk moist or dry, not sticky; flesh yellow fading to white, odor foul and unpleasant, taste slightly disagreeable; cap ¾–2" (2–5 cm) wide, conspicuously conic; dull orange-yellow to yellow-ochre, fading to yellow; gills pale dull yellow, not golden yellow, subdistant to distant; stalk ¼–³⁄₁₆" (5–7 mm) thick at the apex, tapered downward and about ⅛" (2–3 mm) thick at the base, hollow, pale bright yellow overall or white at the base; spores 7–9 × 4–6 μm; on soil or among mosses in swamps and wet areas in conifer or mixed woods; edibility unknown—*H. auratocephalus* (Ellis) Murrill.

12. Cap very small, ¼–¾" (5–20 mm) wide at maturity, convex to plane; surface slimy to sticky, orange-red to reddish orange, fading to orange-yellow; gills orange-buff; stalk orange-yellow to yellow with a white base; spores 7–10 × 4–6 μm; on soil in conifers or hardwoods; edibility unknown—*H. minutulus* Peck.

12. Cap ⅜–2" (1–5 cm) wide, convex to nearly plane; surface orange to yellow-orange; gills brilliant orange; stalk hollow, pale orange-yellow, moist or dry but not sticky to slimy; spores 7–10 × 4–6 μm; on humus and soil in mixed woods; edible—*H. marginatus* var. *marginatus* Peck.

12. Cap ½–2" (1.2–5 cm) wide; gills lavender to purple, fading to orange-lavender or yellowish in age—*H. purpureofolius* Bigelow (see p. 144).

12. Cap ½–1¼" (1.3–3.2 cm) wide; surface and flesh bitter to very bitter—*H. reai* Maire (see p. 144).

12. Not with the above combinations of characters → 13.

13. Cap sharply conic, at least when young; gills not decurrent → 14.
13. Cap not sharply conic; gills decurrent → 15.
13. Cap not sharply conic; gills attached or somewhat decurrent → 18.

    14. Cap color variable (orange-red, orange, ochraceous orange, or yellow to pale yellow, but not brilliant red to scarlet), typically not fading or slowly fading in age, sticky or dry; gills yellow; stalk colored like the cap or paler, sometimes with a whitish base, longitudinally striate or twisted-striate, usually blackening at the base in age, but not blackening when bruised; spores 9–15 × 5–9 μm; on the ground in grassy areas and in woods; edible—*H. acutoconicus* (Clements) Smith.

    14. Cap brilliant red to scarlet overall, fading to orange-red or paler in age, smooth and sticky to slimy when moist; gills orange to yellow; stalk orange-yellow to reddish yellow or yellow with a whitish base, not blackening in age or when bruised—*H. cuspidatus* Peck (see p. 139).

    14. Cap dark orange-red to red, quickly staining black when bruised, typically smooth and dry; stalk staining black when bruised or in age—*H. conicus* (Fries) Fries (see p. 138).

    14. Nearly identical to the previous choice except cap and stalk viscid to slimy, cap translucent-striate when fresh; edibility unknown—*H. singeri* Smith and Hesler.

    14. Cap dark red to violet-brown, becoming black when handled or bruised, silky-fibrillose; gills yellow to grayish orange when young, becoming olive-ochre at maturity, staining black when bruised; stalk reddish orange to yellow with a whitish base, staining black when bruised; spores 8–10 × 5–6.5 μm; on the ground under conifers, especially hemlock; edibility unknown—*H. conicus* var. *atrosanguineus* Grund and Harrison.

15. Cap and stalk dry or moist but not sticky or slimy when fresh → 16.
15. Cap and stalk sticky or slimy when fresh → 17.

    16. Cap ⅜–1⅜" (1–3.5 cm) wide, often deeply depressed in age; surface dry, silky, finely roughened especially on the disc, orange, orange-red, or scarlet, and paler in age; gills subdistant to distant, ivory to creamy yellow or orange-yellow; stalk ⅛–¼" (2–5 mm) thick, colored like the cap or paler; spores 7–13 × 4–7 μm; on rich humus, on decaying wood, or in bogs—*H. cantharellus* (Schweinitz) Fries (see p. 137).

    16. Cap ¾–2¾" (2–7 cm) wide, often depressed to funnel-shaped; surface fibrillose to fibrillose-scaly, dry or moist and slippery but not viscid, bright red, fading to reddish orange to orange-yellow; gills subdistant to distant, whitish at first, becoming pale orange-yellow; stalk ⅛–½" (4–12 mm) thick, red on the upper portion, paler below; spores 11–18 × 7–10 μm; on the ground in mixed woods—*H. appalachianensis* Hesler and Smith (see p. 135).

    16. Cap ¾–2¾" (2–7 cm) wide, obtuse to convex when young, becoming broadly convex, typically with a low broad umbo at maturity, smooth, often cracked on the disc in age, reddish orange to orange when young, fading to pale orange to orange-yellow or tawny in age; gills subdistant to distant, often crossveined, salmon-buff to pale orange; stalk whitish to pale salmon-buff; spores 5–8 × 3–5 μm; in grassy areas or on humus in woods; edible—*H. pratensis* (Fries) Fries.

    16. Cap ⅜–1⅜" (1–3.5 cm) wide, convex, becoming deeply depressed; surface covered with tiny scales with orange-red to orange-yellow tips, orange-red when young, fading to orange and finally orange-yellow in age; gills distant to subdistant, orange to orange-yellow or pale yellow; stalk 1½–4¾" (4–12 cm) long, colored like the cap, pale yellow near the base and whitish where buried in sphagnum

mosses; spores 9–14 × 5–10 μm; cells of the stalk cuticle ellipsoid to cylindrical, less than 120 μm long; among sphagnum mosses in bogs; edibility unknown—*H. turundus* var. *sphagnophilus* (Peck) Hesler and Smith.

16. Nearly identical to the previous choice except that the cells of the stalk cuticle are 120–190 μm long, the spores measure 8–11 × 5.5–7 μm, and the cap scales remain colored like the cap or darken to brown in age; edibility unknown—*H. coccineocrenatus* Orton.

16. Nearly identical to the previous two choices except that the cap scales have smoky gray to dark brown tips and the stalk is shorter, 1⅛–2⅜" (3–6 cm) long; spores 9–14 × 5–8 μm; on moist soil and among sphagnum mosses, typically at high elevations; edibility unknown—*H. turundus* var. *turundus* (Fries) Fries.

17. Growing among sphagnum mosses under larch in bogs and woods; cap and stalk coated with a slimy universal veil; partial veil and ring absent; cap bright orange-red overall when young, becoming orange to orange-yellow, palest toward the margin—*H. speciosus* var. *speciosus* Peck (see p. 145).

17. Growing among sphagnum mosses in bogs or on the ground in conifer woods, usually associated with larch; nearly identical to the previous choice except that below the slimy universal veil, a thin, white partial veil covers the gills of young specimens, leaving a thin, evanescent ring; edible—*H. speciosus* var. *kauffmanii* Hesler and Smith.

17. Growing on the ground in grassy areas, neither in bogs nor associated with larch; cap ¾–2" (2–5 cm) wide, convex to nearly plane or sometimes depressed; surface covered with a thin, slimy coating that soon dries, reddish orange fading to orange and finally pale orange-yellow to dull yellow in age; gills subdistant, whitish at first, becoming pale yellow to pale salmon; stalk covered with a thin, slimy coating when fresh, pale yellow to salmon, whitish near the base; spores 7–10 × 4–5.5 μm; edibility unknown—*H. hypothejus* var. *aureus* (Arrhenia) Imler.

18. Odor (especially several minutes after collecting) sweetish, resembling honey, or unpleasant; cap ⅜–1¾" (1–4.5 cm) wide, hemispheric to convex when young, becoming nearly plane and somewhat depressed on the disc in age, smooth and brilliant scarlet when moist, fading to orange or yellow, becoming dry and fibrillose-scurfy; stalk colored like the cap overall or on the upper two-thirds, becoming yellow toward the base; spores 6–9 × 3.5–5 μm, binucleate; edible—*H. reidii* Kühner.

18. Odor not distinctive; cap ¾–1⅝" (2–4 cm) wide, convex when young; surface smooth and brilliant scarlet when moist, fading to orange or yellow, becoming dry and fibrillose-scurfy; stalk colored like the cap; spores 6–8 × 4–6 μm; uninucleate—*H. miniatus* (Fries) Fries (see p. 141).

18. Odor not distinctive; cap ¾–2¾" (2–7 cm) wide, convex when young, becoming depressed to funnel-shaped; surface fibrillose to fibrillose-scaly, dry or moist and slippery but not viscid, bright red, fading to reddish orange to orange-yellow; stalk red on the upper portion, paler below; spores 11–18 × 7–10 μm—*H. appalachianensis* Hesler and Smith (see p. 135).

18. Odor not distinctive; cap ¾–2⅜" (2–6 cm) wide, hemispheric to convex when young, becoming nearly plane and often somewhat depressed on the disc, or sometimes umbonate; surface fibrillose to fibrillose-scaly, bright scarlet to orange-red with a narrow orange-yellow margin when moist, fading to orange or orange-yellow overall in age; stalk orange-red to orange with a white to pale yellow base; spores 6–8 × 3.5–5 μm, binucleate; among sphagnum mosses in bogs and wet woodlands; edible—*H. squamulosus* Ellis and Everhart.

18. Odor not distinctive; cap ¾–3⅛" (2–8 cm) wide, conic when young; surface smooth and silky when moist or dry, not sticky or slimy, not fibrillose-scurfy, bright red, fading to orange-red in age; spores 7–11 × 4–5 μm—*H. coccineus* (Fries) Fries *sensu* Ricken (see p. 138).
18. Odor not distinctive; cap surface sticky to slimy → 19.
19. Cap sticky; stalk not sticky; cap ¾–3⅛" (2–8 cm) wide; dark red to bright red when young, fading to reddish orange, then orange; stalk ⅜–⅝" (1–1.6 cm) thick, smooth, reddish, soon fading orange to yellow, sometimes with a whitish base; spores 7–12 × 4–6 μm; on the ground in conifer and hardwoods; edibility unknown (listed as edible in many books, but reportedly poisonous to some)—*H. puniceus* (Fries) Fries.
19. Cap and stalk slimy; cap color variable, orange-brown to pinkish brown or brownish orange to reddish orange when young, becoming dull orange to orange-yellow, sometimes with an olive tint; margin pale orange to yellow and translucent-striate—*H. perplexus* Smith and Hesler (see p. 142).
20. Odor of crushed flesh resembling freshly husked green corn; cap ¾–1¾" (1–4.5 cm) wide, sticky and purplish brown to purplish gray when moist, fading to pale gray, sometimes with a tinge of cinnamon-buff on the disc; gills decurrent, close to subdistant, pale purplish drab when young, becoming dingy buff in age; stalk dry, whitish; spores 5–7 × 3.5–5.5 μm; on the ground in mixed woods, especially low, wet areas; edibility unknown—*H. rainierensis* Hesler and Smith. The crushed flesh of *H. fornicatus* Konrad and Maublanc also typically has a faint odor of green corn (see p. 133).
20. Odor of crushed flesh pungent, medicinal, nitrous → 21.
20. Odor of crushed flesh resembling raw potato; taste bitter; cap 1½–4½" (4–11.5 cm) wide, sticky to slimy when fresh, whitish overall in very young specimens, becoming dull yellowish brown over the disc by maturity; gills and stalk whitish; spores 6–9 × 4.5–6 μm; on the ground under conifers; edibility unknown—*H. tennesseensis* Smith and Hesler.
20. Odor of crushed flesh disagreeable and resembling coal tar—*H. camarophyllus* (Fries) Dumée (see p. 137).
20. Odor of crushed flesh resembling almond extract or cherry pits → 22.
20. Not as above; odor fragrant, fruity, or not distinctive → 23.
21. Flesh pale grayish to brownish gray, typically staining pink when cut and rubbed; gills nearly free from the stalk, whitish, staining pink when bruised and sometimes blackening; stalk ⅛–5/16" (3–8 mm) thick, pale brown, staining pinkish brown when bruised and sometimes blackening; cap ¾–2" (2–5 cm) wide, pale yellow-brown, dry; spores 7.5–10 × 4.5–6 μm; on the ground in hardwoods; edibility unknown—*H. nitiosus* Blytt.
21. Nearly identical to the previous choice but lacking pink or black stains on the flesh, gills and stalk—*H. nitratus* Fries (see p. 142).
22. Cap surface slimy to sticky, ashy gray to brownish gray or grayish tan; gills attached to the stalk or slightly decurrent—*H. agathosmus* Fries (see p. 135).
22. Cap surface sticky when young, soon dry, pale yellow-brown to cinnamon-buff, sometimes flushed pale vinaceous red, becoming pale tan in age; gills decurrent, whitish, flushed pale yellow-brown to vinaceous brown in age, often forked, cross-veined; stalk white at first, flushed pale vinaceous brown in age; spores 10–14 × 5.5–7.5 μm; on soil and among mosses in conifer woods; edibility unknown—*H. monticola* Hesler and Smith.
22. Cap surface slimy when fresh, yellow-brown, cinnamon-brown or shades of orange-brown on the disc, becoming whitish to cream toward the margin—*H. bakerensis* Smith and Hesler (see p. 136).

23. Cap sticky to slimy when fresh, often shiny when dry → 24.
23. Cap dry or moist but not sticky or slimy, typically dull when dry → 30.
    24. Cap violet-gray, dark violet to brownish violet; 1–2⅜" (2.5–6 cm) wide; surface moist and sticky when fresh, gills decurrent, whitish at first, soon smoky violaceous; stalk white or tinted pale violet to grayish; flesh pale violet-gray, taste mild at first, sometimes becoming bitter, then acrid; spores 6–8 × 4–6 μm; on the ground and among sphagnum mosses in conifer or hardwoods and swampy areas; edibility unknown—*H. subviolaceous* Peck.
    24. Cap dark brown to black on the disc, becoming paler ashy gray on the margin, conspicuously streaked with gray to black fibrils, slimy when fresh, 1⅛–2½" (3–6.5 cm) wide; gills attached, white to pale grayish; stalk slimy when fresh, sheathed up to a white apex with blackish fibrils beneath the coating of gluten; spores 9–12 × 5–6 μm; on the ground under conifers, especially spruce; edible—*H. olivaceoalbus* (Fries) Fries.
    24. Cap olive-brown, becoming pale olive-brown to dull yellow on the margin and remaining olive-brown on the disc in age, slimy when fresh, ¾–2⅜ (2–6 cm) wide, convex to depressed, sometimes with a small umbo; gills decurrent, white at first, soon pale yellow, often developing orange stains in age; stalk slimy when fresh, whitish when young, becoming yellowish to orangish, with a thin fibrillose band near the apex; spores 7–10 × 4–6 μm; on sandy soil among mosses under mixed pines and oaks—*H. hypothejus* (Fries) Fries (see p. 140).
    24. Cap dark brown to yellow-brown on the disc, pale yellow-brown toward the margin; 1–3" (2.5–8 cm) wide, convex, becoming nearly flat in age with a distinct, broad umbo; margin incurved at first, becoming elevated, arched, and split in age, moist and sticky when fresh, flesh white, odor typically faint of green corn when crushed and rubbed; gills notched or attached, white; stalk white; spores 7–8.5 × 4.5–5.5 μm; on the ground under conifers, usually among mosses; edibility unknown—*H. fornicatus* Konrad and Maublanc.
    24. Not with the above combinations of characters → 25.
25. Cap typically 1⅛–2⅜" (3–6 cm) wide at maturity → 26.
25. Cap typically 2⅜–4¾" (6–12 cm) wide at maturity → 29.
    26. Stalk ⅛–³⁄₁₆" (2–4.5 mm) thick, typically dry, covered with tiny fibrils that are white at first and soon become grayish, at least on the upper portion; cap sticky to slimy when fresh, dark ashy gray to brownish gray on the disc at first, becoming ashy gray to brownish gray overall—*H. tephroleucus* (Fries) Fries (see p. 146).
    26. Stalk ¼–⁵⁄₁₆" (5–8 mm) thick, typically dry, white, covered with tiny dark gray points and scales, at least on the upper portion; cap sticky to slimy when fresh, dark ashy gray to brownish gray on the disc at first, becoming ashy gray to brownish gray overall; spores 7–9 × 4–5 μm; on the ground under conifers, especially fir; edible—*H. pustulatus* (Fries) Fries.
    26. Not with the above combinations of characters → 27.
27. Stalk pale brownish gray to grayish brown, slimy, lacking a superior annular zone of whitish fibrils; cap margin of young specimens lacking cottony fibers; gills white to grayish white, attached or notched, subdistant, crossveined; cap slimy when fresh, shiny when dry, brownish black when young, becoming brown to pale grayish brown in age; margin translucent-striate; stalk ⅛–¼" (2–5 mm) thick; spores 6–10 × 4–6 μm—*H. unguinosus* (Fries) Fries (see p. 146).
27. Stalk white, very slimy up to a superior annular zone of whitish fibrils; cap margin of young specimens with white, cottony fibers; cap surface slimy when fresh, dark purplish brown to fuscous or blackish when young, fading to grayish brown or pale grayish

brown in age; gills attached to decurrent, close, white to whitish, sometimes tinged pinkish buff; spores 8–13 × 5–7 μm; on the ground in conifer and mixed woods, often associated with pine during October and November; edibility unknown—*H. fuscoalbus* (Lasch) Fries.

27. Not with the above combinations of characters; gills yellow to orange-yellow at maturity → 28.

    28. Cap sticky when fresh but soon dry; dark olive-brown at first, developing yellow tones along the margin and finally dull brownish yellow with an olive-brown disc, 1–1½" (2.5–4" cm) wide, conic when young, becoming broadly conic at maturity, flesh pale yellow, odor pungent or not distinctive; gills close, attached, pale yellow to yellow, edges eroded; stalk ¼–⁵⁄₁₆" (5–8 mm) thick, colored like the gills or paler, dry; spores 8–10 × 5–5.5 μm; on the ground under hardwoods or on burned ground associated with blueberry; edibility unknown—*H. spadiceus* (Fries) Fries.

    28. Cap slimy when fresh; color variable, orange-brown to pinkish brown or brownish orange to reddish orange when young, becoming dull orange to orange-yellow, sometimes with an olive tint; margin pale orange to yellow and translucent-striate; stalk slimy—*H. perplexus* Smith and Hesler (see p. 142).

29. Cap slimy when fresh, rusty to cinnamon-brown when young, fading to pale brown, dull orange, pinkish tan, salmon, buff, or yellowish white toward the margin, with the disc remaining reddish brown to shades of orange-brown—*H. subsalmonius* Smith and Hesler (see p. 145).

29. Cap slimy when fresh, blackish to dark brown or olive-brown when young, becoming pale olive-brown to grayish brown or olive-gray and darker on the disc at maturity, convex to nearly plane; margin incurved when young; gills close to subdistant, white to cream; stalk ⅜–1" (1–2.5 cm) thick, silky to scurfy, white, coated with gluten; spores 7–9 × 4–5.5 μm; on the ground in conifer and mixed woods; edibility unknown—*H. fuligineus* Frost.

29. Cap smooth, sticky when wet, soon dry, dark pinkish brown to pinkish cinnamon, paler and often with rosy pink tints toward the margin, paler overall in age; gills attached or slightly decurrent, close, white; stalk ¼–¾" (6–20 mm) thick, densely pruinose to scurfy, at least on the upper half, white; spores 6–9 × 3.5–5 μm; on the ground under conifers and hardwoods, especially oak; edible—*H. roseibrunneus* Murrill.

    30. Cap covered with small brownish olive scales with darker brown to blackish tips over a whitish to honey-yellow ground color; stalk pale yellow; clustered or in groups on the ground under conifers and hardwoods—*H. caespitosus* (Murrill) Murrill (see p. 136).

    30. Cap smooth, not covered with scales; gills bright to dull orange, close to subdistant, sinuate; cap ¾–2" (2–5 cm) wide, bell-shaped to broadly convex with a distinct umbo, smooth, moist, olive-brown on the disc, developing dull orange tones overall at maturity; stalk pale olive-yellow to yellow, often whitish toward the base; spores 6.5–8 × 4–5 μm; on the ground or on very decayed conifer wood; edibility unknown—*H. marginatus* var. *olivaceus* Smith and Hesler.

    30. Cap not covered with scales; surface covered with a thin coating of tiny, flattened, grayish fibrils when young, becoming pale brownish gray in age, often spotted or streaked brown or darker gray, ¾–2" (2–5 cm) wide; gills subdistant to distant, often crossveined, pale gray to pale yellowish gray; stalk pale gray to pale yellowish gray, sometimes whitish at the base; spores 4.5–6 × 4–4.5 μm; on the ground in conifer and mixed woods, usually near hemlock; edibility unknown—*H. canescens* Smith and Hesler.

    30. Cap smooth, not covered with scales; gills pale yellow to yellow, edges eroded;

cap 1–1½" (2.5–4 cm) wide, conic when young, becoming broadly conic at maturity, dark olive-brown at first, developing yellow tones along the margin, and finally dull brownish yellow with an olive-brown disc; stalk colored like the gills or paler; spores 8–10 × 5–5.5 μm; on the ground under hardwoods or on burned ground associated with blueberry; edibility unknown—*H. spadiceus* Fries) Fries.

30. Not with the above combinations of characters → 31.

31. Flesh white, not staining red when cut and rubbed, odor not distinctive; cap ¾–3½" (2–9 cm) wide, smooth, dark pinkish brown to pinkish cinnamon, paler and often with rosy pink tints toward the margin, paler overall in age; gills attached or slightly decurrent, close, white; stalk ¼–¾" (6–20 mm) thick, densely pruinose to scurfy at least on the upper half, white; spores 6–9 × 3.5–5 μm; on the ground under conifers and hardwoods, especially oak; edible—*H. roseibrunneus* Murrill.

31. Flesh pale grayish brown, staining reddish, then black when cut and rubbed, odor of ammonia or not distinctive; cap ¾–2⅜" (2–6 cm) wide, olive-brown to dark brown, often cracked and scurfy on the disc in age; gills notched, subdistant, grayish at first, becoming grayish brown at maturity, staining reddish when bruised; stalk ¼–⅜" (6–10 mm) thick, hollow, often curved, grayish brown to grayish olive, dark gray or blackish; spores 6–10 × 4.5–7 μm, elliptic; on the ground under conifers or hardwoods; edibility unknown—*H. ovinus* (Fries) Fries.

31. Flesh whitish to brownish, not staining reddish when cut and rubbed, odor fragrant; cap ⅝–1⅛" (1.5–3 cm) wide, gray-brown to blackish brown, often cracked and scurfy on the disc in age; gills whitish to grayish, staining reddish when bruised; stalk hollow, grayish to blackish gray; spores 5–7 × 5–6 μm, globose; on the ground in hardwoods; edibility unknown—*H. subovinus* Hesler and Smith.

Hygrophorus agathosmus Fries

COMMON NAME: Gray Almond Waxy Cap.

CAP: ¾–3¼" (2–8 cm) wide, convex, becoming broadly convex to nearly flat; surface smooth, slimy to sticky, ashy gray to brownish gray or grayish tan, darker on the disc and fading toward the margin; margin inrolled and slightly velvety when young; flesh whitish gray to white, soft; odor of almond extract or cherry pits; taste not distinctive.

GILLS: attached to slightly decurrent, subdistant, narrow, sometimes forked, waxy, creamy white.

STALK: 2–3¼" (5–8 cm) long, ¼–¾" (5–20 mm) thick, nearly equal, solid, dry, fibrillose to pruinose when young, smooth in age, whitish to ashy gray; partial veil and ring absent.

SPORE PRINT: white.

MICROSCOPIC FEATURES: spores 7–10 × 4.5–5.5 μm, elliptic, smooth, hyaline, inamyloid.

FRUITING: solitary, scattered, or in groups on the ground under conifers, especially spruce and pine; July–October; occasional.

EDIBILITY: edible.

Hygrophorus appalachianensis Hesler and Smith

CAP: ¾–2¾" (2–7 cm) wide, convex, becoming depressed to funnel-shaped; surface fibrillose to fibrillose-scaly, dry or moist and slippery but not viscid, bright red, fading to reddish orange to orange-yellow; margin incurved when young, becoming uplifted and wavy in age; flesh yellowish, tinged orange; odor and taste not distinctive.

GILLS: attached at first, becoming decurrent, subdistant to distant, whitish at first, becoming pale orange-yellow.

STALK: 1⅛–2¾" (3–7 cm) long, ⅛–½" (4–12 mm) thick, nearly equal or tapered at the base, smooth or slightly scaly, moist or dry, not viscid, hollow, red on the upper portion, paler below.

SPORE PRINT: white.

MICROSCOPIC FEATURES: spores 11–18 × 7–10 μm, elliptic, smooth, hyaline; cheilocystidia and pleurocystidia 32–58 × 7–14 μm, clavate to spathulate, buried.

FRUITING: solitary, scattered, or in groups on the ground in mixed woods; June–September; occasional.

EDIBILITY: edible.

COMMENTS: its size, very large spores, and presence of cystidia are distinctive features that differentiate *H. appalachianensis* from *H. miniatus* (edible), *H. coccineus* (edible), *H. cantharellus* (edible), and *H. turundus* (edibility unknown).

Hygrophorus bakerensis Smith and Hesler

COMMON NAME: Tawny Almond Waxy Cap.

CAP: 1⅝–5⅞" (4–15 cm) wide; rounded when young, becoming nearly plane in age; surface slimy when moist, covered with small, flattened, brownish fibers beneath the gluten, yellow-brown, cinnamon-brown, or shades of orange-brown on the disc, becoming whitish to cream toward the margin; margin incurved and cottony when young, becoming smooth and flat in age; flesh thick, firm, white; odor strongly fragrant, resembling almond extract or cherry pits; taste not distinctive.

GILLS: decurrent, close to subdistant, waxy, whitish to cream.

STALK: 1⅝–5½" (4–14 cm) long, ⅜–1" (8–25 mm) thick, equal or tapering downward, solid, dry, cottony at the apex when young, smooth overall in age, often with clear drops of liquid when moist, white to cream overall; partial veil and ring absent.

SPORE PRINT: white.

MICROSCOPIC FEATURES: spores 7–10 × 4.5–6 μm, elliptic, smooth, hyaline, inamyloid.

FRUITING: scattered or in groups on the ground under conifers; August–November; occasional.

EDIBILITY: edible.

Hygrophorus caespitosus (Murrill) Murrill

CAP: ⅜–2⅜" (1–6 cm) wide, broadly convex to nearly flat, sometimes depressed on the disc; surface moist at first, soon dry, not sticky, covered with small, brownish olive scales with darker brown to blackish tips over a whitish to honey-yellow ground color; margin often uplifted and torn in age; flesh thick, yellowish; odor not distinctive; taste not distinctive or resembling raw potato.

GILLS: attached to slightly decurrent, subdistant to distant, broad, sometimes forked, waxy, whitish to pale yellow.

STALK: ¾–2" (2–5 cm) long, ⅛–¼" (3–7 mm) thick, nearly equal, tapering downward slightly, becoming hollow, pale yellow with olive-yellow to yellow tones near the base, smooth, dry; partial veil and ring absent.

SPORE PRINT: white.

MICROSCOPIC FEATURES: spores 6.5–10 × 4–7 μm, elliptic, smooth, hyaline, inamyloid.

FRUITING: clustered or in groups on the ground under conifers and hardwoods; June–August; occasional.

EDIBILITY: unknown.

Hygrophorus calyptraeformis Berkeley and Broome

CAP: 1–2⅜" (2.5–6 cm) wide, sharply conic when young, becoming broadly convex with a sharply pointed umbo in age; surface smooth, slightly sticky when wet, coral-red to brownish pink or vinaceous pink when young, fading to coral-pink to shell-pink or salmon-pink to pinkish buff in age; margin often split and lobed in age; flesh thin, whitish pink; odor and taste not distinctive.

GILLS: notched, close to subdistant, narrow, waxy, pinkish buff to cream with lavender tints.

STALK: 2¼–5½" (5.5–14 cm) long, ⅛–¼" (3–7 mm) thick, equal, hollow, smooth, moist or dry but not sticky, fragile, whitish overall when young, becoming lavender to pinkish on the upper two-thirds in age; partial veil and ring absent.

SPORE PRINT: white.

MICROSCOPIC FEATURES: spores 6–9 × 4–6 μm, elliptic, smooth, hyaline, inamyloid.

FRUITING: scattered or in groups on the ground under conifers and hardwoods; July–October; occasional.

EDIBILITY: edible.

Hygrophorus camarophyllus (Fries) Dumée

COMMON NAME: Dusky Waxy Cap.

CAP: 1⅝–5⅛" (4–13 cm) wide; convex to broadly convex, occasionally with an umbo; surface dry, sticky when moist, smooth but appearing streaked and minutely downy, tan to brownish gray to coffee-brown; margin downy at first; flesh thick, white; odor of coal tar; taste not distinctive.

GILLS: attached to slightly decurrent, close to subdistant, broad, waxy, white to cream or grayish.

STALK: 1¼–5⅛" (3–13 cm) long, ⅜–¾" (1–2 cm) thick, equal or tapered downward, solid, dry, colored like the cap or paler; partial veil and ring absent.

SPORE PRINT: white.

MICROSCOPIC FEATURES: spores 7–9 × 4–5 μm, elliptic, smooth, hyaline, inamyloid.

FRUITING: scattered or in groups on the ground under conifers; July–November; occasional.

EDIBILITY: edible.

Hygrophorus cantharellus (Schweinitz) Fries

COMMON NAME: Chanterelle Waxy Cap.

CAP: ⅜–1⅜" (1–3.5 cm) wide; convex, becoming flat and finally depressed; surface dry, silky, finely roughened, escpecially on the disc, orange, orange-red or scarlet, becoming paler in age; margin often recurved and wavy or scalloped in age; flesh thin, colored like the cap or yellowish; odor and taste not distinctive.

GILLS: strongly decurrent, subdistant to distant, broad, waxy, ivory to creamy yellow or orange-yellow.

STALK: 1–4¾" (2.5–12 cm) long, ⅛–¼" (2–5 mm) thick, equal or enlarged slightly upward, stuffed or hollow, smooth, dry, fragile, colored like the cap or paler, whitish or yellowish white at the base; partial veil and ring absent.

SPORE PRINT: white.

MICROSCOPIC FEATURES: spores 7–13 × 4–7 μm, elliptic to nearly oval, smooth, hyaline, inamyloid.

FRUITING: in groups or clustered on rich humus, on decaying wood or in bogs; July–October; fairly common.

EDIBILITY: edible.

Hygrophorus chrysodon (Fries) Fries

COMMON NAME: Golden-spotted Waxy Cap.

CAP: 1¼–3⅛" (3–8 cm) wide; convex, becoming broadly convex to nearly flat; surface sticky when moist, shiny when dry, smooth, white, covered with yellow-orange to golden yellow granules over the entire surface when young, often remaining on the margin and on the disc in age; margin shaggy and inrolled initially, expanding in age; flesh white, thick; odor and taste not distinctive.

GILLS: decurrent, distant, moderately broad, white to cream, edges sometimes with yellow powder.

STALK: 1¼–3⅛" (3–8 cm) long, ¼–⅝" (5–15 mm) thick, equal, stuffed, white with golden yellow granules at the apex, often dark yellow at the base; partial veil and ring absent.

SPORE PRINT: white.

MICROSCOPIC FEATURES: spores 7–11 × 3.5–6 μm, elliptic, smooth, hyaline, inamyloid.

FRUITING: solitary, scattered, or in groups on the ground under conifers; July–September; fairly common.

EDIBILITY: edible.

Hygrophorus coccineus (Fries) Fries sensu Ricken

COMMON NAME: Scarlet Waxy Cap.

CAP: ¾–3⅛" (2–8 cm) wide, conic when young, becoming convex to flat with an umbo in age; surface smooth, not sticky or slimy, bright red when young, fading to orange-red in age, not fibrillose-scurfy when dry; flesh thin, fragile, yellow-orange, slowly staining gray in $FeSO_4$; odor and taste not distinctive.

GILLS: attached, close to subdistant, broad, waxy, yellowish, red-orange to creamy peach.

STALK: ¾–3⅛" (2–8 cm) long, ⅛–⅜" (2–10 mm) thick, equal, hollow, sometimes compressed, dry, smooth, yellow-orange to orange-red, paler yellow downward and often coated with white mycelium on the base; partial veil and ring absent.

SPORE PRINT: white.

MICROSCOPIC FEATURES: spores 7–11 × 4–5 μm, elliptic, smooth, hyaline, inamyloid.

FRUITING: solitary, scattered, or in groups on the ground in conifer and hardwood forests; July–September; fairly common.

EDIBILITY: edible.

Hygrophorus conicus (Fries) Fries

COMMON NAME: Witch's Hat.

CAP: ¾–3½" (2–9 cm) wide, sharply conic to bell-shaped, usually with an umbo; surface smooth, slightly sticky when moist, otherwise dry, dark orange-red to red, orange, lighter orange near the margin or sometimes yellow overall, often with olive-green tints, quickly staining black when bruised or in age; flesh thin, fragile, colored like the cap, bruising black; odor and taste not distinctive.

GILLS: free from the stalk, close, broad, waxy, light yellow to greenish orange, bruising black.

STALK: ¾–4" (2–10 cm) long, ⅛–⅜" (3–10 mm) thick, equal, hollow, fragile, smooth, not sticky, often longitudinally striate or twisted-striate, yellow to yellow-orange, pale yellow near the base, staining black when bruised or in age; partial veil and ring absent.

SPORE PRINT: white.

MICROSCOPIC FEATURES: spores 8–14 × 5–7 μm, elliptic, smooth, hyaline, inamyloid.

FRUITING: solitary to scattered on the ground under conifers; July–September; occasional to fairly common.

EDIBILITY: poisonous.

Hygrophorus cuspidatus Peck

CAP: ¾–2¾" (2–7 cm) wide; sharply conic when young, becoming broadly conic to
nearly flat with an umbo in age; surface smooth, sticky to slimy when moist, shiny
when dry, brilliant red to scarlet overall, fading to orange-red or paler in age; margin
often uplifted and split in age; flesh thin, fragile, pale yellow to nearly white; odor and
taste not distinctive.

GILLS: free from the stalk, close to crowded, waxy, orange to yellow-orange to yellow.

STALK: 2–3½" (5–9 cm) long, ¼–⅜" (5–10 mm) thick, equal or enlarged slightly downward, stuffed to hollow, smooth, not sticky, often longitudinally striate or twisted-
striate, orange-yellow, reddish yellow, or yellow, whitish at the base, not blackening
in age or when bruised; partial veil and ring absent.

SPORE PRINT: white.

MICROSCOPIC FEATURES: spores 8–12 × 4–6.5 μm, elliptic, smooth, hyaline, inamyloid.

FRUITING: in groups on the ground in mixed woods; June–August; occasional.

EDIBILITY: unknown.

Hygrophorus eburneus (Fries) Fries

COMMON NAME: White Waxy Cap.

CAP: ¾–2¾" (2–7 cm) wide, obtuse to convex, becoming broadly convex to nearly plane,
sometimes with a low umbo, margin incurved when young; surface glutinous to viscid,
smooth, bright white to whitish, typically duller over the disc; flesh white, unchanging
when exposed; odor and taste not distinctive.

GILLS: subdistant to distant, decurrent, white when young, buff to yellowish in age, with
2 or 3 tiers of lamellulae.

STALK: 1¾–4½" (4.5–11.5 cm) long, ⅛–⅝" (4–15 mm) thick, nearly equal or tapered
downward, sometimes abruptly narrowed at the base, glutinous, hollow in age, white
to whitish.

SPORE PRINT: white.

MICROSCOPIC FEATURES: spores 6–9 × 3.5–5 μm, ellipsoid, smooth, hyaline.

FRUITING: scattered or in groups on the ground in conifer and mixed woodlands;
August–November; occasional.

EDIBILITY: edible.

COMMENTS: *Hygrophorus borealis* (edible) has a moist to dry, but not slimy, whitish cap
with yellow to pale tan on the disc and a translucent-striate margin.

Hygrophorus erubescens (Fries) Fries

CAP: ¾–3½" (2–9 cm) wide, convex, becoming broadly convex with a low, broad umbo;
surface slimy to sticky, color variable, spotted and streaked vinaceous pink to purplish
red over a pale pink to white ground color, typically white along the margin and
darkest on the disc; margin incurved and white when young; flesh white, sometimes
staining yellowish when cut and rubbed; odor and taste not distinctive.

GILLS: attached at first, becoming decurrent, close to subdistant, medium broad, whitish
to pinkish buff when young, becoming spotted pinkish red in age.

STALK: 1½–3½" (4–9 cm) long, ⅜–¾" (1–2 cm) thick, nearly equal down to a tapered
base, dry, scurfy overall or at least on the upper portion, white and beaded with
moisture drops at the apex when fresh, white below with pinkish red fibrils and dots,
sometimes yellow, especially at the base, in age or when bruised.

SPORE PRINT: white.

MICROSCOPIC FEATURES: spores 7–11 × 5–6 μm, elliptic, smooth, hyaline, inamyloid.

FRUITING: scattered or in groups on the ground under conifers, especially pine; July–October; occasional.

EDIBILITY: unknown.

Hygrophorus flavodiscus Frost

COMMON NAME: Yellow-centered Waxy Cap.

CAP: ¾–2¾" (2–7 cm) wide, hemispheric when young, becoming convex to flat, sometimes depressed in age; surface smooth, slimy and sticky, becoming streaked when dry, pale brownish yellow to honey-yellow or reddish yellow, orange-tan or pale yellow, at least on the disc, becoming pale yellow to creamy yellow or whitish at the margin; flesh white, thick; odor and taste not distinctive.

GILLS: decurrent, close to subdistant, moderately broad, thin-edged, waxy, pinkish when young, becoming whitish in age.

STALK: 1–3½" (2.5–9 cm) long, ⅛–⅝" (5–15 mm) thick, nearly equal, solid, covered with a slimy, thick glutinous layer, with tiny scales at the apex, whitish to yellowish, sometimes streaked or stained with yellow in age; flesh pale green in KOH.

SPORE PRINT: white.

MICROSCOPIC FEATURES: spores 6–8.5 × 3.5–5 μm, elliptic, smooth, hyaline, inamyloid.

FRUITING: scattered, in groups, or in clusters on the ground in pine woods; September–November; occasional to fairly common.

EDIBILITY: edible.

Hygrophorus gliocyclus Fries

CAP: 1¾–3½" (2–9 cm) wide, convex, becoming nearly flat, sometimes with a very broad umbo; surface smooth, sticky, slimy, creamy buff to pale yellow on the disc, often paler on the margin; flesh white; odor and taste not distinctive.

GILLS: decurrent, subdistant, broad, waxy, colored like the cap, becoming yellowish in age.

STALK: 1¾–2⅜" (2–6 cm) long, ¼–½" (6–12 mm) thick, equal, tapering sharply at the base, solid, colored like the cap, covered with a hyaline glutinous layer; partial veil and ring absent.

SPORE PRINT: white.

MICROSCOPIC FEATURES: spores 8–11 × 4.5–6 μm, elliptic, smooth, hyaline, inamyloid.

FRUITING: in groups or clusters on the ground under spruce and pine; August–November; occasional.

EDIBILITY: edible.

Hygrophorus hypothejus (Fries) Fries

CAP: ¾–2⅜" (2–6 cm) wide, convex, then depressed at maturity, sometimes with a small umbo; surface glutinous when young and fresh, becoming viscid and finally dry in age, olive-brown on young specimens, fading to pale olive-brown to yellow-brown or dull yellow on the margin on mature specimens and remaining olive-brown on the disc; margin sterile and incurved at first, becoming decurved and remaining so well into maturity, often wavy in age; flesh white, unchanging; odor and taste not distinctive.

GILLS: subdistant, attached to decurrent, with three to four tiers of attenuate lamellulae, white at first, soon pale yellow, often developing orange stains in age, typically veined on the faces and sometimes crossveined.

STALK: 1⅛–2⅛" (3–5.5 cm) long, ½–⅜" (7–11 mm) thick, tapered slightly downward, weakly fibrillose, glutinous when fresh, with a thin fibrillose band near the apex, whitish at first, soon yellowish to pale ochraceous orange on the upper portion and remaining white toward the base.

SPORE PRINT: white.

MICROSCOPIC FEATURES: spores 7–10 × 4–6 μm, ellipsoid, smooth hyaline, inamyloid.

FRUITING: scattered or in groups on sandy soil among mosses under mixed pines and oaks; September–November; occasional.

EDIBILITY: edible.

Hygrophorus laetus (Fries) Fries

CAP: ¼–1⅜" (5–35 mm) wide, hemispheric when young, becoming convex to nearly flat in age, often depressed at the center; surface slimy to sticky when fresh, shiny when dry, smooth, color variable, shell-pink to peach, orange-red to orange to olive-orange, or shades of violet-gray, yellow, or cream; margin translucent-striate; flesh thin, pinkish to buff; odor skunk-like to fishy, not distinctive; taste not distinctive.

GILLS: attached to decurrent, distant to subdistant, narrow, color variable like the cap.

STALK: ⅝–4¾" (1.5–12 cm) long, 1/16–¼" (1.5–6 mm) thick, equal or slightly enlarged at the apex, hollow, colored like the cap, slimy to sticky when fresh; partial veil and ring absent.

SPORE PRINT: white.

MICROSCOPIC FEATURES: spores 5–10 × 3–6 μm, elliptic, smooth, hyaline, inamyloid.

FRUITING: scattered or in groups in sphagnum bogs or on the ground in mixed woods, often among mosses; August–October; fairly common.

EDIBILITY: edible.

Hygrophorus marginatus var. concolor Peck

COMMON NAME: Orange-gilled Waxy Cap.

CAP: ⅝–1½" (1.5–4 cm) wide, conic, becoming bell-shaped to nearly flat; surface smooth, moist, slightly sticky when wet, orange-yellow when young, becoming bright orange-yellow to bright golden yellow to yellow at maturity; flesh thin, yellow; odor and taste not distinctive.

GILLS: attached, distant, broad, waxy, orange-yellow to yellow.

STALK: 1–4" (2.5–10 cm) long, ⅛–⅜" (3–9 mm) thick, nearly equal, becoming hollow in age, smooth, moist, pale yellow to orange-yellow, sometimes white at the base.

SPORE PRINT: white.

MICROSCOPIC FEATURES: spores 7–10 × 4–7 μm, elliptic, smooth, hyaline, inamyloid.

FRUITING: single to scattered on the ground or among sphagnum mosses at the edge of bogs and in mixed woods; July–October; fairly common.

EDIBILITY: edible.

Hygrophorus miniatus (Fries) Fries

CAP: ¾–1⅝" (2–4 cm) wide, convex, becoming broadly convex to nearly flat with a depressed disc in age; surface smooth and brilliant scarlet when moist, fading to orange or yellow, becoming dry and fibrillose-scurfy; flesh thin, colored like the cap or paler; odor and taste not distinctive.

GILLS: attached, sometimes slightly decurrent, close to subdistant, broad, colored like cap when young, becoming paler orange, yellow or orange-pink in age.

STALK: 1–3" (2.5–7.5 cm) long, ⅛–¼" (3–6 mm) thick, equal, stuffed, smooth, colored like the cap when young, fading to orange-yellow or yellow in age; partial veil and ring absent.

SPORE PRINT: white.

MICROSCOPIC FEATURES: spores 6–8 × 4–6 μm, elliptic, smooth, uninucleate, hyaline, inamyloid.

FRUITING: scattered or in groups on the ground, among mosses, or on decaying wood in deciduous and mixed woods; July–October; fairly common.

EDIBILITY: edible.

COMMENTS: *Hygrophorus miniatus* var. *mollis* (edible) has an orange-yellow to yellow cap even when young, and pale yellow to vividly yellow gills. *Hygrophorus calciphilus* (edible) is very similar but has binucleate spores.

Hygrophorus nitratus Fries

CAP: ¾–2¾" (2–7 cm) wide, bell-shaped when young, becoming convex in age; surface smooth, dry, not sticky or slimy, gray-brown, often with ivory or buff streaks; flesh thin, pale to dark brownish gray, not staining when cut and rubbed; odor pungent, medicinal, nitrous; taste acidic.

GILLS: sinuate, subdistant, broad, waxy, ivory with grayish tints, darkening in age.

STALK: 1⅝–4" (4–10 cm) long, ⅛–⅜" (2–10 mm) thick, nearly equal or enlarged at the base, smooth, hollow, pale brown to ivory-brown when young, becoming grayish brown in age, whitish at the base; partial veil and ring absent.

SPORE PRINT: white.

MICROSCOPIC FEATURES: spores 7.5–10 × 4.5–6 μm, oval to elliptic, smooth, hyaline, inamyloid.

FRUITING: scattered or in groups on the ground under conifers and hardwoods; July–September; occasional.

EDIBILITY: unknown.

Hygrophorus parvulus Peck

CAP: ⅜–1⅛" (1–3 cm) wide, convex to broadly convex or nearly flat, sometimes depressed on the disc; surface smooth, sticky to slimy, translucent-striate, orange-yellow or amber-yellow, becoming paler yellow in age; flesh waxy, colored like the cap; odor and taste not distinctive.

GILLS: decurrent, subdistant to distant, crossveined, whitish to cream or pale yellow.

STALK: 1¼–2⅜" (3–6 cm) long, 1/16–⅛" (1.5–3 mm) thick, equal or tapered at the base, hollow, fragile, smooth, not sticky, colored like the cap, lower portion and base sometimes darker yellow-brown to reddish brown; partial veil and ring absent.

SPORE PRINT: white.

MICROSCOPIC FEATURES: spores 5–8.5 × 3.5–5 μm, elliptic, smooth, hyaline, inamyloid.

FRUITING: scattered on the ground under hardwoods, in mixed woods and under rhododendron; July–September; occasional.

EDIBILITY: unknown.

Hygrophorus perplexus Smith and Hesler

CAP: ⅜–1⅛" (1–3 cm) wide; conic, becoming nearly flat, with a broad umbo; surface smooth, slimy, color variable, orange-brown to pinkish brown or brownish orange to reddish orange when young, becoming dull orange to orange-yellow, sometimes with an olive tint; margin incurved when young, translucent-striate when fresh; flesh colored like the cap; odor and taste not distinctive.

GILLS: attached, subdistant, ivory-yellow with faint pink tones when young, becoming yellow to orange-yellow in age.

STALK: 1⅛–2" (3–5 cm) long, ⅛–¼" (2–5 mm) thick, nearly equal or enlarged slightly downward, hollow, smooth, slimy, amber-gold to pale yellow-tan, sometimes with faint pinkish-gray tints toward the apex; partial veil and ring absent.

SPORE PRINT: white.

MICROSCOPIC FEATURES: spores 6–8 × 4–5 μm, oval to elliptic, smooth, hyaline, inamyloid.

FRUITING: in groups or in clusters on the ground under hardwoods, especially beech; July–September; occasional.

EDIBILITY: unknown.

Hygrophorus psittacinus (Fries) Fries

COMMON NAME: Parrot Mushroom.

CAP: ⅜–1¼" (1–3 cm) wide; broadly conic, becoming convex to nearly flat, sometimes with a broad umbo; surface sticky to slimy when fresh, smooth and shiny when dry, green to olive-green initially, becoming amber-orange, orange, pinkish orange or pinkish; margin translucent-striate at first, becoming opaque in age; flesh thin, colored like the cap; odor and taste not distinctive.

GILLS: attached, subdistant, waxy, pale yellow-orange, yellow, greenish or reddish.

STALK: 1¼–2¾" (3–7 cm) long, ⅛–¼" (2–5 mm) thick, smooth, sticky to slimy overall, nearly equal or tapered slightly upward, hollow, green when young, soon becoming amber-orange, yellow, pinkish, or yellowish white; partial veil and ring absent.

SPORE PRINT: white.

MICROSCOPIC FEATURES: spores 6.5–10 × 4–6 μm, elliptic, smooth, hyaline, inamyloid.

FRUITING: scattered or in groups on the ground under conifers and hardwoods and in grassy areas; June–September; occasional.

EDIBILITY: edible.

Hygrophorus pudorinus (Fries) Fries

COMMON NAME: Turpentine Waxy Cap.

CAP: 1½–4⅜" (4–11 cm) wide, obtuse to convex, becoming broadly convex in age, margin incurved when young; surface smooth, sticky when moist, pinkish buff to pale flesh-colored, pale pinkish salmon or pale tan, often cracked, especially on the disc; flesh white or tinged pinkish red; odor not distinctive or fragrant; taste not distinctive or turpentine-like.

GILLS: subdistant, subdecurrent, white to whitish or sometimes with reddish or salmon tinges, occasionally forked and often crossveined.

STALK: 1½–3½" (4–9 cm) long, ⅜–¾" (1–2 cm) thick, nearly equal or tapered in either direction, dry or slightly viscid, solid, white to buff, sometimes tinged pinkish red, with tiny white scales or points near the apex which quickly stain yellow in KOH.

SPORE PRINT: white.

MICROSCOPIC FEATURES: spores 6.5–10 × 4–5.5 μm, ellipsoid, smooth, hyaline.

FRUITING: scattered or in groups on the ground in conifer woods or among sphagnum mosses in bogs; August–October; occasional.

EDIBILITY: edible.

COMMENTS: at least five varieties of this mushroom are recognized and are distinguished by color and odor.

Hygrophorus purpurascens (Fries) Fries

CAP: 1⅛–4¾" (3–12 cm) wide, convex, becoming broadly convex to nearly flat, dry, covered with minute flattened fibers giving a streaked to somewhat scaly appearance, dark purplish brown, red-brown or cinnamon on the disc, pale pink, pinkish tan to whitish toward the margin; margin inrolled when young, incurved to expanded in age, fibrous-cottony; flesh white, thick; odor and taste not distinctive.

GILLS: attached and often decurrent, subdistant, narrow, white when young, becoming pinkish buff, often spotted or streaked purplish to red-brown.

STALK: 1⅛–4" (3–10 cm) long, ⅜–1" (1–2.5 cm) thick, nearly equal, tapered at the base, solid, dry, silky and white at the apex, scurfy and colored like the cap below, with vinaceous brown spots and streaks, staining yellow to yellow-brown in KOH; partial veil white, cottony to fibrillose, leaving a sparse, fibrillose superior ring that becomes purplish red at maturity.

SPORE PRINT: white.

MICROSCOPIC FEATURES: spores 5.5–8 × 3–4.5 μm, elliptic, smooth, hyaline, inamyloid.

FRUITING: in groups or clusters on the ground under conifers, especially spruce and pine; July–October; infrequent.

EDIBILITY: edible.

Hygrophorus purpureofolius Bigelow

COMMON NAME: Lavender-gilled Waxy Cap.

CAP: ½–2" (1.2–5 cm) wide, conic, becoming broadly convex to nearly plane with a broad umbo; surface smooth, moist, not sticky or slimy, dark orange-red overall, fading in age, palest along the margin; flesh thin, white; odor and taste not distinctive.

GILLS: attached to slightly decurrent, close to subdistant, waxy, lavender to purple, fading to orange-lavender or yellowish in age.

STALK: 1–3" (2.5–7.5 cm) long, ⅛–½" (5–12 mm) thick, nearly equal or enlarged downward, hollow, smooth, often compressed with a vertical groove, orange-red to orange-yellow; partial veil and ring absent.

SPORE PRINT: white.

MICROSCOPIC FEATURES: spores 7–11 × 4–5.5 μm, elliptic, smooth, hyaline, inamyloid.

FRUITING: scattered or in groups on the ground in deciduous and mixed woods; July–September; occasional.

EDIBILITY: unknown.

Hygrophorus reai Maire

COMMON NAME: Bitter Waxy Cap.

CAP: ½–1¼" (1.3–3.2 cm) wide, convex, becoming broadly convex; surface smooth, sticky, brilliant orange-red on the disc, becoming reddish orange to bright orange toward the margin; margin slightly translucent-striate; flesh thin, reddish orange; odor not distinctive; taste very bitter.

GILLS: attached, subdistant, broad, waxy, whitish when young, becoming yellow to yellowish orange in age.

STALK: 1–2½" (2.5–6.3 cm) long, ¹⁄₁₆–¼" (2–7 mm) thick, nearly equal, colored like the cap or paler, smooth, sticky, often translucent; partial veil and ring absent.

SPORE PRINT: white.

MICROSCOPIC FEATURES: spores 6.5–9 × 4–5.5 μm, elliptic, smooth, hyaline, inamyloid.

FRUITING: scattered or in groups under conifers, usually among mosses; July–October; occasional.

EDIBILITY: inedible.

COMMENTS: the cap surface also tastes bitter.

Hygrophorus russula (Fries) Quélet
COMMON NAME: Russula-like Waxy Cap.
CAP: 2–4¾" (5–12 cm) wide, hemispheric to convex, becoming broadly convex to nearly
plane in age; margin cottony, white and inrolled when young and remaining so well
into maturity; surface sticky at first but soon dry, coated with tiny appressed fibrils
and scales, color variable, spotted and streaked vinaceous pink to purplish red over
a pale pink to white ground color; flesh white or tinged pinkish; odor and taste not
distinctive.
GILLS: close to crowded, attached at first, becoming decurrent at maturity, white when
young, soon developing purplish red spots and flushed purplish red in age.
STALK: 1⅛–3⅛" (3–8 cm) long, ⅝–1⅜" (1.5–3.5 cm) thick, nearly equal or sometimes
tapered downward, dry, solid, typically pruinose at the apex and smooth below, white
at first, soon colored like the cap.
SPORE PRINT: white.
MICROSCOPIC FEATURES: spores 6–8 × 3–5 μm, ellipsoid, smooth, hyaline.
FRUITING: solitary, scattered or in groups on the ground under oak; August–October;
occasional.
EDIBILITY: edible.

Hygrophorus speciosus var. **speciosus** Peck
COMMON NAME: Larch Waxy Cap.
CAP: ¾–2" (2–5 cm) wide, convex, becoming nearly flat, sometimes with an umbo; surface smooth, sticky to slimy, bright orange-red overall when young, becoming orange
to orange-yellow toward the margin in age; flesh white to pale yellow; odor and taste
not distinctive.
GILLS: attached to slightly decurrent, distant, narrow, waxy, white to yellowish.
STALK: 1⅝–4" (4–10 cm) long, ¼–⅜" (4–10 mm) thick, nearly equal, sometimes enlarged
at the base, stuffed, smooth, sticky to slimy, whitish to pale amber-yellow, often stained
tawny-orange on the upper portion or overall as the gluten dries, white at the apex
above the glutinous covering; partial veil and ring absent.
SPORE PRINT: white.
MICROSCOPIC FEATURES: spores 8–10 × 4.5–6 μm, elliptic, smooth, hyaline, inamyloid.
FRUITING: scattered, in groups or in clusters on the ground or among sphagnum mosses
under larch in woods and in bogs; August–October; occasional.
EDIBILITY: edible.

Hygrophorus subsalmonius Smith and Hesler
CAP: 2⅜–4¾" (6–12 cm) wide, convex, becoming broadly convex to flat, depressed on the
disc in age; surface smooth, sticky to slimy, rusty or cinnamon-brown when young,
fading to pale brown, dull orange, pinkish tan, salmon-buff or yellowish white toward
the margin in age; disc remaining reddish brown, cinnamon, or shades of orange-
brown; margin inrolled when young, with cottony fibers, becoming smooth and somewhat uplifted in age; flesh thick, whitish; odor and taste not distinctive.
GILLS: attached to decurrent, close to subdistant, waxy, cream to faintly pink, becoming
ivory to pinkish tan in age.
STALK: 1⅛–4" (3–10 cm) long, ¼–¾" (5–20 mm) thick, equal or somewhat enlarged at
the apex and tapered at the base, solid, smooth, typically slimy at the base, white to
pale tan, staining yellow, orange-yellow, or brownish in age or when bruised; partial
veil and ring absent.

SPORE PRINT: white.

MICROSCOPIC FEATURES: spores 6.5–8 × 3–5 μm, elliptic, smooth, hyaline, inamyloid.

FRUITING: scattered or in groups on the ground under hardwoods, especially oak and hickory; August–November; occasional.

EDIBILITY: edible.

Hygrophorus tephroleucus (Fries) Fries

COMMON NAME: Spotted-stalk Waxy Cap.

CAP: ½–1¼" (1–3 cm) wide, convex, becoming broadly convex to nearly plane, sometimes depressed; surface smooth, sticky to slimy when fresh, with minute fibers beneath the slime layer, pale to dark ashy gray or brownish gray overall, often minutely scaly in age; margin inrolled when young, often incurved at maturity; flesh white, thin; odor and taste not distinctive.

GILLS: attached to slightly decurrent, subdistant, broad, waxy, white to pale yellow in age.

STALK: 1–2⅜" (2.5–6 cm) long, about ⅛" (2–4 mm) thick, nearly equal, solid, white, covered with minute white scales that become ashy gray to dark gray in age; partial veil and ring absent.

SPORE PRINT: white.

MICROSCOPIC FEATURES: spores 7–10 × 4–5 μm, elliptic, smooth, hyaline, inamyloid.

FRUITING: scattered or in groups on the ground or among mosses in conifer woods or in bogs; August–October; occasional.

EDIBILITY: unknown.

Hygrophorus unguinosus (Fries) Fries

CAP: ¾–2" (2–5 cm) wide, hemispheric when young, becoming convex and nearly plane in age; surface slimy when fresh, shiny when dry, brownish black when young, becoming brown to pale grayish brown in age; margin translucent-striate; flesh grayish to white; odor and taste not distinctive.

GILLS: attached or notched, subdistant, broad, crossveined, waxy, white to grayish white.

STALK: 1⅛–3½" (3–9 cm) long, ⅛–¼" (2–5 mm) thick, nearly equal, slimy when fresh, pale brownish gray to grayish brown; partial veil and ring absent.

SPORE PRINT: white.

MICROSCOPIC FEATURES: spores 6–10 × 4–6 μm, elliptic, smooth, hyaline, inamyloid.

FRUITING: scattered or in groups on the ground, often among mosses in conifer and mixed woods and the borders of swamps; July–October; occasional.

EDIBILITY: unknown.

Genus *Hypholoma*

Species in *Hypholoma*, a fairly small genus, have routinely been included in the genus *Naematoloma* in most field guides. They are small to medium mushrooms that grow on decaying conifer and hardwood logs and stumps or on the surrounding soil, or among sphagnum mosses in bogs, wet areas in woods, swamps, and marshes. Their caps are yellow to orange, reddish or olive-yellow to pale olive-brown. They have whitish to yellowish gills when young that often become purplish gray to purple-brown at maturity. Their spore prints are purple-brown or sometimes dull rusty brown, and their spores are smooth with an apical pore. A fibrous to cortinate partial veil is often present, leaving remnants on the cap margin or a thin annular zone. Some are excellent edibles, but at least one causes gastric upset, and the edibility of several is unknown.

Key to Species of *Hypholoma*

1. Growing on decaying conifer or hardwood logs and stumps or on the surrounding soil → 2.
1. Growing among sphagnum mosses in bogs, wet areas in woods, swamps, or marshes → 3.
    2. Cap yellow-orange to reddish orange or orange-cinnamon, 1–2¾" (2.5–7 cm) wide, moist or dry, typically with white veil remnants on the margin, convex to broadly convex; flesh white, odor and taste not distinctive; gills attached, close, whitish to grayish at first, becoming purple-brown at maturity; stalk 2–4" (5–10 cm) long, ⅛–⅜" (3–10 mm) thick, rusty brown from the base upward, whitish to yellowish near the apex, fibrillose; partial veil fibrous to cortinate, whitish, leaving remnants on the cap margin and a thin annular zone; spores 6–8 × 3.5–4.5 μm; edible—*H. capnoides* (Fries) Karsten.
    2. Cap orange-yellow to sulphur-yellow or greenish yellow; margin usually not rimmed with veil remnants; gills yellow to greenish yellow when young, becoming grayish then tinged pale purple-brown at maturity; taste bitter—*H. fasciculare* (Hudson : Fries) Kummer (see p. 148).
    2. Cap brick-red with yellow-orange at or near the margin, typically not rimmed with veil remnants on the margin; gills whitish to pale greenish yellow when young, becoming purplish gray to purple-brown at maturity; taste not distinctive—*H. sublateritium* (Fries) Quélet (see p. 148).
3. Cap moist, not viscid, honey-yellow to greenish yellow, yellowish at the margin—*H. elongatum* (Persoon : Fries) Ricken (see p. 147).
3. Cap moist, not viscid, pale brownish red to orange-brown or sometimes olive-brown, yellow toward the margin; partial veil absent or very sparse, usually not forming remnants on the cap margin or a ring on the stalk—*H. udum* (Persoon : Fries) Kühner (see p. 149).
3. Cap viscid, olive-yellow to pale olive-brown overall, sometimes with greenish tints; partial veil fibrous, white, leaving remnants on the margin and an evanescent ring, or more typically, a fibrous annular zone—*H. myosotis* (Fries : Fries) M. Lange (see p. 148).

Hypholoma elongatum (Persoon : Fries) Ricken.

CAP: ⅜–1" (1–2.5 cm) wide, conic to bell-shaped when young, becoming broadly convex to flat in age; surface moist, smooth, honey-yellow to greenish yellow, yellowish at the margin, translucent-striate to shallowly sulcate at maturity; flesh whitish; odor not distinctive; taste not distinctive to slightly bitter.

GILLS: attached, subdistant, pale tawny, becoming gray-brown or pinkish brown at maturity.

STALK: 1½–6" (4–15.5 cm) long, ¹⁄₁₆–⅛" (1–3 mm) thick, equal, white-pruinose on a pale straw-yellow ground color, reddish brown toward the base; partial veil and ring absent.

SPORE PRINT: purplish brown.

MICROSCOPIC FEATURES: spores 9–13 × 5.5–7 μm, elliptic with an apical pore, smooth, pale brown.

FRUITING: scattered or in groups among sphagnum mosses in swamps, bogs, and marshes; August–October; fairly common.

EDIBILITY: unknown.

COMMENTS: also known as *Naematoloma elongatum*.

Hypholoma fasciculare (Hudson : Fries) Kummer

COMMON NAME: Sulphur Tuft.

CAP: ¾–3⅛" (2–8 cm) wide, convex, becoming broadly convex to nearly flat, often with an umbo; margin occasionally rimmed with veil remnants; surface smooth, moist or dry, orange-yellow to sulphur-yellow or greenish yellow, with a darker orange to brownish orange disc; flesh pale yellow, bruising brownish; odor not distinctive; taste bitter.

GILLS: attached, close, yellow to greenish yellow when young, becoming grayish, then tinged pale purple-brown at maturity.

STALK: 2–4¾" (5–12 cm) long, ⅛–⅜" (3–10 mm) thick, nearly equal or slightly tapered in either direction, fibrillose, pale yellow to yellow, becoming fulvous from the base upward; partial veil fibrous to cortinate, whitish, usually leaving a thin, superior annular zone.

SPORE PRINT: purple-brown.

MICROSCOPIC FEATURES: spores 6.5–8 × 3.5–4 µm, elliptic with an apical pore, smooth, pale purple-brown.

FRUITING: in clusters on conifer and hardwood logs and stumps or on the surrounding soil; May–November; fairly common.

EDIBILITY: poisonous.

COMMENTS: also known as *Naematoloma fasciculare*. It causes gastric distress.

Hypholoma myosotis (Fries : Fries) M. Lange

CAP: ⅜–1⅜" (1–3.5 cm) wide, broadly conic to convex; surface smooth, viscid, olive-yellow to pale olive-brown, sometimes with greenish tints when fresh, fading to olive-buff in age; margin typically rimmed with tiny patches of veil remnants; flesh thin, olive-buff; odor and taste not distinctive.

GILLS: attached, subdistant, whitish when young, becoming olive-buff and finally pale brown at maturity; edges whitish and finely fringed in age.

STALK: 2¾–6" (7–15.5 cm) long, ⅛–¼" (2–6 mm) thick, nearly equal, hollow, sheathed from the base up to the annular zone by white veil fibrils when young, becoming olive-brown to grayish brown in age; partial veil fibrous, white, leaving remnants on the margin and an evanescent ring, or more typically, a fibrous annular zone.

SPORE PRINT: dull rusty brown.

MICROSCOPIC FEATURES: spores 14–18 × 7–9 µm, obovate to elliptic with an apical pore, smooth, pale brown.

FRUITING: scattered or in groups among sphagnum mosses in bogs or wet areas in woods; July–October; fairly common.

EDIBILITY: unknown.

COMMENTS: also known as *Naematoloma myosotis* and *Pholiota myosotis*.

Hypholoma sublateritium (Fries) Quélet

COMMON NAME: Brick Cap, Brick Tops.

CAP: 1–4" (2.5–10 cm) wide, convex, becoming broadly convex to nearly flat; surface smooth, moist or dry, with scattered yellowish fibrils, brick-red with yellow-orange at or near the margin; flesh dull yellow to pale yellowish brown, thick, firm; odor not distinctive; taste mild or bitter.

GILLS: Attached, close, narrow, whitish to pale greenish yellow, becoming purplish gray to purple-brown at maturity; not staining when cut or bruised.

STALK: 2–4" (5–10 cm) long, ¼–⅝" (6–15 mm) thick, equal, hollow in age, pale yellow to whitish above the ring, dull brown or grayish below, covered with reddish brown fibrils; partial veil fibrous to cortinate, leaving a sparse superior ring or annular zone.

SPORE PRINT: purple-brown.

MICROSCOPIC FEATURES: spores 6–7 × 3.5–4.5 μm, elliptic with an apical pore, smooth, pale brown.

FRUITING: in dense clusters or scattered on hardwood stumps or logs; August–October; common.

EDIBILITY: edible.

COMMENTS: also known as *Naematoloma sublateritium*.

Hypholoma udum (Persoon : Fries) Kühner

CAP: ¾–1½" (1–4 cm) wide, obtusely conic to convex, becoming broadly convex to bell-shaped or nearly flat in age; surface smooth, moist, pale brownish red to orange-brown or sometimes olive-brown, typically yellow toward the margin, fading to dingy yellowish overall in age; flesh yellowish, thin; odor not distinctive; taste somewhat bitter.

GILLS: attached, close, pale yellow at first, becoming grayish and finally purple-brown at maturity.

STALK: 2–4¾" (5–12 cm) long, 1⁄16–⅛" (1.5–4 mm) thick, fragile, nearly equal, slightly fibrillose, rusty brown to rusty red from the base upward and yellow near the apex; partial veil absent or very sparse, usually not forming remnants on the cap margin and not forming a ring.

SPORE PRINT: purple-brown.

MICROSCOPIC FEATURES: spores 13–19 × 5–8 μm, elliptic with an apical pore, smooth, pale purple-brown.

FRUITING: scattered or in groups among sphagnum mosses in bogs or wet areas in woods; July–October; common.

EDIBILITY: unknown.

COMMENTS: also known as *Naematoloma udum*.

Genus *Hypsizygus*

Hypsizygus tessulatus (Bulliard : Fries) Singer

COMMON NAME: Elm Oyster.

CAP: 2–6" (5–15 cm) wide, convex at first, becoming more or less flat, often with a sunken center in age; surface smooth to very finely hairy, white to yellowish buff, with distinct water spots and usually becoming cracked in age; flesh thick, white, firm to hard; odor and taste not distinctive.

GILLS: attached, often appearing notched, sometimes with fine, barely subdecurrent lines, close to subdistant, whitish.

STALK: 2–4" (5–10 cm) long, ⅜–1" (1–2.5 cm) thick, off-center to almost lateral, dry, solid, very tough, smooth to finely hairy, whitish.

SPORE PRINT: white to pale buff.

MICROSCOPIC FEATURES: spores 5–7 μm, globose, smooth, hyaline, inamyloid.

FRUITING: solitary or in groups or clusters of two or three, usually on living deciduous trees, especially elm and box elder; August–December; common.

EDIBILITY: edible, but tough unless collected when young.

COMMENTS: this species has been terribly confused. According to Scott Redhead (1986), it is the same fungus as *Hypsizygus marmoreus, H. ulmarius,* and *H. elongatipes.* The cap's water spots are more important than the cracks. We find it a ubiquitous species on wounds of living box elders during autumn in the Northeast. Sometimes spelled *H. tesselatus.*

Genus *Inocybe*

Inocybe is a fairly large genus of small to medium terrestrial mushrooms commonly found in grassy areas near trees, on soil, and on humus. They typically have an umbonate cap with a fibrillose to scaly surface and a margin that is often lacerated in age. Their spore prints are dull brown to dull yellow-brown, and their spores are smooth and typically elliptic, or angular and warted, and lack an apical pore. A partial veil may be present and cortinate, or absent, and the gill edges are often white and finely fringed. The stalk apex is usually white-pruinose.

This genus is not commonly collected for the table, and the edibility of most of the species is unknown. The identification of most species is often very difficult and requires the use of a microscope. Because the genus *Inocybe* is complex and includes a large number of species, a comprehensive key to species is beyond the scope of this book. We have included a sampling of some of the more common and distinctive species that occur in this region.

Inocybe fuscodisca (Peck) Massee
COMMON NAME: Black-nipple Fiber Head.
CAP: ⅜–1" (1–2.5 cm) wide, conic when young, becoming bell-shaped to broadly convex and finally flat with a prominent, dark brown umbo in age; surface smooth and subviscid over the disc, moist to dry and covered with dark brown to dull brown, radially flattened fibrils over a pale buff to whitish ground color; flesh whitish to pale buff; odor unpleasant, spermatic; taste not distinctive.
GILLS: attached, close, broad, cream at first, becoming pale brown to gray-brown at maturity.
STALK: 1½–3" (4–7.5 cm) long, 1/16–⅛" (1.5–3 mm) thick, equal, somewhat bulbous at the base, pruinose at the apex, sheathed from the base to the annular zone with dark brown to olive-brown fibrils over a whitish ground color; annular zone evanescent.
SPORE PRINT: brown.
MICROSCOPIC FEATURES: spores 7–10 × 4.5–6 μm, elliptic, smooth, pale brown.
FRUITING: solitary, scattered, or in groups on the ground under conifers; August–November; occcasional.
EDIBILITY: poisonous.

Inocybe lacera (Fries) Karsten
CAP: ⅜–1½" (1–4 cm) wide, convex, becoming nearly plane, often with a low umbo; surface dry, fibrillose to fibrillose-scaly and sometimes ragged and torn, especially near the margin, dull brown becoming tawny ochraceous in age; flesh whitish; odor and taste not distinctive.
GILLS: attached or notched, close, whitish at first, becoming grayish brown at maturity.
STALK: 1⅛–2⅜" (3–6 cm) long, ⅛–½" (3–6 mm) thick, fibrillose, dry, whitish near the apex, grayish brown below.
SPORE PRINT: brown.
MICROSCOPIC FEATURES: spores 10–18 × 4–6 μm, elliptic to subcylindric, smooth, pale brown; pleurocystidia and cheilocystidia thin-walled, subcylindric to subventricose or ventricose-subovoid, rounded or obtuse at the apex.
FRUITING: scattered or in groups on the ground, usually in sandy soil, under conifers, especially pines, in hardwoods, or in mixed woods; May–October; fairly common.
EDIBILITY: poisonous.

Inocybe rimosa (Bulliard : Fries) Kummer

COMMON NAME: Straw-colored Fiber Head.

CAP: ¾–3½" (2–9 cm) wide, broadly conic when young, becoming broadly convex to nearly flat with a distinct, conic umbo in age; surface slippery when wet, silky and shiny when dry, covered overall with radial fibers, golden, straw-yellow to honey-brown, often darker over the disc; margin incurved when young, whitish, becoming expanded, cracked, and yellowish in age; flesh white; odor and taste not distinctive.

GILLS: attached, close, narrow, whitish when young, becoming grayish, then coffee-brown at maturity.

STALK: 1½–3½" (4–9 cm) long, ⅛–½" (3–12 mm) thick, nearly equal or tapered slightly downward, smooth, silky-fibrillose, longitudinally striate, whitish, developing yellowish tinges in age; partial veil and ring absent.

SPORE PRINT: brown.

MICROSCOPIC FEATURES: spores 9–15 × 5–8 μm, elliptic, smooth, pale brown.

FRUITING: solitary, scattered or in groups on the ground and among mosses under conifers and hardwoods; July–October; fairly common.

EDIBILITY: poisonous, causing gastric distress.

COMMENTS: also known as *I. fastigiata*. The nearly identical *I. fastigiata* var. *microsperma* (poisonous) has a spermatic odor.

Inocybe sororia Kauffman

COMMON NAME: Pungent Fiber Head, Corn Silk Inocybe.

CAP: ¾–3" (2–7.5 cm) wide, conic when young, becoming broadly convex to nearly flat with a distinct umbo in age; surface dry, silky, streaked with radial fibers, pale buff to yellowish to yellowish brown, often darker over the disc; margin incurved when young, becoming expanded and split in age; flesh cream to pale buff; odor strong of green corn or cornhusks; taste not distinctive.

GILLS: attached, becoming free, close to crowded, narrow, white to cream when young, soon becoming yellowish, then brownish in age.

STALK: 2–4" (2.5–10 cm) long, 1/16–¼" (2–6 mm) thick, equal or enlarged slightly at the base, solid, silky, fibrillose, somewhat scurfy toward the apex, whitish, becoming yellowish to brownish; partial veil and ring absent.

SPORE PRINT: brown.

MICROSCOPIC FEATURES: spores 9–17 × 5–8 μm, elliptic, smooth, pale brown.

FRUITING: solitary, scattered, or in groups on the ground under hardwoods and conifers; July–November; occasional.

EDIBILITY: poisonous.

COMMENTS: *Inocybe lanatodisca* var. *phaeoderma* (edibility unknown) is nearly identical but has a darker, ochraceous brown cap, smaller spores, 9–11 × 5–6 μm, and a complex fragrant odor described as spermatic with a green corn component.

Inocybe subochracea (Peck) Earle

CAP: ¾–2" (2–5 cm) wide, hemispheric when young, becoming broadly convex with a low, broad umbo in age; surface dry, covered with radial fibers nearly overall and flattened scales, especially over the disc, honey-brown to golden orange; margin incurved, with fibrous veil remnants, becoming expanded in age; flesh whitish to pale yellowish; odor and taste not distinctive.

GILLS: notched, close, moderately broad, whitish, becoming pale golden brown to rusty brown.

STALK: 1–2" (2.5–5 cm) long, ⅛–¼" (2–5 mm) thick, equal, solid, white-pruinose at the apex, with white mycelium coating the base, sheathed from the base to the annular zone with a dense coating of dark brown to ochre fibers, often scurfy; partial veil fibrous, whitish to buff, leaving an evanescent annular zone.

SPORE PRINT: brown.

MICROSCOPIC FEATURES: spores 7–10 × 4.5–6 μm, elliptic, smooth, pale brown.

FRUITING: solitary, scattered, or in groups on the ground under conifers and hardwoods; August–October; occasional.

EDIBILITY: unknown.

Inocybe tahquamenonensis Stuntz

CAP: ½–1¼" (1.2–3 cm) wide, bell-shaped when young, becoming broadly convex to nearly flat in age, with or without a low, broad umbo; surface dry, covered overall with tiny erect to flattened scales, dark purplish brown, reddish brown or blackish brown; flesh reddish purple; odor not distinctive or mildly radish-like; taste not distinctive.

GILLS: attached, subdistant, moderately broad to broad, purplish red when young, becoming reddish brown to chocolate-brown in age; edges often whitish at maturity.

STALK: 1¼–3" (3–7.5 cm) long, ⅛–¼" (3–7 mm) thick, nearly equal, solid, dry, covered with tiny, flattened scales, colored like the cap.

SPORE PRINT: brown.

MICROSCOPIC FEATURES: spores 6–8.5 × 5–6 μm, short-oblong, angular-nodulose, pale brown.

FRUITING: solitary, scattered, or in groups on the ground under hardwoods; August–October; uncommon.

EDIBILITY: unknown.

Genus *Laccaria*

Laccaria is a terrestrial genus that lacks a veil and has gills that are typically thick and waxy, as in the Waxy Caps (Genus *Hygrophorus*). Laccarias are typically brownish, pinkish, or purplish, while *Hygrophorus* species tend toward yellow, orange, and red. The stalk is always dry, fibrous, and very tough, quite unlike that of a Waxy Cap. The spore print color varies from white to pale purplish, and the lower stalk is consistently coated with white to purplish mycelium that is sometimes copious and extends well up the stalk.

Key to Species of *Laccaria*

1. Base of stalk coated with mycelium that is white (not distinctly lilac to violet) in young specimens → 2.
1. Base of stalk coated with mycelium that is distinctly lilac to violet in young specimens → 4.
    2. Cap 1¼–3" (3–7.5 cm) wide, grossly scaly (not minutely scaly) overall at maturity, brownish orange, sometimes darker on the disc, margin not striate; gills attached, sometimes notched, close to distant, thick but particularly waxy, pinkish flesh color; stalk 1–4" (2.5–10 cm) long, ⅛–⅝" (4–15 mm) thick, grossly fibrous and longitudinally striate, often scaly at the top in age, brownish orange; usually found at high elevations on mountains; spores 7.5–10.5 × 6.5–8.5 μm, not including the crowded spines which are mostly about 1 μm long, usually broadly elliptic to sub-globose; basidia 4-spored; cheilocystidia absent; edibility unknown—*L. nobilis* Mueller : Smith.

2. Cap ¾–1½" (2–3.5 cm) wide, finely fibrous to minutely scaly overall (only coarsely scaly over the disc or not at all), reddish brown or orangish brown, fading to pinkish buff in age, margin translucent-striate to plicate-striate; gills attached, sometimes notched, close to distant, thin or thick, pinkish flesh-color; stalk ¾–4" (2–10 cm) long, ¹⁄₁₆–⁵⁄₁₆" (1.5–8 mm) thick, sometimes with a slight bulb, smooth, not longitudinally striate, reddish brown or darker; especially frequent in wet moss; spores 7–11 × 7–11 μm, not including the crowded spines, which are 1.5–3 μm long, mostly globose; basidia 4-spored; cheilocystidia absent; edibility unknown—*L. glabripes* McNabb.

2. Cap ³⁄₈–1½" (1–4 cm) wide, smooth to finely fibrous, reddish brown fading to orangish brown and eventually buff, strongly plicate-striate; gills attached, sometimes notched, distant, moderately thick, pinkish flesh color; stalk ⁵⁄₈–1¼" (1.5–3.5 cm) long, ¹⁄₁₆–⅛" (1–4 mm) thick, sometimes with a slight bulb, dry, smooth to finely fibrous, colored like the cap; spores 7.5–10 × 7–9.5 μm, not including the crowded spines, which are mostly 1.5–3 μm long, globose to subglobose; basidia 4-spored; cheilocystidia absent to abundant; edibility unknown—*L. ohiensis* (Montagne) Singer. [Note: *L. striatula* (Peck) Peck is very similar. Primary differences are its usually darker reddish brown and more translucent-striate cap and its smoother, longer, up to 4" (10 cm), stalk. It prefers a habitat of wet, mossy areas in hemlock-beech woods.]

2. Cap ⁵⁄₈–3⅛" (1.6–8 cm) wide, finely fibrillose to smooth, slightly to strongly translucent-striate, orange-brown; gills attached, distant, pale flesh color; stalk 2¾–6¾" (6.5–16.5 cm) long, ⅛–³⁄₈" (3–10 mm) thick; growing among sphagnum mosses, usually in bogs—*L. longipes* Mueller (see p. 155).

2. Not as in the above choices; basidia 2- or 4-spored—*L. laccata* complex (see comments on p. 155 or go on to 3).

3. Cap ³⁄₈–1" (1–2.5 cm) wide, finely fibrous to very finely scaly, margin strongly plicate-striate, dull brownish orange or brick-red at first, fading somewhat; gills attached to subdecurrent, subdistant to distant, broad, moderately thick, pinkish flesh-color; stalk ¾–2" (2–5 cm) long, ¹⁄₁₆–¼" (2–5 mm) thick, nearly smooth or finely fibrous, longitudinally striate, brownish orange to dark brownish red; spore print white; spores 8.5–13 × 7.5–13 μm, not including the crowded spines, which are mostly 1–2 μm long; basidia mostly 2-spored; cheilocystidia absent; edibility unknown—*L. lateritia* Malençon.

3. Cap ³⁄₈–2¼" (1–6 cm) wide, finely fibrous to finely scaly, reddish orange or orangish brown at first, fading to pinkish buff or orangish buff; margin not truly striate but sometimes translucent-striate; gills attached, sometimes notched, sometimes becoming barely subdecurrent, subdistant to distant, broad, thick, pinkish flesh-color; stalk typically 1–3" (2.5–7.5 cm) long, ⅛–³⁄₈" (3–10 mm) thick, sometimes slightly enlarged at the base, fibrous, sometimes slightly longitudinally striate, colored like the cap or darker; spore print white; spores 7–12.5 × 6–9 μm, not including the spines, which are mostly 0.5–1.5 μm long, elliptic to broadly elliptic; basidia 4-spored; cheilocystidia present; edibility unknown—*L. proxima* (Boudier) Patouillard.

3. Cap ³⁄₈–2¼" (1–6 cm) wide, finely fibrous-scaly, orangish brown at first, fading eventually to buff; margin striate or not, sometimes translucent-striate or plicate-striate; gills attached, sometimes notched or subdecurrent to decurrent, close to distant, broad, thin to thick, slightly waxy, pinkish flesh-color, sometimes becoming slightly purplish in age; stalk ⁵⁄₈–4" (1.5–10 cm) long, ⅛–³⁄₈" (3–10 mm) thick, often twisted or flattened, sometimes slightly enlarged downward or with a slight bulb, fibrous, longitudinally striate or not, orangish brown fading eventually to buff; spore print white; spores 6.5–13 × 6–11.5 μm, not including the well-spaced to crowded spines, which are mostly 1–2.5 μm long,

globose to subglobose; basidia mostly 4-spored; cheilocystidia absent to abundant—
L. laccata (Scopoli : Fries) Berkeley and Broome (see p. 155).

4.　　Stalk typically more than ⅝" (1.5 cm) thick overall → 5.

4.　　Stalk typically less than ⅝" (1.5 cm) thick overall → 6.

5.　　Growing in sand or in very sandy soil, especially in dunes; cap 1¼–3" (3.5–7.5 cm) wide, finely fibrous to finely scaly, brown to buff or pinkish buff, margin not striate; flesh thick, pale purple-gray; gills attached to somewhat decurrent, thick and waxy, close to subdistant, purple to dark violet-gray; stalk 1½–3½" (4–9 cm) long, typically ⅜–1" (1–2.5 cm) thick, enlarged at the base, fibrous, longitudinally striate, brown to buff or pinkish buff, covered with sand; spore print white; spores 16–21 × 6–9 μm, smooth to finely roughened—*L. trullisata* (Ellis) Peck (see p. 156).

5.　　Habitat not as above; cap 2–5" (5–12.5 cm) wide, nearly smooth to very finely fibrous or occasionally slightly scaly in age, distinctly purplish (light violet-brown or pinkish buff) at first, fading to buff in age, margin not striate; flesh thick, violet-buff; gills attached to somewhat decurrent, thick and waxy, close to subdistant, dark purple or violet to violet-gray; stalk usually ⅝–1¼" (1.5–3.5 cm) thick, sometimes enlarged at the base, coarsely fibrous, usually with brownish to reddish brown longitudinal striations and/or scales, ground color light violet-brown or pinkish buff at first, fading to buff in age; spore print white to pale violet—*L. ochropurpurea* (Berkeley) Peck (see p. 155).

6.　　Gills distinctly purplish (purplish gray to grayish purple), even in age; cap ⅜–2" (1–5 cm) wide, pruinose or finely fibrous to finely scaly, bright grayish purple at first, fading to buff in age, margin striate or not, sometimes translucent-striate; gills attached, notched to slightly decurrent, subdistant to distant, moderately broad, thick, slightly waxy, grayish purple to purplish gray; stalk ⅜–3" (1–7.5 cm) long, ¹⁄₃₂–¼" (1–7 mm) thick, sometimes slightly enlarged at the base, fibrous, longitudinally striate, bright grayish purple at first, fading to buff in age; spore print white to very pale violet; spores 7–10 × 6.5–10 μm, not including the spines, which are 1.5–3 μm long, globose; edible—*L. amethystina* (Hudson) Cooke.

6.　　Gills only slightly purplish at first, soon fading to light pinkish flesh-color; cap ⅜–3" (1–7.5 cm) wide, finely fibrous to fibrous-scaly, margin not striate, pinkish flesh-color at first, fading to buff in age; gills attached to slightly subdecurrent, subdistant to distant, broad, moderately thick, slightly waxy, light purplish at first, fading to pinkish flesh-color; stalk 1–5" (2.5–12.5 cm) long, ⅛–⅝" (3–15 mm) thick, sometimes somewhat enlarged at the base, fibrous, longitudinally striate, pinkish flesh-color at first, fading to buff in age; spore print white; spores 6–10 × 6–9 μm, not including the spines, which are 1–2 μm long, nearly globose to broadly oval; edibility unknown—*L. bicolor* (Maire) Orton.

6.　　Gills pinkish flesh-color (lacking any purplish tints); cap ⅜–1¼" (1–3 cm) wide, strongly pruinose to fibrous or occasionally finely scaly, brownish orange at first, eventually fading to buff, margin not striate; gills attached, sometimes notched, subdistant to distant, moderately broad, thick, slightly waxy, pinkish flesh-color; stalk 1–2¼" (2.5–6 cm) long, ¹⁄₁₆–¼" (2–7 mm) thick, sometimes slightly enlarged at the base, fibrous, often longitudinally striate, brownish orange at first, eventually fading to buff; spore print white; spores 6–10 × 6–9 μm, not including the crowded spines, which are 0.5–2 μm long, globose to subglobose; edibility unknown—*L. lavendipes* Mueller.

6.　　Cap ⅜–2¼" (1–6 cm) wide, finely fibrous to finely scaly, reddish orange or orangish brown at first, fading to pinkish buff or orangish buff, margin not truly striate but sometimes translucent-striate; gills attached, occasionally notched or becoming barely subdecurrent, subdistant to distant, broad, thick, pinkish flesh-color; stalk typically 1–3" (2.5–7.5 cm) long, ⅛–⅜" (3–10 mm) thick, sometimes slightly enlarged at the base, fibrous, sometimes slightly longitudinally striate, reddish orange or orangish brown at first, fading to pinkish buff or orangish buff, base sometimes darker; spore print white; spores 7–12.5 × 6–9 μm, not including the spines, which are mostly 0.5–1.5 μm long, elliptic to broadly elliptic; basidia 4-spored; cheilocystidia present; edibility unknown—*L. proxima* (Boudier) Patouillard.

Laccaria laccata (Scopoli : Fries) Berkeley and Broome

COMMON NAME: Common Laccaria.

CAP: ⅜–2¼" (1–6 cm) wide, finely fibrous-scaly; surface orangish brown at first, fading eventually to buff, margin striate or not, sometimes translucent-striate or plicate-striate; flesh thin, colored more or less like the cap; odor and taste not distinctive.

GILLS: attached, occasionally notched or subdecurrent to decurrent, close to distant, broad, thin to thick, slightly waxy, pinkish flesh-color, sometimes becoming slightly purplish in age.

STALK: ⅝–4" (1.5–10 cm) long, ⅛–⅜" (3–10 mm) thick, often twisted or flattened, sometimes slightly enlarged downward or with a slight bulb, fibrous, longitudinally striate or not, orangish brown fading eventually to buff.

SPORE PRINT: white.

MICROSCOPIC FEATURES: spores 6.5–13 × 6–11.5 μm, not including the well-spaced to crowded spines, which are mostly 1–2.5 μm long, usually globose to subglobose; hyaline, inamyloid; basidia mostly 4-spored; cheilocystidia absent to abundant.

FRUITING: solitary to gregarious, occasionally in small clusters, on the ground in a wide variety of habitats, sometimes among mosses; June–November; common.

EDIBILITY: edible.

COMMENTS: Gregory Mueller's monograph lists more than two dozen taxonomic synonyms. Sorting these fungi out requires devoting a great deal of time to them.

Laccaria longipes Mueller

CAP: ⅝–3⅛" (1.6–8 cm) wide, convex to broadly convex, sometimes depressed; margin often uplifted; surface finely fibrillose to smooth, slightly to strongly translucent-striate, moist and lubricous when fresh, orange-brown, hygrophanous, fading to pale tan; flesh pale flesh-color; odor and taste not distinctive.

GILLS: attached, distant, pale flesh-color.

STALK: 2¾–6¾" (6.5–16.5 cm) long, ⅛–⅜" (3–10 mm) thick, nearly equal or with a slightly swollen base, dry, somewhat fibrillose, and longitudinally striate, colored like the cap, coated with a white basal mycelium.

SPORE PRINT: white.

MICROSCOPIC FEATURES: spores 7–9 × 6–8.5 μm, subglobose to broadly ellipsoid, echinulate, hyaline.

FRUITING: scattered or in groups among sphagnum mosses, usually in bogs; August–October; common.

EDIBILITY: edible.

Laccaria ochropurpurea (Berkeley) Peck

COMMON NAME: Purple-gilled Laccaria.

CAP: 2–5" (5–12.5 cm) wide, nearly smooth to very finely fibrous or occasionally slightly scaly in age; surface purplish, light violet-brown or pinkish buff at first, fading to buff in age, margin not striate; flesh thick, violet-buff; odor and taste not distinctive.

GILLS: attached to somewhat decurrent, thick and waxy, close to subdistant, dark purple or violet to violet-gray.

STALK: ⅝–1¼" (1.5–3.5 cm) thick, sometimes enlarged at the base, coarsely fibrous, usually with brownish to reddish brown longitudinal striations and/or scales, ground color light violet-brown or pinkish buff at first, fading to buff in age.

SPORE PRINT: white to pale violet.

MICROSCOPIC FEATURES: spores 6.5–11 × 6.5–9.5 μm, not including crowded spines, which are mostly 1–1.5 μm long, mostly globose to subglobose, hyaline, inamyloid.

FRUITING: solitary to scattered, rarely in clusters of two to four, sometimes gregarious, on the ground under conifers or in mixed woods, especially with oaks and pines; July–October; occasional to common.

EDIBILITY: edible and popular.

COMMENTS: an unusually robust and distinctive mushroom for the genus.

Laccaria trullisata (Ellis) Peck

COMMON NAME: Sandy Laccaria.

CAP: 1¼–3" (3.5–7.5 cm) wide, convex to plane, sometimes depressed; surface dry, fibrillose to finely scaly, grayish purple when very young, becoming red-brown to brown or buff to pinkish buff; margin incurved to decurved, not striate; flesh thick, pale purple to pale purple-gray; odor and taste not distinctive.

GILLS: attached to somewhat decurrent, thick and waxy, close to subdistant, purple to dark violet-gray, becoming dull reddish violet in age.

STALK: 1½–3½" (4–9 cm) long, typically ⅜–1" (1–2.5 cm) thick, enlarged at the base, dry, fibrillose, longitudinally striate, brown to buff or pinkish buff, covered with sand on the lower portion or overall.

SPORE PRINT: white.

MICROSCOPIC FEATURES: spores 14–21 × 5.5–8 μm, subfusiform to fusiform-ellipsoid, smooth to very finely roughened, hyaline.

FRUITING: solitary, scattered, or in groups in sand or in very sandy soil, especially in dunes; August–November; fairly common.

EDIBILITY: edible.

COMMENTS: the elliptical, smooth spores separate this mushroom from all other *Laccaria* species.

Genus *Lactarius*

Lactarius is a large genus, with over two hundred species described from North America. They are small to large, mostly terrestrial mushrooms with distinctly brittle flesh; a central stalk; white, cream, yellow, or ochre spore prints; and ornamented amyloid spores. No universal veil, partial veil, or ring is formed by members of this genus. Their gills, cap flesh, or stalk exudes a latex when cut. The latex may be clear and watery, white to cream or variously colored immediately on exposure. The latex of some species changes color and may stain tissues. Latex production varies from scanty to copious depending on the species and on the condition of the mushroom; dry specimens may not exude any latex.

Key to Species of *Lactarius*

1.   Latex clear and watery immediately on exposure → 2.
1.   Latex white to cream immediately on exposure, changing to sulfur-yellow within two minutes → 5.
1.   Latex yellow to dingy yellow or brownish yellow immediately on exposure → 6.
1.   Latex orange, vinaceous red, yellow-brown, reddish brown to wine-brown, or blue immediately on exposure → 7.
1.   Latex white to cream immediately on exposure, not changing color significantly within two minutes (color change occurs slowly, not at all, or only on drying; latex sometimes staining tissues yellow, green, lilac, violet, reddish or brownish) → 10.
      2.   Odor sweet, fragrant, like maple sugar or burnt sugar or resembling anise or coconut → 3.
      2.   Odor not distinctive → 4.
3.   Odor sweet, like maple sugar or burnt sugar; cap dry, orange-brown to pale cinnamon-brown; stalk pale orange-tan to pale pinkish cinnamon; among sphagnum mosses in bogs or in conifer or mixed woods—*L. aquifluus* Peck (see p. 167).
3.   Odor sweet, fragrant, often like burnt sugar; latex mild; cap pale to dark brown, often zoned, disc glaucous; stalk colored like the cap, not darkening significantly from the base up as it ages; spores 7–10 × 6–8 μm; on soil in conifer and mixed woods; edible—*L. mutabilis* Peck.
3.   Odor mildly fragrant, often like burnt sugar or not distinctive; latex slowly and mildly acrid; cap dark purplish brown or purplish gray when young, fading to pinkish brown to pinkish buff; stalk colored like the cap, darkening progressively toward the apex as it matures; latex white in very young specimens, soon watery—*L. quietus* var. *incanus* Hesler and Smith (see p. 176).
3.   Odor fragrant, resembling anise or coconut; cap dry, scurfy, dark pinkish brown to dark pinkish gray, paler in age; latex of young specimens white, becoming watery in age; acrid, sometimes slowly—*L. hibbardae* var. *hibbardae* Peck (see p. 171).
      4.   Cap up to 1⅜" (3.5 cm) wide, convex with a distinct umbo when young, plane or depressed in age, usually with a distinct umbo, dark brown to olive-brown; latex scanty, taste unpleasant, bitter, or metallic, but not acrid; spores 6–8 × 5.5–6.5 μm; on buried wood or among sphagnum mosses in wet areas of hardwoods; edibility unknown—*L. louisii* Homola.
      4.   Cap up to 1¾" (4.5 cm) wide, with a small pointed umbo, dark reddish brown when young, fading to orange-brown; umbo typically retaining the darker reddish brown color well into maturity and resembling an eyespot; latex scanty, watery to whey-like, taste very slowly acrid, then fading—*L. oculatus* (Peck) Burlingham (see p. 174).
5.   Gills whitish, becoming yellowish with age, not spotted vinaceous red or becoming reddish overall; cap up to 2" (5 cm) wide, whitish when young, becoming brownish red in age; spores 6–8 × 5–6 μm; in mixed woods; edibility unknown—*L. colorascens* Peck.
5.   Gills whitish to pinkish buff when young, soon spotted vinaceous red to pinkish brown or dark reddish brown; cap up to 4¾" (12 cm) wide, pale pinkish cinnamon when young, becoming darker pinkish cinnamon to orange-cinnamon in age—*L. vinaceo-rufescens* Smith (see p. 179).
5.   Gills whitish to pale yellow, not spotting vinaceous red or becoming reddish overall; cap up to 3½" (9 cm) wide, whitish to pale yellowish cinnamon; flesh and latex acrid; spores 6–9 × 5.5–6.5 μm; on the ground in hardwoods and mixed woods; poisonous—*L. chrysorheus* Fries.

5. Gills creamy buff to pale yellow-tan, staining sulfur-yellow when cut or bruised; cap 2–4" (5–10 cm) wide, bright orange to saffron-yellow; margin incurved and downy when young, not bearded with coarse hairs; latex bitter to acrid—*L. croceus* Burlingham (see p. 169).

5. Gills whitish to yellow, staining yellow to greenish yellow; edges often finely toothed; cap 2¾–7" (7–18 cm) wide, disc typically deeply depressed, pale yellow to dark orange-yellow; margin long-inrolled and bearded with coarse hairs, becoming smooth in age; stalk ⅝–1⅛" (1.5–3 cm) thick, tawny, scrobiculate; latex acrid; spores 6–9 × 5–7.5 μm; on the ground in coniferous woods; edibility unknown—*L. scrobiculatus* (Fries) Fries complex.

5. Gills white when young, becoming pinkish buff at maturity, staining dingy yellow when bruised; cap 2⅜–6" (6–15.5 cm) wide, deeply depressed to funnel-shaped, milky white overall or with ochraceous zones; margin bearded with coarse hairs at least on young specimens; stalk inconspicuously scrobiculate or lacking scrobiculations; latex slowly acrid—*L. resimus* (Fries) Fries (see p. 177).

    6. Latex at first dingy yellow, becoming yellow-brown, scanty, slightly acrid (often slowly); flesh azure-blue, becoming yellowish near the gills; cap azure-blue with dingy orange-brown areas when young, soon fading to orange-brown with green to olive tints, staining green when bruised; spores 7–9 × 5–7 μm; on the ground or in grass under conifers, especially pine; edibility unknown—*L. chelidonium* var. *chelidonioides* (Smith) Hesler and Smith.

    6. Nearly identical to the previous choice except latex yellow at first, becoming dingy yellow, then yellow-brown, scanty, mild-tasting—*L. chelidonium* var. *chelidonium* Peck.

7. Latex dark blue immediately on exposure, soon dark bluish green; cap, gills, and stalk indigo-blue, fading to bluish gray; spores 7–9 × 5.5–7.5 μm; on the ground in woods; edible—*L. indigo* (Schweinitz) Fries.

7. Latex orange, often scanty → 8.

7. Latex vinaceous red, yellow-brown or reddish brown to wine-brown → 9.

    8. Latex slowly staining tissues green; cap and stalk orange, typically with green spots—*L. deterrimus* Gröger (see p. 170).

    8. Latex slowly staining tissues vinaceous red (especially near the stalk base), not staining tissues green; cap and stalk orange, lacking green spots; spores 9–12 × 7.5–9 μm; on the ground and among sphagnum mosses in cedar woods, swamps, and bogs; edible—*L. thyinos* Smith. (Note: *L. salmonicolor* is a nearly identical edible species that is found in dry upland habitats, not in swamps and bogs).

9. Latex dark vinaceous red, scanty, mild to slightly acrid; cap dark vinaceous red zoned with paler red to pink and gray, fading in age, often with green spots; stalk colored like the cap or duller, scrobiculate; spores 8–11 × 6.5–8 μm; on the ground in conifer and mixed woods, usually near pine or hemlock; edible—*L. subpurpureus* Peck.

9. Latex dark wine-brown, scanty, mild to slowly and slightly acrid, staining tissues green; flesh whitish with greenish to bluish tints; gills pinkish orange; cap zoned with bands of grayish blue, grayish purple, green, and blue, with a silvery sheen when young—*L. paradoxus* Beardslee and Burlingham (see p. 174).

9. Latex yellow-brown, scanty, slightly acrid (often slowly); flesh azure-blue, becoming yellowish near the gills; gills yellowish to yellow-brown; cap azure-blue with dingy orange-brown areas when young, soon fading to orange-brown with green to olive tints, staining green when bruised; spores 7–9 × 5–7 μm; on the ground or in grass

under conifers, especially pine; edibility unknown—*L. chelidonium* var. *chelidonioides* (Smith) Hesler and Smith.

9.   Nearly identical to the previous choice except latex mild-tasting and gills grayish yellow —*L. chelidonium* var. *chelidonium* Peck.

    10.   Exposed latex staining gills or drying reddish orange, salmon, pinkish, or reddish, typically within one hour (check gill surfaces for prior staining reactions) → 11.

    10.   Exposed latex staining gills or drying lilac, violaceous, or purple, typically within one hour (check gill surfaces for prior staining reactions) → 14.

    10.   Exposed latex staining gills or drying pale green, bluish green, greenish olive, olive-gray, olive-brown, tawny-brown, reddish brown, vinaceous brown, or dark brown, typically within one hour (check gill surfaces for prior staining reactions) → 17.

    10.   Exposed latex not staining gills, or if so, staining some shade of yellow; latex sometimes changing color when dry → 26.

11.   Cap margin conspicuously pleated or deeply grooved → 12.

11.   Cap margin lacking conspicuous pleats or deep grooves → 13.

    12.   Cap pale tan to yellowish buff; spore print pinkish buff; gills distant; flesh and latex acrid; cut flesh rosy salmon; typically found in the southern part of the region, usually in hardwoods or mixed woods; edibility unknown—*L. subplinthogalus* Coker (see p. 178).

    12.   Cap blackish brown to yellow-brown when young, usually fading in age, typically with a small, pointed umbo, velvety, sometimes wrinkled; spore print bright ochre; flesh and latex mild to slightly bitter, not acrid; gills close to subdistant; cut flesh staining rosy pink or dull reddish—*L. lignyotus* Fries (see p. 172).

    12.   Cap blackish brown at first, fading to yellow-brown, velvety, wrinkled to radially wrinkled, with or without an umbo; spore print white; flesh and latex mild to slightly acrid; gills subdistant to distant; cut flesh staining vinaceous pink; spores 8–10 × 7–9 μm; on the ground in conifer or mixed woods; edibility unknown— *L. gerardii* var. *subrubescens* (Smith and Hesler) Hesler and Smith.

13.   Cap whitish when young, soon tinged buff, sometimes pale smoky brownish; spore print yellow; latex acrid; gills crowded to close; cut flesh staining pinkish; spores 6.5– 7.5 × 6–7.5 μm; on the ground in hardwoods and mixed woods, typically found in the southern part of the region; edibility unknown—*L. subvernalis* var. *cokeri* (Smith and Hesler) Hesler and Smith.

13.   Cap pale yellow-brown to smoky gray or dingy whitish, wrinkled or smooth; spore print pinkish buff; flesh and latex taste variable, acrid becoming mild or mild becoming acrid; gills close to crowded; cut flesh staining salmon; spores 6–8 × 6–7.5 μm; on the ground in conifer or hardwoods; edibility unknown—*L. fumosus* Peck.

    14.   Cap margin of young specimens bearded with coarse hairs → 15.

    14.   Cap margin of young specimens silky, pruinose, grooved or smooth, not bearded → 16.

15.   Cap azonate to faintly zoned, pale yellow to orange-yellow, staining purplish where bruised, 2⅜–7" (6–18 cm) wide, convex to broadly funnel-shaped; flesh and latex mild to slightly acrid or somewhat bitter; gills close to crowded, cream to pale ochraceous, staining purplish; stalk pale yellow to orange-yellow, scrobiculate, staining purplish; on the ground under conifers, especially spruce—*L. repraesentaneus* Britzelmayr (see p. 176).

15.   Cap strongly zonate, pale cinnamon or orange-brown, zoned with bands of pinkish tan and cinnamon-brown, 1⅛–4" (3–10 cm) wide, convex to broadly funnel-shaped; flesh

and latex mild to slightly bitter or acrid; gills close, cream to pinkish buff, staining dull lilac; stalk whitish to buff with reddish brown and yellowish spots and streaks; on the ground in hardwoods and mixed woods; spores 10–13.5 × 9–11 μm; edibility unknown—*L. speciosus* Burlingham (see p. 177).

15. Nearly identical to the previous choice except: spore print yellowish and spores smaller, 7.5–10 × 6–7.5 μm; edibility unknown—*L. dispersus* (Fries) Fries.

   16. Cap blackish brown to yellow-brown when young, usually fading in age, typically with a small pointed umbo, velvety, sometimes wrinkled; spore print yellow; flesh and latex mild to slightly bitter, not acrid; gills close to subdistant; cut flesh staining dark violet; spores 7–11 × 7–9 μm; on the ground in coniferous woods; edibility unknown—*L. lignyotus* var. *nigroviolascens* (Atkinson) Hesler and Smith.

   16. Very similar to the previous choice, but cut flesh slowly staining paler violet; and gills typically somewhat marginate; spores 9–11 × 8.5–10 μm—*L. lignyotus* var. *marginatus* (Smith and Hesler) Hesler and Smith.

   16. Cap pale yellow, slimy to sticky when fresh, azonate; stalk slimy when fresh, pale yellow; gills crowded, decurrent, pale yellow; flesh and latex bitter, then slowly and slightly acrid, staining gills violaceous; spore print pale pinkish buff; spores 7–10 × 7–8 μm; edibility unknown—*L. aspideoides* Burlingham.

   16. Cap pale yellow, sticky, not slimy when fresh, azonate; stalk dry or sticky, not slimy when fresh, pale yellow; flesh and latex mild, then bitter, staining gills violaceous; spore print yellowish; spores 8–11 × 7–8.5 μm; edibility unknown—*L. aspideus* (Fries) Fries.

   16. Cap pale pinkish lilac to pale lilac-brown with lilac spots, sticky to slimy, typically azonate, 1⅛–4" (3–10 cm) wide; stalk up to ⅝" (1.6 cm) thick, lacking scrobiculations, whitish to pale buff; gills attached, close, creamy white; latex white, soon changing to cream and staining tissues dull lilac; flesh and latex mild then bitter; spores 7.5–11 × 6.5–8.5 μm—*L. uvidus* (Fries) Fries (see p. 179).

   16. Cap grayish buff to dull grayish lilac, brownish lilac or brownish gray, sticky when fresh, concentrically spotted and zoned, especially toward the margin, broadly funnel-shaped at maturity, 3½–7" (9–18 cm) wide; stalk ¾–1½" (2–4 cm) thick, scrobiculate, colored like the cap or paler; gills attached to decurrent, close, whitish to creamy white; latex whitish to creamy white, staining tissues dull lilac; flesh and latex weakly to distinctly acrid; spores 9–12 × 7–11 μm—*L. maculatus* Peck (see p. 173).

17. Exposed latex staining gills or drying pale green, bluish green, dull green, or olive → 18.

17. Exposed latex staining gills or drying some shade of gray or brown → 19.

   18. Cap pale yellow to pale ochre or pale pinkish cinnamon, sticky to slimy when fresh and moist, broadly convex with a depressed disc, azonate, 2–5½" (5–14 cm) wide; latex white, drying pale olive to olive; flesh and latex acrid; spores 8–10 × 6.5–8 μm; edibility unknown—*L. affinis* var. *viridilactis* (Kauffman) Hesler and Smith.

   18. Cap dark brown to gray-brown overall when young, soon fading to pinkish brown with a pinkish buff margin, often darker on the disc, slimy to sticky, with or without a small, pointed umbo, azonate, 1⅛–3⅛" (3–8.5 cm) wide; latex white, drying bluish green to olive; flesh and latex acrid; stalk sticky to slimy; spores 7.5–10 × 6–8 μm; edibility unknown—*L. mucidus* var. *mucidus* Burlingham. (Note: *L. mucidus* var. *mucidioides* Hesler and Smith, edibility unknown, is nearly identical but has a cream spore print; both varieties occur on soil under conifers.)

   18. Cap whitish to cream, often with yellow-brown stains in age, dry, smooth to finely cracked, azonate, 1¾–4¾" (4.5–12 cm) wide; gills crowded, often forked, dull

cream; stalk white to cream; latex whitish, drying pale green; flesh and latex strongly acrid; spores 6.5–9 × 5.5–6.5 μm; edibility unknown—*L. piperatus* var. *glaucescens.*

18. Cap pale pinkish cinnamon to pinkish buff, often with grayish red tints, becoming pale cinnamon to orange-brown and finally dull brick-red in age, often streaked and spotted white, especially on the margin, dry, somewhat velvety, azonate, 2⅜–6" (6–15.5 cm) wide; flesh white, slowly pinkish, rapidly staining vinaceous red in FeSO₄; latex white, staining gills dull green to olive and then slowly dull brown; flesh and latex acrid—*L. allardii* Coker (see p. 166).

18. Cap dark pinkish brown to dark pinkish gray, paler in age, faintly zoned with slightly darker colors, dry, scurfy, ¾–2¾" (2–7 cm) wide; odor resembling anise or coconut; latex white in young specimens, watery in mature specimens, drying pale grayish green; flesh and latex slowly acrid; spores 6.5–8 × 5.5–6.5 μm; edibility unknown—*L. hibbardae* var. *glaucescens* Hesler and Smith.

18. Cap orange-red to brick-red, becoming pale reddish orange in age, distinctly zoned with bay-red to orange, scurfy to nearly smooth, dry, 2–3⅛" (5–8 cm) wide; latex copious, white, drying bluish green; flesh and latex acrid; gills cinnamon-buff, becoming pinkish cinnamon; spores 6–8 × 5.5–7.5 μm; edibility unknown—*L. peckii* var. *glaucescens* Hesler and Smith.

19. Odor of gills and flesh of mature specimens fishy or unpleasant and resembling spoiled crab; flesh and latex mild → 20.

19. Odor fragrant or not distinctive; flesh and latex acrid (sometimes weakly or slowly) → 21.

20. Cap dark orange-brown to cinnamon-brown at the center, paler orange-brown toward the margin, smooth when young, becoming finely wrinkled in age or more so in wet weather; stalk up to 4½" (11.5 cm) long, pale orange-brown to dull orange—*L. volemus* var. *volemus* (Fries) Fries (see p. 180).

20. Cap reddish brown to vinaceous brown, paler and sometimes orange-cinnamon on the margin, often with a whitish bloom, distinctly wrinkled to finely corrugated; stalk up to 4⅜" (11 cm) long, pale grayish cinnamon to pale pinkish cinnamon—*L. corrugis* Peck (see p. 168).

20. Cap whitish to buff, becoming brownish in age, convex at first, often shallowly depressed in age, dry, slightly velvety; flesh white, staining brown; latex copious, watery-white to white, staining gills brown—*L. luteolus* Peck (see p. 173).

21. Cap whitish, darkening to dull brownish ochre in age → 22.

21. Cap pinkish buff, pinkish cinnamon, orange-brown to dull brick-red → 23.

21. Cap pale yellow, yellow to orange-yellow or orange-tawny to dull orange, faintly to distinctly zoned → 24.

21. Cap brownish yellow to olive, dark grayish green, tan, lilac-gray to lilac-brown or gray-brown to dark brown → 25.

22. Cap large, 3–10" (7.5–25.5 cm) wide, convex, becoming broadly funnel-shaped, dry, scaly in age, whitish, darkening to dull brownish ochre; margin distinctly inrolled and cottony when young; gills close to subdistant, whitish to pale ochre; stalk up to 1⅛" (3 cm) thick, white—*L. deceptivus* Peck (see p. 169).

22. Cap medium to large, 2–5⅝" (5–15 cm) wide, convex, becoming nearly flat with a depressed center, dry, felty to somewhat velvety, whitish, developing pinkish buff or yellowish tints in age; margin incurved when young and often remaining so well into maturity, not cottony and inrolled—*L. subvellereus* Peck (see p. 178).

23. Cap small to medium, 1–2¾" (2.5–7 cm) wide, convex with a depressed center,

becoming somewhat funnel-shaped, dark vinaceous buff to pale pinkish cinnamon, fading to pale vinaceous tan to vinaceous gray; flesh pale pinkish buff, odor fragrant; latex white, staining gills vinaceous brown; flesh and latex mild at first, then slowly bitter, and finally weakly acrid; gills slightly decurrent, pale to dark vinaceous buff; stalk pinkish buff near the apex, darker toward the base; spores 7.5–10 × 6–7.5 μm; edibility unknown—*L. subdulcis* (Fries) S. F. Gray.

23. Cap medium to large, 2–6" (5–15.5 cm) wide, faintly to distinctly zoned, color variable, typically dull brick-red or brownish red, but also orange-brown or dull reddish orange with darker bay-red to orange-red zones; gills pale buff at first, soon tinged pinkish cinnamon—*L. peckii* Burlingham (see p. 175).

23. Cap medium to large, 2⅜–6" (6–15.5 cm) wide, azonate, pale pinkish cinnamon to pinkish buff, often with grayish red tints, becoming pale cinnamon to orange-brown, and finally dull brick-red in age, often streaked and spotted white, especially on the margin, dry, somewhat velvety; flesh white, slowly pinkish, rapidly staining vinaceous red in FeSO₄; latex white, staining gills dull green to olive and then slowly dull brown; flesh and latex acrid—*L. allardii* Coker (see p. 166).

24. Cap surface dry, nearly smooth, coated with a thin layer of tiny white silky fibrils at first, becoming pale yellow to orange-yellow over the disc and gradually paler to cream along the margin, streaked and spotted darker orange-tawny, darkening in age to orange-yellow or dull orange, with faint to distinct orange-tawny zones or azonate; flesh white, odor pungent to somewhat fragrant; latex drying bright gray to pale brownish gray on gill edges, gills staining tawny-brown when bruised—*L. zonarius* (St. Amans) Fries *sensu* Neuhoff (see p. 180).

24. Cap sticky to slimy when fresh and moist, scurfy-roughened, somewhat felty, and often with a whitish to silvery sheen when dry, yellow-orange to dull orange, faintly zoned; flesh whitish, odor faintly fragrant; latex white, staining gills rusty brown—*L. agglutinatus* Burlingham (see p. 166).

25. Cap dry to slightly sticky, pale brownish yellow with olive-green tints, darker olive-brown on the disc and paler on the margin, 2–4" (5–10 cm) wide, broadly convex and centrally depressed; flesh white with a pinkish tinge; latex white, staining gills brown; gills white to pale yellow; stalk colored like the cap and scrobiculate; spores 6–8 × 5.5–6.5 μm; edibility unknown—*L. sordidus* Peck.

25. Cap dry, olive to dark grayish green with darker olive to dark green concentrically arranged spots; gill edges typically olive-brown to greenish—*L. atroviridis* Peck (see p. 168).

25. Cap slimy and sticky when fresh, lilac-brown when young, fading to lilac-tan or pale lilac-gray and finally pale tan—*L. argillaceifolius* Hesler and Smith (see p. 167).

25. Cap dry, slightly sticky when wet, dark brown to gray-brown when young, in age fading to pinkish brown or lilac-gray; latex white, changing to olive-gray and staining gills olive-gray to grayish brown—*L. vietus* (Fries) Fries (see p. 179).

26. Cap olive-brown to brown with reddish tints on the disc; very small, ¼–⅝" (6–16 mm) wide, convex and soon nearly flat, often with a tiny umbo, translucent-striate, dry, latex white; flesh and latex acrid; spores 6–8 × 5–6.5 μm; on the ground, usually under alder; edibility unknown—*L. obscuratus* (Lasch) Fries.

26. Cap apricot-orange to orange, often paler along the margin, convex to nearly flat, smooth, sticky, ½–2½" (1.3–6.5 cm) wide; flesh pale buff, odor not distinctive; latex white; flesh and latex mild; gills attached, pale yellow; stalk orange-buff; spores 7–9 × 6–7 μm; under conifers, usually among sphagnum mosses—*L. splendens* Hesler and Smith.

26. Cap dull purple to purple-gray, becoming vinaceous to purple-brown, convex, then soon depressed, covered with tiny, dark scales overall, 1⅛–3⅛" (3–8 cm) wide; flesh pale vinaceous, slightly fragrant or odor not distinctive; latex white; flesh and latex mild to slightly acrid; gills attached to slightly decurrent, subdistant to distant, whitish, then tinted vinaceous; stalk colored like the cap or paler; spores 6.5–8 × 5.5–7 μm; edibility unknown—*L. purpureo-echinatus* Hesler and Smith.

26. Not with the above combinations of characters → 27.

27. Cap margin bearded with coarse, white hairs on young specimens → 28.

27. Cap margin not bearded; odor of gills and flesh sweet, fragrant, often like maple sugar or burnt sugar, anise, or coconut → 29.

27. Cap margin not bearded; odor of gills and flesh not sweet or fragrant; cap and stalk white, cream, pale yellow, or pale ochre-yellow → 32.

27. Not with the above combinations of characters → 33.

28. Cap azonate, white to cream except for the disc, which may be reddish orange to vinaceous, fibrillose except for the disc; marginal hairs white; stalk whitish when young, becoming ochraceous from the base up in age, apex tinged flesh-pink; spores 6–8.5 × 5–6.5 μm—*L. pubescens* Fries (see p. 175).

28. Cap typically zoned, pale pink with a whitish margin, fading to whitish and typically with dull pink bands, fibrillose except on the disc, dry except sticky on the disc, centrally depressed, 2–4¾" (5–12 cm) wide; marginal hairs white; flesh white to pale flesh-pink; latex white; flesh and latex acrid; gills slightly decurrent, whitish tinged vinaceous; stalk dry, smooth, colored like the cap or paler; spores 7.5–10 × 6–7.5 μm; suspected to be poisonous—*L. torminosus* (Fries) S. F. Gray.

29. Odor of gills and flesh sweet, fragrant, often like maple sugar or burnt sugar → 30.

29. Odor of gills and flesh resembling anise or coconut → 31.

30. Cap ¾–1¾" (2–4.5 cm) wide, broadly conic to broadly convex and depressed, usually with a small pointed umbo, smooth, moist to dry, dark red-brown when young, fading to dull brick-red, azonate; flesh fragrant; latex white to whey-like; flesh and latex slightly disagreeable to bitter but typically not acrid; gills pinkish cinnamon to dull vinaceous red; stalk colored like the cap and paler near the apex; spores 7–9 × 6–8 μm; usually with conifers, especially hemlock and pine; edibility unknown—*L. camphoratus* (Fries) Fries.

30. Cap 1⅛–4" (3–10 cm) wide, convex to nearly flat and often shallowly depressed, dry, pale to dark brown, often zoned, fading in age; disc glaucous; flesh sweet, fragrant, often like burnt sugar and strong when dried; latex white or watery; flesh and latex mild; gills whitish, sometimes stained yellowish or reddish; stalk colored like the cap; spores 7–10 × 6–8 μm; in conifer or mixed woods; edible—*L. mutabilis* Peck.

30. Cap 1¼–4⅜" (3–11 cm) wide, surface coated with a layer of tiny silky fibrils and dark purplish brown or purplish gray when young, becoming dark red-brown at the center in age, obscurely zoned or azonate; flesh sweet, fragrant, often like burnt sugar and strong when dried; latex white in young specimens, soon watery as the mushroom matures; flesh and latex slowly and mildly acrid; stalk coated with whitish fibrils when young, becoming colored like the cap and darkening progressively toward the apex as it ages—*L. quietus* var. *incanus* Hesler and Smith (see p. 176).

31. Cap pale vinaceous gray to pale lilac-gray or whitish, convex, becoming plane and shallowly depressed to funnel-shaped; dry, coated with tiny matted fibers, ¾–3⅛" (2–8 cm) wide; KOH on the cuticle staining dull pale yellow; flesh pale buff, odor of coconut;

latex white; flesh and latex slowly and slightly acrid; gills pinkish buff to pinkish cinnamon; stalk whitish to pale vinaceous gray; on the ground under birch and alder or in mixed woods—*L. glyciosmus* (Fries) Fries (see p. 170).

31. Cap dark pinkish brown to dark pinkish gray, fading to paler pinkish brown to pinkish gray in age; flesh whitish, odor fragrant, resembling anise or coconut—*L. hibbardae* var. *hibbardae* Peck (see p. 171).

    32. Cap sticky to slimy when fresh; pale ochre-yellow to pale ochre or pale pinkish cinnamon, lighter toward the margin, azonate, smooth; gills close—*L. affinis* var. *affinis* Peck (see p. 166).

    32. Cap slimy and sticky when fresh; whitish to cream with pale yellow zones and spots, tawny on the disc in age, smooth; margin finely pubescent when young; on the ground under oak; gills crowded—*L. maculatipes* Burlingham (see p. 173).

    32. Cap dry, white to creamy white, often stained dull tan, azonate, smooth, convex, becoming depressed to broadly funnel-shaped, 2–6" (5–15.5 cm) wide; margin not cottony or inrolled; flesh white; latex white; flesh and latex strongly acrid; gills crowded, attached, white; stalk dry, white; spores 5–7 × 5–5.5 μm; on soil in hardwoods; edible—*L. piperatus* var. *piperatus* (Fries) S. F. Gray.

    32. Cap white to whitish, sometimes with pinkish to lavender or brownish stains, usually with narrow darker zones near the margin; convex, becoming depressed to broadly funnel-shaped, typically dry, covered with tiny, matted fibrils, 2⅜–7" (6–18 cm) wide; margin inrolled at first, with tiny matted fibers; flesh white; latex white; flesh and latex slowly and strongly acrid; gills close to crowded, attached to slightly decurrent, gills pale pink to pale vinaceous brown; stalk white, dry; spores 6–8 × 4–5 μm; on the ground under willow and poplar; edibility unknown—*L. controversus* (Fries) Fries.

    32. Cap white, staining ochraceous to tan, dry, azonate, somewhat velvety, convex, becoming depressed, then broadly funnel-shaped, 2–6" (5–15.5 cm) wide; margin inrolled to incurved when young but not cottony; flesh white, taste strongly acrid; latex white, taste mild to slightly bitter but typically not acrid, drying yellowish on the gills; gills close at first, soon becoming subdistant to distant, whitish to cream, developing brown stains in age; stalk tapered toward the base, dry, white; spores 7.5–10 × 6.5–8.5 μm; on the ground under hardwoods; edibility unknown. *L. vellereus* (Fries) Fries *sensu lato*.

33. Flesh and latex mild or bitter at first, remaining mild or slowly slightly to strongly acrid → 34.

33. Flesh and latex quickly acrid to strongly acrid → 38.

    34. Flesh and latex mild and remaining so; cap orange-brown to dull cinnamon, dry, smooth to slightly velvety; flesh white, odor not distinctive; gills distant at maturity, attached to slightly decurrent, white to cream; stalk pale orange-brown to orange-yellow—*L. hygrophoroides* var. *hygrophoroides* Berkeley and Curtis (see p. 172).

    34. Flesh and latex mild or bitter at first, then slowly slightly to strongly acrid → 35.

35. Cap olive-gray to olive-buff with lilac tinges or dark brown to yellow-brown to tan or purple-brown to purple-gray → 36.

35. Cap pale reddish brown, pale cinnamon, brick-red to rusty red, dark chestnut or dull orange-cinnamon to pale dull orange → 37.

    36. Flesh and latex mild, then slighty to distinctly acrid; cap olive-gray to olive-brown when young, becoming pale olive-gray with a lilac tinge, olive-buff to pinkish buff on the margin at maturity, azonate to faintly zoned, sticky to slimy when wet,

becoming dry; stalk pinkish buff to pale pinkish gray, whitish at the apex—
L. cinereus var. *fagetorum* Hesler and Smith (see p. 168).

36. Flesh and latex mild, then slightly acrid; cap brown to yellow-brown to tan,
     wrinkled, velvety when dry; stalk dark brown to yellow-brown, paler and pleated
     at the apex—*L. gerardii* var. *gerardii* Peck (see p. 170).

36. Flesh and latex mild or slowly slightly acrid, especially in young specimens; cap
     dark purple-brown to purple-gray with a small, pointed umbo, fading in age to
     pale purple-brown to red-brown at the center and purplish buff to reddish buff
     on the margin, dry, covered with fibers and tiny scales; on decaying wood, often
     moss-covered, or among sphagnum in bogs—*L. griseus* Peck (see p. 171).

37. Flesh and latex bitter at first, then acrid; latex white, slowly turning sulfur-yellow; cap
     pale cinnamon to pinkish cinnamon or pale reddish brown, typically pitted and stained
     with darker pinkish cinnamon, dry to slightly sticky when wet, broadly convex and
     shallowly depressed with a small umbo, azonate, 1⅛–3" (3–8 cm) wide; gills attached to
     slightly decurrent, close, whitish, soon flushed pinkish cinnamon; stalk dry, pinkish buff
     at first, then pinkish cinnamon, with white basal mycelium; spores 8–11 × 7–8 μm;
     edibility unknown—*L. imperceptus* Beardslee and Burlingham.

37. Flesh and latex mild, then slowly strongly acrid; cap brick-red to rusty red or dark
     chestnut, paler toward the margin, moist or dry, convex, becoming shallowly depressed,
     typically with a small umbo, 1⅛–3½" (3–9 cm) wide; margin often weakly crimped at
     maturity; KOH on cuticle staining dull olive; latex white, unchanging; gills slightly
     decurrent to decurrent, close, whitish, then slowly dull reddish to vinaceous red; stalk
     colored like the cap but paler, with coarse, whitish hairs at the base; spores 8–10 × 7–8
     μm; on the ground usually among mosses in conifer woods; edibility unknown—
     L. hepaticus Plowright.

37. Flesh and latex mild, then slowly acrid; cap dull orange-cinnamon to reddish orange or
     dark apricot, fading to pale dull orange to orange-buff, convex, becoming depressed,
     then funnel-shaped, azonate, or faintly zoned, ¾–2¾" (2–7 cm) wide; flesh whitish,
     soon staining yellow where cut; latex white, slowly turning yellow; gills decurrent, close
     to crowded, pale pinkish buff, becoming spotted pinkish, then darkening to orange-
     cinnamon in age; stalk colored like the cap but paler, with coarse, whitish hairs at the
     base; spores 7–9 × 6–7.5 μm; among sphagnum mosses or on the ground in woods,
     usually near birch; edibility unknown—*L. thejogalus* (Fries) S. F. Gray.

38. Cap dry, pale ashy gray with a darker gray disc, lacking olive tones, fading in age,
     smooth, sticky when fresh, convex at first, then centrally depressed and broadly
     funnel-shaped, ¾–2⅜" (2–6 cm) wide; spore deposit white; latex white, not stain-
     ing the gills; stalk smooth, pale ashy gray; spores 6.5–9 × 5–6.5 μm; on the ground
     in woods; edibility unknown—*L. cinereus* var. *cinereus* Peck.

38. Cap slimy and sticky when fresh, shiny when dry, dark reddish brown on the
     disc, pinkish cinnamon on the margin; gills whitish to buff when young, becom-
     ing orange-yellow at maturity; stalk colored like the cap or paler, scrobiculate—
     L. hysginus Fries (see p. 172).

38. Cap dry, dark brownish red to dark bay-red, zoned with slightly darker bands, 1⅝–
     4¾" (4–12 cm) wide; stalk not scrobiculate; latex white, unchanging—*L. rufus* var.
     rufus (Fries) Fries (see p. 177).

38. Cap dry to moist but not sticky or slimy; pinkish gray to pale gray with gray-
     brown spots and streaks, typically zonate → 39.

39. Gills close to subdistant, attached, crossveined, pale pinkish buff, becoming pale
     pinkish cinnamon; stalk 1⅛–2" (3–5 cm) long, ⅜–1⅛" (1–3 cm) thick, margin with very

inconspicuous matted fibers (use a hand lens) when young—*L. pseudoflexuosus* Hesler and Smith (see p. 175).

39. Gills subdistant to distant, attached to slightly decurrent, crossveined, pale yellowish, then ochre-yellow with a slight pinkish tint; stalk 1⅛–2¾" (3–7 cm) long, ¼–½" (5–12 mm) thick, margin lacking matted fibers when young; cap grayish olive-brown to pale gray, usually with a purplish tint, with watery, gray-brown zones or sometimes azonate; flesh white; latex white; stalk pale pinkish gray to gray; spores 6–7.5 × 5–6 μm; on the ground in hardwoods and mixed woods; edibility unknown—*L. pyrogalus* (Fries) Fries.

Lactarius affinis var. affinis   Peck

CAP: 2–4⅜" (5–11 cm) wide, convex, becoming broadly convex, then shallowly depressed at the center in age; surface smooth, sticky to slimy when fresh, pale ochre-yellow to pale ochre or pale pinkish cinnamon, lighter toward the margin; flesh white; odor not distinctive; latex white; flesh and latex acrid.

GILLS: attached to slightly decurrent, close, broad, white to cream.

STALK: 1⅛–3⅛" (3–8 cm) long, ½–¾" (1.2–2 cm) thick, nearly equal, hollow in age, smooth, sticky when fresh, white.

SPORE PRINT: white.

MICROSCOPIC FEATURES: spores 9–11 × 7–8 μm, elliptic, with warts and ridges, hyaline, amyloid.

FRUITING: scattered or in groups on the ground in conifer or hardwoods; June–September; occasional.

EDIBILITY: unknown.

Lactarius agglutinatus   Burlingham

CAP: 1½–3½" (4–9 cm) wide, convex, becoming broadly depressed and funnel-shaped; margin inrolled and scurfy when young, becoming uplifted in age; surface faintly zoned, scurfy-roughened and slimy when moist, becoming dry and glaucous to dull, yellow-orange to dull orange; flesh whitish, slowly staining rusty brown when cut or bruised; odor faintly fragrant; latex white, unchanging; flesh and latex acrid.

GILLS: attached to slightly decurrent, close, often forked near the stalk, whitish to pinkish cream, staining rusty brown when cut or bruised, with several tiers of lamellulae.

STALK: 1⅝–3" (4–7.5 cm) long, ½–¾" (1.3–2 cm) thick, nearly equal, smooth, dry, white flushed pale orange.

SPORE PRINT: pale yellow.

MICROSCOPIC FEATURES: spores 7–8 × 5–6.5 μm, elliptic, with warts and ridges, hyaline, amyloid.

FRUITING: solitary or scattered on the ground under hardwoods and mixed woods, usually with oak; July–September; occasional.

EDIBILITY: unknown.

Lactarius allardii   Coker

CAP: 2⅜–6" (6–15.5 cm) wide, convex, becoming broadly convex with a depressed center; margin incurved and remaining so well into maturity; surface dry, somewhat velvety to nearly smooth, azonate, pale pinkish cinnamon to pinkish buff, often with grayish red tints, becoming pale cinnamon to orange-brown, and finally dull brick-red in age, typically streaked and spotted white, especially on the margin; flesh firm, compact, white, slowly pinkish, rapidly staining vinaceous red in FeSO$_4$; odor not distinctive; latex white, slowly becoming greenish olive, then brownish, staining gills dull green to olive, then slowly dull brown; flesh and latex acrid.

GILLS: attached to decurrent, close to subdistant at first, becoming nearly distant at maturity, white to ivory-yellow.

STALK: ¾–2" (2–5 cm) long, ⅜–1⅛" (1–3 cm) thick, nearly equal, hollow, dry, smooth, colored like the cap or paler.

SPORE PRINT: white to whitish.

MICROSCOPIC FEATURES: spores 8–11 × 5–8 μm, ellipsoid to subglobose, with warts and ridges, hyaline, amyloid.

FRUITING: scattered or in groups on the ground in hardwoods and mixed woods; June–Septmeber; occasional.

EDIBILITY: unknown.

Lactarius aquifluus Peck

COMMON NAME: Burnt-sugar Milky.

CAP: 1⅛–5⅛" (3–15 cm) wide; broadly convex, becoming flat with a slightly depressed center in age, with or without an umbo; margin incurved when young, becoming expanded in age; surface dry, covered with tiny fibrils, orange-brown to light grayish brown to pale cinnamon-brown, zoned with thin bands of dark grayish brown or water spots; flesh pale whitish to tan, not staining when bruised; odor sweet, fragrant, like maple sugar or burnt sugar, latex watery, unchanging, not staining; flesh and latex mild.

GILLS: attached to the stalk, slightly decurrent, narrow to broad, close, often forked near the stalk, creamy white, becoming pale orange-tan to pale pinkish cinnamon.

STALK: 1⅝–3⅛" (4–8 cm) long, ⅜–¾" (1–2 cm) thick, nearly equal or enlarged at the base, stuffed, becoming hollow, dry, pruinose, colored like the cap, whitish at the base.

SPORE PRINT: white to cream.

MICROSCOPIC FEATURES: spores 6–9 × 5.5–7.5 μm, elliptic, with warts and ridges, hyaline, amyloid.

FRUITING: scattered, in groups, or in clusters on the ground in conifer and mixed woods and among sphagnum mosses in bogs; July–October; fairly common.

EDIBILITY: edible and quite good.

COMMENTS: formerly known as *L. helvus* var. *aquifluus.* Some books list this species as mildly poisonous or poisonous. As the late Alexander Smith pointed out in *The Mushroom Hunter's Field Guide,* comments indicating that this species is poisonous are based on a European species called *L. helvus* (poisonous), which does not occur in North America.

Lactarius argillaceifolius Hesler and Smith

CAP: 1⅝–7" (4–18 cm) wide; convex, becoming broadly convex with a depressed center; margin incurved, often remaining so into maturity; surface downy to finely hairy when young, becoming smooth in age, slimy and sticky when wet, lilac-brown when young, fading to lilac-tan or pale lilac-gray and finally pale tan or pinkish buff at the center; flesh white to buff; odor not distinctive; latex creamy white, staining the gills grayish brown to dark brown or olive-brown; flesh and latex mild to slowly acrid.

GILLS: attached to slightly decurrent, close, cream when young, developing pinkish tints near the margin, staining buff to olive-brown to dark brown and finally dark reddish brown in age or when cut.

STALK: 2⅜–3½" (6–9 cm) long, ⅝–1⅜" (1.5–3.5 cm) thick, nearly equal or tapering downward, slimy or dry, whitish, spotted and stained brownish in age.

SPORE PRINT: pinkish buff.

MICROSCOPIC FEATURES: spores 7–11 × 7–8 μm, nearly round to elliptic, with warts and ridges, hyaline, amyloid.

FRUITING: scattered or in groups on the ground under hardwoods, especially oak; July–October; occasional.

EDIBILITY: unknown.

Lactarius atroviridis Peck

CAP: 2⅜–5⅝" (6–15 cm) wide, broadly convex, becoming flat with a depressed center; margin incurved when young, remaining so into maturity; surface dry, scurfy to minutely scaly, pitted, various shades of green ranging from olive to dark grayish green, usually zoned with dark green spots arranged concentrically; flesh whitish to pinkish brown, thick to thin, not staining; odor not distinctive; latex white, very slowly turning greenish, staining gills dark grayish green to greenish brown; flesh and latex acrid.

GILLS: attached to slightly decurrent, close, cream to pinkish white, occasionally forked near the stalk, staining greenish gray to brownish; edges typically olive-brown to greenish.

STALK: ¾–3⅛" (2–8 cm) long, ⅜–1⅛" (1–3 cm) thick, nearly equal, hollow, dry, colored and streaked like the cap, scrobiculate.

SPORE PRINT: cream to buff.

MICROSCOPIC FEATURES: spores 7–10 × 6–9 μm, nearly round to elliptic, forming a partial reticulum with warts and ridges, hyaline, amyloid.

FRUITING: solitary to scattered on the ground under conifers and hardwoods; July–October; uncommon.

EDIBILITY: unknown.

Lactarius cinereus var. **fagetorum** Hesler and Smith

CAP: 1⅛–2¾" (3–7 cm) wide, convex, becoming nearly plane with a depressed center and broadly funnel-shaped in age; margin incurved when young; surface sticky to slimy when wet, becoming dry, olive-gray to olive-brown when young, becoming pale olive-gray with a lilac tinge, olive-buff to pinkish buff on the margin at maturity, azonate to faintly zoned; flesh white; odor not distinctive; latex white, often scanty; flesh and latex mild, then slightly to distinctly acrid.

GILLS: decurrent, crowded, white, becoming creamy white.

STALK: 1½–3" (4–7.5 cm) long, ⅜–⅝" (1–1.6 cm) thick, pinkish buff to pale pinkish gray, whitish at the apex, dry, hollow in age.

SPORE PRINT: pale yellow.

MICROSCOPIC FEATURES: spores 6–8 × 5–6 μm, elliptic, with warts and ridges, hyaline, amyloid.

FRUITING: solitary, scattered, or in groups on the ground, usually under beech; August–October; fairly common.

EDIBILITY: unknown.

Lactarius corrugis Peck

COMMON NAME: Corrugated-cap Milky.

CAP: 1½–7⅞" (4–20 cm) wide; convex, becoming broadly convex to nearly flat with a depressed center in age; margin often wrinkled; surface velvety, dry, reddish brown to vinaceous brown, paler and sometimes orange-brown on the margin, often with a whitish bloom, especially when young, distinctly wrinkled to finely corrugated; flesh whitish, staining brown; odor slightly to strongly fishy in mature mushrooms, often not distinctive in young specimens; latex copious, white, staining gills and flesh tawny-brown; flesh and latex mild.

GILLS: attached, close, occasionally forked, pale cinnamon to pale golden brown, staining darker brown when bruised.

STALK: 2–4⅜" (5–11 cm) long, ⅝–1⅛" (1.5–3 cm) thick, equal, solid, dry, velvety, pale grayish cinnamon to pale pinkish cinnamon, sometimes with a whitish bloom.

SPORE PRINT: white.

MICROSCOPIC FEATURES: spores 9–12 × 8.5–12 μm, nearly round, forming a partial reticulum with warts and ridges, hyaline, amyloid.

FRUITING: solitary to scattered on the ground in hardwoods and mixed woods; July–October; fairly common.

EDIBILITY: edible.

Lactarius croceus Burlingham

CAP: 2–4" (5–10 cm) wide, convex, becoming broadly convex to nearly flat with a depressed center; margin incurved and downy when young; surface smooth, slimy when wet, shiny when dry, bright orange to saffron-yellow initially, fading to paler yellow-orange or yellowish tan in age, zoned with dark orange-brown bands; flesh whitish, thick, staining sulfur-yellow before the latex changes color; odor not distinctive; latex white, turning sulfur-yellow, staining gills and flesh sulfur-yellow; flesh and latex bitter to acrid.

GILLS: attached to slightly decurrent, broad, close to subdistant, creamy buff to pale yellow-tan, staining sulfur-yellow where cut or bruised.

STALK: 1⅛–2⅜" (3–6 cm) long, ⅜–¾" (1–2 cm) thick, nearly equal, stuffed, becoming hollow, smooth, at times velvety at the base, colored like the cap or paler, sometimes spotted brownish.

SPORE PRINT: yellowish.

MICROSCOPIC FEATURES: spores 7.5–10 × 5.5–7.5 μm, elliptic, forming a partial reticulum with warts and ridges, hyaline, amyloid.

FRUITING: solitary or scattered on the ground under hardwoods, especially oak; July–October; occasional.

EDIBILITY: unknown.

Lactarius deceptivus Peck

COMMON NAME: Deceptive Milky.

CAP: 3–10" (7.5–25.5 cm) wide, convex, becoming broadly funnel-shaped; surface dry, smooth when young, whitish, often with yellowish or brownish stains, becoming coarsely scaly and darkening to dull brownish ochre in age; margin distinctly inrolled and cottony when young; flesh thick, white; odor pungent or not distinctive; latex white, unchanging, staining tissues brownish; flesh and latex strongly acrid.

GILLS: attached, close to subdistant, white at first, then cream to pale ochre.

STALK: 1½–4" (4–10 cm) long, up to 1⅛" (3 cm) thick, nearly equal or tapered downward, dry, scurfy to nearly smooth, white, staining brownish in age.

SPORE PRINT: white to pale ochre-buff.

MICROSCOPIC FEATURES: spores 9–13 × 7–9 μm, broadly ellipsoid, with warts and ridges, hyaline, amyloid.

FRUITING: solitary, scattered, or in groups on the ground in conifer or hardwoods, often under oak and hemlock; June–October; fairly common.

EDIBILITY: edible when thoroughly cooked.

Lactarius deterrimus Gröger

CAP: 1⅛–5⅛" (3–13 cm) wide; convex, becoming broadly convex to nearly flat in age, with a depressed center, often broadly funnel-shaped; margin incurved and pruinose when young, becoming naked and uplifted in age; surface smooth, slippery when wet, pinkish orange to carrot-orange, zoned with distinct bands of rusty orange, fading in age to pale whitish orange at the center, typically developing green spots; flesh whitish to pale yellowish orange, staining bright green in $FeSO_4$, staining green, then grayish or grayish brown when bruised; odor not distinctive; latex orange, slowly staining cut surfaces yellow-orange, then green, and finally grayish green; flesh and latex not distinctive or slightly acrid.

GILLS: attached to slightly decurrent, narrow, close, pinkish orange to carrot-orange, fading to pale salmon-buff in age, bruising green to gray-green.

STALK: 1⅝–3" (4–7.5 cm) long, ⅝–1⅛" (1.5–3 cm) thick, equal or tapered at the base, stuffed, becoming hollow in age, smooth, dry, colored like the cap, often with green spots and stains in age, whitish at the base.

SPORE PRINT: cream to whitish.

MICROSCOPIC FEATURES: spores 7.5–9 × 6–7 μm, broadly ellipsoid, with warts and ridges, hyaline, amyloid.

FRUITING: solitary, scattered, or in groups on the ground under conifers, especially pine; July–October; fairly common.

EDIBILITY: edible.

COMMENTS: often incorrectly called *L. deliciosus,* which is a similar European species.

Lactarius gerardii var. gerardii Peck

COMMON NAME: Gerard's Milky.

CAP: 1¼–5⅛" (3–13 cm) wide; convex, becoming broadly convex to flat, wrinkled, with a depressed center in age, with or without an umbo; margin becoming uplifted and wavy in age; surface dry, slightly sticky when wet, velvety when dry, brown to yellow-brown to tan; flesh white; odor not distinctive; latex white, unchanging, not staining; flesh and latex mild, then slightly acrid.

GILLS: attached to decurrent, broad, subdistant to distant, crossveined, whitish to cream.

STALK: 1⅜–3⅛" (3.5–8 cm) long, ⅜–¾" (8–20 mm) thick, nearly equal or enlarged downward, usually with an abruptly tapered base, stuffed when young, becoming hollow in age, dry, velvety, dark brown to yellow-brown, paler and pleated at the apex.

SPORE PRINT: white.

MICROSCOPIC FEATURES: spores 7–10 × 7.5–9 μm, round to elliptic, with warts and ridges, forming a partial reticulum, hyaline, amyloid.

FRUITING: scattered or in groups on the ground under hardwoods and conifers; July–October; occasional.

EDIBILITY: edible.

Lactarius glyciosmus (Fries) Fries

CAP: ¾–3½" (2–8 cm) wide, convex, becoming plane, then shallowly depressed to funnel-shaped; surface dry, coated with tiny, matted fibers, pale vinaceous gray to pale lilac-gray or whitish, cuticle staining dull pale yellow when KOH is applied; flesh pale buff; odor fragrant like coconut; latex white, unchanging, not staining gills or flesh; flesh and latex slowly and slightly acrid.

GILLS: close, subdecurrent to decurrent, pinkish buff to pinkish cinnamon, typically crossveined.

STALK: ¾–3½" (2–9 cm) long, ⅛–⅜" (3–10 mm) thick, nearly equal, dry, smooth or nearly so, whitish to pale vinaceous gray.

SPORE PRINT: pale cream to pinkish buff.

MICROSCOPIC FEATURES: spores 6–10 × 5–7 μm, ellipsoid to subglobose, with warts and ridges, hyaline, amyloid.

FRUITING: scattered or in groups on the ground or among mosses under birch and alder or in mixed woods; August–November; fairly common.

EDIBILITY: edible.

COMMENTS: compare with *L. hibbardae*.

Lactarius griseus Peck

CAP: ⅝–2" (1.5–5 cm) wide, convex, becoming broadly convex to nearly flat with a depressed center and a small, pointed umbo, often funnel-shaped in age; margin inrolled when young, becoming expanded to uplifted in age; surface dry, covered with fibers and tiny scales, dark purple-brown to purple-gray when young, fading in age to paler purple-brown to red-brown on the disc and purplish buff to reddish buff on the margin; flesh white to yellow; odor not distinctive; latex white, drying yellowish, not staining flesh or gills; flesh and latex mild or slowly acrid, especially in young specimens.

GILLS: decurrent, broad, close to subdistant, whitish, becoming pinkish or purplish buff in age, not staining when cut or bruised.

STALK: ¾–2½" (2–6.5 cm) long, ⅛–⅜" (3–10 mm) thick, equal or enlarged slightly downward, hollow, dry, colored like the gills at the apex, purplish below when young, becoming pale rosy brown in age, base coated with coarse white hairs.

SPORE PRINT: yellowish.

MICROSCOPIC FEATURES: spores 7–9 × 6–7 μm, subglobose to elliptic, with warts and ridges, hyaline, amyloid.

FRUITING: in groups or clusters on decaying wood or among sphagnum mosses in bogs; July–October; occasional.

EDIBILITY: unknown.

Lactarius hibbardae var. hibbardae Peck

CAP: ⅝–4" (1.5–10 cm) wide; convex, becoming broadly convex to flat with a depressed center, sometimes with an umbo; margin incurved, becoming expanded to uplifted and wavy in age; surface fibrillose to scurfy, dry, not sticky or slimy, dark pinkish brown to dark pinkish gray, becoming paler pinkish brown to pinkish gray in age, faintly zoned with slightly darker colors, staining yellowish to orange with KOH and yellow with NH_4OH; flesh whitish with tinges of pinkish gray or pinkish brown; odor fragrant, like anise or coconut; latex white when young, becoming watery; flesh and latex acrid.

GILLS: attached to slightly decurrent, close to crowded, pale pinkish buff to cream, not staining or slowly staining very slightly brownish when cut or bruised.

STALK: ¾–4" (2–10 cm) long, ⅛–⅝" (4–15 mm) thick, nearly equal or tapering downward, hollow, pruinose, dry, colored like the gills or like the cap in age, whitish at the base.

SPORE PRINT: white to cream.

MICROSCOPIC FEATURES: spores 6.5–9 × 5–6.5 μm, elliptic, with warts and ridges, forming a partial reticulum, hyaline, amyloid.

FRUITING: solitary to scattered on the ground under conifers and hardwoods or among sphagnum mosses; July–October; fairly common.

EDIBILITY: unknown.

Lactarius hygrophoroides var. **hygrophoroides** Berkeley and Curtis
COMMON NAME: Hygrophorus Milky.
CAP 1⅛–4" (3–10 cm) wide, broadly convex, becoming nearly flat with a depressed center
in age, sometimes broadly funnel-shaped; margin incurved, becoming uplifted in age;
surface dry, smooth to slightly velvety, sometimes slightly wrinkled toward the margin
in age, orange-brown to dull cinnamon; flesh white; odor not distinctive; latex white,
unchanging, not staining gills or flesh; flesh and latex mild.
GILLS: attached to slightly decurrent, broad, distant at maturity, crossveined, white to
cream to yellowish buff.
STALK: 1⅛–2" (3–5 cm) long, ¼–⅝" (5–15 mm) thick, nearly equal, solid, dry, pale orange-
brown to orange-yellow.
SPORE PRINT: white.
MICROSCOPIC FEATURES: spores 7.5–10.5 × 6–7.5 μm, elliptic, with warts and ridges,
hyaline, amyloid.
FRUITING: solitary, scattered, or in groups on the ground under hardwoods; July–
October; occasional.
EDIBILITY: edible.

Lactarius hysginus var. **hysginus** Fries
CAP: 1⅛–3½" (3–9 cm) wide, convex, becoming broadly convex to flat with a depressed
center, often broadly funnel-shaped; margin incurved when young, becoming
expanded to uplifted in age; surface smooth, sticky to slimy when fresh, shiny when
dry, dark reddish brown at the center, pinkish cinnamon on the margin, fading in age;
flesh whitish, staining pinkish brown in FeSO₄; odor not distinctive; latex white,
unchanging, not staining flesh or gills; flesh and latex acrid.
GILLS: attached to slightly decurrent, narrow, close to crowded, whitish to buff when
young, becoming orange-yellow at maturity, not staining when cut or bruised.
STALK: 1⅛–3⅛" (3–8 cm) long, ⅜–⅝" (1–1.5 cm) thick, equal, hollow, slimy when young,
becoming dry, colored like the cap or paler, whitish at the apex, scrobiculate.
SPORE PRINT: yellowish.
MICROSCOPIC FEATURES: spores 5.5–7.5 × 5.5–7 μm, round to elliptic, with warts and
ridges, forming a partial reticulum, hyaline, amyloid.
FRUITING: scattered or in groups on the ground under conifers; July–October;
occasional.
EDIBILITY: unknown.
COMMENTS: *L. hysginus* var. *americanus* (edibility unknown) has pale yellow gills when
young, becoming tan to dull cinnamon in age.

Lactarius lignyotus Fries
COMMON NAME: Chocolate Milky.
CAP: ¾–4" (2–10 cm) wide, convex with an incurved margin and a small pointed umbo
when young, becoming nearly plane or with a depressed disc and retaining the umbo;
surface velvety, dry, usually wrinkled, blackish brown to yellow-brown when young,
fading in age; flesh white, staining rosy pink to dull reddish when cut; latex white,
staining gills reddish; flesh and latex mild to slightly bitter, not acrid.
GILLS: attached to somewhat decurrent, close to subdistant, white, slowly becoming pale
ochraceous.
STALK: 2–4½" (5–11 cm) long, ¼–½" (5–13 mm) thick, enlarged downward or nearly
equal, solid, with distinct longitudinal ridges at the apex, dry, velvety, blackish brown
to pale yellow-brown with a white base.

SPORE PRINT: bright ochre.

MICROSCOPIC FEATURES: spores 9–11 × 9–10 μm, nearly globose, forming a partial reticulum with warts and ridges, hyaline, amyloid.

FRUITING: scattered or in groups on the ground and among mosses in conifer woods and bogs; August–October; fairly common.

EDIBILITY: unknown.

COMMENTS: several varieties of this species are recognized.

Lactarius luteolus Peck

COMMON NAME: Buff Fishy Milky.

CAP: 1–3⅛" (2.5–8 cm) wide, convex to nearly plane, often shallowly depressed in age; surface dry, slightly velvety, azonate, whitish to buff with a white bloom, becoming brownish in age; flesh white, staining brown; odor not distinctive in young specimens, soon becoming fishy or unpleasant and resembling spoiled crab; latex copious, watery white to white, staining gills brown; flesh and latex mild.

GILLS: attached to slightly decurrent, close, white becoming cream, staining yellow-brown to brown.

STALK: 1–2⅜" (2.5–6 cm) long, ¼–¾" (6–20 mm) thick, tapered downward or nearly equal, solid to stuffed, dry, slightly velvety, whitish to buff, staining brown.

SPORE PRINT: white to cream.

MICROSCOPIC FEATURES: spores 7–9 × 5.5–7 μm, elliptic, with warts and ridges, sometimes forming a partial reticulum, hyaline, amyloid.

FRUITING: solitary, scattered, or in groups on the ground in hardwoods and mixed woods; June–October; occasional.

EDIBILITY: edible.

Lactarius maculatipes Burlingham

CAP: 2–3½" (5–9 cm) wide, broadly convex, becoming nearly flat with a depressed center, often broadly funnel-shaped in age; margin incurved at first, becoming expanded and uplifted; surface finely pubescent on the margin when young, becoming smooth, slimy and sticky when fresh, whitish to cream overall, becoming tawny on the disc in age, with pale yellow zones and spots that darken in age; flesh white; odor not distinctive; latex white, staining gills and flesh yellowish; flesh and latex slowly and sometimes weakly acrid.

GILLS: decurrent, narrow, crowded, whitish, becoming pinkish buff, staining yellowish to pale ochre when cut or bruised.

STALK: 1¼–3⅛" (3–8 cm) long, ⅜–1" (1–2.5 cm) thick, tapering downward, slimy, colored like the cap, spotted and streaked with darker tan, staining yellow when bruised or in age.

SPORE PRINT: pinkish buff to yellowish.

MICROSCOPIC FEATURES: spores 6.5–8 × 6–7.5 μm, nearly round to elliptic, with warts and ridges, hyaline, amyloid.

FRUITING: scattered or in groups on the ground under oak, most commonly encountered in the southern part of the region; July–December; occasional.

EDIBILITY: unknown.

Lactarius maculatus Peck

CAP: 3½–7" (9–18 cm) wide, convex when young, becoming centrally depressed, then broadly funnel-shaped at maturity; surface smooth, sticky when fresh, concentrically spotted and zoned, especially toward the margin, grayish buff to dull grayish lilac, brownish lilac or brownish gray; flesh pale grayish; odor not distinctive; latex whitish to creamy white, staining tissues dull lilac; flesh and latex weakly to distinctly acrid.

GILLS: attached to decurrent, close, whitish to creamy white, darker in age, sometimes forked.

STALK: 2–4¾" (5–12 cm) long, ¾–1½" (2–4 cm) thick, nearly equal, scrobiculate, colored like the cap or paler, smooth but not sticky, hollow in age.

SPORE PRINT: white.

MICROSCOPIC FEATURES: spores 9–12 × 7–11 μm, broadly ellipsoid, with warts and ridges, hyaline, amyloid.

FRUITING: solitary, scattered, or in groups on the ground in hardwoods and mixed woods; August–October; infrequent.

EDIBILITY: unknown.

Lactarius oculatus (Peck) Burlingham

COMMON NAME: Eyespot Milky.

CAP: ⅜–1¾" (1–4.5 cm) wide, broadly convex, becoming nearly flat in age with a slightly depressed center and small, pointed umbo; surface sticky when fresh, shiny when dry, dark reddish brown to orange-brown at the center, paler toward the margin, fading to pale orange-brown but the umbo usually retaining the dark reddish brown color well into maturity and resembling an eyespot; flesh thin, fragile, watery, pinkish brown; odor not distinctive; latex scanty, watery to whey-like, unchanging; flesh and latex very slowly acrid then fading.

GILLS: attached to slightly decurrent, broad, close to subdistant, cream to pinkish buff, staining light pinkish brown to dull cinnamon in age.

STALK: ¾–1¾" (2–4.5 cm) long, ⅛–⅜" (3–10 mm) thick, nearly equal or enlarged slightly downward, hollow, smooth, moist but not sticky, dull pinkish brown to pinkish orange, often with whitish mycelium at the base.

SPORE PRINT: pale yellow.

MICROSCOPIC FEATURES: spores 7.5–9 × 6–7 μm, elliptic, with warts and ridges, hyaline, amyloid.

FRUITING: scattered or in groups in sphagnum mosses in bogs or on the ground under conifers, especially spruce and pine; July–October; fairly common.

EDIBILITY: edible.

Lactarius paradoxus Beardslee and Burlingham

COMMON NAME: Silver-blue Milky.

CAP: 2–3⅛" (5–8 cm) wide, broadly convex, becoming flat with a depressed center, often funnel-shaped in age; margin incurved at first, becoming expanded to uplifted; surface smooth, slimy to sticky when wet, with a silvery sheen when young, zoned with bands of grayish blue, grayish purple, green and blue, staining green when bruised; flesh thick, whitish with greenish to bluish tints, slowly staining greenish when cut or bruised; odor not distinctive; latex scanty, dark wine-brown, staining tissues green; flesh and latex mild or slightly acrid.

GILLS: attached to slightly decurrent, narrow to broad, close, occasionally forked near the stalk, pinkish orange, staining blue-green when bruised.

STALK: 1–1¼" (2–3 cm) long, ⅜–⅝" (1–1.5 cm) thick, nearly equal or tapered downward, hollow, dry, colored like the cap, staining green when bruised or in age.

SPORE PRINT: cream to yellow.

MICROSCOPIC FEATURES: spores 7–9 × 5.5–6.5 μm, elliptic, forming a partial reticulum with warts and ridges, hyaline, amyloid.

FRUITING: solitary or scattered on the ground or in grass under pine, oak, and palmetto palm; August–December; occasional.

EDIBILITY: edible.

Lactarius peckii Burlingham

COMMON NAME: Peck's Milky.

CAP: 2–6" (5–15.5 cm) wide, broadly convex with a depressed center; margin inrolled at first, becoming decurved and remaining so well into maturity; surface dry, velvety when young, then scurfy to nearly smooth; color variable, typically dull brick-red or brownish red but also orange-brown to orange-red or dull reddish orange, often paler toward the margin, faintly to distinctly zoned with darker bay-red to orange-red bands, often fading in age; flesh pale vinaceous brown; latex copious, white, staining tissues reddish brown; flesh and latex extremely acrid.

GILLS: decurrent, close, narrow, pale cinnamon-buff at first, soon tinged pinkish cinnamon and darkening in age, staining rusty brown to reddish brown.

STALK: ¾–2⅜" (2–6 cm) long, ⅜–1" (1–2.5 cm) thick, nearly equal, hollow in age, covered with a whitish bloom when young, colored like the cap but usually paler, often spotted reddish brown or dull orange.

SPORE PRINT: white.

MICROSCOPIC FEATURES: spores 6–7.5 μm, globose to subglobose, with warts and ridges, forming a partial to complete reticulum, hyaline, amyloid.

FRUITING: solitary, scattered, or in groups on the ground in oak woods; July–September; occasional to fairly common.

EDIBILITY: unknown.

COMMENTS: this species has gills that are among the darkest in the genus.

Lactarius pseudoflexuosus Hesler and Smith

CAP: 1½–4" (4–10 cm) wide, convex, becoming nearly flat with a depressed center in age; surface sticky when wet, soon moist to dry, smooth at maturity, pinkish gray to pale gray with gray-brown spots and streaks, faintly to distinctly zoned; margin inrolled with very inconspicuous matted fibers (use a hand lens) when young; flesh whitish; odor not distinctive; latex dull creamy buff; flesh and latex acrid.

GILLS: attached, close to subdistant, crossveined, pale pinkish buff when young, becoming pale pinkish cinnamon at maturity, sometimes forking.

STALK: 1⅛–2" (3–5 cm) long, ⅜–1⅛" (1–3 cm) thick, nearly equal or tapered downward, uneven and often pitted, dry, hollow in age, pale pinkish gray to gray, sometimes with a whitish bloom when young.

SPORE PRINT: pale pinkish buff.

MICROSCOPIC FEATURES: spores 6.5–9 × 6–7 μm, broadly elliptic to subglobose, with warts and ridges, forming a partial reticulum, hyaline, amyloid.

FRUITING: solitary, scattered, or in groups on the ground in mixed woods; July–October; uncommon.

EDIBILITY: unknown.

Lactarius pubescens Fries *sensu lato*

CAP: 1–4" (2.5–10 cm) wide, obtuse to convex, becoming broadly convex with a depressed center; surface dry and fibrillose except for the disc, which is sticky and smooth when fresh, white to cream; disc reddish orange to vinaceous, azonate; margin inrolled and bearded with coarse white hairs when young; flesh white; odor faintly of geraniums; latex white; flesh and latex acrid.

GILLS: attached to slightly decurrent, crowded, whitish to pale yellow with flesh-pink tinges.

STALK: ¾–2½" (2–6.5 cm) long, ¼–¾" (6–20 mm) thick, nearly equal or tapered downward, silky, hollow in age, whitish when young, becoming ochraceous from the base up in age, apex tinged flesh-pink.

SPORE PRINT: cream with a flesh-pink tint.

MICROSCOPIC FEATURES: spores 6–8.5 × 5–6.5 μm, elliptic, with warts and ridges, forming a partial reticulum, hyaline, amyloid.

FRUITING: scattered or in groups on the ground in wet areas under birch and other hardwoods; August–October; occasional.

EDIBILITY: unknown.

Lactarius quietus var. **incanus** Hesler and Smith

CAP: 1¼–4⅜" (3–11 cm) wide, broadly convex, becoming flat with a depressed center in age; margin incurved at first, becoming expanded to uplifted, opaque when young, becoming slightly translucent-striate in age; surface coated with a layer of tiny whitish silky fibrils when young, soon naked and smooth or with scattered small bumps, moist to dry, at times with water spots, dark purplish brown or purplish gray when young, becoming dark red-brown at the center in age, with a paler pinkish brown to pinkish buff margin, staining pale olive in KOH; flesh pale pinkish buff; odor sweet, fragrant, often like burnt sugar or rarely not distinctive; latex white in young specimens, soon watery as specimens mature; flesh and latex slowly and mildly acrid.

GILLS: attached to slightly decurrent, narrow, often forked, close, white to whitish with pink tints, becoming cinnamon, staining orange-cinnamon in age.

STALK: 1⅛–5½" (4–14 cm) long, ¼–⅜" (5–10 mm) thick, equal or slightly enlarged at the base, solid, becoming hollow, dry, covered with a layer of tiny, whitish, silky fibrils when young, becoming colored like the cap and darkening progressively toward the apex as it matures.

SPORE PRINT: pinkish buff.

MICROSCOPIC FEATURES: spores 6.5–9 × 5–7.5 μm, elliptic, with warts and ridges, hyaline, amyloid.

FRUITING: scattered or in groups on the ground under oaks; August–October; occasional.

EDIBILITY: unknown.

Lactarius representaneus Britzelmayer

COMMON NAME: Northern Bearded Milky.

CAP: 2⅜–7" (6–18 cm) wide, convex to broadly funnel-shaped; surface azonate to faintly zoned, with a thin layer of matted fibers, often scurfy in age, dry to somewhat sticky, pale yellow to orange-yellow, sometimes with rusty tints in age; flesh white, staining purplish when cut; latex white to cream, staining all tissues purplish; flesh and latex mild to slightly acrid or somewhat bitter.

GILLS: attached, close to crowded, cream to pale ochraceous, staining purplish.

STALK: 2–4¾" (5–12 cm) long, ⅜–1⅛" (1–3 cm) thick, nearly equal or enlarged downward, hollow at maturity, sticky to dry, scrobiculate, pale yellow to orange-yellow, staining purplish.

SPORE PRINT: yellowish.

MICROSCOPIC FEATURES: spores 8–11 × 6.5–8 μm, elliptic, with warts and ridges, hyaline, amyloid.

FRUITING: scattered or in groups on the ground under conifers, especially spruce; August–September; occasional.

EDIBILITY: poisonous.

Lactarius resimus (Fries) Fries

CAP: 2⅜–6" (6–15.5 cm) wide, deeply depressed with an arched margin, becoming funnel-shaped in age; margin bearded with coarse hairs at least on young specimens; surface viscid to glutinous, finely tomentose to nearly glabrous, azonate to faintly zoned when young, sometimes conspicuously zoned on old specimens, milky white overall or with ochraceous zones; flesh white; odor not distinctive; latex scant and white on exposure, quickly changing to sulfur-yellow; flesh and latex slowly acrid.

GILLS: decurrent, crowded, often forked near the stalk, white when young, becoming pinkish buff at maturity, staining dingy yellow when bruised.

STALK: 2–3½" (5–9 cm) long, ⅜–1" (1–2.5 cm) thick, tapered downward, sticky or dry, hollow, white to whitish or pale yellow, inconspicuously scrobiculate or lacking scrobiculations.

SPORE PRINT: whitish.

MICROSCOPIC FEATURES: spores 6–8 × 5–6 μm, broadly ellipsoid, with warts and ridges, hyaline, faintly amyloid.

FRUITING: scattered on the ground under conifers or hardwoods; August–October; uncommon.

EDIBILITY: unknown.

Lactarius rufus var. **rufus** (Fries) Fries

COMMON NAME: Red-hot Milky.

CAP: 1⅝–4¾" (4–12 cm) wide, convex, becoming broadly convex to flat with a depressed center, broadly funnel-shaped in age; margin incurved at first, becoming uplifted, sometimes ribbed; surface often coated with a whitish bloom when young, becoming dull, dry, dark brownish red to dark bay-red, zoned with slightly darker bands; flesh pale purplish tan, slowly staining olive-gray in $FeSO_4$; odor not distinctive; latex white, copious in young specimens, unchanging; flesh and latex intensely acrid.

GILLS: attached to slightly decurrent, narrow, crowded, whitish to pinkish tan.

STALK: 2–4⅜" (5–11 cm) long, ⅜–1¾" (9–18 mm) thick, equal, stuffed, dry, with a whitish bloom at first, then colored like the cap, whitish at the base.

SPORE PRINT: cream to pale yellow.

MICROSCOPIC FEATURES: spores 8.5–12 × 6–8 μm, elliptic, with warts and ridges, forming a partial reticulum, hyaline, amyloid.

FRUITING: solitary, scattered, or in groups on the ground under pines, or among sphagnum mosses in bogs; August–October; fairly common.

EDIBILITY: poisonous.

Lactarius speciosus Burlingham

CAP: 1⅛–4" (3–10 cm) wide, broadly convex, becoming nearly flat with a depressed center in age, often broadly funnel-shaped; margin inrolled and bearded with coarse hairs when young, becoming expanded to uplifted and smooth in age; surface sticky when young, becoming dry, covered with a thin layer of fibrils, pale cinnamon or orange-brown, zoned with bands of pinkish tan and cinnamon-brown; flesh thick, firm, whitish; odor not distinctive; latex white, staining the gills and flesh dull lilac; flesh and latex mild to slightly bitter or acrid.

GILLS: attached to slightly decurrent, close, cream to pinkish buff, staining dull lilac when cut or bruised.

STALK: 1⅛–2" (3–5 cm) long, ⅜–1" (1–2.5 cm) thick, equal, hollow, dry, whitish to buff, with reddish brown and yellowish spots and streaks.

SPORE PRINT: white.

MICROSCOPIC FEATURES: spores 10–13.5 × 9–11 μm, elliptic, forming a partial reticulum with warts and ridges, hyaline, amyloid.

FRUITING: scattered or in groups in hardwoods and mixed woods; July–September; occasional.

EDIBILITY: unknown.

Lactarius subplinthogalus Coker

CAP: 1¼–2" (3–5 cm) wide, nearly flat, becoming broadly funnel-shaped in age; margin uplifted and conspicuously pleated; surface nearly smooth when young, becoming finely wrinkled in age, pale tan to yellowish buff overall, becoming brownish yellow in age; flesh whitish, staining pinkish orange when cut or bruised; odor not distinctive; latex white, staining gills and flesh rosy salmon; flesh and latex acrid.

GILLS: attached to slightly decurrent, broad, distant, colored like the cap, staining rosy salmon when cut and bruised.

STALK: 1¼–3⅛" (3–8 cm) long, ¼–⅝" (7–15 mm) thick, equal or tapered slightly downward, solid, becoming hollow, smooth, dry, cream with orangish or orange-brown stains in age.

SPORE PRINT: pinkish buff to cinnamon-buff.

MICROSCOPIC FEATURES: spores 7.5–9.5 × 7–8 μm, nearly globose to elliptic, with warts and ridges, hyaline, amyloid.

FRUITING: solitary to scattered on the ground under oak or in oak-pine woods; July–September; occasional.

EDIBILITY: unknown.

Lactarius subvellereus Peck

CAP: 2–5⅝" (5–15 cm) wide; broadly convex, becoming nearly flat with a depressed center in age; margin incurved when young, often remaining so well into maturity; surface dry, felty to somewhat velvety, whitish, developing pinkish buff or yellowish tints in age; flesh whitish; odor not distinctive; latex abundant, whitish to pale creamy yellow, staining gills reddish brown; flesh and latex acrid.

GILLS: attached to slightly decurrent, crowded, narrow, often forked, pinkish buff to pale yellow, staining tawny when bruised, drying to reddish brown.

STALK: ¾–2" (2–5 cm) long, ½–1" (1.2–2.5 cm) thick, short, equal or tapering downward, solid, dry, whitish with pinkish buff tints.

SPORE PRINT: white.

MICROSCOPIC FEATURES: spores 7.5–9 × 5.5–7 μm, elliptic, with warts and ridges, hyaline, amyloid.

FRUITING: scattered or in groups on the ground in mixed woods; July–October; occasional.

EDIBILITY: edible.

Lactarius uvidus (Fries) Fries

CAP: 1⅛–4" (3–10 cm) wide, convex to nearly plane, sometimes with a low umbo or shallowly depressed; margin incurved at first; surface sticky to slimy, smooth, typically azonate, pale pinkish lilac to pale lilac-brown with lilac spots; flesh whitish, staining dull lilac when cut; odor not distinctive; latex white, soon changing to cream and staining tissues dull lilac; flesh and latex mild at first, then bitter.

GILLS: attached to decurrent, close, creamy white, staining dull lilac when bruised.

STALK: 1⅛–3⅛" (3–8 cm) long, ⅜–⅝" (1–1.6 cm) thick, equal, lacking scrobiculations, sticky when fresh, shiny when dry, whitish to pale buff, typically ochraceous near the base in age.

SPORE PRINT: pale yellow.

MICROSCOPIC FEATURES: spores 7.5–11 × 6.5–8.5 μm, broadly ellipsoid to subglobose, with warts and ridges, hyaline, amyloid.

FRUITING: scattered or in groups on the ground, usually in mixed woods with aspen, birch, and pine; August–October; occasional.

EDIBILITY: unknown.

Lactarius vietus (Fries) Fries

CAP: ⅜–3⅛" (1–8 cm) wide, broadly convex, becoming flat in age with a depressed center; surface dry, slightly sticky when wet, smooth, moist, dark brown to gray-brown when young, in age fading to pinkish brown or lilac-gray; margin wavy and sometimes finely scalloped or pleated in age; flesh whitish to pinkish buff, slowly staining grayish in $FeSO_4$; odor not distinctive; latex white, changing to olive-gray, staining gills gray to olive-gray or grayish brown; flesh and latex acrid.

GILLS: attached to slightly decurrent, broad, close, pinkish white to cream, staining gray to olive-gray or olive-brown.

STALK: 1¼–2⅜" (3–6 cm) long, ⅜–¾" (8–18 mm) thick, equal, fragile, dry to moist, pale pinkish brown overall.

SPORE PRINT: white to cream.

MICROSCOPIC FEATURES: spores 7.5–9 × 6–7.5 μm, elliptic, with warts and ridges, hyaline, amyloid.

FRUITING: scattered or in groups or clusters on decaying wood, especially birch; August–October; occasional.

EDIBLILITY: unknown.

Lactarius vinaceorufescens Smith

COMMON NAME: Yellow Latex Milky.

CAP: 1⅝–4¾" (4–12 cm) wide, convex, becoming broadly convex to nearly flat; margin incurved at first, expanding and somewhat uplifted and uneven in age; surface smooth, pale pinkish cinnamon with pinkish buff at the margin when young, becoming darker pinkish cinnamon to orange-cinnamon in age, faintly zoned with bands or water spots of nearly the same color; flesh moderately thick, white to pinkish, staining bright sulfur-yellow when cut; odor not distinctive; latex white, rapidly turning bright sulfur-yellow after exposure; flesh and latex acrid.

GILLS: attached to slightly decurrent, narrow, close, often forked near the stalk, with many tiers of lamellulae, whitish to pinkish buff, soon spotted vinaceous red to pinkish brown or dark reddish brown.

STALK: 1⅝–2¾" (4–7 cm) long, ⅜–1" (1–2.5 cm) thick, nearly equal or enlarged slightly downward, hollow, nearly smooth, with white to brownish stiff hairs at the base, pinkish white overall, darkening in age.

SPORE PRINT: white to yellowish.

MICROSCOPIC FEATURES: spores 6.5–9 × 6–7 μm, nearly round to elliptic, with warts and ridges, hyaline, amyloid.

FRUITING: scattered or in groups on the ground under pine; August–October; fairly common.

EDIBILITY: poisonous.

Lactarius volemus var. volemus (Fries) Fries

COMMON NAME: Voluminous-latex Milky.

CAP: 2–4" (5–10 cm) wide, broadly convex, becoming nearly flat with a depressed center to broadly funnel-shaped in age; margin incurved at first, expanding and becoming uplifted in age; surface dry, pruinose to velvety when young, becoming smooth to finely wrinkled in age, dark orange-brown to cinnamon-brown at the center, paler orange-brown toward the margin, fading in age to pale orange-brown, then honey-yellow; flesh thick, brittle, white, staining brownish when cut or bruised, instantly staining dark blue-green in FeSO$_4$; odor not distinctive in very young specimens, soon becoming fishy as the mushrooms mature; latex copious, white, becoming creamy white, then brownish or grayish and staining gills and flesh tawny-brown; flesh and latex mild.

GILLS: attached to slightly decurrent, broad, close, often forked, whitish to cream, bruising tawny-brown.

STALK: 2–4½" (5–11.5 cm) long, ¼–¾" (5–20 mm) thick, nearly equal or tapered at the base, solid, sometimes hollow in age, nearly smooth, pale orange-brown to dull orange.

SPORE PRINT: white.

MICROSCOPIC FEATURES: spores 7.5–10 × 7.5–9 μm, oval, with warts and ridges, hyaline, amyloid.

FRUITING: solitary, scattered, or in groups on the ground in hardwoods or mixed woods; July–September; fairly common.

EDIBILITY: edible.

COMMENTS: *L. volemus* var. *flavus* (edible) is nearly identical but has a yellow cap and cream to pale yellow stalk and occurs in the southern part of the region.

Lactarius zonarius (St. Amans) Fries *sensu* Neuhoff

CAP: 2–4" (5–10 cm) wide, convex at first, soon depressed, then broadly funnel-shaped; margin inrolled with tiny yellowish fibers when young, becoming decurved and smooth; surface dry, nearly smooth, coated with a thin layer of tiny white silky fibrils at first, becoming pale yellow to orange-yellow over the disc and fading to cream on the margin, streaked and spotted darker orange-tawny, darkening in age to orange-yellow or dull orange, with faint to distinct orange-tawny zones or sometimes azonate; flesh white; odor pungent to somewhat fragrant; latex white, drying bright gray to pale brownish gray on the gill edges; flesh and latex strongly acrid.

GILLS: decurrent, close to crowded, narrow, white, soon creamy yellow, then bright ochre-yellow, staining tawny-brown when bruised.

STALK: ¾–2" (2–5 cm) long, ⅜–¾" (1–2 cm) thick, nearly equal or tapered downward, solid, hollow in age, smooth, white, becoming pale ochre-yellow and bruising tawny-orange.

SPORE PRINT: bright ochre.

MICROSCOPIC FEATURES: spores 7–10 × 6–7.5 μm, broadly elliptic, with ridges forming a partial reticulum, typically lacking warts, hyaline, amyloid.

FRUITING: solitary, scattered, or in groups on the ground under hardwoods, especially oak; July–September; occasional.

EDIBILITY: unknown.

Genus *Lentinellus*

The distinctive feature of the wood-decaying genus *Lentinellus* is a combination of distinctly serrate gills, acrid flesh, and strongly amyloid spores. Species of the genus *Lentinus* also often have serrate gills, but their spores are inamyloid, and the flesh is not acrid.

Key to Species of *Lentinellus*

1. Stalk either absent or lateral and stubby → 2.
1. Not as above; stalk central to lateral and well-developed → 3.
   2. Cap 1–4" (2.5–10 cm) wide, distinctly velvety to fuzzy, at least at the center, dark brown overall, paler toward the margin; stalk absent; usually growing in clusters but sometimes solitary or in groups on dead deciduous logs and stumps—*L. ursinus* (Fries) Kühner (see p. 182).
   2. Cap ¼–1½" (5–40 mm) wide, smooth, whitish to pinkish or brownish, typically darker at the margin; stalk absent or very short and stubby and eccentric to lateral; spores 5–7 × 4–6 μm; in groups or clusters on dead deciduous stumps, logs, and sticks—*L. flabelliformis* (Bolton : Fries) Ito.
   2. Not as in either of the above choices → 3.
3. Growing in clusters, with numerous caps typically arising from a single basal stalk; more or less shoehorn- to vase-shaped overall; cap pinkish to cinnamon, more or less smooth (cap may be somewhat scaly but not at all fuzzy)—*L. cochleatus* (Fries) Karsten (see p. 181).
3. Growing in clusters, but each cap arising from a more or less separate stalk; cap with white hairs on a pale pinkish to pinkish cinnamon ground color; cap and stalk both distinctly velvety to fuzzy—*L. vulpinus* (Fries) Kühner and Maire (see p. 182).
3. Not growing in clusters; cap typically less than 2" (5 cm) wide, sunken in age, smooth or only vaguely velvety in age, pinkish to brownish; stalk central to eccentric—*L. omphalodes* (Fries) Karsten (see p. 182).

Lentinellus cochleatus (Fries) Karsten

CAP: ⅜–2" (1–5 cm) wide, somewhat spatula-shaped at first, becoming convex and rounder, nearly funnel-shaped in age; surface smooth to somewhat scaly, pinkish to brownish, hygrophanous, often with darker radial fibers; flesh tough, wet, dingy pale brownish or pinkish; odor of anise or not distinctive; taste very acrid.

GILLS: very decurrent, close, thickened, pale pinkish to cinnamon, with dingy brownish stains in age.

STALK: ¼–2" (7–50 mm) long, ⅛–⅜" (3–10 mm) thick, deeply furrowed with ridges that are *not* extensions of the decurrent gills, distinctly velvety, central to eccentric, dry, pinkish to cinnamon.

SPORE PRINT: white.

MICROSCOPIC FEATURES: spores 3.5–5.5 × 3.5–4.5 μm, globose to subglobose, minutely spiny, hyaline, strongly amyloid.

FRUITING: in tightly packed clusters on deciduous wood, especially birch, beech, and ash; July–November; occasional to frequent.

EDIBILITY: inedible.

Lentinellus omphalodes (Fries) Karsten

CAP: ¼–2" (5–50 mm) wide, broadly convex to nearly flat with a depressed center in age; surface smooth overall, pinkish to brownish, hygrophanous; flesh soft, whitish to dingy pale yellowish brown; odor not distinctive; taste mild for several moments, then very acrid.

GILLS: attached to almost decurrent, subdistant, pale pinkish to pale cinnamon.

STALK: ¼–2" (7–50 mm) long, about ¹⁄₃₂–⅛" (1–4 mm) thick, ridged, not velvety or fuzzy, central to eccentric, dry, tough, brown overall, more yellowish brown near the apex.

SPORE PRINT: buff.

MICROSCOPIC FEATURES: spores 4.5–6.5 × 3.5–4.5 μm, oval to subglobose, minutely spiny, thin-walled, hyaline, strongly amyloid.

FRUITING: solitary or in small groups on deciduous or coniferous wood but sometimes on moss-covered ground; August–November; infrequent to rare.

EDIBILITY: inedible.

Lentinellus ursinus (Fries) Kühner

COMMON NAME: Bear Lentinus.

CAP: 1–4" (2.5–10 cm) wide, semicircular, convex to nearly flat; surface distinctly velvety to fuzzy, at least at the center, dark brown overall, paler toward the margin; flesh firm, fairly thick, whitish to pale brownish; odor fruity or not distinctive; taste exceedingly acrid.

GILLS: radiating from the point of attachment to the substrate, broad, close to subdistant, color rather variable but typically pinkish to brownish.

STALK: absent.

SPORE PRINT: white.

MICROSCOPIC FEATURES: spores 3–4.5 × 2–3.5 μm, subglobose to oval, with minute amyloid spines, hyaline.

FRUITING: typically in clusters on deciduous wood, especially oak, maple and beech; July–November; fairly common.

EDIBILITY: inedible.

Lentinellus vulpinus (Fries) Kühner and Maire

CAP: 2–4½" (5–11.5 cm) wide, convex, shell-shaped; surface velvety to fuzzy, covered with short white hairs over a pale pinkish to pale cinnamon ground color, dry or moist; margin inrolled at first; flesh watery, dull white with pinkish to brownish tints; odor not distinctive; taste acrid.

GILLS: short decurrent with ridges extending down the stalk, close to crowded, often torn, whitish at first, becoming pale pinkish cinnamon at maturity; edges serrate.

STALK: lateral, short, usually less than 2" (5 cm) long, stout, narrowed downward and fused with others to form a common base, velvety to fuzzy, dry, pinkish brown.

SPORE PRINT: white.

MICROSCOPIC FEATURES: spores 3.5–5 × 2.5–3.5 μm, subglobose, weakly echinulate, amyloid.

FRUITING: in overlapping clusters on decaying hardwood trees, logs, and stumps; August–October; occasional.

EDIBILITY: inedible.

Genus *Lentinus*

This genus *Lentinus* of wood-decaying fungi is most easily recognized by a combination of the following characteristics: a central to eccentric stalk; a whitish, yellowish, or orangish spore print; a fibrous to scaly cap; growth on wood, sometimes buried; and in many species, serrate gill edges. The spores are smooth, elliptic to cylindric, and inamyloid (compare to *Lentinellus*, p. 181). David Pegler, in *The Genus Lentinus: A World Monograph*, has eliminated the genus *Panus* and made it a subgenus of *Lentinus*. Species of *Panus* are now transferred to *Lentinus* and included in the following key. Most of the species are technically edible but are almost too tough to bother with. However, one is a popular and widely cultivated edible: *Lentinus edodes*, better known as Shiitake.

Key to Species of *Lentinus*

1. Gill edges distinctly serrate → 2.
1. Gill edges not serrate → 3.
   2. Cap 1¼–12" (3–30 cm) wide, whitish to tan with brown scales; partial veil present, leaving a ring on the stalk; odor typically of anise—*L. lepideus* (Fries : Fries) Fries (see p. 184).
   2. Cap ⅜–3½" (1–9 cm) wide, convex and dark grayish brown at first, becoming whitish, with numerous dark brown scales, and depressed at the center; spores 6–10 × 2.5–3.5 µm—*L. tigrinus* (Fries) Fries.
   2. Cap ¾–3⅛" (2–8 cm) wide, creamy white, somewhat fibrillose and dry when young, becoming pinkish orange to tawny-brown with reddish orange spots, slightly viscid at maturity; gills decurrent, cream, subdistant, edges typically minutely serrate and bearing brownish droplets when fresh; spores 6–10 × 2.5–3.5 µm; considered very rare in North America reported only in early spring (March–April) on hemlock in Maine and New Hampshire—*L. adhaerens* (Albertini and Schweinitz : Fries) Fries = *Panus adhaerens* (Albertini and Schweinitz : Fries) Corner.
   2. Cap ⅜–2" (1–5 cm) wide, broadly convex, soon depressed, becoming funnel-shaped at maturity, pale yellow to tawny ochraceous; flesh white to pale cream; odor strongly anise-like; growing on decaying wood of poplar and willow—*L. suavissimus* Fries (see p. 184).
3. Mushroom violet overall when young, soon fading to brownish to tan, usually with some violet tints remaining on stalk; cap 1–5" (2.5–13 cm) wide, dry, smooth or only minutely hairy; gills strongly decurrent, sometimes forked near the stalk—*L. torulosus* (Persoon : Fries) Lloyd = *Panus torulosus* (Persoon : Fries) Fries = *Panus conchatus* (Fries) Fries (see p. 185).
3. Mushroom pinkish tan to reddish brown with purplish tints when young, becoming pale reddish brown to pale ochraceous or tan in age; cap densely hairy and velvety over-all; gills strongly decurrent, sometimes forked near the stalk—*L. strigosus* (Schweinitz) Fries = *Panus rudis* Fries (see p. 184).
3. Mushroom white overall when young; cap 1–8" (2.5–21 cm) wide, dry, coated with short erect or matted hairs, at least over the disc, becoming nearly smooth in age, white dis-coloring yellowish to salmon-buff in age; gills colored like the cap; stalk eccentric or central, sometimes nearly lateral, with matted fibers or nearly smooth near the apex, densely coated with white hairs below; flesh white, tough, odor fragrant, citrus-like, resembling grapefruit, or not distinctive; spores 9–16 × 4–6 µm—*L. levis* (Berkeley and Curtis) Murrill = *Panus levis* Berkeley and Curtis.

Lentinus lepideus (Fries : Fries) Fries

COMMON NAME: Train Wrecker.

CAP: 1¼–10" (3–25 cm) wide, convex to nearly plane; surface smooth and whitish only when very young, soon developing coarse, brown scales; flesh thick, firm, white, becoming quite tough in age; odor somewhat fragrant or not distinctive; taste not distinctive or sometimes disagreeable in older specimens.

GILLS: attached, with decurrent lines extending a short distance down the stalk, subdistant, broad, white to yellowish.

STALK: 1–4" (2.5–10 cm) long, ⅜–1" (1–2.5 cm) thick, central to eccentric, dry, whitish, developing brown scales.

SPORE PRINT: white.

MICROSCOPIC FEATURES: spores 9–12 × 4–5 μm, cylindric, smooth, hyaline, inamyloid.

FRUITING: solitary, in groups, or in clusters on wood, especially frequent on fence posts and railroad ties; May–September; occasional to common.

EDIBILITY: edible but tough; recommended only if the specimens are very young.

COMMENTS: this fungus seems far more effective at colonizing chemically treated wood than most mushrooms.

Lentinus strigosus (Schweinitz) Fries

COMMON NAME: Ruddy Panus.

CAP: 1–3" (2.5–7.5 cm) wide, kidney- to fan-shaped or broadly funnel-shaped; surface densely hairy and velvety to fuzzy, pinkish tan to reddish brown with purplish tints when young, becoming pale reddish brown to pale ochraceous or tan in age; flesh tough, thin, white; odor not distinctive; taste slightly bitter.

GILLS: decurrent, close, narrow, white to tan.

STALK: short, stubby, lateral, pinkish tan to light reddish brown.

SPORE PRINT: white.

MICROSCOPIC FEATURES: spores 4.5–7 × 2.5–3 μm, elliptic, smooth, hyaline, inamyloid.

FRUITING: solitary to clustered on deciduous wood; May–November; infrequent to frequent.

EDIBILITY: edible.

COMMENTS: also known as *Panus rudis*.

Lentinus suavissimus Fries

CAP: ⅜–2" (1–5 cm) wide, broadly convex, soon depressed, becoming funnel-shaped at maturity; surface smooth, dry to moist, pale yellow to tawny-ochraceous; margin incurved when young, becoming elevated and wavy at maturity, often translucent-striate when young and moist; flesh white to pale cream; odor strongly anise-like; taste not distinctive.

GILLS: deeply decurrent, close to crowded, white to dull white, with two tiers of lamellulae; edges finely serrate to denticulate.

STALK: ⅜–¾" (1–2 cm) long, ⅛–¼" (3–6 mm) thick, central to eccentric or sometimes lateral, nearly equal down to a slightly swollen base, solid, smooth except woolly at the base, colored like the cap; partial veil and ring absent.

SPORE PRINT: white.

MICROSCOPIC FEATURES: spores 6–9 × 2–4 μm, cylindric, smooth, hyaline, inamyloid.

FRUITING: scattered or in groups on decaying wood of poplar and willow; June–October; rare.

EDIBILITY: unknown.

COMMENTS: the pungent odor of anise, small funnel-shaped stature, growth on decaying wood, and decurrent gills with serrate to denticulate edges are the distinctive features.

Lentinus torulosus (Persoon : Fries) Lloyd

CAP: 1–5" (2.5–13 cm) wide, convex when young, becoming broadly convex, sometimes depressed on the disc or broadly funnel-shaped; margin becoming uplifted and wavy or lobed in age; surface dry, smooth and violet to purplish at first, soon fading to pale yellowish brown to tan, sometimes cracking and forming tiny scales in age; flesh tough, white; odor and taste not distinctive.

GILLS: decurrent, with lines extending well down the stalk, close to crowded, purplish fading to buff or tannish.

STALK: ⅜–1½" (1–4 cm) long, ¼–1" (5–25 mm) thick, central to eccentric, dry, velvety to fuzzy, at least when young, violet at first, fading to grayish white.

SPORE PRINT: white.

MICROSCOPIC FEATURES: spores 5–6.5 × 2.5–3.5 μm, elliptic to cylindric, smooth, hyaline, inamyloid.

FRUITING: typically in clusters on wood or on the ground from roots; May–November; occasional to fairly common.

EDIBILITY: edible, but quite tough and of mediocre flavor.

COMMENTS: this mushroom can be difficult to identify once its distinctive purplish colors have faded.

Genus *Lepiota* and Allies

The genus *Lepiota* and Allies is a large group of mushrooms with over two hundred species known from North America. They are terrestrial species characterized by having a cap that separates cleanly from a central stalk, free white gills (two may be green), and a partial veil that usually leaves a ring on the stalk. Most species have a white spore print, but one is green and another may be white or green. They vary from very small, cap diameter up to 2" (5 cm), to very large, cap diameter up to 10" (25.5 cm). Species of the Lepiota family are usually divided into several genera, including *Lepiota, Macrolepiota, Chlorophyllum, Leucoagaricus,* and *Leucocoprinus,* among others. Some are choice edibles, two species have caused fatalities, and the edibility of others is unknown.

Key to Species of *Lepiota* and Allies

1. Cap 1⅛–6" (3–15.5 cm) wide, obtuse to convex, becoming nearly flat, often with an umbo, smooth and pinkish brown to reddish brown when young, breaking up into coarse pinkish brown to reddish brown scales on a white ground color that reddens in age; flesh in cap and stalk white, staining yellow to yellow-orange when cut; stalk usually enlarged in the middle and tapering above and below, white at first, staining red in age, with a white, membranous superior ring; spores 8–14 × 5–10 μm; in groups or clusters on stumps, wood chips, and sawdust piles; July–October; edible—*Lepiota americana* Peck.

1. Cap ¾–4⅜" (2–11 cm) wide, egg-shaped when young, becoming convex and nearly flat in age, covered with small, erect, pointed, reddish brown to cinnamon-brown scales over a white ground color; scales easily wiped off; partial veil fibrous-cottony, white, leaving a white superior to median ring—*Lepiota acutesquamosa* (Weinmann) Kummer (see p. 188).

1. Cap ¾–2⅜" (2–6 cm) wide, bell-shaped to broadly conic, typically with an umbo, powdery or with tiny scales, bright to pale yellow; margin distinctly striate to the disc; stalk usually enlarged downward, smooth or powdery, colored like the cap; partial veil fibrous-cottony, bright yellow, leaving a movable persistent or evanescent ring; spores 8–13 × 5–8 μm, dextrinoid; on rich soils, on wood chips used for landscaping, in greenhouses, and on soil in potted plants; July–October, year-round indoors; poisonous— *Leucocoprinus birnbaumii* (Corda) Singer = *Lepiota lutea* (Bolton) Quélet.

1. Cap ⅜–1½" (1–4 cm) wide, conic to broadly convex, smooth when young, becoming wrinkled in age, pale to dark brown, fading to pale brown or creamy white along the margin in age; gills bright yellow; stalk ⅛–¼" (3–7 mm) thick, pale to dark yellow on the upper portion, brown below; partial veil fibrous, leaving an appressed, evanescent, fibrillose superior ring; spores 4.5–5 × 2.5–3 μm, dextrinoid; on the ground in hardwoods; July–September; edibility unknown—*Lepiota luteophylla* Sundberg.

1. Not with the above combinations of characters → 2.
   2. Cap very large, 2¾–12" (7–30 cm) wide, stalk ⅜–1" (1–2.5 cm) thick → 3.
   2. Not as above, cap typically smaller → 4.

3. Spore print green; cap 2¾–12" (7–30 cm) wide, white, covered with large pinkish brown to cinnamon patches; gills white, becoming greenish to grayish green in age— *Chlorophyllum molybdites* Massee (see p. 76).

3. Spore print white; cap 2⅜–8" (7–20 cm) wide, rounded when young, smooth at first, soon breaking up and forming coarse scales as the cap expands; scales cinnamon-brown to pinkish brown or grayish, with dingy white to white flesh between the scales; stalk smooth, white, staining yellow-orange or saffron when cut or bruised; partial veil white, membranous, leaving a persistent, thick-edged, movable superior ring—*Macrolepiota rachodes* (Vittadini) Singer (see p. 195).

3. Spore print white; cap 2¾–10" (7–25 cm) wide, egg-shaped when young, becoming bell-shaped to broadly convex, with an umbo, reddish brown and nearly smooth when young, soon breaking up into coarse, reddish brown scales and patches with white flesh between; flesh white or tinged reddish, not staining yellow-orange when cut or bruised, odor and taste not distinctive; stalk 6–12" (15.5–30 cm) long, ⅜–½" (1–1.3 cm) thick, enlarged downward to a bulbous base, coated with tiny, brownish scales on a white ground color; partial veil white, membranous, leaving a persistent thick-edged, movable superior ring; spores 15–20 × 10–13 μm, dextrinoid; on the ground in woods; July–October; edible—*Macrolepiota procera* (Scopoli) Singer.

3. Cap similar to the previous choice but smaller, 2–6" (5–15.5 cm) wide, with smaller scales; stalk 4–6" (10–15.5 cm) long, ⅜–½" (1–1.3 cm) thick, coated with tiny, brownish scales, with a thin, funnel-shaped ring; spores 10–13 × 7–8 μm, dextrinoid; July–October; edible—*Macrolepiota gracilenta* (Fries) Wasser.

3. Cap similar to the previous two choices but 3⅛–6" (8–15.5 cm) wide, with a white cap and scattered, tiny, brown scales; stalk 6–8" (15.5–20.5 cm) long, ⅜–½" (1–1.3 cm) thick, coated with tiny, brownish scales, with a thick, double-edged ring; spores 9–10 × 6–7 μm, dextrinoid; July–October; edible—*Macrolepiota prominens* (Fries) Moser.
   4. Cap and stalk white to pale grayish white → 5.
   4. Not as above; cap or stalk more richly colored → 6.

5. Cap 2–4¾" (5–12 cm) wide, egg-shaped when young, becoming convex to nearly flat in age, white to dull white to grayish; margin not striate; partial veil white, membranous, leaving a thickened, double-edged superior ring that is usually movable; scattered or in groups in grassy areas—*Lepiota naucinoides* Peck (see p. 189).

5. Cap ¾–2¾" (2–7 cm) wide, obtusely conic to bell-shaped when young, becoming broadly conic to nearly flat, typically with an umbo, white, darkening in age, coated

with white, powdery scales, stalk ⅛–¼" (3–6 mm) thick, enlarged near the base, white, slowly staining yellow or brown when bruised; partial veil membranous, white, leaving a persistent superior ring; spores 6–11 × 6–8 μm, dextrinoid; on sawdust, wood chips, compost, and rich soils; June–October; poisonous—*Lepiota cepaestipes* (Sowerby : Fries) Kummer.

6.   Cap with concentrically arranged scales; scales reddish brown to cinnamon-brown, yellowish brown, orange-tawny or yellowish, darkest on the disc, paler toward the margin → 7.

6.   Cap with or without scales; scales, if present, scattered and not concentrically arranged; cap reddish to brownish pink, coral-pink, pinkish orange, purple to gray or blackish, darkest on the disc, paler toward the margin → 9.

7.   Cap ⅜–2¾" (1–7 cm) wide; scales reddish brown; flesh odor unpleasant to spicy or absent; stalk ⅛–¼" (2–6 mm) thick, smooth, white to pale pinkish buff, with a persistent, small, white superior ring; spores 5.5–8 × 3–4.5 μm, wedge-shaped, dextrinoid, with a small corner spur—*Lepiota cristata* (Fries) Kummer (see p. 188).

7.   Cap ¾–2" (2–5 cm) wide; scales cinnamon-brown to pinkish brown; flesh white, often reddish in age, odor faintly musty; stalk ⅛–¼" (2–7 mm) thick, smooth above the ring, coated with cinnamon-brown to pinkish brown scales below the ring; partial veil fibrillose, often leaving remnants on the cap margin and an evanescent superior ring; spores 6–8 × 2.5–4.5 μm; on the ground in woods and grassy areas, often under oak; deadly poisonous—*Lepiota josserandii* Bonorden and Boiffard.

7.   Fruiting body nearly identical to the previous choice except for a somewhat more membranous veil and larger elliptic dextrinoid spores, 7–10 × 4–5 μm; deadly poisonous—*Lepiota helveola* Bresadola.

7.   Not with the above combinations of characters → 8.

8.   Scales tawny to orange-tawny or darker yellowish brown, darkest on the disc; stalk ⅛–⁵⁄₁₆" (3–8 mm) thick, shaggy or with cottony scales, tawny; cap ¾–2⅜" (2–6 cm) wide; margin rimmed with flaps of partial veil; partial veil fibrous-cottony, leaving an evanescent ring; spores 14–20 × 4–6 μm, spindle-shaped, dextrinoid; on the ground in woods; August–October; poisonous—*Lepiota clypeolaria* (Fries) Kummer.

8.   Scales dark brown to reddish brown, darkest on the disc, paler toward the margin and fading in age; stalk ⅛–⁵⁄₁₆" (3–8 mm) thick, not shaggy but streaked with white cottony fibers and patches of dark brown scales; cap ¾–2⅜" (2–6 cm) wide; margin often rimmed with fibrous-cottony flaps of partial veil; partial veil fibrous to cottony, leaving a persistent or evanescent superior ring; spores 6–9 × 4–6 μm, elliptic, dextrinoid; on humus under conifers; August–October; edibility unknown —*Lepiota clypeolarioides* Rea.

8.   Scales tawny to pale reddish brown, darkest on the disc; stalk ¼–¾" (6–20 mm) thick, bulbous, smooth near the apex, coated with tawny to pale reddish brown fibers and scales below the ring; cap 1⅛–4" (3–10 cm) wide, partial veil fibrillose, leaving an evanescent superior ring; spores 7.5–10 × 3–4 μm, elliptic, truncate, dextrinoid; on needles under conifers; August–October; edibility unknown—*Lepiota cortinarius* Lange.

9.   Cap reddish to brownish pink when young, becoming coral-pink to pinkish orange or brownish pink in age, darkest on the disc; 1⅛–3⅛" (3–8 cm) wide, smooth when young, becoming cracked and covered with tiny scales or fibers in age; stalk smooth, white, with a persistent, white, membranous superior ring—*Lepiota rubrotincta* Peck (see p. 189).

9.   Cap purple to grayish purple or blackish purple on a white ground color, darkest on the

disc; ⅜–1½" (1–4 cm) wide, smooth when young, becoming cracked and covered with tiny scales or fibers in age; stalk ⅛–¼" (3–6 mm) thick, enlarged downward, white, with a persistent, white, membranous superior ring; spores 6–9 × 4–5 μm, dextrinoid; on humus in coniferous woods; edibility unknown—*Lepiota roseilivida* Murrill.

9. Cap dark gray to blackish on a white ground color and darkest on the disc; ⅜–1⅜" (1–3.5 cm) wide, smooth when young, becoming cracked and covered with tiny scales or fibers in age; stalk 1/16–⅛" (1.5–3 mm) thick, enlarged downward, white, with a persistent, white, membranous superior ring; spores 5–6 × 3–3.5 μm, dextrinoid; on decaying wood of conifers and the surrounding rich soil; edibility unknown—*Lepiota phaeosticta* Morgan.

9. Not as above; cap ⅜–1¾" (1–4.5 cm) wide, ovate to bell-shaped when young, becoming broadly convex with a small umbo, translucent-striate from the margin to the disc, tawny to ochre-brown on the umbo, pale bright yellow over the remainder or pale yellowish white on the margin; stalk long and extremely fragile—*Leucocoprinus fragilissimus* (Ravenel in Berkeley and Curtis) Patouillard (see p. 191).

Lepiota acutesquamosa (Weinmann) Kummer

COMMON NAME: Sharp-scaled Lepiota.

CAP: ¾–4⅜" (2–11 cm) wide, egg-shaped when young, becoming convex and nearly flat in age, surface covered with small, erect, pointed, reddish brown or cinnamon-brown scales over a white ground color; scales easily wiped off; flesh white; odor and taste not distinctive.

GILLS: free, crowded, white to pale cream.

STALK: 2–4⅜" (5–11 cm) long, ⅜–⅝" (1–1.6 cm) thick, enlarged downward, smooth or nearly so, white; base somewhat bulbous; partial veil fibrous-cottony, white, leaving a white superior to median ring with reddish brown scales on the edge; ring often collapsing or sometimes evanescent.

SPORE PRINT: white.

MICROSCOPIC FEATURES: spores 6.5–8.5 × 2–3 μm, narrowly elliptic, smooth, hyaline, dextrinoid.

FRUITING: solitary, scattered, or in groups on humus in woods; July–October.

EDIBILITY: edible.

Lepiota cristata (Fries) Kummer

COMMON NAME: Malodorous Lepiota.

CAP: ⅜–2¾" (1–7 cm) wide, convex, becoming broadly convex to nearly plane, sometimes with an umbo; surface dry, smooth at first, soon breaking up into concentric rings of tiny, reddish brown scales, reddish brown to yellowish brown over the disc, paler toward the margin, with white flesh showing between the scales; flesh white, odor unpleasant to spicy or absent; taste not distinctive.

GILLS: nearly free, close, narrow, white; edges even to somewhat toothed.

STALK: ¾–3⅛" (2–8 cm) long, ⅛–¼" (2–6 mm) thick, nearly equal, sometimes bulbous at the base, hollow in age, smooth, white to pale pinkish buff, often brownish at the base; partial veil membranous, white, leaving a persistent, small, white superior ring.

SPORE PRINT: white or green.

MICROSCOPIC FEATURES: spores 5.5–8 × 3–4.5 μm, wedge-shaped with a small corner spur, smooth, hyaline, dextrinoid.

FRUITING: scattered or in groups on the ground in woods and grassy areas; June–October; fairly common.

EDIBILITY: unknown.

Lepiota naucinoides Peck

COMMON NAME: Smooth Parasol.

CAP: 2–4¾" (5–12 cm) wide, egg-shaped when young, becoming convex to nearly flat in age; surface dry, smooth to slightly roughened with small scales, white to dull white to grayish, occasionally staining yellow when bruised or in age, especially near the center; flesh white, thick, not staining when cut or bruised; odor and taste not distinctive.

GILLS: free from the stalk, close, broad, white, not staining when cut or bruised, becoming pinkish to grayish or pale brown in age.

STALK: 2–4¾" (5–12 cm) long, ⅜–¾" (1–2 cm) thick, nearly equal or enlarged downward, especially at the base, solid, becoming hollow in age, smooth, dry, white, sometimes staining yellowish when bruised or in age; partial veil white, membranous, leaving a thickened double-edged superior ring that is often movable.

SPORE PRINT: white.

MICROSCOPIC FEATURES: spores 7–9 × 5–6 μm, oval, smooth, with an apical pore, hyaline, dextrinoid.

FRUITING: scattered or in groups in grassy areas; September–November; fairly common.

EDIBILITY: edible.

COMMENTS: also known as *Lepiota naucina*, *Leucoagaricus naucina*, and *Leucoagaricus naucinoides*. Carefully compare with white species of *Amanita*.

Lepiota rubrotincta Peck

COMMON NAME: Red-tinged Lepiota.

CAP: 1⅛–3⅛" (3–8 cm) wide, rounded, becoming convex to flat with an umbo in age; surface smooth and dry when young, becoming cracked and covered with tiny scales or fibers in age, reddish to brownish pink when young, becoming coral-pink to pinkish orange or brownish pink in age with the disc remaining uncracked, smooth, and darker red-brown or coral-pink; margin often torn or cracked; flesh white, thin, not staining when cut or bruised; odor and taste not distinctive.

GILLS: free from the stalk, close, narrow, white, not staining when cut or bruised.

STALK: 2–5⅞" (5–15 cm) long, ⅛–⅜" (3–10 mm) thick, nearly equal or enlarged downward, becoming hollow at maturity, white, sometimes becoming discolored in age; partial veil white, membranous, forming a superior ring.

SPORE PRINT: white.

MICROSCOPIC FEATURES: spores 6–10 × 4–6 μm, elliptic, smooth, hyaline, dextrinoid.

FRUITING: solitary, scattered, or in groups on the ground in hardwood forests, on compost, and on leaf litter; July–October; occasional.

EDIBILITY: unknown.

Genus *Leptonia*

Leptonia incana (Fries) Quélet

CAP: ⅜–1½" (1–4 cm) wide, convex at first, becoming flat or with a raised margin and developing a depressed, scaly center; surface yellow to greenish yellow, becoming brownish in age, with a somewhat translucent-striate margin; flesh yellowish or greenish, staining a distinct teal color when bruised; odor described as similar to mouse feces or rancid buttered popcorn; taste not distinctive.

GILLS: attached to slightly decurrent, close to subdistant, whitish or pale green at first, becoming pinkish-tinted in age.

STALK: ⅜–1½" (1–4 cm) long, 1/16–⅛" (1.5–3 mm) thick, bright greenish yellow when fresh, smooth overall, the base coated with white mycelium that stains teal when bruised.

SPORE PRINT: salmon-pink.

MICROSCOPIC FEATURES: spores 11–14 × 8–9 μm, angular, smooth, hyaline, inamyloid.

FRUITING: solitary or scattered in various habitats, especially with exposed limestone; June–September; rare to occasional.

EDIBILITY: unknown.

Leptonia odorifer (Hesler) Largent

CAP: ⅜–2" (2–5 cm) wide, convex at first, becoming flat or with a raised margin and soon developing a deeply depressed, scaly center; surface pale to rich grayish brown, margin somewhat striate; odor strong, similar to burnt rubber, sulphur, or cat urine; taste strongly unpleasant.

GILLS: subdecurrent to decurrent, close, whitish at first, becoming pale pinkish cinnamon.

STALK: 1½–3½" (4–9 cm) long, ⅛–¼" (3–5 mm) thick, whitish to grayish, mostly smooth, the base coated with white mycelium.

SPORE PRINT: salmon-pink.

MICROSCOPIC FEATURES: spores 9–12 × 7–9 μm, angular, smooth, hyaline, inamyloid.

FRUITING: scattered to gregarious on rich soil or humus in mixed woods; July–September; rare.

EDIBILITY: unknown.

COMMENTS: also known as *Entoloma odorifer*. To date, this mushroom has been reported only from the Great Smoky Mountains area and from Syracuse, New York.

Leptonia rosea Longyear

CAP: ⅝–1½" (1.5–4 cm) wide, convex, developing a depressed, scaly center; surface pale rosy pink beneath a dense layer of rose-colored to rosy brown scales that thin out toward the cap margin; odor fragrant or not distinctive; taste not distinctive.

GILLS: attached, close, whitish at first, becoming fleshy pink.

STALK: 2–3" (5–7.5 cm) long, about ⅛" (2–3 mm) thick, somewhat enlarged downward, brownish rose, smooth overall except pruinose at the apex, dry, the base coated with white mycelium.

SPORE PRINT: salmon-pink.

MICROSCOPIC FEATURES: spores 8–12 × 6.5–8 μm, angular, smooth, hyaline, inamyloid.

FRUITING: solitary to scattered on humus or rotting wood in mixed woods; July–September; rare.

EDIBILITY: unknown.

COMMENTS: also known as *Entoloma roseum*.

Leptonia serrulata (Fries) Kummer

CAP: ⅜–1½" (1–4 cm) wide, convex at first, becoming flat or with a raised margin and developing a depressed, scaly center, dark bluish black at first, fading to grayish purple; odor and taste not distinctive.

GILLS: attached to slightly decurrent, close; surface whitish at first, becoming bluish gray to pinkish gray, with finely toothed, dark bluish black edges.

STALK: 1½–3" (4–7.5 cm) long, about ⅛" (1.5–3 mm) thick, bluish black and smooth overall except scurfy at the apex, dry, the base coated with white mycelium.

SPORE PRINT: salmon-pink.

MICROSCOPIC FEATURES: spores 9–12 × 6–8 μm, angular, smooth, hyaline, inamyloid.

FRUITING: solitary to scattered on humus in mixed woods, among mosses, or sometimes on mossy logs; June–September; occasional.

EDIBILITY: unknown.

COMMENTS: also known as *Entoloma serrulatum*.

Genus *Leucocoprinus*

Leucocoprinus fragilissimus (Ravenel) Patouillard

CAP: ⅜–1¾" (1–4.5 cm) wide, ovate to bell-shaped when young, becoming broadly convex with a small umbo; margin incurved at first, becoming uplifted in age; surface translucent-striate to sulcate from the margin to the disc, tawny to ochre-brown on the umbo, pale bright yellow over the remainder, or pale yellowish white on the margin, fading in age; flesh very thin, whitish; odor and taste not distinctive.

GILLS: attached, becoming nearly free, narrow, white.

STALK: 1½–6" (4–15.5 cm) long, 1/16–⅛" (1–3 mm) thick, enlarged downward, extremely fragile, pale grayish yellow when fresh, darkening in age, coated overall with tiny yellowish scales; partial veil cottony-membranous, whitish to pale yellow, leaving a superior ring.

SPORE PRINT: white.

MICROSCOPIC FEATURES: spores 9–13 × 7–8 μm, elliptic with an apical pore, smooth, hyaline, dextrinoid; cheilocystidia 12–25 × 9–15 μm; pleurocystidia absent.

FRUITING: solitary or scattered on the ground in mixed woods in the most southern part of the region; July–October; occasional.

EDIBILITY: unknown.

COMMENTS: *Leucocoprinus magnicystidiosus* (edibility unknown) is very similar but has a darker brown disc and very large cheilocystidia and pleurocystidia. It is known from Tennessee.

Genus *Leucopaxillus*

Members of the small genus *Leucopaxillus* are medium to large, mostly whitish to tan, sturdy mushrooms with a white spore print. They have dry caps and stalks and grow on the ground in hardwoods and conifer woods. Their gill layers readily separate from the cap flesh, and their stalk base is attached to copious white mycelium that binds leaves, needles, and debris. Most species have a farinaceous or disagreeable odor and a disagreeable, bitter, or farinaceous taste. They produce roughened, amyloid spores. Little is known about edibility for most of the species in this genus.

Key to Species of *Leucopaxillus*

1. Cap very large, 3½–12" (9–30 cm) wide, dry, unpolished, pinkish buff, becoming dull tan over the disc or nearly overall; margin strongly inrolled, typically furrowed; flesh white, odor pungent to farinaceous, taste mild to unpleasant; gills close to crowded, pale yellow, darkening in age; stalk nearly equal down to an abruptly bulbous base, whitish, with copious white mycelium at the base; spores 6–8 × 4–5.5 μm; solitary to scattered on the ground under hardwoods, especially oak; July–October; edible— *L. tricolor* (Peck) Kühner.

1. Cap smaller than the previous choice, 1–7" (2.5–18 cm) wide → 2.

2. Stalk yellow with a white basal mycelium; cap 1⅜–4" (3.5–10 cm) wide, yellow, hygrophanous; flesh yellowish, odor and taste farinaceous or not distinctive; gills close, yellow; spores 4–5 × 3.5–4.5 μm; scattered on the ground under hardwoods, especially beech; September–November; edibility unknown—*L. subzonalis* (Peck) Bigelow.

2. Stalk white to whitish, dull, coated with soft hairs overall or at least on the lower portion at maturity; cap 2–4" (5–10 cm) wide, white to creamy white, usually developing pale yellow-brown tints over the disc and becoming finely cracked in age; margin cottony and ribbed; flesh white, odor pungent and disagreeable or not distinctive, taste farinaceous to disagreeable or sometimes mild; gills white; spores 5–7.5 × 3–5 μm; on the ground under hardwoods; August–October; edibility unknown—*L. paradoxus* (Costantin and Dufour) Boursier.

2. Stalk white, dull, smooth or nearly so, not coated with soft hairs → 3.

3. Cap white overall or white with a dull pinkish tinge over the disc, 1⅜–7" (3.5–18 cm) wide; margin strongly inrolled at first, often ribbed or grooved; flesh white, odor farinaceous, taste strongly farinaceous-bitter; gills white, close to crowded; stalk white to creamy white, dull, with copious white mycelium; spores 3.5–5.5 × 3.5–5 μm, globose to oval; on decaying leaves under hardwoods, especially beech and oak; June–November; edibility unknown—*L. laterarius* (Peck) Singer and Smith.

3. Cap white at first, often becoming pale yellow-brown over the disc, 1–3¾" (2.5–9.5 cm) wide; flesh white, odor fragrant or absent, taste farinaceous-bitter; spores 5–8.5 × 3.5–5 μm, elliptic; on humus, leaves, and conifer needles in woods—*L. albissimus* (Peck) Singer (see p. 192).

Leucopaxillus albissimus (Peck) Singer

COMMON NAME: White Leucopax.

CAP: 1–3¾" (2.5–9.5 cm) wide, convex when young, becoming nearly plane in age; surface dull, nearly smooth when young, soon becoming finely cracked and breaking into tiny irregular patches, white at first, often becoming pale yellow-brown over the disc and paler toward the margin, sometimes remaining white overall; margin incurved and typically remaining so well into maturity, often finely ribbed, sometimes lobed; flesh white, unchanging when cut; odor fragrant or absent; taste farinaceous-bitter.

GILLS: attached to decurrent, close to crowded, white, with several tiers of lamellulae.

STALK: 1⅛–2¾" (3–7 cm) long, ¼–⅝" (6–16 mm) thick, enlarged downward, sometimes tapered below, solid, white, typically coated with copious white mycelium and trapped debris at the base; partial veil and ring absent.

SPORE PRINT: white.

MICROSCOPIC FEATURES: spores 5–8.5 × 3.5–5 μm, elliptic, with sparse strongly amyloid warts, hyaline.

FRUITING: scattered or in groups on humus, leaves, and conifer needles in woods; August–October; fairly common.

EDIBILITY: unknown.

COMMENTS: Singer and Smith recognized 12 varieties and forms of this species.

Genus *Limacella*

Limacella is the only genus except *Amanita* in the familiy Amanitaceae; the two are grouped together because of unique microscopic features that they share. All species of *Limacella* are rare to locally common.

Key to Species of *Limacella*

1. Stalk dry, with a membranous ring; cap ¼–⅝" (5–15 cm) wide, viscid when fresh, pale pinkish cream, often radially streaked and/or split, convex to broadly convex; gills whitish (to pale greenish gray in var. *fischeri* [Kauffman] H. V. Smith), crowded; stalk 4–6" (10–15 cm) long, ⅜–¾" (1–2 cm) thick, with a thin, whitish, membranous ring; spores 4–5 × 3.5–4 μm; most often found near elm or ash trees in swampy woods in the Great Lakes region; edibility unknown—*L. lenticularis* (Lasch) Earle.
1. Stalk dry, lacking a membranous ring; cap 1⅛–3⅛" (3–8 cm) wide, viscid when fresh, reddish brown fading to light pinkish brown; stalk with only a slight ring of fibers at most; odor and taste farinaceous; edibility unknown—*L. glioderma* (Fries) Earle (see p. 193).
1. Stalk dry overall but typically beaded with moisture above a membranous white ring, smooth; cap 1–3" (2.5–7.5 cm) wide, viscid when fresh, whitish to pale pinkish buff, convex to flat; gills barely attached, crowded, narrow, white to creamy; partial veil white, membranous, leaving a persistent ring on the upper stalk; stalk 3–4" (7.5–10 cm) long, ⅜–⅝" (1–1.5 cm) thick; flesh odor and taste farinaceous; spores 4–5 × 3.5–5 μm; growing on the ground in mixed woods; very rare on the East Coast (reported from New York); edibility unknown—*L. solidipes* (Peck) H. V. Smith.
1. Stalk distinctly slimy; mushroom not as in any of the above choices → 2.
    2. Slime on the cap and stalk thick, clear; cap ¾–2¾" (2–7 cm) wide, white to creamy, slime typically dripping from cap edge; gills free or nearly so, close, broad, white; stalk 2–4" (5–10 cm) long, ⅛–⅜" (3–10 mm) thick, whitish staining brownish where bruised, typically lacking a distinct partial veil or ring; found in various habitats, including woods, fields, and sand dunes—*L. illinita* (Fries) Earle (see p. 194).
    2. Slime on the cap and stalk thick, reddish brown; cap ¾–1½" (2–4 cm) wide, convex, whitish to pale yellowish brown; gills free, close, broad, white; stalk 1½–3" (4–7.5 cm) long, about ⅜" (8–12 mm) thick, whitish to slightly brownish, no distinct ring or partial veil evident; flesh white; odor and taste mild or not distinctive; spores 5–6.5 × 4.5–5.5 μm; typically single, on the ground in mixed woods; edibility unknown—*L. glischra* (Morgan) Murrill.

Limacella glioderma (Fries) Earle

CAP: 1⅛–3⅛" (3–8 cm) wide, convex, becoming nearly plane in age; surface slimy when fresh, reddish or orangish brown to chestnut-brown, fading to light pinkish brown; gills barely attached to the stalk, close, broad, whitish; flesh whitish or with a slight purplish tint; odor and taste farinaceous.

GILLS: gills barely attached to the stalk, close, broad, whitish.

STALK: 2–4" (5–10 cm) long, ³⁄₁₆–⅜" (5–10 mm) thick, with only a slight ring of fibers at most.

SPORE PRINT: white.

MICROSCOPIC FEATURES: spores 3–4 μm, globose, smooth, hyaline, inamyloid.

FRUITING: solitary to gregarious on humus or on soil in woods, also reported on humus in greenhouses; August to October; rare.

EDIBILITY: unknown.

Limacella illinita (Fries) Earle

CAP: ¾–2¾" (2–7 cm) wide, convex, becoming nearly plane in age; surface coated with hyaline slime when fresh, smooth, white to creamy; margin sometimes dripping slime during rainy periods; flesh white; odor and taste not distinctive.

GILLS: free or nearly so, close, broad, white.

STALK: 2–4" (5–10 cm) long, ⅛–⅜" (3–10 mm) thick, slimy, whitish, staining brownish where bruised, typically lacking a ring or sometimes with a sparse, fibrillose annular zone.

SPORE PRINT: white.

MICROSCOPIC FEATURES: spores 4–6.5 μm, globose to broadly elliptic, smooth, hyaline, inamyloid.

FRUITING: scattered or in groups in various habitats, including woods, fields, and sand dunes; July–November; rare to locally common.

EDIBILITY: unknown.

Genus *Lyophyllum*

Lyophyllum is a poorly understood genus that needs more research; it is hard to identify some of the species to genus without careful microscopic study. The key presented here treats four species that are relatively distinct, including one popular edible.

Key to Species of *Lyophyllum*

1. Gills and sometimes other parts of the mushroom slowly staining gray to black where bruised → 2.
1. Not as above → 3.
   2. Gills attached to subdecurrent, gray at first, slowly staining yellowish, then bluish, and finally black where bruised; cap 1½–4" (4–10 cm) wide, brown-black to blackish gray, shiny; stalk pale gray, striate to fibrous; odor and taste farinaceous to rancid; spores 7–9 × 4–4.5 μm, elliptic; on the ground in coniferous woods; edibility unknown—*L. semitale* (Fries) Kühner.
   2. Gills attached, whitish at first, slowly staining orangish yellow and finally gray where bruised; cap 1–2¼" (2.5–6 cm) wide, creamy white at first, moist, smooth, slowly staining yellowish and finally grayish where bruised; stalk ¾–1½" (2–4 cm) long, about ⅜" (1 cm) thick near the top, enlarged downward, nearly smooth, whitish at first, staining like the gills where bruised; odor strongly of green corn; spores 6–10 × 3.5–6 μm, fusoid, smooth; edibility unknown—*L. luteogriseascens* Clemenceau and Smith.
3. Growing in large clusters; caps gray, grayish yellow, or grayish brown, 2–5" (5–12.5 cm) wide; gills attached to subdecurrent, close, whitish, not staining appreciably where bruised; stalk 2–4" (5–10 cm) long, ¼–¾" (5–20 mm) thick, smooth, white, becoming brownish from the base upward; spores 4–5.5 × 4–5 μm, nearly globose—*L. decastes* (Fries) Singer (see p. 195).
3. Growing in clusters or not; caps whitish to pale brownish orange at first, fading to dingy gray in age, 1¼–4" (3–10 cm) wide; gills attached, crowded, white to yellowish at first, slowly staining dingy ochre where bruised; spores 5–6 × 3–3.5 μm, oval to elliptic—*L. multiforme* (Peck) Bigelow (see p. 195).

Lyophyllum decastes (Fries) Singer

CAP: 2–5" (5–12.5 cm) wide, convex to nearly flat; surface moist, smooth, margin incurved at first, gray to grayish yellow or yellowish brown; flesh whitish; odor and taste not distinctive.

GILLS: attached to subdecurrent, close, whitish, not substantially changing color where bruised.

STALK: 2–4" (5–10 cm) long, ¼–¾" (5–20 mm) thick, smooth, dry, white, soon becoming brownish below.

SPORE PRINT: white.

MICROSCOPIC FEATURES: spores 4–5.5 × 4–5 μm, globose to subglobose, smooth, hyaline, inamyloid.

FRUITING: usually in large clusters on the ground in various habitats, especially grassy areas, but not typically in forests; July–October; infrequent to fairly common.

EDIBILITY: edible with caution (see comments below).

COMMENTS: this is a popular edible, but it has few distinguishing characters, and reportedly some people are sensitive to it.

Lyophyllum multiforme (Peck) Bigelow

CAP: 1¼–4" (3–10 cm) wide, smooth, slippery, whitish to pale brownish orange at first, fading to dingy gray in age; flesh whitish; odor and taste not distinctive.

GILLS: attached, crowded, narrow, white to yellowish at first, slowly staining dingy ochre where bruised.

STALK: 1½–3" (4–7.5 cm) long, ¼–⅜" (5–10 mm) thick, often eccentric, hollow, becoming fibrous in age.

SPORE PRINT: white.

MICROSCOPIC FEATURES: spores 5–6 × 3–3.5 μm, elliptic, smooth, hyaline, inamyloid.

FRUITING: gregarious or in small clusters on woody soil; August–November; occasional.

EDIBILITY: unknown.

Genus *Macrolepiota*

Macrolepiota rachodes (Vittadini) Singer

COMMON NAME: Shaggy Parasol.

CAP: 2¾–8" (7–20 cm) wide, rounded to convex, becoming nearly flat in age; surface dry, initially smooth, forming coarse scales as the cap expands, cinnamon-brown, pinkish brown, or grayish with dingy white flesh between the scales, disc usually remaining uncracked and brown; margin often shaggy or ragged with veil remnants; flesh white, thick, staining orange, then reddish-brown when cut or bruised; odor and taste not distinctive.

GILLS: free from the stalk, close, broad, white, staining brown when bruised or in age.

STALK: 2–5⅞" (5–15 cm) long, ⅜–¾" (1–2 cm) thick, enlarged downward, occasionally bulbous, smooth, white on the upper portion, shades of brown on the lower portion; stalk flesh staining yellow-orange or saffron when cut or bruised; partial veil white, membranous, leaving a thick-edged superior movable ring.

SPORE PRINT: white.

MICROSCOPIC FEATURES: spores 6–9.5 × 6–7 μm, elliptic, smooth, hyaline, dextrinoid.

FRUITING: solitary or in groups on the ground among leaves, conifer needles, and wood chips, grassy areas, and in gardens; September–November; fairly common.

EDIBILITY: edible and often rated choice.

COMMENTS: also known as *Lepiota rachodes;* the species name is sometimes spelled *rhacodes.*

Genus *Marasmiellus*

Marasmiellus nigripes (Schweinitz) Singer

COMMON NAME: Black-footed Marasmius.

CAP: ⅜–¾" (1–2 cm) wide, convex, becoming nearly flat in age; surface finely powdery (use a hand lens), dry, chalky white, wrinkled to distinctly plicate in age; flesh thin, rubbery-gelatinous, white; odor and taste not distinctive.

GILLS: attached to subdecurrent, subdistant to distant, often forked and crossveined at maturity, white, becoming reddish-stained in age.

STALK: ¾–2" (2–5 cm) long, 1/32–1/16" (0.75–1.5 mm) thick, distinctly tapered toward the base, tough, hollow, whitish when very young, soon becoming black with a coating of minute white hairs (use a hand lens), often with fine rhizomorph-like threads connected to the base.

SPORE PRINT: white.

MICROSCOPIC FEATURES: spores 7–9 μm, triangular, smooth, hyaline.

FRUITING: solitary, scattered, or in groups on decaying twigs, leaves, and grasses; July–September; occasional.

EDIBILITY: unknown.

Marasmiellus praeacutus (Ellis) Halling

COMMON NAME: Pointed-stalked Marasmiellus.

CAP: ¼–¾" (5–20 mm) wide, convex when young, nearly flat in age, often depressed on the disc, dry, smooth, with shallow radial grooves, reddish brown overall when young, fading to grayish brown and finally whitish overall in age; flesh white, thin; odor and taste onion-like or fetid.

GILLS: attached, subdistant to close, white, sometimes forked and crossveined.

STALK: ½–1½" (1.2–4 cm) long, about 1/16" (1–2 mm) thick, nearly equal or slightly enlarged down to the base, which tapers abruptly to a point, reddish brown except for the white pointed base.

SPORE PRINT: white to pale cream.

MICROSCOPIC FEATURES: spores 5.5–8.5 × 2.5–3.5 μm, lacrymoid to narrowly elliptic, smooth, hyaline, inamyloid.

FRUITING: scattered or in groups on hardwood leaves, conifer needles, cone scales, and decaying wood; June–August; occasional.

EDIBILITY: unknown.

Genus *Marasmius*

Though many are beautiful, *Marasmius* and allies are mostly so small and thin-fleshed that they are classified as inedible for that reason alone. Unless otherwise stated, the species included in the key are inedible or their edibility is unknown. Most are decomposers of leaves, pine needles, and other small plant materials, though a few are wood decayers. *Marasmius* species make up for their small size by their resistance to rapid decay, which limits the spore production of most other gilled mushrooms. They dry out and shrivel quite readily but revive and begin producing spores anew when adequate moisture is available.

Key to Species of *Marasmius* and Allies

1. Odor (especially when the cap is crushed), and sometimes taste, distinctly foul or fetid (e.g., like stagnant water) or like garlic → 2.

1. Odor and taste not distinctive; stalk tough, base with long yellow to orange hairs (either bristle-like or somewhat woolly); cap yellowish, orangish, reddish, or brownish, often with a tiny depressed area at the center; gills yellowish, often decurrent; spores 5–7 × 3–4 μm, elliptic to cylindric, amyloid; growing in dense recurved clusters on well-decayed conifer wood—*Xeromphalina campanella* (Batsch : Fries) Kühner and Maire (see p. 270).
1. Odor and taste disagreeable; nearly identical to the previous choice but with a darker reddish brown cap and narrower spores, 5.5–7 × 2.5–3 μm, amyloid; found in the northern part of the region, frequent; inedible—*Xeromphalina brunneola* O. K. Miller, Jr.
1. Odor and taste not distinctive; nearly identical to *X. campanella* (see above) but growing in dense, erect clusters on decaying hardwood, especially oak; distributed throughout the region and common; spores 5–6 × 2.5–3.5 μm; inedible—*Xeromphalina kauffmanii* Smith.
1. Odor and taste not distinctive; nearly identical to the previous choice but growing solitary, scattered, or in small groups on conifer debris or among sphagnum mosses; stalk base conspicuously tufted with orangish to amber hairs; spores 5–7.5 × 3–4.5 μm; edibility unknown—*Xeromphalina cornui* (Quélet) Favre.
1. Odor and taste not distinctive; stalk tough, with orange-brown, velvety, matted hairs over the entire surface, usually lacking conspicuous hairs or rhizomorphs at the base; cap 1–2¾" (2.5–7 cm) wide, convex to nearly plane, dry, velvety, orange-brown; gills attached with a fine decurrent line, close to subdistant, white, becoming pale yellow in age; spores 6–9 × 3–5 μm; on decaying hardwood; edibility unknown—*Xeromphalina tenuipes* (Schweinitz) Smith.
1. Not as in any of the above choices → 3.
    2. Growing on individual conifer needles (one or two mushrooms per needle); odor and taste strongly of stagnant water and/or with a hint of garlic; cap 5⁄16–½" (8–12 mm) wide, dry, convex to nearly flat, somewhat sulcate in age, whitish at first, becoming somewhat reddish, developing a perforation at the center; gills often appearing to be attached to a collarium, close to subdistant, crossveined, whitish, lamellulae present in moderate numbers; stalk scurfy, whitish to yellowish at apex, purplish brown to reddish brown overall, nearly black at the base; spores 5–7 × 3–3.5 μm—*Micromphale perforans* (Hoffmann and Fries) Singer.
    2. Usually growing in clusters on sticks or fallen branches; odor foul; cap ⅜–1⅛" (1–3 cm) wide, reddish brown at first, fading to tan, plicate, convex to nearly flat, with a small sunken area at the center in age; gills attached to a collarium, yellowish at first, becoming somewhat reddish in age; stalk brown, quite velvety, with a cottony base; spores 8–10 × 3.5–4 μm—*Micromphale foetidum* (Fries) Singer.
    2. Gregarious to clustered on decaying wood; odor rather fetid (especially when the cap is crushed); cap ⅛–⅝" (3–15 mm) wide, convex to nearly flat, dry, appearing glabrous, but minutely powdery under a hand lens, reddish brown at first, fading (starting at the margin) to dull grayish brown and eventually whitish, margin becoming slightly sulcate in age; gills attached, close to subdistant, often forked and crossveined in age, narrow, white; stalk pale at the apex, reddish brown overall, tapering abruptly to a white, pointed base—*Marasmiellus praeacutus* (Ellis) Halling (see p. 196).
    2. Not as in any of the above choices → 7.
3. Cap ¼–⅝" (5–15 mm) wide, convex with a distinct, nipple-like umbo, slightly striate, dingy whitish with a pinkish tinge, sometimes minutely hairy; gills subdecurrent, close to subdistant, whitish or yellowish; spores 10–11 × 3–4 μm—*Marasmiellus papillatus* (Peck) Redhead and Halling.

3.   Cap ⅜–¾" (1–2 cm) wide, wrinkled to distinctly plicate in age, lacking a pronounced umbo, finely powdery (use a hand lens), chalky white; gills attached to subdecurrent, subdistant to distant, often forked and crossveined in age, white, becoming reddish-stained in age; stalk about 1⁄16" (1–1.5 mm) thick, distinctly and rather evenly tapered toward the base, tough, hollow, whitish when very young, soon becoming black with a coating of minute, white hairs (use a hand lens); fine rhizomorph-like threads typically connected to the base of the stalk—*Marasmiellus nigripes* (Schweinitz) Singer (see p. 196).

3.   Not as in any of the above choices → 4.
    4.   Cap distinctly sulcate to plicate → 5.
    4.   Cap not distinctly sulcate to plicate → 6.

5.   Mushroom very tiny, cap ⅛–⅜" (3–10 mm) wide; growing on dead or decaying grass blades, especially common on lawns; cap orange to brown or reddish brown, fading to yellowish white; lamellulae absent; stalk about 1⁄100" (0.1–0.5 mm) thick; rhizomorphs absent—*Marasmius graminum* (Libert) Berkeley and Broome (see p. 200).

5.   Cap typically ⅛–¾" (3–20 mm) wide; growing in clusters on dead wood (especially common on beech); cap typically white to yellowish or rarely pale orange; gills attached to a collarium that may be either free from or collapsed onto the upper stalk, lamellulae absent or few; stalk no more than about 1⁄32" (0.3–1 mm) thick; rhizomorphs usually present but minimal, brown to blackish brown, typically wiry, curled or not—*Marasmius rotula* (Scopoli : Fries) Fries (see p. 201).

5.   Mushroom almost identical to previous choice but growing on oak leaves (not wood) and not in clusters; gills attached to a collarium that is rather consistently free from the stalk, lamellulae absent or few; rhizomorphs present but very minute (use a hand lens), black, twisted, and curled—*Marasmius capillaris* Morgan (see *Marasmius rotula*, p. 202).

5.   Cap very small, no more than ⅜" (10 mm) wide, purplish to reddish brown with a dense pruina at first (use a hand lens), distinctly sulcate to plicate at maturity; often shrinking in dry weather and reviving in wet weather; stalk base inserted into the substrate; gregarious to scattered on tree bark, especially on standing trunks—*Mycena corticola* (Fries) S. F. Gray (see p. 209).

5.   Cap ⅛–1" (3–25 mm) wide, orange to brown; substrate variable, usually growing on deciduous leaves or twigs but sometimes on eastern white pine needles, rarely on wood; odor and taste not distinctive; lamellulae absent to few; stalk no more than 1⁄32" (0.2–1 mm) thick, yellowish at first, darkening to brown from the base upward; rhizomorphs absent—*Marasmius siccus* (Schweinitz) Fries (see p. 202).

5.   Not as in any of the above choices → 11.
    6.   Stalk no more than 1⁄32" (1 mm) thick at narrowest portion, often bristle-like or horsehair-like → 7.
    6.   Stalk more than 1⁄32" (1 mm) thick at narrowest portion, not typically bristle-like or horsehair-like → 8.

7.   Cap brownish pink at the center, light yellowish pink at the edge, 1⁄32–⅛" (1–3 mm) wide; growing on spruce, fir, or white cedar needles; odor and taste distinctly of garlic (crush the cap if uncertain); lamellulae absent at first but typically developing a single tier of lamellulae by maturity; stalk about 1⁄100" (0.1–0.2 mm) thick, yellowish white near the apex, yellowish brown below; rhizomorphs rare to abundant, dark brown, curled, extremely thin; spores 8–11 × 2.5–4 µm—*Marasmius thujinus* Peck.

7.   Cap brown overall, 1⁄16–¾" (2–20 mm) wide; typically growing on conifer needles, cone scales, twigs, or other coniferous debris, rarely on deciduous leaves; odor and taste not distinctive; lamellulae abundant; stalk no more than 1⁄32" (0.2–1 mm) thick, brownish at the apex, blackish brown to black below; rhizomorphs rare to abundant, blackish

brown, bristle-like, typically twisted and curled; spores 7–10 × 3–4 μm—*Marasmius androsaceus* (Linnaeus : Fries) Fries.

7. Not as in either of the above choices → 9.

    8. Cap light brown to dull yellowish brown, fading to pale tan or pale yellowish white, with a brownish umbo that does not fade substantially in age, 1⅛–3⅛" (3–8 cm) wide; flesh whitish; odor and taste not distinctive; stalk dry, chalky white, typically longitudinally striate; scattered or in groups or clusters on the ground in mixed woods; spores 6–7 × 3–4.5 μm, elliptic, smooth, hyaline, inamyloid—*Marasmius nigrodiscus* (Peck) Halling.

    8. Cap light yellow, sometimes tinted orangish in spots (especially at the edge), ¾–2½" (2–6.5 cm) wide; growing on deciduous leaves; odor and taste either not distinctive or somewhat radishy; gills yellowish white to pale orangish yellow, lamellulae numerous; stalk up to ⅜" (9 mm) thick, whitish to pale yellow, hollow; rhizomorphs absent, but with an abundant pale yellow basal mycelium that usually creates a mat of several leaves—*Marasmius strictipes* (Peck) Singer (see p. 202).

    8. Cap tan, brownish or yellowish white, ⅜–1¾" (1–4.5 cm) wide; growing in grass on lawns, often in fairy rings or arcs; odor sweet or fragrant like almonds or not distinctive, taste not distinctive; gills pale tannish to yellowish white, lamellulae numerous; stalk 1/16–3/16" (2–4.5 mm) thick, white to yellowish or faintly orangish near the apex, more brownish below; rhizomorphs absent; edible—*Marasmius oreades* (Bolton : Fries) Fries (see p. 200).

    8. Not as in any of the above choices; cap typically no more than 1¼" (3 cm) wide at maturity → 9.

    8. Not as in any of the above choices; cap typically more than 1¼" (3 cm) wide at maturity → 10.

9. Cap reddish brown to yellowish brown but fading (especially at the edge), ¼–1¼" (6–30 mm) wide; typically growing on conifer needles but sometimes on bark or even grass; odor and taste usually of garlic but sometimes not distinctive; gills usually forked, lamellulae abundant; stalk 1/32–⅛" (0.5–3 mm) thick, yellowish at the apex, brown to reddish brown below; rhizomorphs absent; spores 7–10 × 3–5 μm; edible—*Marasmius scorodonius* (Fries) Fries.

9. Cap orangish yellow to brownish orange, ⅛–1" (4–25 mm) wide; usually growing on decaying oak leaves, rarely on wood, especially common in oak-hickory woods; odor and taste not distinctive; lamellulae numerous, usually in two to four tiers; stalk 1/32–3/32" (0.5–2 mm) thick, white to pale yellow at the apex, darkening to dark brown near the base; rhizomorphs absent—*Marasmius pyrrhocephalus* Berkeley (see p. 201).

9. Cap gray or light yellowish brown to brown with an olive tinge, becoming olive-gray in age, ⅜–¾" (1–2 cm) wide; odor and taste not distinctive; gills grayish to brownish, developing two irregular tiers of lamellulae; stalk 1/16–⅛" (2–2.5 mm) thick, whitish near the top, dark brownish below, usually rather long; rhizomorphs absent; spores 6.5–11 × 4–7 μm—*Marasmius chordalis* Fries.

9. Not as in any of the above choices; lamellulae numerous → 10.

9. Not as in any of the above choices; lamellulae absent to single-tiered → 11.

    10. Cap yellowish, lacking distinct brown tones, ¼–1½" (7–40 mm) wide; growing on deciduous leaves and twigs, especially oak; odor usually mildly fragrant but sometimes strong and with a fetid or spermatic odor, taste not distinctive; gills yellowish, lamellulae numerous; stalk 1/32–⅛" (1–3 mm) thick, yellowish overall at first, becoming brownish to nearly black from the base upward; rhizomorphs absent; spores 5.5–9 × 3–5 μm—*Marasmius delectans* Morgan.

    10. Cap yellowish brown to brown, ⅜–1½" (1–3.5 cm) wide; growing on deciduous

leaves, twigs, and occasionally logs; odor mild to pungent or earthy, taste mild or
with a bitter aftertaste; gills yellowish white to pinkish yellow, lamellulae numerous; stalk ⅓₂–⅛" (1–3 mm) thick, yellowish, becoming brownish to nearly black
from the base upward; rhizomorphs absent, but with a white to yellowish brown
basal mycelium; spores 7–10 × 3–5.5 μm—*Marasmius cohaerens* var. *cohaerens*
(Persoon : Fries) Cooke and Quélet.

10. Not as in either of the above choices → 11.

11. Cap brown to orangish brown or reddish brown, minutely velvety (use a hand lens),
¾–1¾" (2–4.5 cm) wide; more common in the southern part of the region; odor and
taste mildly farinaceous; gills typically fewer than 30, yellowish white, lamellulae absent;
stalk slightly less than 1⁄16" (1–1.25 mm) thick, pinkish near the apex, shading to blackish
brown or black toward the base; rhizomorphs absent, but with a conspicuous tuft of
cottony, white mycelium at the base; spores 15–18 × 3–4.5 μm—*Marasmius fulvo-
ferrugineus* Gilliam.

11. Cap pink to reddish but often pale yellowish brown when young, ⅛–⅝" (3–17 mm)
wide; growing on conifer needles or deciduous leaves or other woody debris; odor not
distinctive, taste mild to faintly radishy or bitter; lamellulae absent or few; stalk about
1⁄100" (0.2–0.6 mm) thick, pinkish overall, darker below; rhizomorphs absent; spores
10–15 × 2.5–4.5 μm—*Marasmius pulcherripes* Peck.

11. Cap yellowish brown to yellow-brown, ¼–1" (5–25 mm) wide; growing on deciduous
leaves; odor and taste mildly spermatic; gills yellowish, lamellulae typically single-tiered;
stalk 1⁄16" (2 mm) thick at most, white to yellowish above, deep brown toward the base;
rhizomorphs absent, but with white to pale orangish yellow basal mycelium;
spores 7–12 × 3–6 μm—*Marasmius glabellus* Peck.

Marasmius graminum (Libert) Berkeley and Broome
COMMON NAME: Grass Marasmius.
CAP: ⅛–⅜" (3–10 mm) wide, convex, sulcate, typically with a tiny depression at the center
of the cap; surface orange to brown or reddish brown, fading to yellowish white; flesh
indistinguishably thin; odor and taste not distinctive.
GILLS: attached to a collarium or to the stalk, distant, narrow, white to yellowish, sometimes with brownish edges; lamellulae typically absent but sometimes one can be
found.
STALK: ⅝–1¼" (1.5–3.5 cm) long, about 1⁄100" (0.1–0.5 mm) thick, shiny; yellowish white
at the extreme apex, reddish brown to blackish brown below; rhizomorphs absent.
SPORE PRINT: white (but very difficult to obtain).
MICROSCOPIC FEATURES: spores 6.5–11 × 3–5.5 μm, roughly appleseed-shaped, smooth,
hyaline, inamyloid.
FRUITING: in groups on dead grass blades; June–September; very common but easily
overlooked.
EDIBILITY: inedible.
COMMENTS: even knowledgeable, experienced mycophiles miss this rather ubiquitous but
miniscule mushroom. It is easy to find if, shortly after a rain or early in the morning
while the grass is still wet with dew, you get down on your hands and knees on a lawn
with plenty of dead grass blades and look around awhile.

Marasmius oreades (Bolton : Fries) Fries
COMMON NAME: Fairy Ring Mushroom.
CAP: ⅜–1¾" (1–4.5 cm) wide, typically broadly bell-shaped, with a broad umbo; surface
tan, brownish, or yellowish white, shiny and smooth when moist, dull and minutely

velvety when dry; flesh moderately thick, white to yellowish white; odor sweet or fragrant, like almonds or not distinctive; taste not distinctive.

GILLS: free to attached, close to distant, narrow at first, becoming quite broad, rather thick, pale tannish to yellowish white, tough; lamellulae numerous.

STALK: ⅜–2¾" (1–7 cm) long, about ⅛" (2–5 mm) thick, white to yellowish or faintly orangish near the apex, more brownish below, flexible but very tough (easily bent without breaking); rhizomorphs absent.

SPORE PRINT: white.

MICROSCOPIC FEATURES: spores 6.5–10 × 3.5–6 μm, typically roughly lemon-shaped with a small rounded apical bump, smooth, hyaline, inamyloid.

FRUITING: in grass, especially on lawns, often in fairy rings or arcs; May–October; frequent to common.

EDIBILITY: edible with caution (see comments below).

COMMENTS: a fine edible, but you must be sure not to misidentify the poisonous Sweating Mushroom, *Clitocybe dealbata*, which has comparatively fragile, decurrent gills, does not revive if dried and then rehydrated, and has a stalk that cannot be bent and twisted without breaking.

Marasmius pyrrhocephalus Berkeley

CAP: ⅛–1" (4–24 mm) wide, convex to rounded when young, becoming broadly convex and often depressed; margin typically uplifted in age; surface moist or dry, translucent-striate, smooth when young, becoming weakly sulcate in age, orangish yellow to brownish orange, darkest on the disc; flesh whitish; odor and taste not distinctive.

GILLS: attached to notched, close to subdistant, white at first, becoming dull pale yellow, often spotted brownish when bruised or in age; lamellulae numerous, usually in two to four tiers.

STALK: 1⅜–4" (3.5–10 cm) long, 1⁄32–3⁄32" (0.5–2 mm) thick, dry, white to pale yellow at the apex, progressively darkening to blackish brown at the base, coated with matted fibrils on the lower portion; rhizomorphs absent.

SPORE PRINT: white to whitish.

MICROSCOPIC FEATURES: spores 6–10 × 3–5 μm, elliptic, obovate, or somewhat fusiform, smooth, hyaline, inamyloid.

FRUITING: scattered or in groups on decaying oak leaves and humus under hardwoods or mixed woods; May–November; occasional.

EDIBILITY: unknown.

Marasmius rotula (Scopoli : Fries) Fries

COMMON NAME: Pinwheel Marasmius.

CAP: ⅛–¾" (3–20 mm) wide, convex with a small depression at the center; surface typically white to yellowish or rarely pale orange; flesh very thin at the edge, visibly white to yellowish at the center; odor not distinctive; taste mild or with a slight bitter aftertaste.

GILLS: attached to a collarium that may be either free from or collapsed onto the upper stalk, distant or nearly so, narrow to rather broad, nearly white to pale yellow; lamellulae typically absent but infrequently with a single short tier.

STALK: ⅝–3½" (1.5–8.5 cm) long, less than 1⁄32" (0.3–1 mm) thick, yellowish white to brownish at the apex, reddish brown to dark brown to blackish brown overall at maturity, sometimes with a tiny bulb at the base; rhizomorphs usually present but minimal, brown to blackish brown, typically wiry, curled or not.

SPORE PRINT: white.

MICROSCOPIC FEATURES: spores 6–9.5 × 3–4.5 μm, more or less narrowly lacrymoid, smooth, hyaline, inamyloid.

FRUITING: growing in clusters on dead wood, especially common on beech; May–October; common.

EDIBILITY: inedible.

COMMENTS: compare the very similar *Marasmius capillaris* Morgan (edibility unknown), which grows on oak leaves and is never clustered.

Marasmius siccus (Schweinitz) Fries

COMMON NAME: Orange Pinwheel Marasmius.

CAP: ⅛–1" (3–25 mm) wide, cushion- to bell-shaped at first, becoming convex, depressed, or umbonate on the disc; surface somewhat roughened, strongly plicate to sulcate at maturity, orange to reddish orange or yellowish brown, often fading in age to orange-yellow or brownish orange; flesh thin, white, dextrinoid; odor and taste not distinctive.

GILLS: notched, sinuate or free, distant, white to pale yellow; lamellulae absent to few.

STALK: ¾–2¾" (2–7 cm) long, up to ¹⁄₃₂" (0.2–1 mm) thick, nearly equal, cartilaginous, hollow, yellowish at first, darkening to brown from the base upward; rhizomorphs absent.

SPORE PRINT: white.

MICROSCOPIC FEATURES: spores 15–22 × 3–5 μm, club- to spindle-shaped, often curved, smooth, hyaline.

FRUITING: scattered or in groups on decaying leaves, twigs, and conifer needles, especially white pine; June–October; fairly common.

EDIBILITY: unknown.

Marasmius strictipes (Peck) Singer

COMMON NAME: Orange-yellow Marasmius.

CAP: ¾–2½" (2–6.5 cm) wide, convex to broadly umbonate; surface light yellow, sometimes tinted orangish in spots (especially at the edge); flesh yellowish white to white, dextrinoid; odor and taste mild to somewhat radishy.

GILLS: attached but often seceding, close to crowded, narrow, yellowish white to pale orangish yellow, occasionally forked, lamellulae numerous.

STALK: 1–3½" (2.5–8.5 cm) long, ⅛–⅜" (3–9 mm) thick, whitish to pale yellow, hollow; rhizomorphs absent, but with an abundant pale yellow basal mycelium that usually binds together a mat of several leaves.

SPORE PRINT: white.

MICROSCOPIC FEATURES: spores 6–10.5 × 3–4.5 μm, roughly elliptic to lacrymoid, smooth, hyaline, inamyloid.

FRUITING: scattered to gregarious on deciduous leaves; July–September; infrequent to frequent.

EDIBILITY: inedible.

COMMENTS: it is easily mistaken for a *Collybia*; indeed, it was originally described as *Collybia strictipes*.

Genus *Megacollybia*

Megacollybia platyphylla (Persoon : Fries) Kotlaba and Pouzar
CAP: 1–6" (2.5–15 cm) wide, convex to nearly flat, often with a broad umbo in age; surface smooth, dry to moist, dull brownish gray to grayish brown, streaked with dark radial fibers; flesh white, moderately thick at the center; odor not distinctive; taste mild to somewhat bitter.
GILLS: attached, subdistant, very broad, white to grayish.
STALK: 3–5" (7.5–12.5 cm) long, ⅜–¾" (1–2 cm) thick, dry, white, with thick, white rhizomorphs.
SPORE PRINT: white.
MICROSCOPIC FEATURES: spores 7–9 × 5–7 μm, oval, smooth, hyaline, inamyloid.
FRUITING: solitary, in groups or in clusters on or about decaying deciduous wood; May–October; common as far west as Ohio, rarer westward.
EDIBILITY: edible with caution; some people get upset stomachs from it, and it must be thoroughly cooked.
COMMENTS: this is one of the first large, conspicuous gilled mushrooms to appear late each spring, and it is rather ubiquitous by early summer. Formerly classified as *Tricholomopsis platyphylla*.

Genus *Melanoleuca*

Melanoleuca is a small and fairly distinctive genus typically found growing in humus but sometimes on lawns, and never on decaying wood. A fairly reliable set of field characters for *Melanoleuca* includes a smooth, convex to flat, white to yellowish to dark brown, often hygrophanous cap with a leather-like surface texture and an umbo; very crowded white gills that are always attached but never decurrent; a tall, slender stalk that is usually longitudinally striate; and a spore print that is usually white but in some species ranges to pale yellowish cream or orangish yellow. The spores are warty, with a plage, and amyloid. The two species keyed out here have white spore prints; both are reportedly edible, but the genus has not been studied well enough for us to recommend them without reservation.

Key to Species of *Melanoleuca*

1. Cap dark brown at first, fading to pale brownish, 1–3" (2.5–7.5 cm) wide, convex at first, becoming plane with a low umbo; gills close, white; stalk 1–3⅛" (2.5–8 cm) long, ⅛–½" (3–12 mm) thick, white, longitudinally striate with dark fibers; spores 7–9 × 5–6 μm; often on lawns and other grassy areas; reportedly edible—*Melanoleuca melaleuca* (Persoon : Fries) Murrill complex.
1. Cap yellowish brown, fading to yellowish white—*Melanoleuca alboflavida* (Peck) Murrill (see p. 203).

Melanoleuca alboflavida (Peck) Murrill
CAP: 2–4" (5–10 cm) wide, convex to nearly flat or slightly depressed, with a low umbo; surface moist, smooth, yellowish brown fading to yellowish white, but the umbo remaining somewhat darker; flesh white; odor and taste not distinctive.
GILLS: attached, typically sinuate, crowded, narrow, whitish.
STALK: 1¼–6" (3–15 cm) long, ¼–⅜" (5–10 mm) thick, slightly enlarged at the base, dry, whitish, typically slightly twisted, longitudinally striate with dark fibers.

SPORE PRINT: white.

MICROSCOPIC FEATURES: spores 7–9 × 4–5.5 μm, oval to elliptic, finely warty, hyaline, amyloid.

FRUITING: solitary, scattered, or in groups on the ground in deciduous or mixed forests; June–September; occasional to common.

EDIBILITY: edible with caution (see comments).

COMMENTS: this mushroom is quite tasty, according to several sources, but accurate identification is necessary to be safe.

Genus *Melanophyllum*

Melanophyllum echinatum (Roth : Fries) Singer

COMMON NAME: Red-gilled Agaricus.

CAP: ½–2" (1.2–5 cm) wide, conic to convex, becoming nearly plane in age; surface dry, coated with short, erect, brown scales over a grayish brown to yellowish ground color; scales fragile, soon powdery and easily removed; flesh whitish; odor and taste not distinctive; margin sometimes rimmed with torn partial veil, especially on young specimens.

GILLS: free, close, bright to dark red, becoming brown at maturity.

STALK: ¾–3½" (2–9 cm) long, 1/16–3/16" (1.5–4.5 mm) thick, nearly equal down to a bulbous base, scurfy to nearly smooth, pinkish brown, darkening in age; partial veil membranous, grayish, leaving veil remnants or a sparse superior ring.

SPORE PRINT: dull red when fresh, drying purplish brown.

MICROSCOPIC FEATURES: spores 5–7 × 2–3 μm, elliptic, finely punctate to nearly smooth, hyaline.

FRUITING: scattered, in groups or clusters on rich humus, leaf litter, and very well-decayed wood under hardwoods; June–October; occasional.

EDIBILITY: unknown.

Genus *Mycena*

Mycena is a large genus of small mushrooms, with caps usually less than two inches wide and stalks less than 1/8" (3 mm) in diameter. Though a few species frequently appear terrestrial, Mycenae are saprobes, decaying wood, leaves, and other dead plant materials. Some have rather peculiar habitat tendencies *(M. osmundicola,* for example, is frequently collected on dead fern stalks) or even strict requirements *(M. luteopallens,* exclusively on decaying walnut or hickory nut hulls). Identification of *Mycena* species varies from simple to daunting. There are a good number of very distinctive species that are easy to key out but many others, especially species of Section Typicae, are nondescript and require careful microscopic study for accurate identification. We know of no worthwhile edible species and have therefore omitted information on edibility from the key. They should all be considered inedible.

Key to Species of *Mycena* and Allies

1. Stalk with a tiny bulb that flattens into a disc by maturity; cap 1/8–5/8" (3–15 mm) wide, almost conic at first, becoming almost flat, off-white with a paler margin, translucent, surface smooth (if the mushroom is young, the center of the cap may be minutely spiny, moist); gills slightly attached, whitish to slightly grayish, with one or two tiers of lamellulae; stalk 3/8–2¼" (1–6 cm) long, little more than 1/64" (0.5 mm) thick, cleanly

separable from the cap; spores 6–10 × 3.5–4.5 μm; growing on oak leaves or conifer needles—*M. stylobates* (Fries) Quélet.

1. Not as in the above choice; cap white-pruinose and generally remaining somewhat so even in age; cap very small, ⅛–¼" (3–6 mm) wide, convex to conic, cuticle grayish to whitish, moist and striate; gills free to barely attached, subdistant to distant, narrow; stalk ¾–1¼" (2–3 cm) long, about 1/64" (0.5 mm) thick, slightly thicker toward the base, finely hairy at first, becoming coated with powder; spores 7–10 × 4–5 μm; growing on conifer needles, deciduous leaves, or especially on fern debris—*M. osmundicola* Lange.

1. Not as in either of the above choices; cap and stalk distinctly viscid to glutinous (not merely slippery-smooth), though sometimes covered by a white, pruinose layer → 2.

1. Not as in any of the above choices; stalk neither viscid nor glutinous (though it may be slippery-smooth) → 4.

    2. Gills bright lilac when young, fading to pale, dull lilac in age; cap grayish lavender when very young, soon becoming dingy yellow with a paler margin; scattered or in groups on decaying conifer wood, especially balsam fir and hemlock—*Chromosera cyanophylla* (Fries) Redhead, Ammirati and Norvell = *M. lilacifolia* (Peck) Smith (see p. 76).

    2. Gills pinkish orange to orange, gill edges deep reddish orange; cap reddish orange at first, fading to yellowish orange; growing in clusters on decaying deciduous wood, especially frequent on beech—*M. leaiana* (Berkeley) Saccardo (see p. 212).

    2. Not as in either of the above choices → 3.

3. Growing on humus or among mosses in pine or pine-oak woods; odor farinaceous, taste farinaceous to somewhat disagreeable; gills whitish, often with a greenish tinge at first, sometimes developing reddish brown spots in age; cap and stalk both distinctly white-pruinose at first; cap ⅜–1¼" (1–3 cm) wide, deep olive to greenish brown, becoming nearly black in age and often covered with a hoary sheen; stalk 1½–3½" (4–9 cm) long, 1/32–⅛" (0.75–3 mm) thick, yellow to greenish yellow to grayish, becoming purplish brown near the base—*M. griseoviridis* Smith (see p. 211).

3. Growing among mosses or on humus in pine or pine-oak woods; odor and taste farinaceous; gills pale grayish to whitish, sometimes developing reddish brown spots in age; cap and stalk both distinctly white-pruinose at first; cap ⅜–1" (1–2.5 cm) wide, olive-gray, not fading to whitish in age, sometimes covered with a hoary sheen; stalk 1½–3" (4–7.5 cm) long, about 1/16" (1–2 mm) thick, pale greenish yellow, often becoming reddish at the base—*M. epipterygia* var. *epipterygioides* (Pearson) Kühner (see p. 210).

3. Growing on humus or among mosses in pine or pine-oak woods; odor not distinctive or mildly fragrant, taste not distinctive; gills whitish to faintly yellowish, not developing reddish brown spots; cap and stalk not distinctly pruinose at first; cap up to ¾" (2 cm) wide, yellow, becoming gray-brown to grayish or sometimes fading to whitish; stalk yellow, not becoming reddish brown near the base; spores 8–11 × 5–6 μm—*M. epipterygia* var. *epipterygia* (Scopoli : Fries) S. F. Gray.

3. Growing on conifer wood, often at the base of standing trees; odor farinaceous to pumpkin-like, taste unpleasant and often somewhat bitter; gills yellowish to greenish yellow or sometimes whitish, developing reddish brown spots in age; flesh yellowish, becoming reddish in age; cap and stalk both distinctly white-pruinose at first; cap up to 1⅛" (3 cm) wide, typically lacking a distinct umbo, color variable, greenish yellow, grayish yellow, or greenish gray to yellowish gray when young, becoming sordid brownish in age, usually developing reddish brown spots at maturity—*M. epipterygia* var. *viscosa* (Maire) Ricken (see p. 211).

3. Growing on conifer wood, often moss-covered; odor and taste farinaceous; gills white

to pale yellow, not developing reddish brown spots; cap white-pruinose at first, ¼–⅝"
(7–16 mm) wide, distinctly umbonate, greenish yellow to yellow with a paler margin—
M. epipterygia var. *lignicola* Smith (see p. 210).

    4.   Latex not exuding from the stalk when cut; cap pale blue to greenish blue at first,
often becoming grayish or brownish overall in age but typically retaining at least a
hint of blue at the margin; cap ¼–¾" (5–20 mm) wide, shape variable but essentially convex, sticky to slippery, smooth overall but often translucent- striate and
somewhat granular at the margin; cuticle entirely removable; odor and taste not
distinctive; gills close to crowded, variously attached, whitish to grayish, edges
minutely hairy (use a hand lens), with abundant lamellulae; stalk bluish to bluish
green overall and densely pruinose or pubescent when young, 1–3" (2.5–7.5 cm)
long, ¹⁄₃₂–³⁄₃₂" (0.75–2.5 mm) thick, becoming grayish to brownish in age, elastic,
hollow, with a distinct greenish coating of mycelium on the base when young;
growing on humus, decaying woody hardwood debris (bark, twigs, etc.), or
leaves, not clustered—*M. subcaerulea* (Peck) Saccardo (see p. 213).

    4.   Latex orangish yellow to blood-red → 5.

    4.   Latex copious, clear and watery → 6.

    4.   Not as in any of the above choices; latex, if any, scant and watery → 7.

5.   Latex orangish yellow in cap, dull orange to dull reddish brown in stalk, gills distinctly
marginate; cap ¼–1" (5–25 mm) wide, shape variable but essentially convex, very finely
pruinose when young but typically appearing shiny, reddish brown; gills attached, close
to almost crowded, pale yellow when young, edges maroon; stalk ¾–1½" (2–4 cm)
long, ¹⁄₁₆–⅛" (1–3 mm) thick, often flattened, typically rooting, pale at the apex, purplish brown below, smooth overall but scantily covered with brownish fibers, and with
a distinctly hairy base; spores 7–9 × 4–6 μm; typically gregarious to clustered on deciduous leaves, especially beech—*M. atkinsoniana* Smith.

5.   Latex reddish in cap and stalk; gills distinctly marginate; cap ⅛–¾" (3–20 mm) wide,
conic to convex, densely grayish pruinose when young, soon becoming shiny and moist
and sulcate, reddish brown overall, redder at the center and more yellowish at the margin; odor and taste not distinctive; gills attached and typically with a fine subdecurrent
"tooth," distant or nearly so, dirty reddish to grayish, the edges a very dark reddish
brown; stalk ¾–2¼" (2–6 cm) long, ¹⁄₆₄–¹⁄₃₂" (0.5–0.75 mm) thick, fragile, more or less
smooth and the same color as the cap but finely pruinose at first and with a finely hairy
base even in mature specimens; spores 8–11 × 4–6 μm; typically gregarious on dead
leaves or conifer needles but also in beds of moss—*M. sanguinolenta* (Fries) Quélet.

5.   Latex dark blood-red in cap and stalk; gills not marginate or only vaguely so; cap ⅜–2"
(1–5 cm) wide, conic to convex to flat in age, dry and densely pruinose at first but soon
becoming polished and moist, dark reddish brown at the center and fading toward the
margin, which is typically torn at maturity; odor not distinctive, taste mild to bitter
(latex typically bitterish); gills narrowly attached, whitish to grayish purple, becoming
dingy reddish brown with pale minutely powdery edges and with abundant lamellulae;
stalk ¹⁄₃₂–¹⁄₁₆" (0.75–1.5 mm) thick, consistently with a distinctly reddish tint, fragile,
hollow, with a distinctly hairy base; typically growing in clusters but sometimes solitary
on decaying wood—*M. haematopus* (Fries) Quélet (see p. 211).

    6.   Gills distinctly marginate (edges dirty brownish); cap ⅜–1¼" (1–3 cm) wide,
shape variable but essentially convex, glabrous, brownish gray to grayish brown,
paler at the margin, even to slightly wrinkled, appearing dry and velvety; flesh
thin, oozing drops of clear latex when cut; gills attached to subdecurrent, close to
crowded, narrow, with numerous lamellulae; stalk ⅝–1¼" (1.5–3 cm) long,
¹⁄₃₂–¹⁄₁₆" (0.75–1.5 mm) thick, dark gray to blackish brown with a pruina at first,

fading to grayish brown or translucent-gray and becoming glabrous; odor and taste not distinctive; scattered to gregarious on conifer wood; spores 6–7.5 × 3.5–4 μm, elliptic with an apiculus—*M. marginella* var. *marginella* (Fries) Quélet.

6. Gills not marginate; mushroom otherwise nearly identical to previous choice—*M. marginella* var. *rugosodisca* (Peck) Smith.

6. Gills not marginate; entire mushroom staining dark blue where bruised; with copious clear latex that stains fingers or white paper quickly purple, then dark blue; cap ⅜–1¼" (1–3 cm) wide, shape variable but essentially convex, brownish overall, finely fibrous; flesh whitish, staining bluish black; gills whitish, staining bluish black; stalk ¾–2" (2–5 cm) long, about ³⁄₃₂" (2 mm) thick, dry, white-pruinose over a brownish ground color; odor and taste not distinctive; scattered to clustered on dead conifer wood, especially hemlock; spores 3–4 μm, subglobose to globose, small, with an apiculus—*M. fuliginaria* (Fries) *sensu* Kühner.

7. Cap very small, ⅛–⅜" (3–10 mm) wide, shape variable but essentially convex, deep red with a fine pruina at first, fading to orangish yellow with a yellow margin, smooth, fairly glabrous at maturity; gills attached, pale orange to whitish, the edges often paler than the sides, with numerous lamellulae; stalk ⅜–2¼" (1–6 cm) long, very thin, almost hair-like in some specimens, up to about ¹⁄₃₂" (0.75 mm) thick in others, yellow to orangish yellow, at first with a dense white pruina, base with white, strigose hairs; odor and taste not distinctive; spores 9–11 × 3.5–4.5 μm; single to gregarious on hardwood debris—*M. acicula* (Fries) Quélet.

7. Cap very small, no more than ⅜" (1 cm) wide, purplish to reddish brown with a dense pruina at first (use a hand lens), distinctly sulcate to plicate at maturity; often shrinking in dry weather and reviving in wet weather; stalk base inserted into the substrate; gregarious to scattered on tree bark, especially on standing trunks—*M. corticola* (Fries) S. F. Gray (see p. 209).

7. Cap very small, ⅛–⅜" (3–10 mm) wide, somewhat conic to convex, becoming flat, chalky white with a fine white pruina (use hand lens); gills free or nearly so, narrow, white; stalk ⅜–1¼" (1–3 cm) long, about ¹⁄₃₂" (.075 mm) thick, white with a fine pruina, base with a dense mat of hairs; scattered to gregarious on conifer needles and twigs; spores 7–12 × 2.5–3.5 μm, somewhat cylindric, with a tapered and often eccentric apiculus; cheilocystidia and pleurocystidia present but difficult to find, many rather bowling-pin-shaped—*M. delicatella* (Peck) Smith.

7. Cap small, ¼–⅝" (5–15 mm) wide, convex, becoming sunken at the center, white, dry to sticky; gills attached at first, becoming decurrent, close, white, edges often uneven (use a hand lens), with numerous lamellulae; stalk ½–2" (1.2–5 cm) long, ¹⁄₆₄–¹⁄₁₆" (0.5–1.5 mm) thick, white, sticky, with a tuft of white mycelium at the base; scattered to grouped or in small clusters on deciduous leaves or twigs—*Resinomycena rhododendri* (Peck) Redhead and Singer (see p. 241).

7. Cap small, ⅛–⅝" (3–15 mm) wide; yellowish orange to orangish yellow when young, fading somewhat, smooth, glabrous; margin inrolled at first, radially lined; gills strongly decurrent, whitish; stalk ⅜–2" (1–5 cm) long, about ¹⁄₁₆" (1.5 mm) thick, minutely hairy at first (use a hand lens), becoming glabrous, fragile, orangish, hollow; spores 4–5 × 2–2.5 μm, elliptic, hyaline, inamyloid; growing among mosses—*Rickenella fibula* (Bulliard : Fries) Raithelhuber (see p. 243).

7. Not as in any of the above choices; gills not marginate; cap surface white or with tints of yellow, orange, pink, red, or purple when young → 8.

7. Not as in any of the above choices; gills marginate; cap surface yellow, orange, pink, red, purple, or darker brownish gray to grayish brown → 9.

7. Not as in any of the above choices → 10.

8. Cap, gills, and stalk usually purplish or lilac when young; odor and taste distinctly of radishes—*M. pura* (Fries) Quélet (see p. 213).

8. Growing on walnut or hickory nut hulls; cap orange-yellow to yellow at first—*M. luteopallens* (Peck) Saccardo (see p. 212).

8. Mushroom lacking any yellow tints, even in age; cap pinkish red at first; gills and stalk white to pink or reddish; typically growing in moss (especially sphagnum) but also on beds of conifer needles; cheilocystidia with long, slender apical necks—*M. amabilissima* (Peck) Saccardo (see p. 209).

8. Cap 5/16–1/2" (8–12 mm) wide, shape variable but essentially conic to convex, fleshy pink, becoming yellowish at the center and whitish near the margin, glabrous, moist; gills attached with a fine subdecurrent line, pinkish at first, fading to whitish; stalk typically about 2" (5 cm) long, 1/32–1/16" (0.75–1.5 mm) thick, grayish pink overall, sometimes with a paler apex; apparently restricted to beds of white pine needles; spores 7–10 × 4–5 μm; cheilocystidia essentially narrowly clavate with several small rounded projections on the swollen portion—*M. subincarnata* (Peck) Saccardo.

9. Odor distinctly of radishes, taste bitter or of radishes; cap 5/8–3" (1.5–7.5 cm) wide, shape variable but essentially convex, typically purplish brown but fading markedly to brown, sometimes with a dirty yellow tint, glabrous, moist, slippery, with a striate margin when moist; flesh fairly thick, yellowish to whitish; gills attached but variously so and often seceding from the stalk, quite broad, purplish brown with reddish purple edges, with at least one tier of lamellulae; stalk 1 1/4–3" (3–8 cm) long, 1/8–3/8" (3–10 mm) thick, hollow, longitudinally striate, grayish white with a yellowish apex, often quite roughened; spores 8–10 × 3.5–5 μm; scattered to gregarious on humus and debris, especially in oak and hickory forests—*M. rutilantiformis* Murrill.

9. Odor and taste not distinctive; cap reddish orange, fading to yellow or yellowish white; gills yellow to orange with reddish edges; stalk orangish; cap 3/8–3/4" (1–2 cm) wide, shape variable but essentially convex, moist, smooth, slippery, with a somewhat translucent-striate margin when moist; gills finely attached with a subdecurrent line, subdistant, with at least one tier of lamellulae; stalk 1–1 1/2" (2.5–4 cm) long, 1/32–1/16" (0.75–1.5 mm) thick, with an orange pruina at first, becoming glabrous; spores 7–9 × 4–5.5 μm; gregarious on beds of conifer needles, especially pine, typically at higher elevations on mountains; uncommon in the Northeast—*M. strobilinoides* Peck.

9. Odor and taste not distinctive; cap pinkish to purplish, fading to somewhat yellowish, gills pinkish to reddish with dirty reddish edges, stalk pinkish to reddish overall when young; cap 1/4–1 1/4" (6–30 mm) wide, conic to convex to plane, moist, slippery, margin translucent-striate when moist; gills attached but variously so, with one or two tiers of lamellulae; stalk 1–2 3/4" (2.5–7 cm) long, 1/32–1/16" (0.75–1.5 mm) thick, glabrous and slippery but not slimy, base with white strigose hairs or coated with white mycelium; spores 7–9 × 4–5 μm; growing on conifer needle beds—*M. rosella* (Fries) Quélet.

9. Odor and taste not distinctive; cap dark purple at first; gills pale gray with dark purple edges; stalk dark purple, often with a paler apex; cap 1/4–1" (5–25 mm) wide, shape variable but essentially convex, surface hoary-pruinose at first, becoming more glabrous, opaque at first, becoming markedly translucent-striate when moist and mature; gills finely attached, narrow; stalk 1 1/4–4" (3–10 cm) long, 1/32–1/16" (0.75–1.5 mm) thick, relatively tough and cartilaginous, with white, strigose hairs at the base; spores 8–14 × 6–8.5 μm; on conifer wood and debris—*M. purpureofusca* (Peck) Saccardo.

NOTE: *The following species belong to* Section Typicae, *which includes at least fifty rather non-descript gray, brown, and blackish species from northeastern North America. Identifications made with this portion of the key should be considered tentative at best.*

10. Cap and gills developing dirty reddish brown spots in age; cap blackish brown to nearly black at first; flesh slowly staining reddish brown when cut; odor not distinctive, taste mild to faintly farinaceous; spores 7–10 × 4–6 µm; gregarious to clustered on deciduous or coniferous wood—*M. maculata* Karsten.

10. Mushroom with a bleach-like odor; cap blackish at first, fading to dirty yellowish brown or grayish at maturity, with a whitish margin; taste slightly acid or not distinctive; gills sometimes developing reddish brown stains; spores 8–11 × 4.5–6 µm; gregarious to almost clustered on decaying conifer wood or humus, especially of larch (tamarack)—*M. alcalina* (Fries) Quélet.

10. Cap dark gray at the center, whitish at the margin, becoming markedly crenate at maturity; lower stalk at first coated with white myclium that breaks up into fibrous flecks; stalk base densely strigose, the hairs becoming pale orangish brown at maturity; odor and taste slightly farinaceous; spores 7–10 × 5–7 µm; sometimes gregarious but typically in large clusters on deciduous wood—*M. inclinata* (Fries) Quélet.

10. Cap slightly radially wrinkled, more or less brown, fading to tan or dirty cinnamon; odor and taste slightly to strongly farinaceous; gills rather broad and cross-veined; stalk base naked; spores 8–11 × 5.5–7 µm; scattered to gregarious or almost clustered on deciduous wood or debris—*M. galericulata* (Fries) S. F. Gray.

Mycena amabilissima (Peck) Saccardo

CAP: ⅛–¾" (3–20 mm) wide; conic to convex at first, becoming broadly conic or somewhat bell-shaped; surface glabrous, slippery when moist; brilliant pinkish red at first, fading to nearly white in age, lacking yellow tints; margin initially pressed against the stalk, opaque to distinctly translucent-striate; flesh thin, pinkish, fading to white; odor and taste not distinctive.

GILLS: attached, subdistant to distant, crossveined, rather narrow at first; white or reddish, with numerous lamellulae.

STALK: 1¼–2" (3–5 cm) long, 1/64–1/16" (0.5–1.5 mm) thick, hollow, fragile, pruinose at first, becoming glabrous and moist; reddish fading to white, in some specimens white at first; base becoming slightly yellowish.

SPORE PRINT: white.

MICROSCOPIC FEATURES: spores 7–9 × 3–4 µm, narrowly elliptic, tapering to an apiculus, smooth, hyaline, inamyloid; cheilocystidia with long, slender apical necks.

FRUITING: typically gregarious among mosses, especially sphagnum, but sometimes scattered on beds of conifer needles; May–November; occasional to common.

EDIBILITY: inedible.

Mycena corticola (Fries) S. F. Gray

CAP: ⅛–⅜" (3–10 mm) wide, shape variable but essentially convex; surface purplish to reddish brown with a dense pruina at first (use a hand lens), fading to pale grayish brown, distinctly sulcate to plicate and fairly glabrous at maturity; shriveling and shrinking in dry conditions and reviving like a *Marasmius* in rainy weather; odor and taste not distinctive.

GILLS: attached to slightly decurrent, broad, distant, colored similar to the cap but paler, with one or two tiers of lamellulae.

STALK: ³⁄₁₆–³⁄₈" (5–10 mm) long, about ¹⁄₆₄" (0.5 mm) thick, colored like the cap or gills but covered with a whitish pruina at first, base with only a few white hairs, stalk inserted into the substrate.

SPORE PRINT: white.

MICROSCOPIC FEATURES: spores 9–11 μm, globose with a small apiculus, hyaline, amyloid; pleurocystidia absent; cheilocystidia clavate and decorated with numerous fingerlike projections that are often contorted.

FRUITING: gregarious, but sometimes scattered, on coniferous or deciduous bark, especially on standing trunks; August–November, but also collected through winter and into early spring; common.

EDIBILITY: inedible.

COMMENTS: this minute Mycena is easily overlooked.

Mycena epipterygia var. **epipterygioides** (Pearson) Kühner

CAP: ³⁄₈–1" (1–2.5 cm) wide, ovoid to obtusely conic, becoming broadly conic to convex or nearly plane in age, often with a low, broad umbo or flattened disc; surface pruinose or covered with a hoary sheen at first, becoming bald at maturity, viscid, sulcate-striate in age, olive-gray to olive-buff, not fading to whitish in age; flesh dark olive brownish; odor and taste farinaceous.

GILLS: bluntly attached and often forming a collar, subdistant to close, pale grayish to whitish, sometimes developing reddish brown spots in age, with two tiers of attenuate lamellulae.

STALK: 1½–3⅛" (4–8 cm) long, ¹⁄₁₆–⅛" (1.5–3 mm) thick, nearly equal, viscid, pruinose, at least near the apex, and smooth below, pale greenish yellow, base often coated with scattered fibrils and becoming reddish in age.

SPORE PRINT: white.

MICROSCOPIC FEATURES: spores 8–12 × 4–6 μm, ellipsoid, smooth, hyaline, amyloid.

FRUITING: scattered or in groups among mosses or on humus in pine or oak woods; September–November; occasional to fairly common.

EDIBILITY: unknown.

Mycena epipterygia var. **lignicola** Smith

CAP: ¼–⅝" (7–16 mm) wide, oval at first, soon developing a somewhat conic umbo; surface pruinose at first, viscid, with a removable cuticle; greenish yellow to rich yellow with a whitish margin; flesh yellowish, not staining reddish; odor and taste slightly to decidedly farinaceous.

GILLS: finely attached, subdistant to distant, narrow, white to pale yellow, not developing reddish brown spots.

STALK: 1½–2¼" (4–6 cm) long, ¹⁄₆₄–¹⁄₃₂" (0.5–1.5 mm) thick, viscid, glabrous, bright yellow fading to pale yellow, with a moderately strigose base.

SPORE PRINT: white.

MICROSCOPIC FEATURES: spores 9–12 × 5.5–8 μm, oval to subelliptic with an often eccentric apiculus, smooth, hyaline, amyloid.

FRUITING: gregarious on moss-covered, conifer wood; season, distribution, and frequency not well known.

EDIBILITY: inedible.

Mycena epipterygia var. **viscosa** (Maire) Ricken

CAP: 5/16–3/8" (8–10 mm) wide; surface beneath powdery layer yellowish or yellowish gray or greenish gray, becoming brownish in age, distinctly coated with a fine, white, powdery layer at first; odor strongly farinaceous to pumpkin-like; taste strongly rancid-farinaceous and very unpleasant.

GILLS: attached, yellowish to greenish yellow or sometimes whitish, often developing reddish brown spots; edges pale.

STALK: 1¼–3" (3–8 cm) long, 1/32–1/8" (0.75–3 mm) thick, lemon yellow to greenish yellow, becoming reddish near the base; distinctly white-pruinose at first.

SPORE PRINT: white.

MICROSCOPIC FEATURES: spores 8–11 × 5–8 μm, lacrymoid to elliptic, smooth, hyaline, amyloid.

FRUITING: scattered, in groups or clusters on conifer wood, often at the base of standing trees; August–November; occasional to fairly common.

EDIBILITY: unknown.

Mycena griseoviridis Smith

CAP: 3/8–1¼" (1–3.5 cm) wide, ovoid when young, soon becoming broadly conic to convex or bell-shaped; margin typically sulcate, often flaring; surface viscid, sometimes striate or translucent-striate, conspicuously white-pruinose when young, soon bald, deep olive to greenish brown, becoming nearly black in age; flesh thin, whitish; odor and taste farinaceous to somewhat disagreeable.

GILLS: attached with a decurrent tooth, close to subdistant, white with a greenish tinge at first, becoming greenish gray and often developing reddish brown spots; edges whitish.

STALK: 1½–3½" (4–9 cm) long, 1/32–1/8" (0.75–3 mm) thick, nearly equal, yellow to greenish yellow to grayish, becoming purplish brown near the base, which is typically somewhat white-strigose.

SPORE PRINT: white.

MICROSCOPIC FEATURES: spores 9–11 × 5–6.5 μm, ellipsoid, smooth, hyaline, weakly amyloid; cheilocystidia with one to several long needle-like apical projections.

FRUITING: scattered or in groups on humus in oak or pine woods; September–November; common.

EDIBILITY: unknown.

Mycena haematopus (Fries) Quélet

COMMON NAME: Bleeding Mycena.

CAP: 3/8–2" (1–5 cm) wide, more or less oval at first, typically becoming bell-shaped to conic but sometimes merely convex and finally nearly flat but typically with an umbo; margin initially pressed against the stalk, later becoming somewhat torn; surface dry and densely pruinose at first, becoming moist and smooth, opaque to translucent-striate, sometimes sulcate in age; dark reddish brown at the center, fading toward the margin; flesh thin, exuding a dark, blood-red latex when cut; odor not distinctive; taste mild or slowly somewhat bitter.

GILLS: finely attached, close to subdistant, narrow to moderately broad, whitish or grayish at first, becoming dirty reddish brown, edges finely fringed or powdery in appearance (use a hand lens); lamellulae numerous.

STALK: 1–4" (2.5–10 cm) long or sometimes longer, 1/32–1/8" (0.75–3 mm) thick, hollow,

reddish brown, the upper portion at first densely pruinose but soon becoming glabrous; exuding a dark, blood-red latex when cut or broken; base hairy-strigose.

SPORE PRINT: white.

MICROSCOPIC FEATURES: spores 8–11 × 5–7 μm, oval to elliptic with a small eccentric
apiculus, smooth, hyaline, amyloid.

FRUITING: typically in clusters but sometimes solitary on wood; typically March–
November but year-round in mild weather; very common.

EDIBILITY: edible.

COMMENTS: without question the commonest and one of the most easily identified of all
Mycenae. The photograph also shows *Spinellus fusiger,* a common mold parasite of this
species.

Mycena leaiana (Berkeley) Saccardo

COMMON NAME: Orange Mycena.

CAP: ⅜–2" (1–5 cm) wide, convex, sometimes slightly depressed at the center in age; surface glabrous or minutely pruinose, viscid to slightly glutinous, reddish orange at first,
fading to yellowish orange or sometimes losing virtually all color in age; margin typically translucent-striate when young and moist; flesh thick, white to orangish; odor
vaguely farinaceous; taste not distinctive.

GILLS: attached to sinuate, close to crowded, broad, often rather thickened; pinkish
orange to orange, staining yellowish orange where bruised, edges brilliant reddish
orange to orangish red; with numerous lamellulae.

STALK: 1–3" (2.5–7.5 cm) long, ¹⁄₁₆–⅛" (1.5–4 mm) thick, sometimes slightly swollen at the
apex; hollow, rather tough; initially somewhat powdery and orange, becoming glabrous and almost glutinous in age; orange overall, sometimes yellowish at the apex;
base densely hairy, the hairs typically orangish; exuding a scant orangish watery latex
when cut.

SPORE PRINT: white.

MICROSCOPIC FEATURES: spores 7–10 × 5–6 μm, elliptic to almost ovoid with a small
eccentric apiculus, smooth, hyaline, amyloid.

FRUITING: in clusters on deciduous wood, especially beech; April–November; very
common.

EDIBILITY: unknown.

COMMENTS: a distinctive and very common mushroom.

Mycena luteopallens (Peck) Saccardo

COMMON NAME: Walnut Mycena.

CAP: ¼–⅝" (8–15 mm) wide, oval, becoming convex to nearly flat, sometimes with a
slight umbo in age; surface smooth and glabrous except when very young; margin
translucent-striate in age; brilliant orange-yellow to rich yellow, fading as it dries to yellowish white; flesh thin, yellowish to whitish; odor and taste not distinctive.

GILLS: variously attached, subdistant, fairly broad, yellowish to slightly pinkish, with
whitish edges.

STALK: 2–4" (5–10 cm) long, ¹⁄₃₂–¹⁄₁₆" (0.75–1.5 mm) thick, hollow, base with strigose
hairs; orangish near the apex, yellowish below.

SPORE PRINT: white.

MICROSCOPIC FEATURES: spores 7–9 × 4–5.5 μm, oval with an eccentric apiculus, very
slightly roughened, hyaline, lightly amyloid.

FRUITING: in small groups growing from decaying walnut or hickory nut hulls; August–
November; occasional to common.

EDIBILITY: inedible.

COMMENTS: easily identified if you note the peculiar substrate, but often overlooked.

Mycena pura (Fries) Quélet

COMMON NAME: Pink Mycena.

CAP: ¾–3" (2–7.5 cm) wide, convex to flat, glabrous, moist; margin translucent-striate; surface color quite variable, pink, red, purple, grayish lilac, or sometimes whitish at first, but invariably with some pink to purple tint; hygrophanous but usually retaining some hint of the original pigments; flesh whitish to purplish or bluish, rather thick at the center, thinning abruptly about halfway toward the margin; odor and taste distinctly and strongly of radishes.

GILLS: variously attached, close to subdistant, very broad for the cap size, interveined, whitish to bluish or purplish but often with a strong grayish tint, edges whitish; lamellulae numerous.

STALK: 1–4" (2.5–10 cm) long, ¹⁄₁₆–¼" (1.5–6 mm) thick, often flattened or twisted, sometimes enlarged near the base, hollow, tough, glabrous to scaly; colored more or less like the cap or paler, sometimes simply whitish.

SPORE PRINT: white.

MICROSCOPIC FEATURES: spores 6–10 × 3–3.5 μm, narrowly elliptic to subcylindric with an eccentric apiculus, smooth, hyaline, amyloid.

FRUITING: solitary, scattered, grouped, or gregarious on humus in coniferous or deciduous woods; April–November; occasional to common.

EDIBILITY: reportedly edible but neither prized nor recommended.

COMMENTS: this mushroom's remarkable variability in color can complicate its identification; the radish-like odor and taste provide the most important clues.

Mycena subcaerulea (Peck) Saccardo

COMMON NAME: Blue Mycena.

CAP: ¼–¾" (5–20 mm) wide, bell-shaped to convex; surface smooth, slightly viscid, translucent-striate, pale blue to greenish blue at first, often becoming grayish or brownish with some bluish tints in age; flesh very thin, whitish; odor and taste not distinctive.

GILLS: attached to nearly free, close, whitish to grayish; edges finely fringed at maturity.

STALK: 1⅛–3½" (3–9 cm) long, ¹⁄₃₂–³⁄₃₂" (0.75–2 mm) thick, nearly equal, pruinose to pubescent, hollow, bluish to bluish green, often with a greenish basal mycelium when young.

SPORE PRINT: white.

MICROSCOPIC FEATURES: spores 6–8 μm, globose, smooth, hyaline, amyloid.

FRUITING: scattered or in small groups on the ground, on decaying hardwood debris or leaves; June–September; occasional.

EDIBILITY: unknown.

Genus *Nolanea*

Nolanea is a very difficult genus; a key to species here would be beyond the scope of this work. They can be generally identified to genus in the field by the combination of a conic cap when young; gills that are barely attached to the stalk (sometimes appearing free); and a slender, straight, fragile stalk. *N. quadrata* and *N. murraii* are distinctive; *N. verna* and *N. conica* have several look-alikes in the genus and require careful microscopic examination for positive identification.

Nolanea conica (Peck) Saccardo

CAP: ⅜–¾" (1–2 cm) wide, acutely conic at first, becoming broadly conic in age with a minute umbo; surface smooth, dull watery cinnamon when moist, fading to pale grayish cinnamon and appearing silky-shiny when dry; margin extending slightly beyond the gills, appearing slightly striate only when moist; flesh brownish; odor and taste not distinctive.

GILLS: barely attached or sometimes appearing free, close to subdistant, narrow, bright flesh color when mature.

STALK: 1½–2¼" (4–6 cm) long, about 1/16" (1–2 mm) thick, brown, straight, hollow, the base coated with white mycelium.

SPORE PRINT: salmon-pink.

MICROSCOPIC FEATURES: spores 8–11 × 6–8 μm, angular, 5-sided, hyaline, inamyloid.

FRUITING: in groups or scattered among mosses or on humus in mixed woods, rarely on decaying wood; July–September; occasional.

EDIBILITY: unknown.

Nolanea murraii (Berkeley and Curtis) Saccardo

COMMON NAME: Yellow Unicorn Entoloma.

CAP: ⅜–1¼" (1–3 cm) wide, bell-shaped to conic at first, becoming more convex but retaining a small, pointed umbo in most specimens; surface moist, silky-smooth, yellow to orangish yellow, fading somewhat in age; flesh yellowish; odor and taste pleasant or not distinctive.

GILLS: narrowly attached to the stalk, subdistant, broad, yellow, becoming pinkish tinged in age.

STALK: 2–4" (5–10 cm) long, 1/16–¼" (2–5 mm) thick, dry, yellow, the base usually slightly coated with white mycelium.

SPORE PRINT: salmon-pink.

MICROSCOPIC FEATURES: spores 9–12 × 8–10 μm, angular, typically 4-sided, appearing somewhat square, hyaline, inamyloid.

FRUITING: grouped or scattered on the ground in woods, especially in swamps or moist areas; July–October; occasional to common.

EDIBILITY: unknown.

COMMENTS: also known as *Entoloma cuspidatum*. Compare to *Hygrophorus marginatus* (see p. 141), which has a less conic cap and lacks a pointed umbo.

Nolanea quadrata Berkeley and Curtis

COMMON NAME: Salmon Unicorn Entoloma.

CAP: ⅝–2" (1.5–5 cm) wide, convex, bell-shaped to acutely conic, with a small, pointed umbo in most specimens; surface smooth, moist, salmon-orange, fading somewhat in age; flesh orangish; odor and taste not distinctive.

GILLS: slightly attached, subdistant, broad, salmon-orange.

STALK: 2–4" (5–10 cm) long, 1/16–¼" (2–6 mm) thick, fragile, salmon-orange, sometimes developing a greenish tinge in age, hollow, the base coated with white mycelium.

SPORE PRINT: salmon-pink.

MICROSCOPIC FEATURES: spores 10–12 μm, angular, typically 4-sided, appearing somewhat square, hyaline, inamyloid.

FRUITING: scattered on humus or among mosses in woods; July–October; occasional to common.

EDIBILITY: unknown, but believed to be toxic.

COMMENTS: also known as *Entoloma salmoneum* and *Nolanea salmonea*.

Nolanea verna (Lundell) Kotlaba and Pouzar

COMMON NAME: Spring Entoloma.

CAP: 1–2" (2.5–5 cm) wide, conic to bell-shaped at first, often becoming more convex in age but retaining a pointed umbo in most specimens; surface dark brown at first, fading to tan; flesh pale brown; odor and taste not distinctive.

GILLS: attached, subdistant, broad, grayish to pinkish brown.

STALK: 1½–3⅛" (4–8 cm) long, ⅛–⅜" (3–10 mm) thick, scurfy-fibrous, brown, the base coated with white mycelium.

SPORE PRINT: salmon-pink.

MICROSCOPIC FEATURES: spores 8–11 × 7–8 μm, angular, typically 6-sided, hyaline, inamyloid.

FRUITING: solitary to scattered on the ground in mixed woods; April–June; occasional to frequent.

EDIBILITY: poisonous.

COMMENTS: the springtime fruiting habit of this mushroom is shared by several nearly identical species.

Genus *Omphalina*

Omphalina ectypoides (Peck) Bigelow

COMMON NAME: Wood Clitocybe.

CAP: 1–2⅜" (2.5–6 cm) wide, broadly convex when young, soon becoming flat with a sunken center or funnel-shaped; surface moist, brownish yellow to yellow-brown, covered with minute blackish brown to reddish brown matted fibers and scales that often disappear in age; margin uplifted in age; flesh yellowish; odor and taste not distinctive.

GILLS: strongly decurrent, subdistant, narrow, occasionally forked, yellowish, sometimes with reddish brown stains in age.

STALK: 1–2½" (2.5–6.5 cm) long, ⅛–⅜" (3–9 mm) thick, solid, smooth to slightly scurfy, enlarged downward, honey-yellow, staining brownish when handled.

SPORE PRINT: white.

MICROSCOPIC FEATURES: spores 6.5–8 × 3.5–5 μm, elliptic, smooth, hyaline, amyloid.

FRUITING: scattered or in groups on decaying conifer wood, especially hemlock; July–September; occasional.

EDIBILITY: unknown.

COMMENTS: formerly known as *Clitocybe ectypoides*.

Omphalina epichysium (Persoon) Quélet

CAP: ⅜–1⅛" (1–3 cm) wide, convex when young, becoming depressed on the disc and finally broadly funnel-shaped; surface smooth, hygrophanous, grayish brown, translucent-striate; flesh grayish; odor and taste not distinctive.

GILLS: decurrent, subdistant, pale grayish brown.

STALK: ⅝–1" (1.6–2.5 cm) long, about ⅛" (2–4 mm) thick, slightly enlarged downward to nearly equal, smooth, dark brown to grayish brown.

SPORE PRINT: white.

MICROSCOPIC FEATURES: spores 6–9 × 3.5–5 μm, elliptic, smooth, hyaline, inamyloid.

FRUITING: scattered on decaying wood in coniferous and mixed woods; July–October; occasional.

EDIBILITY: unknown.

Genus *Omphalotus*

Omphalotus olearius (De Candolle : Fries) Singer

COMMON NAME: Jack O' Lantern.

CAP: 2¾–7" (7–18 cm) wide, convex when young, becoming nearly flat and shallowly depressed at the disc in age, often with a small umbo; surface dry, smooth, streaked with tiny fibrils, bright orange to yellow-orange, often stained reddish brown in age; margin uplifted and wavy in age; flesh white with an orange tint; odor not distinctive or somewhat unpleasant; taste not distinctive.

GILLS: decurrent, close, narrow, thin, yellow-orange.

STALK: 2–7" (5–18 cm) long, ¼–⅞" (5–22 mm) thick, nearly equal, tapered at the base, solid, dry, smooth, becoming scurfy in age, yellow-orange.

SPORE PRINT: whitish cream.

MICROSCOPIC FEATURES: spores 3–5 μm, round, smooth, hyaline, inamyloid.

FRUITING: in clusters at the base of hardwood trees and stumps, especially oak, or on the ground attached to buried wood; July–November; fairly common.

EDIBILITY: poisonous, causing gastrointestinal upset.

COMMENTS: fresh specimens often glow green in the dark.

Genus *Panaeolus*

Panaeolus is a small genus with fewer than twenty-five species in North America. They are small to medium mushrooms with dark brown to black spore prints, mostly slender stalks, and grayish to brownish conic to bell-shaped caps. Their spores are smooth and truncate with an apical pore and do not discolor when mounted in concentrated H_2SO_4. A partial veil is present on the button stage of some species, usually leaving remnants on the cap margin and typically not forming a ring on the stalk. Maturing gills are often gray or brown, mottled with black. At maturity, the gills are dark brown to black and the gill edges are usually white. A few species are edible, some are weakly to moderately hallucinogenic, and the edibility of many is unknown. We do not recommend this group for the table.

Key to Species of *Panaeolus*

1. Partial veil present on immature specimens, leaving remnants on the cap margin and typically not forming a ring → 2.
1. Partial veil absent → 3.
    2. Veil remnants on the cap margin distinctly and regularly tooth-like, whitish; cap smooth or nearly so, brownish, olive-gray to grayish tan, conic to bell-shaped, ⅜–1½" (1–4 cm) wide; flesh brownish; odor not distinctive; taste not distinctive or unpleasant; gills grayish, becoming mottled with black and finally black at maturity; edges whitish and finely fringed; stalk 2⅜–5½" (6–14 cm) long, 1⁄16–¼" (1.5–5 mm) thick, grayish white, becoming brown, whitish pruinose; on horse or cow dung—*P. campanulatus* (Fries) Quélet (see p. 217).
    2. Veil remnants on the cap margin uneven and irregular, whitish; cap wrinkled and pitted, sometimes reticulate; otherwise nearly identical to the previous choice; sometimes hallucinogenic—*P. retirugis* (Fries) Gillet.
3. Cap white to buff, sometimes yellowish in age, dry, smooth to wrinkled, often cracked in age, 1½–4" (4–10 cm) wide; flesh whitish; odor and taste not distinctive; gills pale grayish, becoming mottled with black and finally black at maturity; edges whitish; stalk

2¾–7" (7–18 cm) long, ¼–⅝" (5–16 mm) thick, dry, white to grayish white, longitudinally twisted-striate, sometimes with moisture beads at the apex when fresh; spores 14–20 × 9–12 μm; on horse or cow dung or on wood shavings, sawdust, or straw mixed with dung; edible—*P. solidipes* (Peck) Saccardo = *Anellaria sepulchralis* (Berkeley) Singer.

3. Cap grayish tan to grayish brown, cinnamon-brown to dark reddish brown or brown → 4.

    4. Cap 1⅛–2" (3–5 cm) wide, cinnamon-brown to dark reddish brown or brown, fading to pinkish tan and retaining a darker reddish brown band on the margin, then fading overall, often grayish or blackish when coated with spores, smooth or slightly wrinkled, becoming cracked and scaly in age or when dry; flesh brownish; odor and taste not distinctive; gills brown, becoming mottled with black and finally black at maturity; edges white; stalk 1½–4" (4–10 cm) long, ⅛–⅜" (3–10 mm) thick, nearly equal, white-pruinose over a reddish brown ground color, paler at the apex, often longitudinally striate; spores 10–14 × 7–9 μm; on dung, on manured grassy areas, and in gardens; mildly to moderately hallucinogenic— *P. subbalteatus* (Berkeley and Broome) Saccardo.

    4. Cap ⅜–1⅛" (1–3 cm) wide, reddish brown to grayish brown, often fading to grayish tan; on lawns and other grassy areas, not on dung—*P. foenisecii* (Persoon : Fries) Kühner (see p. 217).

Panaeolus campanulatus (Bulliard : Fries) Quélet

COMMON NAME: Bell-cap Panaeolus.

CAP: ⅜–1½" (1–4 cm) wide, conic, becoming bell-shaped; margin rimmed with distinctly tooth-like, whitish veil remnants; surface smooth or nearly so, brownish or olive-gray to grayish tan; flesh brownish; odor not distinctive; taste not distinctive or unpleasant.

GILLS: notched, subdistant, broad, grayish at first, becoming mottled with black and finally black at maturity; edges whitish and finely fringed.

STALK: 2⅜–5½" (6–14 cm) long, 1/16–¼" (1.5–5 mm) thick, grayish white, becoming brown, whitish pruinose; partial veil white, not leaving a ring but leaving remnants on the cap margin.

SPORE PRINT: blackish.

MICROSCOPIC FEATURES: spores 13–16 × 8–11 μm, elliptic with an apical pore, smooth, blackish.

FRUITING: solitary, scattered, or in groups on horse dung; June–September; occasional.

EDIBILITY: sometimes hallucinogenic if eaten in sufficient quantity.

Panaeolus foenisecii (Persoon : Fries) Kühner

COMMON NAME: Haymaker's Mushroom.

CAP: ¼–1⅛" (5–30 mm) wide, convex to bell-shaped; surface dry, smooth or sometimes finely cracked when dry, reddish brown to grayish brown, often fading to grayish tan; margin slightly striate when fresh and moist; flesh thin, grayish brown to tan; odor and taste not distinctive.

GILLS: attached, close, brown, becoming purple-brown to blackish brown.

STALK: 1–4" (2.5–10 cm) long, 1/16–⅛" (1.5–3 mm) thick, nearly equal, smooth, white to pinkish brown, darkening in age; partial veil and ring absent.

SPORE PRINT: dark brown to purple-brown.

MICROSCOPIC FEATURES: spores 11–18 × 6–9 μm, elliptic with an apical pore, roughened, pale purple-brown.

FRUITING: scattered or in groups on lawns and other grassy areas; May–October; very common.

EDIBILITY: hallucinogenic (see comments).

COMMENTS: this mushroom contains variable amounts of psilocybin (usually trace levels). Although it is edible in small quantities, it is hallucinogenic when eaten in large quantities. We do not recommend eating this mushroom.

Genus *Panellus*

The mushrooms of the genus *Panellus* are best characterized by their combination of white to yellowish spores, rather tough texture, lateral stalk (if a stalk is present at all), nonserrate gills, and occurrence on wood. The spores are smooth, cylindric to sausage-shaped, and amyloid. Only one, *P. serotinus,* is commonly collected for the table.

Key to Species of *Panellus*

1. Spore print yellowish; cap smooth to finely velvety in age, up to 4" (10 cm) wide, variously yellow to green or purple in color; gills yellow, neither forked nor crossveined; mushroom rather tough, not decaying readily; found only in autumn after frosts on decaying wood; spores smooth, sausage-shaped, amyloid—*Panellus serotinus* (Fries) Kühner (see p. 218).
1. Spore print white; cap hairy to scaly, tan to pale brown, less than 1½" (4 cm) wide; mushroom rather tough, not decaying readily; taste quite acrid; normally luminescing green when fresh (view in complete darkness for five to ten minutes); spores smooth, sausage-shaped, amyloid—*Panellus stipticus* (Bulliard : Fries) Karsten (see p. 219).
1. Not as in either of the above choices → 2.
    2. Cap white, less than ½" (13 mm) wide; found on conifer wood; spores 3.5–6 × 1–1.5 μm, smooth, cylindric to sausage-shaped, amyloid—*Panellus mitis* (Persoon : Fries) Singer.
    2. Cap brown with a distinct violet tint, sometimes with a very fine, white, velvety coating, usually nearly 1" (2.5 cm) wide at maturity; gills subdistant, often crossveined; spores as in previous choice except 6.5–10 × 2–3.5 μm; compare carefully with next choice—*Panellus violaceofulvus* (Batsch : Fries) Singer (see p. 219).
    2. Cap purplish or lilac overall; stalk with a dense white fuzzy coating; gills not interveined; spores as in the previous two choices except 5–7 × 1–2 μm—*Panellus ringens* (Fries) Romagnesi.

Panellus serotinus (Fries) Kühner

COMMON NAME: Late Fall Oyster Mushroom.

CAP: 1–4" (2.5–10 cm) wide, semicircular; surface smooth and slippery to sticky when young, typically becoming somewhat velvety or fuzzy in age, yellowish green to greenish yellow, usually with a purplish tint, sometimes purplish overall; margin inrolled at first; odor not distinctive; taste not distinctive or bitter.

GILLS: attached to decurrent, close, yellowish to orangish yellow.

STALK: short, stubby, hairy, lateral, yellowish to brownish.

SPORE PRINT: yellowish.

MICROSCOPIC FEATURES: spores 4–6 × 1.5–2 μm, sausage-shaped, smooth, hyaline, amyloid.

FRUITING: solitary, in groups, or in overlapping clusters, most frequent on dead deciduous wood, especially beech, but also on coniferous wood; August–December, rather exclusively after autumn's first frosts; frequent to common.

EDIBILITY: edible, but rather tough; it must be cooked a long time, over low heat.

Panellus stipticus (Bulliard : Fries) Karsten

COMMON NAME: Luminescent Panellus.

CAP: ¼–1¼" (5–30 mm) wide, semicircular to kidney-shaped; surface minutely scaly to hairy or fuzzy, dingy white to pale brownish; odor not distinctive; taste rather acrid.

GILLS: attached to decurrent, close to crowded, narrow, pinkish to pale brownish.

STALK: short, stubby, hairy, lateral, whitish to brownish.

SPORE PRINT: white.

MICROSCOPIC FEATURES: spores 3–6 × 2–3 μm, oblong to sausage-shaped, smooth, hyaline, amyloid.

FRUITING: usually gregarious or in clusters on dead deciduous logs and sticks; May–December; frequent to common.

EDIBILITY: inedible.

COMMENTS: when fresh, this mushroom will give off a greenish glow if viewed in total darkness for several minutes. The species name, *stipticus,* refers to its reputed value in stopping bleeding.

Panellus violaceofulvus (Batsch : Fries) Singer

CAP: ⅛–⅝" (3–15 mm) wide, kidney-shaped; surface finely coated with white hairs, deep violet-brown; flesh thin, violet-brown; odor not distinctive; taste not determined.

GILLS: attached, subdistant, typically crossveined, pale violet to violet-brown.

STALK: short, stubby, violet-brown, finely coated with white hairs.

SPORE PRINT: white.

MICROSCOPIC FEATURES: spores 6.5–10 × 2–3.5 μm, cylindric to sausage-shaped, smooth, hyaline, amyloid.

FRUITING: scattered, in groups or clusters on decaying conifer wood; May–July; infrequent to occasional.

EDIBILITY: unknown.

Genus *Paxillus*

Species of the small genus *Paxillus* are medium to large mushrooms that grow on the ground or on decaying wood. Some have a central to eccentric stalk; others have a rudimentary stalk or are stalkless. They typically have decurrent gills that easily separate from the cap flesh and have inrolled cap margins when young. Their spore print colors vary from yellowish buff to olive-yellow or yellow-brown, and their spores are smooth. None is recommended for the table.

Key to Species of *Paxillus*

1. Stalk velvety, with a dense covering of dark brown or blackish brown matted hairs, eccentric to nearly lateral, ¾–4" (2–10 cm) long, ⅜–1¼" (1–3 cm) thick—*P. atrotomentosus* (Batsch : Fries) Fries (see p. 220).

1. Stalk not covered with dark brown hairs, usually central, yellow-brown, often with darker brown stains, ¾–4" (2–10 cm) long, ¼–¾" (6–20 mm) thick—*P. involutus* (Batsch : Fries) Fries (see p. 220).

1. Stalk rudimentary or absent → 2.

2. Gills conspicuously wavy and corrugated, blunt, frequently forked and anastomosing, orange-yellow, staining darker orange; flesh pale yellow, odor not distinctive when fresh, typically disagreeable when specimens have been dried; spores 3–3.5 × 1.5–2 μm—*P. corrugatus* Atkinson (see p. 220).

2. Gills often conspicuously wavy and corrugated or sometimes nearly straight, thin, occasionally forked or anastomosing, radiating from the point of attachment to the substrate; flesh whitish, odor not distinctive when specimens are fresh or dried; spores 4–6 × 3–4 μm—*P. panuoides* Fries (see p. 221).

Paxillus atrotomentosus (Batsch : Fries) Fries (see photo, p. 10)

COMMON NAME: Velvet-footed Pax.

CAP: 1½–5⅞" (4–15 cm) wide, convex, becoming flat, sometimes depressed at the center; surface dry, felty to smooth, covered with matted hairs, dull olive-brown to rusty brown or yellowish brown to blackish brown; margin inrolled when young; flesh whitish, thick, tough, not staining when cut or bruised; odor and taste not distinctive.

GILLS: decurrent, close, often forked or pore-like near the stalk, tan to yellow-brown or dull yellow, not staining when cut or bruised.

STALK: ¾–4" (2–10 cm) long, ⅜–1¼" (1–3 cm) thick, equal, eccentric to nearly lateral, solid, velvety with a covering of densely matted dark brown or blackish brown hairs, apex often lighter; partial veil and ring absent.

SPORE PRINT: dull yellow to pale brownish yellow.

MICROSCOPIC FEATURES: spores 5–7 × 3–4 μm, elliptic, smooth, hyaline to pale brown.

FRUITING: solitary, in groups, or in clusters on decaying conifer stumps and logs or partially buried wood; July–October; occasional to fairly common.

EDIBILITY: unknown.

Paxillus corrugatus Atkinson

CAP: ¾–2" (2–5 cm) wide, convex, becoming nearly flat in age, fan- to petal-shaped or semicircular, narrowed at the point of attachment; surface dry, smooth to felty or suede-like, pale brownish yellow to olive-yellow, often with reddish brown stains, especially along the margin and the base; margin incurved when young, expanding in age; flesh pale yellow; odor not distinctive when fresh, typically disagreeable when specimens have been dried; taste somewhat bitter or mild.

GILLS: close to crowded, blunt, frequently forked and anastomosing, usually crossveined, conspicuously wavy and corrugated, orange-yellow, staining darker orange, easily separated from the cap.

STALK: absent or rudimentary.

SPORE PRINT: yellowish olive to olive-yellow when fresh, drying yellowish cinnamon.

MICROSCOPIC FEATURES: spores 3–3.5 × 1.5–2 μm, broadly elliptic to oval, smooth, hyaline to pale yellow, dextrinoid.

FRUITING: solitary, scattered, or in overlapping clusters on decaying conifer wood, especially hemlock and pine; August–October; uncommon.

EDIBILITY: unknown.

COMMENTS: compare with *P. panuoides* and *Phyllotopsis nidulans*. Singer, Garcia and Gomez have recently published this species as a bolete and assigned the name *Meiorganum curtisii*.

Paxillus involutus (Batsch : Fries) Fries

COMMON NAME: Poison Paxillus.

CAP: 1⅝–4¾" (4–12 cm) wide, convex, becoming flat, often depressed on the disc; surface dry, smooth, somewhat sticky when moist, occasionally finely cracked in age, covered with matted hairs, dull brown, yellow-brown or red-brown, sometimes with olive tints; margin inrolled until maturity; flesh dull yellow to pale tan, staining reddish brown when cut or bruised; odor and taste not distinctive.

GILLS: decurrent, crowded, broad, forked and often pore-like near the stalk, tan to dull yellow or olive-yellow, staining reddish brown when bruised or in age.

STALK: ¾–4" (2–10 cm) long, ¼–¾" (6–20 mm) thick, nearly equal or enlarged downward, solid, usually central, yellow-brown, often with darker brown stains, smooth; partial veil and ring absent.

SPORE PRINT: pale to dark yellow-brown.

MICROSCOPIC FEATURES: spores 7–9 × 4–6 μm, elliptic, smooth, pale brown.

FRUITING: solitary, scattered or in groups on the ground or on decaying wood in coniferous or mixed woods; July–October; fairly common.

EDIBILITY: poisonous; see comments below.

COMMENTS: despite its reputation, we have encountered several Europeans collecting this mushroom for the table. Their preparation technique includes boiling for several minutes in two or more changes of water. However, we do not recommend this because there are reports that this mushroom may produce a gradually acquired hypersensitivity that causes kidney failure.

Paxillus panuoides Fries

CAP: 1–4⅜" (2.5–11 cm) wide, fan- to petal-shaped; surface dry, smooth to felty, olive-yellow to yellow-brown; margin incurved when young, becoming uplifted, thin and wavy at maturity; flesh whitish to pale cream; odor and taste not distinctive or taste somewhat bitter.

GILLS: thin, radiating from the point of attachment to the substrate, sometimes forked or anastomosing, occasionally crossveined, often conspicuously wavy and corrugated but sometimes straight, pale yellow to pale yellow-orange, easily separated from the cap.

STALK: absent or rudimentary.

SPORE PRINT: yellowish.

MICROSCOPIC FEATURES: spores 4–6 × 3–4 μm, elliptic, smooth, hyaline, weakly amyloid, not dextrinoid.

FRUITING: scattered or in overlapping clusters on decaying conifer wood; May–November; occasional.

EDIBILITY: unknown.

Genus *Phaeocollybia*

Phaeocollybia species are small to medium mushrooms with smooth, sticky caps that are various shades of brown. Their stalks are cartilaginous with a long tapering root that is mostly buried in humus. They grow on the ground in coniferous woods and have pale cinnamon to rusty brown spore prints. Species of this small genus have free to deeply notched gills and roughened to finely wrinkled spores. Very little is known about the edibility of these mushrooms.

Key to Species of *Phaeocollybia*

1. Cap dark to pale olive to olive-brown, ⅜–2" (1–5 cm) wide, conic, becoming nearly flat, with a conspicuous, pointed umbo, slimy to sticky, dark to pale olive to olive-brown, hygrophanous and fading to pale yellow-green, typically developing dark spots; flesh

olive-buff to olive-green; odor radish-like, taste not distinctive; stalk 3–6" (8–15.5 cm) long, ⅛–⅜" (3–10 mm) thick, tapered downward and rooting, hollow, pale yellow on the upper portion, becoming orange-brown below; spores 7–9 × 4.5–6 μm; growing on the ground under conifers; July–September; edibility unknown—*P. fallax* Smith.

1. Cap apricot-orange to brownish orange → 2.
   2. Cap ⅜–1⅛" (1–3 cm) wide, narrowly and sharply conic when young, becoming broadly conic and tipped with a sharp point at maturity; apricot-orange to brownish orange; flesh colored like the cap, odor and taste somewhat like radishes or crushed almonds; stalk colored like the cap on the upper portion, becoming reddish brown at the base—*P. christinae* (Fries) Heim (see p. 222).
   2. Cap ⅝–2⅛" (1.6–5.5 cm) wide, conic with a conspicuous umbo and straight margin, smooth, shiny, brownish orange, hygrophanous; flesh whitish in cap; odor like potato or unpleasant, taste slightly bitter; stalk 4¾–8⅝" (12–22 cm) long, ⅛–¼" (2–5 mm) thick, tapered downward and rooting, orange-yellow at the apex, darkening to reddish brown at the base; spores 4–6 × 2.5–4.5 μm; growing on soil under conifers and plants of the heath family; July–September; edibility unknown—*P. jennyae* (Karsten) Heim.

Phaeocollybia christinae (Fries) Heim

CAP: ⅜–1⅛" (1–3 cm) wide, narrowly and sharply conic when young, becoming broadly conic and tipped with a sharp point at maturity; surface smooth and sticky when fresh, sometimes spotted, apricot-orange to brownish orange, often reddish brown in age; flesh colored like the cap; odor and taste somewhat like radishes or almonds.

GILLS: notched, close to crowded, pale yellow to orange-yellow, darkening in age.

STALK: 2–7" (5–18 cm) long, ⅛–¼" (3–6 mm) thick, equal down to ground level, then tapered below and rooting, apricot-orange to brownish orange, reddish brown at the base, smooth, sticky; partial veil and ring absent.

SPORE PRINT: cinnamon to rusty brown.

MICROSCOPIC FEATURES: spores 7–10 × 4–5 μm, elliptic, finely roughened, hyaline to pale brown.

FRUITING: scattered or in groups and deeply rooted in sandy soil under conifers, especially spruce; August–October; occasional.

EDIBILITY: unknown.

COMMENTS: also known as *P. rufipes*.

Genus *Phaeomarasmius*

Phaeomarasmius erinaceellus (Peck : Singer)

COMMON NAME: Powder-scale Pholiota.

CAP: ⅜–1½" (1–4 cm) wide, convex to nearly flat; surface granular to finely scaly, dry, rust-brown at first, soon fading to more or less yellowish brown; odor not distinctive; taste metallic to somewhat bitter.

GILLS: notched, crowded, broad, whitish or yellowish at first, becoming slightly to deeply rust-colored in age; partial veil fibrous, whitish to pale yellowish, usually leaving only an inconspicuous zone of fibers on the upper stalk.

STALK: 1¼–2¼" (3–6 cm) long, ⅛–¼" (3–6 mm) thick, dry, with rust-colored scales overall but buff to yellowish near the top.

SPORE PRINT: cinnamon-brown.

MICROSCOPIC FEATURES: spores 6–8 × 4–4.5 μm, elliptic, smooth, pale brownish.

FRUITING: solitary to almost gregarious on decaying deciduous wood; June–October; occasional to common.

EDIBILITY: unknown.

COMMENTS: it would be wise to compare the description of this mushroom to the key description of *Pholiota granulosa* (p. 224), especially if you are not using a microscope. Also known as *Pholiota erinaceella*.

Genus *Pholiota*

Pholiota is a conspicuous genus of brown- to rusty-spored species that are fairly easy to recognize in the field. The gills are always attached; the caps are yellow or orangish to red, brown, or blue to greenish blue and are usually scaly; a partial and/or universal veil is always evident in young specimens; and most species grow in clusters on wood. The spores are smooth, and an apical pore is usually evident. If your specimen doesn't key out here and the spore print is orange to bright rusty brown, try *Gymnopilus* (p. 120). In their 1968 monograph, *The North American Species of Pholiota*, Smith and Hesler reported 205 species. Microscopic examination is required to accurately identify the majority of species in this genus. We have included 18 of the most distinctive species common to northeastern North America.

Determining whether the cap cuticle is truly dry or has a gelatinized layer is tricky but important. If a mature cap appears dry, a drop of clean water on the cap may revive a slimy gelatinized layer. If the cap is covered with scales, gently pull them off. If the cuticle tears, it can be interpreted as dry. If the cuticle does not tear, then it probably has a gelatinized layer. Also, if bits of debris are tightly stuck to the cap, the cuticle probably has a gelatinized layer.

Several Pholiotas cause gastrointestinal upset, and identification can be difficult. For these reasons, Pholiotas are not ideal choices for novice mycophagists.

Key to Species of *Pholiota*

1. Growing on burned soil or charred wood; cap ¾–2¼" (2–6 cm) wide, sticky or slippery to slimy or glutinous, yellowish brown to reddish brown, fading in age; flesh yellowish to brownish, odor and taste not distinctive; gills whitish to yellowish at first; stalk ¹⁄₁₆–³⁄₁₆" (2–6 mm) thick at the top; partial veil cortinate, leaving, at most, a zone of fibers ringing the upper stalk; spores 6–8 × 4–4.5 μm, elliptic to oval, with a distinct apiculus and apical pore; inedible—*P. highlandensis* (Peck) Smith and Hesler.

1. Not as in the above choice; cap flesh yellow or with distinct yellowish tones → 2.

1. Not as in either of the above choices; flesh in cap more or less white or whitish (lacking distinct yellowish tones) → 3.

1. Not as in any of the above choices → 5.

   2. Cap up to 6" (15 cm) wide, sticky or slippery to slimy, smooth, orangish yellow to yellowish orange, with large, flattened scales; flesh yellow; gills pale yellowish at first; veil whitish, cortinate, leaving only a zone of fibers; stalk dry, yellowish or yellowish brown, covered with fibrous scales below the annular zone; in clusters on deciduous or coniferous wood; spores 7–11 × 4.5–6 μm, broadly elliptic, lacking a distinct apiculus, with a distinct apical pore but not appearing truncate—*P. aurivella* (Fries) Kummer (see p. 227).

   2. Very similar to previous choice except cap smaller, up to 4" (10 cm) wide; cap, gills, and stalk bright yellow, generally lacking orange tones except at center of cap; gill edges soon staining brown where rubbed; spores 4–5 × 2.5–3 μm, oblong to

elliptic, usually with a distinct apiculus, apical pore not evident—*P. flammans* (Fries) Kummer (see p. 228).

2. Similar to the key descriptions for *P. aurivella* and *P. flammans* above except cap only up to 2½" (6 cm) wide, lacking scales, with distinct greenish tones and a darker brownish center, appearing streaked with fibers, shiny; odor sometimes of green corn; gills also with a distinct greenish tinge; stalk rather smooth, yellow, becoming dirty brown from the base upward; spores 7–9 × 4–4.5 μm, mostly with a fairly acute apiculus, with a distinct apical pore but not appearing truncate— *P. spumosa* (Fries) Singer (see p. 229).

2. Not as in any of the above choices → 4.

3. Cap cuticle on mature specimens dry, not moist, sticky, slippery, slimy, or glutinous → 4.

3. Cap cuticle on mature specimens moist, sticky, slippery, slimy, or glutinous → 6.

4. Cap 1¼–5" (3–12.5 cm) wide, cuticle entirely broken up into dry, tan to pinkish tan or brownish scales, flesh pale yellow between the scales; odor usually strong of garlic or onions; veil fibrous to membranous, yellowish to tan, leaving a distinct ring and/or bits of tissue adhering to the edge of the cap; gills pale yellow at first, soon developing distinct green tones; stalk dry, scaly like the cap; in large clusters on living or dead deciduous or coniferous trees, stumps, or logs; spores 5–8 × 3.5–4.5 μm, elliptic to oval, with a distinct apical pore but not appearing truncate—*P. squarrosa* (Fries) Kummer (see p. 229).

4. Not as in the above choice → 7.

5. Cap dry, densely covered by tiny, erect, bright yellowish brown scales that are somewhat granular near the cap margin, ground color more yellow between the scales, ⅜– 1¼" (1–3.5 cm) wide; margin even at maturity somewhat incurved and usually decorated with tiny yellow bits of the partial veil; flesh thin, watery brown; odor and taste not distinctive; partial veil yellow to yellowish brown, fibrous, leaving a thin, fibrous ring on the upper stalk; gills attached, close to crowded, moderately narrow, pale yellow when young, brownish yellow in age; stalk 1–2" (2.5–5 cm) long, 1⁄16–⅛" (2–3.5 mm) thick, dry, yellow at the top, becoming dark rusty brown from the base upward, scurfy-fibrous; solitary to gregarious on well-decayed deciduous or coniferous wood; spores 7.5–9 × 4–6 μm, with a very slight apiculus and only a minute apical pore; edibility unknown—*P. granulosa* (Peck) Smith and Hesler.

5. Cap slimy to sticky, not densely covered by scales; blue to greenish blue when young, developing pale cinnamon blotches and finally that color overall but usually retaining a bluish to greenish tint, ¾–1½" (2–4 cm) wide; flesh thin, bluish, odor and taste not distinctive; veil white, leaving a fibrous ring halfway up the stalk or disappearing; gills attached to subdecurrent, close, moderately broad, pale brown at first, cinnamon-brown in age; stalk 1¼–2¼" (3–6 cm) long, 1⁄16–¼" (1.5–6 mm) thick, dry, colored similar to the cap but paler, silky to fibrous, with tiny white flecks, base with numerous white rhizomorphs, sometimes slightly enlarged; solitary or in small clusters on soil or debris; very rare (reported from Idaho, Oregon, Washington, and Maine); spores 7–9 × 4–4.5 μm, with a distinct apiculus and a minute apical pore, not truncate or only slightly truncate, not dextrinoid; edibility unknown—*P. subcaerulea* Smith and Hesler.

5. Not as in either of the above choices → 7.

6. Gill edges not finely fringed and whitish, not beaded with tiny white droplets; cap 2¼–7½" (6–20 cm) wide, sticky to slimy, whitish to dark yellowish brown at the center or sometimes overall, with cottony whitish to pale brownish scales that may adhere to the cap or be washed off by rain in age, margin with hanging bits of veil

tissue; flesh thick, white, odor and taste not distinctive; partial veil thick, white, cottony-membranous, disintegrating and decorating much of the stalk, sometimes leaving a thin ring; gills attached, close, broad, white at first, slowly becoming rusty brown; stalk 2–7" (5–18 cm) long, ⅜–1¼" (1–3 cm) thick at the apex, usually enlarged downward to 2–3" (5–7.5 cm) thick at the base, white at first, often becoming brownish in age, especially at the base, cottony-scurfy, decorated as explained above; usually in clusters of 3 to 8 specimens on stumps or logs of cottonwood, aspen, or poplar, especially frequent at the ends of cut logs; spores 7–9.5 × 4–5.5 μm, elliptic to oval, mostly with a distinct apiculus and a distinct apical pore, sometimes appearing slightly truncate; edible — *P. destruens* (Brondeau) Gillet.

6.   Gill edges finely fringed, whitish and usually beaded with tiny white droplets; cap 1–4" (2.5–10 cm) wide, sticky to glutinous, orangish brown to dark reddish brown, bald, smooth, and usually somewhat conic at the center, decorated with numerous brownish scales near the margin, which is also rimmed with bits of veil tissue; flesh thick, whitish, odor and taste not distinctive; partial veil whitish to pale brown, fibrous, not leaving a distinct ring; gills attached to subdecurrent or with fine decurrent lines, close, broad, whitish at first, becoming grayish and finally rusty brown; stalk 1¼–6" (3–15 cm) long, ¼–⅝" (5–15 mm) thick, dry, fibrous, with brown cottony scales, whitish to grayish above a slight fibrous ring (ring often not evident), dark brown near the base; solitary to several on decaying deciduous trunks, logs, and stumps, especially maple, also reported on hemlock; spores 10–18 × 5.5–8.5 μm, oval with an apiculus at one end, to rather football-shaped (apiculate at both ends), with only a minute apical pore, thick-walled; edibility unknown — *P. albocrenulata* (Peck) Saccardo.

6.   Not as in either of the above choices → 7.

7.   Cap more or less scaly overall → 8.

7.   Cap not scaly overall; scales absent or restricted to an area near the cap margin; cap more or less brownish overall or with distinct brownish to cinnamon tones at first or in age → 9.

7.   As in the previous choice, but cap not brownish overall, nor with distinct brownish tones → 11.

8.   Growing on trunks and stumps of deciduous trees, especially maple, birch, and beech, usually in large clusters but sometimes solitary; cap 1–4" (2.5–10 cm) wide, sticky to slimy, whitish at first, soon becoming orangish to brownish beneath a dense covering of scales; scales erect, pointed, tan to pale brownish; flesh thick, whitish; odor and taste not distinctive; partial veil whitish, cottony-fibrous to almost membranous, usually leaving a torn ring on the upper stalk and remnants hanging from the edge of the cap; gills attached, usually deeply notched, close to crowded, moderately broad, whitish at first, becoming dull rusty brown in age; stalk 2–6" (5–15 cm) long, ¼–⅝" (5–15 mm) thick, dry, covered except for the upper portion by brownish to orangish scales, whitish and silky at the apex; spores 4–6 × 2.5–3.5 μm, elliptic to oval, with a slight apiculus, without a distinct apical pore; edible but not recommended — *P. squarrosoides* (Peck) Saccardo.

8.   Growing on the ground, usually in large clusters; cap ¾–4" (2–10 cm) wide, slippery to slimy, grayish brown to dirty brownish yellow and usually streaked with fibers beneath a layer of brown, fibrous scales that often wash away by maturity; flesh moderately thick, buff to brownish; odor and taste not distinctive; partial veil whitish, fibrous, leaving a ring and/or bits of tissue hanging from the edge of the

cap; gills attached, crowded, narrow, whitish at first, soon becoming more or less brownish and sometimes developing a greenish tint, edges finely roughened; stalk 1–4" (2.5–10 cm) long, ⅛–⅜" (3–10 mm) thick, dry, covered except above the annular zone by dark yellowish brown scales that are larger and more abundant upward; spores 4.5–7 × 3.5–4.5 μm, elliptic to broadly elliptic, with a distinct apiculus and a minute but distinct apical pore; edible but not recommended— *P. terrestris* Overholts.

9. Cap small, ¾–1¼" (2–3.5 cm) wide at maturity, with a broad, almost conic umbo at first; found in springtime (April–June); sticky to slippery, smooth, pale to dark yellow-brown, margin translucent-striate, often hung with bits of veil tissue; flesh watery, pinkish buff to cinnamon- buff; odor and taste not distinctive; partial veil pale brown, usually leaving a fibrous-membranous ring on the upper stalk; gills crowded, quite narrow, pale dingy tan, edges often becoming finely fringed in age; stalk ¾–2¼" (2–6 cm) long, 1/16–¼" (2–5 mm) thick, dry, silky-fibrous, grayish or buff to pale yellow-brown, soon staining dull rusty brown from the base upward; usually in groups or clusters on well-decayed deciduous or coniferous wood; spores 5.5–7.5 × 3–4.5 μm, oval, with a slight but distinct apiculus, appearing barely truncate from a distinct apical pore; edibility unknown—*P. vernalis* (Peck) Smith and Hesler = *Kuehneromyces vernalis* (Peck) Singer and Smith.

9. Cap larger than in previous choice, 1½–2¼" (4–6 cm) wide at maturity → 10.

    10. Stalk moderately scurfy only on the middle to upper portion, watery whitish or light tan to pale cinnamon at first, not darkening substantially in age; cap ¾–2¼" (2–6 cm) wide, moist to sticky or slippery, smooth overall but with fine fibers near the margin at first, pale cinnamon to yellowish cinnamon at first, fading to pale yellowish tan or nearly whitish from the center outward in age, margin finely translucent-striate in age; flesh colored about the same as the cap surface; odor and taste not distinctive; partial veil yellowish cinnamon, usually leaving a thin, brownish, membranous ring on the upper stalk; gills close to crowded, pale pinkish tan at first, darkening to nearly cinnamon in age; stalk 1½–3½" (4–9 cm) long, ⅛–½" (4–12 mm) thick, dry, color similar to but paler than the cap, nearly smooth overall in age, with a coating of white mycelium at the base; usually in large clusters on well-decayed deciduous wood or sawdust, April–June; spores 5.5–7.5 × 3.5–5 μm, elliptic to oval, with or without a distinct apiculus, appearing slightly to distinctly truncate from a distinct to indistinct apical pore—*P. veris* (Singer and Smith) Smith and Hesler (see p. 230).

    10. Stalk distinctly scaly below the ring, whitish at first, becoming dark brown to blackish brown from the base upward in age; cap ⅝–2¼" (1.5–6 cm) wide, sticky to slippery, smooth overall but sometimes coated with inconspicuous white fibers, dull reddish cinnamon at first, soon fading to yellowish or yellowish pink at the center or in a circular zone between the center and the margin; flesh thin, moist to watery, dingy whitish; odor weak but pleasantly spicy, taste not distinctive to slightly unpleasant, not bitter; partial veil membranous, whitish, leaving either a membranous ring or merely a fibrous zone near the top of the stalk; gills close to crowded, whitish at first, soon becoming dull cinnamon; stalk 1½–4" (4–10 cm) long, 1/16–½" (2–12 mm) thick, dry; usually in large clusters on well-decayed deciduous wood; spores 5.5–7.5 × 3.5–6 μm, oval, with a slight but distinct apiculus and a distinct apical pore making the spores appear truncate; edible **but not recommended because of possible confusion with the Deadly Galerina** *Galerina autumnalis* (see p. 118)—*P. mutabilis* (Fries) Kummer.

11. Taste not bitter, odor slightly radish-like or not distinctive; cap whitish to grayish, sometimes with a pinkish tinge, 1–4" (2.5–10 cm) wide, sticky to glutinous, smooth, with scattered white cottony flecks; partial veil white, cortinate; stalk whitish above, brownish below; in small clusters on woody soil or humus; spores 5.5–7 × 3.5–4.5 μm, with a slight to fairly acute apiculus and a minute apical pore, slightly dextrinoid— *P. lenta* (Fries) Singer (see p. 228).

11. Taste bitter, odor fragrant or not distinctive; cap yellow at first, becoming deeper yellow to reddish at the center, often with a greenish tinge near the margin, 1¼–2½" (3–6 cm) wide, moist to sticky or slippery, smooth, margin at first scurfy or fibrous, often rimmed with bits of veil tissue in age; flesh pale yellow; partial veil thin, fibrous to cortinate, whitish; gills close, broad, whitish to pale yellowish at first, becoming brownish; stalk 1½–4" (4–10 cm) long, ⅛–½" (4–12 mm) thick, dry, pale yellow at first, becoming brownish from the base upward, fibrous; in clusters on deciduous or coniferous wood; spores 8–10 × 4–5.5 μm, with an almost acute apiculus and a usually inconspicuous apical pore; edibility unknown—*P. alnicola* (Fries) Singer.

11. Not as in either of the above choices → 12.

  12. Odor and taste not distinctive or rarely bleach-like, odor not like green corn; cap 1¼–5" (3–12.5 cm) wide, moist to sticky or slippery at first but soon dry, rather bell-shaped when young with a large, somewhat conic umbo, smooth, decorated at first with patches of brownish fibrils, yellow to orangish yellow, sometimes with a greenish tinge, margin opaque; flesh whitish, rather thick at the center; partial veil whitish to yellowish, fibrous to membranous, usually leaving a thin, membranous ring on the upper stalk and bits of tissue adhering to the edge of the cap; gills close to crowded, narrow, yellow at first, becoming reddish cinnamon in age, edges finely whitish-fringed; stalk 1½–4½" (4–12 cm) long, ⅛–½" (4–12 mm) thick, sometimes swollen at the base, dry, yellow overall, darkening to orangish brown near the base, often whitish at the top, fibrous; in clusters, usually at the base of deciduous or coniferous trees or stumps but sometimes on woody debris or soil; spores 8.5–12 × 4.5–6 μm, elliptic to oval with slightly thickened walls, an obscure to prominent apiculus and a small apical pore, somewhat dextrinoid; inedible— *P. malicola* var. *malicola* (Kauffman) Smith.

  12. As in the previous choice except odor faintly fragrant, usually of green corn; cap usually not so bell-shaped; gills sometimes slowly staining orange where bruised, edges not white-fringed—*P. malicola* var. *macropoda* Smith and Hesler (see p. 228).

Pholiota aurivella (Fries) Kummer

COMMON NAME: Golden Pholiota.

CAP: 1½–6" (4–15 cm) wide, convex to nearly plane; surface sticky or slippery to slimy, smooth beneath a dispersed layer of large flattened scales that are typically arranged in a concentric pattern, cuticle orangish yellow to yellowish orange; flesh yellow; odor not distinctive or somewhat sweetish; taste not distinctive.

GILLS: attached, close, pale yellowish at first, reddish brown in age; veil whitish, cortinate, leaving only a zone of fibers near the top of the stalk.

STALK: 2–3" (5–7.5 cm) long, ¼–⅝" (5–15 mm) thick, dry, yellowish or yellowish brown, covered with fibrous scales below the annular zone.

SPORE PRINT: brown to rusty brown.

MICROSCOPIC FEATURES: spores 7–11 × 4.5–6 μm, broadly elliptic, lacking a distinct apiculus, with a distinct apical pore but not appearing truncate, smooth, brownish, somewhat dextrinoid.

FRUITING: in clusters on deciduous or coniferous trunks or stumps; July–November; occasional to frequent.

EDIBILITY: edible.

COMMENTS: several names have been given to this mushroom, including *P. adiposa* and *P. squarrosa-adiposa*.

Pholiota flammans (Fries) Kummer

COMMON NAME: Yellow Pholiota.

CAP: 1¼–4" (3–10 cm) wide, convex to nearly plane; surface sticky to slippery, smooth, bright yellow, to orangish yellow or yellowish orange only at the center, with curved, fibrous scales; flesh yellow; odor and taste not distinctive.

GILLS: attached and sharply notched, close to crowded, rather broad, bright yellow at first, soon staining brown on the edges where bruised; partial veil thin, yellow, fibrous to cortinate, usually leaving only a zone of fibers but sometimes an almost membranous ring on the upper stalk.

STALK: dry, bright yellow, becoming orangish yellow near the base in age, densely covered with cottony-fibrous scales below the annular zone.

SPORE PRINT: rusty brown.

MICROSCOPIC FEATURES: spores 4–5 × 2.5–3 μm, oblong to elliptic, usually with a distinct apiculus, apical pore not evident, smooth, brownish.

FRUITING: solitary or more typically in small clusters on coniferous logs and stumps; August–October; occasional.

EDIBILITY: inedible.

COMMENTS: a beautiful, distinctive species.

Pholiota lenta (Fries) Singer

CAP: 1–4" (2.5–10 cm) wide, convex to nearly plane; surface sticky to glutinous, with scattered, white, fibrous scales, whitish to grayish, sometimes with a pinkish tinge, becoming slightly darker at the center, margin rimmed with bits of white veil tissue; flesh white; odor not distinctive or slightly radish-like; taste not distinctive.

GILLS: attached, sometimes with fine decurrent lines, close, narrow to moderately broad, whitish at first, grayish brown in age, edges sometimes finely fringed; partial veil white, thick, cortinate, sometimes leaving a zone of fibers on the upper stalk.

STALK: 1¼–4" (3–10 cm) long, ⅛–½" (4–12 mm) thick, sometimes swollen at the base, fibrous, white toward the top, brownish near the base.

SPORE PRINT: brown.

MICROSCOPIC FEATURES: spores 5.5–7 × 3.5–4.5 μm, with a slight to fairly acute apiculus and a minute apical pore, smooth, pale brownish, slightly dextrinoid.

FRUITING: in small clusters on woody soil or humus; July–November; infrequent to occasional.

EDIBILITY: inedible.

Pholiota malicola var. macropoda Smith and Hesler

CAP: 1¼–5" (3–12.5 cm) wide, convex to nearly flat; surface moist to sticky or slippery at first but soon dry, smooth, yellow to orangish yellow, sometimes with a greenish tinge, decorated at most with faint whitish to buff fibers near the opaque margin; flesh yellowish, rather thick; odor faintly fragrant (often like green corn); taste not distinctive.

GILLS: attached or notched, close, narrow to moderately broad, yellowish at first, becoming pale rusty brown in age, sometimes slowly staining orange where bruised, edges even (not white-fringed); partial veil whitish to buff, fibrous to cortinate, usually leaving a slight zone of fibers on the upper stalk and cap margin.

STALK: 1½–7" (4–18 cm) long, ⅛–1" (4–25 mm) thick, nearly equal or sometimes tapered in either direction, dry, whitish to yellowish and silky at the top, lower portion fibrous to striate, becoming dark rusty brown from the base upward, with only a slight zone of fibers on the upper stalk.

SPORE PRINT: brown.

MICROSCOPIC FEATURES: spores 7.5–11 × 4.5–5.5 μm, with a distinct apical pore but apex not truncate, smooth, pale brownish, somewhat dextrinoid.

FRUITING: in clusters, usually at the base of deciduous or coniferous trees or stumps but sometimes on woody debris or soil; August–November; infrequent.

EDIBILITY: inedible.

COMMENTS: this variety is probably more frequent than known, as collections are often simply identified as *P. malicola*.

Pholiota spumosa (Fries) Singer

CAP: ¾–2½" (2–6 cm) wide, convex to nearly flat, usually with an almost conic umbo at first; surface sticky or slippery to slimy or glutinous, shiny, without scales, yellowish brown with distinct greenish tones, especially near the margin, and a darker brownish center, streaked with fibers; margin paler yellow to greenish yellow in age; flesh yellow to greenish yellow; odor faintly of green corn or not distinctive; taste not distinctive.

GILLS: attached or notched, sometimes with fine decurrent lines, close, moderately broad, yellow or, more typically, greenish yellow at first, cinnamon-brown in age but usually still with a greenish tint; partial veil yellow, fibrous to cortinate.

STALK: 1¼–4" (3–10 cm) long, ⅛–⅜" (3–8 mm) thick, with a thin coating of yellow fibers over most of its length but appearing more or less smooth and lacking a distinct ring, greenish yellow and appearing finely powdery at the apex, yellow to greenish yellow below, becoming dingy brown from the base upward.

SPORE PRINT: brown.

MICROSCOPIC FEATURES: spores 7–9 × 4–4.5 μm, mostly with a fairly acute apiculus, with a distinct apical pore but not truncate, smooth, pale brownish.

FRUITING: in small clusters, usually on decaying coniferous and sometimes deciduous stumps, logs, humus, sawdust, woodchips, or soil; June–November; occasional to frequent.

EDIBILITY: edibility unknown.

COMMENTS: several closely related species are difficult to distinguish.

Pholiota squarrosa (Fries) Kummer

COMMON NAME: Scaly Pholiota.

CAP: 1¼–5" (3–12.5 cm) wide, convex to nearly plane; surface entirely broken up into dry, tan to pinkish tan or brownish scales; margin usually decorated with yellowish to tan remnants from the partial veil at maturity; flesh pale yellow; odor usually strong of garlic, in some collections reportedly absent or more similar to onions; taste unpleasant or not distinctive.

GILLS: attached, usually with fine decurrent lines, close to crowded, moderately narrow, pale yellow at first, soon developing distinct green tones, finally dirty rusty brown but often with greenish tones remaining evident, covered at first by a fibrous-membranous yellowish or tan partial veil.

STALK: 1½–5" (4–12.5 cm) long, ⅛–⅝" (4–15 mm) thick, sometimes tapered toward the base, dry, scaly like the cap, often with a somewhat membranous ring.

SPORE PRINT: brown.

MICROSCOPIC FEATURES: spores 5–8 × 3.5–4.5 μm, elliptic to oval, with a distinct apical pore but not truncate, smooth, brownish, weakly dextrinoid.

FRUITING: usually in large clusters on or at the base of deciduous or coniferous trees, stumps, or logs; July–November; occasional to common.

EDIBILITY: poisonous, causes gastrointestinal upset.

COMMENTS: this is one of the most easily identified Pholiota species, especially when the odor is detected.

Pholiota veris (Singer and Smith) Smith and Hesler

CAP: ¾–2¼" (2–6 cm) wide, convex to nearly plane; surface moist to sticky or slippery, smooth overall but with fine fibers near the margin at first, pale cinnamon to yellowish cinnamon at first, fading to pale yellowish tan or whitish from the center outward in age, margin finely translucent-striate in age; flesh colored like the cap or paler; odor and taste not distinctive.

GILLS: close to crowded, pale pinkish tan at first, darkening to nearly cinnamon in age; partial veil yellowish cinnamon, usually leaving a thin, brownish, membranous ring on the upper stalk.

STALK: 1½–3½" (4–9 cm) long, ⅛–½" (4–12 mm) thick, dry, color similar to but paler than the cap, moderately scurfy only on the middle to upper portion, watery-whitish or light tan to pale cinnamon at first, not darkening substantially in age; usually with a thin, brownish, membranous ring on the upper stalk at maturity, nearly smooth overall in age, with a coating of white mycelium at the base.

SPORE PRINT: brown.

MICROSCOPIC FEATURES: spores 5.5–7.5 × 3.5–5 μm, elliptic to oval, with or without a distinct apiculus, slightly to distinctly truncate from a distinct to indistinct apical pore, smooth, brownish.

FRUITING: scattered, in groups or clusters on well-decayed deciduous wood or sawdust; April–June; infrequent to occasional.

EDIBILITY: unknown.

COMMENTS: one of the earliest agarics to fruit in spring.

Genus *Phyllotopsis*

Phyllotopsis nidulans (Fries) Singer (see photo, p. 10).

COMMON NAME: Orange Mock Oyster.

CAP: 1–3" (2.5–7.5 cm) wide, sometimes wider, convex to nearly plane; surface yellowish orange, sometimes becoming somewhat brownish or fading to orangish yellow, densely fuzzy; flesh orangish; odor offensive, often compared to rotten cabbage; taste very disagreeable.

GILLS: close to crowded, fairly narrow, yellow to orange.

STALK: absent.

SPORE PRINT: pale pink.

MICROSCOPIC FEATURES: spores 6–8 × 3–4 μm, sausage-shaped, smooth, hyaline, inamyloid.

FRUITING: solitary to clustered on dead wood, almost always of deciduous trees; typically August–November but also sometimes during winter thaws; fairly common.

EDIBILITY: inedible.

COMMENTS: a very identifiable fungus: orange, stalkless, nasty-smelling fuzzy caps with distinctive pink spores.

Genus *Pleurocybella*

Pleurocybella porrigens (Persoon : Fries) Singer
COMMON NAME: Angel's Wings.
CAP: 1⅜–4" (3.5–10 cm) wide, fan- to shell-shaped; surface smooth, white; flesh very thin, white; odor and taste not distinctive.
GILLS: crowded, narrow, white.
STALK: absent.
SPORE PRINT: white.
MICROSCOPIC FEATURES: spores 6–7 × 5–6 μm, subglobose to oval, smooth, hyaline, inamyloid.
FRUITING: solitary to clustered on dead conifer wood, especially eastern hemlock; August–October; common.
EDIBILITY: a good edible when fresh and abundant.

Genus *Pleurotus*

Pleurotus is a small genus of medium to large, wood-inhabiting mushrooms that usually grow in overlapping clusters. The stalk is usually eccentric to lateral or rudimentary to absent, but is sometimes central. Except for a single species, no veils or rings are present. Their spore prints are white, buff, cream, or grayish lilac, and the spores are inamyloid. All species are edible, and some are rated as choice.

Key to Species of *Pleurotus*

1. Veil present on young specimens, leaving a ring on the stalk or remnants on the cap margin; cap coated with tiny, matted, grayish fibrils on a whitish ground color— *P. dryinus* (Fries) Kummer (see p. 231).
1. Veil and ring absent on all stages; cap color variable, dark brown, yellowish brown to grayish brown, pale gray, tan to yellowish buff, creamy white or white—*P. ostreatus* (Jacquin : Fries) Kummer complex (see p. 232).

Pleurotus dryinus (Persoon : Fries) Kummer
COMMON NAME: Veiled Oyster.
CAP: 2–5" (5–12.5 cm) wide, convex, becoming broadly convex; surface dry, coated with tiny, matted, grayish fibrils on a whitish ground color, becoming slightly scurfy and whitish to dull yellowish tan overall in age; margin inrolled when young, becoming decurved at maturity, often rimmed with flaps of white veil; flesh thick, firm, white; odor fragrant to slightly pungent; taste not distinctive.
GILLS: decurrent, close to subdistant, sometimes forked near the stalk or crossveined, white, discoloring yellowish, covered by a white, membranous veil in the button stage.
STALK: 1½–4" (4–10 cm) long, ¾–1⅛" (2–3 cm) thick, eccentric to central, nearly equal or tapered downward, fibrous-tough, solid, whitish, sometimes with a white, membranous, superior ring.
SPORE PRINT: white.
MICROSCOPIC FEATURES: spores 9–14 × 3.5–5 μm, elliptic, smooth, hyaline, inamyloid.

FRUITING: solitary, scattered, or in clusters on decaying hardwoods; July–October; occasional.

EDIBILITY: edible but tough; it requires thorough cooking.

Pleurotus ostreatus (Jacquin : Fries) Kummer complex

COMMON NAME: Oyster Mushroom.

CAP: 1½–7⅛" (4–18 cm) wide, convex, oyster shell– to fan-shaped; surface smooth, moist or dry but not viscid; color variable, dark brown, yellowish brown to grayish brown, pale gray, tan to yellowish buff, creamy white, or white; flesh white; odor anise-like, fragrant, fruity, or not distinctive; taste not distinctive.

GILLS: decurrent, close to subdistant, white, grayish white or pale cream, typically with two or more tiers of lamellulae.

STALK: eccentric, lateral, rudimentary or absent; when present, up to 1½" (4 cm) long and up to 1⅜" (3.5 cm) thick, dry, solid, enlarged in either direction or nearly equal, often coated, at least near the base, with downy white hairs, white to dingy yellow.

SPORE PRINT: white, buff, cream, or grayish lilac (see comments below).

MICROSCOPIC FEATURES: variable (see comments below).

FRUITING: typically growing in overlapping clusters, sometimes scattered, on logs, stumps, and standing trees, usually hardwoods; April–November, or year-round when conditions allow; very common.

EDIBILITY: edible and often rated as choice.

COMMENTS: species in this complex are highly variable and often difficult to separate, even using microscopic features. The following information may assist you if you have access to a microscope. The spores of *P. ostreatus* = *P. sapidus* measure 7–12 × 3.5–5 μm, and the spore print is white to grayish lilac. *P. populinus* (edible) grows on various hardwoods, especially cottonwood, and has a buff spore print and larger spores (9–15 × 3–5 μm). *P. cystidiosus* (edible) has a white spore print, larger spores (11–17 × 4–5 μm), abundant clavate pileocystidia, and short pyriform cheilocystidia.

Genus *Pluteus*

Members of the genus *Pluteus* are small to medium mushrooms with pink to salmon or brownish pink spore prints and smooth, inamyloid spores. They grow on decaying wood, sawdust, wood chips, and other woody debris. They have free gills, a central stalk that separates cleanly from the cap, and flesh that is typically soft and putrescent. Their gills are typically white when young and pale pink to salmon or brownish pink in age. They are easily separated from species of *Volvariella* because they lack a volva. Some species are edible, but none is highly rated.

Key to Species of *Pluteus*

1. Cap 1–1¾" (2.5–4.5 cm) wide, wrinkled to reticulate on the disc when young, often becoming nearly smooth at maturity, pink to pinkish buff or pinkish gray when young, hygrophanous and fading to white or whitish in age; flesh whitish, odor not distinctive, taste disagreeable to nauseous; stalk whitish and silvery streaked, watery brownish at the base; spores 5–7 × 4.5–6 μm; on decaying hardwood logs; July–September; edibility unknown—*P. pallidus* Homola.
1. Not as above; cap white to whitish, with or without brownish fibrils and scales → **2**.
1. Not as above; cap some shade of yellow, orange, or red → **3**.
1. Not as above; cap some shade of brown, gray, or black → **4**.

2. Cap covered with tiny soft hairs, white overall, 1⅛–4" (3–10 cm) wide; spores 5–7 × 4.5–6 μm; pleurocystidia lacking horns; growing on decaying conifer wood or hardwood; July–October; edibility unknown—*P. tomentosulus* Peck.

2. Cap smooth, white to dull white overall, 1⅛–4" (3–10 cm) wide; spores 7–8 × 5–6 μm; pleurocystidia with prominent horns; growing on decaying hardwoods; July–October; edible—*P. pellitus* (Fries) Kummer.

2. Cap white, soon streaked with brownish fibrils and scales and becoming finely cracked over the disc, 1⅛–4" (3–10 cm) wide; spores 6–7.5 × 4.5–5 μm; pleurocystidia with prominent horns; growing on sawdust piles and decaying wood; July–October; edible—*P. petasatus* (Fries) Gillet.

3. Cap brilliant scarlet to orange-red when young, fading to orange or yellow at maturity, smooth to finely wrinkled, ⅜–3⅛" (1–8 cm) wide—*P. aurantiorugosus* (Trog) Saccardo (see p. 234).

3. Cap bright yellow and translucent-striate on the margin when young, hygrophanous and becoming bright yellow-brown in age, smooth, moist, ⅜–1⅛" (1–3 cm) wide; flesh white to yellowish, odor and taste not distinctive; gills whitish at first, then yellow, and finally brownish pink; stalk 1⁄16–⅛" (1.5–3 mm) thick, yellow; growing on decaying hardwood; July–October—*P. admirabilis* Peck (see p. 234).

3. Cap dull yellow to olive-yellow, often dark brownish yellow on the disc, smooth, moist or dry, ¾–3½" (2–9 cm) wide; margin lacking striations or only slightly striate; smooth, moist or dry, flesh whitish to yellowish, odor and taste not distinctive; stalk whitish to pale pinkish, becoming dull yellow; spores 6–7 × 4.5–5.5 μm; growing on decaying hardwood; July–October; edible—*P. flavofuligineus* Atkinson.

4. Cap finely granular (use a hand lens) to velvety, dark reddish brown, ⅜–1¼" (1–3.2 cm) wide; margin finely striate; flesh watery tan to whitish, odor and taste not distinctive; stalk 1⁄16–⅛" (1.5–3 mm) thick, fragile, whitish on the upper portion, grayish to dull brown toward the base; spores 5–6 × 4–4.5 μm; caulocystidia in clusters, elongate-clavate to slender fusoid-ventricose, containing dark brown pigment when revived in KOH; on decaying hardwood; July–October; edibility unknown—*P. seticeps* (Atkinson) Singer.

4. Cap slightly to distinctly wrinkled to veined, at least over the disc, dark brown when young, hygrophanous and fading to yellow-brown in age, ⅜–1⅛" (1–3 cm) wide; margin finely striate; flesh white to pale watery brown, odor and taste strongly farinaceous or not distinctive; stalk ⅛–¼" (2–6 mm) thick, white to whitish olive, becoming yellow to yellowish brown from the base upward, longitudinally streaked grayish to silvery; spores 5–7 × 5–6 μm; caulocystidia absent; on decaying hardwood; July–October; edibility unknown—*P. chrysophaeus* (Fries) Quélet.

4. Cap wrinkled to somewhat granular, olive-brown to yellowish olive, ⅜–2" (1–5 cm) wide; flesh whitish to pale yellow, odor and taste not distinctive; stalk 1⁄16–⅛" (1.5–3 mm) thick, dull yellow, brighter yellow near the base; spores 6–7 × 5–6 μm; on soil and decaying hardwood; July–October; edibility unknown—*P. lutescens* (Fries) Bresadola.

4. Not with the above combinations of characters → 5.

5. Cap surface smooth, dark grayish brown to bluish gray, ¾–2" (2–5 cm) wide; margin striate or not; flesh odor disagreeable, taste unpleasant; stalk whitish, usually stained blue at the base; spores 7–9 × 5–6 μm; pleurocystidia with prominent horns; on decaying hardwood; July–October; hallucinogenic—*P. salicinus* (Fries) Kummer.

5. Cap striate to furrowed from the margin to the disc, sometimes with granules or tiny

scales on the disc, fragile and often collapsing as it matures, grayish brown to pale red-dish brown, ⅜–2" (1–5 cm) wide; flesh whitish to watery brown, odor and taste not dis-tinctive; stalk ¹⁄₁₆–⅛" (1.5–3 mm) thick, white, with inconspicuous longitudinal stria-tions; spores 6–7.5 × 5–5.5 µm; on decaying hardwood; August–September; edible — *P. longistriatus* Peck.

5.   Cap coarsely wrinkled and ridged from the disc to the margin, granular to velvety plush to the unaided eye, brown on the disc with yellow between the ridges, ¾–2⅜" (2–6 cm) wide; flesh odor not distinctive, taste somewhat disagreeable; stalk velvety, brown; on decaying hardwoods — *P. granularis* Peck (see p. 235).

5.   Not with the above combination of characters → 6.

    6.   Cap surface conspicuously reticulate-veined on the disc, especially of young speci-mens, blackish brown fading to dull brown, ⅜–1⅜" (1–3.5 cm) wide — *P. thompsonii* (Berkeley and Broome) Dennis (see p. 235).

    6.   Cap surface streaked with flattened black fibrils, blackish brown, 1⅛–4" (3–10 cm) wide; flesh white, odor and taste not distinctive; gills white at first, edges blackish brown; stalk brownish, coated with blackish fibrils, paler than the cap; spores 6.5–8 × 4.5–5 µm; on decaying conifer wood; August–October; edible — *P. atromargina-tus* (Konrad) Kühner.

    6.   Cap surface smooth or streaked with tiny fibers, often wrinkled when young, dull brown to grayish brown or pale cinnamon-brown, 1⅛–4¾" (3–12 cm) wide; stalk ¼–¾" (6–20 mm) thick at the apex — *P. cervinus* (Schaeffer : Fries) Kummer (see p. 235).

Pluteus admirabilis (Peck) Peck

COMMON NAME: Yellow Pluteus.

CAP: ⅜–1⅛" (1–3 cm) wide, bell-shaped to broadly convex with a low umbo; surface moist when fresh, wrinkled on the disc or sometimes nearly smooth overall; margin translucent-striate; bright yellow when young, hygrophanous and fading to bright yellow-brown in age; flesh white to yellowish; odor and taste not distinctive.

GILLS: free, close, whitish at first, becoming yellow, and finally brownish pink at maturity.

STALK: 1–2⅜" (2.5–6 cm) long, ¹⁄₁₆–⅛" (1.5–3 mm) thick, nearly equal, moist, smooth, yellow.

SPORE PRINT: salmon.

MICROSCOPIC FEATURES: spores 5–7 × 4.5–6 µm, subglobose, smooth, hyaline.

FRUITING: solitary, scattered, or in groups on decaying hardwood; June–September; fairly common.

EDIBILITY: edible.

COMMENTS: *P. leoninus* (edible) is similar but has a white stalk and a smooth cap.

Pluteus aurantiorugosus (Trog) Saccardo

CAP: ⅜–3⅛" (1–8 cm) wide, obtusely conic to convex, becoming nearly plane and fre-quently umbonate in age; surface smooth to slightly velvety when fresh and moist, hygrophanous, becoming fibrillose to finely wrinkled in age or when dry, brilliant scar-let to orange-red when young, fading to orange or yellow and remaining darkest over the disc at maturity; margin incurved at first, becoming lobed and slightly uplifted in age, somewhat striate to finely cracked at maturity; flesh yellowish, watery, fragile; odor and taste not distinctive.

GILLS: notched to nearly free when young, free in age, moderately broad, white at first, becoming pink at maturity, edges fimbriate, crowded to close, with several tiers of attenuate lamellulae.

STALK: 1⅛–3½" (3–9 cm) long, ⅛–⅜" (3–10 mm) thick, nearly equal or tapered in either direction, striate, fibrillose to fibrillose-scaly when young, nearly smooth at maturity; white to pale yellow, becoming orange-red toward the base and paler toward the apex at maturity, sometimes with a white mycelium; partial veil and ring absent.

MICROSCOPIC FEATURES: spores 6–7.5 × 4.5–5 μm, broadly oval, smooth, hyaline.

SPORE PRINT: pinkish.

FRUITING: solitary, scattered, or in groups on decaying hardwood; July–September; uncommon.

EDIBILITY: unknown.

Pluteus cervinus (Schaeffer : Fries) Kummer

COMMON NAME: Fawn Mushroom, Deer Mushroom.

CAP: 1⅛–4¾" (3–12 cm) wide, convex, becoming broadly convex to nearly flat in age; surface smooth, sometimes streaked with tiny fibers, often wrinkled when young, dull brown to grayish brown or pale cinnamon-brown; flesh soft, white, thick; odor and taste not distinctive.

GILLS: free from the stalk, close to crowded, broad, white when young, becoming pale pink to salmon in age.

STALK: 2–4" (5–10 cm) long, ¼–¾" (6–20 mm) thick, nearly equal or enlarged downward, solid, dry, white, often with dull brown to grayish fibers; partial veil and ring absent.

SPORE PRINT: pink to salmon to brownish pink.

MICROSCOPIC FEATURES: spores 5.5–7 × 4–6 μm, elliptic, smooth, hyaline.

FRUITING: solitary or in groups on decaying hardwoods and conifers, on the ground over decaying wood, and on sawdust piles; June–October; fairly common.

EDIBILITY: edible.

COMMENTS: also known as *P. atricapillus*. Some authors recognize specimens with a blackish brown wrinkled cap when young and a thicker stalk (up to ¾" [2 cm] wide) as a separate species called *P. magnus* (edible).

Pluteus granularis Peck

CAP: ¾–2⅜" (2–6 cm) wide, convex, becoming broadly convex to nearly plane, umbonate; surface wrinkled and ridged from the disc to the margin, granular to velvety plush to the unaided eye, brown on the disc with yellow between the ridges; flesh yellowish; odor not distinctive; taste somewhat disagreeable.

GILLS: free, crowded, white at first, becoming pinkish.

STALK: 1⅛–2¾" (3–7 cm) long, ⅛–¼" (2–6 mm) thick, nearly equal, solid, velvety, dry, brown or paler.

SPORE PRINT: dull pink.

MICROSCOPIC FEATURES: spores 5–6.5 × 4–5 μm, broadly elliptic, smooth, hyaline.

FRUITING: solitary or scattered on decaying wood in conifers or hardwoods; July–October; occasional.

EDIBILITY: unknown.

Pluteus thompsonii (Berkeley and Broome) Dennis

CAP: ⅜–1⅜" (1–3.5 cm) wide, obtuse to convex or bell-shaped when young, becoming nearly plane in age; surface conspicuously reticulate-veined on the disc, especially on young specimens, becoming smoother in age; blackish brown when young, hygrophanous and dull brown in age; margin typically translucent-striate; flesh white; odor and taste not determined.

GILLS: free or nearly so, close to subdistant, white to grayish when young, becoming pale pink at maturity.

STALK: ¾–2¾" (2–7 cm) long, ⅛–¼" (2–6 mm) thick, nearly equal, hollow, white to silvery gray, darkest at the base, longitudinally streaked, whitish pruinose when young.

SPORE PRINT: dull pink.

MICROSCOPIC FEATURES: spores 6–8 × 5.5–6 μm, broadly oval to nearly round, smooth, hyaline.

FRUITING: solitary or in groups on decaying hardwood or on the ground, probably attached to buried wood; August–October; uncommon.

EDIBILITY: unknown.

Genus *Porpoloma*

Porpoloma umbrosum (Smith and Wallroth) Singer

COMMON NAME: Amyloid Tricholoma.

CAP: 2–4" (5–10 cm) wide, convex; surface dull gray to grayish brown, sometimes with slight reddish stains, smooth, but often becoming roughened or cracked in age, dry; flesh grayish, staining reddish when bruised; odor and taste farinaceous.

GILLS: attached to sinuate, crowded, broad, grayish, often staining pinkish to reddish.

STALK: 1–2" (2.5–5 cm) long, ½–¾" (1.5–2 cm) thick, grayish brown, often roughened.

SPORE PRINT: white.

MICROSCOPIC FEATURES: spores 7–9 × 3–4 μm, elliptic, smooth, hyaline, amyloid.

FRUITING: solitary to several under various conifers; July–September; infrequent to occasional.

EDIBILITY: unknown.

COMMENTS: a unique species likely to be keyed out as a *Tricholoma*, but species of that genus have inamyloid spores. Also known as *Pseudotricholoma umbrosum*.

Genus *Psathyrella*

Psathyrella is a very large genus with nearly four hundred known species throughout North America. They are small to medium, with a dark brown to blackish or sometimes dull reddish to pinkish gray spore print. *Psathyrella* species have a slender central stalk that is brittle and can be easily snapped in half. The gills are usually attached, not decurrent, and are often mottled. Some species are known to be edible, but the edibility of most species is unknown. Although none has been reported to cause fatalities, the edible species are usually considered mediocre.

Key to Species of *Psathyrella*

1. Growing on the decaying remains of the Shaggy Mane mushroom, *Coprinus comatus* (see p. 103); cap ¾–2⅜" (2–6 cm) wide, rimmed with flaps of partial veil when young, silky, white, becoming dingy white in age; stalk white; ring evanescent; spores 7–9 × 4–5 μm; August–September; edibility unknown—*P. epimyces* (Peck) Smith.
1. Growing on decaying leaves and stems of cattails and sedges near the water line; cap ¼–¾" (6–20 mm) wide, cinnamon-brown, quickly fading to pale yellowish brown, distinctly striate; stalk white, typically lacking a ring; spores 10–12 × 5–6.5 μm; May–August; edibility unknown—*P. typhae* (Kalchbrenner) Pearson and Dennis.

1. Growing on burned ground; cap ⅝–2⅜" (1.5–6 cm) wide; at first covered with white patches and strands of partial veil, chocolate to dark brown, hygrophanous and fading to pale cinnamon-buff to pinkish buff or pale grayish brown; stalk white, becoming pale brown in age, often whitish pruinose near the apex; ring absent or evanescent; spores 6–8 × 3–4 µm; July–October; edibility unknown—*P. carbonicola* Smith.

1. Growing on soil, organic debris, in grass, among mosses, or on wood → 2.

    2. Cap 1⅛–4" (3–10 cm) wide, smooth, pale to dark honey-yellow; stalk ⅛–⅜" (3–10 mm) thick, white, usually not forming a ring; typically around stumps in woods or in lawns—*P. candolleana* (Fries) Maire (see p. 238).

    2. Cap ⅝–2" (1.5–5 cm) wide, smooth when young, distinctly wrinkled in age, cinnamon-brown, fading to pale pinkish brown and finally pale pink; stalk very long and slender, 2⅜–4¾" (6–12 cm) long, 1/16–⅛" (1.5–3 mm) thick, whitish, smooth; partial veil and ring absent or rudimentary; spores 11–15 × 6.5–8 µm; August–October; edibility unknown—*P. gracilis* (Fries) Quélet.

    2. Not as above; cap surface covered with a dense layer of tiny flattened orange-brown to dark brown fibers or conspicuous coarse dark brown fibers → 3.

    2. Not as above; cap surface greater than 2" (5 cm) wide at maturity, conspicuously radially grooved and wrinkled → 4.

    2. Not as above; cap surface reddish brown to dark brown, smooth to somewhat wrinkled but not conspicuously radially grooved and wrinkled → 5.

3. Cap surface covered with a dense layer of tiny silky fibers, orange-brown to dark yellow-brown, ¾–4¾" (2–12 cm) wide; usually in grassy areas—*P. velutina* (Fries) Singer (see p. 239).

3. Cap surface ranging from brownish orange to tawny orange on the disc to pale tawny on the margin, covered with conspicuous, coarse, dark brown fibers, 1⅛–4" (3–10 cm) wide—*P. echiniceps* (Atkinson) Smith (see p. 238).

    4. Cap 1⅛–4" (3–10 cm) wide, obtuse when young, broadly convex to nearly plane in age, sometimes with an umbo; surface coated with silky, white fibrils when young, at first dark reddish brown to dark rusty brown, fading to dull orange-brown or dingy tan in age; margin sometimes rimmed with flaps of partial veil; stalk white and silky-pruinose near the apex; partial veil fibrillose-membranous, not leaving a ring; spores 6.5–9 × 4.5–5.5 µm; on decaying hardwood—*P. delineata* (Peck) Smith (see p. 238).

    4. Fruiting body nearly identical to the previous choice; watery brown to tawny cap fading to dingy tan; spores 9–11 × 6–8 µm, finely warted, with a protruding blunt pore, pale brown; edibility unknown—*P. rugocephala* (Atkinson) Smith.

5. Gills pale brown when young, becoming vinaceous red at maturity; cap margin not rimmed with white, tooth-like flaps of partial veil; cap ¾–2" (2–5 cm) wide; dark brown, fading to reddish buff; stalk white to reddish buff, lacking a ring; spore print pinkish red; spores 6.5–8 × 3.5–5 µm; usually in dense clusters on humus at the base of trees and stumps; August–October; edibility uknown—*P. conissans* (Peck) Smith.

5. Gills brown at maturity; cap margin not rimmed with white, tooth-like flaps of partial veil; cap ¾–1⅜" (2–3.5 cm) wide, smooth, moist, dark cinnamon-brown to reddish brown, fading to dingy tan or grayish in age, translucent-striate on the margin; stalk 2–4" (5–10 cm) long, about ⅛" (2–3.5 mm) thick, smooth, white, lacking a ring; spores 9–12 × 5–6 µm; on wet soil; August–October; edibility unknown—*P. baileyi* Smith.

5. Gills brown at maturity; cap margin rimmed with white tooth-like flaps of partial veil; cap ⅝–2" (1.5–5 cm) wide, smooth, sparsely coated with silky, whitish fibrils when young, reddish brown at first, becoming pale reddish brown to cinnamon-brown or

sometimes darker reddish brown in age; stalk white, lacking a ring; spores 7–10 × 4.5–5 μm; on decaying hardwood logs and stumps—*P. septentrionalis* Smith (see p. 239).

5. Fruiting body nearly identical to the previous choice except lacking tooth-like flaps of partial veil on the margin and with smaller spores, 4.5–7 × 3–4 μm; edibility unknown— *P. hydrophila* (Fries) Maire.

Psathyrella candolleana (Fries) Maire

COMMON NAME: Common Psathyrella, Suburban Psathyrella.

CAP: 1⅛–4" (3–10 cm) wide, obtusely conic to convex when young, becoming nearly plane in age, sometimes with an umbo; surface smooth, pale to dark honey-yellow, sometimes darkening to purplish brown near the margin, fading to whitish in age, coated with silky, white fibrils when young; margin striate, rimmed with flaps of partial veil; flesh pale yellow to whitish; odor and taste not distinctive.

GILLS: attached, close to crowded, white when young, becoming grayish brown, sometimes tinged with violet; edges whitish and very finely scalloped.

STALK: 2–4⅜" (5–11 cm) long, ⅛–⅜" (3–10 mm) thick, equal, hollow, white, coated with white fibrils; partial veil thin, fragile, fibrillose-membranous, not forming a ring or sometimes leaving an evanescent superior ring.

SPORE PRINT: purplish brown.

MICROSCOPIC FEATURES: spores 7–10 × 4–5 μm, elliptic, truncate, smooth, with an apical pore, pale brown.

FRUITING: scattered or in groups around decaying hardwood stumps, on wood chips, and in lawns; May–September; fairly common.

EDIBILITY: edible.

Psathyrella delineata (Peck) Smith

CAP: 1⅛–4" (3–10 cm) wide, obtuse when young, becoming convex and finally broadly convex to nearly plane in age, sometimes with a low broad umbo; surface coated with silky, white fibrils when very young, conspicuously radially grooved and wrinkled, dark reddish brown to dark rusty brown when young, fading to dull orange-brown and finally dingy tan, sometimes forming a darker zone along the margin; margin sometimes rimmed with flaps of partial veil; flesh watery brown; odor and taste not distinctive.

GILLS: attached, close, dull white when young, becoming pale brown and finally purplish brown to dark rusty brown at maturity; edges minutely fringed in age, sometimes whitish.

STALK: 1½–4" (4–10 cm) long, ⅜–⅝" (1–1.6 cm) thick, enlarging slightly downward, hollow, white and silky-pruinose near the apex, whitish to pale brown below, coated with silky fibers that darken in age, with conspicuous white fibers near the base; partial veil fibrillose-membranous, not leaving a ring.

SPORE PRINT: purple-brown.

MICROSCOPIC FEATURES: spores 6.5–9 × 4.5–5.5 μm, oval to elliptic, smooth, with an apical pore, pale brown.

FRUITING: scattered or in groups on decaying hardwood; July–September; occasional.

EDIBILITY: unknown.

Psathyrella echiniceps (Atkinson) Smith

CAP: 1⅛–4" (3–10 cm) wide, obtuse to convex when young, becoming nearly flat in age; surface brownish orange to tawny-orange on the disc, fading to pale tawny on the margin, covered with conspicuous, coarse, dark brown fibers and tiny scales; margin often rimmed with flaps of partial veil; flesh pale to dark buff; odor and taste somewhat disagreeable.

GILLS: attached, close, pale brown when young, becoming dark reddish brown to purplish brown in age; edges white and minutely fringed.

STALK: 2¾–4¾" (7–12 cm) long, ⅜–⅝" (1–1.6 cm) thick, enlarging slightly downward, hollow, dull white to pale yellow, sparsely coated on the lower two-thirds with coarse, dark brown fibers and tiny scales, pruinose at the apex; partial veil fibrous, typically not leaving a ring.

SPORE PRINT: purplish brown.

MICROSCOPIC FEATURES: spores 7–9 × 4–5 μm, elliptic, smooth, with an apical pore, pale brown.

FRUITING: scattered, in groups, or in small clusters near hardwood stumps; September–November; uncommon.

EDIBILITY: unknown.

Psathyrella septentrionalis Smith

CAP: ⅝–2" (1.5–5 cm) wide, obtuse to convex when young, becoming nearly flat in age; surface smooth, moist, reddish brown, and sparsely coated with silky whitish fibrils when young, becoming pale reddish brown to cinnamon-brown or sometimes dark reddish brown in age; margin rimmed with white, tooth-like flaps of partial veil when young, striate in age; flesh watery brown to tan; odor and taste not distinctive.

GILLS: attached, close, pale brown when young, becoming dark brown in age; edges even.

STALK: ⅝–2⅜" (1.5–6 cm) long, ¹⁄₁₆–¼" (1.5–5 mm) thick, nearly equal, white, pruinose near the apex, silky fibrillose to nearly smooth below, becoming pale brown from the base upward in age; partial veil fibrillose-membranous, white, not leaving a ring.

SPORE PRINT: brownish black.

MICROSCOPIC FEATURES: spores 7–10 × 4.5–5 μm, elliptic, smooth, with an apical pore, pale brown.

FRUITING: in clusters and groups on decaying hardwood logs and stumps; August–October; occasional.

EDIBILITY: unknown.

COMMENTS: compare with *P. hydrophila* (edibility unknown), which lacks flaps of partial veil on the cap margin and has smaller spores.

Psathyrella velutina (Fries) Singer

COMMON NAME: Velvety Psathyrella.

CAP: ¾–4¾" (2–12 cm) wide, oval to hemispheric when young, becoming convex and finally plane, often with an umbo; surface covered with a dense layer of tiny, silky, flattened fibers, many of which aggregate into tiny scales, orange-brown to dark yellow-brown, darkest on the disc; margin often rimmed with flaps of partial veil; flesh yellowish brown, staining dark rusty brown in KOH; odor and taste not distinctive.

GILLS: attached, close, pale to dark brown, mottled; edges white, often beaded with moisture drops when fresh.

STALK: 1⅛–4¾" (3–12 cm) long, ⅛–⅝" (3–16 mm) thick, equal, pale yellowish brown with dull orange-brown fibers and scales below the ring, smooth and whitish above; partial veil cottony-fibrous, leaving a thin superior zone of fibrils.

SPORE PRINT: blackish brown.

MICROSCOPIC FEATURES: spores 8–12 × 6–8 μm, elliptic, smooth, with an apical pore, pale
  brown.
FRUITING: solitary, scattered, or in small clusters in grassy areas and organic debris;
  July–October; fairly common.
EDIBILTY: edible.
COMMENTS: also known as *Lacrymaria velutina*.

Genus *Psilocybe*

Psilocybe species are small to medium mushrooms with smooth caps, slender stalks, and dark
purplish brown spore prints. They grow on a variety of substrates, including wood, soil, and
dung, and among mosses. Their spores are smooth with a truncate apical pore. The cap and/or
stalk of some species stains greenish blue in age or when bruised. This fairly large genus
contains many hallucinogenic species.

Key to Species of *Psilocybe*

1.   Growing on decaying wood, wood chips, or decaying peat moss → 2.
1.   Growing on dung, manured soil, or a mixture of dung and straw → 3.
1.   Not with the above combinations of characters → 4.
   2.   Cap cinnamon-brown to reddish brown, hygrophanous and fading to ochraceous
        buff to brownish yellow, often with greenish stains, slowly blueing when bruised,
        ⅜–1½" (1–4 cm) wide, conic to convex when young, becoming nearly plane, often
        with a small umbo, sticky to dry, gills close to crowded, brown with whitish edges;
        stalk whitish to reddish brown, staining greenish blue when bruised, lacking a
        ring; flesh blueing when cut; odor and taste farinaceous; spores 7–11 × 4–6 μm;
        August–October; hallucinogenic—*P. caerulipes* (Peck) Saccardo.
   2.   Cap chestnut-brown to olive-brown or brownish orange, hygrophanous and
        fading to yellowish or whitish or remaining dark olive-brown, often with greenish
        stains, slowly blueing when bruised, ⅜–2¼" (1–5.5 cm) wide, conic to convex
        when young, becoming nearly plane, sometimes with a small umbo, sticky to dry;
        gills close, grayish to violet-gray, sometimes with whitish edges; stalk whitish,
        staining blue-green when bruised, lacking a ring; flesh blueing when cut; odor and
        taste farinaceous; spores 10–14 × 5.5–7 μm; September–October; hallucinogenic—
        P. baeocystis Singer and Smith.
3.   Stalk lacking a ring, whitish to brownish, not blueing; cap ⅜–1⅛" (1–3 cm) wide, sticky
     to dry, dark reddish brown to orangish brown, hygrophanous and fading to pale gray-
     ish brown or straw-color, not blueing; gills attached, subdistant, pale brown to purplish
     brown with whitish edges; flesh not blueing, odor and taste not distinctive; spores 10–
     14 × 7–9 μm—*P. coprophila* (Bulliard : Fries) Kummer (see p. 241).
3.   Nearly identical to the previous choice except that the stalk has a white, membranous
     ring; hallucinogenic—*P. merdaria* (Fries) Ricken.
   4.   Growing among sphagnum mosses in bogs, swamps, and wet woodlands; cap
        dark reddish brown, sticky, lacking blue stains; flesh dark brown, not blueing, taste
        farinaceous or not distinctive—*P. atrobrunnea* (Lasch) Gillet (see p. 241).
   4.   Growing among Haircap Moss *(Polytrichum)* on sandy soil; cap ⅛–¾" (3–20 mm)
        wide, convex to bell-shaped, dull rusty brown, not blueing when bruised; stalk
        very thin, 1/32–1/16" (0.5–1.5 mm) thick, dull rusty brown; spores 6.5–8.5 × 4–5.5 μm;
        May–June; edibility unknown—*P. polytrichophila* Peck = *P. montana* (Persoon :
        Fries) Kummer.

4. Growing on hardwood leaves; cap ⅛–¾" (3–20 mm) wide, dark reddish brown fading to dingy tan; stalk brown, sometimes forming a thin flat disc at the base; not blueing on any part; spores 5.5–8 × 4–5 μm, rhomboid; July–September; edibility unknown—*P. phyllogena* Peck.

4. Growing on the decaying remains of Fireweed *(Epilobium)*; cap ⅜–1" (1–2.5 cm) wide, sticky, dark rust-brown, hygrophanous and fading to ochraceous or beige; flesh brownish, odor and taste not distinctive, not blueing on any part; spores 5.5–7.5 × 4–6.5 μm, subrhomboid; July–September; edibility unknown—*P. acadiensis* Smith.

Psilocybe atrobrunnea (Lasch) Gillet

CAP: ⅜–2⅜" (1–6 cm) wide, convex to bell-shaped when young, becoming broadly convex at maturity, often with an umbo; surface dark reddish brown, moist and sticky when fresh, fading in age to pale reddish brown to ochraceous buff; margin inrolled at first, lacking veil remnants, lacking greenish stains; flesh dark brown, not blueing when cut; odor and taste not distinctive.

GILLS: attached or notched, subdistant to close, pale cinnamon-brown when young, becoming dark purplish brown in age, usually with white minutely fringed edges.

STALK: 3⅛–7" (8–18 cm) long, ⅛–¼" (3–6 mm) thick, equal, hollow at maturity, cartilaginous and tough, pale reddish brown with white fibrils, pruinose at the apex; partial veil present as white fibrils on very young specimens, not forming a ring.

SPORE PRINT: dark purplish brown.

MICROSCOPIC FEATURES: spores 8–14 × 5–8 μm, elliptic, smooth, with a narrow apical pore, pale brown.

FRUITING: scattered or in groups among sphagnum mosses in bogs, swamps, and wet woodlands; August–October; occasional.

EDIBILITY: unknown.

Psilocybe coprophila (Bulliard : Fries) Kummer

COMMON NAME: Dung-loving Psilocybe.

CAP: ⅜–1⅛" (1–3 cm) wide, convex to bell-shaped; surface sticky to dry, smooth, shiny, dark reddish brown to orangish brown, hygrophanous and fading to pale grayish brown or straw-color; margin translucent-striate to the disc, often coated with tiny white patches when young; flesh brownish, not blueing; odor and taste not distinctive.

GILLS: attached, subdistant, pale brown to purplish brown; edges whitish.

STALK: ¾–2" (2–5 cm) long, ¹⁄₁₆–⅛" (1.5–4 mm) thick, nearly equal, whitish when young, becoming brown at maturity, coated with tiny white scales and flecks, not blueing; partial veil absent, or if present, sparse, white, evanescent, not forming a ring.

SPORE PRINT: purplish brown.

MICROSCOPIC FEATURES: spores 10–14 × 7–9 μm, elliptic with an apical pore, smooth, pale brown.

FRUITING: scattered or in groups on horse or cow dung; June–October; fairly common.

EDIBILITY: hallucinogenic.

Genus *Resinomycena*

Resinomycena rhododendri (Peck) Redhead and Singer

CAP: ⅛–⅝" (3–15 mm) wide, convex at first, becoming nearly flat with a depressed center; surface white, usually very sticky, smooth to finely coated with crystalline granules, white to vaguely yellowish; flesh very thin, whitish; odor and taste not distinctive.

GILLS: attached, soon becoming strongly decurrent, close, narrow, whitish, the edges often finely jagged and usually beaded with tiny resinous droplets.

STALK: ⅜–2" (1–5 cm) long, about ¹⁄₁₆" (1–2 mm) thick, typically very resinous or sticky, white, with a tuft of white mycelium at the base.

SPORE PRINT: white.

MICROSCOPIC FEATURES: spores 5.5–8.5 × 2.5–4 μm, elliptic to subcylindric with a prominent apiculus, smooth, hyaline, amyloid.

FRUITING: scattered, in groups or clusters on leaves, twigs, or woody soil; June–September; occasional to common, more common eastward.

EDIBILITY: unknown.

COMMENTS: distinctive by virtue of its stickiness: it will readily adhere to your fingers.

Genus *Resupinatus*

Resupinatus applicatus (Bataille : Fries) S. F. Gray

COMMON NAME: Black Jelly Oyster.

CAP: ⅛–¼" (2–6 mm) wide, convex to cup-shaped; surface dry, coated with tiny fibers, dark bluish gray to grayish black, stalkless; flesh thin, rubbery-gelatinous, grayish black; odor and taste not distinctive.

GILLS: arising from the point of attachment, subdistant, pale grayish at first, becoming dark gray to blackish.

SPORE PRINT: white.

MICROSCOPIC FEATURES: spores 4–6 μm, globose, smooth, hyaline, inamyloid.

FRUITING: scattered, in groups, or in dense clusters on the underside of decaying hardwood logs; June–November; fairly common.

EDIBILITY: unknown.

COMMENTS: this mushroom is not easily discovered unless you search the underside of decaying logs.

Genus *Rhodocybe*

Rhodocybe mundula (Lasch) Singer

COMMON NAME: Cracked-cap Rhodocybe.

CAP: 1–2" (2.5–5 cm) wide, convex to nearly flat; surface dingy grayish white, dry, smooth, soon becoming cracked or lined in a concentric pattern; margin inrolled at first; flesh whitish; odor farinaceous; taste bitter.

GILLS: strongly decurrent, crowded, narrow, whitish to grayish at first, becoming pinkish.

STALK: 1–2" (2.5–5 cm) long, ⅛–¼" (2.5–7 mm) thick, solid, finely scurfy-hairy, rather flexible, off-white, the base somewhat coated with white mycelium.

SPORE PRINT: salmon-pink.

MICROSCOPIC FEATURES: spores 4–6 × 4–5 μm, globose to subglobose, smooth to slightly bumpy, ends angular, hyaline, inamyloid.

FRUITING: scattered or in small clusters on the ground in mixed woods; July–October; occasional to common.

EDIBILITY: unknown.

Genus *Rhodotus*

Rhodotus palmatus (Bulliard : Fries) Maire

COMMON NAME: Netted Rhodotus.

CAP: 1–2″ (2.5–5 cm) wide, convex, becoming broadly convex; surface reddish to pinkish at first, fading to more or less orangish, conspicuously netted (see photo); flesh gelatinous; odor and taste not distinctive.

GILLS: attached, close, moderately broad and crossveined, whitish, becoming pinkish from spores.

STALK: 1–2″ (2.5–5 cm) long, ⅛–¼″ (3–5 mm) thick, reddish to pinkish, eccentric to almost lateral, dry, tough.

SPORE PRINT: pinkish.

MICROSCOPIC FEATURES: spores 6–8 μm, globose, warted, hyaline, inamyloid.

FRUITING: scattered or in groups on dead hardwood trees, logs, or stumps, especially maple; June–September; infrequent to occasional.

EDIBILITY: unknown.

COMMENTS: a distinctive mushroom, endemic to northeastern North America.

Genus *Rickenella*

Rickenella fibula (Bulliard : Fries) Raithelhuber

CAP: ⅛–⅜″ (3–15 mm) wide, dry to moist, convex with a depressed center; surface margin inrolled at first, becoming striate to sulcate, smooth, orangish brown to reddish brown, fading to yellowish orange or orangish yellow; flesh extremely thin; odor and taste not distinctive.

GILLS: strongly decurrent, distant, whitish.

STALK: ⅜–2″ (1–5 cm) long, about 1/16″ (1–2 mm) thick, fragile, dry to moist, very finely hairy at first, becoming smooth, yellowish orange to orangish yellow, hollow.

SPORE PRINT: white.

MICROSCOPIC FEATURES: spores 4–5 × 2–2.5 μm, elliptic, smooth, hyaline, inamyloid.

FRUITING: usually scattered or in groups among mosses; June–November; fairly common.

EDIBILITY: inedible.

Genus *Rozites*

Rozites caperata (Fries) Karsten

COMMON NAME: Gypsy.

CAP: 2–6″ (5–15.5 cm) wide, egg-shaped when young, becoming broadly convex, often with a low, broad umbo; surface wrinkled and fibrillose to scurfy, sometimes nearly smooth in age, coated with a thin, hoary sheen when young, moist to dry, not sticky, brownish orange to brownish yellow or ochre-orange, becoming paler in age; flesh whitish to yellowish; odor and taste not distinctive.

GILLS: attached, close, pale yellow at first, becoming dull orange-brown at maturity.

STALK: 2–5″ (5–10 cm) long, ⅜–¾″ (1–2 cm) thick, enlarged downward or nearly equal; dry, solid, white when young, soon pale yellow, then pale yellowish brown in age; partial veil membranous, white, leaving a membranous superior to median ring on the stalk; base often sheathed with remnants of a fibrous, matted, white universal veil.

SPORE PRINT: rusty brown.

MICROSCOPIC FEATURES: spores 11–14 × 7–9 μm, almond-shaped, slightly roughened,
  pale brown.
FRUITING: solitary, scattered, or in groups on the ground in conifer or hardwoods;
  July–November; fairly common.
EDIBILITY: edible.
COMMENTS: compare with *Cortinarius claricolor* (see p. 107) and *C. torvus* (see p. 112).

Genus *Russula*

Russula is a large genus estimated to have nearly two hundred species in North America. They
are small to large, typically terrestrial mushrooms, with brittle flesh that easily crumbles. Their
spore prints range in color from white to cream, yellow, or ochre. They have globose to broadly
elliptic spores that are amyloid and roughened with warts and ridges that often form a reticulum. The stalk is central and lacks a ring.

  This is a fairly safe group of mushrooms to collect for the table. Only a few species are
known to be poisonous, and no fatalities have been reported from eating species in this genus.
Because the genus is complex and includes a large number of species, a comprehensive key to
species is beyond the scope of this book. We have included a sampling of some of the more
common and distinctive species that occur in this region. For additonal information, we highly
recommend *Keys to the Species of Russula in Northeastern North America* by Geoffrey Kibby and
Raymond Fatto.

Russula brevipes Peck
COMMON NAME: Short-stalked White Russula.
CAP: 3½–8" (9–20 cm) wide, convex when young, becoming nearly flat to funnel-shaped
  in age; surface dry, coated with tiny, matted fibrils, white to cream, staining yellowish
  to brown, often coated with dirt; margin inrolled when young, becoming uplifted in
  age; flesh white, brittle; odor not distinctive or unpleasant; taste not distinctive to
  slowly slightly acrid (very acrid in var. *acrior;* see comments below).
GILLS: attached, crowded, narrow, white to cream, staining reddish brown.
STALK: 1–3" (2.5–7.5 cm) long, 1–1½" (2.5–4 cm) thick, equal or tapering slightly down-
  ward, hollow in age, dry, smooth, white to cream, staining brownish.
SPORE PRINT: white to cream.
MICROSCOPIC FEATURES: spores 8–11 × 6–8 μm, elliptic to oval, with warts and a partial
  to complete reticulum, hyaline, amyloid.
FRUITING: solitary, scattered, or in groups on the ground in mixed woods;
  July–September; fairly common.
EDIBILITY: edible.
COMMENTS: transformed into a choice edible when attacked by the Lobster Mushroom,
  Hypomyces lactifluorum. The photograph shows *R. brevipes* var. *acrior,* which has bluish
  green tints on the gills and stalk apex and very acrid flesh.

Russula compacta Frost
COMMON NAME: Firm Russula.
CAP: 2¾–5⅞" (7–15 cm) wide, convex when young, becoming broadly convex to nearly
  flat in age, often depressed over the disc or funnel-shaped; surface dry, sticky when
  wet, dull, often cracked toward the center in age, whitish to cream, staining cinnamon
  to yellowish brown or orangish brown; flesh firm, white, with yellowish tones in age,
  staining gray-green with $FeSO_4$; odor fishy; taste not distinctive to slightly acrid.

GILLS: attached, close to subdistant, white, staining reddish brown when bruised.

STALK: ¾–4¾" (2–12 cm) long, ⅜–1½" (1–4 cm) thick, equal, dry, smooth, solid, becoming hollow in age, white, staining reddish brown when bruised.

SPORE PRINT: white.

MICROSCOPIC FEATURES: spores 7–10 × 6–8 μm, elliptic to oval, with warts and reticulation, hyaline, amyloid.

FRUITING: solitary, scattered, or in groups on the ground in mixed woods; July–September; fairly common.

EDIBILITY: edible.

Russula laurocerasi Melzer

COMMON NAME: Almond-scented Russula.

CAP: 1½–4" (4–10 cm) wide, hemispheric when young, becoming broadly convex to nearly flat in age; surface smooth, shiny and sticky when wet, dull when dry, ivory when very young, becoming yellowish to tawny to yellow-brown; margin tuberculate-striate; flesh white, brittle; odor fragrant, like almonds or marzipan; taste not distinctive to slightly acrid.

GILLS: attached, crowded to subdistant, broad, grayish cream to grayish yellow to yellowish white.

STALK: 1–4" (2.5–10 cm) long, ⅜–1⅛" (1–3 cm) thick, equal, dry, smooth, whitish to yellowish white, staining yellowish brown to brown.

SPORE PRINT: cream to pale yellowish.

MICROSCOPIC FEATURES: spores 7.5–10.5 × 7.5–9 μm, elliptic to oval, with warts, ridges, and a partial to complete reticulum, hyaline, amyloid.

FRUITING: solitary, scattered, or in groups on the ground under hardwoods and in mixed woods; July–September; fairly common.

EDIBILITY: unknown.

COMMENTS: *R. fragrantissima* (edibility unknown) is nearly identical but has a larger cap, 2¾–8" (7–21 cm) wide, a stouter stalk, and flesh that is moderately to strongly acrid.

Russula lutea (Hudson : Fries) S. F. Gray

CAP: 1–2¾" (2.5–7 cm) wide, convex when young, becoming broadly convex to nearly flat with a depressed center in age; surface viscid when fresh, becoming dry, smooth, bright yellow to yellow-orange or apricot; margin striate in age; flesh white; odor not distinctive or sweet, especially in age; taste not distinctive.

GILLS: attached to nearly free, subdistant, broad, whitish when young, becoming yellow to deep ochre at maturity.

STALK: 1⅛–2" (3–5 cm) long, ¼–¾" (5–20 mm) thick, equal, smooth, dry, white.

SPORE PRINT: yellow to deep ochre.

MICROSCOPIC FEATURES: spores 7–9 × 6–8 μm, elliptic, with warts, hyaline, amyloid.

FRUITING: solitary, scattered, or in groups on the ground under hardwoods; July–September; occasional.

EDIBILITY: edible.

COMMENTS: *R. claroflava* (edibility unknown) has a white stalk that slowly stains grayish to blackish when bruised.

Russula nigricans Fries

CAP: 2–5⅞" (5–15 cm) wide, convex when young, becoming broadly convex to nearly flat with a sunken center in age; surface tacky when young, becoming dry, felty, whitish when young, becoming brownish to blackish brown at maturity, often cracked and fissured in age; flesh white, staining bright orange to red, then turning brownish black when bruised; odor not distinctive; taste slowly acrid.

GILLS: attached, distant, broad, whitish, staining bright reddish, then black when bruised.

STALK: 1⅛–3⅛" (3–8 cm) long, ⅜–1½" (1–4 cm) thick, equal, whitish, staining like the cap when bruised.

SPORE PRINT: white to pale cream.

MICROSCOPIC FEATURES: spores 6–8 × 5–7 μm, elliptic, with warts, finely reticulate, hyaline, amyloid.

FRUITING: solitary, scattered, or in groups on the ground under hardwoods; July–September; infrequent.

EDIBILITY: unknown.

COMMENTS: *R. densifolia* (edibility unknown) and *R. dissimulans* (edibility unknown) have close to subdistant gills. The cuticle of *R. densifolia* peels from the margin, halfway to the disc, but the cuticle of *R. dissimulans* does not peel.

Russula pulchra Burlingham

CAP: 1½–4" (4–10 cm) wide, convex when young, becoming broadly convex to nearly flat with a shallowly depressed center in age; surface dry, minutely fibrillose when young, soon becoming smooth, often finely cracked in age, reddish orange when fresh, becoming pale peach in age, the disc sometimes cream to pinkish yellow; margin finely striate; flesh white; odor and taste not distinctive.

GILLS: attached, subdistant, broad, forked toward the stalk, white to cream.

STALK: 1⅛–4" (3–10 cm) long, ⅝–1" (1.5–2.5 cm) thick, equal, stuffed, dry, smooth, white, occasionally with pink tones.

SPORE PRINT: cream.

MICROSCOPIC FEATURES: spores 6.5–9 × 5.5–7.5 μm, elliptic, with warts, hyaline, amyloid.

FRUITING: solitary, scattered, or in groups on the ground under hardwoods, especially beech; July–September; occasional.

EDIBILITY: edible.

Russula subgraminicolor Murrill

CAP: 2⅜–3⅛" (6–8 cm) wide, convex when young, becoming broadly convex to flat with a depressed center in age; surface dry, minutely fibrillose when young, soon becoming smooth; color various shades of blue-green to grass-green overall, fading toward the margin in age; margin faintly to distinctly striate; flesh white; odor and taste not distinctive.

GILLS: attached, subdistant, broad, forked near the stalk, whitish to cream.

STALK: 1⅛–4" (3–10 cm) long, ⅝–1" (1.6–2.5 cm) thick, equal, dry, smooth, white.

SPORE PRINT: cream to pale ochre.

MICROSCOPIC FEATURES: spores 8.5–11 × 6–8.5, elliptic, with warts, hyaline, amyloid.

FRUITING: solitary, scattered, or in groups on the ground under oaks; July–September; occasional.

EDIBILITY: edible.

Russula tenuiceps Kauffman

CAP: 2¾–4¾" (7–12 cm) wide, convex when young, becoming broadly convex to flat with a shallowly depressed center in age; surface smooth, sometimes sticky; color variable, shades of peach-red, rose-red, or crimson, often with cream or yellow-orange blotches, especially over the disc; margin striate; flesh white; odor not distinctive; taste strongly acrid.

GILLS: attached, becoming nearly free, crowded, narrow, creamy yellow.

STALK: 1½–4⅜" (4–11 cm) long, ½–¾" (1.6–2 cm) thick, equal, smooth, dry, white with red to pink tones.

SPORE PRINT: yellow to deep ochre.

MICROSCOPIC FEATURES: spores 7–9 × 6–7.5 μm, elliptic, with warts and fine lines, forming a partial reticulum, hyaline, amyloid.

FRUITING: solitary, scattered, or in groups on the ground under hardwoods, especially oaks and maples; July–September; occasional.

EDIBILITY: unknown.

Russula ventricosipes Peck

CAP: 2–5½" (5–14 cm) wide, convex, becoming nearly plane; surface viscid when moist, often coated with sand, needle litter, and debris, tawny-yellow to dull brown; margin typically tuberculate-striate; flesh whitish to pale yellow; odor weakly or strongly of marzipan or bitter almonds, or somewhat fetid; taste weakly to strongly acrid.

GILLS: attached, close, often forked near the stalk, dull white to creamy yellow.

STALK: ¾–4" (2–10 cm) long, ⅜–1½" (1–4 cm) thick, nearly equal down to a tapered base, reddish pruinose overall or at least near the base, ground color whitish.

SPORE PRINT: dark cream.

MICROSCOPIC FEATURES: spores 7–10 × 4–6 μm, lacrymoid, with very tiny projections, appearing nearly smooth, hyaline, amyloid.

FRUITING: solitary, scattered, or in groups, often partially buried in sandy soil near pines; July–October; fairly common.

EDIBILITY: unknown.

Genus *Simocybe*

Simocybe centunculus (Fries) Karsten

COMMON NAME: American Simocybe.

CAP: ⅜–1⅛" (1–3 cm) wide, convex to nearly flat in age; surface velvety to silky, brownish olive; margin translucent-striate when fresh and moist; flesh thin, brown; odor and taste not distinctive.

GILLS: attached, close to crowded, grayish brown to yellowish brown; edges whitish and finely serrate.

STALK: ⅜–1½" (1–4 cm) long, about ⅛" (2–4 mm) thick, nearly equal, frequently curved, central to eccentric, pruinose to slightly scurfy, especially near the apex when young, nearly smooth in age, dingy whitish, becoming brownish olive; base often coated with white mycelium.

SPORE PRINT: olive-brown to dark ochre.

MICROSCOPIC FEATURES: 6–7.5 × 4–5 μm, elliptic, smooth, pale brown.

FRUITING: solitary, scattered, or in groups on decaying hardwood; July–October; occasional.

EDIBILITY: unknown.

COMMENTS: also known as *Ramicola centuncula* and *Naucoria centuncula*.

Genus *Squamanita*

Squamanita umbonata (Sumstine) Bas

COMMON NAME: Knobbed Squamanita.

CAP: 1¼–2¼" (3–6 cm) wide, conic at first, becoming nearly flat but retaining a conic umbo; surface orangish buff to orangish brown, dry, fibrous to scaly, the cuticle often splitting near the edge to reveal thick, white flesh; odor and taste not distinctive.

GILLS: attached, close to crowded, fairly broad, white; partial and universal veils evident on young specimens.

STALK: 1–3" (2.5–7.5 cm) long, ¼–¾" (5–20 mm) thick, whitish and cottony above, usually with fibers or scales like the cap below, with rings of coarse, brownish scales near the ground, arising from an enlarged cylindric tuber.

SPORE PRINT: white.

MICROSCOPIC FEATURES: spores 6–9 × 3.5–5.5 μm, elliptic, smooth, hyaline, inamyloid; chlamydospores sometimes present on lower stalk.

FRUITING: solitary, scattered, or in groups on the ground in mixed woods; August–September; rare; most reports are from Pennsylvania, New York, and New England.

EDIBILITY: unknown.

COMMENTS: formerly called *Armillaria umbonata*.

Genus *Strobilurus*

Strobilurus conigenoides (Ellis) Singer

COMMON NAME: Magnolia-cone Mushroom.

CAP: ¼–¾" (5–20 mm) wide, convex to flat; surface white, dry, covered with a dense layer of minute hairs; margin incurved at first; flesh very thin; odor and taste not distinctive.

GILLS: attached, close to almost crowded, broad, white.

STALK: 1–2" (2.5–5 cm) long, ¹⁄₃₂–¹⁄₁₆" (0.75–2 mm) thick, dry, covered with a dense layer of minute hairs.

SPORE PRINT: white.

MICROSCOPIC FEATURES: spores 6–7 × 3–3.5 μm, elliptic, smooth, hyaline, inamyloid.

FRUITING: in groups on magnolia cones and sweetgum fruits; September–November; rare in the northern states and Canada, common in the South.

EDIBILITY: unknown.

COMMENTS: this mushroom's range coincides with the natural ranges of its hosts.

Genus *Stropharia*

Stropharia is a fairly small genus of small to large mushrooms that grow on wood chips, sawdust, soil, and dung. Their spore prints are dark purple-brown to violaceous fuscous, and their spores are smooth, with a truncate apical pore. The caps are usually viscid when fresh, and they have a central stalk with rhizomorphs often attached. A partial veil is usually present, leaving a ring on the stalk or remnants on the cap margin. The gill edges are often whitish and finely fringed at maturity. Some are excellent edibles, at least one species is poisonous, and the edibility of some is unknown.

Key to Species of *Stropharia*

1.   Cap green to bluish green, fading to yellowish green and finally yellowish, convex to nearly plane, smooth to slightly scaly, 1–3" (2.5–8 cm) wide; margin typically rimmed

with white veil remnants; flesh whitish, odor and taste not distinctive; gills close, whitish at first, becoming purple-brown with a white edge at maturity; stalk 1½–4" (4–10 cm) long, up to ⅝" (1.6 cm) thick, sheathed from the base up to the ring with white, soft, cottony patches and scales; partial veil whitish, leaving a delicate superior ring or annular zone; spores 7–9 × 4–5 μm; on woody debris, leaf litter, humus, and grassy areas; poisonous according to some reports—*S. aeruginosa* (Curtis : Fries) Quélet.

1. Cap orange to brick-red, ochre to brownish ochre, reddish brown, grayish brown to brown or reddish purple → 2.
1. Cap white to cream or pale yellow → 3.
   2. Cap moist or dry but not viscid, reddish brown to reddish purple when young, fading to pale brown, buff, or pale gray in age; underside of partial veil and ring conspicuously and regularly segmented; ring thick, cottony-membranous, with radial striations or grooves on the upper surface—*S. rugosoannulata* Farlow : Murrill (see p. 250).
   2. Cap subviscid, becoming dry, ochre to brownish ochre, fading to ochre-yellow in age; ring thick, cottony-membranous, not conspicuously and regularly segmented on the underside and typically not radially striate or grooved on the upper surface—*S. hardii* Atkinson (see p. 250).
   2. Cap viscid when fresh, pale purplish brown, reddish brown or grayish brown, fading to pale reddish brown in age; ring thin, delicate, membranous, and often tearing and falling away; stalk sheathed from the base up to the ring with white, soft, cottony patches and scales—*S. hornemannii* (Fries) Lundell and Nanfeldt (see p. 250).
   2. Cap viscid when fresh, orange to brick-red; ring conspicuous, membranous; stalk sheathed from the base up to the ring with whitish scales over a dull ochre ground color—*S. thrausta* (Schulzer) Saccardo (see p. 251).
3. Growing on horse dung in grassy areas, in gardens, and along trails in woods—*S. semiglobata* (Batsch : Fries) Quélet (see p. 251).
3. Growing on grassy areas; cap yellow, viscid when fresh, smooth, ¾–2" (2–5 cm) wide; flesh white, odor and taste not distinctive; gills close, whitish when young, becoming purple-brown at maturity; stalk 1⅛–2" (3–5 cm) long, up to ⅜" (1 cm) thick, dry, fibril-lose to nearly smooth, white, usually with white rhizomorphs at the base; partial veil cottony-membranous, white, leaving a persistent, white, superior to median ring that is radially striate on the upper surface; spores 7–10 × 4–5.5 μm; reportedly poisonous—*S. coronilla* (Bulliard : Fries) Quélet.
3. Growing on grassy areas; cap white, dry to moist but not viscid, smooth, 1⅛–2⅜" (3–6 cm) wide, margin often rimmed with torn partial veil remnants; flesh white, odor and taste not distinctive; gills attached, close, white when young, becoming gray, then purple-black at maturity; stalk 1⅛–3⅛" (3–8 cm) long, ⅜–⁵⁄₁₆" (4–8 mm) thick, dry, nearly smooth, white; partial veil cottony-membranous, white, leaving a fragile, white, superior ring that is radially striate on the upper surface; spores 10–13 × 6–8 μm; edible —*S. melanosperma* (Fries) Quélet.
3. Growing on humus in woods; cap white to whitish, becoming tinged pale yellow on the disc or overall in age, typically umbonate, viscid when fresh, shiny, smooth, ¾–2" (2–5 cm) wide; flesh white, odor and taste not distinctive; gills subdistant, whitish at first, becoming pale violet-brown and darkening to purple-brown in age; stalk 1⅜–3" (3–7.5 cm) long, up to ¼" (6 mm) thick, smooth or nearly so, white to yellowish; partial veil thin, membranous, white, leaving a persistent superior ring that is usually radially striate on the upper surface; spores 7–9 × 4–5 μm; edibility unknown—*S. albonitens* (Fries) Karsten.

3. Growing on wood chips; cap buff at first, becoming pale yellow-brown, at least over the disc, becoming cracked in age or when dry, 1⅝–5⅞" (4–15 cm) wide, underside of partial veil and ring conspicuously and regularly segmented; ring thick, cottony-membranous, with radial striations or grooves on the upper surface—*S. rugosoannulata* Farlow : Murrill (see p. 250).

Stropharia hardii Atkinson

COMMON NAME: Hard's Stropharia.

CAP: 1–4" (2.5–10 cm) wide, convex when young, becoming nearly flat in age; surface subviscid, becoming dry, nearly smooth when young, coated with tiny matted fibrils and scales, especially near the margin in age, ochre to brownish ochre, fading to ochre-yellow with darker fibrils and scales, often with darker tones near the center; flesh whitish; odor not distinctive to slightly farinaceous; taste not distinctive.

GILLS: attached, close, whitish when very young, soon becoming grayish brown then purplish brown at maturity.

STALK: 1½–4¾" (4–12 cm) long, ¼–¾" (6–20 mm) thick, nearly equal, often with a slightly bulbous base, sticky when young and fresh, becoming dry, slightly scaly to nearly smooth, whitish to pale ochre-yellow; base with whitish rhizomorphs; partial veil white, membranous, leaving a persistent superior ring.

SPORE PRINT: purplish brown.

MICROSCOPIC FEATURES: spores 6–9 × 3–5 μm, elliptic with apical pore, smooth, pale purplish brown.

FRUITING: solitary, scattered, or in groups on the ground under hardwoods; July–October; occasional.

EDIBILITY: unknown.

Stropharia hornemannii (Fries) Lundell and Nannfeldt

COMMON NAME: Lacerated Stropharia.

CAP: 1½–6" (4–15.5 cm) wide, hemispheric to convex when young, becoming broadly convex to flat with a low broad umbo in age; surface smooth, viscid when fresh, pale purplish brown, reddish brown or grayish brown, fading in age; margin with white, cottony veil remnants; flesh white to pale yellowish near the stalk; odor not distinctive; taste not distinctive or mildly unpleasant.

GILLS: attached, close, broad, pale grayish buff, becoming grayish purple to purplish brown in age.

STALK: 2–5⅞" (5–15 cm) long, ¼–1" (7–25 mm) thick, equal, smooth, sheathed from the base up to the ring with white, soft, cottony patches and scales over a yellowish ground color, silky and white above the ring; base with white rhizomorphs; partial veil white, leaving a thin, delicate, membranous ring on the upper portion of the stalk.

SPORE PRINT: purple-brown.

MICROSCOPIC FEATURES: spores 10.5–13 × 5.5–7 μm, elliptic, with an apical pore, smooth, pale purplish brown.

FRUITING: solitary, scattered, or in groups on the ground under conifers or on well-decayed conifer wood; August–November; occasional.

EDIBILITY: unknown.

Stropharia rugosoannulata Farlow : Murrill

COMMON NAME: Wine-cap Stropharia, Wine-red Stropharia.

CAP: 1⅝–5⅞" (4–15 cm) wide, broadly convex or bell-shaped, becoming nearly flat in age;

surface dry, smooth, becoming cracked in age, reddish brown to reddish purple, fading to pale brown, buff, or pale gray in age; flesh white, thick, not staining when cut or bruised; odor and taste not distinctive.

GILLS: attached or notched, crowded, broad, white, soon becoming gray to purple-gray, then grayish black to purple-black, not staining when bruised.

STALK: 2¾–7⅞" (7–20 cm) long, ⅜–1¼" (1–3 cm) thick, nearly equal or enlarged and bulbous at the base, solid, white, fibrous, staining yellow to brown in age, with white mycelium and rhizomorphs at the base; partial veil white, conspicuously and regularly segmented on the underside, leaving a persistent superior ring; ring thick, cottony-membranous, segmented on the underside, radially striate or grooved on the upper surface.

SPORE PRINT: dark purple-brown to purple-black.

MICROSCOPIC FEATURES: spores 10–14 × 7–9 μm, elliptic, smooth, pale purplish brown.

FRUITING: scattered or in groups on wood chips, mulch, straw, and in gardens or lawns; June–October; fairly common.

EDIBILITY: edible.

COMMENTS: cracked, buff to pale brown caps may be encountered.

Stropharia semiglobata (Batsch : Fries) Quélet

COMMON NAME: Round Stropharia.

CAP: ⅜–2⅜" (1–6 cm) wide, hemispheric when young, becoming broadly convex; surface viscid when fresh, smooth, pale yellow to yellow; margin sometimes rimmed with veil remnants; flesh whitish; odor and taste not distinctive.

GILLS: attached, close to subdistant, grayish when young, becoming dark purple-brown at maturity.

STALK: 1½–4" (4–10 cm) long, ⅛–¼" (2–5 mm) thick, nearly equal or slightly enlarged at the base, viscid, white to yellowish, usually lacking rhizomorphs on the base; partial veil thin, fibrous, viscid, leaving a superior annular zone that soon becomes dark purple-brown from falling spores.

SPORE PRINT: dark purple-brown.

MICROSCOPIC FEATURES: spores 15–19 × 7–10 μm, elliptic with an apical pore, smooth, purplish brown.

FRUITING: solitary, scattered, or in groups on horse dung in grassy areas, gardens, and along horse trails in woods; June–September; fairly common.

EDIBILITY: edible, but of poor quality.

Stropharia thrausta (Schulzer) Saccardo

CAP: 1⅛–3⅛" (3–8 cm) wide, convex, becoming broadly convex, sometimes with a low umbo; surface viscid when fresh, orange to brick-red, with scattered, pointed, flattened, whitish scales, especially near the margin; flesh white; odor and taste not distinctive.

GILLS: attached, sometimes with a decurrent tooth, close to subdistant, grayish, soon becoming violet-gray, then purple-brown in age; edges whitish.

STALK: 2⅜–4¾" (6–12 cm) long, ⅛–⅜" (3–10 mm) thick, nearly equal, densely coated from the base up to a ring by whitish scales over a dull ochre ground color, white pruinose above the ring; ring superior, whitish, membranous, flaring.

SPORE PRINT: purple-brown.

MICROSCOPIC FEATURES: spores 12–14 × 6–7.5 μm, elliptic with an apical pore, smooth; brown.

FRUITING: scattered, in groups or clusters on buried wood and on organic debris under conifers or hardwoods; September–November; uncommon to rare.

EDIBILITY: unknown.

COMMENTS: also known as *Stropharia squamosa* var. *thrausta* and *Psilocybe thrausta*.

Genus *Tectella*

Tectella patellaris (Fries) Murrill

COMMON NAME: Veiled Panus.

CAP: ¼–¾" (7–20 mm) wide, convex, more or less shell–shaped, often drooping; surface brownish, slimy to sticky when young, becoming dry and fibrous; margin inrolled, decorated with bits of veil tissue; flesh tough; odor and taste not distinctive.

GILLS: attached, distant, narrow, brownish; partial veil present on young specimens.

STALK: usually absent; if present, then lateral and tiny.

SPORE PRINT: white.

MICROSCOPIC FEATURES: spores 3–5 × 1–1.5 μm, cylindric, smooth, hyaline, weakly amyloid.

FRUITING: in groups or clusters on deciduous wood, especially beech, birch, and willow; July–October; infrequent to occasional.

EDIBILITY: unknown.

COMMENTS: this is the only species in its genus; no other stalkless agarics have partial veils. Previously known as *Panus operculatus*.

Genus *Tephrocybe*

Tephrocybe palustris (Peck) Donk

CAP: ⅜–¾" (1–2 cm) wide, convex to flat, typically with a slight, rounded umbo, sometimes becoming slightly depressed at the center in age; surface gray to dark brown at first, fading to grayish tan, moist, translucent-striate or sulcate-striate about two-thirds of the way toward the disc; flesh thin, pale buff; odor somewhat rancid-farinaceous; taste not distinctive.

GILLS: attached, with a slight decurrent tooth, close, whitish to yellowish.

STALK: 2–4" (5–10 cm) long, 1/16–⅛" (2–3 mm) thick, moist, colorless-translucent to gray or dark brown, fragile, finely pruinose when young.

SPORE PRINT: white.

MICROSCOPIC FEATURES: spores 5.5–8.5 × 3–5 μm, elliptic, smooth, hyaline, inamyloid.

FRUITING: in groups among sphagnum mosses; July–September; occasional to frequent.

EDIBILITY: unknown.

COMMENTS: also known as *Lyophyllum palustre*.

Genus *Tricholoma*

Members of the genus *Tricholoma* are medium to large terrestrial mushrooms with white to pale cream spore prints and inamyloid spores. They have a central stalk, and most species lack a partial veil and ring. A few species have a sparse fibrous partial veil that is evident only in the button stage, and a few others have a well-formed veil that leaves a prominent ring. More than one hundred species are known from North America. Some species are edible, some are poisonous, and the edibility of the majority is unknown. *Porpoloma umbrosum*, a species that macroscopically resembles some *Tricholoma* species, is also included in the key.

Key to Species of *Tricholoma* and Ally

1. Cap mostly white to cream, buff, or tan, sometimes developing yellow to reddish brown or grayish tinges or fibrous scales, especially on the disc, as the mushroom matures → 2.
1. Cap yellow, greenish yellow, yellowish green, olive, or orange-yellow to orange → 3.
1. Cap reddish, red-brown, golden brown to yellow-brown, olive-brown or grayish brown to blackish brown → 7.
1. Cap pale to dark gray to black → 14.
    2. Odor pungent, disagreeable, like coal tar; taste disagreeable but not acrid; stalk white to cream, brownish near the base, not staining yellow when bruised or in age; cap 1⅛–2⅜" (3–6 cm) wide—*T. inamoenum* (Fries) Quélet (see p. 261).
    2. Odor pungent, disagreeable, like coal tar (sometimes faint); taste not distinctive; stalk white at first, becoming creamy white to yellow, staining yellow when bruised or in age; cap 2–4¾" (5–12 cm) wide—*T. sulphurescens* Bresadola (see p. 264).
    2. Odor not distinctive; taste acrid; cap opaque, white, becoming buff in age, discoloring dingy ochraceous when handled, coated with flattened radiating fibrils; gills white, slowly becoming yellowish in age, with an evanescent, cortinate partial veil on the button stage; stalk white, soon becoming yellow on the base; spores 5–7 × 3–4 μm; on the ground under conifers; edibility unknown—*T. albatum* (Quélet) Maublanc and d'Astis.
    2. Odor and taste farinaceous—*T. subresplendens* (Murrill) Ovrebo (see p. 264).
    2. Odor and taste not distinctive; cap ¾–2¾" (2–7 cm) wide, smooth, viscid when wet, soon dry and shiny, slightly silky on the disc; white, becoming yellowish on the disc at maturity; gills white, sinuate to nearly free, close to crowded; stalk dry, white; spores 6–7.5 × 4–4.5 μm; edibility unknown—*T. resplendens* (Fries) Quélet.
    2. Odor not distinctive, taste not distinctive or bitter; flesh white to grayish; cap whitish, typically with pale gray, brownish gray, or tan tints on the disc, dry, with radiating fibrils and sometimes with tiny, grayish or brownish scales, 1⅛–3⅛" (3–8 cm) wide; gills white, edges often serrate or eroded and sometimes blackening; stalk white, often dingy white or pinkish at the base; spores 6–8 × 5–6 μm; on the ground in hardwoods; edibility unknown—*T. serratifolium* Peck.
    2. Odor like dirty gym socks but usually described as spicy-sweet, aromatic, fruity, or fragrant; taste not distinctive; cap robust, 2–8" (5–20.5 cm) wide, dry, white to whitish and nearly smooth at first, soon developing yellow, flattened fibers that aggregate into fibrous scales; fibrous scales becoming yellowish brown to reddish brown, especially on the disc; gills of young specimens covered by a thick, white, cottony-membranous partial veil; stalk sheathed from the base up to a thick, cottony ring by a white veil that breaks into patches and becomes reddish brown; ring white on the upper surface, reddish brown below; spores 5–7 × 4.5–5.5 μm; on the ground under conifers—*T. magnivelare* (Peck) Redhead = *Armillaria ponderosa* (Peck) Saccardo (see p. 261).
3. Odor pungent, disagreeable, like coal tar; cap typically dull sulphur-yellow to olivaceous yellow or dull pale yellow, often tinged brownish to grayish on the disc, ¾–3⅛" (2–8 cm) wide; flesh yellow, taste farinaceous to unpleasant; gills and stalk yellow; spores 7–11 × 5–6.5 μm; inedible—*T. sulphureum* (Fries) Kummer.
3. Odor pungent, disagreeable, like coal tar; cap dry, shiny and smooth at first, becoming soft and somewhat suede-like at maturity, yellow when young, fading to yellowish buff

to yellowish tan in age, ¾–3⅛" (2–8 cm) wide; flesh pale yellow, taste farinaceous to disagreeable; stalk pale yellow, darkening somewhat in age or when bruised; spores 7–9 × 5–6 μm; edibility unknown—*T. odorum*.

3.   Odor distinctly soapy; taste not distinctive or slightly soapy; gills white; cap color highly variable, grayish olive, lead-gray, greenish gray, or grayish brown to brown; stalk flesh pinkish to pinkish orange—*T. saponaceum* (Fries) Kummer (see p. 263).

3.   Odor and taste farinaceous or unpleasant; cap coated with conspicuous, radiating, blackish fibrils over a yellow to golden yellow ground color—*T. sejunctum* (Sowerby : Fries) Quélet (see p. 263).

3.   Odor farinaceous or not distinctive; taste farinaceous, bitter, or acrid; cap yellowish green to greenish yellow → 4.

3.   Odor and taste farinaceous or not distinctive; cap yellow, sulphur-yellow, golden yellow, smoky yellow, orange-yellow, or dull orange; gills white or yellow → 5.

4.   Growing under conifers or hardwoods; odor not distinctive; taste bitter at first, then strongly acrid; cap yellowish green to greenish yellow with grayish fibrils; stalk pale yellowish white to yellowish green—*T. aestuans* (Fries) Gillet (see p. 258).

4.   Growing under hardwoods; odor not distinctive; taste bitter; cap pale yellowish green, often tinged gray or brown, especially on the disc, sometimes streaked with darker fibrils, 1⅛–3¾" (3–9.5 cm) wide; gills pale greenish yellow to pale yellowish green; stalk pale yellowish green; spores 6–8 × 5–6 μm—*T. palustre* (edible) Smith.

4.   Growing under conifers; odor and taste farinaceous; stalk white overall; cap greenish yellow, typically with a low, broad umbo, smooth, dry or slightly viscid when moist, 2–3¾" (5–9.5 cm) wide; flesh white, gills crowded, free or slightly notched, white; stalk white overall, not yellow on the upper portion; spores 5–6 × 3.5–4 μm, elliptic; edibility unknown—*T. intermedium* Peck.

4.   Growing under conifers; odor and taste farinaceous; cap with a conspicuous, sharp umbo, dry, fibrillose, yellowish green, often dark green on the umbo, sometimes streaked with dark fibrils, 1½–4⅜" (4–11 cm) wide; flesh pale green to greenish buff; gills close, pale yellowish green, sometimes staining reddish; stalk silky to slightly scurfy, pale greenish yellow, often tinged reddish at the base; spores 6–9 × 4–5 μm; edibility unknown—*T. davisiae* Peck.

5.   Gills bright sulphur-yellow; cap bright to dull sulphur-yellow, becoming reddish brown on the disc—*T. flavovirens* (Persoon : Fries) Lundell (see p. 260).

5.   Gills white, whitish, or tinged yellowish → 6.

6.   Stalk white to dull white overall, becoming yellowish to dingy yellowish in age; cap yellow overall or with pale, smoky brown over the disc, moist or dry, smooth, slightly hygrophanous, sometimes with water spots, 1½–3½" (4–9 cm) wide; flesh whitish to yellowish, odor and taste not distinctive; gills close to subdistant, broadly sinuate at first, nearly free in age, whitish to yellowish; spores 6–8 × 5–6 μm; growing under conifers or among sphagnum mosses in bogs; edibility unknown—*T. fumosoluteum* (Peck) Saccardo.

6.   Stalk white to whitish overall; cap coated with tiny, golden brown to reddish brown or yellow-brown, flattened scales and fibers over a yellowish ground color, 1½–4" (4–10 cm) wide; flesh white, odor and taste farinaceous; gills close, sinuate, white to whitish; on the ground under conifers—*T. leucophyllum* Ovrebo and Tylutki (see p. 261).

6.   Stalk yellow on the upper portion, becoming white toward the base; cap color variable when young, dull orange or golden yellow to yellow, sometimes with

obscure dark fibrils, becoming pale yellow at maturity, dry or slightly viscid when moist; flesh white, odor and taste farinaceous; growing under conifers — *T. subluteum* Peck (see p. 263).

7. Ring present, membranous and distinct or an obscure fibrous zone → 8.

7. Ring absent → 9.

    8. Odor like dirty gym socks but usually described as spicy-sweet, aromatic, fruity, or fragrant; taste not distinctive; cap robust, 2–8" (5–20.5 cm) wide, dry, white to whitish and nearly smooth at first, soon developing yellow, flattened fibers that aggregate into fibrous scales; fibrous scales becoming yellowish brown to reddish brown, especially on the disc; gills of young specimens covered by a thick, white, cottony-membranous partial veil; stalk sheathed from the base up to a thick cottony ring by a white veil that breaks into patches and becomes reddish brown; ring white on the upper surface, reddish brown below; spores $5–7 \times 4.5–5.5$ μm; on the ground under conifers; edible — *T. magnivelare* (Peck) Redhead = *Armillaria ponderosa* (Peck) Saccardo (see p. 261).

    8. Odor not distinctive or fragrant to spicy; taste nutty, bitter, or not distinctive; surface dry, reddish brown to brown, breaking up into rings and patches of coarse flattened scales on a whitish to pale pinkish buff ground color — *T. caligatum* (Viviani) Ricken complex (see p. 259).

    8. Odor and taste strongly farinaceous or unpleasant; ring membranous, white on the upper surface and orange-brown below — *T. zelleri* (Stuntz and Smith) Ovrebo and Tylutki (see p. 265).

    8. Odor and taste strongly farinaceous or unpleasant; ring an obscure fibrous zone — *T. aurantium* (Schaeffer : Fries) Ricken (see p. 259).

9. Cap dry, reddish brown; finely to distinctly cracked and/or scaly; odor not distinctive or farinaceous to disagreeable → 10.

9. Cap dry, buffy brown or pale golden brown to dark olive-brown on disc, dull yellow to dull yellowish buff on the margin, with flattened, radiating fibrils except smooth on the disc, sometimes finely cracked, 1–2¾" (2.5–7 cm) wide; flesh white, odor and taste strongly farinaceous; gills close to subdistant, pale buff to pale yellow, often discoloring brown or olivaceous brown on the edge; stalk nearly equal with a rounded base, pale buff overall and often dingy buff at the base, slightly scurfy; spores $6–8 \times 5–6$ μm; on the ground in beech-maple woods; edibility unknown — *T. subaureum* Ovrebo.

9. Cap dry, blackish brown to grayish brown or dull yellow-brown, densely matted-fibrillose and often finely cracked on the disc, 1½–4¾" (4–12 cm) wide; margin pale grayish brown and often tinged dingy yellow or yellowish green; flesh pale dull gray, odor and taste farinaceous; gills close, pale buff to pale gray, discoloring yellow near the cap margin; stalk solid at maturity, nearly equal and often with a slightly bulbous base, slightly scurfy, white to pale gray, sometimes tinged yellowish brown; spores $7–8 \times 5–6$ μm; on the ground in beech-maple or oak-hickory woods; edibility unknown — *T. luteomaculosum* Smith.

9. Cap sticky to slimy when fresh and moist, soon dry; surface coated with tiny, flattened scales and fibers or nearly smooth; odor and taste farinaceous → 11.

    10. Stalk stuffed, then hollow at maturity, reddish brown or paler, fibrillose or fibrillose-scaly; surface of young and mature caps coated with small, fibrous scales, typically not coarsely cracked on the disc at maturity; margin inrolled and cottony-fibrous at first, becoming expanded and remaining somewhat cottony-fibrous in age; cap 1–3⅛" (2.5–8 cm) wide, umbonate; flesh white, developing reddish tinges

when cut; gills notched to sinuate, close, whitish to buff when young, staining reddish brown when bruised or in age; partial veil cortinate, not leaving a ring— *T. vaccinum* (Persoon : Fries) Kummer (see p. 264).

10. Stalk solid at maturity, white to whitish, slowly staining reddish brown from the base up; margin thin, incurved, and coated with soft downy hairs at first, nearly naked at maturity; cap coarsely cracked and scaly on the disc at maturity with whitish flesh showing in the cracks— *T. imbricatum* (Fries : Fries) Kummer (see p. 260).

11. Growing on the ground under cottonwood → 12.

11. Growing on the ground in mixed hardwoods, mixed woods, or sometimes with northern scrub pine → 13.

11. Growing on the ground or among mosses in conifer woods or sometimes with oak; stalk whitish when young, becoming reddish brown in age, fibrous and scurfy, nearly equal or enlarged downward; flesh white, staining reddish; cap chestnut-brown over the disc, paler golden brown toward the margin— *T. pessundatum* (Fries) Quélet (see p. 262).

11. Growing on the ground under conifers; stalk white to whitish, often stained brownish, especially toward the base; flesh white, not staining reddish; cap coated with tiny, golden brown to yellow-brown or reddish brown, flattened scales and fibers over a yellowish ground color; margin smooth, yellow to brownish yellow— *T. leucophyllum* Ovrebo and Tylutki (see p. 261).

12. Gill edges uniformly colored orange-brown to reddish brown; gills close, whitish with small brownish spots; stalk nearly equal to enlarged downward, white at the apex, unevenly colored like the cap over a whitish ground color; cap viscid then dry, smooth or nearly so, orange-brown to reddish brown when young, becoming paler orange-brown to reddish brown in age, margin paler, often yellowish tan, 1⅛–3¾" (3–9.5 cm) wide; spores 4.5–6 × 3–4 μm; edibility unknown— *T. fulvimarginatum* Ovrebo and Halling.

12. Gill edges white or unevenly stained reddish brown; gills close, white, staining reddish brown; stalk nearly equal or enlarged downward, white, staining reddish brown; cap viscid then dry, with radiating fibers and water spots, pale to dark cinnamon-brown with a whitish margin, 2–5½" (5–14 cm) wide; spores 5–6 × 3.5–4 μm; edible— *T. populinum* Lange.

13. Cap reddish brown to yellow-brown, slightly viscid when fresh, often shiny when dry, smooth, pale yellow-brown to dull yellow on the margin; stalk pale yellow to yellowish buff with reddish brown longitudinal fibrils; gills close, yellowish buff to buff bruising brownish; on the ground under hardwoods or mixed woods, often with birch— *T. flavobrunneum* (see p. 260).

13. Cap dark brown or sometimes brownish gray, viscid, wrinkled or sometimes smooth, 1⅛–3⅛" (3–8 cm) wide; stalk white to pale buff, often dingy whitish at the base, slightly scurfy; gills close, white to pale buff, discoloring brownish to brownish gray on the edge; spores 6–9 × 4–6 μm; on the ground in beech-maple or oak-hickory woods or sometimes under northern scrub pine; edibility unknown— *T. olivaceobrunneum* Ovrebo.

14. Flesh grayish, staining reddish when bruised, odor and taste farinaceous; cap dull gray to grayish brown; on the ground under conifers; spores amyloid— *Porpoloma umbrosum* (Smith and Wallroth) Singer (see p. 236).

14. Flesh whitish to pale grayish, not staining, odor and taste farinaceous; stalk white with a persistent, flaring, white, cottony-membranous ring; partial veil present on the button stage; cap dry, radially fibrillose, pale gray to grayish buff, ¾–2½"

(2–6.5 cm) wide; spores 5–6 × 3–4 μm; on the ground under willow; edibility unknown—*T. cingulatum* (Fries) Jacobasch.

14. Flesh white to pale gray, not staining, odor not distinctive, taste not distinctive or bitter; cap whitish, typically with pale gray, brownish gray, or tan tints on the disc, dry, with radiating fibrils and sometimes with tiny, grayish or brownish scales, 1⅛–3⅛" (3–8 cm) wide; gills white, edges often serrate or eroded and sometimes blackening; stalk white, often dingy white or pinkish at the base; spores 6–8 × 5–6 μm; on the ground in hardwoods; edibility unknown—*T. serratifolium* Peck.

14. Flesh white to pale gray, not staining, odor and taste not distinctive; cap gray to brownish gray, especially on the disc, whitish toward the margin, moist when fresh, becoming dry and somewhat shiny, smooth, 2⅜–4" (6–10 cm) wide; gills white at first, becoming grayish at maturity, subdistant, notched to sinuate, edges entire; stalk up to ¾" (2 cm) thick, solid, smooth, white, becoming grayish in age; spores 5–6 × 2–3 μm; on the ground among fallen leaves in hardwoods; edibility unknown—*T. tumidum* Fries.

14. Not with the above combinations of characters → 15.

15. Cap conspicuously scaly to fibrillose-scaly overall or at least over the center at maturity → 16.

15. Cap not conspicuously scaly to fibrillose-scaly; surface fibrillose or with scattered, tiny, fibrillose scales or nearly smooth, sometimes virgate → 17.

16. Growing under hardwoods; flesh pale grayish white to grayish brown, odor and taste farinaceous; cortinate partial veil absent on the button stage; stalk ¼–⅝" (5–16 mm) thick, coated with tiny grayish to brownish fibrillose scales on a pale gray ground color; cap dry, dark brownish gray, 1⅛–4" (3–10 cm) wide; gills pale gray, bruising dark gray on the edge, sometimes bruising orange-pink; spores 6–7 × 4–5 μm; edibility unknown—*T. squarrulosum* Bresadola.

16. Growing under hardwoods; flesh white to grayish, odor and taste weakly farinaceous; cortinate partial veil absent on the button stage; stalk slender, ¼–⅜" (5–10 mm) thick, smooth, dry, white to grayish white, cap dry, at first fibrillose, becoming fibrillose-scaly at maturity, gray to dark gray or grayish brown when young, becoming silvery gray and paler toward the margin, ¾–2⅜" (2–6 cm) wide; gills close, white to pale gray, sometimes becoming yellowish in age; spores 5–6 × 3–4 μm; edibility unknown—*T. terreum* (Shaeffer : Fries) Kummer.

16. Growing under conifers; flesh whitish, odor and taste farinaceous; cortinate partial veil present on the button stage; cap dry, brownish gray or pale to dark gray, ¾–2¾" (2–7 cm) wide; gills pale gray when young, becoming white overall or retaining the gray on the edge at maturity; stalk silky, white to whitish; spores 5–7 × 3–4 μm; edibility unknown—*T. sculpturatum* (Fries) Quélet.

16. Growing under conifers or in mixed woods; flesh grayish, odor not distinctive, taste quickly bitter; cortinate partial veil absent on the button stage; cap dry, dark gray, 1⅛–2½" (3–6.5 cm) wide; gills pale buff when young, becoming buff to pale gray and blackening on the edge in age; stalk silky to slightly scurfy, grayish buff; spores 7.5–9 × 5–6 μm; edibility unknown—*T. atrodiscum* Ovrebo.

16. Growing under conifers; flesh pale gray, odor and taste not distinctive; cortinate partial veil present on the button stage; cap dry, gray, dark gray or dark brownish black, ¾–2⅜" (2–6 cm) wide; gills pale gray when young, fading to whitish in age; stalk silky to slightly scurfy, whitish; spores from specimens with 4-spored basidia 6.5–7.5 × 4–5 μm; spores from specimens with 2-spored basidia 8.5–11.5 × 4–6 μm—*T. myomyces* (Persoon : Fries) Lange (see p. 262).

17. Growing in hardwoods → 18.

17. Growing in conifer woods or near conifers in mixed woods → 19.
    18. Flesh buff, sometimes with pinkish orange tinges, odor and taste farinaceous; stalk and gills buff at first, discoloring dull orange; cap dry, pale gray to reddish gray on a reddish orange ground color when young, becoming brownish gray to dark reddish brown on the disc and buff to pale reddish buff on the margin, often with a pinkish tone overall, 1⅜–4" (3.5–10 cm) wide; spores 6–7 × 4–5 μm; reported only from Michigan to date; edibility unknown—*T. insigne* Ovrebo.
    18. Flesh grayish white, odor not distinctive, taste bitter or acrid; gills pale pinkish gray with dark gray coloration on portions of the edges and faces; cap dry, virgate, dark gray, paler on the margin, 1¾–6" (4.5–15.5 cm) wide; stalk smooth, whitish; spores 6.5–8 × 5–6 μm; edibility unknown—*T. pullum* Ovrebo.
    18. Flesh pale gray, odor not distinctive, taste bitter; gills pale grayish overall or with fuscous-black coloration on the edges, sometimes spotted fuscous-black on the faces; cap dry, fibrillose or with scattered tiny fibrillose scales, not virgate, gray overall or darker fuscous-gray on the disc in age, 1–4⅜" (2.5–11 cm) wide; stalk silky, sometimes pruinose at the apex, whitish; spores 6.5–7.5 × 5–6 μm; edibility unknown—*T. acre* Peck.
19. Flesh white, odor and taste farinaceous or not distinctive; cap surface smooth, viscid when moist, soon dry, gray overall, sometimes with faint yellow or purplish tints, virgate with dark gray fibrils, 2⅜–4⅝" (6–12 cm) wide; gills close, white, becoming yellowish in age; stalk whitish, sometimes with yellow tints in age; spores 6–7 × 3–4 μm; on the ground under pines; edible—*T. portentosum* (Fries : Fries) Quélet (see p. 262).
19. Nearly identical to the previous choice but lacking any yellowish coloration on the gills, stalk, or cap; gills white to pale gray; spores 7–9 × 3–4 μm; on the ground under pitch pine and northern scrub pine; edible—*T. niveipes* Ovrebo, Halling and Baroni.
19. Flesh white to grayish, odor musty, earthy, or not distinctive, taste acrid or bitter; cap dry, not viscid, with a pointed umbo, silvery gray to gray, darkest on the disc, paler toward the margin; virgate with dark gray fibrils that are most evident on young specimens, 1½–4" (4–10 cm) wide; gills white, often tinged grayish; stalk nearly smooth, white; spores 6–7.5 × 5–6 μm; edibility unknown—*T. virgatum* (Fries : Fries) Kummer.
19. Flesh white to grayish, odor not distinctive, taste bitter; cap lacking an umbo or with a low, rounded umbo; pale gray to silvery gray, typically not virgate or slightly virgate with dark gray fibrils, 1½–2½" (4–6.5 cm) wide; gills white, discoloring dark brown on the edge in age; stalk silky to slightly scurfy, white; spores 7.5–8.5 × 5–6 μm; edibility unknown—*T. argenteum* Ovrebo.
19. Flesh white to watery gray; odor and taste strongly farinaceous; cap often with a low umbo, viscid when fresh, becoming dry in age, often wrinkled between the disc and the margin or smooth overall, light medium gray with a whitish margin; gills close, white; stalk smooth to slightly scurfy, white; spores 7–8 × 5–6 μm; edibility unknown—*T. marquettense* Ovrebo.

Tricholoma aestuans (Fries) Gillet

CAP: 1⅜–3½" (3.5–9 cm) wide, hemispheric or convex when young, becoming broadly convex or nearly flat in age, often with a low, broad umbo; surface dry, yellowish green to grayish yellow, brownish on the disc, radially streaked with grayish fibrils; margin incurved when young, expanded to slightly uplifted in age, sometimes lobed or wavy; flesh grayish white to grayish green; odor not distinctive; taste bitter or not distinctive at first, then strongly acrid.

GILLS: attached, close, broad, honey-yellow to greenish yellow, bruising pale brownish gray.

STALK: 1⅛–2¾" (3–7 cm) long, ⅜–¾" (1–2 cm) thick, nearly equal or tapering slightly downward, hollow, silky, pale yellowish white to yellowish green.

SPORE PRINT: white.

MICROSCOPIC FEATURES: spores 6.5–8 × 4–5 μm, elliptic, smooth, hyaline, inamyloid.

FRUITING: solitary, scattered, or in groups on the ground under pine or sometimes oak; September–November; occasional.

EDIBILITY: unknown.

Tricholoma aurantium (Schaeffer : Fries) Ricken

COMMON NAME: Veiled Trich.

CAP: 1⅝–3⅛" (4–8 cm) wide, convex, becoming broadly convex to nearly flat in age; surface slimy and sticky when fresh, shiny when dry, covered with tiny, flattened scales or fibers, brownish gold to orange-brown or tawny ochre, often darker on the disc; margin incurved when young, becoming expanded in age; flesh white; odor and taste strongly farinaceous or unpleasant.

GILLS: notched, close, narrow, white, staining brownish red to brownish orange.

STALK: 1½–3⅛" (4–8 cm) long, ⅜–¾" (1–2 cm) thick, nearly equal or tapered downward, solid, dry, white and smooth at the apex, sheathed from the base up to an obscure fibrous zone with bright orange-brown to reddish orange, tiny, flattened scales; partial veil thin, cortinate.

SPORE PRINT: white.

MICROSCOPIC FEATURES: spores 4–6 × 3–4 μm, elliptic, smooth, hyaline, inamyloid.

FRUITING: solitary, scattered, or in groups on the ground under conifers and in mixed woods; August–October; occasional.

EDIBILITY: inedible.

Tricholoma caligatum (Viviani) Ricken complex

CAP: 2⅜–4¾" (6–12 cm) wide, hemispheric when young, becoming broadly convex to nearly flat in age, sometimes shallowly depressed; surface dry, reddish brown to pale cinnamon-brown, breaking up into coarse flattened scales and fibers on a whitish to pinkish buff ground color; margin inrolled and cottony-membranous when young, becoming expanded to uplifted in age, sometimes rimmed with veil remnants; flesh white to cream; odor not distinctive or fragrant to spicy; taste nutty, bitter, or not distinctive.

GILLS: attached, close, broad, white.

STALK: 1⅝–4" (4–10 cm) long, ⅝–1⅛" (1.5–3 cm) thick, nearly equal or enlarged in the middle and tapered in both directions, solid, dry, sheathed from the base up to the ring by a cinnamon-brown veil that breaks into patches on a white ground color, white above the ring, sometimes pruinose; ring white, cottony-membranous, often flaring upward, median to superior.

SPORE PRINT: white.

MICROSCOPIC FEATURES: spores 6–8 × 4.5–5.5 μm, elliptic, smooth, hyaline, inamyloid.

FRUITING: solitary, scattered, or in groups on the ground under conifers, especially hemlock, or under hardwoods, especially oak; July–October; occasional.

EDIBILITY: edible.

COMMENTS: because this is a complex of several forms, there is a wide range of odor and taste.

Tricholoma flavobrunneum (Fries) Kummer

CAP: 1⅛–3⅛" (3–8 cm) wide, convex when young, becoming broadly convex to nearly flat in age, sometimes shallowly depressed, with or without a small umbo; surface smooth, slightly viscid when fresh, often shiny when dry, with radiating fibrils, reddish brown to yellow-brown, darker over the disc and paler yellow-brown to dull yellow on the margin; flesh white to yellowish; odor and taste farinaceous.

GILLS: notched, close, broad, yellowish buff to buff, bruising brownish.

STALK: 2¾–5⅞" (7–15 cm) long, ⅜–¾" (1–2 cm) thick, nearly equal or slightly enlarged downward, slippery to viscid when young and fresh, becoming dry, solid, hollow in age, pale yellow to yellowish buff with reddish brown longitudinal fibrils, pale yellowish to whitish at the apex.

SPORE PRINT: white.

MICROSCOPIC FEATURES: spores 5.5–7 × 4–6 μm, elliptic, smooth, hyaline, inamyloid.

FRUITING: solitary, scattered, or in groups on the ground under hardwoods or mixed woods; August–October; occasional.

EDIBILITY: unknown.

Tricholoma flavovirens (Persoon : Fries) Lundell

COMMON NAME: Canary Trich.

CAP: 2–4" (5–10 cm) wide, convex, becoming broadly convex to nearly flat in age; surface sticky when fresh, pale yellow to sulphur-yellow, becoming reddish brown on the disc; flesh white; odor farinaceous; taste farinaceous or not distinctive.

GILLS: notched, close, broad, pale yellow to bright sulphur-yellow.

STALK: 1⅛–2¾" (3–7 cm) long, ⅜–¾" (1–2 cm) thick, nearly equal, occasionally enlarged at the base, solid, dry, pale yellow to sulphur-yellow.

SPORE PRINT: white.

MICROSCOPIC FEATURES: spores 6–7 × 4–5 μm, elliptic, smooth, hyaline, inamyloid.

FRUITING: solitary to scattered or in groups on the ground under conifers and in mixed woods; August–October; fairly common.

EDIBILITY: edible and choice.

Tricholoma imbricatum (Fries : Fries) Kummer

COMMON NAME: Shingled Trich.

CAP: 1–3½" (2.5–9 cm) wide, convex, becoming broadly convex, usually with a low, broad umbo; surface dry, fibrillose to fibrillose-scaly at first, becoming coarsely cracked and scaly on the disc at maturity with whitish flesh showing in the cracks, reddish brown on a whitish ground color; scales near the disc often arranged like shingles; margin thin, incurved, and coated with soft, downy hairs at first, expanded in age; flesh white, staining reddish when bruised; odor and taste not distinctive.

GILLS: sinuate to nearly free in age, close, moderately broad, white when young, spotted or flushed reddish brown in age.

STALK: 2–3½" (5–9 cm) long, ⅜–⅝" (1–1.6 cm) thick, nearly equal down to a tapered base, solid, dry, fibrillose, white to whitish, slowly staining reddish brown from the base up; apex often white-pruinose.

SPORE PRINT: white.

MICROSCOPIC FEATURES: spores 5–6.5 × 3.5–4.5 μm, broadly elliptic, smooth, hyaline, inamyloid.

FRUITING: scattered or in groups on the ground under conifers; September–November; occasional.

EDIBILITY: edible.

Tricholoma inamoenum (Fries) Quélet

CAP: 1⅛–2⅜" (3–6 cm) wide, convex when young, becoming broadly convex to nearly
flat in age, often with a low, broad umbo; surface smooth, dull, pale buff to whitish,
often developing grayish to brownish tints on the disc; flesh white; odor pungent,
disagreeable, like coal tar; taste disagreeable.

GILLS: notched, subdistant to distant, broad, often crossveined, white, with three or four
tiers of lamellulae.

STALK: 1½–3½" (4–9 cm) long, ⅛–⅜" (4–10 mm) thick, nearly equal or enlarging slightly
downward to a swollen base that often tapers abruptly below, smooth, dry, pruinose at
the apex, white to cream, brownish near the base.

SPORE PRINT: white.

MICROSCOPIC FEATURES: spores 10–12 × 6.5–8 μm, elliptic, smooth, hyaline, inamyloid.

FRUITING: solitary or scattered on the ground under conifers; August–November;
occasional.

EDIBILITY: unknown.

COMMENTS: also known as *T. platyphyllum*.

Tricholoma leucophyllum Ovrebo and Tylutki

CAP: 1½–4" (4–10 cm) wide, convex, becoming nearly plane in age, sometimes with a
broad umbo; surface viscid when fresh and moist, soon dry, coated with tiny golden
brown to yellow-brown or reddish brown flattened scales and fibers over a yellowish
ground color; margin inrolled at first, decurved well into maturity, smooth, yellow to
brownish yellow; flesh white; odor and taste farinaceous.

GILLS: sinuate, close, white to whitish; edges often eroded at maturity; with many tiers of
lamellulae.

STALK: 1⅛–3½" (3–9 cm) long, ⅜–1" (1–2.5 cm) thick, nearly equal, scurfy, white to
whitish, often stained brownish, especially toward the base.

SPORE PRINT: white.

MICROSCOPIC FEATURES: spores 5–7 × 3.5–4.5 μm, elliptic, smooth, hyaline.

FRUITING: scattered or in groups on the ground under conifers; August–October;
occasional.

EDIBILITY: unknown.

Tricholoma magnivelare (Peck) Redhead

COMMON NAME: Matsutake, White Matsutake.

CAP: 2–8" (5–20.5 cm) wide, convex, becoming broadly convex to nearly plane in age;
margin cottony and inrolled at first; surface dry, white to whitish and nearly smooth
at first, soon developing yellow, flattened fibers which aggregate into fibrous scales;
fibrous scales becoming yellowish brown to reddish brown, especially on the disc;
flesh white, very firm; odor like dirty gym socks but usually described as spicy-sweet,
aromatic, fruity, or fragrant; taste not distinctive.

GILLS: crowded to close, attached to notched, white to whitish, staining brownish in age,
covered by a thick, white, cottony-membranous partial veil when young.

STALK: 1½–6" (4–15.5 cm) long, ½–2" (1.3–5 cm) thick, nearly equal or tapered downward,
typically with a narrowed base, dry, solid, sheathed from the base up to a thick cottony
ring by a white veil that breaks into patches and becomes reddish brown; ring white on
the upper surface, reddish brown below.

SPORE PRINT: white.

MICROSCOPIC FEATURES: spores 5–7 × 4.5–5.5 μm, broadly elliptic, smooth, hyaline,
inamyloid.

FRUITING: solitary, scattered or in groups on the ground under conifers, especially hemlock; August–October; uncommon.

EDIBILITY: edible and choice.

COMMENTS: compare with *T. caligatum* (see p. 259).

Tricholoma myomyces (Persoon : Fries) Lange

CAP: ¾–2⅜" (2–6 cm) wide, bluntly conic with an incurved margin when young, becoming convex to nearly plane, sometimes shallowly depressed or with a low, broad umbo; surface dry, covered with matted fibrils, often forming fibrillose scales at maturity, gray to dark gray, fuscous or brownish black; flesh pale gray; odor and taste not distinctive.

GILLS: covered by a cortinate partial veil on the button stage, soon evanescent; sinuate, close and pale gray when young, becoming nearly free and whitish or pale gray in age.

STALK: 1½–2¾" (4–7 cm) long, ¼–⅜" (5–10 mm) thick, nearly equal, silky to slightly scurfy, whitish.

SPORE PRINT: white.

MICROSCOPIC FEATURES: spores from specimens with 4-spored basidia 6.5–7.5 × 4–5 µm; spores from specimens with 2-spored basidia 8.5–11.5 × 4–6 µm; all spores elliptic to ovate, smooth, hyaline, inamyoid.

FRUITING: scattered or in groups under conifers; August–October; fairly common.

EDIBILITY: unknown.

Tricholoma pessundatum (Fries) Quélet

COMMON NAME: Red-brown Trich.

CAP: 2–5⅞" (5–15 cm) wide, convex when young, becoming broadly convex to flat in age; surface viscid when moist, smooth, chestnut-brown over the disc, paler golden brown toward the margin; margin incurved when young, uplifted and faintly striate in age; flesh white, staining reddish; odor and taste farinaceous.

GILLS: notched, close, broad, whitish to buff, staining reddish.

STALK: 2¾–4" (7–10 cm) long, ⅜–¾" (1–2 cm) thick, nearly equal or enlarged downward, solid, smooth, whitish when young, becoming reddish brown in age, fibrous and scurfy; flesh white, staining reddish.

SPORE PRINT: white.

MICROSCOPIC FEATURES: spores 4–6 × 2.5–3 µm, elliptic, smooth, hyaline, inamyloid.

FRUITING: scattered, in groups, or in clusters on the ground or among mosses under conifers or sometimes oak; September–November; occasional and most commonly collected in the northern part of the region.

EDIBILITY: poisonous.

Tricholoma portentosum (Fries : Fries) Quélet

COMMON NAME: Sticky Gray Trich.

CAP: 2⅜–4⅝" (6–12 cm) wide, broadly conic when young, becoming broadly convex to nearly flat in age; surface smooth, viscid when moist, gray overall, sometimes with faint yellow or purplish tints, virgate with dark gray fibrils; margin incurved at first, becoming expanded in age; flesh white, fragile; odor and taste farinaceous or not distinctive.

GILLS: notched, close, broad, white, becoming yellowish in age, occasionally tinted yellow or gray.

STALK: 2–4" (5–10 cm) long, ⅜–¾" (1–2 cm) thick, nearly equal, often curved, solid, coated with tiny fibrils, whitish, sometimes with yellow tints in age.

SPORE PRINT: white.

MICROSCOPIC FEATURES: spores 6–7 × 3–4 μm, elliptic to oval, smooth, hyaline, inamyloid.

FRUITING: in groups or scattered among mosses or on the ground under pines; September–November; fairly common.

EDIBILITY: edible.

COMMENTS: compare with *T. niveipes*, and with *T. virgatum*, which has a dry cap.

Tricholoma saponaceum (Fries) Kummer

COMMON NAME: Soapy Trich.

CAP: 1–3⅛" (2.5–8 cm) wide, convex when young, becoming broadly convex to nearly flat in age; surface smooth at first, often cracked over the disc at maturity, moist and slippery when fresh, becoming slimy when wet; cap color highly variable, grayish olive, lead-gray, greenish gray, or grayish brown to brown; margin inrolled when young, becoming expanded in age; flesh white; odor distinctly soapy; taste not distinctive or slightly soapy.

GILLS: attached or notched, subdistant, broad, white.

STALK: 1⅜–3⅛" (3.5–8 cm) long, ⅜–1" (1–2.5 cm) thick, nearly equal, solid, dry, smooth, white, staining brownish in age; flesh pinkish to pinkish orange.

SPORE PRINT: white.

MICROSCOPIC FEATURES: spores 5–7 × 3.5–5 μm, oval, smooth, hyaline, inamyloid.

FRUITING: solitary, scattered, or in groups on the ground under conifers and hardwoods; July–October; fairly common.

EDIBILITY: inedible.

Tricholoma sejunctum (Sowerby : Fries) Quélet

CAP: 1⅝–3⅛" (4–8 cm) wide, convex, becoming broadly convex to nearly flat in age; surface dry, slightly sticky when moist, coated with conspicuous radiating blackish fibrils over a yellow to golden yellow ground color; flesh white to yellowish, fragile; odor and taste farinaceous to unpleasant.

GILLS: notched, subdistant to close, broad, white, sometimes tinged yellowish in age.

STALK: 2–3⅛" (5–9 cm) long, ⅜–¾" (1–2 cm) thick, nearly equal, sometimes enlarged at the base, solid, smooth, dry, whitish to yellowish.

SPORE PRINT: white.

MICROSCOPIC FEATURES: spores 6–7 × 4–5.5 μm, oval, smooth, hyaline, inamyloid.

FRUITING: solitary, scattered, or in groups on the ground under conifers, especially pine, or in mixed woods; August–November; fairly common.

EDIBILITY: unknown.

COMMENTS: some authors consider *T. intermedium* (edibility unknown) to be a variety of *T. sejunctum*.

Tricholoma subluteum Peck

CAP: 1½–4⅜" (4–11 cm) wide, obtusely conic at first, becoming convex, then broadly convex with a low, broad umbo; surface dry to slightly viscid when moist; color variable when young, dull orange or golden yellow to yellow, sometimes with obscure dark fibrils, becoming pale yellow at maturity; flesh white; odor and taste farinaceous.

GILLS: notched to nearly free, close, white.

STALK: 3–4¾" (7.5–12 cm) long, ⅜–⅝" (1–1.6 cm) thick, nearly equal or slightly enlarged downward, dry, solid at first, becoming hollow in age, yellow on the upper portion, becoming white toward the base.

SPORE PRINT: white.

MICROSCOPIC FEATURES: spores 4.5–6 × 4–5.5 μm, globose to subglobose, smooth, hyaline, inamyloid.

FRUITING: solitary, scattered, or in groups on the ground under conifers; August–October; occasional.

EDIBILITY: unknown.

Tricholoma subresplendens (Murrill) Ovrebo

CAP: 1¾–4⅜" (4.5–11 cm) wide, convex, becoming broadly convex to nearly plane, sometimes with a low, broad umbo; margin incurved when young, sometimes wavy and lobed in age; surface dry to slightly viscid when moist, smooth, often slightly silky, white to cream, becoming tinged or spotted yellowish tan to tan or pale pinkish cinnamon, especially on the disc; flesh white; odor and taste farinaceous to weakly farinaceous.

GILLS: sinuate to notched, becoming nearly free at maturity, broad, close to subdistant, white to whitish; edges uneven, sometimes finely scalloped in age; lamellulae numerous but not arranged in distinct tiers.

STALK: 2–4" (5–10 cm) long, ⅜–1" (1–2.5 cm) thick, nearly equal or slightly tapered downward, smooth to silky-fibrillose or slightly scurfy, dry, whitish, often stained brownish especially near the base.

SPORE PRINT: white.

MICROSCOPIC FEATURES: spores 6–8 × 4–5 μm, elliptic, smooth, hyaline, inamyloid.

FRUITING: solitary, scattered, or in groups on the ground in hardwoods and mixed woods, sometimes near hemlock; September–November; occasional.

EDIBILITY: unknown.

COMMENTS: *Tricholoma resplendens* (edibility unknown) is very similar but lacks the farinaceous odor and taste and has a shiny cap when dry.

Tricholoma sulphurescens Bresadola

CAP: 2–4¾" (5–12 cm) wide, hemispheric to convex when young, becoming nearly plane, sometimes with an umbo; surface dry, smooth, white at first, becoming creamy white, staining yellow to dull yellow when bruised or in age; flesh white; odor pungent, disagreeable, like coal tar (sometimes faint); taste not distinctive.

GILLS: attached or notched, close, white to creamy white, becoming yellowish, especially near the edges in age.

STALK: 1⅛–4" (3–10 cm) long, ⅜–¾" (1–2 cm) wide, club-shaped when young, becoming nearly equal at maturity, dry, smooth to somewhat fibrillose, white at first, becoming creamy white to yellow, staining yellow when bruised or in age.

SPORE PRINT: white.

MICROSCOPIC FEATURES: spores 4.5–6 × 3.5–5 μm, broadly ellipsoid to subglobose, smooth, hyaline.

FRUITING: solitary, scattered, or in groups on the ground in hardwoods, especially oak; August–October; infrequent.

EDIBILITY: unknown.

Tricholoma vaccinum (Persoon : Fries) Kummer

COMMON NAME: Russet–scaly Trich.

CAP: 1–3⅛" (2.5–8 cm) wide, conic to convex when young, becoming broadly convex to nearly plane, with an umbo; surface coated with small, fibrous scales, typically not coarsely cracked on the disc at maturity; fibrous scales reddish brown on a white ground color; margin inrolled and cottony-fibrous at first, becoming expanded and remaining cottony-fibrous and somewhat thickened in age; flesh white, developing reddish tinges when cut.

GILLS: notched to sinuate, close, whitish buff when young, staining reddish brown when bruised or in age.

STALK: 1–3⅛" (2.5–8 cm) long, ⅜–⅝" (1–1.6 cm) thick, nearly equal or slightly enlarged downward, hollow at maturity, fibrillose, reddish brown or paler; partial veil cottony-fibrous, not leaving a ring.

SPORE PRINT: white.

MICROSCOPIC FEATURES: spores 5.5–7 × 4–5 μm, oval to elliptic, smooth, hyaline.

FRUITING: scattered, in groups, or in clusters under conifers; July–November; occasional to fairly common.

EDIBILITY: unknown.

Tricholoma zelleri (Stuntz and Smith) Ovrebo and Tylutki

CAP: 2–5⅞" (5–15 cm) wide, broadly conic when young, becoming broadly convex to nearly flat with a low, broad umbo in age; surface slimy and sticky when fresh, dark orange when young, becoming orange-brown, paler or yellowish on the margin, covered with minute fibrils, breaking up and becoming scaly in age; margin incurved when young and often rimmed with veil remnants; flesh whitish, bruising brownish; odor and taste strongly farinaceous to unpleasant.

GILLS: attached, crowded, narrow to broad, whitish, staining rusty brown in age.

STALK: 2–5⅛" (5–13 cm) long, ⅜–1⅛" (1–3 cm) thick, tapered downward to a pointed base, dry, smooth and white above the ring, sheathed from the base up to the ring with orange-brown fibers and tiny scales; partial veil cottony-membranous, leaving a cottony-membranous superior ring that flares upward initially; ring white on the upper surface and orange-brown below.

SPORE PRINT: white.

MICROSCOPIC FEATURES: spores 4–5.5 × 3–4 μm, elliptic, smooth, hyaline, inamyloid.

FRUITING: scattered or in groups on the ground under conifers; July–September; occasional.

EDIBILITY: edible.

Genus *Tricholomopsis*

Tricholomopsis, a small but interesting group of mushrooms that decay coniferous wood, is characterized by the combination of a white spore print, central to eccentric (not lateral) stalk, and typically yellowish gills and flesh. Also keyed out here is *Megacollybia platyphylla*, which was classified as a *Tricholomopsis* until quite recently.

Key to Species of *Tricholomopsis* and *Megacollybia*

1. Cap gray to brownish gray, with dark radial fibers, 1–6" (2.5–15 cm) wide; flesh white; rhizomorphs thick, white—*Megacollybia platyphylla* (Persoon : Fries) Kotlaba and Pouzar (see p. 203).

1. Cap yellowish with reddish fibrous scales, 2–4" (5–10 cm) wide; flesh yellowish; gills attached, close, narrow, yellowish; stalk yellowish with reddish fibrous scales — *T. rutilans* (Schaeffer : Fries) Singer (see p. 266).
1. Not as in either of the above choices → 2.
    2. Cap yellowish overall, covered with small, blackish, fibrous scales; flesh yellowish; gills attached, crowded, narrow, yellow; stalk yellowish, covered with tiny, blackish scales — *T. decora* (Fries) Singer (see p. 266).
    2. Mushroom very similar to previous choice but lacking blackish, fibrous scales on cap and stalk — *T. sulfureoides* (Peck) Singer.
    2. Cap covered with ascending to recurved rusty brown to tawny scales; stalk fibrillose-scaly, colored like the cap or paler — *T. formosa* (Murrill) Singer (see p. 266).

Tricholomopsis decora (Fries) Singer

COMMON NAME: Decorated Mop.
CAP: 1–2¼" (2.5–6 cm) wide, convex to nearly flat, sometimes with a slight central depression; surface yellowish with blackish, fibrous scales; flesh yellowish; odor and taste not distinctive.
GILLS: attached, crowded, narrow, yellow.
STALK: 1–2¼" (2.5–6 cm) long, ⅛–⅜" (3–10 mm) thick, yellow with tiny, blackish scales.
SPORE PRINT: white.
MICROSCOPIC FEATURES: spores 6–7.5 × 4–5.5 μm, elliptic, smooth, hyaline, inamyloid.
FRUITING: solitary or in small groups on coniferous wood; June–October; occasional to frequent.
EDIBILITY: inedible.

Tricholomopsis formosa (Murrill) Singer

CAP: 1⅛–3⅛" (3–8 cm) wide, convex to nearly plane; margin incurved when young, often wavy in age; surface covered with ascending to recurved rusty brown to tawny scales over a cinnamon-buff ground color, dry; flesh whitish; odor and taste not distinctive or slightly disagreeable.
GILLS: attached, close to crowded, whitish to pinkish cream.
STALK: 1½–3" (4–7.5 cm) long, ¼–⅜" (5–10 mm) thick, nearly equal or tapered slightly downward, fibrillose-scaly, dry, colored like the cap or paler; partial veil and ring absent.
SPORE PRINT: white.
MICROSCOPIC FEATURES: spores 5–7 × 5–6 μm, ovoid, smooth, hyaline, inamyloid.
FRUITING: solitary, scattered, or in groups on rich humus attached to buried wood or roots, on sawdust and litter debris, or on decaying wood; July–October; occasional.
EDIBILITY: unknown.

Tricholomopsis rutilans (Schaeffer : Fries) Singer

COMMON NAME: Plums and Custard.
CAP: 2–4" (5–10 cm) wide, convex, becoming nearly plane in age; surface dry, covered with red to purplish red scales and fibers over a yellowish ground color; flesh pale yellow; odor and taste not distinctive.
GILLS: attached or notched, close, yellow, with numerous tiers of lamellulae.
STALK: 2–4⅜" (5–11 cm) long, ⅜–1" (1–2.5 cm) thick, nearly equal, frequently curved, dry, coated with red to purplish red scales and fibers over a yellowish ground color, often hollow in age.

SPORE PRINT: white.

MICROSCOPIC FEATURES: spores 5–7 × 3–5 μm, elliptic, smooth, hyaline.

FRUITING: scattered or in groups on decaying conifer wood, especially pine, and on rich humus in conifer woods; May–November; occasional.

EDIBILITY: edible, but of poor quality.

Genus *Tubaria*

Tubaria is a fairly small genus of approximately twenty-five species in North America. They are small to medium mushrooms with pale pinkish brown, cinnamon, or reddish brown caps that often have tiny patches of veil remnants on the margin. The stalk is fairly slender and typically has a ring or at least an annular zone. They grow on soil, logs, stumps, and wood chips and sometimes on the ground. Their spore prints are ochre to brownish ochre or dark reddish cinnamon. The spores are smooth, thin-walled, and lack an apical pore; the spores often collapse, especially when mounted in KOH.

The edibility of species in this genus is unknown, and none is recommended for the table. The genus is not well understood and additional work is needed before a comprehensive key is available. We have included the two most common species in this region.

Key to Species of *Tubaria*

1. Ring conspicuous, membranous, persistent; cap reddish brown to dark reddish cinnamon, coated with conspicuous whitish fibrils—*T. confragosa* (Fries) Kühner (see p. 267).
1. Ring inconspicuous, fibrous, often merely a fibrous annular zone; cap reddish brown, coated with inconspicuous whitish fibrils—*T. furfuracea* (Persoon : Fries) Gillet (see p. 268).

Tubaria confragosa (Fries) Kühner

COMMON NAME: Ringed Tubaria.

CAP: ⅜–2" (1–5 cm) wide, convex, becoming broadly convex; margin incurved at first, becoming decurved and remaining so well into maturity, sometimes wavy in age; surface moist, coated with conspicuous whitish fibrils over a reddish brown to dark reddish cinnamon ground color, hygrophanous and fading to cinnamon-buff, sometimes cracked and finely scaly when dry; flesh thin, reddish brown; odor and taste not distinctive.

GILLS: attached, close, moderately broad, yellowish buff at first, becoming reddish cinnamon to reddish brown; edges sometimes finely fringed.

STALK: 1½–3" (4–7.5 cm) long, ⅛–¼" (3–7 mm) thick, nearly equal, hollow in age, coated with whitish fibrils over a reddish cinnamon to reddish brown ground color; partial veil white, leaving a persistent, membranous, superior ring; silky to pruinose above the ring.

SPORE PRINT: dark reddish cinnamon.

MICROSCOPIC FEATURES: spores 6–9 × 4–6 μm, elliptic, smooth, thin-walled, often collapsing, especially when mounted in 3% KOH, pale brown.

FRUITING: in groups or clusters on decaying conifer or hardwood logs, stumps, and wood chips; July–October; occasional.

EDIBILITY: unknown.

Tubaria furfuracea (Persoon : Fries) Gillet

COMMON NAME: Fringed Tubaria.

CAP: ⅜–2" (1–5 cm) wide, convex, becoming nearly flat; margin incurved at first, becoming uplifted and wavy in age, sometimes with tiny, white patches of veil; surface moist and slippery when fresh, smooth, coated with tiny, whitish, matted fibrils, translucent-striate when fresh, reddish brown when young, hygrophanous and fading to tan or pinkish buff; flesh thin, watery, yellow-brown; odor and taste not distinctive.

GILLS: attached, close, moderately broad, veined on the faces, pale yellowish buff at first, becoming pale yellow-brown.

STALK: ¾–2" (2–5 cm) long, ⅛–¼" (2–6 mm) thick, nearly equal, coated with whitish fibrils at first, becoming pale brown, fibrous, becoming hollow and often splitting longitudinally in age, with copious white mycelium at the base; partial veil white, fibrous, sometimes leaving a fibrous superior ring or annular zone; silky to pruinose above the ring.

SPORE PRINT: ochre to brownish ochre.

MICROSCOPIC FEATURES: spores 6–9 × 4–7 μm, obovate to elliptic, smooth, thin-walled, often collapsing, especially when mounted in 3% KOH, pale yellow.

FRUITING: scattered, in groups, or sometimes in clusters on wood chips and other woody debris or on the ground attached to buried wood; April–November; occasional to fairly common.

EDIBILITY: unknown.

Genus *Volvariella*

Members of the genus *Volvariella* are small to large mushrooms with free gills at maturity, a saccate volva, and a pinkish spore print. They have a central stalk that is cleanly separable from the cap, their spores are smooth, and they lack a partial veil and ring. They grow on a variety of substrates, including wood, sawdust, compost, soil, and other mushrooms. A few species are excellent edibles.

Key to Species of *Volvariella*

1. Growing on other mushrooms, especially *Clitocybe nebularis;* cap 1–3⅛" (2.5–8 cm) wide, silky-fibrillose, white to grayish white, sometimes yellowish or brownish over the disc in age; stalk white to grayish white; spores 5–7.5 × 3–5 μm; September–October; edibility unknown—*V. surrecta* (Knapp) Singer.

1. Growing on soil, on straw compost, and in greenhouses; cap pale brown with a blackish brown disc, sometimes streaked blackish brown, 1½–4" (4–10 cm) wide, often radially cracked; margin usually toothed and often splitting in age; gill edges minutely fringed; stalk dull white to pale brown, surrounded by a large, pale brown volva; spores 7–10.5 × 4.5–7 μm; June–September; edible—*V. volvacea* (Bulliard : Fries) Singer.

1. Not with the above combinations of characters → 2.

    2. Caps large, 2–8" (5–20 cm) wide, white to creamy white → 3.

    2. Caps small to medium, ¼–2" (6–50 mm) wide, white to grayish white → 4.

    2. Caps small to medium, ¾–2⅜" (2–6.5 cm) wide, pinkish gray to gray → 5.

3. Growing on soil in woods, in gardens, on lawns, and on dung; cap 2–5⅞" (5–15 cm) wide, smooth, sticky when fresh, shiny when dry; stalk ⅜–¾" (1–2 cm) thick, white, surrounded by a deep, white volva; spores 12–20 × 7–12 μm; June–August; edible—*V. speciosa* (Fries) Singer.

3. Growing on standing hardwood trees, stumps, and logs; cap 2–8" (5–20 cm) wide, silky, dry—*V. bombycina* (Schaeffer : Fries) Singer (see p. 269).
    4. Cap margin striate; cap ¼–1⅛" (6–30 mm) wide, silky, white overall or sometimes grayish on the disc; stalk white to grayish white, smooth, surrounded by a small, white to grayish white volva; in gardens and greenhouses, on lawns, and occasionally in woods—*V. pusilla* (Fries) Singer (see p. 269).
    4. Cap margin not striate; cap ½–2" (1.3–5 cm) wide, silky, white overall or pale yellow to pale brown over the disc in age; stalk white, densely to sparsely pubescent or nearly smooth, surrounded by a small, white volva; spores 6–8 × 3.5–6 μm; scattered in sandy soil in mixed woods and along roadsides; August–October; edibility unknown—*V. hypopithys* (Fries) Shaffer.
5. Volva deep, white, covered with conspicuous long hairs; cap ¾–1⅜" (2–3.5 cm) wide; dull gray to brownish gray with a darker gray disc; margin not striate; stalk smooth, white; spores 5.5–7.5 × 3.5–4.5 μm; attached to decaying leaves; July–September; edibility unknown—*V. villosavolva* (Lloyd) Singer.
5. Volva lobed, white on the inner surface and grayish or brownish on the outer surface, lacking conspicuous long hairs; cap ¾–2⅜" (2–6 cm) wide; pale gray to pinkish gray, silky; margin not striate but often torn in age; gill edges minutely fringed; stalk smooth, white; spores 5.5–9 × 4–6 μm; on soil or humus in woods; July–September; edibility unknown—*V. taylori* (Berkeley and Broome) Singer.

Volvariella bombycina (Schaeffer : Fries) Singer

COMMON NAME: Tree Volvariella.
CAP: 2–8" (5–20 cm) wide, oval, becoming convex to nearly plane in age; surface silky, dry, white to pale yellowish white; flesh white; odor and taste not distinctive.
GILLS: free from the stalk, close to crowded, white when young, becoming pink to brownish pink in age.
STALK: 2⅜–8" (6–20 cm) long, ⅜–¾" (1–2 cm) thick, enlarging downward, white, smooth, dry, surrounded at the base by a white to pale brown, saccate volva.
SPORE PRINT: pink to brownish pink.
MICROSCOPIC FEATURES: spores 6.5–10.5 × 4.5–6.5 μm, elliptic, smooth, hyaline.
FRUITING: solitary or in groups on standing hardwood trees, stumps, and logs; July–October; occasional.
EDIBILITY: edible.

Volvariella pusilla (Fries) Singer

CAP: ¼–1⅛" (6–30 mm) wide, egg-shaped when young, becoming convex, then broadly convex to plane in age; surface dry, silky, coated with tiny, matted fibrils and typically scurfy, white, sometimes grayish on the disc; margin striate; flesh white; odor and taste not distinctive.
GILLS: free, close to subdistant, white, becoming pink.
STALK: ⅜–2" (1–5 cm) long, 1/16–3/16" (1.5–5 mm) thick, nearly equal, white to grayish white; base surrounded by a small, white to grayish white volva.
SPORE PRINT: white.
MICROSCOPIC FEATURES: spores 5.5–8 × 4–6 μm, oval, smooth, hyaline.
FRUITING: solitary, scattered, or in groups on lawns or in gardens, greenhouses, and woods; July–September; occasional.
EDIBILITY: unknown.

Genus *Xeromphalina*

Xeromphalina campanella (Bataille : Fries) Kühner and Maire
  CAP: ⅛–1" (3–25 mm) wide, convex to broadly convex with a depressed center; surface smooth, moist, with a striate margin, yellowish orange to orangish brown; odor and taste not distinctive.
  GILLS: decurrent, subdistant to distant, pale yellow to pale orange.
  STALK: ⅜–2" (1–5 cm) long, ¹⁄₃₂–⅛" (0.5–3 mm) thick, dry, yellow at the apex, shading below to dark reddish brown, base with a dense tuft of long, orangish hairs.
  SPORE PRINT: pale buff.
  MICROSCOPIC FEATURES: spores 5–7 × 3–4 μm, elliptic to cylindric, smooth, hyaline, amyloid.
  FRUITING: in dense clusters, typically recurved, on well-decayed conifer wood; May–November; common.
  EDIBILITY: inedible.
  COMMENTS: the densely hairy base and growth on conifer wood are very important field characters.

Genus *Xerula*

Until recently, these mushrooms have been lumped together and called *Collybia radicata* or *Oudemansiella radicata*. The key shared character of the four species is the long rooting stalk, which is an unusual feature for a white-spored gilled mushroom. They can be difficult to identify to genus if the collector misses the "root," which often snaps off when the mushroom is picked. These fungi are being studied for possible medicinal value: they contain chemicals which have antibiotic properties.

Key to Species of *Xerula*

1. Gills with rusty marginate edges; cap rusty brown; aboveground portion of the stalk whitish, readily staining rust where bruised; odor and taste not distinctive; edibility unknown—*X. rubrobrunnescens* Redhead, Ginns and Shoemaker
1. Gills not marginate; cap whitish to hazel; aboveground portion of stalk smooth, white, not staining where bruised; odor and taste of carrots or geraniums; spores 18–23 × 10–14 μm; edible—*X. megalospora* (Clements) Redhead, Ginns and Shoemaker.
1. Nearly identical to the previous choice, except odor and taste not distinctive; spores smaller, 15–18.5 × 10–12 μm—*X. radicata* (Rehlan : Fries) Dörfelt var. *radicata*.
1. Gills not marginate; aboveground portion of stalk distinctly roughened with grayish to brownish scales or fibers over a white ground color; odor and taste not distinctive—*X. furfuracea* (Peck) Redhead, Ginns and Shoemaker (see p. 270).

Xerula furfuracea (Peck) Redhead, Ginns and Shoemaker
  COMMON NAME: Rooted Collybia.
  CAP: ¾–4" (2–10 cm) wide, sometimes wider, convex to nearly flat, with a thick, rubbery cuticle, grossly wrinkled around a distinct umbo, finely velvety at first, becoming nearly smooth in age, dry to slippery, light brown to dark grayish brown; flesh white; odor and taste not distinctive.
  GILLS: attached but typically seceding from the stalk in age, subdistant, fairly broad, white.

STALK: aboveground portion 3–8" (7.5–20 cm) long, ⅛–¾" (3–20 mm) thick, usually enlarged toward the base, white beneath a covering of grayish to brownish scales or fibers.

SPORE PRINT: white.

MICROSCOPIC FEATURES: spores 14–17 × 9.5–12 μm, broadly oval to elliptic, appearing smooth, with slightly thickened walls and a small rounded apiculus, hyaline, inamyloid.

FRUITING: solitary, scattered, or in groups on the ground, but also infrequently on decaying deciduous logs; May–October; very common.

EDIBILITY: the caps are edible, but the stalks are tough.

COMMENTS: a ubiquitous species; *X. megalospora* and *X. radicata* var. *radicata* are also fairly common, but *X. rubrobrunnescens* is rare.

Agaricus arvensis

Agaricus campestris

Agaricus fuscofibrillosus

Agaricus placomyces

Agaricus silvicola

Agaricus subrufescens

Agrocybe acericola

Agrocybe erebia

Agrocybe firma

Agrocybe praecox

Amanita brunnescens var. *brunnescens*

Amanita caesarea

Amanita citrina f. *lavendula*

Amanita flavorubescens

Amanita flavoconia

Amanita longipes

Amanita muscaria var. *formosa*

Amanita onusta

Amanita pantherina var. *velatipes*

Amanita phalloides

Amanita porphyria

Amanita rubescens

Amanita virosa

Amanita sinicoflava

Armillaria mellea

Armillaria ostoyae

Armillaria tabescens

Asterophora lycoperdoides

Asterophora parasitica

Baeospora myosura

Baeospora myriadophylla

Bolbitius variicolor

Callistosporium purpureomarginatum

Calocybe persicolor

Cantharellula umbonata

Catathelasma ventricosa

Cheimonophyllum candidissimus

Chlorophyllum molybdites

Chromosera cyanophylla

Chroogomphus rutilus

Chroogomphus vinicolor

Chrysomphalina chrysophylla

Claudopus parasiticus

Clitocybe candicans

Clitocybe clavipes

Clitocybe gibba

Clitocybe irina

Clitocybe nuda

Clitocybe odora

Clitocybe robusta

Clitocybe squamulosa

Clitocybe subclavipes

Clitocybe subditopoda

Clitocybe tarda

Clitocybe trullaeformis

Clitocybula lacerata

Clitopilus prunulus

Collybia acervata

Collybia alkalivirens

Collybia confluens

Collybia dichrous

Collybia dryophila

Collybia luxurians

Collybia maculata

Collybia polyphylla

Collybia semihirtipes

Collybia subnuda

Collybia tuberosa

Conocybe lactea

Conocybe tenera

Coprinus atramentarius

Coprinus cinereus

Coprinus comatus

Coprinus dilectus

Coprinus lagopus

Coprinus micaceus

Coprinus plicatilis

Coprinus variegatus

Cortinarius alboviolaceus

Cortinarius armillatus

Cortinarius bolaris

Cortinarius camphoratus

Cortinarius claricolor

Cortinarius corrugatus

Cortinarius croceofolius

Cortinarius distans

Cortinarius evernius

Cortinarius iodes

Cortinarius luteus

Cortinarius marylandensis

Cortinarius paleiferus

Cortinarius pholideus

Cortinarius pyriodorus

Cortinarius semisanguineus

Cortinarius torvus

Cortinarius traganus

Cortinarius violaceus

Crepidotus applanatus var. *applanatus*

Crepidotus cinnabarinus

Crepidotus mollis

Crinipellis zonata

Cyptotrama asprata

Cystoderma amianthinum

Cystoderma granulosum

Entoloma abortivum

Flammulina velutipes

Galerina autumnalis var. *autumnalis*

Galerina tibiicystis

Gomphidius glutinosus

Gymnopilus luteofolius

Gymnopilus picreus

Gymnopilus sapineus

Gymnopilus spectabilis

Hebeloma crustuliniforme

Hebeloma mesophaeum

Hebeloma radicosum

Hebeloma velatum

Hohenbuehelia petaloides

Hygrophoropsis aurantiaca

Hygrophorus agathosmus

Hygrophorus appalachianensis

Hygrophorus bakerensis

Hygrophorus caespitosus

Hygrophorus calyptraeformis

Hygrophorus camarophyllus

Hygrophorus cantharellus

Hygrophorus chrysodon

Hygrophorus coccineus

Hygrophorus conicus

Hygrophorus cuspidatus

Hygrophorus eburneus

Hygrophorus erubescens

Hygrophorus flavodiscus

Hygrophorus gliocyclus

Hygrophorus hypothejus

Hygrophorus laetus

Hygrophorus marginatus var. *concolor*

Hygrophorus miniatus

Hygrophorus nitratus

Hygrophorus parvulus

Hygrophorus perplexus

Hygrophorus psittacinus

Hygrophorus pudorinus

Hygrophorus purpurascens

Hygrophorus purpureofolius

Hygrophorus reai

Hygrophorus russula

Hygrophorus speciosus var. *speciosus*

Hygrophorus subsalmonius

Hygrophorus tephroleucus

Hygrophorus unguinosus

Hypholoma elongatum

Hypholoma fasciculare

Hypholoma myosotis

Hypholoma sublateritium

Hypholoma udum

Hypsizygus tessulatus

Inocybe fuscodisca

Inocybe lacera

Inocybe rimosa

Inocybe sororia

Inocybe subochracea

Inocybe tahquamenonensis

Laccaria laccata

Laccaria longipes

Laccaria ochropurpurea

Laccaria trullisata

Lactarius affinis var. *affinis*

Lactarius agglutinatus

Lactarius allardii

Lactarius aquifluus

Lactarius argillaceifolius

Lactarius atroviridis

Lactarius cinereus var. *fagetorum*

Lactarius corrugis

Lactarius croceus

Lactarius deceptivus

Lactarius deterrimus

Lactarius gerardii var. *gerardii*

Lactarius glyciosmus

Lactarius griseus

Lactarius hibbardae var. *hibbardae*

Lactarius hygrophoroides var. *hygrophoroides*

Lactarius hysginus var. *hysginus*

Lactarius lignyotus

Lactarius luteolus

Lactarius maculatipes

Lactarius maculatus

Lactarius oculatus

Lactarius paradoxus

Lactarius peckii

Lactarius pseudoflexuosus

Lactarius pubescens

Lactarius quietus var. *incanus*

Lactarius representaneus

Lactarius resimus

Lactarius rufus var. *rufus*

Lactarius speciosus

Lactarius subplinthogalus

Lactarius subvellereus

Lactarius uvidus

Lactarius vietus

Lactarius vinaceorufescens

Lactarius volemus var. *volemus*

Lactarius zonarius

Lentinellus cochleatus

Lentinellus omphalodes

Lentinellus ursinus

Lentinellus vulpinus

Lentinus lepidius

Lentinus strigosus

Lentinus suavissimus

Lentinus torulosus

Lepiota acutesquamosa

Lepiota cristata

Lepiota naucinoides

Lepiota rubrotincta

Leptonia incana

Leptonia odorifer

Leptonia rosea

Leptonia serrulata

Leucocoprinus fragilissimus

Leucopaxillus albissimus

Limacella glioderma

Limacella illinita

Lyophyllum decastes

Lyophyllum multiforme

Macrolepiota rachodes

Marasmiellus nigripes

Marasmiellus praeacutus

Marasmius graminum

Marasmius oreades

Marasmius pyrrhocephalus

Marasmius rotula

Marasmius siccus

Marasmius strictipes

Megacollybia platyphylla

Melanoleuca alboflavida

Melanophyllum echinatum

Mycena amabilissima

Mycena corticola

Mycena epipterygia var. *epipterygioides*

Mycena epipterygia var. *lignicola*

Mycena epipterygia var. *viscosa*

Mycena griseoviridis

Mycena haematopus and *Spinellus fusiger*

Mycena leaiana

Mycena luteopallens

Mycena pura

Mycena subcaerulea

Nolanea conica

Nolanea murraii

Nolanea quadrata

Nolanea verna

Omphalina ectypoides

Omphalina epichysium

Omphalotus olearius

Panaeolus campanulatus

Panaeolus foenisecii

Panellus serotinus

Panellus stipticus

Panellus violaceofulvus

Paxillus corrugatus

Paxillus involutus

Paxillus panuoides

Phaeomarasmius erinaceellus

Phaeocollybia christinae

Pholiota aurivella

Pholiota flammans

Pholiota lenta

Pholiota malicola var. *macropoda*

Pholiota spumosa

Pholiota squarrosa

Pholiota veris

Pleurocybella porrigens

Pleurotus dryinus

Pleurotus ostreatus complex

Pluteus admirabilis

Pluteus aurantiorugosus

Pluteus cervinus

Pluteus granularis

Pluteus thompsonii

Porpoloma umbrosum

Psathyrella candolleana

Psathyrella delineata

Psathyrella echiniceps

Psathyrella septentrionalis

Psathyrella velutina

Psilocybe atrobrunnea

Psilocybe coprophila

Resinomycena rhododendri

Resupinatus applicatus

Rhodocybe mundula

Rhodotus palmatus

Rickenella fibula

Rozites caperata

Russula brevipes var. *acrior*

Russula compacta

Russula laurocerasi

Russula lutea

Russula nigricans

Russula pulchra

Russula subgraminicolor

Russula tenuiceps

Russula ventricosipes

Simocybe centunculus

Squamanita umbonata

Strobilurus conigenoides

Stropharia hardii

Stropharia hornemannii

Stropharia thrausta

Stropharia rugosoannulata

Stropharia semiglobata

Tectella patellaris

Tephrocybe palustris

Tricholoma æstuans

Tricholoma aurantium

Tricholoma caligatum complex

Tricholoma flavobrunneum

Tricholoma flavovirens

Tricholoma imbricatum

Tricholoma inamoenum

Tricholoma leucophyllum

Tricholoma magnivelare

Tricholoma myomyces

Tricholoma pessundatum

Tricholoma portentosum

Tricholoma saponaceum

Tricholoma sejunctum

Tricholoma subluteum

Tricholoma subresplendens

Tricholoma sulphurescens

Tricholoma vaccinum

Tricholoma zelleri

Tricholomopsis decora

Tricholomopsis formosa

Tricholomopsis rutilans

Tubaria confragosa

Tubaria furfuracea

Volvariella pusilla

Volvariella bombycina

Xeromphalina campanella

Xerula furfuracea

Boletes

Boletes, also known as fleshy pore fungi, are among the most fascinating and highly prized mushrooms. Their beautiful colors, distinctive features, and relative abundance makes them one of the most popular groups collected. Boletes are a relatively safe group to collect for the table and are immensely popular among mycophagists. Most boletes grow on the ground and are soft and fleshy. They have a cap, a stalk, and a sponge-like layer of tubes on the undersurface of the cap. Except for genus *Gastroboletus,* species of which have enclosed and irregularly arranged tubes, boletes have vertically arranged tubes, each of which terminates in a pore. The tube layer is easily detached and typically separates cleanly from the cap flesh. (Polypores also have tubes but can easily be differentiated from boletes because most grow on wood. Moreover, their fruiting bodies are typically tough and leathery to woody, and their tube layers usually do not separate cleanly from the cap flesh.)

The majority of boletes are mycorrhizal with trees, and only a few can be collected in open fields or grassy areas. One of the most important steps in bolete identification is obtaining a spore print. Although some can be very difficult to identify even with the aid of chemical tests and the microscope, many boletes are easily identified using only macroscopic features.

KEY TO THE GENERA OF BOLETES

1. Pores elongated and radially arranged, sometimes weakly gill-like; pore surface pale to dull yellow, decurrent, usually slowly blueing when injured; growing on the ground, usually near ash trees; cap dry, smooth, yellowish brown to reddish brown; stalk typically eccentric, lacking glandular dots, scabers, and reticulation, solid at maturity; spore print olive-brown—*Gyrodon merulioides* (Schweinitz) Singer = *Boletinellus merulioides* (Schweinitz) Murrill (see p. 349).
1. Pores elongated and radially arranged to strongly gill-like, thickened, decurrent; stalk lacking glandular dots, scabers, and reticulation; veil and ring absent; spore print yellowish ochraceous to olive-brown—Genus *Phylloporus* (see p. 332).
1. Pores small, angular to circular, not elongated and radially arranged; cap bright sulfur-yellow to reddish orange, dry, powdery or coated with tiny matted fibers and scales; stalk bright yellow, lacking glandular dots, scabers, and reticulation, solid at maturity; veil bright yellow, membranous to powdery, fragile, sometimes forming a ring on the stalk or remnants on the cap margin; pore surface yellow when young, becoming grayish brown at maturity; spore print olive-gray to olive-brown—*Pulveroboletus ravenelii* (Berkeley and Curtis) Murrill (see p. 353).

1. Not with the above combinations of characters → 2.
    2. Stalk solid at maturity; cap covered with coarse, dry, grayish pink, pinkish tan, pinkish brown, or gray to blackish scales; pore surface white when young, becoming gray to blackish gray in age; flesh whitish, quickly staining reddish orange when exposed and rubbed; spore print blackish brown to black—Genus *Strobilomyces* (see p. 332).
    2. Stalk hollow at maturity; spore print pale to bright yellow; cap dry, color variable, dingy yellow, orange, yellow-brown or dark vinaceous red; pore surface white to yellow; pores very small and circular; stalk lacking glandular dots, scabers, and reticulation; veil and ring absent—Genus *Gyroporus* (see p. 329).
    2. Fruiting body not with the above combinations of characters → 3.
3. Stalk ornamented by scabers that usually darken in age; scabers at maturity are darker than the stalk surface color; stalk lacking glandular dots and reticulation, solid at maturity; pore surface white to grayish white when young, becoming tan to dingy yellow or yellowish brown at maturity; pores circular to angular, not elongated and radially arranged; veil and ring absent; spore print yellow-brown to cinnamon-brown or rusty brown to vinaceous-brown—Genus *Leccinum* (see p. 329).
3. Stalk not ornamented by scabers and not reticulate, usually glandular dotted, solid at maturity; veil and ring present or absent; spore print olive-yellow, yellow-brown, olive-brown, cinnamon-brown, or dark brown; cap usually sticky to slimy or sometimes dry; pore surface white or yellow; pores circular to angular, infrequently elongated and radially arranged to somewhat gill-like—Genus *Suillus* (see p. 332).
3. Stalk not ornamented with glandular dots or scabers but occasionally reticulate, solid at maturity; veil and ring usually present; spore print reddish to vinaceous, vinaceous-brown to purple-brown, or grayish brown to chocolate-brown; cap usually sticky to slimy but sometimes scaly and dry; pore surface white or yellow when young, becoming grayish brown to reddish brown, or brownish yellow at maturity; pores often elongated and radially arranged to somewhat gill-like—Genus *Fuscoboletinus* (see p. 328).
3. Fruiting body not with the above combinations of characters → 4.
    4. Spore print flesh-pink, vinaceous, pinkish brown to purplish brown, rusty brown, or gray-brown; cap dry, smooth, rarely sticky to slimy; pore surface typically white when young, becoming pinkish to pinkish brown at maturity (three species have a dark chocolate-brown to purple-brown or blackish pore surface when young), pores circular to angular, not elongated and radially arranged; stalk frequently reticulate, at least near the apex, or scurfy, lacking glandular dots, usually without scabers (two species have distinct scabers), solid at maturity, sometimes unusually long in relation to the cap diameter; veil and ring absent—Genus *Austroboletus* and Genus *Tylopilus* (see p. 316).
    4. Spore print olive to olive-brown, yellow-brown, cinnamon-brown or dark brown; cap usually dry; pore surface white, yellow, orange, red, brown, or grayish, sometimes blueing when bruised, pores circular to angular, not elongated and radially arranged; stalk reticulate or not, sometimes strongly lacerated, lacking glandular dots and scabers, solid at maturity; veil and ring absent—Genus *Boletellus,* Genus *Boletus,* and Genus *Chalciporus* (see p. 319).

Key to Species of *Austroboletus* and *Tylopilus*

1. Stalk distinctly reticulate overall or at least on the upper third → 2.
1. Stalk not reticulate to obscurely reticulate or distinctly reticulate only at the apex → 3.
    2. Cap dark red, reddish orange, bright yellow, or yellow-brown to red-brown,

sometimes with a yellow margin; flesh greenish yellow to orange-yellow, not blue-ing when cut or bruised; stalk yellow to dark red or dull red, coarsely reticulate with raised yellow ribs that may redden in age—*A. betula* (Schweinitz) Horak = *Boletellus betula* (Schweinitz) Gilbert (see p. 335).

2.   Cap some shade of brown or tan; pore surface white when young, becoming pinkish or pinkish tan; flesh very bitter; reticulation coarse and dark brown—*T. felleus* (Bulliard : Fries) Karsten (see p. 361).

2.   Cap yellow-brown to orange-brown or brown; pore surface yellow-brown to brown or at least with brown tints; flesh mild to slightly bitter; stalk typically equal at maturity; flesh white, usually staining slowly purplish buff to pinkish buff when cut and rubbed; growing in oak-pine woods; July–September; occasional, especially in the southern part of the region; edibility unknown—*T. tabacinus* (Peck) Singer.

2.   Cap dark brown to dark gray or blackish; pore surface black when young, becoming gray in age, staining grayish orange or darker gray when mature; stalk blackish brown to gray; flesh grayish, staining orange to orange-red when cut and rubbed, eventually blackening; taste mild, growing on the ground in oak-pine woods; July–September; edibility unknown—*T. griseocarneus* Wolfe and Halling.

2.   Cap white to buff or pale yellowish; pore surface white to pale gray when young, eventually becoming pinkish, not staining blue or brown when bruised; stalk colored like the cap, with very coarse reticulation formed by very prominent raised ribs that often give the stalk a pitted appearance; flesh white, typically bitter; on sandy soils in oak-pine woods; June–October; inedible, bitter—*A. subflavidus* (Murrill) Wolfe = *T. subflavidus* Murrill.

3.   Cap bright orange to bright orange-red, fading to dull orange, cinnamon, or tan in age—*T. ballouii* (Peck) Singer (see p. 359).

3.   Cap pink to rose-colored; stalk predominantly white or pinkish except for a bright yellow base—*T. chromapes* (Frost) Smith and Thiers (see p. 360).

3.   Cap black to dark grayish brown, sometimes with a whitish bloom, velvety, at least when young; stalk colored like the cap or paler, especially near the apex, flesh staining pinkish to reddish gray, then blackening when cut and rubbed—*T. alboater* (Schweinitz) Murrill (see p. 359).

3.   Not with the above combinations of characters → 4.

4.   Cap white or whitish when young, occasionally with a pinkish tinge, sometimes becoming pale tan, developing brownish stains in age; cap surface uneven and often wrinkled like parchment; stalk white or whitish; flesh white, bitter—*T. intermedius* Smith and Thiers (see p. 361).

4.   Cap purple-brown to grayish brown or reddish brown; pore surface dark brown to purple-brown to dark grayish brown; stalk colored like the cap or paler; stalk surface scurfy from a dense coating of tiny, purplish to purplish brown scabers—*T. eximius* (Peck) Singer (see p. 360).

4.   Cap dark brown to greyish brown, olive-brown, or vinaceous brown; pore surface dark coffee-brown, dark reddish brown, or blackish brown; flesh slowly staining blue at first, then becoming reddish brown to brown; stalk smooth, often white at the base; spores 12–18 × 6–7.5 μm; edibility unknown—*T. pseudoscaber* (Secretan) Smith and Thiers (see comments section of *T. sordidus*, p. 363).

4.   Nearly identical to the previous choice except the flesh slowly stains fuscous but does not stain blue; edibility unknown—*T. porphyrosporus* (Fries) Smith and Thiers (see comments section of *T. sordidus*, p. 363).

4.   Cap violet or purple when young, soon fading to purple-brown to purple-gray or

dingy cinnamon; pore surface white when young, becoming pinkish or pinkish tan in age, not staining blue or brown when bruised; stalk violet or purple when young, retaining a purplish tint even in age; flesh very bitter—*T. plumbeoviolaceous* (Snell and Dick) Singer (see p. 362).

4.   Cap dark to bright purple when young, becoming purple-brown or reddish brown and cinnamon to dull brown in age; pore surface white when young, becoming pinkish to pinkish brown, staining brown when bruised, not blueing; stalk developing conspicuous olive to olive-brown stains from the base upward when handled or in age; stalk white to brown, lacking any purplish tints; flesh very bitter— *T. rubrobrunneus* Mazzer and Smith (see p. 362).

4.   Not with the above combinations of characters → 5.

5.   Cap maroon to reddish brown or cinnamon, at times tawny to yellow-brown, margin not beveled; pore surface white when young, becoming pinkish to pinkish brown at maturity, not staining blue or brown when bruised; flesh white to pinkish, not staining blue or brown when cut and rubbed, taste mild to tart, not bitter; stalk long and slender in relation to the cap diameter, 3–7" (7.5–18 cm) long, ¼–⅜" (6–10 mm) thick at the apex, solid, with elevated, anastomosing lines that sometimes form an obscure, narrow reticulation overall or at least on the upper half—*A. gracilis* (Peck) Wolf (see p. 335).

5.   Cap maroon and velvety when young and fresh, soon purplish brown to dark reddish brown, then duller and smooth in age, margin typically beveled; pore surface white for a long time, then dingy white to brownish; pore surface not blueing, often staining brownish when bruised, not pinkish at maturity; stalk not developing olive stains when handled, lacking elevated, anastomosing lines and not forming an obscure narrow reticulation over the surface; flesh mild-tasting, not bitter; spores 6.5–10.5 × 2.5–4 μm; July– October; edible—*T. badiceps* (Peck) Smith and Thiers.

5.   Cap dark brown to brown when young, becoming yellow-brown to dull cinnamon in age, margin not beveled; pore surface white when young, becoming pinkish or pinkish brown at maturity, staining brown when bruised, not blueing; flesh mild-tasting, not bitter, slowly staining brown or pinkish when cut and rubbed; lacking elevated, anastomosing lines and not forming an obscure narrow reticulation over the stalk surface— *T. indecisus* (Peck) Murrill (see p. 361).

5.   Cap gray-brown to olive-brown, or dark brown, surface finely tomentose to somewhat velvety-tomentose, becoming finely to conspicuously rimose areolate in age, 1¾–5⅛" (4.5–13 cm) wide; flesh whitish, staining greenish blue; pore surface whitish to grayish buff when young, becoming pinkish brown then reddish brown to yellow-brown in age, staining dark blue to dark blue-green, then dark brownish red when bruised; pores angular to irregular and 1–2 mm wide at maturity; stalk brownish with much darker longitudinal streaks, with a greenish or bluish green tint or zone near the apex; spores mostly 10–14 × 4–6 μm—*T. sordidus* (Frost) Smith and Thiers (see p. 363).

5.   Cap dull olive gray to brownish, olive-brown to gray-brown or dark brown, 2–6" (5–15 cm) wide; flesh whitish, not blueing, slowly staining red-brown then fuscous; odor disagreeable; pore surface grayish to pale yellowish gray at first, becoming gray-brown to reddish brown; stalk lacking greenish or bluish tints near the apex; spores 11–20 × 5–8 μm; July–October; edibility unknown—*T. nebulosus* (Peck) Wolf.

Key to Species of *Boletellus,* *Boletus,* and *Chalciporus*

1.   Pore surface whitish to grayish when fresh, not yellow when young or at maturity, not blueing when bruised—*Boletus griseus* Frost (see p. 340).

1.   Pore surface dark brown to dark yellow-brown, dark maroon-red or red-brown to dark reddish brown when fresh, quickly blueing when bruised → 2.

1.   Pore surface brilliant golden yellow when fresh, not blueing when bruised → 3.

1.   Pore surface orange, reddish orange, orange-red, red, or orange-brown when fresh → 4.

1.   Pore surface white becoming yellow, or pale yellow, yellow, lemon-yellow, greenish yellow, or olive-yellow but not brilliant golden yellow when fresh → 10.

    2.   Cap surface instantly staining dark vinaceous with KOH, dark brown, gray-brown, yellow-brown, or reddish brown when young, fading to dull cinnamon in age; flesh whitish, blueing when cut, staining pinkish orange with KOH; stalk brownish pruinose over a dull yellow ground color, lacking olive tints; spores 11–15 × 4–6 μm—*Boletus vermiculosus* Peck (see p. 348).

    2.   Cap surface staining very dark brown with KOH, yellow when young, becoming brown as it matures; flesh yellow, fading to pale yellow in age, blueing when cut, blued surface staining yellow with KOH; stalk brownish pruinose over a dull whitish or pale yellow ground color with olive tints; spores 9–12 × 3–4 μm; on the ground under oak; July–September—*Boletus vermiculosoides* Smith and Thiers (see p. 347).

    2.   Cap brownish olive, olive, or yellowish olive, bruising dark blue to blackish, 1⅛– 4" (3–10 cm) wide, dry, fibrillose-scurfy, becoming finely cracked and somewhat scaly, especially over the disc in age; flesh ochraceous, instantly staining blue when cut, odor and taste not distinctive; stalk 3⅛–4¾" (8–12 cm) long, ⅜–1⅛" (1–3 cm) thick, enlarged downward, punctate with reddish dots and points that form a reticulum over the upper half, with a yellowish ground color, reddish toward the base; flesh yellow, instantly staining blue when cut, rusty rose in the base; spores 13–16 × 5–7 μm; cap and stalk tissues amyloid; under hardwoods or mixed woods, especially oak and beech; edibility unknown—*Boletus pseudo-olivaceus* Smith and Thiers.

3.   Stalk short and stout, 1⅛–2⅜" (3–6 cm) long, ⅜–⅝" (1–1.6 cm) thick at the apex, often enlarged up to 1" (2.5 cm) downward, typically club-shaped and distinctly swollen above a tapered base, yellowish, streaked with dark brown tones, with yellow mycelium sometimes visible at the base; cap surface dry, becoming somewhat viscid when wet, often finely cracked in age, dull reddish brown, yellow-brown, or grayish brown, often dull purplish red on the margin, becoming dull cinnamon when dry; flesh white to pale yellow, vinaceous under the cuticle, with pale brownish tinges in the stalk; odor musty, similar to a sectioned fruiting body of *Scleroderma citrinum,* taste not distinctive; spores 8–11 × 3–5 μm; growing in clusters fused at the stalk bases or solitary, scattered, or in groups on the ground under hardwoods, especially oak; edible—*Boletus innixus* Frost = *Boletus caespitosus* Peck (see p. 340).

3.   Stalk long and slender, 1½–4½" (4–11.5 cm) long, ¼–⅝" (6–17 mm) thick, slightly enlarged downward or nearly equal, narrowed abruptly and usually covered with copious white mycelium at the base, pale yellow at the apex, streaked and flushed pale pinkish brown downward; cap moist and sticky when fresh, pinkish cinnamon to pinkish brown or vinaceous brown, often fading in age or when dry; flesh white to pale yellow except vinaceous under the cuticle, odor and taste not distinctive; spores 11–16 × 4–6 μm; scattered or in groups on the ground under hardwoods, especially oak—*Boletus auriporus* Peck (see p. 337).

4.   Cap at maturity usually less than 2" (5 cm) wide; pore surface bright or dull rose-red to rose-pink when fresh → 5.

4.   Cap at maturity 1½–4" (4–10 cm) wide; pore surface with brown tones, cinnamon, reddish cinnamon, cinnamon-brown to reddish brown when fresh → 6.

4.   Cap at maturity 2–5⅛" (6–13 cm) wide; pore surface orange, orange-red, red, or bright red to dark red when fresh, lacking brown tones → 7.

5.   Cap red when young, fading to yellow, at least over the margin, in age; pore surface not blueing when bruised; flesh bright yellow, not blueing when cut and rubbed, taste mild; stalk colored like the cap or paler, with yellow tints in age; spores 12–15 × 3–5 μm, strongly dextrinoid—*Chalciporus rubinellus* (Peck) Singer = *Boletus rubinellus* Peck (see p. 349).

5.   Cap yellowish to pinkish red, becoming pinkish cinnamon in age, pore surface bright rose-red, not blueing when bruised; flesh pale yellow, not blueing when cut and rubbed, taste mild; stalk rose-pink on the upper portion, yellowish near the base; spores 9–13 × 3–4 μm, weakly dextrinoid to inamyloid; edibility unknown—*Chalciporus pseudorubinellus* (A. H. Smith and Thiers) A. E. Bessette, A. R. Bessette, and D. W. Fischer, *comb. nov.* Basionym: *Boletus pseudorubinellus* Smith & Thiers. *Boletes of Michigan,* 300.

5.   Cap cinnamon-red with yellow streaks; pore surface not blueing when bruised; flesh whitish tinged yellow, not blueing when cut and rubbed; taste mild; stalk dingy apricot-yellow; spores 9–13 × 3–4 μm, weakly dextrinoid to inamyloid; edibility unknown—*Chalciporus rubritubifer* (Kauffman) Singer = *Boletus rubritubifer* Kauffman.

6.   Cap typically less than 3" (7.5 cm) wide at maturity, buff to orangish brown·or reddish brown; pore surface cinnamon to reddish cinnamon or cinnamon-brown to reddish brown, not blueing when bruised—*C. piperatus* (Bulliard : Fries) Bataille (see p. 348).

6.   Fruiting body nearly identical to the previous choice except the pore surface stains blue when bruised; edibility unknown—*C. piperatoides* Smith and Thiers.

6.   Cap up to 4" (10 cm) wide at maturity, bright sulfur-yellow to orange-yellow or golden yellow; pore surface yellow when very young, soon brick-red to reddish brown; stalk bright yellow to brick-red or dark red, base coated with bright yellow mycelium; growing on or near stumps, decaying wood, or sawdust; edibility unknown—*Boletus hemichrysus* Berkeley and Curtis.

7.   Cap whitish to tan, grayish, pinkish tan, or olive-gray; pore surface red to red-orange, blueing when bruised; stalk colored like the cap, developing olive stains when handled, with a slight, red reticulation at the apex or not reticulate at all; inedible, bitter—*Boletus firmus* Frost = *Boletus piedmontensis* Grand and Smith.

7.   Cap pinkish orange to reddish orange when young, becoming orange-yellow to yellow and typically developing brownish or olive tones, especially over the disc, in age, blueing when bruised; flesh yellow, blueing when bruised; pore surface orange-red, becoming orange to orange-yellow at maturity, darkest near the stalk and palest near the margin, blueing when bruised; stalk ground color pale yellow, punctate with reddish dots and points, becoming reddish brown in age, blueing when bruised, lacking short, stiff, dark red hairs at the base of mature specimens; under hardwoods, especially beech and oak, or in mixed woods; edibility unknown—*Boletus discolor* (Quélet) Bigeard and Guillemin.

7.   Cap dark to bright red, sticky when fresh; pore surface dark red, quickly blueing when bruised; flesh yellow, quickly blueing when cut; stalk dark red, deeply and coarsely reticulate—*Boletus frostii* Russell (see p. 339).

7. Not with the above combinations of characters; stalk typically reticulate, at least over the upper portion → 8.

7. Not with the above combinations of characters; stalk lacking obvious reticulation, often punctate with tiny red or orange-cinnamon to brown dots and points → 9.

    8. Cap dark red to deep wine-red overall; flesh yellow, quickly staining blue when cut; pore surface dark red when fresh, blueing when bruised; stalk dark red except yellow at the apex, quickly blueing when bruised, with dark red reticulation over at least the upper portion; growing under hardwoods—*Boletus rubroflammeus* Smith and Thiers (see p. 344).

    8. Fruiting body very similar to the previous choice except for more variable cap color, ranging from dark red to brick-red or brownish red, a yellow stalk base, and growth in conifer woods; edibility unknown—*Boletus flammans* Snell and Dick.

    8. Cap yellowish to olive-brown, brown, or reddish brown; pore surface dark red to orange-red or brownish red, blueing when bruised; stalk yellow, with red reticulation and often punctate with reddish dots and points; flesh yellow, quickly blueing when cut; poisonous—*Boletus luridus* Schaeffer : Fries.

    8. Not with the above combinations of characters → 9.

9. Flesh bright yellow, quickly staining dark blue when cut; cap color variable, cinnamon-brown to yellow-brown, reddish brown, or reddish orange to orange-yellow, quickly staining blue to blue-black when bruised; pore surface red to orange, quickly blueing when bruised; stalk punctate with reddish dots and points, quickly blueing when bruised; with short, stiff, dark red hairs at the base of mature specimens—*Boletus subvelutipes* Peck (see p. 347).

9. Flesh yellow, quickly staining dark blue when cut; cap dark brown to blackish brown, sometimes with olive tones or reddish brown; pore surface bright red to red-orange, quickly blueing when bruised; stalk yellow, punctate with reddish dots and points, quickly blueing when bruised, lacking short, stiff, dark red hairs at the base of mature specimens; edibility unknown—*Boletus luridiformis* Rostkovius = *B. erythropus* (Fries) Krombholz.

9. Flesh bright yellow, quickly staining dark blue when cut, blued flesh staining yellow, then orange with $FeSO_4$; cap red to orange-red, instantly staining dark violet when bruised; margin with a narrow sterile band extending beyond the pore surface; pore surface pinkish red to dark red, becoming orange-red, quickly blueing when bruised; stalk pale yellow, faintly pruinose-scurfy overall, darkening to brownish from the base upward in age or where handled, flesh reddish in the base, lacking short, stiff, dark red hairs at the base; growing under oak; edibility unknown—*Boletus subluridellus* Smith and Thiers.

9. Flesh yellow, instantly staining blue when cut, blued flesh staining yellow, then slowly olive with the addition of $FeSO_4$; cap dry, somewhat velvety, dull brick-red to reddish brown, with ochraceous to yellowish brown overtones, becoming dingy cinnamon in age; pore surface maroon-red when young, orange-red at maturity, quickly blueing when bruised; stalk yellow, punctate on the lower half with orange-cinnamon to brown dots and points, lacking short, stiff, dark red hairs at the base; growing on the ground under hardwoods; edibility unknown—*Boletus rufocinnamomeus* Smith and Thiers.

9. Flesh yellow to pale yellow; slowly staining blue when cut; cap dark red to bright apple-red; pore surface orange-red to red, staining greenish blue when bruised; stalk dark red to bright apple-red, lacking reticulation or sometimes reticulate only at the very apex; spores 11–15 × 4–5 μm; growing on the ground under hardwoods, recorded from Michigan east to New England; edibility unknown—*Boletus bicolor* var. *borealis* Smith and

Thiers. (A similar undescribed species, soon to be published, is found in the southern part of the range and has yellowish flesh that does not stain blue when exposed, a reticulate stalk, and smaller spores that measure 8–11 × 3–4 μm.)

10. Pore surface of fresh specimens staining dark blue, blue, or blue-green instantly, or within 30 seconds after bruising; stalk distinctly reticulate over the upper half or at least at the top, sometimes overall → 11.

10. Pore surface of fresh specimens staining dark blue, blue, or blue-green instantly, or within 30 seconds after bruising; stalk lacking distinct reticulation or only finely so at the apex → 13.

10. Pore surface of fresh specimens typically not blueing when bruised (an occasional weak blueing reaction may be observed) → 20.

11. Cap bright rose-red to rose-pink, retaining its color well into maturity, becoming pinkish brown in age; stalk yellow, often with red tints, especially toward the base; cap flesh quickly blueing when cut; edible—*Boletus speciosus* Frost.

11. Cap whitish to grayish white, darkening to olivaceous brown or pale grayish brown, surface often finely to conspicuously cracked in age; stalk yellowish on the upper portion, red to purplish red near the base, reticulation and upper portion sometimes pink; cap flesh white to pale yellow, blueing when cut; taste distinctly bitter; growing on the ground under oak and other hardwoods, or under spruce and hemlock; inedible—*Boletus inedulis* (Murrill) Murrill.

11. Cap yellow-brown, olive-brown, reddish brown, or brown → 12.

12. Cap reddish brown or yellow-brown to olive-brown; flesh pale yellow, quickly blueing when cut, odor and taste not distinctive; stalk yellow on the upper portion, pinkish red to purplish red on the lower portion or at least tinged reddish, especially near the base, reticulate overall or at least over the upper half, blueing when bruised; growing on the ground under beech, maple or conifers, especially hemlock—*Boletus speciosus* var. *brunneus* Peck (see p. 346).

12. Cap brown to yellow-brown; cap flesh yellow, quickly blueing when cut, taste mild, not bitter; stalk yellow on the upper portion, brownish red and coated with tiny, brown flecks on the lower portion; growing on the ground in oak-pine woods in the southern part of the region; edibility unknown—*Boletus luridellus* (Murrill) Murrill.

12. Cap olive-brown, becoming dark yellow-brown then fading to paler brown to grayish brown or tan in age; surface often finely to conspicuously cracked in age; stalk red on the lower portion or overall, typically yellow on the upper portion; cap flesh whitish to pale yellow, blueing when cut; taste distinctly bitter; growing on the ground under conifers; inedible—*Boletus calopus* Fries.

13. Cap whitish to buff or pale gray-brown when young, darkening in age; stalk whitish; growing on the ground under hardwoods, especially oak—*Boletus pallidus* Frost (see p. 342).

13. Cap bright lemon-yellow, becoming duller yellow in age, sometimes with bright brownish yellow tints on the disc, occasionally tinged brownish red, dry, velvety when young, dull or shiny at maturity, quickly staining bluish black when bruised; flesh and pore surface bright lemon-yellow, quickly bruising blue; stalk lemon-yellow, bruising blue, then brown; KOH quickly staining orange on all blueing surfaces; FeSO$_4$ quickly staining yellow on all blueing surfaces; growing on the ground in hardwoods or mixed woods, especially with oak, beech, maple, and yellow birch; July–September; edibility unknown—*Boletus pseudosulphureus* Kallenbach.

13. Cap with distinct brown tones when fresh, rusty golden yellow, olive brownish, yellow-

brown, orange-brown, pinkish brown, reddish brown, bay brown, rusty brown, or blackish brown → 14.

13. Cap pink, rose-red, red, dark red, brick-red to orange-yellow, lacking distinct brown tones when fresh → 17.

    14. Cap and stalk reddish brown to yellow-brown or rusty golden yellow, often finely cracked in age; pore surface at least slightly decurrent; flesh blueing slowly or not at all when cut and rubbed; stalk typically eccentric; growing on or near conifer stumps or sometimes sawdust; edibility unknown—*Boletus lignicola* Kallenbach.

    14. Cap blackish brown and velvety when young, becoming chocolate-brown to chestnut-brown and often finely cracked in age; stalk reddish brown to blackish brown, decorated with tiny hairs, points, or scales, staining blue, then slowly reddish when bruised; growing on decaying wood, often at the base of trees; spores longitudinally striate—*Boletellus chrysenteroides* (Snell) Singer (see p. 336).

    14. Cap dark olive to olive-brown, becoming deeply cracked in age, with reddish tinges in the cracks; stalk yellow at the apex, purplish red at the base, staining blue-green when bruised; spores not truncate—*Boletus chrysenteron* (Bulliard) Fries (see p. 338).

    14. Fruiting body nearly identical to the previous choice except that many of the spores are truncate to slightly notched, 10–15 × 4.5–7 μm; edible—*Boletus truncatus* (Singer, Snell and Dick) Pouzar.

    14. Fruiting body similar to the previous two choices; cap 1⅛–2" (3–5 cm) wide, dry, somewhat velvety, finely cracked toward the margin at maturity, dingy olive brownish at first, becoming dingy yellow-brown in age; flesh pale buff, blueing when bruised; stalk yellowish brown with brown pruina; spores 9–12 × 4.5–5 μm, truncate to slightly notched; on the ground in hardwood forests, edibility unknown—*Boletus subdepauperatus* Smith and Thiers.

    14. Not with the above combinations of characters → 15.

15. Cap reddish brown to chestnut-brown or yellow-brown, staining green to blue in NH_4OH; stalk colored like the cap; flesh whitish when fresh, usually staining yellow, then blueing near the tubes when cut and rubbed, remainder of the flesh usually not blueing; pore surface pale yellow, becoming greenish yellow in age; growing on the ground or on decaying wood, especially conifer, in conifer woods or beech-maple forests; edible—*Boletus badius* Fries.

15. Cap rusty brown to reddish brown, fading to dull cinnamon or dingy yellow-brown in age, staining blue, then dull purplish with NH_4OH; flesh bright yellow, quickly blueing when cut or bruised; stalk yellow at the apex, flushed rusty red, pinkish or purplish below—*Boletus pseudosensibilis* Smith and Thiers (see p. 343).

15. Cap pale to dark brown or reddish brown, staining green to dark green with NH_4OH; stalk yellow except at the base, which is reddish brown; all parts instantly blueing when cut or bruised; growing on the ground in mixed woods, most often under oak; edible—*Boletus pulverulentus* Opatowski.

15. Not with the above combinations of characters → 16.

    16. Cap dull yellow-brown to dull cinnamon-brown, very large, 2–10" (5–25.5 cm) wide; stalk often conspicuously enlarged downward, pale yellow on the upper portion, dull yellow to pale brown below, staining darker brown where handled, sometimes with reddish stains near the base; flesh pale yellow, slowly and weakly blueing when cut; growing on the ground under conifers, especially hemlock, or in mixed hardwoods, especially beech and oak; edible—*Boletus huronensis* Smith and Thiers.

16. Cap reddish brown to dark orange-brown, dull brick-red, or dull brown, instantly staining blue-black when bruised; stalk yellow on the upper portion, often with red tints, pale brown below, develping olive-brown to dark brown tones from the base upward as it matures, typically staining blue-black when bruised; flesh blueing when cut and rubbed; growing on sandy soil in oak-pine woods; edibility unknown—*Boletus oliveisporus* (Murrill) Murrill.

16. Cap pinkish brown, rosy brown, dull red, reddish brown, or grayish brown; cap flesh yellowish except for a thin reddish zone just beneath the cuticle, staining reddish then blueing erratically when cut; stalk color variable, whitish to yellowish, sometimes splashed with red; flesh yellow; growing on sandy soils in woods in the southern part of the region; edible—*Boletus communis* Bulliard. (Note: There is considerable confusion regarding the name of this entity. Quite likely, the species name *communis* will soon be changed.)

17. Cap 1½–4" (4–10 cm) wide, red to rose-red when young, developing olive, olive-brown, or olive-gray tones as it matures, typically olive-gray and cracked in age, usually showing red or pink tints in the cracks in age; stalk yellow at the apex, punctate with rhubarb-red dots and points below, with a white to pale yellow basal mycelium, usually with faint longitudinal ribs; flesh typically blueing when cut and rubbed; growing on the ground in hardwoods, especially oak; spores longitudinally striate; edibility unknown—*Boletellus intermedius* Smith and Thiers.

17. Cap 2–6" (5–15.5 cm) wide, dark red, red, rose-red, purple-red, or pink when fresh, becoming yellow in age; flesh pale yellow, slowly staining blue when cut; stalk yellow at the apex, red or rosy red on the lower two-thirds or more; growing on the ground under oaks—*Boletus bicolor* var. *bicolor* Peck (see p. 337).

17. Cap ¾–3⅛" (2–8 cm) wide, velvety, dry, often becoming cracked in age, dark red when young, fading to brick-red in age; flesh yellow, slowly staining bluish green, odor and taste not distinctive; stalk up to ½" (1.3 cm) thick, tapered down to a somewhat pointed base, yellow at the apex, punctate with dull orange dots and points from the base up to the apex, darkening to brown where handled; stalk flesh yellow, slowly staining bluish green, reddish orange in the base; pore surface typically depressed near the stalk; cap cuticle staining dull orange with KOH; flesh staining olive-green and cap cuticle blackish with $FeSO_4$; growing on the ground in hardwoods and mixed woods, especially under oak and beech—*Boletus rubellus* Krombholz (see p. 344).

17. Cap up to 2" (5 cm) wide at maturity → 18.

17. Not with the above combinations of characters; cap diameter more than 2" (5 cm) wide at maturity → 19.

18. Cap red or dark red, becoming pinker or duller in age; surface soon developing conspicuous cracks and fissures with yellow to whitish flesh showing in the cracks; pores angular, 1–2 per mm, not radially elongated to gill-like near the stalk; flesh typically blueing when cut; stalk reddish pruinose over a yellow ground color; basal mycelium white to pale yellow or not evident; growing on the ground in woods—*Boletus fraternus* Peck (see p. 339).

18. Cap dull rose-red to bright orange-red when fresh, darkening in age; surface not conspicuously cracked and fissured; pores 1–2 mm wide, irregular to angular, becoming radially elongated to gill-like near the stalk; flesh pale yellow, typically blueing when cut; stalk colored like the cap or slightly paler; basal mycelium white or not evident; growing on the ground in woods; edibility unknown—*Boletus subfraternus* Coker and Beers.

18. Cap rose-red when fresh, becoming paler pinkish red and finely cracked in age;

stalk predominantly yellow, typically reddish pruinose especially toward the base; stalk base coated with yellow mycelium; pores 1–2 per mm, circular to angular, not radially elongated to gill-like near the stalk; flesh typically blueing when cut; growing on the ground in woods and grassy areas; edible—*Boletus campestris* Smith and Thiers.

19. Cap dark to pale brick-red, fading to dull rose; stalk yellow, often tinged pink or red, especially near the base; flesh bright yellow; all parts quickly blueing when cut or bruised—*Boletus sensibilis* Peck (see p. 345).

19. Cap bright brick-red to bright rose-red, becoming olivaceous in age; stalk pale yellow, often tinged red to reddish brown at the base; flesh yellow, blueing when cut; growing on the ground in forests, especially under oak; poisonous—*Boletus miniato-olivaceus* Frost.

19. Cap red to brick-red, soon fading to reddish orange or orange-yellow; flesh pale yellow, blueing when cut; stalk yellow above or overall when young, developing orange to reddish or reddish brown tones or streaks in age—*Boletus miniato-pallescens* Smith and Thiers (see p. 341).

  20. Stalk distinctly to weakly reticulate overall or at least on the upper portion, or deeply grooved and shaggy to lacerated → 21.

  20. Stalk smooth or slightly roughened, lacking distinct reticulation, sometimes with raised rib-like lines forming a partial reticulum at the apex or nearly overall → 29.

21. Cap color variable, pale gray, yellow, olive, yellow-brown, or olive-brown; stalk yellow, with prominent, raised reticulation overall or at least on the upper two-thirds; flesh yellow, only slightly bitter-tasting—*Boletus ornatipes* Peck (see p. 342).

21. Cap white to dull white, yellow, golden yellow, orange, or red → 22.

21. Cap dark to pale purplish brown, brownish lilac, or dull pink when young, becoming pinkish brown to pale yellow-brown in age; stalk with fine, whitish reticulation → 23.

21. Not with the above combinations of characters → 24.

  22. Cap white to dull white; stalk white to pale yellow; flesh white, not staining blue when cut, taste mild; growing on the ground under oaks; edibility unknown—*Boletus albisulphureus* (Murrill) Murrill.

  22. Cap and stalk bright orange to orange-yellow or golden yellow; cap surface finely velvety to powdery when young and fresh; stalk typically with raised lines to strongly reticulate on the upper portion; flesh not blueing when cut; pore surface often developing reddish orange tints; growing on the ground in oak-pine woods; edibility unknown—*Boletus auriflammeus* Berkeley and Curtis.

  22. Cap dark red, reddish or orange, sometimes with a yellow margin, becoming bright yellow to yellow in age; stalk yellow to dark red or dull red, coarsely reticulate with raised yellow ribs—*Austroboletus betula* (Schweinitz) Horak = *Boletellus betula* (Schweinitz) Gilbert (see p. 335).

23. Cap dark purple to purplish brown when young, becoming pinkish brown to yellow-brown in age; stalk whitish with lilac tinges when young, becoming mostly dark lilac in age; cap or stalk surface staining green to blue-green with KOH; flesh white, not blueing when cut; growing on the ground under hardwoods, especially oak; edible—*Boletus pseudoseparans* Grand and Smith (see Comments for *B. separans,* p. 345).

23. Cap brownish lilac to purplish brown, or pale pinkish brown to dull pink, becoming pale cinnamon-brown to yellow-brown in age; stalk colored like the cap or paler when young, usually with a pinkish, lilac, or wine-colored tinge when fresh; cap or stalk surface staining green to blue-green with KOH—*Boletus separans* Peck (see p. 345).

  24. Cap brown to yellow-brown, dry; stalk typically bulbous when young, more than

½" (1.3 cm) thick at maturity, golden yellow to yellow; flesh yellow, not blueing when cut, taste mild; growing on the ground under hardwoods, especially oak; edible—*Boletus auripes* Peck.

24. Cap blackish brown, becoming dark yellow-brown; stalk blackish brown to grayish brown, with a coarse white reticulum; flesh white, not blueing when cut; growing on the ground in beech-maple woods; edible—*Boletus variipes* var. *fagicola* Smith and Thiers.

24. Cap pale to dark cinnamon-brown, reddish brown, yellow-brown, or olive-gray; stalk relatively long in relation to the cap, typically 3" (7.5 cm) or more in length, pinkish tan to reddish brown, coarsely reticulate to shaggy → 25.

24. Not with the above combinations of characters → 26.

25. Stalk reddish brown to pinkish tan, deeply grooved and ridged for most or all of its length; ridges branched or torn to create a honeycomb or shaggy-bark effect—*Boletellus russellii* (Frost) Gilbert (see p. 336).

25. Stalk pale to dark cinnamon-brown to reddish brown, with prominent reticulation overall or at least on the upper two-thirds—*Boletus projectellus* Murrill (see p. 343).

26. Cap 1⅛–3½" (3–9 cm) wide at maturity, pale yellow-brown to pinkish cinnamon or dull brick-red to dark reddish brown; stalk yellow to pale yellow, weakly to coarsely reticulate; reticulation yellowish to dark brown, flesh whitish to pale yellow, not blueing when cut → 27.

26. Cap 3½–7" (9–18 cm) wide or more at maturity, various shades of brown; stalk reticulation fine, white to whitish, sometimes inconspicuous; flesh white, not blueing when cut → 28.

27. Cap dull brick-red to reddish brown with an olive tint; surface staining green, then blue, and finally fuscous with NH_4OH; pore surface yellow to dingy yellow; pores angular, 1–3 mm wide; stalk acutely tapered downward; reticulum brown to dark brown, coarse, distinctly wide-meshed; growing on the ground under oak; edibility unknown—*Boletus tenax* Smith and Thiers.

27. Cap pale brownish yellow when young, becoming pale yellow-brown to pinkish cinnamon, dry, somewhat velvety, producing a fleeting brilliant green reaction with NH_4OH, instantly staining dark brown with KOH; pore surface lemon-yellow, not blueing when bruised; pores angular, 1–2 mm wide on mature specimens; stalk yellow, tapered downward, with longitudinal rib-like lines that form a partial reticulum; growing on the ground under oak; edible—*Boletus illudens* Peck (see p. 340).

28. Cap pale lemon-yellow splashed irregularly with brick to bright rusty red, finally brick-red overall; flesh white, changing to olive-gray with the addition of $FeSO_4$; stalk with a distinct, whitish reticulum on the upper one-third or more—*Boletus chippewaensis* Smith and Thiers (see p. 338).

28. Cap pale brown to reddish brown, pale cinnamon-brown, rusty red or yellowish tan; stalk with a distinct, whitish reticulum on the upper one-third or more—*Boletus edulis* Bulliard : Fries (see p. 339).

28. Cap grayish buff to grayish brown, often pale yellow-brown, sometimes with prominent cracks and patches of tiny scales at maturity; stalk whitish to brown, distinctly reticulate overall or at least on the upper portion; growing on the ground under oak or beech, sometimes with maple and hemlock; edible—*Boletus variipes* Peck.

29. Cap olive to tawny-olive; growing solitary or in small groups on fruiting bodies of the Common Earthball, *Scleroderma citrinum*—*Boletus parasiticus* Bulliard : Fries (see p. 343).

29. Cap rusty red to dark orange-yellow, surface dry, granular-scaly when young, becoming smooth at maturity; pore surface staining pale cinnamon when bruised; stalk pale orange-yellow, typically with a distinct orange zone at the apex, longitudinally striate, at least on the upper portion; flesh whitish to pale yellow, taste unpleasant; growing on the ground under oak; edibility unknown—*Boletus roxanae* Frost.

29. Cap bright yellow or sometimes orange-yellow, typically slimy to sticky when fresh; stalk yellow, sticky or slimy, sheathed at the base with cottony, white mycelium; flesh white to pale yellow, not blueing; growing on the ground in pine woods; edibility unknown—*Boletus curtisii* Berkeley.

29. Cap pale brownish yellow when young, becoming pale yellow-brown to pinkish cinnamon, dry, somewhat velvety, producing a fleeting brilliant green reaction with NH_4OH, instantly staining dark brown with KOH; pore surface lemon-yellow, not blueing when bruised; pores angular, 1–2 mm wide on mature specimens; stalk yellow, tapered downward, with longitudinal rib-like lines that form a partial reticulum; growing on the ground under oak; edible—*Boletus illudens* Peck (see p. 340).

29. Not with the above combinations of characters → 30.

    30. Cap dry, reddish brown, dark brown, brown, orange-brown, yellow-brown, or yellowish tan; cap cuticle not staining blue with NH_4OH; pore surface whitish to yellowish, becoming ochre to rusty ochre in age, slowly staining yellow-brown when bruised; stalk whitish, tinged yellowish, pinkish tan or brown when bruised or in age; flesh white, not blueing when cut, taste mild; growing on the ground under hardwoods and conifers; edible—*Boletus affinis* Peck.

    30. Fruiting body nearly identical to the previous choice except that the cap is mottled with few to many white to pale yellow spots—*Boletus affinis* var. *maculosus* Peck (see p. 336).

    30. Fruiting body similar to the previous two choices but cap dark red to purple-red or maroon; cap cuticle staining blue with NH_4OH (pers. comm. with Ernst Both); growing on the ground, usually with oak; edible—*Boletus purpureofuscus* Smith = *Xanthoconium purpureum* Snell and Dick.

    30. Not with the above combinations of characters → 31.

31. Cap surface smooth to slightly wrinkled, chestnut, ochre, cinnamon, or reddish brown; flesh pale yellow to whitish, rarely blueing slightly when cut, taste mild to slightly acidic; stalk yellow with occasional reddish or reddish brown tinges on the lower portion, minutely scurfy with a thin coating of tiny, yellow scales, dots, or points—*Boletus subglabripes* Peck (see p. 346).

31. Fruiting body nearly identical to the previous choice except that the cap is strongly wrinkled and pitted; edible—*Boletus hortonii* Smith and Thiers = *Boletus subglabripes* var. *corrugis* Peck.

31. Not with the above combinations of characters → 32.

    32. Cap dry, dark olive-brown to dark grayish brown or reddish brown at the center, often yellow to olive-gold on the margin; pore surface dull yellow at first, becoming orange to reddish orange then brownish orange to brick-red at maturity; stalk bright yellow to greenish yellow beneath a coating of small, red to reddish brown scales; flesh pale yellow, not blueing, but reddening when cut; growing on the ground in mixed woods and hardwoods; edibility unknown—*Boletus morrisii* Peck.

    32. Cap sticky to slimy when fresh, smooth to slightly pitted, butterscotch-yellow to yellow-brown, yellow-orange, orange, orange-brown or reddish orange, often becoming olive-green with an orange center in age, staining cherry-red with KOH; stalk often curved at the base, whitish to yellowish or with reddish to pink-

ish tan tones, scurfy from tiny scales—*Boletus longicurvipes* Snell and Smith (see p. 341).

32. Cap dry to slightly sticky, distinctly wrinkled and pitted, reddish brown to dull red or dingy reddish orange, pore surface developing reddish brown stains in age; stalk often curved near the base, ground color yellow, streaked and punctate with reddish dots and points; flesh yellowish, not blueing and not reddening when cut; growing in mixed woods with oak or chestnut; edibility unknown—*Boletus rubropunctus* Peck.

32. Cap dry, somewhat velvety, often finely cracked in age, olive-brown to yellow-brown, instantly staining reddish brown in NH$_4$OH; pore surface rarely blueing slightly and slowly when bruised; stalk predominantly yellow with reddish brown streaks and stains, scurfy, sometimes with raised, rib-like, longitudinal lines that may form a partial reticulum at the apex; flesh whitish, unchanging or blueing slightly when cut; growing on the ground under hardwoods and conifers; edible—*Boletus subtomentosus* Fries.

32. Cap dry, somewhat velvety, sometimes cracked in age, dark olive to olive-yellow with reddish tints, sometimes reddish brown in age, giving a fleeting blue or blue-green reaction changing to reddish brown in NH$_4$OH—*Boletus spadiceus* Fries (see p. 345).

Key to Species of *Fuscoboletinus*

1. Pore surface pale to bright yellow when young, becoming dull yellow to yellowish brown in age → 2.

1. Pore surface whitish to grayish white when young, becoming reddish brown to grayish brown in age → 4.

2. Cap surface dry, not sticky or slimy, covered with matted fibers or scales, pale pinkish purple to reddish purple; pores large, angular, and radially arranged when young, becoming gill-like and crossveined at maturity; stalk yellow at the apex, pale pinkish purple to reddish purple below; ring absent or present as a faint annular zone; on sphagnum mosses under larch, usually in bogs; August–November; edible—*F. paluster* (Peck) Pomerleau and Smith.

2. Cap surface sticky to slimy → 3.

3. Cap surface covered with coarse, pinkish gray to reddish brown scales or patches; growing among sphagnum mosses under larch trees—*F. spectabilis* (Peck) Pomerleau and Smith (see p. 349).

3. Cap surface lacking scales, red when young, becoming mahogany-red to chestnut-brown in age, red in KOH or NH$_4$OH, blue in FeSO$_4$; growing on the ground under conifers, especially balsam fir and hemlock; August–October; edible—*F. glandulosus* (Peck) Pomerleau and Smith.

4. Cap surface covered by a dark reddish brown slime layer; margin rimmed with cottony pieces of grayish white partial veil; stalk dingy white above the ring, whitish with pinkish brown to yellowish brown streaks below; ring superior; on the ground under larch, white pine, and other conifers, especially near bogs; August–October; edible—*F. serotinus* (Frost) Smith and Thiers.

4. Cap surface covered by a colorless slime layer; grayish to greenish gray or pale grayish brown; flesh white, staining greenish blue when exposed; growing on the ground under larch; August–October; edible—*F. viscidus* (Linnaeus) Grund and Harrison = *F. aeruginascens* (Secretan) Pomerleau and Smith.

4. Cap surface slightly sticky to nearly dry, whitish to pale olive or olive-gray; flesh not staining when cut or bruised; growing on the ground or among sphagnum mosses under larch; August–October; edible—*F. grisellus* (Peck) Pomerleau and Smith.

Key to Species of *Gyroporus*

1. Pore surface and flesh instantly staining dark lilaceous to indigo and finally deep blue when cut or bruised—*G. cyanescens* var. *violaceotinctus* Watling (see p. 350).

1. Pore surface and flesh staining greenish yellow then greenish blue to blue when cut or bruised, otherwise nearly identical to the above choice; edible—*G. cyanescens* (Bulliard : Fries) Quélet.

1. Pore surface and flesh not blueing when cut or bruised → 2.
    2. Cap and stalk vinaceous red to burgundy; cap up to 3⅛" (8 cm) wide, dry, somewhat velvety; spores 8–11 × 5–7 μm; growing on the ground, usually under pine and oak; July–October; edible—*G. purpurinus* (Snell) Singer.
    2. Cap and stalk white to yellowish or some shade of brown → 3.
3. Cap orange-brown to chestnut-brown or yellow-brown—*G. castaneus* (Fries) Quélet (see p. 350).
3. Cap and stalk dull white to dingy yellow, often tinged pink or apricot, darkening to cinnamon or brownish when handled or in age, typically coated with sand; cap large, up to 4¾" (12 cm) wide; flesh whitish, odor and taste not distinctive; spores 9–14 × 4.5–6 μm; growing in sandy soil in oak-pine woods; July–October; edible—*G. subalbellus* Murrill.

Key to Species of *Leccinum*

1. Cap orange to yellowish orange, rusty orange or brick-red when young and fresh; flesh staining purple-gray to blackish when cut and rubbed, with or without a reddish intermediate phase → 2.

1. Cap bright to dull yellow or orange-yellow when young and fresh, darkening to pale yellow-brown to olive-brown in age → 3.

1. Cap whitish to buff, pinkish buff to pale cinnamon-buff, tan to grayish tan or pale gray when young and fresh, darkening at maturity, sometimes with a pinkish or greenish tinge; scabers whitish to buff when young, darkening in age → 4.

1. Cap yellow-brown to grayish brown, dark brown, or black when young and fresh; scabers dark brown to black on mature specimens → 7.
    2. Flesh white, staining rose-vinaceous, then vinaceous gray to purplish gray when cut and rubbed; scabers hazel to vinaceous buff; usually partially buried and coated with sand, growing with beach grass, beach heather, and sedges in coastal sand dunes; New Brunswick to Cape Cod (the exact distribution is unknown); cap orange to yellowish orange, fading to dull cinnamon, 2–6" (5–15.5 cm) wide; July–September; edible—*L. arenicola* Redhead and Watling.
    2. Flesh white, staining pinkish to wine-red, then gradually darkening to purple-gray or blackish when cut and rubbed; cap orange to brick-red, rusty red, or reddish brown; stalk whitish; scabers whitish to buff on young specimens, becoming orange-brown to reddish brown, and finally blackish brown on mature specimens; growing on the ground in mixed hardwoods and conifer forests—*L. aurantiacum* (Bulliard) S. F. Gray (see p. 351).
    2. Flesh white, staining pinkish, then purple-gray to blackish; cap dull orange to tan

or brownish, with tiny fibers that break up to form downy patches or small scales in age; stalk whitish; scabers coal-black on both young and mature specimens; growing on the ground under hardwoods, especially birch—*L. atrostipitatum* Smith, Thiers and Watling (see p. 351).

2.   Flesh white, staining purplish gray, then blackish when cut and rubbed, lacking a distinct reddish stage; scabers brownish on young specimens, becoming blackish at maturity; growing on the ground under hardwoods, especially aspen and birch, rarely under conifers; cap rusty orange to rusty cinnamon or rusty brown, 1½–6" (4–15.5 cm) wide; July–September; edible—*L. insigne* complex Smith, Thiers and Watling.

3.   Cap 1⅛–2¾" (3–7 cm) wide, pale yellow when young, darkening to yellow and finally olive-brown in age, typically wrinkled and pitted and finely cracked at maturity; cap flesh whitish to pale yellow, staining pinkish gray to pinkish brown when cut and rubbed, staining olive in FeSO₄; scabers grayish to blackish; growing on the ground under hardwoods, especially blue beech; June–September; edible—*L. luteum* Smith, Thiers and Watling.

3.   Cap 1¾–5½" (4.5–14 cm) wide, orange-yellow to yellow-ochre, becoming yellow-brown in age, distinctly wrinkled and pitted, at least at maturity, staining cherry red in KOH; flesh white to pale yellow, staining reddish or burgundy when cut and rubbed; scabers pale brown, darkening in age; growing on the ground under hardwoods, especially oak—*L. rugosiceps* (Peck) Singer (see p. 352).

3.   Cap 2–4" (5–10 cm) wide, yellow to orange-yellow, dry, coated with a thin layer of grayish fibrils; cap flesh whitish, staining grayish, then fuscous when cut and rubbed; scabers whitish to pale brown when young, becoming blackish brown at maturity; growing on the ground under aspen and birch; August–October; edible—*L. insigne* f. *ochraceum* Smith and Thiers.

4.   Flesh in cap and stalk whitish, slowly staining pinkish gray to purple-gray or grayish black when cut and rubbed → 5.

4.   Flesh in cap and stalk whitish, not staining pinkish gray to purple-gray or grayish black when cut and rubbed → 6.

5.   Cap 2–6" (5–15.5 cm) wide, pale grayish cinnamon-buff, dry, coated with tiny, matted fibrils, sometimes finely cracked and darkening somewhat in age; margin rimmed with flaps of sterile tissue; pores white when young, becoming olive-buff, then pale yellow-brown in age, staining yellow-brown when bruised; stalk white when young, coated with whitish scabers that become coarse and grayish brown to blackish brown in age; flesh in cap and stalk whitish, staining blue in FeSO₄, staining pinkish gray to purple-gray or blackish gray when cut and rubbed, sometimes staining reddish at the base; growing on the ground under birch and aspen; July–September; edible—*L. insolens* var. *insolens* Smith, Thiers and Watling.

5.   Cap 2–4" (5–10 cm) wide, dull white to pale pinkish buff, dry, coated with tiny matted fibrils, staining dull brown when bruised; margin rimmed with flaps of sterile tissue; pores white when young, becoming olive-buff, then pale yellow-brown in age, staining yellow-brown when bruised; stalk white, coated with whitish scabers that become coarse and dark brown in age; flesh in cap and stalk white, staining blue in FeSO₄, slowly staining purple-gray to blackish gray when cut and rubbed, sometimes staining blue near the base; growing on the ground under birch and aspen; July–September; edible—*L. insolens* var. *brunneo-maculatum* Smith, Thiers and Watling.

6.   Cap 1⅛–4" (3–10 cm) wide, white to whitish, occasionally with gray, buff, tan, or pinkish tints, often darkening and developing a greenish tinge; growing on

the ground in and around bogs, cedar swamps, or wet birch woods—*L. holopus* (Rostkovius) Watling (see p. 352).

6.   Cap ¾–2⅜" (2–6 cm) wide, whitish, pale tan, or pale gray; stalk slender, usually about ¼" (6 mm) thick; growing on the ground under hardwoods, especially oak; July–September; edible—*L. albellum* (Peck) Singer.

6.   Cap ¾–3⅛ (2–8 cm) wide, whitish to tan or grayish tan, darkening in age; typically wrinkled and irregularly depressed; cap and stalk flesh white, instantly staining dark blue in FeSO$_4$; growing on the ground under birch; July–September; edible—*L. rotundifoliae* (Singer) Smith, Thiers and Watling.

6.   Cap ¾–3⅛" (2–8 cm) wide, color variable, often whitish when young, sometimes tan or brown, usually darkening in age; stalk whitish; scabers whitish when young, brown to blackish at maturity; flesh white, slowly staining pinkish or reddish when cut and rubbed but not darkening to gray or grayish black; growing on the ground under birch; July–September; edible—*L. oxydabile* (Singer) Singer.

7.   Cap surface typically coarsely wrinkled and cracked or pitted at maturity → 8.

7.   Cap surface not coarsely wrinkled and cracked or pitted at maturity → 9.

8.   Pore surface whitish to grayish when fresh, typically staining greenish then slowly dingy yellow when bruised or in age; cap yellow-brown to brown when young, becoming olive-brown to grayish brown at maturity, often coarsely wrinkled and cracked at maturity; flesh whitish, staining gray, often slowly, when cut and rubbed, not staining red or blue-green; growing on the ground under hardwoods, especially oak; July–September; edible—*L. griseum* (Quélet) Singer.

8.   Pore surface yellow to dingy yellow when fresh, becoming dingy olive-brown in age; cap dark brown to blackish brown when young, becoming pale yellow-brown in age, 1½–3" (4–7.5 cm) wide; stalk often swollen at or below the middle and tapered in either direction, but sometimes nearly equal, roughened by a dense coating of brown to blackish brown scabers; flesh pale yellow, typically staining pinkish gray, burgundy, or darker when cut and rubbed, odor and taste not distinctive; growing on the ground under oak; edible—*L. nigrescens* (Richon and Roze) Singer = *L. crocipodium* (Letellier) Watling.

9.   Cap covered with tiny dark brown to black fibers when young, fading to yellowish brown in age; flesh white, slowly staining gray to grayish black when cut and rubbed, with or without a reddish intermediate phase, slowly staining bluish gray in FeSO$_4$; pore surface whitish to grayish when fresh; growing on the ground under hardwoods, especially birch; July–October; edible—*L. subleucophaeum* Dick and Snell.

9.   Cap covered with tiny, dark brown to black fibers when young, fading to dark yellowish brown in age; flesh white, staining pinkish to reddish when cut and rubbed, especially at the juncture of the cap and stalk, not darkening to gray or black, staining blue in FeSO$_4$, usually with blue-green stains on the lower portion; pore surface whitish to grayish when fresh; growing on the ground under hardwoods, especially yellow birch—*L. snellii* Smith, Thiers, and Watling (see p. 352).

9.   Cap grayish brown to yellow-brown when young, often developing olive tones in age; flesh white, not staining when cut and rubbed or slowly staining pinkish on flesh near the stalk surface, not darkening to gray or black, sometimes with blue-green stains near the base; pore surface whitish to grayish when fresh; growing on the ground under hardwoods, especially birch; July–October; edible—*L. scabrum* (Fries) S. F. Gray.

9.   Cap color variable, often whitish when young, sometimes tan or brown, usually darkening in age, ¾–3⅛" (2–8 cm) wide, stalk whitish; scabers whitish when young, brown to blackish at maturity; flesh white, slowly staining pinkish or reddish when cut and

rubbed but not darkening to gray or grayish black; pore surface whitish to grayish when fresh; growing on the ground under birch; July–September; edible—*L. oxydabile* (Singer) Singer.

Key to Species of *Phylloporus*

1. Cap color variable (dark red, dull red, reddish yellow to reddish brown, or olive-brown), dry, somewhat velvety, sometimes finely cracked; pore surface bright yellow to ochre, not blueing when bruised, distinctly gill-like; basal mycelium yellow— *P. rhodoxanthus* (Schweinitz) Bresadola (see p. 353).
1. Cap reddish brown to chestnut, usually paler on the disc, dry, somewhat velvety, often concentrically cracked at maturity; pore surface gill-like, dull yellow, becoming olivaceous in age, not blueing when bruised; stalk pale brown, punctate, with a white basal mycelium; growing on sandy soil in hardwoods, especially under beech and oak; July–September; edible—*P. leucomycelinus* Singer.
1. Cap cinnamon to dark pinkish brown, dry, velvety to nearly smooth; pores elongated and radially arranged to gill-like, pale olive-buff when young, becoming dark olive buff at maturity, sometimes staining dark blue to bluish green; growing on sandy soil in oak-pine woods; July–September; edible—*P. boletinoides* Smith and Thiers.

Key to Species of *Strobilomyces*

1. Cap surface covered with coarse, woolly or cottony, flattened or erect, gray, purplish gray, or blackish scales; spores covered by a distinct and complete reticulum—*S. floccopus* (Vahl : Fries) Karsten (see p. 354).
1. Cap surface covered with small, erect, often stiff and pointed, gray, purplish gray or blackish scales; spores 9–12 × 9–10 μm, ornamented with irregular and short ridges, sometimes forming a partial reticulum; July–October; edible—*S. confusus* Singer.
1. Cap surface covered with coarse, woolly or cottony, flattened or erect, grayish pink, pinkish tan, or pinkish brown scales; stalk shaggy to scaly, dry; scales colored like the cap, darkest on mature specimens; spores 9–12 × 7–9 μm, covered by a distinct and complete reticulum; growing on the ground in sandy soil under oak; most common in the southern part of the region but extending as far north as Cape Cod, Massachusetts; July–October; occasional to frequent; edible—*S. dryophilus* Cibula and Weber.

Key to Species of *Suillus*

1. Partial veil present, at least on young specimens, at first covering the pore surface, then tearing and adhering to the margin or forming a ring on the stalk → 2.
1. Partial veil absent on all stages → 7.
    2. Cap surface dry, not sticky or slimy when fresh → 3.
    2. Cap surface sticky or slimy when fresh → 4.
3. Cap surface covered with conspicuous pink to red or dark red scales or patches, fading to dull yellow in age—*S. pictus* (Peck) Kuntze (see p. 357).
3. Cap surface covered with dark brown to yellow-brown fibers and scales; pore surface pale yellow to olive-yellow; stalk typically hollow, at least on the lower half—*S. cavipes* (Opatowski) Smith and Thiers (see p. 355).
3. Cap surface covered with tiny fibers or scales, orangish to dull yellow, tan or pale reddish brown; pore surface orange-yellow to yellow, becoming brownish yellow in age;

stalk solid, orangish to dull yellow, lacking glandular dots; partial veil whitish to yellowish; scattered on the ground under pines; July–September; edible—*S. decipiens* (Berkeley and Curtis) Kuntze.

4.   Partial veil at first thick and distinctly baggy, sometimes flaring from the stalk on the lower portion, forming a prominent ring on the stalk; cap color variable, yellow, orange-yellow, yellow-brown, olive-brown, or dark brown; pore surface dull yellow to salmon orange—*S. salmonicolor* (Frost) Halling (see p. 358).

4.   Partial veil conspicuous but not distinctly baggy; underside of veil and ring dull purple to purplish gray; cap dark reddish brown to cinnamon-brown, yellow-brown or ochre—*S. luteus* (Linne : Fries) S. F. Gray (see p. 356).

4.   Partial veil not baggy and not dull purple to purplish gray on the underside; cap surface pale yellow when young, becoming tan, ochre-yellow to yellow-brown in age, covered with sticky gluten that tastes acidic; stalk pale yellow to ochre with dingy glandular dots and smears and a superior, gelatinous, band-like ring at maturity; stalk flesh salmon-ochraceous in the lower portion—*S. intermedius* (Smith and Thiers) Smith and Thiers (see p. 356).

4.   Partial veil not baggy and not dull purple to purplish gray on the underside; cap pinkish buff to vinaceous buff or pale pinkish cinnamon, covered with sticky gluten that tastes mild, not acidic; margin often rimmed with remnants of the partial veil; flesh whitish near the cuticle, yellowish above the tubes, slowly staining vinaceous brownish, odor and taste not distinctive; stalk dull white to yellow, often yellow in both the interior and exterior at the apex, surface coated with vinaceous cinnamon glandular dots both above and below the ring, with a superior, gelatinous, band-like ring at maturity; spores 8–11 × 3–3.5 μm; on the ground under mixed stands of red and white pine; edible—*S. subalutaceus* (Smith and Thiers) Smith and Thiers.

4.   Partial veil not baggy and not dull purple to purplish gray on the underside; cap surface pale yellow to pale ochraceous when young, darkening to dull yellow to vinaceous cinnamon at maturity, covered with sticky gluten when fresh, taste of the gluten mild, not acidic; margin of immature specimens with white, cottony veil remnants; flesh white, slowly becoming yellow at maturity, odor and taste not distinctive, staining pink, then lilac-gray when KOH is added, staining olive-blue when FeSO$_4$ is applied; pore surface whitish at first, becoming yellow at maturity; pores very small, circular; stalk white and lacking glandular dots at first, becoming yellowish near the apex and brownish toward the base, with brownish glandular dots, lacking a ring; spores 6–9 × 2.5–3 μm; on the ground under pines; edible— *S. neoalbidipes* Palm and Stewart.

4.   Not with the above combinations of characters → 5.

5. Cap bright yellow to ochre-yellow, with cinnamon to reddish patches or streaks; margin rimmed with white to yellow or pale brown cottony veil tissue, not forming a ring on the stalk; stalk typically less than ⅜" (1 cm) thick; growing on the ground under white pine—*S. americanus* (Peck) Snell (see p. 354).

5. Fruiting body nearly identical to *S. americanus;* partial veil sometimes forming a slight ring on the stalk; stalk ¼–⅝" (6–16 mm) thick; growing on the ground under mixed conifers; July–October; edible—*S. sibiricus* (Singer) Singer.

5. Not with the above combinations of characters; partial veil forming a prominent ring on the stalk → 6.

6.   Cap surface yellow, yellow-orange or pale orange-brown with pale pinkish cinnamon streaks; stalk base of fresh specimens quickly staining distinctly green when

cut; growing on the ground or among sphagnum mosses under larch—*S. proximus* Smith and Thiers (see p. 357).

6.   Cap color variable, orange-yellow, dull red, red-brown, reddish brown, dark reddish brown, or dark chestnut-brown; stalk base of fresh specimens not staining distinctly green when cut; growing on the ground or among mosses under larch— *S. grevillei* (Klotzsch : Fries) Singer (see p. 355).

7.   Cap surface sticky, white or whitish when young, gradually becoming pale yellow; pore surface whitish when young but soon becoming yellowish; stalk whitish, becoming pale yellow at the apex, conspicuously coated with pinkish tan glandular dots and smears; growing on the ground under white pine; July–October; edible—*S. placidus* (Bonorden) Singer.

7.   Cap surface velvety to felted when young, becoming nearly smooth in age, reddish brown to burgundy-brown or dark brown; pore surface pinkish brown to tan or buff, darkening in age to yellowish brown, not yellow; stalk colored like the cap or paler, lacking glandular dots; growing on the ground, usually under oak; July–October; edibility unknown—*S. castanellus* (Peck) Smith and Thiers = *Boletinus squarrosoides* Snell and Dick.

7.   Cap surface covered with tufts of tiny, matted fibers when young, soon becoming smooth, sticky, and dull ochre-orange; stalk dull orangish brown when young, becoming dull ochre-orange to ochre-yellow and densely covered with brown to dark brown glandular dots and smears; flesh pale yellow; odor fragrant, like almond extract—*S. punctipes* (Peck) Singer (see p. 358).

7.   Not with the above combinations of characters → 8.

8.   Cap surface smooth, sticky to slimy when fresh, dark brown to cinnamon-brown, fading in age; stalk white to pale yellow, lacking conspicuous glandular dots; pore surface of fresh specimens not beaded with a milk-like latex; growing on the ground under pines; July–October; edible—*S. brevipes* (Peck) Kuntze.

8.   Cap surface smooth, sticky to slimy when fresh, pinkish buff to pinkish cinnamon, becoming darker reddish cinnamon in age; stalk pale yellow, with inconspicuous pale yellow glandular dots that do no darken in age, staining brownish in age or when handled; pores pale yellow, duller in age, beaded with a milk-like latex when fresh; growing on the ground under white pine; edible—*S. lactifluus* (Withering : S. F. Gray) Smith and Thiers.

8.   Cap surface sticky or slimy when fresh, color variable, some shade of tan, brown, cinnamon, or orangish cinnamon; stalk whitish when young, becoming yellowish in age, especially at the apex, covered with conspicuous pinkish tan to brownish glandular dots and smears; growing on the ground under pines—*S. granulatus* (Linne : Fries) Kuntze (see p. 355).

8.   Not with the above combinations of characters → 9.

9.   Pore surface dark cinnamon to pinkish brown when young, slowly becoming dingy yellow in age, staining blue to greenish blue when bruised, at least on mature specimens; cap coated with tiny, yellow to orange-yellow fibers and scales when young, becoming nearly smooth in age, sticky beneath the fibers and scales when fresh; flesh pale yellow to yellow, slowly staining greenish blue when cut; stalk colored like the cap, with glandular dots; growing on soil and humus under pine; August–October; edible— *S. tomentosus* (Kauffman) Singer, Snell, and Dick.

9.   Pore surface pale yellow to yellowish orange when young, not staining blue to greenish blue when cut or bruised → 10.

10.   Cap surface with scattered, flattened, brownish to reddish fibers over an apricot-orange to yellow ground color; pore surface pale yellowish orange when fresh,

becoming dull ochre to dingy yellow in age; pores angular, mostly 2 per mm; growing on the ground in hardwoods, especially oak, or in mixed woods— *S. subaureus* (Peck) Snell (see p. 358).

10. Cap surface covered with scattered tufts of reddish, brownish, or grayish fibers on a yellowish ground color; pore surface pale yellow when young, becoming orange-buff in age; pores angular, mostly 1 per mm; growing on the ground under pines; July–September; edible—*S. hirtellus* (Peck) Kuntze.

Austroboletus betula (Schweinitz) Horak

COMMON NAME: Shaggy-stalked Bolete.

CAP: 1⅛–3½" (3–9 cm) wide, convex, becoming broadly convex; surface smooth, sticky or slimy when moist, dark red, reddish or orange, sometimes with a yellow margin, becoming bright yellow to yellow in age; flesh greenish yellow to orange-yellow, not blueing when cut or bruised; odor and taste not distinctive.

PORE SURFACE: yellow to greenish yellow, not staining blue or brown when bruised; pores circular, 1 mm wide.

STALK: 4–8" (10–20 cm) long, ¼–¾" (6–20 mm) thick, nearly equal, solid, yellow to dark red or dull red, coarsely reticulate with raised yellow ribs that may redden in age, often with white mycelium at the base; partial veil and ring absent.

SPORE PRINT: dark olive to olive-brown.

MICROSCOPIC FEATURES: spores 15–19 × 6–10 μm; narrowly elliptic, ornamented with a loose reticulum and scattered minute pits, typically with a distinct apical pore, pale brown.

FRUITING: solitary to scattered on the ground in mixed oak-pine and beech forests; July–September; occasional.

EDIBILITY: edible.

COMMENTS: also known as *Boletellus betula*.

Austroboletus gracilis (Peck) Wolfe

COMMON NAME: Graceful Bolete.

CAP: 1⅛–4" (3–10 cm) wide, convex to broadly convex; surface dry, finely velvety when young, sometimes finely cracked in age, maroon to reddish brown or cinnamon, at times tawny to yellow-brown; flesh white or tinged pink, not staining blue or brown; odor not distinctive; taste mild or slightly tart.

PORE SURFACE: white when young, becoming pinkish to pinkish brown or burgundy-tinged in age, not staining blue or brown when bruised; pores circular, 1–2 per mm.

STALK: long and slender in relation to the cap diameter, 3–7" (7.5–18 cm) long, ¼–⅜" (6–10 mm) thick at the apex, enlarging downward or nearly equal, colored like the cap or paler, whitish at the base, solid, with elevated, anastomosing lines that sometimes form an obscure, narrow reticulation overall or at least on the upper half; partial veil and ring absent.

SPORE PRINT: pinkish brown to reddish brown.

MICROSCOPIC FEATURES: spores 10–17 × 5–8 μm, narrowly ovoid to subelliptic, pitted, pale brown.

FRUITING: solitary, scattered, or in groups on the ground or decaying wood in conifer and hardwoods; June–October; frequent.

EDIBILITY: edible.

COMMENTS: also known as *Tylopilus gracilis* and *Porphyrellus gracilis*. *Gyroporus* species have a hollow stalk at maturity, a white to yellow pore surface, and a pale to bright yellow spore print. *Tylopilus* species have smooth (not pitted) spores.

Boletellus chrysenteroides (Snell) Singer

CAP: 1⅛–2⅜" (3–6 cm) wide, convex, becoming broadly convex; surface dry, not sticky or slimy, blackish brown and velvety when young, becoming chocolate-brown to chestnut-brown and often finely cracked in age; cracks when present white to pale yellow without red tints; flesh pale yellow, staining blue when cut or bruised; odor and taste not distinctive.

PORE SURFACE: pale yellow, becoming yellow or greenish yellow, staining blue when cut or bruised; pores angular, 1 mm wide.

STALK: ¾–1½" (2–4 cm) long, ¼–½" (6–12 mm) thick, equal or enlarging slightly downward, solid, reddish brown to blackish brown, often yellowish at the apex, not reticulate but often decorated with tiny hairs, points, or scales that may form a reticulate pattern; staining blue, then slowly reddish when bruised or handled; partial veil and ring absent.

SPORE PRINT: olive-brown to dark brown.

MICROSCOPIC FEATURES: spores 10–16 × 5–8 μm, narrowly ovate to nearly oblong, longitudinally striate, yellowish to brownish.

FRUITING: solitary or in groups, on decaying wood, often at the bases of trees, sometimes on the ground; June–September; fairly common.

EDIBILITY: edible.

Boletellus russellii (Frost) Gilbert

COMMON NAME: Russell's Bolete.

CAP: 1⅛–5⅛" (3–13 cm) wide, hemispheric to convex, becoming broadly convex; margin strongly incurved; surface velvety when young, dry, becoming finely cracked or forming scale-like patches, yellow-brown to reddish brown, cinnamon-brown, or olive-gray; flesh pale yellow to yellow, often brownish around larval tunnels, not blueing when cut or bruised; odor and taste not distinctive.

PORE SURFACE: yellow to greenish yellow, not bruising blue but sometimes changing to brighter yellow when cut or rubbed; pores angular, 1 mm or more wide.

STALK: 4–8" (10–20 cm) long, ⅜–¾" (1–2 cm) thick, equal or enlarging slightly downward, often curved, frequently sticky at the base when moist, solid; reddish brown to pinkish tan, deeply grooved and ridged for most or all of its length; ridges branched or torn to create a honeycomb or shaggy-bark effect; partial veil and ring absent.

SPORE PRINT: dark olive to olive-brown.

MICROSCOPIC FEATURES: spores 15–20 × 7–11 μm, elliptic, longitudinally striate with deep grooves or wrinkled with a cleft in the wall at the apex, pale brown.

FRUITING: solitary to scattered on the ground and on humus under oak, hemlock, and pine; July–September; uncommon.

EDIBILITY: edible.

Boletus affinis var. maculosus Peck

COMMON NAME: Spotted Bolete.

CAP: 1⅜–4" (3.5–10 cm) wide, convex, becoming broadly convex to nearly flat; surface finely velvety, smooth to finely wrinkled, sometimes pitted and cracked in age, dry, red-brown varying to dark brown, brown, yellow-brown, and orange-brown, mottled with white to pale yellow spots; flesh white, often pale yellow around larval tunnels, not blueing when cut or bruised; odor and taste not distinctive.

PORE SURFACE: white to pinkish white when young, becoming yellow in age, bruising olive-ochre; pores circular, 1–2 per mm.

STALK: 1⅛–4" (3–10 cm) long, ⅜–¾" (1–2 cm) thick, solid, often enlarged at the base, whitish, tinged yellowish, pinkish tan or brown when bruised or in age, not reticulate or slightly so at the apex; partial veil and ring absent.

SPORE PRINT: yellow-brown to amber.

MICROSCOPIC FEATURES: spores 9–16 × 3–5 μm, nearly oblong, smooth, pale yellow.

FRUITING: scattered to caespitose on the ground in mixed forests, particularly beech; June–September; fairly common.

EDIBILITY: edible.

COMMENTS: also known as *Xanthoconium affine. Boletus affinis* var. *affinis* (edible) is nearly identical but lacks the numerous white to pale yellow spots.

Boletus auriporus Peck

CAP: ¾–3⅛" (2–8 cm) wide, convex, becoming broadly convex to nearly plane; surface coated with tiny, matted fibrils, smooth, moist and sticky when fresh, becoming dull when dry, pinkish cinnamon to pinkish brown or vinaceous brown when fresh, often fading in age or when dry; taste of the cuticle usually acidic when moist and sticky; margin whitish when young, brownish at maturity, projecting beyond the pores; flesh white to pale yellow except vinaceous under the cuticle, not blueing when cut or bruised; odor and taste not distinctive.

PORE SURFACE: brilliant golden yellow when young and fresh, becoming dull yellow in age, slowly staining dull brick-red when bruised; pores angular, 1–2 per mm.

STALK: 1½–4½" (4–11.5 cm) long, ¼–⅝" (6–17 mm) thick, solid, slightly enlarged downward or nearly equal, typically narrowed abruptly at the base, slippery to sticky when fresh, pale yellow at the apex, streaked and flushed pale pinkish brown downward, with copious white mycelium at the base; partial veil and ring absent.

SPORE PRINT: olive-brown.

MICROSCOPIC FEATURES: spores 11–16 × 4–6 μm, fusiform-elliptic, smooth, pale brown.

FRUITING: scattered or in groups on the ground under oak; July–October; fairly common in the southern part of the region.

EDIBILITY: edible.

COMMENTS: *Boletus viridiflavus* is a synonym. *Boletus innixus* (see p. 340) often grows in clusters, has a distinctly swollen stalk above a tapered base, a yellow basal mycelium, and smaller spores that measure 8–11 × 3–5 μm.

Boletus bicolor var. bicolor Peck

COMMON NAME: Two-colored Bolete, Red-and-Yellow Bolete.

CAP: 2–6" (5–15.5 cm) wide, convex, becoming flat or irregular; surface dry, finely velvety when young, sometimes finely cracked in age, dark red, red, rose-red, purple-red, or pink when fresh, turning yellow in age; flesh pale yellow, *slowly* staining blue when bruised or cut; odor and taste not distinctive.

PORE SURFACE: yellow when fresh, dingy yellow or olive-tinged in age, staining greenish blue (sometimes weakly), when bruised; pores angular, 1–2 per mm.

STALK: 2–4" (5–10 cm) long, ⅜–1⅛" (1–3 cm) thick, nearly equal or club-shaped, solid, yellow at the apex, red or rosy-red on the lower two-thirds or more, unchanging or slowly staining blue when bruised or cut; partial veil and ring absent.

SPORE PRINT: olive-brown.

MICROSCOPIC FEATURES: spores 8–12 × 3.5–5 μm, oblong to slightly ventricose, smooth, pale brown.

FRUITING: solitary or in groups on the ground under oaks; July–October; fairly common.

EDIBILITY: edible, choice.

COMMENTS: *Boletus sensibilis* (poisonous) (see p. 345) has a dark to pale brick- red cap that fades to dull rose or dingy cinnamon in age, a predominantly yellow stalk lightly flushed with red, pale yellow flesh and a yellow pore surface, both of which stain blue quickly.

Boletus chippewaensis Smith and Thiers

CAP: 2⅜–6" (6–15.5 cm) wide, convex to broadly convex, becoming nearly plane in age; surface smooth, dry, slightly sticky when wet, pale lemon-yellow splashed irregularly with brick to bright rusty red, finally brick-red overall; flesh white, not blueing when bruised or cut, changing to olive-gray with the addition of $FeSO_4$; odor and taste not distinctive.

PORE SURFACE: white when very young, soon turning creamy lemon-yellow, finally olive-brown in age, staining pinkish cinnamon when bruised; pores circular, 2–3 per mm.

STALK: 2–6" (5–15.5 cm) long, 1–1⅜" (2.5–3.5 cm) thick, equal or enlarging downward, sometimes bulbous, pinkish tan to cinnamon-brown beneath a white reticulum on the upper one-third or more, base whitish, solid; partial veil and ring absent.

SPORE PRINT: yellow-brown.

MICROSCOPIC FEATURES: spores 11–16 × 5–7 μm, elliptic to spindle-shaped, smooth, pale brown.

FRUITING: solitary to scattered under mixed conifers and hardwoods; July–October; uncommon to rare.

EDIBILITY: edible, choice.

COMMENTS: *B. edulis* var. *aurantio-ruber* (edible, choice) is very similar, with a cap that is rusty red overall or ochraceous on the margin, is often wrinkled, lacks the pale lemon-yellow ground color, and has a pale yellow pore surface that becomes olive-yellow, then brownish yellow to brown in age. Other varieties of *B. edulis* (see p. 339) (edible, choice) are usually larger, have reddish brown to yellowish tan caps, and also lack the pale lemon-yellow ground color and creamy lemon-yellow pore surface. (The illustrations of *B. edulis* shown in *Taming the Wild Mushroom* and *Edible Wild Mushrooms of North America* are actually *B. chippewaensis.*)

Boletus chrysenteron (Bulliard) Fries

COMMON NAME: Red-cracked Bolete.

CAP: 1⅛–3⅛" (3–8 cm) wide, convex, becoming broadly convex to nearly flat; surface dry, smooth, velvety when young, becoming cracked in age, dark olive to olive-brown with reddish tinges in the cracks and often with a red marginal zone in age; flesh white when young, turning yellow in age, slowly blueing when bruised or cut, staining lemon-yellow in $FeSO_4$, buff in KOH; odor and taste not distinctive.

PORE SURFACE: yellow to bright yellow when young, often olive-yellow or brownish in age, slowly blueing when cut or bruised; pores irregular, 1 mm wide.

STALK: 1⅝–2⅜" (4–6 cm) long, ¼–⅜" (6–10 mm) thick, equal or tapered downward, solid, yellow at the apex, purplish red at the base, staining blue-green when bruised, finely granular to scurfy overall, sometimes with ridges but not reticulate, mycelium at base white; partial veil and ring absent.

SPORE PRINT: olive-brown.

MICROSCOPIC FEATURES: spores 9–13 × 3.5–4.5 μm, oblong to boat-shaped, smooth, pale brown.

FRUITING: scattered to gregarious on the ground in hardwoods and mixed forests, along roadcuts, and on mossy banks; June–September; fairly common.

EDIBILITY: edible.

Boletus edulis Bulliard : Fries

COMMON NAME: King Bolete, Cep, Porcini.

CAP: 1¾–10" (4.5–25.5 cm) wide, convex to nearly flat; surface smooth to slightly wrinkled, dry, sticky when wet, brown to reddish brown, pale cinnamon-brown, rusty red, or yellowish tan; flesh white, not blueing when bruised; odor and taste not distinctive.

PORE SURFACE: white when young, becoming yellow to olive-yellow, then brownish yellow in age, staining yellowish olive to dull orange-cinnamon or pale yellowish brown when bruised; pores small, circular, 2–3 per mm.

STALK: 2–10" (5–25.5 cm) long, ¾–3" (2–7.5 cm) thick, enlarging downward or nearly equal, sometimes bulbous, white or pale brown, with a distinct, whitish reticulum on the upper one-third or more, solid; partial veil and ring absent.

SPORE PRINT: olive-brown.

MICROSCOPIC FEATURES: spores 13–19 × 4–6.5 μm, elliptic, smooth, pale yellowish brown.

FRUITING: solitary, scattered, or in groups on the ground in woods, especially under conifers; June–October; fairly common.

EDIBILITY: edible and choice; one of the most highly prized edible mushrooms.

COMMENTS: *Tylopilus felleus* (see p. 361) has a brown stalk with coarse brown reticulation, a vinaceous spore print, and very bitter-tasting flesh. *Boletus chippewaensis* (see p. 338) has a pale lemon-yellow cap splashed irregularly with brick to bright rusty red and a creamy lemon-yellow pore surface.

Boletus fraternus Peck

CAP: ¾–2" (2–5 cm) wide, convex to broadly convex; surface dry, finely velvety when young, becoming conspicuously cracked and fissured, with yellow flesh showing in the cracks at maturity, red to dark red when young, pinker or dull brick red in age; flesh yellow to whitish, typically blueing when cut; odor and taste not distinctive.

PORE SURFACE: yellow when young, becoming olive-yellow to brownish yellow in age, typically blueing when bruised; pores angular, 1–2 per mm.

STALK: ¾–2" (2–5 cm) long, ¼–½" (6–12 mm) thick, nearly equal down to a tapered base, solid, reddish pruinose over a yellow ground color, with or without white to pale yellow basal mycelium; partial veil and ring absent.

SPORE PRINT: olive-brown.

MICROSCOPIC FEATURES: spores 12–15 × 4.5–7 μm, fusiform, smooth, pale brown.

FRUITING: solitary, scattered, or in groups on the ground in grassy areas, in woods, along roadcuts, and in gardens; July–September; occasional.

EDIBILITY: edible.

Boletus frostii Russell

COMMON NAME: Frost's Bolete, Apple Bolete.

CAP: 2–6" (5–15.5 cm) wide, hemispheric to convex, becoming broadly convex to flat; surface with a whitish bloom when young, quickly becoming smooth and sticky when moist, initially dark blackish red to bright red, fading with age to blood-red with yellowish areas; margin incurved when young, becoming upturned in age; flesh pale to lemon-yellow, rapidly staining blue when cut or bruised; odor and taste not distinctive.

PORE SURFACE: dark red when fresh, paler in age, often beaded with yellow droplets when young and moist, quickly blueing when bruised; pores circular, 2–3 per mm.

STALK: 1⅝–4¾" (4–12 cm) long, ⅜–1" (1–2.5 cm) thick, nearly equal to enlarging downward, solid, deeply and coarsely reticulate, dark red, often yellow or whitish at the base, slowly staining blue when cut or bruised; partial veil and ring absent.

SPORE PRINT: olive-brown.

MICROSCOPIC FEATURES: spores 11–17 × 4–5 μm, elliptic, smooth, pale brown.

FRUITING: scattered or in groups on the ground under hardwoods, especially oak; July–October; occasional to fairly common.

EDIBILITY: edible.

Boletus griseus Frost

CAP: 2–5¾" (5–14 cm) wide, convex, becoming broadly convex to flat or slightly depressed; surface dry, with flattened dark grayish fibers, not sticky or slimy, often scaly in age, pale to dark gray or brownish gray when young, sometimes developing yellowish or ochre tints in age; flesh whitish with dark yellow-brown around larval tunnels, unchanging or staining dingy red or brown when cut or bruised, staining bluish gray in $FeSO_4$; odor and taste not distinctive.

PORE SURFACE: whitish to grayish or dingy gray-brown, not yellow, unchanging or staining brownish or gray when bruised; pores circular, 1–2 per mm.

STALK: 1⅝–5¾" (4–14.5 cm) long, ⅜–1⅜" (1–3.5 cm) thick, equal or tapering downward, often curved near the base, solid, whitish or grayish when young, developing yellow tones from the base upwards as it matures, sometimes with reddish stains; surface covered overall with a coarse pale to yellowish reticulum that becomes brownish to blackish in age; partial veil and ring absent.

SPORE PRINT: olive-brown to yellow-brown.

MICROSCOPIC FEATURES: spores 9–13 × 3–5 μm, oblong, smooth, pale brown.

FRUITING: solitary to scattered on the ground in mixed hardwoods, especially under oaks; June–September; occasional.

EDIBILITY: edible.

Boletus illudens Peck

CAP: 1⅛–3½" (3–9 cm) wide, convex, becoming nearly plane, surface dry, somewhat velvety, pale brownish yellow when young, becoming yellow-brown to pinkish cinnamon, producing a fleeting brilliant green reaction with NH_4OH, instantly staining dark brown with KOH, slowly staining bluish gray with $FeSO_4$; flesh pale yellow, not blueing when cut, slowly staining bluish gray with $FeSO_4$; odor and taste not distinctive.

PORE SURFACE: lemon-yellow, not blueing when bruised; pores angular, 1–2 mm wide or more on mature specimens.

STALK: 1⅛–3½" (3–9 cm) long, ¼–½" (5–13 mm) thick, tapered downward, dry, yellow, typically marked with longitudinal rib-like lines that form a partial reticulum.

SPORE PRINT: olive to olive-brown.

MICROSCOPIC FEATURES: spores 10–14 × 4–5 μm, elliptic to nearly spindle-shaped, smooth, pale brown.

FRUITING: solitary, scattered, or in groups on the ground under oaks in oak or oak-pine woods; July–October; occasional.

EDIBILITY: edible.

Boletus innixus Frost

CAP: 1⅛–3⅛" (3–8 cm) wide, convex, becoming broadly convex; surface dry, becoming somewhat sticky when wet, often finely cracked in age, dull reddish brown, yellow-brown, or grayish brown, often dull purplish red on the margin, fading to dull cinnamon when dry; flesh white to pale yellow, vinaceous under the cuticle, with pale brownish tinges in the stalk; odor musty, similar to a sectioned fruiting body of *Scleroderma citrinum;* taste not distinctive.

PORE SURFACE: brilliant golden yellow when young and fresh, becoming dull yellow in age; pores circular to angular, 1–3 per mm on young specimens, up to 2 mm wide on mature specimens.

STALK: short and stout, 1⅛–2⅜" (3–6 cm) long, ⅜–⅝" (1–1.6 cm) thick at the apex, often enlarged up to 1" (2.5 cm) downward, typically club-shaped and distinctly swollen above a tapered base, yellowish, streaked with dark brown tones, with yellow mycelium sometimes visible at the base; partial veil and ring absent.

SPORE PRINT: olive-brown.

MICROSCOPIC FEATURES: spores 8–11 × 3–5 μm, elliptic, smooth, pale brown.

FRUITING: growing in clusters fused at the stalk bases, or solitary, scattered, or in groups on the ground under hardwoods, especially oak; June–October; occasional.

EDIBILITY: edible.

COMMENTS: often misnamed *B. auriporus;* carefully compare its description (p. 337) to that of *B. innixus.* There is much confusion about these two species and the name *B. caespitosus,* and many authors who have included illustrations of these species have not examined the type collections and determined spore measurements. We are grateful to Ernst Both, who has studied the types and generously shared his findings.

Boletus longicurvipes Snell and Smith

CAP: ⅝–2⅜" (1.6–6 cm) wide, convex, becoming broadly convex; surface smooth to slightly wrinkled or pitted, sticky to slimy when fresh, butterscotch-yellow to yellow-brown, yellow-orange, orange, orange-brown, or reddish orange, often becoming olive-green with an orange center in age, staining cherry-red when a drop of KOH is applied; flesh white, becoming pale yellowish, not blueing when cut or bruised; odor and taste not distinctive.

PORE SURFACE: pale yellow to dingy yellow, olive-yellow or greenish gray, not blueing when cut or bruised; pores circular, 2 per mm.

STALK: 2–3½" (5–9 cm) long, ⅜–¾" (1–2 cm) thick, nearly equal or enlarging downward, often curved above the base, solid, whitish to yellowish or with reddish or pinkish tan tones, scurfy from tiny scales that are yellow when young, reddish to red-brown in age; partial veil and ring absent.

SPORE PRINT: olive-brown.

MICROSCOPIC FEATURES: spores 13–17 × 4–5 μm, narrowly subfusiform to oblong, smooth, pale brown.

FRUITING: solitary, scattered, or in groups on the ground in oak-pine forests; July–September; fairly common.

EDIBILITY: edible.

Boletus miniato-pallescens Smith and Thiers

CAP: 3⅛–7" (8–18 cm) wide, convex, becoming broadly convex to nearly flat in age; surface dry, smooth or finely velvety, often finely cracked in age, not sticky or slimy, red to brick-red, soon fading to reddish orange or orange-yellow; flesh pale yellow, blueing when cut or bruised; odor and taste not distinctive.

PORE SURFACE: yellow, sometimes with a pale orange tinge when young, becoming dingy yellow or tinged reddish orange in age, blueing quickly when bruised; pores circular, 1–2 per mm.

STALK: 2⅜–5½" (6–14 cm) long, ⅜–1¾" (1–4.5 cm) thick, equal or tapering downward, solid, yellow above or overall when young, developing orange to reddish or reddish brown tones or streaks in age, especially over the lower portion, sometimes blueing when bruised, not reticulate or only finely so at the apex; partial veil and ring absent.

SPORE PRINT: olive-brown.

MICROSCOPIC FEATURES: spores 11–17 × 3–5 μm, narrowly fusiform, smooth, pale brown.

FRUITING: solitary, scattered, or in groups on the ground, especially under oaks; July–September; occasional.

EDIBILITY: unknown.

COMMENTS: *B. sensibilis* (poisonous) has a dark to pale brick-red cap that fades to dull rose or cinnamon in age, not reddish orange, and tends to have less red on its stalk.

Boletus ornatipes Peck

COMMON NAME: Ornate-stalked Bolete.

CAP: 1⅝–6¼" (4–16 cm) wide, convex, becoming nearly flat; surface dry, velvety, color variable, pale gray to purplish gray, yellow, olive, yellow-brown, or olive-brown, often darkest at the center with yellow at the margin; flesh yellow, not blueing when cut or bruised; odor not distinctive; taste slightly to very bitter or sometimes mild.

PORE SURFACE: lemon-yellow, staining brighter yellow, orange-yellow, or orange-brown when cut or bruised; pores circular, 1–2 per mm.

STALK: 3⅛–6" (8–15.5 cm) long, ⅜–1" (1–2.5 cm) thick, equal to slightly enlarged downward, solid, bright yellow, often developing brown tones in age, with coarse, raised reticulation often extending to the base, staining darker orange-yellow when bruised; partial veil and ring absent.

SPORE PRINT: olive-brown to dark yellow-brown.

MICROSCOPIC FEATURES: spores 9–13 × 3–4 μm, oblong to slightly ventricose with apex obtuse, smooth, pale brown.

FRUITING: solitary to caespitose on the ground near or under oak, beech, and other hardwoods, and along roadbanks; June–September; fairly common.

EDIBILITY: edible to inedible; some collections may be very bitter.

COMMENTS: *Boletus retipes* has been used as a synonym by some authors, while others reserve this name for smaller specimens with a powdery, yellow cap.

Boletus pallidus Frost

CAP: 1¾–6" (4.5–15.5 cm) wide, convex when young, becoming broadly convex to nearly flat or slightly depressed in age; surface dry, often finely cracked in age, slightly sticky when moist, whitish to buff or pale gray-brown when young, dingy brown with occasional rose tints in age; flesh whitish or pale yellow, sometimes slowly staining blueish or pinkish when cut or bruised; odor not distinctive; taste mild or slightly bitter.

PORE SURFACE: whitish to pale yellow when young, becoming yellow to greenish yellow in age, staining greenish blue, then grayish brown when bruised; pores circular, becoming nearly angular in age, 1–2 per mm.

STALK: 2–4¾" (5–12 cm) long, ⅜–1" (1–2.5 cm) thick, nearly equal or enlarging downward, solid, whitish when young, sometimes yellow at the apex, with occasional reddish flushes near the base in age, smooth to slightly reticulate near the apex, often with white mycelium at the base, sometimes blueing slightly when cut or bruised; partial veil and ring absent.

SPORE PRINT: olive or olive-brown.

MICROSCOPIC FEATURES: spores 9–15 × 3–5 μm, narrowly oval to subfusoid, smooth, pale brown.

FRUITING: solitary, in groups, or caespitose on the ground under mixed hardwoods, especially oaks; July–September; occasional to fairly common.

EDIBILITY: edible.

Boletus parasiticus Bulliard : Fries

COMMON NAME: Parasitic Bolete.

CAP: ¾–3⅛" (2–8 cm) wide, convex, becoming broadly convex; surface dry, nearly smooth, evenly olive to tawny-olive; margin incurved when young; flesh pale lemon-yellow, not blueing when bruised, instantly orange-ochre in KOH; odor and taste not distinctive.

PORE SURFACE: yellowish to ochre or olive-brown, not staining blue or brown when bruised, but sometimes staining ochraceous to reddish; pores angular, 1–2 mm wide.

STALK: 1⅛–2⅜" (3–6 cm) long, ¼–½" (6–13 mm) thick, nearly equal, yellowish or colored like the cap, usually curved, solid; partial veil and ring absent.

SPORE PRINT: olive-brown.

MICROSCOPIC FEATURES: spores 12–18.5 × 3.5–5 μm, elliptic, smooth, pale brown.

FRUITING: solitary or in small groups on fruiting bodies of the earthball, *Scleroderma citrinum;* July–September; occasional to fairly common.

EDIBILITY: edible.

COMMENTS: this is an easy bolete to identify because it fruits only on the Common Earth-ball. The bolete is edible, but its host is poisonous.

Boletus projectellus Murrill

CAP: 1½–8" (4–20 cm) wide, convex, becoming flat; surface finely velvety when young, often finely cracked in age, dry, not sticky or slimy, pale to dark cinnamon-brown to dull reddish or dark reddish brown, occasionally with gray or olive shades, especially when young; margin extending beyond the pore surface; flesh whitish, often with a rosy tinge, not blueing when cut or bruised but slowly changing to yellow-brown; odor not distinctive; taste acidic.

PORE SURFACE: pale yellow to olive-yellow when fresh, becoming brownish olive in age, not blueing but staining lemon-yellow when cut or bruised; pores circular, 0.5–2 mm wide.

STALK: 2¾–4¾" (7–12 cm) long, ⅜–¾" (1–2 cm) thick, equal or enlarging downward, solid, colored like the cap or somewhat paler, with prominent reticulation overall or at least on the upper two-thirds, typically sticky or slimy at the base in wet weather; partial veil and ring absent.

SPORE PRINT: olive-brown.

MICROSCOPIC FEATURES: spores 18–33 × 7.5–12 μm, oval to ventricose, smooth, pale brown.

FRUITING: solitary to scattered on the ground under pine; July–September; occasional.

EDIBILITY: edible.

COMMENTS: also known as *Boletellus projectellus.* This species has the largest spores of all boletes in North America. The sterile projecting margin explains its species name.

Boletus pseudosensibilis Smith and Thiers

CAP: 2⅜–5½" (6–14 cm) wide, convex, becoming flat in age; surface dry, smooth, not velvety and not sticky or slimy, often becoming cracked in age, dull rusty brown to dull reddish brown, fading to dull cinnamon or dingy yellow-brown; cap surface staining blue, then dull purplish with NH₄OH; cracks, if present, showing yellow flesh; flesh bright yellow, staining quickly blue when cut or bruised; odor and taste not distinctive.

PORE SURFACE: yellow to ochre-yellow when young, olive-ochre in age, rapidly staining blue, then changing to brown when bruised; pores circular to angular, 1–3 per mm.

STALK: 3⅛–6⅜" (8–16 cm) long, ⅝–1⅛" (1.5–3 cm) thick, equal or flared slightly at the apex, solid, typically yellow at the apex and flushed rusty red, pinkish, or purplish below, not reticulate; partial veil and ring absent.

SPORE PRINT: olive-brown.

MICROSCOPIC FEATURES: spores 9–12 × 3–4 μm; subfusiform to oblong, smooth, pale brown.

FRUITING: gregarious on the ground in hardwoods, especially under oak; July–September; occasional.

EDIBILITY: edible, but not recommended because of possible confusion with similar-appearing poisonous species.

Boletus rubellus Krombholz

CAP: ¾–3⅛" (2–8 cm) wide, convex, becoming nearly flat in age; surface velvety, dry, often becoming cracked at maturity, dark red when young, fading to brick-red in age; flesh bright yellow, slowly staining bluish green on exposure, staining dull orange with KOH, staining olive-green with FeSO₄; odor and taste not distinctive.

PORE SURFACE: yellow, staining bluish green when bruised, typically depressed on mature specimens; pores angular, thick-walled, 1–2 per mm.

STALK: 1⅛–3⅛" (3–8 cm) long, ¼–½" (5–13 mm) thick, tapered downward or nearly equal, often with a narrowed base, yellow at the apex, punctate with dull reddish dots and points from the base to the apex, darkening to brown where handled; stalk flesh yellow, slowly staining bluish green, reddish orange in the base.

SPORE PRINT: olive-brown.

MICROSCOPIC FEATURES: spores 10–13 × 4–5 μm, elliptic, smooth, pale brown.

FRUITING: solitary, scattered, or in groups on the ground in hardwoods and mixed woods, especially under oak and beech; July–October; occasional to fairly common.

EDIBILITY: unknown.

Boletus rubroflammeus Smith and Thiers

CAP: 2⅜–4¾" (6–12 cm) wide, convex, becoming broadly convex; surface dry, with finely matted grayish hairs when young, finely cracked at the center in age, dark red to deep wine-red; margin projecting slighty beyond the pores; flesh yellow, quickly blueing when bruised or cut; odor and taste not distinctive.

PORE SURFACE: dark red when young, becoming dingier in age, blueing when bruised; pores small, circular, 1–2 per mm.

STALK: 2⅜–3⅛" (6–8 cm) long, ⅜–1⅛" (1–3 cm) thick, equal or club-shaped, predominantly dark red like the cap, often yellow at the apex, quickly blueing when bruised, with dark red reticulation over at least the upper portion, solid; partial veil and ring absent.

SPORE PRINT: olive-brown.

MICROSCOPIC FEATURES: spores 10–14 × 4–5 μm, suboblong to nearly ventricose, smooth, pale brown.

FRUITING: gregarious to scattered on the ground under hardwoods; July–October; frequency uncertain due to confusion between it and *B. flammans* (see comments below).

EDIBILITY: unknown.

COMMENTS: *B. flammans* (edibility unknown) is nearly identical but has more variable cap color, ranging from dark red to brick-red or brownish red, a yellow stalk base, and grows under conifers.

Boletus sensibilis Peck

CAP: 2–6¼" (5–16 cm) wide, convex, becoming broadly convex to nearly flat; surface finely velvety when young, dry, not sticky, dark to pale brick-red, fading to dull rose or sometimes dingy cinnamon in age, quickly staining blue when bruised; flesh pale yellow, blueing instantly when cut or bruised, instantly yellow with KOH, turning yellow-brown with FeSO₄; odor and taste not distinctive.

PORE SURFACE: yellow when young, becoming duller or browner in age, blueing instantly when bruised; pores circular, 1–2 per mm.

STALK: 3⅛–4¾" (8–12 cm) long, ⅜–1⅜" (1–3.5 cm) thick, equal or enlarging slightly downward, solid, mostly yellow but often tinged pink or red near the base, occasionally finely reticulate only at the very top, quickly staining blue when bruised; flesh bright yellow, quickly staining blue when exposed; partial veil and ring absent.

SPORE PRINT: olive-brown.

MICROSCOPIC FEATURES: spores 10–13 × 3.5–4.5 μm, suboblong to slightly ventricose, smooth, pale brown.

FRUITING: scattered or in groups on the ground in woods, usually hardwoods; July–September; occasional to fairly common.

EDIBILITY: poisonous.

COMMENTS: can cause severe gastrointestinal upset. Compare with *Boletus miniato-olivaceus* (poisonous), which has a bright brick-red to bright rose-red cap that becomes olivaceous in age. Also compare with *B. miniato-pallescens* (edibility unknown), which has a red to brick-red cap that fades to reddish orange or orange-yellow (see p. 341).

Boletus separans Peck

CAP: 2⅜–6" (6–15.5 cm) wide, convex, becoming broadly convex; surface dry, brownish lilac to purplish brown, or pale pinkish brown to dull pink, becoming brown to yellow-brown in age, staining green to blue-green in KOH; flesh white, not blueing when cut or bruised; odor and taste not distinctive.

PORE SURFACE: white when young, becoming yellowish or yellow-olive in age, not blueing when bruised; pores circular, 1–2 per mm.

STALK: 2⅜–6" (6–15.5 cm) long, ⅜–1" (1–2.5 cm) thick, equal or enlarging slightly downward, solid, colored like the cap or paler when young, usually with a pinkish, lilac, or wine-colored tinge when fresh, becoming pale pinkish brown in age, reticulate, at least over the upper half, not blueing when cut or bruised, staining green to blue-green in KOH; partial veil and ring absent.

SPORE PRINT: brownish ochraceous.

MICROSCOPIC FEATURES: spores 12.5–16 × 3.5–4.5 μm, narrowly subfusiform, smooth, pale brown.

FRUITING: solitary, scattered, or in groups on the ground under hardwoods, especially oak, occasionally under pine or mixed conifers; June–October; occasional to frequent.

EDIBILITY: edible, choice.

COMMENTS: *Boletus separans* and *B. pseudoseparans* have great variation both in cap and stalk color and in the extent and color of their reticulation. They are also nearly identical in their response to various reagents as well as microscopic features. It is quite likely that they are variations of the same species.

Boletus spadiceus Fries

CAP: 2–4⅜" (5–11 cm) wide, convex, becoming broadly convex, sometimes nearly flat; surface dry and somewhat velvety, sometimes cracked in age, dark olive to olive-yellow with reddish tints, sometimes reddish brown in age; flesh pale yellow with a reddish line beneath the cuticle, initially bright yellow, then pinkish around larval tunnels, unchanging or blueing slightly when bruised; surface of cap giving a fleeting blue or blue-green reaction, then changing to reddish brown with NH_4OH; odor mild to slightly pungent; taste not distinctive.

PORE SURFACE: yellow to olive-yellow, often but not always staining blue or blue-green when bruised; pores angular, 1–2 mm wide.

STALK: 1½–4" (4–10 cm) long, ⅜–1" (1–2.5 cm) thick, nearly equal but often narrowed downward, solid, mostly yellow or paler, sometimes with brownish stains but never red, typically whitish and narrowed at the base, often with a yellow basal mycelium, sometimes with raised longitudinal lines forming a partial reticulum at the apex or nearly overall; partial veil and ring absent.

SPORE PRINT: olive to olive-brown.

MICROSCOPIC FEATURES: spores 10–14 × 4.5–5 μm, oblong to ventricose, smooth, pale brown.

FRUITING: in groups on the ground in mixed woods and under conifers, along roadbanks and trails; July–September; infrequent to occasional.

EDIBILITY: edible.

Boletus speciosus var. brunneus Peck

CAP: 1½–5½" (4–14 cm) wide, convex, becoming broadly convex to nearly plane in age; surface dry, smooth, lacking scales or conspicuous fibers, reddish brown or yellow-brown to olive-brown; flesh pale yellow, quickly blueing when exposed, staining orange with the addition of KOH and grayish with $FeSO_4$; odor and taste not distinctive.

PORE SURFACE: bright yellow to yellow when fresh, staining blue when bruised; pores circular to angular, 2–3 per mm.

STALK: 1½–4¾" (4–12 cm) long, ⅜–1½" (1–4 cm) thick, nearly equal or enlarged downward, solid, yellow on the upper portion, pinkish red to purplish red on the lower portion, or at least tinged reddish, especially near the base, reticulate overall or at least over the upper half, blueing when bruised; partial veil and ring absent.

SPORE PRINT: olive-brown.

MICROSCOPIC FEATURES: spores 10–15 × 3–4 μm, narrowly oblong to subfusoid, smooth, ochraceous.

FRUITING: solitary, scattered or in groups on the ground under beech, maples or conifers, especially hemlock; June–September; occasional.

EDIBILITY: edible and very good.

COMMENTS: this bolete has often been mistakenly labeled as *B. pseudopeckii*. *Boletus pseudopeckii* has pale yellow flesh that slowly stains pale blue when exposed. Its stalk is initially yellow overall, often with a red zone at the apex, soon developing reddish tones at the base and typically becoming yellow again. It has a reticulum that is so fine that a hand lens is required to observe it on fresh, mature specimens.

Boletus subglabripes Peck

CAP: 1¾–4" (4.5–10 cm) wide, convex, becoming broadly convex to nearly flat, sometimes broadly umbonate; surface smooth to slightly wrinkled, chestnut, ochre, cinnamon, or reddish brown; flesh pale yellow to whitish, rarely blueing slightly when cut or bruised; odor not distinctive; taste mild to slightly acidic.

PORE SURFACE: yellow when fresh, duller or slightly greenish in age, rarely staining slightly blue when cut or bruised; pores circular, 2 per mm.

STALK: 2–4" (5–10 cm) long, ⅜–¾" (1–2 cm) thick, nearly equal, solid, yellow with occasional reddish or reddish brown tinges on the lower portion, minutely scurfy with a thin coating of tiny yellow scales, dots, or points, not reticulate; partial veil and ring absent.

SPORE PRINT: olive-brown.

MICROSCOPIC FEATURES: spores 11–14 × 3–5 μm, narrowly fusoid, smooth, pale brown.

FRUITING: scattered or in groups on the ground under mixed hardwoods, occasionally under conifers; July–October; fairly common.

EDIBILITY: edible.

Boletus subvelutipes Peck

COMMON NAME: Red-mouth Bolete.

CAP: 2⅜–5⅛" (6–13 cm) wide, convex, becoming broadly convex to nearly flat; surface dry, finely velvety when young, occasionally finely cracked with age, color variable, cinnamon-brown to yellow-brown, reddish brown, or reddish orange to orange-yellow, quickly staining blue to blue-black when bruised; flesh bright yellow, quickly staining dark blue when cut or bruised; odor not distinctive; taste mild to slightly acidic.

PORE SURFACE: red or red-orange to orange when fresh, duller in age, quickly staining dark blue to blackish when cut or bruised; pores circular, 2 per mm.

STALK: 1⅛–4" (3–10 cm) long, ⅜–¾" (1–2 cm) thick, nearly equal, solid, punctate with red dots and points, not reticulate, quickly staining dark blue to blackish when bruised; with short, stiff, dark red hairs at the base on mature specimens (immature specimens often have yellow hairs that become dark red in age); partial veil and ring absent.

SPORE PRINT: dark olive-brown.

MICROSCOPIC FEATURES: spores 13–18 × 5–6.5 μm, fusoid-subventricose, smooth, pale brown.

FRUITING: solitary to scattered on the ground under hardwoods, especially oak; June–September; fairly common.

EDIBILITY: poisonous.

COMMENTS: this mushroom causes mild to severe gastrointestinal upset when eaten. Peck's original description of this species states that the flesh is whitish. Although Smith and Thiers report this description in *The Boletes of Michigan*, they then describe Michigan specimens as having bright yellow flesh. Several field guides state that the flesh of *B. subvelutipes* is yellow. In his recent compendium, *The Boletes of North America*, Ernst Both notes Peck's description of the flesh of *B. subvelutipes* as whitish. According to him, a possible explanation for this confusion is that the flesh, which is initially yellow and stains blue instantly, slowly fades to white.

Boletus vermiculosoides Smith and Thiers

CAP: 1½–4¾" (4–12 cm) wide, convex, becoming broadly convex; surface dry, dull, yellow when young, becoming brown as it matures, staining very dark brown with KOH; flesh yellow, fading to pale yellow in age, blueing when cut; blued surface staining yellow with KOH; odor and taste not distinctive.

PORE SURFACE: dark brown when young, paler in age, staining blackish blue when bruised, very slowly fading to dull brownish orange; pores very small, circular to angular, 2–3 per mm.

STALK: 1½–4" (4–10 cm) long, ⅜–¾" (1–2 cm) thick, nearly equal, solid, brownish pruinose over a dull whitish or pale yellow ground color with olive tints, becoming dark brown when handled.

SPORE PRINT: olive-brown.

MICROSCOPIC FEATURES: spores 9–12 × 3–4 μm, elliptic, smooth, pale brown.

FRUITING: solitary, scattered, or in groups on the ground under oak; July–October; occasional.

EDIBILITY: unknown.

Boletus vermiculosus Peck

CAP: 1½–4¾" (4–12 cm) wide, convex, becoming broadly convex to nearly flat; surface dry and velvety when young, occasionally finely cracked in age, dark brown or gray-brown, yellow-brown or reddish brown when young, becoming dull cinnamon-brown in age, staining vinaceous with NH_4OH or KOH; flesh whitish to pale yellow, blueing quickly when cut; odor and taste not distinctive.

PORE SURFACE: typically dark brown when young, becoming reddish brown or brownish orange in age, quickly staining blue or blue-black when cut or bruised; pores circular, 2–3 per mm.

STALK: 1½–3½" (4–9 cm) long, ⅜–¾" (1–2 cm) thick, equal or enlarging slightly downward, solid, brownish pruinose over a dull yellow ground color, staining blue-black when bruised; partial veil and ring absent.

SPORE PRINT: olive-brown.

MICROSCOPIC FEATURES: spores 11–15 × 4–6 μm, suboblong to nearly fusiform, smooth, pale brown.

FRUITING: scattered or in groups on the ground, usually under hardwoods; July–September; infrequent.

EDIBILITY: unknown.

COMMENTS: *Boletus vermiculosoides* (edibility unknown) is very similar but has a yellow cap when young that becomes brown at maturity and stains dark brown with KOH, a whitish to yellow or olive-tinged stalk, yellow flesh that bruises blue and stains yellow with KOH, and smaller spores, 9–12 × 3–4 μm.

Chalciporus piperatus (Bulliard : Fries) Bataille

COMMON NAME: Peppery Bolete.

CAP: ⅝–3½" (1.6–9 cm) wide, convex, becoming nearly flat; surface smooth or with a few fibers, dry or sticky, sometimes finely cracked, buff to yellow-brown, orange-brown, or reddish brown; flesh pale yellow or tinged reddish, becoming dingy purplish brown in age, not blueing when cut or bruised, staining violet-gray with NH_4OH; odor not distinctive; taste distinctly hot and peppery.

PORE SURFACE: dull cinnamon, reddish cinnamon, or cinnamon-brown, becoming darker reddish brown in age, not blueing when cut or bruised but sometimes staining brown; pores angular, sometimes appearing to radiate from the stalk, 0.5–2 mm wide.

STALK: 1½–3¾" (4–9.5 cm) long, ¼–½" (6–12 mm) thick, equal or tapering downward, solid, colored like the cap or paler, base with bright yellow mycelium; flesh lemon-yellow, not blueing; partial veil and ring absent.

SPORE PRINT: brown to cinnamon-brown.

MICROSCOPIC FEATURES: spores 9–12 × 4–5 μm, narrowly fusiform, smooth, pale brown.

FRUITING: solitary, scattered, or in groups on the ground under conifers or hardwoods; July–September; fairly common.

EDIBILITY: unknown.

COMMENTS: also known as *Boletus piperatus*.

Chalciporus rubinellus (Peck) Singer

COMMON NAME: Purple-red Bolete.

CAP: ¾–1¾" (2–4.5 cm) wide, broadly conic when young, becoming convex in age; surface dry, sticky when moist, slightly velvety when young, becoming finely cracked in age, red or reddish when young, yellower in age; flesh bright yellow, not blueing when cut or bruised; odor and taste not distinctive.

PORE SURFACE: bright rose-red when young, becoming dull rose-red to rose-pink in age, not blueing when cut or bruised; pores angular, 1–2 per mm.

STALK: ¾–1⅜" (2–3.5 cm) long, ¼–½" (6–12 mm) thick, nearly equal, solid, red or reddish initially, sometimes mixed with yellow, not reticulate, lacking yellow mycelium at the base; partial veil and ring absent.

SPORE PRINT: brown.

MICROSCOPIC FEATURES: spores 12–15 × 3–5 μm, subfusoid, smooth, pale brown, strongly dextrinoid.

FRUITING: solitary, scattered, or in groups on the ground in mixed woods and under conifers; July–September; occasional.

EDIBILITY: unknown.

COMMENTS: also known as *Boletus rubinellus.*

Fuscoboletinus spectabilis (Peck) Pomerleau and Smith

CAP: 1½–4" (4–10 cm) wide, convex to nearly plane, with or without an umbo; surface covered with coarse, pinkish gray to reddish brown scales or patches, sticky to slimy beneath the scales, pinkish red to orange-red, darkening in age; margin entire, thin, typically supporting gelatinous remnants of the partial veil; flesh yellow, slowly staining pinkish, then brown when exposed; odor disagreeable or somewhat pungent; taste astringent to acidic.

PORE SURFACE: yellow, becoming dull yellowish brown in age, usually staining pinkish when injured; pores angular, elongated, and somewhat radially arranged, 1 mm or more wide.

STALK: 1½–4" (4–10 cm) long, ⅜–¾" (1–2 cm) thick, enlarging downward or nearly equal, solid, yellow, and smooth above the ring, white immediately under, sheathed below by sticky pinkish red to pinkish gray fibers above a yellowish ground color; partial veil gelatinous, pale red to yellowish red, forming a gelatinous, reddish to reddish brown superior ring.

SPORE PRINT: purplish brown.

MICROSCOPIC FEATURES: spores 9–15 × 4–6.5 μm, elliptic, smooth, pale yellowish brown.

FRUITING: solitary, scattered, or in groups among sphagnum mosses under larch trees, especially in bogs; August–October; infrequent.

EDIBILITY: edible.

Gyrodon merulioides (Schweinitz) Singer (see photo, p. 10)

COMMON NAME: Ash-tree Bolete.

CAP: 2–4¾" (5–12 cm) wide, convex with an incurved margin when young, soon becoming flat to concave or wavy; surface dry to slightly sticky, smooth or with tiny fibers, yellowish brown to reddish brown, bruising dull yellow-brown; flesh yellow, unchanging or sometimes slowly bruising blue-green when cut; odor and taste not distinctive.

PORE SURFACE: pale yellow to dull gold or olive, decurrent, usually blueing slowly when bruised, gradually discoloring reddish brown; pores elongated and radially arranged, sometimes weakly gill-like, 1 mm or more wide.

STALK: ¾–1⅝" (2–4 cm) long, ¼–1" (6–25 mm) thick, nearly equal, often curved, off-center, lacking glandular dots, not reticulate, solid, apex colored like the pore surface, lower portion colored like the cap surface, bruising red-brown; partial veil and ring absent.

SPORE PRINT: olive-brown.

MICROSCOPIC FEATURES: spores 7–10 × 6–7.5 μm, oval to nearly round, smooth, pale yellow.

FRUITING: scattered or in groups on the ground near or under ash trees; July–October; fairly common.

EDIBILITY: edible.

COMMENTS: also known as *Boletinellus merulioides.*

Gyroporus castaneus (Fries) Quélet

COMMON NAME: Chestnut Bolete.

CAP: 1⅛–4" (3–10 cm) wide, rounded to broadly convex, becoming flat or slightly depressed; surface finely velvety to nearly smooth, dry, not sticky or slimy, chestnut-brown to yellow-brown or orange-brown; margin often split and flaring in age; flesh brittle, white, not staining blue when cut or bruised, staining brownish in KOH and FeSO$_4$; odor and taste not distinctive.

PORE SURFACE: whitish to buff or yellowish, never pinkish or flesh colored, not blueing when cut or bruised; pores circular, 1–3 per mm.

STALK: 1⅛–3½" (3–9 cm) long, ¼–⅝" (6–16 mm) thick, equal or often swollen in the middle or below, often constricted at the apex and base, brittle, stuffed with a soft pith, developing several cavities or becoming hollow in age, surface uneven, not reticulate, colored like the cap or slightly paler toward the apex; partial veil and ring absent.

SPORE PRINT: pale yellow to buff.

MICROSCOPIC FEATURES: spores 8–13 × 5–6 μm, elliptic to ovoid, smooth, hyaline.

FRUITING: solitary, scattered, or in groups on the ground under mixed conifers and hardwoods; late June–October; fairly common.

EDIBILITY: edible.

Gyroporus cyanescens var. violaceotinctus Watling

CAP: 1½–4¾" (4–12 cm) wide, convex to broadly convex, occasionally nearly flat; surface dry, coarsely tomentose to floccose-scaly, buff or straw-colored to pale olive, or tan to yellowish, often with darker streaks, instantly blueing when bruised; flesh brittle, whitish to pale yellow, instantly staining dark lilaceous to indigo and finally deep blue when cut or bruised; odor and taste not distinctive.

PORE SURFACE: white to yellowish or pale tan, instantly staining dark lilaceous to indigo when bruised; pores circular, 1–2 per mm.

STALK: 1½–4" (4–10 cm) long, ⅜–1" (1–2.5 cm) thick, equal or swollen in the middle or below, brittle, stuffed with a soft pith, becoming hollow or developing several cavities in age; surface coarsely tomentose to fibrillose-scaly when young, often smoother in age, not reticulate, colored like the cap or paler, instantly blueing when cut or bruised; partial veil and ring absent.

SPORE PRINT: pale yellow.

MICROSCOPIC FEATURES: spores 8–10 × 5–6 μm, elliptic, smooth, hyaline.

FRUITING: scattered or in groups among mosses or on sandy soil under birch and maple or in mixed woods; July–September; occasional.

EDIBILITY: edible.

COMMENTS: *Gyroporus cyanescens* (edible) is nearly identical, but the flesh stains greenish blue, then blue when cut. It grows in sandy soil in hardwoods, mixed woods, or along roadcuts.

Leccinum atrostipitatum Smith, Thiers, and Watling

CAP: 2⅜–6¼" (6–16 cm) wide, convex, becoming broadly convex; surface dry to slightly sticky, with tiny fibers that break up to form downy patches or small scales in age, dull orange to tan or brownish; margin rimmed with thin flaps of tissue, at least when young; flesh white, staining pinkish, then purple-gray to blackish when cut and rubbed, especially at the juncture of the cap and stalk; staining bluish gray in $FeSO_4$ and bluish in NH_4OH; odor and taste not distinctive.

PORE SURFACE: whitish to buff or pale gray when young, becoming dingier in age, not blueing when bruised but sometimes staining olive or brownish; pores circular, 2–3 per mm.

STALK: 3⅛–6" (8–15.5 cm) long, ⅜–1" (1–2.5 cm) thick, nearly equal or enlarging downward, solid, surface covered with a dense layer of scabers that are coal-black when young and at maturity, whitish to dingy tan beneath the scabers, not reticulate, with blue or blue-green stains occasionally on the lower portion; flesh white, staining pinkish, then purple-gray to black when cut and rubbed; partial veil and ring absent.

SPORE PRINT: brown.

MICROSCOPIC FEATURES: spores 13–17 × 4–5 μm, subfusiform, smooth, pale brown.

FRUITING: solitary to scattered on the ground in hardwood forests, usually under birch; July–September; occasional.

EDIBILITY: edible.

COMMENTS: *Leccinum testaceoscabrum* (edible) is nearly identical and is distinguished by having a brighter orange and somewhat less scaly cap surface, a paler pore surface when young, and flesh that stains bright green with $FeSO_4$.

Leccinum aurantiacum (Bulliard) S. F. Gray

COMMON NAME: Red-capped Scaber Stalk, Orange-capped Bolete.

CAP: 2–8" (5–20.5 cm) wide, convex to broadly convex, becoming flat in age; surface dry, with tiny, matted fibers when young that often disappear in age, sometimes finely cracked in age, orange to brick-red, rusty red, or reddish brown; margin rimmed with thin flaps of tissue when young; flesh white, slowly staining pinkish to wine-red, then gradually darkening to purple-gray or blackish when cut and rubbed, especially at the juncture of the cap and stalk, staining very pale blue in $FeSO_4$; odor and taste not distinctive.

PORE SURFACE: whitish when young, becoming dingier and grayish to olive-buff to pale olive-brown in age, not blueing when cut or bruised but sometimes staining olive, brownish, or burgundy-brown; pores circular, 2–3 per mm.

STALK: 4–6¼" (10–16 cm) long, ¾–1⅛" (2–3 cm) thick, equal or enlarging downward, solid, whitish, with whitish to buff scabers that darken to orange-brown to reddish brown and finally blackish brown, at least over the lower portion, occasionally with blue-green or yellowish stains on the lower portion, not reticulate; flesh white, slowly staining pinkish to wine-red, then gradually darkening to purple-gray or black; partial veil and ring absent.

SPORE PRINT: brown.

MICROSCOPIC FEATURES: spores 13–18 × 3.5–5 μm, subfusiform, smooth, pale brown.

FRUITING: scattered or in groups on the ground under aspen or pine; July–September; fairly common.

EDIBILITY: edible.

Leccinum holopus (Rostkovius) Watling

COMMON NAME: White Birch Bolete, White Bog Bolete.

CAP: 1⅛–4" (3–10 cm) wide, acutely convex, becoming broadly convex to nearly flat; surface smooth or nearly so, often sticky when moist or in age, predominantly white or whitish when young, occasionally with gray, buff, tan, or pinkish tints, often darkening with age and developing a greenish tinge; flesh white, staining pinkish when cut or rubbed or scarcely staining at all, occasionally brownish; odor and taste not distinctive.

PORE SURFACE: whitish to slightly grayish or pale dingy brown, not blueing when cut or bruised, sometimes staining yellowish or brownish; pores circular, 2–3 per mm.

STALK: 3⅛–5½" (8–14 cm) long, ⅜–¾" (1–2 cm) thick, equal or enlarging slightly downward, solid, whitish beneath scabers that are whitish when young and darken to tan or darker in age, not reticulate, occasionally with greenish stains on the lower portion; partial veil and ring absent.

SPORE PRINT: brown.

MICROSCOPIC FEATURES: spores 14–20 × 5–6.5 μm, subfusoid, smooth, pale brown.

FRUITING: solitary to scattered on the ground in and around bogs, cedar swamps, or wet birch woods; August–October; fairly common.

EDIBILITY: edible.

Leccinum rugosiceps (Peck) Singer

COMMON NAME: Wrinkled Leccinum.

CAP: 2–5⅞" (5–15 cm) wide, convex; surface dry, distinctly wrinkled and pitted at maturity, often finely cracked in age, with pale yellow flesh showing through cracks, not sticky or shiny, orange-yellow becoming yellow-brown in age, staining cherry-red in KOH; flesh white or pale yellow, slowly staining reddish or burgundy when cut or bruised, with stains most visible at the juncture of the cap and stalk; odor and taste not distinctive.

PORE SURFACE: dull yellow to dingy yellowish when young, sometimes dingy olive-brown in age, not blueing when cut or bruised but occasionally having natural blue-green stains; pores circular, less than 1 mm wide.

STALK: 1⅛–4" (3–10 cm) long, ⅜–1⅛" (1–3 cm) thick, nearly equal or tapered at apex or base, solid, pale yellow to brownish beneath pale brown scabers that darken in age, not reticulate, not staining blue when cut or bruised; partial veil and ring absent.

SPORE PRINT: olive-brown.

MICROSCOPIC FEATURES: spores 15–21 × 5–6 μm, fusiform, smooth, pale brown.

FRUITING: solitary or in groups on the ground under oaks in forests and on lawns; July–September; occasional to frequent.

EDIBILITY: edible.

Leccinum snellii Smith, Thiers, and Watling

CAP: 1⅛–3½" (3–9 cm) wide, rounded to convex, becoming broadly convex to nearly flat; surface dry, covered with tiny dark brown to black fibers when young, often fading to dark yellowish brown in age as the fibers erode, staining yellow in KOH; margin not rimmed with flaps of tissue, even when young; flesh white, staining pinkish to reddish, sometimes slowly, when cut and rubbed, especially at the juncture of the cap and stalk, not darkening to purple-gray or black, staining blue in $FeSO_4$; odor and taste not distinctive.

PORE SURFACE: whitish when young, becoming grayish to dingy grayish brown in age, not blueing when cut or bruised but sometimes staining yellowish or brownish; pores circular, 2–3 per mm.

STALK: 1⅝–4⅜" (4–11 cm) long, ⅜–¾" (1–2 cm) thick, nearly equal or enlarging slightly downward, solid, whitish beneath gray to black scabers, not reticulate, often with blue-green stains on the lower portion; stalk flesh typically staining blue-green, at least near the base, and often reddish above; partial veil and ring absent.

SPORE PRINT: brown.

MICROSCOPIC FEATURES: spores 16–22 × 5–7.5 μm, fusoid, smooth, pale brown.

FRUITING: scattered or in groups on the ground in mixed hardwoods, especially under yellow birch; July–October; fairly common.

EDIBILITY: edible.

Phylloporus rhodoxanthus (Schweinitz) Bresadola

COMMON NAME: Gilled Bolete.

CAP: 1–4" (2.5–10 cm) wide, convex, becoming nearly flat or uplifted and funnel-shaped, sometimes with a slightly depressed center; surface dry, somewhat velvety, sometimes finely cracked, dark red, dull red to reddish brown, reddish yellow, or olive-brown, staining blue in NH_4OH; flesh yellow; odor and taste not distinctive.

PORE SURFACE: bright yellow to ochre, not blueing when bruised, decurrent, distinctly gill-like, widely spaced, often with crossveins, separating cleanly from the cap.

STALK: 1⅛–3½" (3–9 cm) long, ¼–½" (6–12 mm) thick, tapering toward the base, solid, dull yellow to reddish brown, nearly smooth; basal mycelium yellow; partial veil and ring absent.

SPORE PRINT: yellowish ochraceous.

MICROSCOPIC FEATURES: spores 9–14 × 3–6 μm, narrowly elliptic to fusiform, smooth, pale yellow.

FRUITING: solitary, scattered, or in groups on the ground in hardwoods or conifers; June–October; occasional.

EDIBILITY: edible.

Pulveroboletus ravenelii (Berkeley and Curtis) Murrill

COMMON NAME: Powdery Sulfur Bolete.

CAP: ⅜–4" (1–10 cm) wide, bluntly rounded to convex, becoming nearly flat in age; surface dry and powdery at first, then coated with tiny, matted fibers and scales, often slightly wrinkled or finely cracked in age, bright sulfur-yellow, becoming orange-red to brownish red from the center toward the margin; margin incurved when young, typically rimmed with pieces of torn partial veil; flesh white to pale yellow, slowly staining pale blue, then dingy yellow to pale brown when cut; odor and taste not distinctive.

PORE SURFACE: bright yellow, becoming dingy yellow to grayish brown at maturity, staining greenish blue, then grayish brown when bruised; pores angular to nearly circular, 1–3 per mm.

STALK: 1½–5¾" (4–14.5 cm) long, ¼–⅝" (6–16 mm) thick, equal or enlarging downward, solid, sheathed from the base upward with tiny, matted fibers, bright sulfur-yellow, bright yellow and smooth above the ring; partial veil membranous to powdery, bright sulfur-yellow, typically leaving a prominent, but sometimes inconspicuous, superior ring.

SPORE PRINT: olive-gray to olive-brown.

MICROSCOPIC FEATURES: spores 8–10.5 × 4–5 μm, elliptic to oval, smooth, pale brown.

FRUITING: solitary, scattered, or in groups on the ground in woods; July–October; occasional.

EDIBILITY: edible.

Strobilomyces floccopus (Vahl : Fries) Karsten (see photo, p. 10)

COMMON NAME: Old Man of the Woods.

CAP: 1⅛–6" (3–15.5 cm) wide, convex, becoming broadly convex and finally flattened in age; surface dry, with a whitish to grayish ground color, covered with coarse, woolly or cottony, flattened or erect, gray to purplish gray or blackish scales; margin bearing cottony pieces of grayish, torn partial veil; flesh nearly white, quickly staining orange-red to orange, then black when cut or bruised; odor and taste not distinctive.

PORE SURFACE: white when young, soon gray, finally black, staining reddish, then black when bruised; pores angular, 1–2 mm wide.

STALK: 1⅝–4¾" (4–12 cm) long, ⅜–1" (1–2.5 cm) thick, nearly equal, sometimes enlarged at the base, dry, solid; flesh whitish quickly staining reddish orange, then black when exposed; reticulate at the apex above the ring, shaggy below the ring; partial veil cottony to woolly, grayish, leaving one or more rings or shaggy zones on the stalk.

SPORE PRINT: black.

MICROSCOPIC FEATURES: spores 9.5–15 × 8.5–12 μm, short elliptic to globose, covered by a distinct reticulum, grayish.

FRUITING: solitary to scattered on the ground in mixed woods; June–October; fairly common.

EDIBILITY: edible.

COMMENTS: *Strobilomyces confusus* (edible) has a slightly smaller cap, up to 4¾" (12 cm) wide, with smaller, erect, often stiff and pointed scales and spores with irregular projections and short ridges that resemble a partial reticulum.

Suillus americanus (Peck) Snell

COMMON NAME: Chicken-fat Suillus, American Slippery Jack.

CAP: 1⅛–4" (3–10 cm) wide, rounded with an incurved margin when young, becoming broadly convex in age, occasionally with an umbo; surface sticky or slimy when moist, bright yellow to ochre-yellow, with cinnamon to reddish patches or streaks; margin rimmed with white to yellow or pale brown, cottony veil tissue; flesh yellow, staining purplish brown when cut; odor and taste not distinctive.

PORE SURFACE: yellow when young, slightly browner in age, slowly staining reddish brown when cut or bruised; pores angular, occasionally radially elongated, often decurrent, 1–2 mm wide.

STALK: 1⅛–3½" (3–9 cm) long, ⅛–⅜" (3–10 mm) thick, nearly equal, becoming hollow, often crooked, not reticulate, yellow, speckled with reddish to dark brown glandular dots and smears, often developing wine-red or wine-brown stains when bruised or in age; veil present, covering most or all of the pore surface when young, but typically remaining attached to the edge of the cap and not leaving a ring on the stalk.

SPORE PRINT: brown.

MICROSCOPIC FEATURES: spores 8–11 × 3–4 μm, nearly fusiform, smooth, pale brown.

FRUITING: solitary, in groups, or caespitose on the ground under white pine; July–October; fairly common to common.

EDIBILITY: edible.

COMMENTS: when handled, this mushroom stains fingers brownish.

Suillus cavipes (Opatowski) Smith and Thiers

COMMON NAME: Hollow-stalked Larch Suillus, Hollow-stemmed Tamarack Jack.

CAP: 1⅛–4" (3–10 cm) wide, rounded to convex, becoming nearly flat, occasionally with a rounded unbo; surface dry, not sticky or slimy, covered with dark brown to yellow-brown fibers and scales with an almost suede-like texture; margin often rimmed with whitish, dry veil remnants; flesh yellow, not staining blue when cut or bruised; odor and taste not distinctive.

PORE SURFACE: pale yellow to olive-yellow, not staining blue when cut or bruised; pores angular, sometimes radiating from the stalk, 0.5–1.5 mm or more wide.

STALK: 1⅛–3½" (3–9 cm) long, ⅜–1" (1–2.5 cm) thick, nearly equal or enlarging slightly downward, typically hollow in the lower half at maturity, yellow on the upper portion, colored like the cap below, not reticulate and without glandular dots or smears, not staining blue or green when cut or bruised; veil present, often leaving a slight ring on the stalk.

SPORE PRINT: brown.

MICROSCOPIC FEATURES: spores 7–10 × 3.5–4 μm, narrowly oval to ventricose, smooth, pale brown.

FRUITING: scattered or in groups on the ground under larch in forests and bogs; September–October; fairly common.

EDIBILITY: edible.

Suillus granulatus (Linne : Fries) Kuntze

COMMON NAME: Dotted-stalked Suillus, Granulated Slippery Jack.

CAP: 2–4¾" (5–12 cm) wide, broadly convex; surface sticky or slimy when fresh, color variable, some shade of tan, brown, cinnamon, or orangish cinnamon, staining olive-gray in KOH; margin lacking veil remnants even when young; flesh white to pale yellow, not staining blue when cut or bruised, staining olive-gray in $FeSO_4$; odor and taste not distinctive.

PORE SURFACE: whitish when young, soon becoming yellowish, not staining blue when cut or bruised; pores irregular to nearly circular, 1 per mm.

STALK: 1⅝–3⅛" (4–8 cm) long, ⅜–1" (1–2.5 cm) thick, nearly equal, solid, whitish when young, becoming yellowish in age, especially at the apex, covered with conspicuous pinkish tan to brownish glandular dots and smears; partial veil and ring absent.

SPORE PRINT: brown.

MICROSCOPIC FEATURES: spores 7–10 × 2.5–3.5 μm, oblong or tapered slightly to the apex, pale brown.

FRUITING: scattered or in groups on the ground under pines; June–November; fairly common.

EDIBILITY: edible.

Suillus grevillei (Klotzsch : Fries) Singer

COMMON NAME: Larch Suillus.

CAP: 1⅜–5½" (3.5–14 cm) wide, convex to nearly plane; surface smooth, shiny, sticky and slimy, color variable (orange-yellow, dull red, red-brown, reddish brown, dark reddish brown, or dark chestnut-brown); flesh pale orange-yellow, bruising pinkish brown; odor and taste not distinctive.

PORE SURFACE: yellow when young, darkening to olive-yellow or olive-brown in age, staining brownish when bruised; pores angular, 1–3 per mm.

STALK: 1½–5½" (4–14 cm) long, ⅜–1⅛" (1–3 cm) thick, nearly equal or enlarging slightly downward, solid, yellow and smooth above the ring, streaked reddish brown to brown below and often whitish near the base, sticky and slimy; partial veil yellowish with reddish brown streaks, cottony with a gelatinous covering, forming a gelatinous superior ring; lacking glandular dots; flesh yellowish, not staining distinctly green at the base when cut.

SPORE PRINT: olive-brown to dull cinnamon.

MICROSCOPIC FEATURES: spores 8–10 × 2.5–3.5 μm, ellipsoid, smooth, pale straw-color to nearly hyaline.

FRUITING: scattered or in groups on the ground or among sphagnum mosses under larch; September–November; fairly common.

EDIBILITY: edible.

COMMENTS: some authors consider the darker colored form to be a variety, *S. grevillei* var. *clintonianus,* or a distinct species, *S. clintonianus.*

Suillus intermedius (Smith and Thiers) Smith and Thiers

CAP: 2–6¼" (5–16 cm) wide, convex to nearly plane; surface smooth, sticky to slimy when fresh, shiny when dry, pale yellow when young, becoming tan or ochre-yellow to yellow-brown, often streaked or spotted in age, covered with a sticky gluten that tastes acidic; margin often rimmed with soft, yellow remnants of partial veil coated with gluten; flesh pale yellow to orange-yellow, not staining when exposed; odor and taste not distinctive.

PORE SURFACE: pale yellow, with beads of fluid when fresh, dingy yellow in age, slowly staining pale reddish brown when bruised or unchanging; pores angular, mostly 2 per mm.

STALK: 1½–4" (4–10 cm) long, ¼–½" (6–12 mm) thick, nearly equal, solid, pale yellow to ochre-yellow, coated at maturity with reddish to brownish glandular dots and smears that darken in age; flesh salmon-ochraceous in the lower portion, paler above; partial veil soft, yellow, cottony, covered by gluten, forming a gelatinous, superior, band-like ring.

SPORE PRINT: dull cinnamon.

MICROSCOPIC FEATURES: spores 8–11 × 3–5 μm, elliptic to spindle-shaped, smooth, pale brown.

FRUITING: scattered or in groups on the ground or among mosses under pine; July–October; fairly common.

EDIBILITY: edible.

COMMENTS: also known as *S. acidus* var. *intermedius.*

Suillus luteus (Linne : Fries) S. F. Gray

COMMON NAME: Slippery Jack.

CAP: 2–4¾" (5–12 cm) wide, rounded when young, becoming broadly convex to flat in age; surface smooth, sticky or slimy when moist, often shiny when dry, dark reddish brown to cinnamon-brown, yellow-brown or ochre; margin rimmed with partial veil remnants; flesh white to pale yellow, not staining blue when cut or bruised; odor and taste not distinctive.

PORE SURFACE: whitish to pale yellow when very young, yellow to dark yellow or olive-yellow in age, not staining blue when bruised; pores angular, 1–2 per mm.

STALK: 1⅛–3⅛" (3–8 cm) long, ⅜–1" (1–2.5 cm) thick, nearly equal, solid, white when young, becoming pale yellow at the apex and often developing dingy purplish or brownish tones toward the base in age, speckled with glandular dots and smears, at least above the ring; partial veil white, with dull purple to purplish-gray tones on the underside, forming a ring; ring often large and flaring or sleeve-like, white with purple tones or dark purple, often gelatinous.

SPORE PRINT: brown.

MICROSCOPIC FEATURES: spores 7–9 × 2.5–3 μm, nearly oblong, smooth, pale brown.

FRUITING: scattered or in groups on the ground near or under pine, spruce, and mixed conifers; August–October; fairly common.

EDIBILITY: edible.

Suillus pictus (Peck) Kuntze

COMMON NAME: Painted Suillus.

CAP: 1⅛–4¾" (3–12 cm) wide, convex, becoming broadly convex to nearly flat, sometimes depressed; surface dry, not sticky or slimy, covered with conspicuous pink to red or dark red scales or patches, fading to dull yellow in age; margin incurved, becoming decurved, frequently rimmed with fibrous, whitish veil remnants; flesh yellow, not staining blue when cut or bruised, sometimes staining slightly reddish; odor and taste not distinctive.

PORE SURFACE: yellow to dingy yellow, not staining blue when bruised, occasionally staining reddish to brownish, often slightly decurrent; pores angular, somewhat radially arranged, 0.5–5 mm wide.

STALK: 1⅝–4¾" (4–12 cm) long, ⅜–1" (1–2.5 cm) thick, equal or enlarging downward, solid, covered with reddish scales, fibers, or patches below the veil, lacking glandular dots or smears, not staining blue when cut or bruised; veil present on young specimens, white, fibrous, often forming a fibrous grayish ring on the stalk.

SPORE PRINT: brown.

MICROSCOPIC FEATURES: spores 8–12 × 3.5–5 μm, narrowly oval, smooth, pale brown.

FRUITING: scattered or in groups on the ground under white pine, in forests and at the edges of bogs; June–October; fairly common.

EDIBILITY: edible.

COMMENTS: also known as *S. spraguei*.

Suillus proximus Smith and Thiers

COMMON NAME: Tamarack Jill.

CAP: 2⅜–3½" (6–9 cm) wide, rounded, becoming convex to broadly convex; surface smooth and sticky or slimy when moist, often shiny when dry, yellow, yellow-orange, or pale orange-brown with pale pinkish cinnamon streaks, staining instantly olive with KOH and bluish gray with $FeSO_4$; flesh yellow or discoloring reddish brown, often slowly staining green when cut or bruised, staining bluish gray with KOH and $FeSO_4$; odor and taste not distinctive.

PORE SURFACE: yellow or dingy yellow when young, becoming olive-yellow to yellow-brown in age, staining reddish or brownish when bruised, not staining blue; pores angular, about 1 per mm.

STALK: 1⅛–3½" (3–9 cm) long, ⅜–⅝" (10–15 mm) thick, nearly equal, solid, yellow at the apex, often streaked with red, brown, or chestnut on the lower portion, surface sticky or slimy when moist, not reticulate, quickly staining distinctly green when cut at the base; partial veil present, yellow to whitish, cottony and covered by a gelatinous layer, forming a thick gelatinous, often brownish, superior ring on the stalk.

SPORE PRINT: brown.

MICROSCOPIC FEATURES: spores 7–10 × 4–4.5 μm, nearly oblong to elliptic, smooth, pale brown.

FRUITING: scattered or in groups on the ground or among sphagnum mosses under larch in forests and bogs; August–October; frequent.

EDIBILITY: edible.

Suillus punctipes (Peck) Singer

CAP: 1⅛–4" (3–10 cm) wide, convex, becoming broadly convex; surface covered with tufts of tiny, gray or brown, matted fibers when young, soon becoming smooth, sticky, and dull ochre-orange; margin without conspicuous veil remnants; flesh pale yellow to yellow-orange, not blueing when cut or bruised, staining purplish to wine-red in KOH; odor fragrant, like almond extract; taste not distinctive.

PORE SURFACE: brown when young, becoming orange-brown to dark or dingy yellow in age, not staining blue when cut or bruised; pores circular to angular, 2 per mm.

STALK: 1⅝–3½" (4–9 cm) long, ⅜–⅝" (1–1.6 cm) thick, equal or enlarging downward, often curved, solid, dull orangish brown when young, becoming dull ochre-orange to ochre-yellow, densely covered with brown to dark brown glandular dots and smears, not reticulate, sometimes with reddish stains at the base; partial veil and ring absent.

SPORE PRINT: brown.

MICROSCOPIC FEATURES: spores 7.5–12 × 3–4 μm, elliptic to subfusiform, smooth, pale brown.

FRUITING: scattered or in groups on the ground under spruce, balsam fir, and other conifers in forests and bogs; July–September; occasional.

EDIBILITY: edible.

COMMENTS: when handled, this mushroom stains fingers brownish.

Suillus salmonicolor (Frost) Halling

COMMON NAME: Slippery Jill.

CAP: 1⅛–3¾" (3–9.5 cm) wide, bluntly rounded or convex to nearly plane; surface sticky or slimy when moist, shiny when dry, color variable, dingy yellow, yellowish orange to ochraceous-salmon, cinnamon-brown or olive-brown to yellow-brown; flesh pale orange-yellow to orange-buff or orange, not staining when exposed; odor and taste not distinctive.

PORE SURFACE: yellow to dingy yellow, or yellowish orange to salmon, darkening to brownish in age, not staining when bruised; pores circular to angular, 1–2 per mm.

STALK: 1–4" (2.5–10 cm) long, ¼–⅝" (6–16 mm) thick, equal or enlarged downward, whitish to yellowish or pinkish ochraceous, coated with reddish brown to dark brown glandular dots and smears; flesh ochraceous to yellowish; partial veil at first thick and distinctly baggy, sometimes flaring from the stalk on the lower portion, forming a gelatinous superior to median ring.

SPORE PRINT: cinnamon-brown to brown.

MICROSCOPIC FEATURES: spores 6–11 × 2.5–4 μm, elliptic, smooth, pale brown.

FRUITING: scattered or in groups on the ground under pine; August–November; fairly common.

EDIBILITY: edible.

COMMENTS: also known as *S. subluteus* and *S. pinorigidus.*

Suillus subaureus (Peck) Snell

CAP: 1⅛–5½" (3–14 cm) wide, convex, becoming broadly convex to nearly flat; surface with scattered, flattened, brownish to reddish fibers over an apricot-orange to yellow ground color, often slightly sticky when moist; margin incurved to inrolled and remaining so for a long time, lacking conspicuous veil remnants even when young; flesh yellow, often reddening slightly when cut or bruised, not staining blue or green; odor not distinctive; taste mild to slightly acidic.

PORE SURFACE: pale yellowish orange when young, becoming dull ochre to dingy yellow in age, not staining blue when cut or bruised; pores angular, mostly 2 per mm.

STALK: 1⅝–3⅛" (4–8 cm) long, ⅜–¾" (1–2 cm) thick, nearly equal, solid, yellow or yellowish, not white, typically spotted with glandular dots that are yellow like the stalk when young but darken with age, not reticulate; partial veil and ring absent.

SPORE PRINT: brown.

MICROSCOPIC FEATURES: spores 7–10 × 2.7–3.5 μm, narrowly elliptic to subfusiform, smooth, pale brown.

FRUITING: scattered or in groups on the ground in hardwoods, especially oak, or in mixed woods; July–October; occasional.

EDIBILITY: edible.

COMMENTS: this mushroom is unusual among species of *Suillus* because it appears not to associate exclusively with conifers.

Tylopilus alboater (Schweinitz) Murrill

COMMON NAME: Black Velvet Bolete.

CAP: 1⅛–5⅞" (3–15 cm) wide, convex, becoming broadly convex or flat; surface dry, finely velvety, occasionally finely cracked in age, black to dark grayish brown, often covered with a thin whitish bloom when young; margin extending slightly beyond the pore surface; flesh white or tinged gray, staining pinkish to reddish gray when cut or bruised, eventually blackening; odor and taste not distinctive.

PORE SURFACE: white or with a tinge of gray when young, becoming dull pinkish or flesh-colored in age, not dark gray or black, usually staining reddish, then slowly black when bruised; pores angular to irregular, about 2 per mm.

STALK: 1½–4" (4–10 cm) long, ⅝–1½" (1.5–4 cm) thick, equal or enlarging downward, solid, colored like the cap or paler, especially near the apex, often covered with a thin, whitish bloom, not reticulate or only slightly so at the apex; partial veil and ring absent.

SPORE PRINT: pinkish to deep flesh color.

MICROSCOPIC FEATURES: spores 7–11 × 3.5–5 μm, narrowly oval, smooth, hyaline.

FRUITING: solitary to scattered on the ground under hardwoods, especially oak; June–September; occasional.

EDIBLE: edible.

Tylopilus ballouii (Peck) Singer

COMMON NAME: Burnt-orange Bolete.

CAP: 2–4¾" (5–12 cm) wide, convex, becoming nearly flat, often irregular; surface dry, not sticky or slimy, bright orange to bright orange-red, fading to dull orange, cinnamon or tan in age; flesh white, staining pinkish tan to violet-brown when cut or bruised; odor not distinctive; taste mild to bitter.

PORE SURFACE: white to dingy white, becoming tan or slightly pinkish in age, not yellow, staining brown when bruised; pores somewhat angular, 1–2 per mm.

STALK: 1–4¾" (2.5–12 cm) long, ¼–1" (6–25 mm) thick, equal or swollen at or above the base, solid, not reticulate or only finely so at the apex, whitish or tinged yellow to orange, staining brownish when cut or bruised or in age, surface smooth or scurfy; partial veil and ring absent.

SPORE PRINT: pale brown, tan or reddish brown.

MICROSCOPIC FEATURES: spores 5–11 × 3–5 μm, elliptic, smooth, hyaline to pale brown.

FRUITING: solitary, scattered, or in groups on the ground on lawns under trees or in woods, especially near oak, beech, and pine; August–September; occasional.

EDIBILITY: edible, sometimes bitter.

Tylopilus chromapes (Frost) Smith and Thiers

COMMON NAME: Chrome-footed Bolete.

CAP: 1⅛–6" (3–15.5 cm) wide, convex, becoming broadly convex to flat, sometimes slightly depressed in age; surface dry or very slightly sticky, pink to rose-colored when young, fading to tan, pinkish tan, or paler as it ages; flesh white, not staining blue when cut or bruised; odor and taste not distinctive.

PORE SURFACE: white when young, becoming pinkish to flesh-colored or dingy pinkish tan in age, not staining blue when cut or bruised; pores circular to angular, 2–3 per mm.

STALK: 1½–5½" (4–14 cm) long, ⅜–1" (1–2.5 cm) thick, nearly equal or tapered slightly in either direction, solid, predominantly white or pinkish except the base, which is bright yellow inside and out; surface usually scurfy from white, pink, or reddish scabers, not reticulate; partial veil and ring absent.

SPORE PRINT: pinkish to pinkish brown.

MICROSCOPIC FEATURES: spores 11–17 × 4–5.5 μm, nearly oblong to narrowly oval, smooth, hyaline to pale brown.

FRUITING: solitary to scattered on the ground under both hardwoods and conifers; June–October; fairly common.

EDIBILITY: edible.

COMMENTS: also known as *Leccinum chromapes*.

Tylopilus eximius (Peck) Singer

COMMON NAME: Lilac-brown Bolete.

CAP: 2–4¾" (5–12 cm) wide, convex, becoming broadly convex to nearly flat; surface dry or slightly sticky, purple-brown to grayish brown, occasionally reddish brown, often with a fine whitish bloom when very young; flesh whitish, not staining blue when cut or bruised, slowly staining gray-brown, reddish, or purplish; odor not distinctive; taste mild to slightly bitter.

PORE SURFACE: dark chocolate-brown to purple-brown, occasionally dark gray to blackish, sometimes becoming yellow-brown in age but not pinkish, not staining blue when bruised; pores nearly circular, up to 3 per mm.

STALK: 1¾–3½" (4.5–9 cm) long, ⅜–1½" (1–4 cm) thick, equal or tapered at either end, solid, colored nearly like the cap, surface scurfy from a dense coating of purplish or purple-brown scabers, not reticulate; partial veil and ring absent.

SPORE PRINT: pinkish to reddish brown to amber-brown.

MICROSCOPIC FEATURES: spores 11–17 × 3.5–5 μm, narrowly subfusoid, smooth, hyaline to pale brown.

FRUITING: solitary to scattered on the ground near conifers, especially hemlock, and in mixed woods; July–October; occasional.

EDIBILITY: not recommended.

COMMENTS: although listed as edible in several field guides, this mushroom has caused severe gastric distress for some individuals.

Tylopilus felleus (Bulliard : Fries) Karsten

COMMON NAME: Bitter Bolete.

CAP: 2–11¾" (5–30 cm) wide, rounded to convex, becoming broadly convex to flat; surface dry, smooth, sometimes sticky when moist, pinkish to reddish purple when young, becoming some shade of brown, buff, or tan; flesh white, not staining blue when cut or bruised; odor not distinctive; taste very bitter.

PORE SURFACE: white when young, becoming pinkish or pinkish tan in age, not staining blue when cut or bruised, often staining brown; pores nearly circular, 1–2 per mm.

STALK: 1½–7⅞" (4–20 cm) long, ⅜–1¼" (1–3 cm) thick, enlarging downward, bulbous, solid, entirely brown or white toward the apex and brown below, often developing olive or olive-brown stains when bruised, prominently reticulate, at least over the upper third; reticulation brown; partial veil and ring absent.

SPORE PRINT: pinkish brown, reddish brown, or rosy brown.

MICROSCOPIC FEATURES: spores 11–17 × 3–5 μm, subfusoid, smooth, pale brown.

FRUITING: solitary or in groups on the ground or on decaying wood under conifers or mixed woods; June–October; common.

EDIBILITY: inedible, bitter.

COMMENTS: sometimes confused with *Boletus edulis,* which has a whitish stalk reticulation and mild-tasting flesh. It is edible and choice for persons who lack the gene for sensing bitter tastes.

Tylopilus indecisus (Peck) Murrill

CAP: 2–6¾" (5–17 cm) wide, convex, becoming broadly convex to nearly flat; surface dry, finely velvety, ochraceous brown to pale brown when young, becoming dull cinnamon in age; flesh white, slowly staining brownish or pinkish when cut or bruised; odor and taste not distinctive.

PORE SURFACE: white when young, becoming pinkish or sometimes brownish in age, staining brown when cut or bruised; pores angular, 2–2.5 per mm.

STALK: 1⅝–4" (4–10 cm) long, ⅜–1⅛" (1–3 cm) thick, thicker at the base when young, nearly equal in age, solid, whitish when young, becoming pale brown below at maturity, staining brown when cut or bruised, often reticulate toward the apex; partial veil and ring absent.

SPORE PRINT: pinkish brown to reddish brown.

MICROSCOPIC FEATURES: spores 10–15 × 3–5 μm, narrowly subfusiform, smooth, pale brown.

FRUITING: scattered or in groups on the ground in oak, pine, and mixed hardwoods; July–September; occasional.

EDIBILITY: edible.

COMMENTS: *Tylopilus ferrugineus* is similar but has a dark reddish brown cap, a dark brown stalk, and spores that measure 8–13 × 3–5 μm.

Tylopilus intermedius Smith and Thiers

CAP: 2⅜–5⅞" (6–15 cm) wide, broadly convex, becoming nearly flat in age; surface uneven, often wrinkled like parchment, occasionally pruinose, white or whitish when young, occasionally with a pinkish tinge or sometimes tan, developing brownish stains in age, staining pinkish in FeSO₄ when young; flesh white, not staining blue when bruised; odor not distinctive; taste bitter.

PORE SURFACE: white when young, pinkish in age, slowly staining brownish when bruised; pores circular or nearly so, 1–2 per mm.

STALK: 3⅛–5½" (8–14 cm) long, ⅜–1⅝" (1–4 cm) thick, enlarging downward, solid, white or whitish, sometimes developing dingy brownish or yellow-brown stains, weakly to distinctly reticulate; partial veil and ring absent.

SPORE PRINT: pinkish to pinkish brown.

MICROSCOPIC FEATURES: spores 10–15 × 3–5 μm, nearly oblong, smooth, hyaline to pale brown.

FRUITING: scattered or in groups on the ground under oak or pine; August–October; infrequent.

EDIBILITY: inedible, bitter.

COMMENTS: most often collected in the southern part of the region.

Tylopilus plumbeoviolaceus (Snell and Dick) Singer

COMMON NAME: Violet-gray Bolete.

CAP: 1½–5⅞" (4–15 cm) wide, convex, becoming broadly convex to nearly flat; surface finely velvety when young, dry, smooth, violet or purple when young, fading to purple-brown, purple-gray, brown, dull cinnamon, or tan, occasionally overlaid with a whitish bloom; flesh white, not staining blue when cut or bruised; odor not distinctive; taste very bitter.

PORE SURFACE: white when young, becoming pinkish or dull pinkish-tan in age, not staining blue when bruised; pores nearly circular, 1–2 per mm.

STALK: 3⅛–4¾" (8–12 cm) long, ⅜–¾" (1–2 cm) thick, equal or enlarging slightly downward, solid, predominantly violet or purple, sometimes marbled with white, fading to dull purple, purple-gray, purple-brown, or brown, occasionally developing olive or olive-brown stains when bruised, sometimes slightly reticulate at the apex; partial veil and ring absent.

SPORE PRINT: pinkish brown.

MICROSCOPIC FEATURES: spores 10–13 × 3–4 μm, elliptic, smooth, pinkish brown.

FRUITING: scattered or in groups on the ground under hardwoods, especially oak; June–September; occasional to frequent.

EDIBILITY: inedible, bitter.

Tylopilus rubrobrunneus Mazzer and Smith

CAP: 3⅛–11¾" (8–30 cm) wide, broadly convex, becoming nearly flat, occasionally slightly depressed in age; surface dry, sometimes finely cracked in age, not sticky or slimy, dark to bright purple when young, becoming purple-brown or dark reddish brown, dull brown, or cinnamon in age; flesh white, not staining blue when cut or bruised, staining purplish-buff with $FeSO_4$; odor not distinctive; taste very bitter.

PORE SURFACE: white at first, becoming pinkish, then dingy pinkish brown, or occasionally brown at maturity, usually staining brown when bruised; pores circular, 1–2 per mm.

STALK: 2⅜–7⅞" (6–20 cm) long, ⅜–2" (1–5 cm) thick, equal or enlarging downward, solid, white to brown, developing olive or olive-brown stains from the base upward as specimens mature or overall when handled, finely reticulate at the apex, smooth below; partial veil and ring absent.

SPORE PRINT: reddish brown or dull pinkish brown.

MICROSCOPIC FEATURES: spores 10–14 × 3–4.5, suboblong to nearly fusoid, smooth, pale brown.

FRUITING: scattered, in groups, or caespitose on the ground under hardwoods, especially beech and maple or in oak-pine woods, also under hemlock; July–September; fairly common.

EDIBILITY: inedible, bitter.

Tylopilus sordidus (Frost) Smith and Thiers

CAP: 1¾–5⅛" (4.5–13 cm) wide, convex, becoming nearly plane in age, margin even; surface dry, finely tomentose to somewhat velvety-tomentose, becoming finely to conspicuously rimose-aerolate, gray-brown to olive-brown, or dark brown, often with dark greenish or bluish tints along the margin, staining purplish to reddish with the addition of KOH, and dark purple with NH_4OH; flesh whitish, staining blue-green when exposed, sometimes with reddish tints, staining bluish gray to greenish gray with the addition of $FeSO_4$, and red to vinaceous or vinaceous tawny with KOH; odor slightly pungent or not distinctive; taste unpleasant or not distinctive.

PORE SURFACE: whitish to grayish buff when young, becoming pinkish brown then reddish brown to yellow-brown in age, staining dark blue to dark blue-green, then dark brownish red when bruised; pores circular at first, becoming angular to irregular at maturity, 1–2 mm wide in age.

STALK: 1⅝–4" (4–10 cm) long, ⅜–¾" (1–2 cm) thick, slightly enlarged downward or nearly equal, dry, solid, brownish with much darker longitudinal streaks, minutely scurfy, typically paler toward the apex and whitish at the base, with a greenish or bluish green tint near the apex, not reticulate; partial veil and ring absent.

SPORE PRINT: reddish brown.

MICROSCOPIC FEATURES: spores 10–14 × 4–6 μm, subelliptic, smooth, pale brown; pleurocystidia fusoid-ventricose, often with an elongated neck.

FRUITING: solitary, scattered, or in groups on the ground under hardwoods, especially oak and under various conifers; July–October; occasional.

EDIBILITY: unknown.

COMMENTS: when wrapped in waxed paper for several minutes, the exposed flesh of this bolete imparts a dark blue-green color. It is also known as *Porphyrellus sordidus. Tylopilus fumosipes = Porphyrellus fumosipes, Porphyrellus pseudoscaber* ssp. *cyaneotinctus*, and *Tylopilus cyaneotinctus = Porphyrellus cyaneotinctus* are all synonyms. This species was first described by Frost as *Boletus sordidus* in 1874, and therefore has priority. Much confusion and contradiction occurs in the literature concerning the identification of two very similar boletes, *T. porphyrosporus* and *T. pseudoscaber.* Some authors have placed them in synonymy. Others have stated that the flesh of *T. pseudoscaber* stains blue then reddish when exposed, whereas the flesh of *T. porphyrosporus* slowly becomes fuscous but does not stain blue when cut.

Austroboletus betula

Austroboletus gracilis

Boletellus chrysenteroides

Boletellus russellii

Boletus affinis var. *maculosus*

Boletus auriporus

Boletus bicolor var. *bicolor*

Boletus chippewaensis

Boletus chrysenteron

Boletus edulis

Boletus fraternus

Boletus frostii

Boletus griseus

Boletus illudens

Boletus innixus

Boletus longicurvipes

Boletus miniato-pallescens

Boletus ornatipes

Boletus pallidus

Boletus parasiticus

Boletus projectellus

Boletus pseudosensibilis

Boletus rubellus

Boletus rubroflammeus

Boletus sensibilis

Boletus separans

Boletus spadiceus

Boletus speciosus var. *brunneus*

Boletus subglabripes

Boletus subvelutipes

Boletus vermiculosoides

Boletus vermiculosus

Chalciporus piperatus

Chalciporus rubinellus

Fuscoboletinus spectabilis

Gyroporus castaneus

Gyroporus cyanescens var. *violaceotinctus*

Leccinum atrostipitatum

Leccinum aurantiacum

Leccinum holopus

Leccinum rugosiceps

Leccinum snellii

Phylloporus rhodoxanthus

Pulveroboletus ravenelii

Suillus americanus

Suillus cavipes

Suillus granulatus

Suillus grevillei

Suillus intermedius

Suillus luteus

Suillus pictus

Suillus proximus

Suillus punctipes

Suillus salmonicolor

Suillus subaureus

Tylopilus alboater

Tylopilus ballouii

Tylopilus chromapes

Tylopilus eximius

Tylopilus felleus

Tylopilus indecisus

Tylopilus intermedius

Tylopilus plumbeoviolaceus

Tylopilus rubrobrunneus

Tylopilus sordidus

Polypores

Members of a very large group of fungi called polypores form fruiting bodies with small, cylindric tubes on the underside of the cap, in which spores are produced. Spores are discharged through a tiny mouth-like opening at the end of each tube, called a pore. Each fruiting body forms many tubes, each with a pore: thus the name *polypore*. Several other common names have been used to describe various species in this diverse group, including *bracket fungi*, *shelf fungi*, and *conks*. Some polypores have a central to eccentric or lateral stalk, while others are stalkless. Most species grow on wood, but a few grow on soil or humus. Several wood-inhabiting polypores are destructive pathogens; others live on decaying or dead material and play an important role in the recycling of nutrients. Many polypores are hard and woody or corky to leathery, but some are fleshy to fibrous. The tube layer of a polypore usually does not separate cleanly and easily from the supporting cap tissue. Several species are excellent edibles if collected when they are young and tender. Although no polypores are known to have caused fatalities, most are inedible.

A similar group called boletes, which resemble polypores that grow on the ground, also produce their spores in tubes. Most boletes grow on the ground, are soft and fleshy, and have tube layers that are usually cleanly and easily separated from the supporting cap tissue. The key to the Crust and Parchment Fungi and Allies should also be consulted if an unknown specimen has characteristics of that group.

Because a comprehensive key to the majority of polypores that occur in this region is beyond the scope of this work, we have included only the most distinctive and the most commonly encountered species.

Key to Groups and Species of Polypores

1. Fruiting body 2–8" (5–20.5 cm) wide and long, a nearly round to cylindric or hoof-shaped, stalkless mass of small, overlapping caps arising from a solid central core; caps laterally fused with petal-shaped projecting margins; radially wrinkled, dull yellow-brown with a tan margin, becoming dark brown to grayish black in age; pore surface purplish gray to dark grayish brown; on trunks and logs of hardwoods—*Globifomes graveolens* (Schweinitz) Murrill (see p. 376).
1. Not as above → 2.
    2. Fruiting body ⅝–3⅜" (1.5–8.5 cm) wide, nearly round to hoof-shaped, stalkless; upper surface whitish to tan or pale yellow-brown; lower surface a thick, membranous veil concealing an interior white to pinkish or pale brown pore surface;

growing on conifer trunks and stumps—*Cryptoporus volvatus* (Peck) Hubbard (see p. 375).

2.   Not as above → 3.

3.   Fruiting body 1⅛–10" (3–25.5 cm) wide, shell- to kidney-shaped, with a short lateral stalk or stalkless, fibrous-tough to corky when fresh; upper surface smooth, pale brown with darker brown streaks, often tearing or breaking up into scales or patches; lower surface depressed from the margin, white and smooth when young, soon developing tiny pores, darkening to pale brown; tubes and pores becoming torn and tooth-like or jagged in age; margin inrolled at maturity; growing on birch trunks and branches— *Piptoporus betulinus* (Fries) Karsten (see p. 377).

3.   Not as above → 4.

4.   Fruiting body 2¾–10" (7–25.5 cm) wide, fan-to spoon-shaped, with a short lateral to eccentric stalk or stalkless; upper surface smooth to velvety, gelatinous, reddish orange to pinkish red or dark red to purplish brown; pore surface whitish to pinkish yellow, becoming reddish brown in age or when bruised; tubes crowded but distinctly separate when viewed with a hand lens; growing on oak trunks and stumps—*Fistulina hepatica* Schaeffer : Fries (see p. 375).

4.   Not as above → 5.

5.   Fruiting body 4–15" (10–38.5 cm) wide, irregularly shaped, stalkless, black to dark brown or reddish brown, hard and brittle, deeply cracked, resembling charred wood or a canker-like growth, usually on standing birch or sometimes ironwood, elm, alder, or beech—*Inonotus obliquus* (Persoon : Fries) Pilát (see p. 376).

5.   Not as above → 6.

6.   Fruiting body 1⅜–5½" (3.5–14 cm) wide, consisting of a circular to fan-shaped cap and a central stalk; cap yellow-brown to dark rusty brown, velvety to fibrous-scaly; lower surface with conspicuous, concentric, gill-like plates and irregular pores, yellow-brown to rusty brown; stalk reddish brown to dark brown; scattered or in groups on the ground, usually under hardwoods—*Coltricia montagnei* (Fries) Murrill (see p. 375).

6.   Fruiting body not with the above combinations of characters → 7.

7.   Fruiting body appearing varnished and shiny, brownish orange to reddish brown or rarely blue, paler toward the margin; laterally stalked or stalkless; growing on wood —Varnish Conks (see p. 377).

7.   Not as above → 8.

8.   Fruiting body with broken pore mouths and tubes; pore surface appearing tooth-like or jagged but pores often evident near the cap edge (use hand lens); growing on decaying wood—Toothed Polypores (see p. 378).

8.   Not as above → 9.

9.   Pore surface weakly to distinctly labyrinthine or radially gill-like; growing on wood— Maze and Gilled Polypores (see p. 379).

9.   Not as above → 10.

10.   Fruiting body with a conspicuous central to eccentric stalk; stalk not branched, usually with a single cap or sometimes with fused caps; cap usually less than 6" (15.5 cm) wide; growing on the ground or on wood—Small, Centrally Stalked Polypores (see p. 381).

10.   Not as above → 11.

11.   Fruiting body fleshy to fibrous-tough, not woody; stalk distinct, repeatedly branched with multiple caps, or unbranched with one or more caps, sometimes forming a rosette or overlapping clusters; caps large, typically 6–12" (15.5–31 cm) wide or more when mature—Large, Centrally to Laterally Stalked Polypores (see p. 385).

11. Not as above → 12.
    12. Fruiting body distinctly woody, stalkless, hoof-shaped to shelf-like; conspicuously thickened at the central point of attachment; growing on wood—Woody, Shelf-like Polypores (see p. 388).
    12. Not as above → 13.
13. Fruiting body not with the above combinations of characters, fleshy to fibrous-tough; shelf-like or irregular; stalkless or with a rudimentary stalk; single-capped to overlapping clusters or rosettes; caps usually small to medium, typically less than 6" (15.5 cm) wide but occasionally up to 12" (31 cm) or more—Fleshy to Fibrous-tough, Shelf-like Polypores (see p. 389).

Coltricia montagnei (Fries) Murrill
COMMON NAME: Green's Polypore.
CAP: 1⅜–5½" (3.5–14 cm) wide, circular to fan-shaped, convex to flat, sometimes depressed; surface velvety to hairy or fibrous-scaly; yellow-brown to dark rusty brown; margin wavy, sometimes torn.
FLESH: up to ¾" (2 cm) thick, soft and spongy, becoming fibrous in age, cinnamon to rusty brown.
PORE SURFACE: yellow-brown to rusty brown, darkening in age; pores angular and radially elongated near the stalk, becoming conspicuously concentric gill-like plates toward the margin.
STALK: 1⅛–2¾" (3–7 cm) long, tapering downward, central to eccentric, velvety, reddish brown to dark brown.
SPORE PRINT: pale brown.
MICROSCOPIC FEATURES: spores 9–15 × 5–7.5 μm, elliptic to oblong, smooth, pale brown.
FRUITING: scattered or in groups on the ground, usually under hardwoods; July–October; uncommon.
EDIBILITY: inedible.
COMMENTS: all parts stain black with KOH. Previously known as *Coltricia montagnei* var. *greenei* and *Cyclomyces greenei*.

Cryptoporus volvatus (Peck) Hubbard
COMMON NAME: Veiled Polypore, Cryptic Globe Fungus.
CAP: ⅝–3⅜" (1.5–8.5 cm) wide, nearly round to hoof-shaped, stalkless; upper surface whitish to tan or pale yellow-brown.
FLESH: up to ¾" (2 cm) thick, soft-corky, white.
PORE SURFACE: white to pinkish or pale brown, covered by a thick, whitish to yellowish membranous veil, tearing open at maturity; pores circular, 3–5 per mm.
SPORE PRINT: pinkish buff.
MICROSCOPIC FEATURES: spores 9–16 × 3–5 μm, cylindric, smooth, hyaline.
FRUITING: scattered or in groups on trunks and stumps of conifers; May–September; occasional.
EDIBILITY: inedible.
COMMENTS: the breakdown of the veil is often assisted by various insects and other tiny arthropods, which bore holes through the membrane.

Fistulina hepatica Schaeffer : Fries
COMMON NAME: Beefsteak Polypore.
CAP: 2¾–10" (7–25.5 cm) wide, fan- to spoon-shaped; surface smooth to velvety, gelatinous, often sticky to slimy; reddish orange to pinkish red or dark red to purplish

brown; often exuding a red juice when squeezed; margin rounded or sharp, often wavy or lobed.

FLESH: ¾–2" (2–5 cm) thick, fleshy and juicy when fresh, becoming fibrous in age, dingy white to pinkish or reddish, zoned with darker and paler areas, slowly darkening when exposed; taste sour to acidic.

PORE SURFACE: whitish to pinkish yellow, becoming reddish brown in age or when bruised; pores circular, 1–3 per mm; tubes crowded but distinctly separate when viewed with a hand lens.

SPORE PRINT: pinkish salmon.

STALK: up to 3⅛" (8 cm) long, lateral to eccentric or sometimes absent, colored like the cap.

MICROSCOPIC FEATURES: spores 4–6 × 2.5–4 μm, oval to lacrymoid, smooth, hyaline.

FRUITING: solitary or in groups on oak trunks and stumps; July–October; infrequent.

EDIBILITY: edible.

COMMENTS: *Fistulina radicata* = *Pseudofistulina radicata* (inedible) has a smaller, up to 3" (7.5 cm) wide, pale yellowish brown cap, a whitish pore surface and a rooting stalk up to 4" (10 cm) long.

Globifomes graveolens (Schweinitz) Murrill

COMMON NAME: Sweet Knot.

FRUITING BODY: 2–8" (5–20.5 cm) wide and long, a nearly round to cylindric or hoof-shaped, stalkless mass of small, overlapping caps, arising from a solid central core.

CAP: laterally fused, with petal-shaped projecting margins, leathery to rigid; surface slightly velvety, radially wrinkled, dull yellow-brown with a tan margin when young, becoming dark brown to grayish black in age.

FLESH: up to ¼" (6 mm) thick, fibrous-tough, yellowish brown.

PORE SURFACE: at first purplish gray, becoming dark grayish brown; pores circular, 3–5 per mm.

SPORE PRINT: brown.

MICROSCOPIC FEATURES: spores 9–14 × 3–4.5 μm, cylindric, smooth, hyaline to pale brown.

FRUITING: solitary or in groups on trunks and logs of hardwoods, especially oak; July–November; infrequent.

EDIBILITY: inedible.

COMMENTS: also known as *Polyporus graveolens*. The common name is a reference to the sweet odor of some fruiting bodies when cut, variously described as odor of apples or green fodder. Most fruiting bodies, however, lack the sweet odor.

Inonotus obliquus (Persoon : Fries) Pilát

COMMON NAME: Clinker Polypore, Birch Canker.

FRUITING BODY: a sterile conk, 4–15" (10–38.5 cm) wide, irregularly shaped, resembling charred wood or a canker-like growth, stalkless; outer portion black to dark brown or reddish brown, hard and brittle, deeply cracked; inner portion corky, bright yellow-brown to rusty brown.

FRUITING: solitary or in groups, usually on standing birch or sometimes on ironwood, elm, alder, or beech; year-round; common.

EDIBILITY: can be ground and steeped to make a refreshing hot or cold drink called Chaga Tea.

COMMENTS: the entire conk is a sterile mass of fungal tissue. The actual fruiting body is seldom observed. It is closely attached to the wood and produces spores that measure $8–10 \times 5–7$ μm and are broadly elliptic, smooth, and hyaline to pale yellow.

Piptoporus betulinus (Fries) Karsten

COMMON NAME: Birch Polypore.

CAP: 1⅛–10" (3–25.5 cm) wide, shell-to kidney-shaped, fibrous-tough to corky when fresh; surface smooth when young, often tearing or breaking up into scales or patches in age; pale brown with darker brown streaks; margin inrolled, most obviously so at maturity.

FLESH: ⅜–2" (1–5 cm) thick, fibrous-tough, white.

PORE SURFACE: depressed from the margin, white and smooth when young, becoming pale brown to yellowish brown in age; pores circular to angular, 3–5 per mm; tubes and pores becoming torn and tooth-like or jagged in age.

STALK: up to 2⅜" (6 cm) long and thick, sometimes rudimentary or lacking, lateral, white to reddish brown or brown.

SPORE PRINT: white.

MICROSCOPIC FEATURES: spores $5–6 \times 1.5–1.7$ μm, cylindric to sausage-shaped, smooth, hyaline.

FRUITING: solitary, scattered, or in groups on birch trees; year-round; common.

EDIBILITY: although listed as edible and eaten by some, most people find it too fibrous-tough and somewhat bitter.

Varnish Conks

Ganoderma tsugae Murrill

COMMON NAME: Hemlock Varnish Shelf.

CAP: 2⅜–12" (6–31 cm) wide, fan- to kidney-shaped; soft and corky when fresh, covered with a thin crust; surface smooth to wrinkled, shiny, appearing varnished or dull and powdery when covered with spores, concentrically zoned and shallowly furrowed; brownish red to mahogany near the center or overall, brownish orange to reddish orange outward, and bright whitish on the margin or rarely blue to bluish green (see photos).

FLESH: up to 1⅛" (3 cm) thick, soft and corky to fibrous-tough, whitish.

PORE SURFACE: white to creamy white, becoming brown in age or when bruised; pores circular to angular. 4–6 per mm.

STALK: 1⅛–6" (3–15.5 cm) long, ⅜–1½" (1–4 cm) thick, typically lateral but sometimes eccentric to central or absent, shiny, appearing varnished, brownish red to mahogany or blackish brown.

SPORE PRINT: brown.

MICROSCOPIC FEATURES: spores $9–11 \times 6–8$ μm, elliptic, truncate, with a thick double wall, appearing rough, pale brown.

FRUITING: solitary or in groups on decaying conifer wood, especially hemlock; May–December; fairly common.

EDIBILITY: inedible.

COMMENTS: *Ganoderma lucidum* (edible when prepared as a tea) has a dark reddish brown cap with a creamy white margin and grows on decaying hardwoods, especially maple.

Toothed Polypores

Irpex lacteus (Fries) Fries

COMMON NAME: Milk-white Toothed Polypore.

CAP: ⅜–1⅛" (1–3 cm) wide, shell- to petal-shaped or semicircular, convex to nearly plane, overlapping and often laterally fused, projecting from a spreading, crust-like mass, leathery to stiff; surface azonate to faintly zoned, densely tomentose or hairy, white to creamy white, dry, stalkless.

FLESH: up to ¹⁄₁₆" (1.5 mm) thick, fibrous-tough, white to pale tan.

PORE SURFACE: white to cream; pores angular, 2–3 per mm; tubes and pores splitting and becoming tooth-like or jagged in age.

SPORE PRINT: white.

MICROSCOPIC FEATURES: spores 5–7 × 2–3 μm, oblong to cylindric, straight to curved, smooth, hyaline.

FRUITING: in spreading, overlapping clusters on decaying hardwood; year-round; common.

EDIBILITY: inedible.

COMMENTS: this is the only species in this genus.

Pycnoporellus alboluteus (Ellis and Everhart) Kotlaba and Pouzar

COMMON NAME: Orange Sponge Polypore.

FRUITING BODY: flattened and crust-like to slightly elevated and cushion-shaped, spreading, stalkless, soft and spongy when fresh; surface loosely hairy to fibrous-scaly, orange.

FLESH: up to ⅛" (3 mm) thick, soft, pale orange.

PORE SURFACE: bright orange to yellow-orange; pores angular, usually 1 mm or wider; tubes and pores becoming torn and tooth-like or jagged in age.

SPORE PRINT: white.

MICROSCOPIC FEATURES: spores 6–9 × 2.5–4 μm, cylindric, smooth, hyaline.

FRUITING: spreading along the lower sides of decaying conifer logs; July–November; rare.

EDIBILITY: inedible.

COMMENTS: all parts quickly stain cherry-red in KOH.

Trichaptum biforme (Fries) Ryvarden

COMMON NAME: Violet Toothed Polypore.

CAP: ⅜–2⅜" (1–6 cm) wide, shell- to petal-shaped or semicircular, convex to nearly plane, leathery to stiff, sometimes smooth in age; distinctly zoned and variously colored, white to grayish, reddish brown; green if covered with algae; margin often violet and wavy; stalkless or with a rudimentary stalk.

FLESH: up to ¹⁄₁₆" (1.5 mm) thick, fibrous-tough, white to ochre.

PORE SURFACE: violet to purple-brown or sometimes fading to buff; pores angular, 2–5 per mm; tubes and pores splitting and becoming tooth-like or jagged in age.

SPORE PRINT: white.

MICROSCOPIC FEATURES: spores 5–8 × 2–2.5 μm, cylindric, smooth, hyaline.

FRUITING: in overlapping clusters on decaying hardwoods; year-round; common.

EDIBILITY: inedible.

COMMENTS: sometimes incorrectly spelled *T. biformis*. *Trichaptum abietinum* (inedible) is smaller, has stiff white hairs on the cap, and grows on conifer wood. *Irpex lacteus* (inedible) has a white to creamy white cap and a tooth-like pore surface and grows on decaying hardwoods. *Spongipellis pachyodon* (inedible) is much larger, 2–8" (5–20.5 cm) wide, has a white to ochraceous cap and a tooth-like pore surface and grows on hardwoods.

Maze and Gilled Polypores

Daedalea quercina Fries

COMMON NAME: Thick-maze Oak Polypore.

CAP: 2–8" (5–20.5 cm) wide, semicircular to kidney-shaped, stalkless, plane to slightly convex, leathery to corky or woody; surface felted to somewhat velvety when young, becoming smooth, then cracked or furrowed in age, often concentrically zoned; color variable, usually brownish yellow to brownish orange, tan, brown, or black; margin whitish and blunt.

FLESH: up to ⅜" (1 cm) thick, fibrous-tough, dull white to pale brown.

PORE SURFACE: whitish to pale yellow-brown or grayish brown, conspicuously labyrinthine, occasionally with elongated pores near the margin; fibrous-tough.

SPORE PRINT: white.

MICROSCOPIC FEATURES: spores 5–6 × 2–3.5 μm, cylindric, smooth, hyaline.

FRUITING: solitary or in groups on decaying hardwoods, especially oak; year-round; occasional.

EDIBILITY: inedible.

COMMENTS: *Cerrena unicolor* (inedible) has a smaller cap, up to 3 ⅛" (8 cm) wide, with a dense layer of short stiff hairs, and is variably colored, often green when coated with algae, with a white to gray labyrinthine pore surface. *Schizopora paradoxa* (inedible) is a white to creamy white or grayish brown, flattened, spreading mass, with a labyrinthine pore surface, which grows on decaying hardwood logs and branches. Also compare with *Daedaleopsis confragosa* (see p. 379), which is smaller and thinner.

Daedaleopsis confragosa (Bolton : Fries) Schroeter

COMMON NAME: Thin-maze Flat Polypore.

CAP: 1⅛–6" (3–15.5 cm) wide, semicircular to kidney-shaped, flat to slightly convex, stalkless, fibrous-tough; surface coarsely wrinkled, finely velvety to smooth, with distinct concentric zones; variously colored, creamy white, tan, grayish, or pale brown; margin sharp, thin.

FLESH: up to ½" (1.2 cm) thick, fibrous-tough, whitish to pale brown.

PORE SURFACE: whitish to pale brown, usually labyrinthine, often with gill-like and elongated pores; fibrous-tough.

SPORE PRINT: white.

MICROSCOPIC FEATURES: spores 7–11 × 2–3 μm, cylindric to sausage-shaped, smooth, hyaline.

FRUITING: solitary, scattered, or in groups on decaying wood, especially hardwood; year-round; common.

EDIBILITY: inedible.

COMMENTS: also known as *Daedalia confragosa*. Compare with *Daedalia quercina* (see p. 379), which is larger and thicker, and has smaller spores.

Gloeophyllum sepiarium (Fries) Karsten

COMMON NAME: Yellow-red Gill Polypore.

CAP: 1–4" (2.5–10 cm) wide, semicircular to kidney-shaped, flat or slightly convex, stalkless, fibrous-tough; surface covered with short stiff hairs, becoming matted and felty or nearly smooth in age, with distinct, concentric zones and furrows, bright yellowish red to reddish brown; margin whitish to orange-yellow or brownish yellow, uneven, with tufts of tiny hairs.

FLESH: up to ¼" (6 mm) thick, fibrous-tough, yellow-brown to rusty brown, black in KOH.

PORE SURFACE: golden brown to rusty brown, gill-like to labyrinthine (often both), and sometimes with elongated pores; pores 1–2 per mm.

SPORE PRINT: white.

MICROSCOPIC FEATURES: spores 9–13 × 3–5 μm, cylindric, smooth, hyaline.

FRUITING: solitary, in groups, or in rosette-like clusters on decaying wood, usually conifer; year-round; common.

EDIBILITY: inedible.

COMMENTS: *Lenzites betulina* (inedible) has white flesh and usually grows on decaying hardwood. *Gloeophyllum trabeum* (inedible) has crowded gills and narrow pores, up to 4 per mm along the margin.

Lenzites betulina (Fries) Fries (see photo, p. 11)

COMMON NAME: Multicolor Gill Polypore.

CAP: 1⅛–4" (3–10 cm) wide, semicircular to kidney-shaped, nearly plane, stalkless, fibrous-tough; surface velvety to hairy, with distinct multicolored concentric zones; colors variable, often white, pink, gray, yellow, orange, or brown, sometimes green when covered with algae.

FLESH: up to 1/16" (1.5 mm) thick, fibrous-tough, white.

PORE SURFACE: white to creamy white, conspicuously gill-like, sometimes forking, occasionally with elongated pores near the margin.

SPORE PRINT: white.

MICROSCOPIC FEATURES: spores 4–6 × 2–3 μm, cylindric to sausage-shaped, smooth, hyaline.

FRUITING: solitary or in groups on decaying wood, especially hardwood; July–December; common.

EDIBILITY: inedible.

COMMENTS: *Gloeophyllum sepiarium* (inedible) has yellow-brown to rusty brown flesh and usually grows on conifer wood.

Trametes cervina (Schweinitz) Bresadola

CAP: 1⅛–3½" (3–9 cm) wide, shell– to fan–shaped or hoof–shaped when young, leathery to stiff, upper surface covered with coarse hairs, distinctly to faintly zoned or sometimes azonate; zones white to pinkish buff and pinkish cinnamon to pale cinnamon; margin uneven and wavy in age; stalkless.

FLESH: up to ⅜" (9 mm) thick, fibrous-tough, white to pale buff.

PORE SURFACE: white when young, becoming cinnamon-buff and darker in age; pores angular to irregular when young, becoming conspicuously labyrinthine, sometimes thin and tooth-like in age.

SPORE PRINT: whitish.

MICROSCOPIC FEATURES: spores 7–10 × 2.5–3 μm, elliptic, slightly curved, smooth, hyaline.

FRUITING: typically in large overlapping clusters on decaying hardwood or rarely conifers; year-round; fairly common.

EDIBILITY: inedible.

COMMENTS: *Trametes trogii* (inedible) has a densely hairy, buff to ochraceous buff pore surface and labyrinthine pores. It typically grows on cottonwood and willow.

Trametes elegans (Sprengel : Fries) Fries

CAP: ¾–5½" (2–14 cm) wide, fan- to shell-shaped or nearly circular, convex to nearly plane, leathery to corky when fresh, becoming stiff when dry; upper surface very finely tomentose to smooth and silky, often concentrically sulcate or warted in age, dry, white to cream or grayish to pale ochraceous; margin thin, stalkless or with a rudimentary stalk.

FLESH: up to ⅝" (1.5 cm) thick, fibrous-tough, white to pale cream.

PORE SURFACE: white at first, becoming creamy white, then pale ochraceous in age, highly variable on each specimen and typically including portions that are labyrinthine, gill-like, and poroid and often split; pores round to angular or irregular, 1–2 per mm.

SPORE PRINT: whitish.

MICROSCOPIC FEATURES: spores 5–7 × 2–3 µm, cylindric to oblong-ellipsoid, smooth, hyaline.

FRUITING: scattered or in groups, often overlapping, on decaying hardwoods; July–December; fairly common.

EDIBILITY: inedible.

COMMENTS: this polypore is easy to recognize because its pore surface changes shape from the base to the margin.

Small, Centrally Stalked Polypores

Albatrellus caeruleoporus (Peck) Pouzar

COMMON NAME: Blue-pored Polypore.

CAP: 1–4¾" (2.5–12 cm) wide, convex to flat and circular to kidney-shaped, sometimes depressed; surface grayish blue to gray, becoming brownish in age; smooth to slightly felty, becoming cracked at maturity; margin incurved and wavy.

FLESH: up to ⅜" (1 cm) thick, firm, whitish; odor and taste not distinctive.

PORE SURFACE: blue to grayish blue; pores angular, 2–3 per mm, decurrent on the stalk.

STALK: 1–2¾" (2.5–7 cm) long, ¼–1" (6–25 mm) thick, central to eccentric, equal or tapering downward; blue to grayish blue, pitted on the upper half, smooth below.

SPORE PRINT: white.

MICROSCOPIC FEATURES: spores 4–6 × 3–5 µm, nearly round, smooth, hyaline.

FRUITING: solitary or in groups on the ground, usually under hemlock; August–October; infrequent.

EDIBILITY: edible.

COMMENTS: *Albatrellus pes-caprae* (edible) has a scaly, grayish brown to blackish brown cap, a decurrent white to greenish yellow pore surface, and large angular pores.

Albatrellus confluens (Albertini and Schweinitz : Fries) Kotlaba and Pouzar

CAP: 2–5½" (5–14 cm) wide, convex and circular to fan-shaped or lobed and irregular; surface pinkish buff to pale orange, smooth when young, sometimes cracked in age; margin incurved and entire, becoming uplifted, wavy, and partially eroded at maturity.

FLESH: ¼–¾" (6–20 mm) thick, soft and fleshy when fresh, creamy white, drying pinkish buff; odor not distinctive; taste cabbage-like or bitter.

PORE SURFACE: white to pale yellow when fresh, becoming salmon-pink when dry; pores circular to angular, 3–5 per mm.

STALK: 1–3⅛" (2.5–8 cm) long, ⅜–1" (1–2.5 cm) thick, tapered downward or nearly equal, central or lateral, creamy white to pinkish buff, becoming pinkish when dry, smooth.

SPORE PRINT: white.

MICROSCOPIC FEATURES: spores 4–5 × 2.5–3.5 μm, ovoid to elliptic, smooth, hyaline, weakly amyloid.

FRUITING: typically in fused clusters but also solitary, on the ground near conifers; July–November; infrequent.

EDIBILITY: edible, but sometimes bitter.

COMMENTS: *Albatrellus ovinus* (edible) forms whitish to creamy yellow or pale grayish fruiting bodies that grow solitary, scattered, or with several fused together on the ground near conifers. *Albatrellus cristatus* (edibility unknown) has a yellowish brown to olive-brown or yellowish green cracked cap and a white to greenish yellow pore surface. *Albatrellus peckianus* (edibility unknown) has a smooth yellow to yellowish tan cap.

Boletopsis subsquamosa (Fries) Kotlaba and Pouzar

COMMON NAME: Kurotake.

CAP: 2–5½" (5–14 cm) wide, circular, convex to broadly convex, usually somewhat depressed; surface dry, smooth, slightly fibrillose-scaly over the disc in age; color variable, dingy white, grayish, brownish or black, darkening when bruised or in age; margin slightly inrolled at first, becoming elevated, wavy, and sometimes furrowed or split in age.

FLESH: firm, thick, white to pale grayish; odor not distinctive, becoming fragrant or spicy when dry; taste somewhat bitter.

PORE SURFACE: subdecurrent, white at first, becoming grayish or pale brownish when dry; pores circular, 1–3 per mm.

STALK: 1⅜–3⅛" (3–8 cm) long, ¾–1⅛" (2–3 cm) thick, central to slightly eccentric, nearly equal, fibrillose-scaly, grayish to olive-brown, solid.

SPORE PRINT: whitish to pale brown.

MICROSCOPIC FEATURES: spores 5–7 × 4–5 μm, angular and irregular, warted, hyaline.

FRUITING: solitary, scattered, or in groups on the ground, usually partially buried in needle litter under conifers, occasionally under hardwoods; September–November; occasional to fairly common.

EDIBILITY: edible, but sometimes bitter.

COMMENTS: this polypore resembles a bolete, but the tube layer does not cleanly separate from the cap, the flesh is too tough, and the spores are not typical of boletes.

Coltricia cinnamomea (Persoon) Murrill

COMMON NAME: Shiny Cinnamon Polypore.

CAP: ½–2" (1.2–5 cm) wide, circular to irregular, plane to depressed, sometimes laterally fused, fibrous-tough; surface concentrically zoned, silky, shiny, bright reddish cinnamon to amber-brown and dark rusty brown; margin faintly striate, often torn, thin, and sharp.

FLESH: up to 1⁄32" (1 mm) thick, fibrous-tough, rusty brown, black in KOH.

PORE SURFACE: yellowish brown to reddish brown, not decurrent; pores angular, 2–4 per mm.

STALK: ⅜–1½" (1–4 cm) long, 1⁄16–¼" (1.5–6 mm) thick, central, nearly equal or tapering downward, velvety, dark reddish brown.

SPORE PRINT: yellowish brown.

MICROSCOPIC FEATURES: spores 6–10 × 4.5–7 μm, elliptic, smooth, hyaline.

FRUITING: solitary, in groups, or fused together on the ground, usually along roadsides and on paths in woods; June–November; fairly common.

EDIBILITY: inedible.

COMMENTS: *Coltricia perennis* (inedible) is larger, has a dull cap, brownish orange to pale cinnamon-brown, and a decurrent pore surface (see p. 383).

Coltricia perennis (Fries) Murrill

CAP: ¾–4¾" (2–11 cm) wide, circular to irregular, plane to depressed or funnel-shaped, often fused with adjacent specimens, fibrous-tough; surface concentrically zoned, dull, velvety to felted, brownish orange to pale cinnamon brown, becoming dark brown or grayish in age; margin thin, sharp, wavy.

FLESH: up to ¹⁄₁₆" (1.5 mm) thick, fibrous-tough, rusty brown, black in KOH.

PORE SURFACE: grayish, staining brown when bruised, yellow-brown to dark brown in age, decurrent; pores angular, 2–4 per mm.

STALK: ⅝–2¾" (1.5–7 cm) long, ⅛–⅜" (3–10 mm) thick, central, nearly equal, velvety, brown.

SPORE PRINT: pale yellowish brown.

MICROSCOPIC FEATURES: spores 6–9 × 3–5 μm, elliptic, smooth, pale yellowish brown.

FRUITING: solitary, in groups, or fused together on the ground, usually under conifers; June–November; fairly common.

EDIBILITY: inedible.

COMMENTS: *Coltricia cinnamomea* (inedible) is smaller, has a bright reddish cinnamon to dark rusty brown, shiny cap and a nondecurrent pore surface (see p. 382).

Inonotus tomentosus (Fries) Teng

COMMON NAME: Woolly Velvet Polypore.

CAP: 1⅜–6½" (3.5–16.5 cm) wide, circular to fan-shaped or irregular, sometimes lobed or fused with adjacent specimens, fibrous-tough; surface dry, velvety, or with matted hairs, coarsely wrinkled and uneven or smooth, tan to ochraceous or rusty brown; margin blunt, wavy to irregular at maturity.

FLESH: up to ¼" (6 mm) thick, fibrous-tough, yellowish brown to rusty brown, black in KOH.

PORE SURFACE: buff at first, becoming grayish brown to dark brown in age; pores angular, 2–4 per mm.

STALK: ¾–2" (2–5 cm) long, ¼–¾" (6–20 mm) thick, sometimes rudimentary, central to eccentric or lateral, nearly equal or irregular, velvety to felty, ochraceous to dark rusty brown.

SPORE PRINT: pale yellow to pale brown.

MICROSCOPIC FEATURES: spores 5–6 × 3–4 μm, elliptic, smooth, yellowish; setae present.

FRUITING: solitary, in groups, or fused together on the ground, duff, or decaying wood under conifers; August–October; fairly common.

EDIBILITY: inedible.

COMMENTS: specimens of *Phaeolus schweinitzii* (inedible) with a dull orange to ochre central stalk are similar, but have dull orange to ochre or rusty brown larger caps, up to 10" (25.5 cm) wide, thicker flesh, up to 1⅛" (3 cm), and a yellowish to greenish yellow pore surface that bruises and darkens brown in age.

Polyporus arcularius Bataille : Fries

COMMON NAME: Spring Polypore.

CAP: ⅜–1⅜" (1–8 cm) wide, circular, convex to depressed or funnel-shaped, fibrous-tough; surface dry, scaly, dark brown when young, becoming golden brown to yellowish brown in age; margin incurved, lined with conspicuous hairs.

FLESH: up to ¹⁄₁₆" (1.5 mm) thick, fibrous-tough, white.

PORE SURFACE: white to pale yellow; pores elongated, hexagonal to angular, often radially aligned, 0.5–1 per mm.

STALK: ¾–2⅜" (2–6 cm) long, ¹⁄₁₆–³⁄₁₆" (1.5–4.5 mm) thick, central, nearly equal, scaly or sometimes nearly smooth, yellowish brown to dark brown.

SPORE PRINT: white.

MICROSCOPIC FEATURES: spores 7–11 × 2–3 μm, cylindric, smooth, hyaline.

FRUITING: solitary or in groups on decaying hardwood or attached to buried wood; April–June; occasional.

EDIBILITY: inedible.

COMMENTS: *Polyporus brumalis* (inedible) has a yellowish brown to blackish brown cap with short, dense, often matted hairs and smaller circular to angular pores that are not radially aligned. *Polyporus craterellus* = *P. fagicola* (edible) is fleshy-fibrous and larger, up to 5½" (14 cm) wide, and has an ochraceous to tan cap with small reddish brown scales, a yellowish white to ochraceous stalk, and angular pores. *Polyporus mori* = *Polyporus alveolaris* = *Favolus alveolaris* (inedible) has a reddish-yellow to reddish orange cap, hexagonal pores, and a short, eccentric to lateral white stalk.

Polyporus badius (Persoon) Schweinitz

COMMON NAME: Black-footed Polypore.

CAP: 1½–8" (4–20.5 cm) wide, circular to irregular, typically funnel-shaped or convex to slightly depressed, fibrous-tough; surface smooth, shiny or dull, pale reddish brown with a darker center when young, soon chestnut-brown to reddish brown with a blackish brown center; margin pale brownish yellow to pale reddish brown and incurved when young, becoming uplifted, thin, wavy, or lobed in age.

FLESH: up to ⅝" (1.6 cm) thick, fibrous-tough, white.

PORE SURFACE: white to pale buff; pores circular to angular, 5–7 per mm.

STALK: ⅜–2" (1–2.5 cm) long, ⅛–⅝" (3–16 mm) thick, central or eccentric, nearly equal or tapering downward, smooth, reddish brown near the apex, black below.

SPORE PRINT: white.

MICROSCOPIC FEATURES: spores 6–10 × 3–5 μm, cylindric, smooth, hyaline; generative hyphae lacking clamp connections.

FRUITING: solitary, scattered, or in groups on decaying hardwood; August–December; fairly common.

EDIBILITY: inedible.

COMMENTS: also known as *P. picipes*. *Polyporus varius* (inedible) is smaller, up to 3⅛" (8 cm) wide, and has a pale buff to tan cap with pinkish brown to grayish radial striations and a black stalk at maturity. *Polyporus elegans* (inedible) is smaller, up to 2⅜" (6 cm) wide, and has a nonstriate tan to chestnut-brown cap and a stalk that is colored like the cap on the upper portion and black below. *Polyporus melanopus* (inedible) has a smaller cap, up to 4" (10 cm) wide, which is pale yellow-brown to gray-brown with radial fibrils, a velvety stalk that is longitudinally wrinkled and black on the lower half or overall, and generative hyphae with clamp connections.

Polyporus radicatus Schweinitz

COMMON NAME: Rooting Polypore.

CAP: 1⅜–8" (3–20.5 cm) wide, circular, convex to nearly plane or slightly depressed, fibrous-tough; surface velvety to finely scaly, sometimes finely cracked and roughened, dry, yellowish brown to reddish brown; margin incurved when young, then uplifted and wavy in age.

FLESH: up to ⅜" (1 cm) thick, fibrous-tough, white.

PORE SURFACE: white to creamy yellow; pores angular, 2–3 per mm.

STALK: 2–6" (5–15.5 cm) long, ¼–1" (6–25 mm) thick, central, yellowish brown, scurfy, enlarging downward, then quickly tapering below as a black underground root.

SPORE PRINT: white.

MICROSCOPIC FEATURES: spores 12–15 × 6–8 μm, oblong to elliptic, smooth, hyaline.

FRUITING: solitary, scattered, or in groups on the ground attached to buried roots, usually under hardwoods; August–October; fairly common.

EDIBILITY: inedible.

COMMENTS: *Polyporus badius* (inedible) and *P. melanopus* (inedible) are both similar but lack the black underground root. See the description and comments section of *P. badius* for additional information (see p. 384).

Large, Centrally to Laterally Stalked Polypores

Bondarzewia berkeleyi (Fries) Bondarzew and Singer (see photo, p. 11)

COMMON NAME: Berkeley's Polypore.

FRUITING BODY: a large overlapping cluster of flattened fused caps, or sometimes a solitary cap, attached to a solid central stalk arising from an underground sclerotium.

CAP: 2⅜–10" (6–25.5 cm) wide, fan-shaped, laterally fused and typically forming a rosette, sometimes lobed, fibrous-tough; surface densely matted and wooly or nearly smooth, dry, radially wrinkled and pitted, obscurely to conspicuously zoned, variously colored whitish to grayish, pale yellow to yellow-brown; margin blunt, wavy.

FLESH: up to 1⅛" (3 cm) thick, corky to fibrous-tough, white; odor not distinctive when fresh; taste mild when very young, bitter in age.

PORE SURFACE: white to creamy white; pores angular, frequently torn and irregular, 0.5–2 per mm.

STALK: aboveground portion 1¾–4¾" (4.5–12 cm) long, 1⅛–2" (3–5 cm) thick, central, roughened, dingy yellow to yellowish brown.

SPORE PRINT: white.

MICROSCOPIC FEATURES: spores 7–9 × 6–8 μm, round, with prominent amyloid ridges, hyaline.

FRUITING: solitary or scattered on the ground at the base of hardwood trees, especially oak; June–October; occasional.

EDIBILITY: edible when young, and easily sectioned, becoming fibrous-tough and bitter in age.

COMMENTS: mature specimens may be more than 40" (102 cm) wide.

Grifola frondosa (Dickson : Fries) S. F. Gray

COMMON NAME: Hen of the Woods.

FRUITING BODY: a large, dense cluster of overlapping caps attached to branches arising from a short, thick common stalk.

CAP: ¾–3⅛" (2–8 cm) wide, fan- to petal-shaped, sometimes lobed, fleshy-fibrous, laterally attached to the stalk branch; surface smooth to finely matted and woolly, faintly to distinctly zoned, gray to brownish gray, becoming dull dark brown in age; margin thin, wavy.

FLESH: up to ¼" (6 mm) thick, fleshy-fibrous, white; odor nutty or not distinctive; taste not distinctive.

PORE SURFACE: white to creamy white; pores angular, 1–3 per mm.

STALK: ¾–1¾" (2–4.5 cm) long, up to 4" (10 cm) or more thick, repeatedly branched, white.

SPORE PRINT: white.

MICROSCOPIC FEATURES: spores 5–7 × 3.5–5 μm, oval to elliptic, smooth, hyaline.

FRUITING: solitary or in groups on the ground at the base of trees, especially oak and maple; August–November; fairly common.

EDIBILITY: edible and choice.

COMMENTS: fruiting bodies may attain a diameter of 24" (60 cm) or more. *Dendropolyporus umbellatus* = *Polyporus umbellatus* (edible, choice) forms a large, overlapping cluster of circular, whitish to pale brown, depressed caps with centrally attached branches on the ground near hardwoods. Also compare with *Meripilus sumstinei* (edible), which has thicker, shelf-like caps and a white pore surface that stains black (see p. 387).

Laetiporus persicinus (Berkeley and Curtis) Gilbertson

FRUITING BODY: a large rosette of flattened to convex, laterally fused and lobed caps attached to a solid, central branching stalk.

CAP: 1⅛–10" (3–25.5 cm) wide, petal- to fan-shaped, soft, fleshy when young, fibrous-tough in age; surface velvety to densely matted and woolly, dry, radially wrinkled, pinkish orange to pinkish brown; margin pale pinkish cream to brownish, blunt, wavy, sometimes lobed.

FLESH: up to ¾" (2 cm) thick, soft, fleshy-fibrous, whitish to pale pinkish yellow; odor nutty or meaty; taste not distinctive.

PORE SURFACE: white to pinkish cream, bruising brownish; pores circular, 3–4 per mm.

STALK: 1½–3½" (4–9 cm) long, up to 2" (5 cm) thick, whitish to pinkish yellow.

SPORE PRINT: white.

MICROSCOPIC FEATURES: spores 6.5–8 × 4–5 μm, oval to elliptic, smooth, hyaline.

FRUITING: solitary or scattered on the ground attached to roots at the base of oak and pine trees; July–October; occasional.

EDIBILITY: edible and choice.

COMMENTS: also known as *Laetiporus sulphureus* var. *semialbinus*. Compare with *L. sulphureus* (see next description), which has bright to dull orange caps and a bright sulphur-yellow pore surface when fresh.

Laetiporus sulphureus (Bulliard : Fries) Murrill

COMMON NAME: Chicken Mushroom, Sulphur Shelf.

FRUITING BODY: a large, overlapping cluster of flattened, laterally fused, and lobed caps, sometimes forming rosettes or a solitary cap, stalkless or with a rudimentary stalk.

CAP: 2–12" (5–31 cm) wide, fan- to petal-shaped, soft, fleshy when young, fibrous-tough in age; surface velvety to densely matted and woolly, dry, radially wrinkled and roughened, bright to dull orange, fading to orange-yellow, then whitish in age; margin pale orange, blunt, wavy, often lobed.

FLESH: up to ¾" (2 cm) thick, fleshy-fibrous, white; odor nutty or not distinctive; taste not distinctive.

PORE SURFACE: bright sulphur-yellow; pores angular, 3–4 per mm.

SPORE PRINT: white.

MICROSCOPIC FEATURES: spores 5–8 × 3.5–5 μm, oval to elliptic, smooth, hyaline.

FRUITING: solitary, overlapping clusters, or rosettes on hardwoods, especially oak and cherry, occasionally on conifers, especially hemlock; May–November; fairly common.

EDIBILITY: edible and choice when collected on hardwoods; may cause gasterointestinal upset when gathered from conifer wood.

COMMENTS: the flesh of this mushroom has the consistency and flavor of white chicken meat. Compare with *L. persicinus* (edible, choice), which has a pinkish orange cap, a white pore surface, and forms rosettes (see p. 386).

Meripilus sumstinei (Murrill) Larsen in Lombard

COMMON NAME: Black-staining Polypore.

FRUITING BODY: a large, dense cluster of overlapping, shelf-like caps attached to a short, thick common stalk.

CAP: 2⅜–8" (6–20.5 cm) wide, fan- to spoon-shaped, fleshy-fibrous; surface yellowish tan to grayish yellow or sometimes yellow-orange to yellow when young, becoming pale brownish yellow to grayish yellow, then dark ochraceous to grayish brown at maturity; margin thin, sharp, wavy, often lobed, blackening when bruised or in age.

FLESH: up to ⅝" (1.6 cm) thick, fleshy-fibrous to fibrous-tough, white.

PORE SURFACE: white to creamy white, bruising black; pores angular, 3–6 per mm.

STALK: ⅜–1⅜" (1–3 cm) long, up to 4½" (11.5 cm) thick, ochre to reddish brown.

SPORE PRINT: white.

MICROSCOPIC FEATURES: spores 6–7 × 4.5–6 μm, oval to nearly round, smooth, hyaline.

FRUITING: solitary or in groups on the ground at the base of trees or stumps of hardwoods, especially oak; July–November; occasional.

EDIBILITY: edible.

COMMENTS: fruiting bodies may attain a diameter of 16" (41 cm) or more. It is commonly and incorrectly labeled *M. giganteus* in many field guides; *M. giganteus* does not occur in North America. Compare with *Grifola frondosa* (see p. 385), which has many gray to brownish gray caps with thinner flesh and a white pore surface that does not stain black.

Phaeolus schweinitzii (Fries) Patouillard

COMMON NAME: Dye Polypore.

FRUITING BODY: a large, overlapping cluster of flattened, fused caps, or sometimes a solitary cap, attached to a solid central stalk.

CAP: 1½–10" (4–25.5 cm) wide, fan- to petal-shaped or circular, fibrous-tough; surface densely matted and woolly or hairy, faintly to distinctly zoned, dull orange to ochre when young, rusty brown to dark brown in age; margin yellow-orange to brownish orange, sharp, wavy, sometimes lobed.

FLESH: up to 1⅛" (3 cm) thick, fibrous-tough, yellowish to reddish brown.

PORE SURFACE: yellow to greenish yellow or orange when young, bruising brown and becoming yellowish brown to dark rusty brown in age; pores angular, 0.5–3 per mm.

STALK: ¾–2¾" (2–7 cm) long, up to 2" (5 cm) thick, branched or unbranched, enlarging upward, pale to dark brown.

SPORE PRINT: whitish to yellowish.

MICROSCOPIC FEATURES: spores 5–9 × 3–5 μm, elliptic, smooth, hyaline.

FRUITING: solitary, overlapping clusters or rosettes on roots at the base of trees or on decaying wood, especially conifers; June–November; fairly common.

EDIBILITY: inedible.

COMMENTS: this mushroom is sometimes used to dye wool. Compare with *Inonotus tomentosus* (inedible), which has a smaller, up to 6½" (16.5 cm) wide cap, thinner flesh, and a brown pore surface (see p. 383).

Polyporus admirabilis Peck

FRUITING BODY: a large, overlapping cluster of laterally fused and lobed caps, sometimes forming rosettes or a solitary cap, laterally to centrally stalked or with a rudimentary stalk.

CAP: 4–14" (10–36 cm) wide, fan-shaped to circular and slightly depressed to funnel-shaped, fibrous-tough; surface dry, smooth, even, bright white to creamy white; margin blunt, wavy, sometimes lobed.

FLESH: up to 2" (5 cm) thick, fibrous-tough, white to creamy white; odor strong and unpleasant or sickeningly sweet; taste not distinctive.

PORE SURFACE: white; pores circular to angular, 3–5 per mm.

STALK: 1–3½" (2.5–9 cm) long, up to 1¾" (4.5 cm) thick, white to yellowish brown, with a narrowed base.

MICROSCOPIC FEATURES: spores 7–9 × 3–3.5 μm, cylindric, smooth, hyaline.

SPORE PRINT: white.

FRUITING: solitary, overlapping clusters, or rosettes on decaying hardwood stumps and trunks; July–November; rare.

EDIBILITY: inedible.

COMMENTS: *Polyporus albiceps* (inedible) is probably a synonym.

Woody, Shelf-like Polypores

Fomes fomentarius (Linnaeus : Fries) Kickx

COMMON NAME: Tinder Polypore.

CAP: 2⅜–8" (6–20.5 cm) wide, hoof-shaped stalkless, woody; surface concentrically furrowed and zoned, hard, thick, crusty, thickened at the central point of attachment, finely cracked and roughened or smooth, pale to dark gray or sometimes pale to dark brown; margin blunt, extending beyond the pore surface.

FLESH: up to 1⅛" (3 cm) thick, fibrous-tough to woody, yellowish brown.

PORE SURFACE: depressed, pale brown; pores circular, 3–5 per mm.

SPORE PRINT: white.

MICROSCOPIC FEATURES: spores 12–20 × 4–7 μm, cylindric, smooth, hyaline.

FRUITING: solitary, in groups, or in clusters on decaying hardwood; year-round; common.

EDIBILITY: inedible.

COMMENTS: *Phellinus igniarius* (inedible) has orange-brown flesh mixed with white mycelial threads, much smaller spores, 5–7 × 4.5–6 μm, and setae present. *Phellinus everhartii* (inedible) has a yellowish brown to black cracked cap, typically grows on oak, and has brown spores and setae. *Phellinus rimosus* (inedible) has a dark brown to blackish deeply cracked cap, typically grows on black locust, has brown spores, and lacks setae.

Fomitopsis pinicola (Swartz : Fries) Karsten (see photo, p. 11)

COMMON NAME: Red-belted Polypore.

CAP: 2¾–15" (7–38 cm) wide, shelf-like to hoof-shaped, woody; surface concentrically furrowed, thickened at the central point of attachment, hard, finely cracked or smooth, variously colored greenish gray to grayish brown or blackish brown, typically reddish near the margin; margin thick, rounded, creamy yellow, usually covered with a sticky reddish brown resinous layer.

FLESH: up to 1½" (4 cm) thick, corky to woody, creamy white to pale brown.

PORE SURFACE: creamy white, staining yellow when bruised; pores circular, 3–5 per mm.

SPORE PRINT: whitish.

MICROSCOPIC FEATURES: spores 6–9 × 3.5–4.5 μm, cylindric-elliptic, smooth, hyaline.

FRUITING: solitary or in groups on decaying wood or living trees, especially conifers; year-round; fairly common.

EDIBILITY: inedible.

COMMENTS: *Ganoderma applanatum* (see p. 389) is typically larger and thicker, lacks the resinous layer and reddish belt, and has a white pore surface that stains brown when bruised. *Phellinus chrysoloma* (inedible) is smaller, up to 3" (7.5 cm) wide, has woody, felted, yellowish brown to reddish brown caps with bright yellowish brown margins in overlapping clusters that arise from a bright ochre crust-like spreading mass with setae present. *Phellinus pini* (inedible) is very similar to *P. chrysoloma* but forms individual caps or rows of caps, not overlapping clusters, has setae, but lacks the crust-like spreading mass.

Ganoderma applanatum (Persoon) Patouillard

COMMON NAME: Artist's Conk.

CAP: 2–26" (5–65 cm) wide, shelf-like to somewhat hoof-shaped, stalkless, woody; surface hard, thick, crusty, concentrically furrowed, thickened at the central point of attachment, finely cracked and roughened, gray to grayish black or brown, dull; margin thin, often white.

FLESH: up to 2⅜" (6 cm) thick, corky to woody, brown.

PORE SURFACE: white, staining brown when bruised; pores circular, 4–6 per mm.

SPORE PRINT: brown.

MICROSCOPIC FEATURES: spores 7–11 × 5–7.5 μm, broadly elliptic, truncate, with a thick double wall, pale brown.

FRUITING: solitary, scattered, or in overlapping clusters on decaying wood, especially hardwood; year-round; common.

EDIBILITY: inedible.

COMMENTS: this species is often used by artists, who etch pictures on the fertile surface with knives or other sharp objects. *Ganoderma lobatum* (inedible) is smaller, up to 8" (20.5 cm) wide, with a yellowish margin, a thin, easily damaged, crusty surface and a creamy white to yellowish buff pore surface. *Heterobasidion annosum* (inedible) is also smaller, up to 10" (25.5 cm) wide, has a grayish white cap that becomes gray-brown to blackish brown in age, a very rough, furrowed surface, and white to pale cream flesh, and grows on conifers.

Fleshy to Fibrous-tough, Shelf-like Polypores

Fomitopsis cajanderi (Karsten) Kotlaba and Pouzar

COMMON NAME: Rosy Polypore.

CAP: 1–5½" (2.5–14 cm) wide, fan-shaped to semicircular or irregular, convex to nearly flat, stalkless, fibrous-tough, becoming corky to brittle when dry; surface coated with tiny, matted hairs, smooth or somewhat roughened but not cracked, typically faintly zoned, pinkish brown to reddish brown or grayish brown; margin fairly sharp, typically whitish when young.

FLESH: fibrous-tough, rosy pink, becoming pinkish brown in age; odor and taste not distinctive.

PORE SURFACE: rosy pink to pinkish brown; pores circular to angular, 3–5 per mm.

SPORE PRINT: whitish.

MICROSCOPIC FEATURES: spores 4–7 × 1.5–2 μm, sausage-shaped, smooth, hyaline.

FRUITING: solitary, scattered, or in overlapping clusters on decaying conifer wood, rarely on hardwood; year-round; fairly common.

EDIBILITY: inedible.

COMMENTS: *Fomitopsis roseus* (inedible) has a hoof-shaped cap that is pale rose pink when young and becomes brownish black, crusty, and cracked in age; a rose-pink to pinkish brown pore surface; and straight, cylindric spores.

Hapalopilus nidulans (Fries) Karsten

COMMON NAME: Tender Nesting Polypore.

CAP: 1–4¾" (2.5–12 cm) wide, fan-shaped to semicircular, convex, stalkless, soft and watery when fresh, becoming corky to brittle when dry; surface coated with tiny, matted hairs, becoming smooth in age, often with one or more shallow concentric furrows, dull brownish orange to cinnamon; margin sharp, curved.

FLESH: up to 1⅛" (3 cm) thick at the base, soft and watery when fresh, pale cinnamon.

PORE SURFACE: ochraceous to cinnamon-brown; pores angular, 2–4 per mm.

SPORE PRINT: white.

MICROSCOPIC FEATURES: spores 3.5–5 × 2–3 μm, elliptic to cylindric, smooth, hyaline.

FRUITING: solitary, in groups, or in overlapping clusters on decaying hardwood; June–November; occasional.

EDIBILITY: inedible.

COMMENTS: all parts instantly stain bright violet with KOH (see photo). This mushroom is very popular with those who dye wool. *Hapalopilus croceus* = *Aurantioporus croceus* (inedible) is larger, up to 8" (20.5 cm) wide, has a bright reddish orange pore surface when fresh, and stains red with KOH. *Phellinus gilvus* (inedible) has an ochre to bright rusty yellow fibrous-tough cap, yellowish brown flesh that stains black in KOH, a grayish brown to dark brown pore surface, and setae.

Inonotus rheades (Persoon) Bondarzew and Singer

CAP: 2–4⅜" (5–11 cm) wide; fan-shaped, semicircular, circular, and irregular, broadly convex, stalkless, moist, fibrous-tough and somewhat flexible when fresh, becoming dry and rigid in age; surface coated with a dense layer of fibrils that becomed matted in age; fibrils pale yellow-brown at first, becoming stained rusty brown from spores, darkening in age; margin incurved when young, often wavy at maturity.

FLESH: bright yellowish brown when young, becoming dark rusty brown, shiny; odor and taste not distinctive.

PORE SURFACE: pale yellowish brown when young, becoming dark reddish brown in age; pores angular, often lacerated, 1–3 per mm.

SPORE PRINT: rusty brown.

MICROSCOPIC FEATURES: spores 5–6 × 3–4 μm, oval to broadly elliptic, often flattened on one end, smooth, pale brown; setae absent.

FRUITING: in groups or overlapping clusters on decaying trunks, logs, and stumps of poplar; June–October; infrequent.

EDIBILITY: inedible.

COMMENTS: all parts stain blackish brown when KOH is applied. A copious deposit of rusty brown spores is often observed on fallen leaves and plants beneath the fruiting bodies.

Ischnoderma resinosum (Fries) Karsten

COMMON NAME: Resinous Polypore.

CAP: 3–10" (7.5–25.5 cm) wide, semicircular to fan-shaped, flattened to convex, stalkless, fleshy-soft when young and fresh, becoming fibrous-tough to brittle in age; surface concentrically and radially furrowed, faintly to distinctly zoned, velvety when young, later covered with a thin, glossy resinous crust, dull brownish orange to dark brown; margin thick, rounded, whitish to ochre, exuding drops of water when fresh.

FLESH: up to ¾" (2 cm) thick, soft, becoming fibrous in age, whitish to pale yellow.

PORE SURFACE: white, bruising brown, becoming pale brown in age; pores angular to circular, 4–6 per mm.

SPORE PRINT: whitish.

MICROSCOPIC FEATURES: spores 4.5–7 × 1.5–2.5 μm, cylindric to sausage-shaped, smooth, hyaline.

FRUITING: solitary or in overlapping clusters on decaying wood; September–November; fairly common.

EDIBILITY: inedible.

COMMENTS: some authors consider *I. resinosum* to be a species that grows only on hardwoods and recognize *I. benzoinum* (inedible) as a similar species that grows only on conifers. Other authors consider these two species to be synonymous. *Coriolopsis gallica* = *Trametes hispida* (inedible) is smaller, up to 4" (10 cm) wide; has a densely hairy, sometimes tufted, yellowish brown to grayish cap, a grayish brown pore surface, brown flesh, spores that measure 10–16 × 3–5 μm, and often grows in dense overlapping clusters.

Oligoporus fragilis (Fries) Gilbertson and Ryvarden

COMMON NAME: Staining Cheese Polypore.

CAPS: ⅜–4" (1–10 cm) wide, fan-shaped to semicircular, sometimes laterally fused, flattened to convex, stalkless, soft when fresh, fibrous-tough in age; surface covered with soft fibers that become matted in age, white to buff, staining reddish brown when bruised or on drying; margin thin, sharp, sometimes wavy.

FLESH: up to ⅝" (1.6 cm) thick, soft to fibrous, white, staining reddish yellow when exposed.

PORE SURFACE: white, staining yellowish, then reddish brown when bruised; pores angular to irregular, sometimes jagged, 2–4 per mm.

SPORE PRINT: whitish.

MICROSCOPIC FEATURES: spores 4–5 × 1–2 μm, cylindric to sausage-shaped, smooth, hyaline.

FRUITING: solitary, in overlapping clusters, or laterally fused in rows on decaying conifer wood; September–November; occasional.

EDIBILITY: inedible.

COMMENTS: *Oligoporus caesius* = *Tyromyces caesius* (inedible) has a white cap and pore surface with grayish blue to dark blue tints, especially on the margin. *Tyromyces chioneus* (inedible) has a white cap and pore surface that lacks marginal tints and does not stain when bruised or on drying. Several other similar species may be collected and should be confirmed microscopically.

Oxyporus populinus (Schumacher : Fries) Donk

COMMON NAME: Mossy Maple Polypore.

CAP: 1⅛–8" (3–20.5 cm) wide, fan- to kidney-shaped, convex, stalkless, fibrous-tough; surface covered with dense matted hairs, smooth in age, white to grayish white or creamy white, often extensively covered with moss; margin sharp, uneven.

FLESH: up to ¾" (2 cm) thick, soft-corky to fibrous-tough, white to yellowish brown.

PORE SURFACE: white to creamy white; pores circular to angular, 4–7 per mm.

SPORE PRINT: white.

MICROSCOPIC FEATURES: spores 3.5–5.5 μm, nearly round, smooth, hyaline.

FRUITING: solitary, in groups, or in overlapping clusters on standing hardwood, especially maple; year-round; fairly common.

EDIBILITY: inedible.

COMMENTS: the species name, *populinus,* which is a reference to poplar, is a misnomer. It is also known as *Fomes connatus. Gloeoporus dichrous = Caloporus dichrous* (inedible) is much smaller, up to 2" (5 cm) wide, has elongated, leathery, white, velvety caps in overlapping clusters and a rubbery-gelatinous, separable, pale reddish to purplish brown pore surface. *Bjerkandera adusta* (inedible) is also much smaller, up to 3" (7.5 cm) wide, has white to grayish velvety caps with a blackish margin, a grayish pore surface, and grows in overlapping clusters.

Polyporus squamosus Hudson : Fries (see photo, p. 11)

COMMON NAME: Pheasant's-back Polypore, Dryad's Saddle.

CAP: 2–12" (5–30.5 cm) wide, kidney- to fan-shaped, circular or funnel-shaped, fleshy-fibrous when young, becoming fibrous-tough in age; surface creamy white to dingy yellowish or pale yellow-brown, with large, flattened, reddish brown scales that are often concentrically arranged; margin thin, entire, or sometimes conspicuously lobed.

FLESH: up to 1½" (4 cm) thick, soft-corky to fibrous-tough, white; odor and taste of watermelon rind.

PORE SURFACE: white to yellowish, decurrent; pores angular, large, 0.25–1 per mm.

STALK: ⅜–4¾" (1–12 cm) long, ⅜–2" (1–5 cm) thick, lateral, eccentric, or sometimes rudimentary, white on the upper portion, brown to brownish black at the base.

SPORE PRINT: white.

MICROSCOPIC FEATURES: spores 10–18 × 4–7 μm, cylindric, smooth, hyaline.

FRUITING: solitary, in groups, or in overlapping clusters on decaying hardwood; May–November; common.

EDIBILITY: edible when young and tender.

COMMENTS: *Polyporus craterellus = P. fagicola* (edible) is smaller, up to 5½" (14 cm) wide, and has fewer and smaller scales and a central yellowish white to ochraceous stalk. *Polyporus mori = Polyporus alveolaris = Favolus alveolaris* (inedible) is much smaller, up to 4" (10 cm) wide, and has a reddish yellow to reddish orange cap, hexagonal pores, and a short, eccentric to lateral, white stalk.

Pycnoporellus fulgens (Fries) Donk

CAP: 1½–3½" (4–9 cm) wide, fan-shaped to semicircular, flattened to slightly convex, stalkless, fibrous-tough; surface coated with tiny matted hairs, becoming nearly smooth in age, pale to dark orange to rusty.

FLESH: up to ¼" (6 mm) thick, corky to fibrous-tough, pale orange, staining red in KOH.

PORE SURFACE: pale to dark orange; pores circular to angular, 2–3 per mm, often lacerated in age.

SPORE PRINT: whitish.

MICROSCOPIC FEATURES: spores 6–9 × 2.5–4 μm, smooth, hyaline, inamyloid.

FRUITING: solitary, scattered, in groups, or sometimes in overlapping clusters on decaying conifer wood or hardwood; year-round; occasional.

EDIBILITY: inedible.

COMMENTS: carefully compare with *P. alboluteus* (see p. 378) and *Pycnoporus cinnabarinus* (see p. 393).

Pycnoporus cinnabarinus (Jacquin : Fries) Karsten

COMMON NAME: Cinnabar-red Polypore.

CAP: 1⅛–5½" (3–14 cm) wide, up to ⅝" (1.6 cm) thick, fan- to kidney-shaped, flattened to convex, stalkless, fibrous-tough; surface wrinkled or smooth, orange to reddish orange, sometimes mixed with other colors in age; margin rounded or sharp.

FLESH: up to ⅝" (1.6 cm) thick, corky to fibrous-tough, reddish orange, black with KOH.

PORE SURFACE: dark to pale orange-red; pores circular to angular, 2–4 per mm.

SPORE PRINT: whitish.

MICROSCOPIC FEATURES: spores 4.5–8 × 2.5–4 μm, cylindric to sausage-shaped, smooth, hyaline.

FRUITING: solitary, in groups, or in overlapping clusters or rosettes on decaying wood; year-round; fairly common.

EDIBILITY: inedible.

COMMENTS: *Pycnoporus sanguineus* (inedible) is much less common, has a smaller, up to 2⅜" (6 cm) wide cap, is thinner, up to ³⁄₁₆" (4.5 mm) thick, and has a dark red pore surface.

Trametes conchifer (Schweinitz : Fries) Pilát

COMMON NAME: Little Nest Polypore.

FRUITING BODY: at first a small cup- to saucer-shaped sterile structure, followed by and typically attached to a cap.

CUP: up to ⅝" (1.6 cm) wide, smooth, concentrically zoned, brown at first, then white and brown.

CAP: ⅜–2" (1–5 cm) wide, fan-to kidney-shaped, flattened, stalkless, fibrous-tough; surface smooth to slightly velvety, radially wrinkled and concentrically zoned; zones of contrasting colors white to grayish white, yellowish or pale brown; sometimes forming secondary cups; margin thin, wavy.

FLESH: less than ¹⁄₃₂" (1 mm) thick, fibrous-tough, white.

PORE SURFACE: white to yellowish; pores angular, 2–4 per mm.

SPORE PRINT: white.

MICROSCOPIC FEATURES: spores 5–7 × 1.5–2.5 μm, cylindric, smooth, hyaline.

FRUITING: in groups on decaying hardwood; June–December; occasional.

EDIBILITY: inedible.

COMMENTS: also known as *Poronidulus conchifer.*

Trametes versicolor (Linnaeus : Fries) Pilát

COMMON NAME: Turkey-tail.

CAP: ¾–4" (2–10 cm) wide, fan- to kidney-shaped, flattened, sometimes laterally fused and forming extensive rows, stalkless, fibrous-tough, thin; surface velvety to silky, with conspicuous concentric zones; zones contrasting and variously colored, often with shades of brown, blue, gray, orange, and green; margin thin, sharp, wavy, sometimes folded or lobed.

FLESH: up to ⅛" (3 mm) thick, fibrous-tough, white to creamy white.

PORE SURFACE: white to grayish; pores angular to circular, 3–5 per mm.

SPORE PRINT: white.

MICROSCOPIC FEATURES: spores 5–6 × 1.5–2 μm, cylindric to sausage-shaped, smooth, hyaline.

FRUITING: solitary, or in overlapping clusters, rows, or rosettes; year-round; very common.

EDIBILITY: inedible.

COMMENTS: *Stereum* species lack pores on their lower surfaces. *Trametes hirsuta* (inedible) has a grayish to yellowish or brownish zoned cap, usually with a brown margin. *Trametes pubescens* (inedible) has a finely hairy to smooth, creamy white to yellowish buff, azonate or faintly zoned cap.

Tyromyces chioneus (Fries) Karsten

COMMON NAME: White Cheese Polypore.

CAP: ¾–4" (2–10 cm) wide, semicircular to nearly circular or petal- to tongue-shaped, convex to nearly plane; surface smooth to slightly tomentose, azonate, dry, soft, white at first, becoming pale yellow to pale grayish in age, stalkless.

FLESH: up to ¾" (2 cm) thick, soft, spongy, white; odor fragrant when fresh, often disappearing after picking; taste not distinctive.

PORE SURFACE: white to pale cream, darkening when dry; pores angular to circular, 3–5 per mm.

SPORE PRINT: white.

MICROSCOPIC FEATURES: spores 4–5 × 1.5–2 μm, cylindric, and typically curved, smooth, hyaline.

FRUITING: scattered or in groups, sometimes overlapping, on decaying hardwoods; July–December; fairly common.

EDIBILITY: inedible.

COMMENTS: *Oligoporus caesius* (inedible) is similar but its upper surface has a tint of blue, it often bruises intensely blue especially on the margin, and it lacks the fragrant odor. *Oligoporus fragilis* (inedible) has a whitish cap that stains reddish brown when bruised or on drying. *Oligoporus tephroleucus* (inedible) has a white, coarsely strigose cap and lacks the fragrant odor. *Tyromyces fissilis* (inedible) forms large, sappy to waxy and tough, stalkless fruiting bodies with a white upper surface that stains cream to ochraceous and thick, white flesh that has a pungent cheese-like odor and a bitter to astringent taste.

Albatrellus caeruleoporus

Albatrellus confluens

Boletopsis subsquamosa

Coltricia cinnamomea

Coltricia montagnei

Coltricia perennis

Cryptoporus volvatus

Daedalea quercina

Daedaleopsis confragosa

Fistulina hepatica

Fomes fomentarius

Fomitopsis cajanderi

Ganoderma applanatum

Ganoderma tsugae

Gloeophyllum sepiarium

Globifomes graveolens

Grifola frondosa

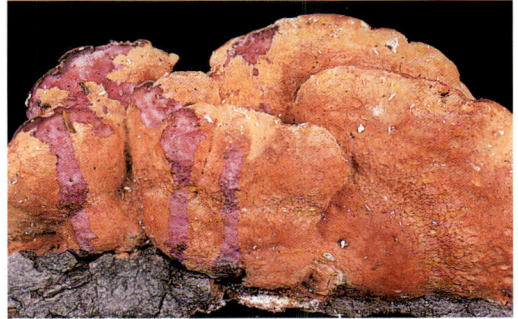

Hapalopilus nidulans

Inonotus obliquus

Inonotus rheades

Inonotus tomentosus

Irpex lacteus

Ischnoderma resinosum

Laetiporus persicinus

Laetiporus sulphureus

Meripilus sumstinei

Oligoporus fragilis

Oxyporus populinus

Phaeolus schweinitzii

Piptoporus betulinus

Polyporus admirabilis

Polyporus arcularius

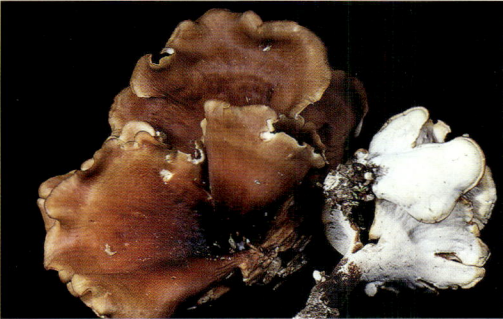

Polyporus badius

Polyporus radicatus

Pycnoporellus alboluteus

Pycnoporellus fulgens

Pycnoporus cinnabarinus

Trametes cervina

Trametes conchifer

Trametes elegans

Trametes versicolor

Trichaptum biforme

Tyromyces chioneus

Tooth Fungi

Tooth Fungi have downward pointing, spine-like teeth on which they produce spores. Although one species of this group grows on fallen pine cones, most Tooth Fungi grow on the ground and form teeth on the underside of their caps. Other members grow on wood and have teeth along branches or at the tips of branches. One species, usually found on standing trees, resembles a satyr's beard and has long spines hanging from a stalkless, solid mass of tissue. They differ from similar tooth-like Polypores by forming conic teeth instead of elongated, flattened, irregular tubes. The spine-like tips of similar Branched and Clustered Coral Fungi do not point downward. Several species are excellent edibles, but many are much too tough to be eaten.

Key to Species of Tooth Fungi

1. Fruiting body growing on the ground; shoehorn- to tongue-shaped or hemispheric with a cap and eccentric stalk, rubbery-gelatinous; gray-translucent, yellowish, or brown; lower surface densely coated with tiny, soft whitish spines; on decaying wood— *Pseudohydnum gelatinosum* (Fries) Karsten (see Jelly Fungi, p. 432).
1. Fruiting body growing on pine cones; with a distinct cap and stalk; cap semicircular to kidney-shaped, often with a hairy fringe; undersurface with white to dull yellow spines —*Auriscalpium vulgare* S. F. Gray (see p. 406).
1. Fruiting body growing on wood → 2.
1. Fruiting body growing on the ground; with a distinct cap and central stalk; cap smooth or cracked when young, becoming scaly at maturity; flesh thick, firm, and brittle → 5.
1. Fruiting body growing on the ground; with a distinct cap and central stalk; cap smooth to slightly roughened, sometimes felty, not developing distinct scales, sometimes with tiny scales over the disc at maturity; flesh thick, firm, and brittle → 6.
1. Fruiting body growing on the ground or on leaf litter, often enveloping needles and usually under conifers; with a cap or several caps fused together and a stalk; flesh of the stalk extremely fibrous-tough and typically flexible → 9.
    2. Fruiting body a bright yellow, spreading crust with crowded, waxy spines that slowly develop reddish stains when bruised; with an intense unpleasant odor; on logs and branches of hardwoods—*Sarcodontia setosa* (Persoon) Donk.
    2. Fruiting body a bright orange-yellow, spreading crust with a whitish margin; with crowded, yellow to orange, rounded spines and orange yarn-like runners; on the underside of decaying hardwood logs—*Phanerochaete chrysorhiza* (Torrey) Gilbert.
    2. Fruiting body an olive-brown to reddish brown, leathery to stiff, spreading crust with uneven flattened teeth; on decaying hardwood branches—*Hydnochaete olivaceus* (Schweinitz) Banker.

2. Fruiting body a whitish, spreading crust with tubes that break into white to creamy white flattened teeth; on decaying hardwood branches—*Irpex lacteus* (Fries) Fries (see Polypores, p. 373).

2. Fruiting body not with the above combinations of characters → 3.

3. Fruiting body consisting of fan-shaped caps growing in overlapping clusters from a common base, 4–10" (10–25.5 cm) wide, 1–2" (2.5–5 cm) thick, 5½–15" (14–38 cm) high, stalkless; whitish to creamy yellow when young, becoming yellow-brown in age; under-surface composed of whitish, crowded spines; odor and taste not distinctive when young, becoming like spoiled ham and bitter in age; on standing hardwood trunks, especially maple—*Climacodon septentrionale* (Fries) Karsten (see p. 407).

3. Fruiting body a cluster of spreading branches with spines arranged in rows along the branches like teeth on a comb, arising from a common base; 2¾–10" (7–25 cm) wide, 2¾–8" (7–20 cm) high, spines ⅛–⅜" (3–10 mm) long, rather evenly distributed along the branches; white to salmon or pinkish; growing on decaying hardwoods—*Hericium coralloides* (Scopoli : Fries) S. F. Gray (see p. 407).

3. Fruiting body a cluster of compact, forking branches with spines arranged primarily in bundles at the tips of the branches, arising from a common base; 4–12" (10–30.5 cm) wide, 6–15" (15.5–36 cm) high, spines ¾–1½" (2–4 cm) long, white; spores 5–6 × 4.5–5.5 μm; growing on decaying hardwoods; edible—*Hericium americanum* Ginns.

3. Fruiting body a stalkless, solid, cushion-shaped mass resembling a beard, giving rise to long spines; 2¾–8" (7–20.5 cm) wide, 3¾–8" (9.5–20.5 cm) high; spines ¾–2¾" (2–7 cm) long, whitish to yellowish; growing on decaying hardwoods—*Hericium erinaceus* (Fries) Persoon (see p. 408).

3. Fruiting body not with the above combinations of characters → 4.

4. Flesh not exuding a creamy white sticky sap when squeezed; fruiting body consisting of a cap, without a stalk or with a rudimentary lateral stalk; cap white when young, cinnamon-brown in age; 1–2½" (2.5–6.5 cm) wide, fan-shaped to circular, overlapping or fused; somewhat velvety, with shallow concentric furrows, whitish to grayish when rubbed, usually zoned at the margin; spines on the undersurface ⅟₃₂–⅛" (1–3 mm) long, fused and appearing forked at their tips; spores 3–4 × 1–1.2 μm; growing on decaying hardwood; July–November; inedible—*Steccherinum adustum* (Schweinitz) Banker = *Mycorrhaphium adustum* (Schweinitz) Maas Geesteranus.

4. Flesh not exuding a creamy white, sticky sap when squeezed; fruiting body a stalkless, spreading, crust-like mass with projecting caps; cap orange-yellow with a whitish margin; ¾–1¾" (2–4.5 cm) wide, semicircular to fan-shaped, sometimes overlapping or fused, velvety to hairy; spines on the undersurface ⅟₃₂–⅟₁₆" (1–1.5 mm) long, usually forked, orange-yellow; spores 3–4 × 2–2.5 μm; growing on decaying hardwoods; June–November; inedible—*Steccherinum ochraceum* (Persoon : Fries) S. F. Gray.

4. Flesh exuding a creamy white, sticky sap when squeezed; fruiting body consisting of a cap, with or without a stalk; cap whitish to pale tan or pinkish tan; 1⅜–4" (3.5–10 cm) wide, semicircular to fan-shaped, typically overlapping or fused, arising from a confluent, spreading base; upper surface densely hairy; spines on the undersurface ⅛–³⁄₁₆" (2.5–5 mm) long, crowded, whitish then pinkish tan in age; stalk absent or a short lateral projection; growing on decaying hardwoods; July–November—*Steccherinum pulcherrimum* (Berkeley and Curtis) Banker (see p. 411).

5. Flesh white to pale brown, mild or slightly bitter, not strongly bitter; cap 2–8" (5–20.5 cm) wide, cracked and conspicuously scaly when immature and mature; scales

dry, dark brown, erect, more or less concentric; spines on the undersurface pale brown tinted grayish when young, becoming reddish brown to dark brown in age; spore print brown; on the ground under conifers or hardwoods—*Sarcodon imbricatus* (Linnaeus : Fries) (see p. 411).

5. Flesh white, odor farinaceous, taste slowly bitter or farinaceous and unpleasant; cap 2–5½" (5–14 cm) wide, cracked when young and soon developing small, flattened, reddish brown scales that darken in age and become partially erect, especially over the center; spines on the undersurface ⅛–⅜" (3–10 mm) long, white to buff or tan with paler tips; stalk pale pinkish brown at first, becoming dark brown in age, with a grayish olive to greenish black base; spore print brown; spores 6–7.5 × 5–6 μm; on the ground in conifer or hardwoods; July–October; uncommon; inedible—*Sarcodon scabrosus* (Fries) Karsten.

5. Flesh white in the cap to pale brown in the stalk, darkest at the stalk base, odor fragrant or farinaceous, taste instantly very bitter; cap 2–5½" (5–14 cm) wide, cracked when young and soon developing small, flattened, pale reddish brown scales that darken in age and become partially erect, especially over the center; spines on the undersurface ¹⁄₁₆–⅛" (1–3 mm) long, white when young, becoming brown with grayish tips; stalk pale pinkish brown to pale reddish brown, tapering downward to an abrupt, white, pointed base; spore print brown; on the ground in hardwoods—*Sarcodon underwoodii* Banker (see p. 411).

5. Not as above → 6.

    6. Growing on the ground under conifers; stalk tapering downward, pale to dark brown with a white basal mycelium; cap 2–6¾" (5–17.5 cm) wide, smooth, color variable, pale pinkish brown to brown with a vinaceous tint, often with dark brown spots in age; flesh yellowish; odor slightly medicinal to farinaceous; taste mild to farinaceous at first, then slowly bitter, or bitter at once; staining dark green, then blackish when a drop of KOH is applied; spines on the undersurface ¼–½" (6–12 mm) long, pale grayish brown with white tips; spore print brown; spores 5–6 × 4–5.5 μm, warted; edibility unknown—*Sarcodon subfelleus* (Harrison) Harrison.

    6. Growing on the ground under conifers; stalk tapering downward to a radicating base, colored like the cap or slightly paler, olive-black or blackish near the base; cap 1–4¾" (2.5–12 cm) wide; pinkish brown to purplish brown or grayish purple, staining dark purple-brown to gray-brown when bruised or handled; flesh bluish gray to dingy pink, instantly dark green, then blackish when a drop of KOH is applied; odor fragrant; taste farinaceous to slowly acrid or disagreeable; spines on the undersurface short, ¹⁄₁₆–⅛" (1–3 mm) long, decurrent, pinkish brown to reddish brown; spore print brown—*Sarcodon fuligineo-violaceus* (Kalchbrenner) Patouillard (see p. 410).

    6. Growing on the ground under conifers or mixed woods; stalk nearly equal to tapered downward, colored like the cap but usually darker brown, often whitish near the apex; cap 1⅛–5" (3–12.5 cm) wide, pale pinkish brown to pale purplish brown or pale grayish brown, finely tomentose to fibrillose, not readily bruising when handled; flesh dull white to pale pinkish buff, instantly staining dark olive-green when a drop of KOH is applied; odor fragrant like maple syrup when fresh, soon disappearing after being picked; taste not distinctive; spines on the under-surface up to ¼" (6 mm) long, decurrent, pale gray; spore print white; spores 4.5–6 × 4–5 μm, subglobose, finely echinulate, hyaline; August–October—*Bankera violascens* (Albertini and Schweinitz) Pouzar (see p. 407).

    6. Growing on the ground under hardwoods, especially oak and beech; stalk nearly

equal down to an abruptly tapering base, purple-brown, olive-green near the base; cap 1⅜–4" (3.5–10 cm) wide, smooth when young, becoming finely cracked and developing tiny, flattened scales at maturity; orange-yellow to reddish brown; flesh whitish, immediately staining pinkish lilac on exposure, then darkening to violet; staining dark green and not blackening when a drop of KOH is applied; odor farinaceous; taste somewhat acrid; spines on the undersurface short, 1/16–⅛" (1.5–3 mm) long, pale pinkish brown when young, purplish brown in age, strongly decurrent; spore print brown; spores 5.5–6.5 × 4.5–5 μm, warted; August–October; rare; inedible—*Sarcodon joeides* (Passerini) Patouillard.

6. Fruiting body not with the above combinations of characters → 7.

7. Cap color variable, pale yellowish tan to grayish tan with olive-green and black tints when young, becoming grayish black at the center and grayish yellow toward the margin, staining black when rubbed or bruised; smooth and somewhat felty, 1–3⅛" (2.5–8 cm) wide; flesh pale grayish white; odor not distinctive; taste bitter; spines on the undersurface long and stout, ¼–⅝" (6–16 mm) long, grayish brown to dark brown with paler tips; stalk colored and staining like the cap; spore print brown; growing on the ground in hardwoods and mixed woods—*Sarcodon atroviridis* (Morgan) Banker (see p. 410).

7. Cap yellow-orange, pale brownish orange to apricot-orange or reddish orange, staining dark orange when bruised, ¾–6" (2–15.5 cm) wide, smooth, convex to nearly plane, sometimes slightly depressed; flesh white, staining orange-yellow when cut and rubbed; odor somewhat nutty to sweet or not distinctive; taste mild or peppery; stalk 1–4" (2.5–10 cm) long, ⅜–1⅜" (1–3.5 cm) thick, white with orange tints or colored like the cap; spore print white—*Hydnum repandum* Linnaeus : Fries (see p. 409).

7. Cap reddish orange to brownish orange or pale dull orange; 1–2" (2.5–5 cm) wide, convex to nearly plane, with a deep central depression, smooth; flesh white to pale orange, staining dark orange when cut and rubbed; odor and taste not distinctive; stalk ¾–2¾" (2–7 cm) long, 3/16–⅜" (5–10 mm) thick, white with orange tints or colored like the cap; spore print white—*Hydnum umbilicatum* Peck (see p. 409).

7. Fruiting body not with the above combinations of characters; cap white to creamy white → 8.

8. Flesh taste mild or slowly bitter; odor not distinctive; white, staining orange when cut and rubbed; cap ¾–4¾" (2–12 cm) wide, white to creamy white, staining orange when bruised; stalk 1⅛–3" (3–7.5 cm) long, white to creamy white, staining orange when bruised; spore print white—*Hydnum repandum* var. *album* (Quélet) Rea (see p. 409).

8. Flesh taste slightly to distinctly acrid or rarely mild; odor not distinctive; white, staining orange when cut and rubbed; cap ⅜–2¾" (1–7 cm) wide, white to creamy white, staining orange when bruised; spines on the undersurface ⅛–¼" (3–6 mm) long, white to creamy white, darkening when bruised, irregularly decurrent or adnate; stalk ¾–2" (2–5 cm) long, white, staining orange when bruised; spore print white; spores 4–5.5 × 3.5–4 μm, oval, smooth, hyaline; growing on the ground under conifers or hardwoods; July–October; edible—*Hydnum albidum* Peck.

9. Cap, stalk, and tips of spines bright sulphur-yellow; flesh homogeneous, not zoned, yellow-brown, brighter yellow at margins; stalk base coated with sulphur-yellow mycelium; spore print brown; spores 4–5 × 3–3.5 μm, warted; under conifers; August–November; inedible—*Hydnellum geogineum* (Fries) Banker.

9. Cap, stalk, and flesh orange-salmon to dark rusty orange or rusty red; cap surface often

white and velvety when young and actively growing; flesh distinctly zoned when longitudinally sectioned; spore print brown; spores 5.5–6.5 × 4.5–5 μm, warted; under conifers; August–November; inedible—*Hydnellum aurantiacum* (Fries) Karsten.

9. Not as above; fruiting body with blue to violet colors present on the cap, stalk, spines, or flesh; flesh distinctly zoned when longitudinally sectioned; cap surface often white and velvety when young and actively growing; spore print brown; spores warted → 10.

9. Not as above; fruiting body yellowish brown, reddish brown to rusty brown or purplish brown; caps commonly fused together, forming large masses; cap surface often white and velvety when young and actively growing, sometimes exuding red droplets when fresh and moist; flesh distinctly zoned when longitudinally sectioned; spore print brown; spores warted → 11.

9. Not as above; spore print white; spores round to oval, weakly to strongly echinulate; caps often fused together into large masses → 14.

    10. Odor pleasant, strongly fragrant-sweet; cap white at first, becoming brownish with violet tints; flesh in cap distinctly zoned with white, blue, and brown; flesh in stalk dark blue to violet-black; stalk blue to blue-black, coated with dark blue mycelium; spores 3.5–5 × 2–3.5 μm, warted; on the ground under conifers; August–November; inedible—*Hydnellum suaveolens* (Fries) Karsten.

    10. Odor variously described as medicinal, menthol-like, or aromatic; cap white to pale brown at first and beaded with red droplets when fresh and moist, becoming brown in age; stalk white with blue tints; spines lilac to dark blue, becoming brownish at maturity; flesh in cap pale gray to pale brown with tints of lilac, darker downwards in the stalk, zonate with bluish bands; spores 4–4.5 × 3.5–4.5 μm, warted; on the ground under conifers; August–November; inedible—*Hydnellum cruentum* Harrison.

    10. Odor not distinctive or slightly farinaceous; cap white at first, soon tinted blue, often entirely blue, changing to brown in the center; cap flesh distinctly zoned blue to purple with brown; stalk flesh orange to orange-red; stalk pale brown at first, becoming darker brown at maturity; spines whitish at first, becoming brown with pale tips in age—*Hydnellum caeruleum* (Hornemann) Karsten (see p. 408).

11. Stalk base conspicuously swollen to bulbous and very spongy → 12.

11. Stalk base not swollen or slightly swollen, not very spongy; flesh taste extremely acrid; flesh in cap and stalk faintly zoned, pale pinkish brown to dark reddish brown; odor fragrant to pungent or not distinctive; cap white to pink and velvety when young or actively growing, exuding red droplets when fresh and moist; becoming brown to vinaceous-brown or dark brown at maturity; surface irregular with pits, horn-like projections, and ridges; spores 5–6 × 4–4.5 μm, warted; on the ground under conifers; August–October; inedible—*Hydnellum peckii* Banker.

11. Stalk base not swollen or slightly swollen, not very spongy; flesh taste farinaceous or not distinctive → 13.

    12. Growing on the ground under hardwoods, especially oak; cap and stalk cinnamon-brown to rusty brown and finely hairy to velvety; stalk base conspicuously swollen to bulbous and very spongy; stalk flesh two-layered, consisting of a hard, zoned central core surrounded by a thick spongy layer, dark brown; odor and taste not distinctive; spines pale to dark brown, darkening when bruised; spores 5.5–7 μm, warted; August–October; inedible—*Hydnellum spongiosipes* (Peck) Pouzar.

    12. Growing on the ground under pines; fruiting body with the characteristics listed in the previous choice except for the following: cap and stalk purplish brown to

rusty brown; cap exuding red droplets when fresh and moist; flesh odor faintly fragrant to slightly acidic; taste slowly disagreeable; spores 4.5–6 × 4–5 μm, warted; inedible—*Hydnellum pineticola* Harrison.

13. Cap with conspicuous concentric zones; surface roughened and irregular with pits, horn-like projections, and ridges; marginal zone white with pinkish tints, staining purple-brown to brown-black when bruised; inner zones of various colors including dull pink, vinaceous-brown, dark rusty cinnamon, and grayish black; flesh orange-brown to dark reddish brown; odor and taste farinaceous or not distinctive; on the ground under conifers or hardwoods—*Hydnellum scrobiculatum* var. *zonatum* (Batsch : Fries) Harrison (see p. 408).

13. Cap lacking conspicuous concentric zones or zoned only along the margin; surface roughened and irregular with pits, horn-like projections, and ridges, dull pink to reddish brown or dark rusty cinnamon; margin sometimes zoned pale brown and pale brownish pink, staining brown-black when bruised; odor and taste farinaceous or not distinctive; spores 4.8–5.5 × 4–5 μm, warted; on the ground under conifers and hardwoods; August–October; inedible—*Hydnellum scrobiculatum* (Fries) Karsten.

    14. Cap surface with conspicuous concentric zones; marginal zone whitish when actively growing; inner zones pale brown to reddish brown or dark brown, darkest at the center; spines on the undersurface up to ⅛" (3 mm) long, white; spores 3.5–4.5 μm, minutely spiny; on the ground under conifers; August–October; inedible—*Phellodon tomentosus* (Fries) Banker.

    14. Cap surface lacking conspicuous concentric zones; spines short, not more than ¹⁄₁₆" (2 mm) long on mature specimens; cap irregular, with pits and ridges, often forming secondary caps; gray to gray-brown or blackish brown and velvety; margin entire or lobed, white, staining gray to dark brown when bruised; flesh reddish gray to blackish brown; odor somewhat spicy-fragrant; taste not distinctive; spores 3–4 μm, spiny; on the ground under hardwoods or mixed woods; August–October; inedible—*Phellodon confluens* (Persoon) Pouzar.

    14. Cap surface lacking conspicuous concentric zones; spines up to ³⁄₁₆" (4 mm) long on mature specimens → 15.

15. Cap surface blue-black to gray when young, becoming black-brown to violet-black at maturity, velvety; margin entire or lobed, grayish white to bluish gray; spines about ⅛" (3 mm) long, pale gray becoming grayish brown; stalk brown-black; flesh brown-black; odor somewhat spicy-fragrant; taste not distinctive; spores 3.5–4.5 × 2.5–3.5 μm, spiny; on the ground under conifers or hardwoods; August–October; inedible—*Phellodon niger* (Fries : Fries) Karsten.

15. Fruiting body nearly identical to the above choice but the cap grayish white when young and dark brown at maturity; spores 4.5–5.5 μm, spiny; August–October; inedible—*Phellodon niger* var. *alboniger* (Peck) Harrison.

Auriscalpium vulgare S. F. Gray (see photo, p. 12)

COMMON NAME: Pine Cone Fungus.

CAP: ⅜–1½" (1–4 cm) wide, kidney-shaped to semicircular, broadly convex; surface dry, densely hairy, pale brown to reddish brown; margin entire or eroded, wavy, pale yellow-brown to pale reddish brown, often with a hairy fringe, notched at the stalk; undersurface covered with spines, about ⅛" (2–3 mm) long, whitish to pinkish, becoming brown at maturity.

FLESH: thin, leathery, white to pale brown.

STALK: 1–3" (2.5–7.5 cm) long, ¹⁄₁₆–⅛" (1.5–3 mm) thick, enlarging downward, pliant, fibrous-tough, densely hairy, dark reddish brown to brown, attached at a notch on the cap margin.

MICROSCOPIC FEATURES: spores 5–6 × 4–5.2 μm, nearly round, slightly roughened, hyaline.

FRUITING: solitary or in groups on pine cones and cone debris on the ground; August–November; occasional.

EDIBILITY: inedible.

Bankera violascens (Albertini and Schweinitz) Pouzar

CAP: 1⅛–5" (3–12.5 cm) wide, convex at first, becoming plane to somewhat depressed at maturity; margin typically lobed or wavy in age; surface finely tomentose to fibrillose, typically lacerate-scaly, at least over the disc in age, pale pinkish brown to pale purplish brown or pale grayish brown, not readily bruising when handled, instantly staining dark olive-green when KOH is applied; undersurface covered with spines up to ¼" (6 mm) long, strongly decurrent, pale gray.

FLESH: dull white to pale pinkish buff, instantly staining dark olive-green when KOH is applied; odor fragrant like maple syrup when fresh, soon disappearing after being picked; taste not distinctive.

STALK: 1⅛–2¾" (3–7 cm) long, ⅜–1" (1–2.5 cm) thick, nearly equal to tapered downward, smooth, sometimes lacerate-scaly, colored like the cap but usually darker brown, often whitish near the apex and dark brown at the base in age.

MICROSCOPIC FEATURES: spores 4.5–6 × 4–5 μm, subglobose, finely echinulate, hyaline.

FRUITING: solitary, scattered, or in groups on the ground under conifers or in mixed woods; August–October; rare.

EDIBILITY: unknown.

COMMENTS: formerly known as *Bankera carnosa*. *Bankera fuligineo-alba* (edibility unknown) is very similar but has a slightly larger cap, up to 6" (15.5 cm) wide, with a conspicuous layer of pine needles adhering to it and does not become scaly.

Climacodon septentrionale (Fries) Karsten

COMMON NAME: Northern Tooth.

FRUITING BODY: 4–10" (10–25.5 cm) wide, 1–2" (2.5–5 cm) thick, 5½–15" (14–38 cm) high, consisting of fan-shaped caps growing in overlapping clusters from a common base, stalkless; upper surface hairy to roughened, whitish to creamy yellow when young, becoming yellow-brown in age; flesh whitish, zoned, fibrous-tough, flexible; odor and taste not distinctive when young, becoming like spoiled ham and bitter in age; under-surface composed of crowded, flexible spines, ¼–¾" (6–20 mm) long, whitish, with lacerated tips.

MICROSCOPIC FEATURES: spores 4–5.5 × 2.5–3 μm, elliptic, thick-walled, smooth, hyaline.

FRUITING: in dense, overlapping clusters on wounds of standing hardwood trunks, especially maple; July–October; fairly common.

EDIBILITY: inedible.

COMMENTS: also known as *Steccherinum septentrionale*.

Hericium coralloides (Scopoli : Fries) S. F. Gray (see photo, p. 12)

COMMON NAME: Comb Tooth.

FRUITING BODY: 2¾–10" (7–25 cm) wide, 2¾–8" (7–20 cm) high, a cluster of spreading branches with spines arranged in rows along the branches like teeth on a comb, arising from a common base; white to salmon or pinkish; flesh thick, white, soft; odor and taste not distinctive; spines ⅛–⅜" (3–10 mm) long, rather evenly distributed along the branches.

MICROSCOPIC FEATURES: spores 3–5 × 3–4 μm, oval to round, slightly roughened, hyaline.

FRUITING: solitary or in groups hanging on decaying hardwood logs and stumps; July–October; fairly common.

EDIBILITY: edible.

COMMENTS: previously known as *H. ramosum*. The Bear's-head Tooth (*H. americanum;* formerly known in North America as *H. coralloides*) is similar but has spines arranged in clusters at the branch tips.

Hericium erinaceus (Fries) Persoon

COMMON NAME: Bearded Tooth, Satyr's Beard.

FRUITING BODY: 2¾–8" (7–20.5 cm) wide, 3¾–8" (9.5–20 cm) high, a whitish to yellow-ish, solid, cushion-shaped, branchless mass resembling a beard, giving rise to long spines; flesh, thick, white, soft; odor and taste not distinctive when young, sour and unpleasant in age; spines ¾–2¾" (2–7 cm) long, tapering to a point.

MICROSCOPIC FEATURES: spores 5–6.5 × 4–5.6 μm, oval to round, smooth to slightly roughened, hyaline.

FRUITING: solitary on standing trunks or fallen logs of hardwoods; August–November; fairly common.

EDIBILITY: edible.

Hydnellum caeruleum (Hornemann) Karsten

COMMON NAME: Bluish Tooth.

CAP: 1³⁄₁₆–4⅜" (3–11 cm) wide, convex to nearly plane; surface dry, soft and velvety on young growing portion, lacking zones; whitish overall, soon tinted blue, often entirely blue, gradually changing to brown or dark brown in the center; undersurface covered with decurrent spines, ⅛–⅜" (3–10 mm) long, crowded, whitish at first, becoming brown with pale tips at maturity.

FLESH: in cap zoned, blue to purple with brown; in stalk homogeneous or somewhat zoned, orange to orange-red, sometimes with blue, especially in young fruiting bodies, thick and extremely fibrous-tough; odor not distinctive to slightly farinaceous or slightly of cooked meat; taste mild or slightly acidic.

STALK: ¾–2⅜" (2–6 cm) long, ⁵⁄₁₆–¾" (7–20 mm) thick, enlarging downward and typically bulbous or tapering downward; pale brown, becoming darker brown in age.

MICROSCOPIC FEATURES: spores 4.5–6 × 3.5–5.5 μm, nearly round to oblong, distinctly warted, pale brown.

FRUITING: scattered or in groups, often with fused caps, on the ground under conifers; August–October; frequent.

EDIBILITY: inedible.

Hydnellum scrobiculatum var. zonatum (Batsch : Fries) Harrison

CAP: 1⅜–5½" (3.5–14 cm) wide, convex to nearly plane, often depressed to funnel-shaped; surface roughened and irregular with pits, horn-like projections and ridges, with con-spicuously concentric zones; marginal zone white with pinkish tints, staining purple-brown to brown-black when bruised; inner zones of various colors, including dull pink, vinaceous-brown, dark rusty cinnamon, and grayish black; undersurface covered with decurrent spines up to ⅛" (3 mm) long, dark vinaceous-cinnamon with paler tips.

FLESH: thick, fibrous-tough, orange-brown to dark reddish brown; odor and taste farinaceous or not distinctive.

STALK: ⅜–1⅜" (1–3.5 cm) long, ⅜–1½" (1–4 cm) thick, typically enlarging downward; colored like the cap.

MICROSCOPIC FEATURES: spores 4.5–5.5 × 4–4.5 μm, somewhat angular, coarsely warted, pale brown.

FRUITING: scattered, in groups, or fused together into masses on the ground under conifers and hardwoods; August–October; fairly common.

EDIBILITY: inedible.

Hydnum repandum Linnaeus : Fries

COMMON NAME: Sweet Tooth, Hedgehog.

CAP: ¾–6" (2–15.5 cm) wide, convex to nearly plane, sometimes slightly depressed; surface dry, felty, becoming somewhat wrinkled and pitted in age, yellow-orange, pale brownish orange to apricot-orange or reddish orange, staining dark orange when bruised; margin wavy, sometimes deeply indented or lobed; undersurface covered with spines, ⅛–⅜" (3–10 mm) long, not decurrent or irregularly so, creamy white to orange-yellow, darkening when bruised.

FLESH: thick, firm, brittle, white, staining orange-yellow when cut and rubbed; odor pleasant, somewhat nutty to sweet or not distinctive; taste mild or sometimes peppery.

STALK: 1–4" (2.5–10 cm) long, ⅜–1⅜" (1–3.5 cm) thick, nearly equal, solid, white with orange tints or colored like the cap, bruising orange-yellow.

MICROSCOPIC FEATURES: spores 7–8.5 × 6–7 μm nearly round, smooth, hyaline.

FRUITING: solitary, scattered, or in groups on the ground under conifers and hardwoods; July–November; common.

EDIBILITY: edible.

COMMENTS: also known as *Dentinum repandum*.

Hydnum repandum var. **album** (Quélet) Rea

CAP: ¾–4¾" (2–12 cm) wide, convex to nearly plane, sometimes slightly depressed; surface dry, felty, occasionally finely wrinkled or cracked in age; white to creamy white, staining orange when bruised; margin entire or wavy, sometimes deeply indented or lobed; undersurface covered with spines, ⅛–¼" (3–6 mm) long, not decurrent or irregularly so; white to creamy white, darkening when bruised.

FLESH: thick, firm, brittle, white, staining orange when cut and rubbed; odor not distinctive; taste mild or slowly bitter.

STALK: 1⅛–3" (3–7.5 cm) long, ¼–1" (1–2.5 cm) thick, nearly equal or enlarged in either direction, often crooked; white to creamy white, staining orange when bruised.

MICROSCOPIC FEATURES: spores 7–8.5 × 5.5–7 μm, oval, smooth, hyaline.

FRUITING: solitary, scattered, or in groups on the ground under conifers and mixed woods; July–October; infrequent.

EDIBILITY: edible.

COMMENTS: also known as *Dentinum repandum* var. *album*.

Hydnum umbilicatum Peck (see photo, p. 12)

CAP: 1–2" (2.5–5 cm) wide, convex to nearly plane, with a deep, central depression that often continues into the stalk; surface dry, felty, becoming somewhat wrinkled and pitted in age, reddish orange to brownish orange or pale dull orange; margin wavy to slightly irregular; undersurface covered with spines, ⅛–¼" (3–6 mm) long, not decurrent; creamy white to pale orange-yellow, darkening when bruised.

FLESH: thick, firm, brittle, white to pale orange, staining darker orange when cut and rubbed; odor and taste not distinctive.

STALK: ¾–2¾" (2–7 cm) long, ³⁄₁₆–⅜" (5–10 mm) thick, enlarging slightly downward or nearly equal, white with orange tints or colored like the cap, staining darker orange when bruised.

MICROSCOPIC FEATURES: spores 7.5–9.5 × 6–7.5 μm, oval, smooth, hyaline.

FRUITING: solitary, scattered, or in groups on the ground or among mosses under conifers and hardwoods, most often found in wet areas; August–October; fairly common.

EDIBILITY: edible.

COMMENTS: also known as *Dentinum umbilicatum*.

Sarcodon atroviridis (Morgan) Banker

CAP: 1–3⅛" (2.5–8 cm) wide, convex to nearly plane; surface dry, smooth, felty; color variable, pale yellowish tan to grayish tan with olive-green and black tints when young, becoming grayish black at the center and grayish yellow toward the margin, sometimes grayish black overall, staining black when rubbed or bruised; margin entire, yellow to yellow-brown with grayish black stains, strongly incurved when young; undersurface covered with long, stout spines, ¼–⅝" (6–16 mm) long, grayish brown when young, dark brown with paler tips at maturity.

FLESH: thick, firm, brittle, pale grayish white; odor not distinctive; taste bitter.

STALK: 2–4" (5–10 cm) long, ¼–⅝" (6–16 mm) thick, nearly equal, smooth, colored like the cap and staining black when rubbed or bruised; hollow in age.

MICROSCOPIC FEATURES: spores 6–8.5 μm, oval to round, strongly warted, pale brown.

FRUITING: solitary, scattered, or in groups on the ground in hardwoods or mixed woods; July–October; infrequent.

EDIBILITY: edible, but of poor quality, according to Roger Phillips.

COMMENTS: also known as *Hydnum atroviride*.

Sarcodon fuligineo-violaceus (Kalchbrenner) Patouillard

CAP: 1–4¾" (2.5–12 cm) wide, convex to broadly convex or nearly plane with a depressed center; surface dry, smooth, or with tiny flattened scales; pinkish brown to purplish brown or grayish purple, staining dark purple-brown to gray-brown when bruised or handled; undersurface covered with short, very crowded spines, ¹⁄₃₂–⅛" (1–3 mm) long, decurrent, pinkish brown to reddish brown with paler tips.

FLESH: thick, brittle, bluish gray to dingy pink, instantly dark green, then blackish when a drop of KOH is applied, slowly staining bluish gray when a drop of $FeSO_4$ is applied; odor fragrant; taste farinaceous to slowly acrid or disagreeable.

STALK: 1⅛–3⅛" (3–8 cm) long, ⅜–¾" (1–2 cm) thick, tapering downward to a radicating base or swollen in the middle and tapering in both directions, colored like the cap or slightly paler, olive-black or blackish near the base, dry, smooth, flesh reddish.

MICROSCOPIC FEATURES: spores 5–7 × 4.5–5.5 μm, broadly elliptic, finely warted, pale brown.

FRUITING: solitary or scattered on the ground under conifers, especially balsam fir and spruce; July–October; infrequent.

EDIBILITY: unknown.

COMMENTS: also known as *Hydnum fuligineo-violaceum*.

Sarcodon imbricatus (Linnaeus : Fries) Karsten

COMMON NAME: Scaly Tooth.

CAP: 2–8" (5–20.5 cm) wide, cracked and conspicuously scaly when immature or mature; convex with a depressed center when young, becoming expanded and deeply depressed at maturity; surface pale brown becoming dark brown, dry, covered with brown to dark brown, erect, pointed scales arranged more or less concentrically and less erect toward the margin; margin incurved and smooth when young, becoming plane and cracked or torn at maturity; undersurface covered with decurrent spines, ¼–⅜" (6–10 mm) long, pale brown tinted grayish when young, becoming reddish brown to dark brown in age.

FLESH: white to pale brown, firm; odor not distinctive; taste mild or slightly bitter but not strongly bitter.

STALK: 1½–4" (4–10 cm) long, ⅝–1⅜" (1.5–3.5 cm) thick, enlarging downward, pale brown when young, darkening in age, smooth, dry, typically hollow at maturity.

MICROSCOPIC FEATURES: spores 6–8 × 5–7.5 μm, nearly round, with large, irregular warts, pale brown.

FRUITING: solitary, scattered, or in groups on the ground in conifers or hardwoods; June–October; infrequent.

EDIBILITY: edible.

COMMENTS: also known as *Hydnum imbricatum*.

Sarcodon underwoodii Banker

CAP: 2–5½" (5–14 cm) wide, convex to nearly plane, often somewhat depressed at the center at maturity; surface pale reddish brown, dry, cracked when young and soon developing small, flattened, more or less concentrically arranged, pale reddish brown scales that darken in age and become partially erect, especially over the center; margin incurved and smooth when young, becoming plane, cracked and wavy at maturity; typically free and extending up to 1⁄16" (2 mm) beyond the spines; undersurface covered with decurrent, short spines, 1⁄16–⅛" (1–3 mm) long, crowded, white when young, becoming brown with grayish tips at maturity.

FLESH: in cap white; in stalk pale brown; firm; odor fragrant or farinaceous, or not distinctive; taste instantly very bitter.

STALK: 1⅜–2¾" (3.5–7 cm) long, ⅜–½" (8–12 mm) thick, tapering downward, usually bent at the ground, with an abrupt, white, pointed base; brown to dark brown, dry.

MICROSCOPIC FEATURES: spores 6–7.5 × 5.5–6.5 μm, oval to nearly round, strongly warted, pale brown.

FRUITING: solitary, scattered, or in groups on the ground in hardwoods; July–October; uncommon.

EDIBILITY: inedible.

Steccherinum pulcherrimum (Berkeley and Curtis) Banker (see photo p. 12)

CAP: 1⅜–4" (3.5–10 cm) wide, semicircular to fan-shaped, arising from a confluent, spreading base; upper surface densely hairy, whitish to pale tan or pinkish tan; undersurface covered with crowded short spines, ⅛–3⁄16" (2.5–5 mm) long, whitish, becoming pinkish tan in age.

FLESH: white, fibrous, pliable, filled with a creamy white, sticky sap that exudes when squeezed; odor of wood or preserved figs; taste not distinctive.

STALK: absent or a short, lateral projection that is sometimes longitudinally furrowed.

MICROSCOPIC FEATURES: spores 4–5 × 2–2.5 μm, elliptic, smooth, hyaline.
FRUITING: in overlapping or fused clusters on decaying hardwoods; July–November; uncommon.
EDIBILITY: inedible.

Bankera violascens

Climacodon septentrionale

Hericium erinaceus

Hydnellum caeruleum

Hydnellum scrobiculatum var. *zonatum*

Hydnum repandum

Hydnum repandum var. *album*

Sarcodon atroviridis

Sarcodon fuligineo-violaceus

Sarcodon imbricatus

Sarcodon underwoodii

Cauliflower Mushrooms

Cauliflower Mushrooms form large, rounded, cauliflower- or lettuce-like clusters that usually grow on the ground at the base of trees. Members of this group are choice edibles and are highly prized for their firm texture and excellent flavor.

Key to Species of Cauliflower Mushrooms

1. Fruiting body with a thick, dark brown to black, deeply rooting stalk; individual branches typically darkest at their tips—*Sparassis crispa* Wulfen : Fries (see p. 415).
1. Fruiting body lacking a thick, dark brown to black, deeply rooting stalk; individual branches typically palest at their tips—*Sparassis herbstii* Peck (see p. 415).

Sparassis crispa Wulfen : Fries (see photo, p. 13)
 COMMON NAME: Rooting Cauliflower Mushroom.
 FRUITING BODY: 4¾–18" (12–46 cm) wide and high, a densely packed, rounded, cauli-flower- or lettuce-like cluster of whitish to pale yellow or tan, flattened, leafy, folded lobes, often resembling ribbon candy; individual branches leaf-like, flexible, darkening upward, typically darkest and broadest at their tips, tapering downward and uniting to form a thick, solid, basal mass above the stalk.
 STALK: 2–5½" (5–14 cm) long, ¾–2" (2–5 cm) thick, tapering toward the base, dark brown to black, deeply rooting.
 FLESH: white, thin, fibrous; odor fragrant ot not distinctive; taste not distinctive.
 SPORE PRINT: white.
 MICROSCOPIC FEATURES: spores 5–6.5 × 3–3.5 μm, oval, smooth, hyaline.
 FRUITING: solitary or in groups on the ground or decaying wood in conifer forests; August–October; occasional.
 EDIBILITY: edible and choice.
 COMMENTS: this mushroom is also known as *S. radicata*. The leaf-like branches are typically smaller than those of *S. herbstii*.

Sparassis herbstii Peck (see photo, p. 13)
 COMMON NAME: Cauliflower Mushroom.
 FRUITING BODY: 6–15" (15–38 cm) wide and high, rounded, a cauliflower- or lettuce-like cluster of cream to pale yellow, flattened branches that are wrinkled, curled, and folded, often resembling ribbon candy; individual branches leaf-like, somewhat stiff, typically palest and broadest at their tips, tapering downward and uniting to form a thick, solid, basal mass; lacking a long, rooting stalk.
 FLESH: white, thin, fibrous; odor pleasant; taste mild.

SPORE PRINT: white.

MICROSCOPIC FEATURES: spores 4–7 × 3–4 μm, oval, smooth, hyaline.

FRUITING: solitary or in groups on the ground, usually at the base of oak or pine trees; July–October; occasional.

EDIBILITY: edible and choice.

COMMENTS: also known as *S. spathulata* and sometimes incorrectly called *S. crispa*. The leaf-like branches are typically larger than those of *S. crispa*.

Branched and Clustered Corals

Branched and Clustered Corals is a group that includes species with two different kinds of fruiting bodies. The first type includes species with spindle-shaped to worm-like, erect, unbranched or infrequently branched stalks, which are often fused at their bases and typically grow in clusters. The second type includes species with erect, repeatedly branched, coral-like stalks that grow solitary or in groups. Most species have brittle flesh and are easily broken, but some are fibrous to tough and flexible. Spores are produced on portions of the smooth to wrinkled outer surface of the stalks and branches. Most species grow on the ground, but some occur on decaying wood or on the bark of standing trees.

Several species are edible, some are poisonous, and the others are inedible or of unknown edibility. A few cases of severe gastrointestinal upset have been caused by some members of this group. However, no fatalities have been reported. Similar species that are erect and unbranched are included in the key to the Earth Tongues, Earth Clubs, and Allies. A number of *Ramaria* species and numerous varieties are not included in the keys because they are difficult to identify, even with a microscope.

Key to Species of Branched and Clustered Corals

1. Fruiting body hard, carbonaceous, with a branched and antler-like, flattened to oval, or sometimes rounded stalk, gray to whitish on the upper portion, black toward the base; on decaying wood—*Xylaria hypoxylon* (Linnaeus : Hooker) Greville (see Carbon and Cushion Fungi, p. 523).
1. Fruiting body rubbery-gelatinous, up to 4" (10 cm) high, consisting of erect, pointed, repeatedly forked, antler-like tines; golden yellow to orange-yellow; on decaying conifer wood—*Calocera viscosa* (Persoon : Fries) Fries (see Jelly Fungi, p. 431).
1. Fruiting body fibrous and leathery, up to 2¾" (7 cm) high, erect, composed of narrow, flattened, pointed to spoon-shaped, forked branches; brown overall or with whitish tips; growing on the ground—Genus *Thelephora* (see Fiber Fans and Vases, p. 425).
1. Fruiting body leathery to fibrous-tough, up to 3¾" (9.5 cm) high, erect, coral-like; branches fused together at their bases; branches broadly flattened to cylindric, with free or fused apices; white to dark buff; growing on the ground—*Tremellodendron* species (see Jelly Fungi, p. 427).
1. Fruiting body not with the above combinations of characters → **2**.
    2. Fruiting body spindle-shaped to worm-like, unbranched or sometimes branched near the apex, growing on the ground in clusters, often with fused bases → **3**.
    2. Fruiting body coral-like, repeatedly branched, growing on the ground or on wood, solitary or in groups → **7**.

3.   Fruiting body yellow to reddish orange → 4.
3.   Fruiting body white to grayish white or pale ochre → 5.
3.   Fruiting body pinkish buff to pale grayish pink or dark reddish purple to dull brown with a purplish tint → 6.

    4.   Fruiting body bright to dull yellow; up to 5½" (14 cm) high, ¹⁄₁₆–⅜" (1.5–10 mm) thick, cylindric to worm-like or somewhat flattened—*Clavulinopsis fusiformis* (Fries) Corner (see p. 421).

    4.   Fruiting body reddish orange to pale orange, yellow to whitish near the base, up to 6" (15.5 cm) high, ¹⁄₁₆–⅜" (1.5–10 mm) thick, spindle-shaped to worm-like, usually flattened, with a conspicuous longitudinal channel, hollow, apex usually pointed; surface smooth; spores 5–6.5 × 5–5.5 μm; growing in clusters; July–October; edible—*Clavulinopsis aurantio-cinnabarina* (Schweinitz) Corner.

5.   Fruiting body up to 4½" (11.5 cm) high, ¹⁄₁₆–¼" (1.5–6 mm) thick, cylindric, worm-like, apex usually pointed, typically solid; surface smooth, translucent-white to bright white, yellowing in age; spores 4–6 × 2.5–3.5 μm; growing in clusters; July–September; edible—*Clavaria vermicularis* Fries.

5.   Fruiting body up to 3⅜" (8.5 cm) high, ⅛–½" (3–12 mm) thick, spindle-shaped to compressed and irregular, sometimes branched near the apex; surface slightly to distinctly wrinkled, white to grayish white or pale ochre; spores 9–14 × 7–10 μm; usually in groups or clusters, but sometimes solitary; June–October; edible—*Clavulina rugosa* (Fries) Schroeter (see comments under *Clavulina cinerea*, p. 421).

    6.   Fruiting body pinkish buff to pale grayish pink, up to 4¾" (12 cm) high, ¹⁄₁₆–¼" (1.5–6 mm) thick, cylindric, worm-like, apex rounded or somewhat pointed, hollow in age; surface smooth—*Clavaria rubicundula* Leathers (see p. 420).

    6.   Fruiting body purple when young, becoming pinkish purple to lavender-gray or grayish brown in age, often whitish at the base; up to 4¾" (12 cm) high, ¹⁄₁₆–¼" (1.5–6 mm) thick, cylindric, worm-like, apex rounded or somewhat pointed, hollow in age; surface smooth; spores 8–12 × 3.5–4.5 μm; growing in clusters; July–September; edible—*Clavaria purpurea* Fries.

7.   Fruiting body growing on decaying wood, sometimes on cones or leaves, but usually not on the ground → 8.

7.   Fruiting body growing on the ground, usually not on wood → 11.

    8.   Fruiting body small, up to 2⅜" (6 cm) high, fibrous-tough and flexible, not brittle → 9.

    8.   Fruiting body large, 2–5½" (5–14 cm) high, brittle or fibrous-tough and flexible → 10.

9.   Fruiting body up to 1½" (4 cm) high, with a short stalk and numerous branches; stalk ¼–1" (6–25 mm) long; branches repeatedly forked with several tips; tips pointed and tine-like; surface smooth, ochre to tan, often with a pinkish tint; surface of stalk paler upward, often whitish on the branch tips; flesh bitter; spores 7–9 × 2.5–4.5 μm, elliptic, smooth, hyaline; on decaying hardwoods or standing trees, especially beech and oak; June–October; inedible—*Lentaria micheneri* (Berkeley and Curtis) Corner.

9.   Fruiting body up to 2½" (6.5 cm) high, with a short stalk and numerous branches; stalk ⅛–¾" (3–20 mm) long, whitish, surrounded at the base by a whitish, cottony mycelium; branches repeatedly forked, with several slender, tapering tips; surface pale, rosy pink to pinkish tan, staining brownish when bruised, darkening to pinkish brown at maturity, with whitish tips that become pinkish brown in age; flesh mild or sometimes bitter; spores 11–16.5 × 3–5.5 μm, cylindric to oblong, smooth, hyaline; on decaying wood or sometimes leaves and cones; July–October; inedible—*Lentaria byssiseda* (Fries) Corner.

10. Odor not distinctive; taste hot and peppery; branch tips typically whitish, short, tine-like; fruiting body up to 3⅛" (8 cm) high, with a short stalk arising from a whitish basal mycelium; branches pinkish buff to pale pinkish cinnamon, staining greenish in FeSO$_4$; flesh whitish, tough; —*Ramaria rubella* var. *blanda* Petersen (see p. 422).

10. Odor strongly aromatic or anise-like; taste bitter or sometimes hot and peppery; branch tips creamy tan to beige; fruiting body up to 5½" (14 cm) high; branches moderately compact, vertical, and parallel; pale yellowish tan to pale cinnamon, sometimes with ochre tints, staining greenish in FeSO$_4$; on decaying hardwoods or infrequently on conifers—*Ramaria concolor* (Corner) Petersen (see p. 422).

10. Odor fragrant, often anise-like; taste bitter; branch tips yellow to greenish yellow; fruiting body up to 5½" (14 cm) high, branches compact, vertical, and parallel; grayish orange to ochre-orange near the base, becoming pale yellow upward, staining greenish in FeSO$_4$; surface of the branches sometimes slowly staining purplish brown when bruised; spores 7–9 × 3.5–5 μm; on decaying conifer wood or sometimes on hardwoods; July–October; inedible—*Ramaria stricta* (Fries) Quélet.

10. Odor usually not distinctive; taste somewhat peppery; branch tips distinctly crown-like; fruiting body up to 5⅛" (13 cm) high; white to pale creamy white when young, becoming ochre-yellow or tan in age; on decaying hardwoods— *Clavicorona pyxidata* (Fries) Doty (see p. 420).

11. Fruiting body up to 6" (15.5 cm) high, cauliflower- to coral-like; stalk up to 1½" (4 cm) thick at the apex, tapering downward, solid, white; branch tips reddish to purplish, staining greenish in FeSO$_4$; flesh white, odor and taste not distinctive; growing in groups or scattered in coniferous or mixed woods—*Ramaria botrytis* (Fries) Ricken (see p. 421).

11. Not as in the previous choice; fruiting body pinkish orange to pale red or reddish purple → 12.

11. Not as in the previous two choices; fruiting body white to pinkish white, ashy gray, yellowish, or ochre → 13.

12. Branches pinkish orange to pale red, darkest at the tips, whitish toward the base, staining greenish in FeSO$_4$; fruiting body up to 4¾" (12 cm) high; flesh white; odor and taste not distinctive; spore print pale yellow-brown; spores 8–13 × 3–5 μm, nearly cylindric, warted, hyaline, inamyloid; growing under conifers; July– October; edible—*Ramaria araiospora* var. *araiospora* Marr and Stuntz.

12. Branches yellowish salmon to pinkish orange, staining greenish in FeSO$_4$; branch tips yellow; fruiting body up to 6½" (16.5 cm) high; flesh whitish, odor not distinctive; taste bitter to astringent; spores 8–12 × 4–5.5 μm; growing under conifers; July–October; poisonous—*Ramaria formosa* (Fries) Quélet.

12. Branches reddish purple, darkest toward the tips; fruiting body up to 3⅛" (8 cm) high, sparingly branched on the upper one-third; flesh reddish purple; in mixed woods—*Clavaria zollingeri* Léveille (see p. 420).

13. Fruiting body up to 3⅛" (8 cm) high, branches white to buff, usually becoming pinkish to pale apricot tinged in age; branch tips short, blunt or somewhat pointed; flesh white; odor and taste not distinctive; surface of the branches not staining greenish within two minutes in FeSO$_4$—*Ramariopsis kunzei* (Fries) Donk (see p. 422).

13. Fruiting body very similar to the previous choice but more robust; branches not developing pinkish to pale apricot tinges, branch tips white at first, becoming vinaceous-buff at maturity; staining distinctly greenish within two minutes in FeSO$_4$; spore print pale tan; July–October—*Ramariopsis lentofragilis* (Atkinson) Corner (see p. 423).

13. Fruiting body up to 3½" (9 cm) high, extremely variable in shape, typically sparingly branched except near the apex, which is crested with numerous pointed or sometimes blunt tips; branches smooth to wrinkled or longitudinally grooved, white to yellowish or grayish ochre, sometimes blackened from the base upward when attacked by a parasitic fungus; flesh soft and fragile; odor and taste not distinctive; spores 7–10 × 6–7.5 μm; growing in woods, especially under conifers; July–October; edible—*Clavulina cristata* (Fries) Schroeter.

13. Fruiting body up to 4⅜" (11 cm) high, extremely variable in shape, typically richly branched from the base upward; apex not crested; branch tips pointed or blunt; branches smooth to wrinkled or longitudinally grooved, white when young, becoming ashy gray, sometimes blackened from the base upward when attacked by a parasitic fungus; flesh white; odor and taste not distinctive—*Clavulina cinerea* (Fries) Schroeter (see p. 421).

Clavaria rubicundula Leathers
COMMON NAME: Smoky Worm Coral.
FRUITING BODY: up to 4¾" (12 cm) high, ¹⁄₁₆–¼" (1.5–6 mm) thick, cylindric to worm-like; apex rounded or somewhat pointed, hollow in age; surface smooth, pinkish buff to pale grayish pink; flesh whitish, brittle; odor weak, resembling tincture of iodine or not distinctive; taste not distinctive.
SPORE PRINT: white.
MICROSCOPIC FEATURES: spores 5.5–8.5 × 3–4 μm, elliptic, smooth, hyaline, inamyloid.
FRUITING: in dense clusters on the ground in hardwoods and mixed woods; June–September; occasional.
EDIBILITY: unknown.

Clavaria zollingeri Léveille (see photo, p. 13)
COMMON NAME: Magenta Coral.
FRUITING BODY: up to 3½" (9 cm) high, erect and coral-like, with repeatedly forked branches arising from a short stalk; branches sparingly branched on the upper one-third; surface smooth, reddish purple, darkest towards the tips; branch tips rounded to blunt or tapering; flesh reddish purple, very brittle; odor not distinctive; taste somewhat radish-like.
SPORE PRINT: white.
MICROSCOPIC FEATURES: spores 5–7.5 × 3–4.5 μm, elliptic to oval, smooth, hyaline, inamyloid.
FRUITING: solitary, scattered, or in groups on the ground in mixed woods; July–October; occasional.
EDIBILITY: edible.
COMMENTS: *Clavaria amethystina* (edible) is rare, has lilac-purple branches when young, brittle, lilac-purple flesh that lacks a distinctive odor and taste, and larger spores, 7–10 × 6–8 μm.

Clavicorona pyxidata (Fries) Doty
COMMON NAME: Crown-tipped Coral.
FRUITING BODY: up to 5⅛" (13 cm) high, erect and coral-like, with numerous, repeatedly forked branches arising from a short stalk; branch tips distinctly crown-like; surface smooth, white to pale creamy white when young, becoming ochre-yellow or tan

in age; flesh tough to brittle, whitish; odor usually not distinctive; taste somewhat peppery.

SPORE PRINT: white.

MICROSCOPIC FEATURES: spores 4–5 × 2–3 μm, elliptic, smooth, hyaline, amyloid.

FRUITING: solitary, scattered, or in groups on decaying hardwoods; June–September; fairly common.

EDIBILITY: edible.

Clavulina cinerea (Fries) Schroeter

COMMON NAME: Gray Coral.

FRUITING BODY: up to 4⅜" (11 cm) high, erect and coral-like, extremely variable in shape, typically richly branched from the base upward; apex not crested; branch tips pointed or blunt; surface smooth to wrinkled or longitudinally grooved, white when young, becoming ashy gray, sometimes blackened from the base upward when attacked by *Spadicioides clavariae,* a parasitic fungus; flesh white, fibrous to brittle; odor and taste not distinctive.

SPORE PRINT: white.

MICROSCOPIC FEATURES: spores 6.5–11 × 5–10 μm, oval to nearly round, smooth, hyaline, inamyloid.

FRUITING: solitary, scattered, or in groups on the ground in woods, especially under conifers; June–October; common.

EDIBILITY: edible.

COMMENTS: compare with *Clavulina cristata* (edible), which is typically sparingly branched except near the apex, which is crested and usually lacks the gray tones. *Clavulina rugosa* (edible) is white to grayish white or pale ochre, spindle-shaped to compressed and irregular, and typically unbranched or sparingly so near the apex. Some authors have suggested that *C. rugosa* may be an atypical form of *C. cristata.*

Clavulinopsis fusiformis (Fries) Corner (see photo, p. 13)

COMMON NAME: Spindle-shaped Yellow Coral.

FRUITING BODY: up to 5½" (14 cm) high, 1/16–⅜" (1.5–10 mm) thick, cylindric to worm-like or somewhat flattened, usually unbranched but sometimes branching near the apex; apex pointed to rounded; surface typically smooth, but sometimes wrinkled or grooved, bright to dull yellow; flesh thin, brittle to fibrous, yellowish.

SPORE PRINT: white to pale yellow.

MICROSCOPIC FEATURES: spores 5–9 × 4–9 μm, broadly oval to globose, smooth, hyaline, inamyloid.

FRUITING: in dense clusters on soil or among grasses in woods and pastures; July–October; fairly common.

EDIBILITY: edible.

Ramaria botrytis (Fries) Ricken

FRUITING BODY: up to 6" (15.5 cm) high, erect and cauliflower- to coral-like, with numerous repeatedly forked branches arising from a large, fleshy stalk; stalk up to 1½" (4 cm) thick at the apex, tapering downward, solid, white; branch tips reddish to purplish, staining greenish in $FeSO_4$; flesh white; odor and taste not distinctive.

SPORE PRINT: pale ochre.

MICROSCOPIC FEATURES: spores 11–20 × 4–6 μm, nearly cylindric, longitudinally striate, hyaline, inamyloid.

FRUITING: solitary, scattered, or in small groups on the ground under conifers or sometimes hardwoods; July–October; occasional.

EDIBILITY: edible and often rated as choice.

Ramaria concolor (Corner) Petersen

FRUITING BODY: up to 5½" (14 cm) high, erect and coral-like, with numerous repeatedly forked branches arising from a short stalk; branches moderately compact, vertical, and parallel; surface smooth, pale yellowish tan to pale cinnamon, sometimes with ochre tints, greenish in $FeSO_4$; branch tips pointed, creamy tan to beige; flesh creamy white to pale tan, fibrous to brittle; odor strongly aromatic or anise-like; taste bitter or sometimes hot and peppery.

SPORE PRINT: yellow.

MICROSCOPIC FEATURES: spores 8–10 × 4–5 μm, elliptic with distinct warts and ridges, hyaline, inamyloid.

FRUITING: scattered or in groups on decaying hardwoods or infrequently on conifers; July–October; fairly common.

EDIBILITY: inedible.

COMMENTS: compare with *Ramaria stricta,* which has yellow to greenish yellow branch tips and typically grows on conifers.

Ramaria rubella var. blanda Petersen

FRUITING BODY: up to 5½" (14 cm) high, erect and coral-like, with numerous, repeatedly forked branches arising on a short stalk from a whitish basal mycelium; surface smooth, pinkish buff to pale pinkish cinnamon, greenish in $FeSO_4$; branch tips short, tine-like, often whitish; flesh whitish, tough; odor not distinctive; taste hot and peppery.

SPORE PRINT: cinnamon-buff.

MICROSCOPIC FEATURES: spores 6–8 × 4–6 μm, broadly ovoid to broadly elliptic with distinct warts and ridges, hyaline, inamyloid.

FRUITING: scattered or in groups on decaying wood; July–September; rare to uncommon.

EDIBILITY: unknown.

Ramariopsis kunzei (Fries) Donk

COMMON NAME: White Coral.

FRUITING BODY: up to 3⅛" (8 cm) high, erect and coral-like, with repeatedly forked branches arising from a short stalk; surface smooth, white, usually developing a distinct pinkish to pale apricot tinge in age, not staining greenish in $FeSO_4$; flesh white, brittle to somewhat flexible; odor and taste not distinctive.

SPORE PRINT: white.

MICROSCOPIC FEATURES: spores 3–5.5 × 2.5–4.5 μm, broadly elliptic to nearly round with minute spines, hyaline, inamyloid.

FRUITING: solitary or scattered on the ground in woods; July–October; fairly common.

EDIBILITY: edible with caution; see comments.

COMMENTS: *Ramariopsis lentofragilis* (poisonous) is very similar but is more robust, does *not* develop pinkish tinges on its branches, has a pale tan spore print, and stains distinctly greenish within two minutes with $FeSO_4$. See comments under *R. lentofragilis* for additional information (see p. 423).

Ramariopsis lentofragilis (Atkinson) Corner

FRUITING BODY: up to 5½" (14 cm) high, erect and coral-like, with repeatedly forked branches arising from a short, pale yellow stalk; surface smooth, creamy white to grayish white, not developing pinkish to pale apricot tinges, staining distinctly green to blackish green within two minutes in FeSO$_4$; branch tips fragile, white at first, becoming vinaceous-buff at maturity; flesh white, brittle to somewhat flexible; odor and taste not distinctive.

SPORE PRINT: pale tan.

MICROSCOPIC FEATURES: spores 4–7 × 3.5–5 μm, ovoid to subglobose, minutely roughened, hyaline, inamyloid.

FRUITING: solitary or scattered on the ground or very decayed wood; August–October; frequency unknown because of possible confusion with *Ramariopsis kunzei* (see comments below).

EDIBILITY: poisonous (see comments below).

COMMENTS: In addition to the specimens shown in the photograph, which were collected in Maine, three additional collections of *R. lentofragilis* have been reported: one from Long Island and two from Nova Scotia. Sam Ristich has reported three cases of poisoning caused by ingestion of this mushroom. Symptoms included severe abdominal pain without diarrhea (except for one individual), general body weakness lasting up to ten days, and severe pain just beneath the sternum that lasted up to two weeks. The toxin(s) causing these symptoms remains unknown. The individuals consumed what they believed to be *R. kunzei*, an edible species. Currie Marr, who devoted considerable time to solving the mystery, determined that the victims had consumed a *R. kunzei* look-alike. The species identification was kindly provided by Ronald Petersen.

Clavaria rubicundula

Clavicorona pyxidata

Clavulina cinerea

Ramaria concolor

Ramaria botrytis

Ramaria rubella var. *blanda*

Ramariopsis kunzei

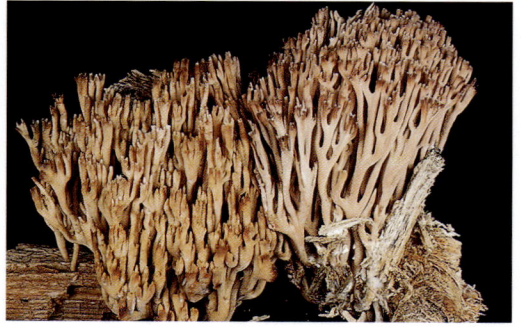

Ramariopsis lentofragilis

Fiber Fans and Vases

Fiber Fans and Vases, members of the genus *Thelephora*, are leathery, fibrous-tough fungi with split or torn margins and brown spores that are usually warted or spiny. Their fruiting bodies are typically some shade of brown at maturity. Some members have distinct caps with or without stalks; others produce coral-like tufts of forking, spoon-shaped to flattened branches. All species are inedible.

Key to Species of Fiber Fans and Vases

1. Fruiting body erect, composed of very narrow to narrow, flattened, pointed to spoon-shaped, forked branches → 2.
1. Fruiting body erect or flattened; branches rounded or absent; caps cup-shaped to fan-shaped, sometimes enveloping roots, stems, branches, or mosses → 3.
    2. Odor and taste not distinctive; fruiting body 1⅛–2¾" (3–7 cm) tall, a coral-like tuft, repeatedly branched from a common stalk, dilating toward the apex and white-fringed or toothed; olive-brown to gray-brown or lilac-brown; dull, leathery; spores 8–10 × 7–8 μm—*Thelephora anthocephala* Fries.
    2. Odor intensely unpleasant; fruiting body 1½–2¾" (4–7 cm) tall, a coral-like tuft, repeatedly branched from a common stalk; branches ending in spoon-shaped to fan-shaped tips; tips often fringed or toothed and vertically grooved; branches whitish when young, becoming gray to lilac-brown in age, with tips remaining whitish; dull, leathery; spores 9–10 × 7–8 μm—*Thelephora palmata* Scopoli : Fries.
3. Fruiting body erect, small, ⅜–1⅛" (1–3 cm) tall, highly variable, typically forming a cluster of funnel-shaped caps that arise on branched stalks from a common base; caps often deeply split or many-parted into irregular lobes—*Thelephora regularis* var. *multipartita* (Schweinitz) Corner (see p. 425).
3. Fruiting body erect, large, 1–4" (2.5–10 cm) tall, highly variable, typically composed of ascending lobes or caps arising from a common central stalk; upper surface whitish to yellowish; lower surface wrinkled, dingy yellow, then grayish brown; spores 4.5–7 × 4.5–5 μm—*Thelephora vialis* Schweinitz.
3. Fruiting body partially erect and spreading, medium-sized, ¾–2" (2–5 cm) high, composed of circular to fan-shaped, stalkless caps in overlapping clusters, often laterally fused, frequently enveloping roots, branches, seedlings, or mosses—*Thelephora terrestris* Fries (see p. 426).

Thelephora regularis var. **multipartita** (Schweinitz) Corner (see photo, p. 14)
  COMMON NAME: Many-parted Thelephora.
  FRUITING BODY: erect, 1–2⅜" (2.5–6 cm) high, 1–2¾" (2.5–7 cm) wide, highly variable,

typically forming a cluster of funnel-shaped caps that arise on stalks from a common base.

CAP: usually funnel-shaped, deeply split into irregular lobes; inner (upper) surface pinkish brown to grayish brown, paler toward the margin, nearly smooth; margin wavy, irregular, divided, whitish at maturity; outer (lower) surface pinkish brown to orange-brown, becoming grayish brown in age, smooth.

FLESH: thin, leathery, watery brown; odor and taste not distinctive.

STALK: erect, nearly equal or enlarging downward, frequently branched, reddish brown, darkening in age, coated with very tiny, soft hairs.

SPORE PRINT: dark brown.

MICROSCOPIC FEATURES: spores 7–10 × 5–6 μm, angularly oval to elliptic, warted, brownish.

FRUITING: solitary, scattered, or in clusters on soil in mixed woods and hardwoods, especially under oak; July–November; infrequent.

EDIBILITY: inedible.

COMMENTS: the small size and many-parted, funnel-shaped caps are distinctive features that separate it from other *Thelephora* species.

Thelephora terrestris Fries (see photo, p. 14)

COMMON NAME: Common Fiber Vase or Earth Fan.

FRUITING BODY: partially erect and spreading, ¾–2" (2–5 cm) high, ½–2¾" (1.3–7 cm) wide, composed of circular to fan-shaped stalkless caps in overlapping clusters, often laterally fused and forming patches 12" (30.5 cm) or more in diameter; upper surface covered with short, stiff hairs, often matted and woolly or scaly, somewhat concentrically zoned, rusty brown to dark brown or grayish brown, becoming blackish brown in age; margin white or brown, woolly, coarsely torn, often with small, fan-shaped outgrowths; lower surface somewhat wrinkled and finely warted, pinkish brown to brown.

FLESH: thin, leathery, watery brown; odor somewhat earthy or absent; taste not distinctive.

SPORE PRINT: purple-brown.

MICROSCOPIC FEATURES: spores 8–12 × 6–9 μm, angularly oval to elliptic, nearly smooth to warted or spiny, purplish.

FRUITING: solitary or in overlapping clusters attached to roots, branches, seedlings, or mosses (especially sphagnum moss), or on the ground in conifer woods; year-round; common.

EDIBILITY: inedible.

COMMENTS: this is the most common *Thelephora* species in northeastern North America. *Thelephora griseozonata* (inedible) has grayish white concentric zones on the upper surface of the cap. *Thelephora intybacea* (inedible), variously described in mycological literature, has a paler cap and a whitish fringed margin that becomes entire and reddish brown at maturity. Both are probably variants of *T. terrestris*.

Jelly Fungi

Jelly Fungi typically produce fruiting bodies with a gelatinous consistency. Some species are firm and rubbery-gelatinous; others have a soft-gelatinous texture. In many cases, young fruiting bodies are rubbery-gelatinous and become soft-gelatinous in age. One species, *Dacrymyces stillatus* Nees : Fries, deliquesces in age to a slimy mass. Another member of this group, *Phleogena faginea* (Fries) Link, forms fruiting bodies that are tiny and dry; although it lacks a gelatinous consistency, it has microscopic features consistent with other members of this group.

During dry periods, Jelly Fungi shrink and become hard and horny as they loose moisture. Considerable variation in shape and color can be seen within this group, and identification based on macroscopic features is usually possible. In some cases, however, microscopic examination of specimens is needed for accurate identification. A few species of Cup Fungi and Earth Clubs that resemble Jelly Fungi have been included in the key. Unlike the Jelly Fungi, which form spores on basidia that are segmented or shaped like a tuning fork, Cup Fungi and Earth Clubs produce spores enclosed in sac-like containers called asci.

Key to Species of Jelly Fungi

1. Fruiting body 1–3⅛" (2.5–8 cm) high, funnel-shaped with a split side or spoon-shaped to tongue-shaped; pinkish red to apricot—*Phlogiotis helvelloides* (Fries) Martin (see p. 432).
1. Fruiting body 1–2¾" (2.5–7 cm) wide, ¾–2" (2–5 cm) high, shoehorn- to tongue-shaped, or hemispheric with a cap and eccentric stalk; upper surface of cap minutely velvety to smooth, gray-translucent, yellowish, or brown; lower cap surface densely coated with tiny, soft, whitish spines; on decaying wood—*Pseudohydnum gelatinosum* (Fries) Karsten (see p. 432).
1. Fruiting body 1⅛–5⅞" (3–15 cm) wide, ear-shaped to irregularly cup-shaped, stalkless, rubbery-gelatinous; upper (outer) surface wrinkled, minutely hairy to velvety, reddish brown; lower (inner) surface smooth, yellowish brown to reddish brown, or grayish brown to purplish brown; solitary, in groups, or in fused clusters on decaying wood, especially conifers—*Auricularia auricula* (Hooker) Underwood (see p. 430).
1. Fruiting body 2–10" (5–25.5 cm) wide, 2–4" (5–10 cm) high, a lettuce-like cluster composed of numerous leaf-like folds, pale to dark reddish brown or rarely pale brownish yellow; on decaying wood—*Tremella foliacea* Person : Fries (see p. 433).
1. Fruiting body not with the above combinations of characters → 2.
    2. Fruiting body a soft, whitish to grayish, irregular, membrane-like, spreading mass

growing on the ground and soon enveloping plant stems and leaves—*Tremella concrescens* (Fries) Burt (see p. 433).

2. Fruiting body a thin, fibrous-tough to leathery, whitish to pale tan, irregular, sheet-like mass forming small, lateral projections; growing on the ground and soon encrusting plant stems, leaves, branches, cones, and debris—*Sebacina incrustans* (Fries) Tulasne (see p. 433).

2. Fruiting body upright, with a central stalk, club-shaped, antler-shaped, densely clustered and coral-like, with leathery, somewhat flattened branches or with erect, finger-like hollow lobes → 3.

2. Fruiting body stalkless or with a rudimentary stalk, brain-like, cup-like, cushion-shaped, or irregularly lobed → 6.

3. Fruiting body rubbery-gelatinous, 2¾–7½" (7–19 cm) wide, 1¾–4½" (4.5–11.5 cm) high, composed of erect, finger-like, hollow lobes that often fuse together; lobes white, irregular, with folds and crests when young, becoming elongated and branched in age; dingy white to yellowish or brownish and darkest on the tips at maturity; spores 9–11 × 5–6 μm; on the ground or decaying wood, usually under hardwoods; July–October; edibility unknown—*Tremella reticulata* (Berkeley) Farlow.

3. Fruiting body leathery to fibrous-tough, 2–5½" (5–14 cm) wide, 1–3¾" (2.5–9.5 cm) high, coral-like, densely clustered, composed of numerous broadly flattened to somewhat cylindric branches; branches fused together at their bases, with their apices frequently united; white to buff—*Tremellodendron pallidum* (Schweinitz) Burt (see p. 434).

3. Fruiting body leathery to fibrous-tough, 1–5⅛" (2.5–13 cm) wide, 1⅜–3½" (3.5–9 cm) high, coral-like, densely clustered, composed of numerous rounded to somewhat flattened branches that are fused together at their bases and repeatedly divided above; apices typically free; whitish to dark buff; spores 7.5–10 × 4.5–5.5 μm; scattered or in clusters on the ground in woods; July–November; inedible—*Tremellodendron candidum* (Schweinitz) Atkinson.

3. Fruiting body not with the above combinations of characters → 4.

4. Fruiting body up to 5⁄16" (8 mm) high, consisting of a tiny head and stalk; head 1⁄16–⅛" (1.5–3 mm) wide, nearly round to flattened or somewhat irregular, finely cracked and roughened, dry, white when young, becoming grayish white to pale brown at maturity; stalk up to ¼" (6 mm) long, equal or enlarged in either direction; in groups or clustered on decaying hardwoods—*Phleogena faginea* (Fries) Link (see p. 432).

4. Fruiting body 1⅜–4" (3.5–10 cm) high, consisting of a cap and stalk; cap irregularly rounded and flattened, smooth and even to distinctly furrowed or brain-like, with a strongly inrolled margin; pale yellow to olive-brown or green; stalk smooth to finely granular; solitary, scattered, or in fused clusters on the ground in woods—*Leotia* species (see key to Earth Tongues and Earth Clubs, p. 503).

4. Fruiting body ⅜–1" (1–2.5 cm) high, 3⁄16–⅜" (5–10 mm) wide, spatula-shaped to shoehorn-shaped with a wavy margin or fan-shaped with deeply cut lobes; yellow-orange to orange; undersurface longitudinally ribbed; stalk round at the base, becoming flattened upward, yellow-orange to orange, darkening in age; in clusters on decaying wood—*Dacryopinax spathularia* (Schweinitz) Martin (see p. 431).

4. Fruiting body not with the above combinations of characters → 5.

5. Fruiting body up to ⅛" (3 mm) wide, ⅝" (1.5 cm) high, consisting of erect, pointed spikes that resemble antler tines, simple or sometimes forked; yellow to orange-yellow; solitary, scattered, or in fused clusters on decaying wood, especially hardwoods—*Calocera cornea* (Batsch : Fries) Fries (see p. 430).

5. Fruiting body ⅜–1" (1–2.5 cm) wide, 1⅜–4" (3.5–10 cm) high, consisting of erect, pointed, repeatedly forked, anter-like tines; golden yellow to orange-yellow; scattered or in groups on decaying conifer wood—*Calocera viscosa* (Persoon : Fries) Fries (see p. 431).

5. Fruiting body ¹⁄₃₂–¹⁄₁₆" (1–1.5 mm) wide, ¼–⅝" (6–15 mm) high, consisting of erect, spindle-shaped to cylindric or pointed spikes, usually simple but sometimes forking; tough, waxy, white to creamy white; scattered or in groups on decaying wood or on soil—*Multiclavula mucida* (Fries) Petersen (see p. 511).

    6. Fruiting body growing in clusters on the cap, gills, and stalk of *Collybia dryophila* (Bulliard : Fries) Kummer ⅛–1" (3–25 mm) wide or larger, brain-like, cup-shaped, or tumor-like, often fused, pale yellow—*Syzygospora mycetophila* (Peck) Ginns (see p. 433).

    6. Fruiting body growing on decaying wood, usually conifers, solitary or in groups; ⅛–¾" (3–20 mm) wide, ⅛–⅜" (3–10 mm) high, brain-like to cushion-shaped or nearly round, whitish to tan or yellow-brown, with a white, firm center when sectioned; spores 9–11 × 7–9 μm, oval to nearly round, smooth, hyaline; edible—*Tremella encephala* Persoon.

    6. Fruiting body growing on decaying wood, especially barkless hardwood, in dense clusters; ¼–1⅛" (7–30 mm) wide, up to ¾" (2 cm) high, brain-like to deeply convoluted, forming extensive irregular masses up to 6" (15.5 cm) or more long; firm and waxy when young, becoming soft-gelatinous in age; white to dingy yellow; spores 9–12 × 4.5–7 μm, sausage-shaped to oval, smooth, hyaline; July–October; edibility unknown—*Ductifera pululahuana* (Patouillard) Donk = *Exidia alba* (Lloyd) Burt.

    6. Fruiting body not with the above combinations of characters → 7.

7. Fruiting body hyaline to whitish at first, becoming pale yellow to brownish yellow, and finally reddish brown in age; ¼–¾" (6–20 mm) wide, up to ⅝" (1.5 cm) high, brain-like to somewhat lobed, fusing together and forming extensive irregular masses up to 4" (10 cm) or more long; containing numerous hard, whitish, seed-like granules up to ¹⁄₃₂" (0.5 mm) in diameter; in groups or clusters on decaying hardwood—*Exidia nucleata* (Schweinitz) Burt (see p. 431).

7. Not as above; fruiting body pale pink, reddish purple, pinkish brown, cinnamon-brown to vinaceous-brown, or blackish brown to black → 8.

7. Not as above; fruiting body pale yellow to orange-yellow, orange or orange-red → 9.

    8. Fruiting body yellowish brown to purplish brown or cinnamon-brown, ⅝–1⅜" (1.5–3.5 cm) wide, up to ¾" (2 cm) high, cushion-shaped to brain-like or irregularly lobed, somewhat erect with a stalk-like base, forming extensive irregular clusters up to 4" (10 cm) or more long; rubbery-gelatinous, shiny, smooth to somewhat roughened and often contorted; margin somewhat thickened and irregular; upper surface typically with scattered, tiny, blackish, wart-like projections (use a hand lens); in groups or clusters on decaying hardwood—*Exidia recisa* Fries (see p. 432).

    8. Fruiting body dark reddish brown to blackish brown or black; ⅜–¾" (1–2 cm) wide, up to ½" (1.3 cm) high, gland- to blister- or brain-like, fusing together to form extensive irregular masses up to 5" (18 cm) or more long; soft, gelatinous, smooth or somewhat warty, shiny; spores 10–16 × 4–5 μm; clustered on decaying hardwood; May–November; edible—*Exidia glandulosa* Bulliard : Fries.

    8. Fruiting body black to blackish brown, often shiny; ⅜–1½" (2–4 cm) wide, ½–1⅜" (1.6–3.5 cm) high, nearly round to top-shaped when young, becoming saucer-shaped in age; rubbery-gelatinous; ascospores 9–16 × 6–7 μm, kidney-

shaped to elliptic, smooth, of two color forms in each ascus: upper four (mature) brown, lower four (immature) hyaline and typically very poorly developed; in groups or clusters on decaying hardwood, especially oak; August–September—*Bulgaria inquinans* Fries (see Cup and Saucer Fungi, p. 491).

8. Fruiting body pale pink to pinkish brown or reddish purple; ¾–1½" (2–4 cm) wide, ⅜–1" (1–2.5 cm) high, forming dense clusters up to 4" (10 cm) long; brain-like, rubbery-gelatinous, smooth, shiny; spores 7–8.5 × 4–5 μm, elliptic to spindle-shaped, often truncate, smooth, hyaline, typically uniseriate in an ascus; on decaying hardwood, especially beech—*Ascotremella faginea* (Peck) Seaver (see Cup and Saucer Fungi, p. 490).

8. Fruiting body pale pink to pinkish brown or yellowish brown; ⅜–1⅜" (1–3 cm) wide, ⅜–1" (1–2.5 cm) high, forming dense clusters up to 4" (10 cm) in diameter; cushion-shaped to top-shaped or irregularly convoluted, rubbery-gelatinous, smooth, shiny; spores 7–9 × 3.5–4.5 μm, elliptic, smooth, uniseriate in an ascus; on decaying hardwoods, especially beech—*Neobulgaria pura* (Fries) Petrak (see Cup and Saucer Fungi, p. 494).

9. Fruiting body yellowish orange to orange or reddish orange, whitish near the point of attachment; ⅜–2⅜" (1–6 cm) wide, up to 1" (2.5 cm) high, a brain-like to multilobed spreading mass; in clusters on decaying conifer branches and logs; year-round—*Dacrymyces palmatus* (Schweinitz) Bresadola (see p. 431).

9. Fruiting body pale yellow to yellow or orange-yellow; ⅝–2¾" (1.5–7 cm) wide, up to 1" (2.5 cm) high, a brain-like to multilobed, spreading mass; on decaying hardwood branches and logs; year-round—*Tremella lutescens* Fries (see p. 433).

9. Fruiting body orange to yellow-orange or yellow; ⅛–⅝" (3–15 mm) wide, cushion-shaped to brain-like or somewhat flattened and blister-like; forming irregular spreading masses up to 2" (5 cm) or more long; smooth to slightly wrinkled or convoluted, deliquescing in age to a slimy mass; spores usually of two types: basidiospores 14–16 × 5–6 μm, sausage-shaped and 3-septate, thick-walled, smooth, hyaline; and conidia 9–15 × 3–4 μm, elliptic and 1-septate, thick-walled, smooth, hyaline; clustered on decaying wood; year-round; fairly common; edible—*Dacrymyces stillatus* Nees : Fries.

Auricularia auricula (Hooker) Underwood
COMMON NAME: Tree-Ear.
FRUITING BODY: 1⅛–5⅞" (3–15 cm) wide, ear-shaped to irregularly cup-shaped, stalkless, rubbery-gelatinous; upper (outer) surface wrinkled, minutely hairy to velvety, reddish brown; lower (inner) surface smooth, yellowish brown to reddish brown or grayish brown to purplish brown.
MICROSCOPIC FEATURES: spores 12–15 × 4–6 μm, sausage-shaped, smooth, hyaline.
FRUITING: solitary, in groups, or in fused clusters on decaying wood, especially conifers; May–November; fairly common.
EDIBILITY: edible.

Calocera cornea (Batsch : Fries) Fries
COMMON NAME: Club-like Tuning Fork.
FRUITING BODY: up to ⅛" (3 mm) wide and ⅝" (1.6 cm) high, consisting of erect, pointed spikes that resemble antler tines, simple or sometimes forked; yellow to orange-yellow, rubbery-gelatinous.
MICROSCOPIC FEATURES: spores 7–11 × 3–4.5 μm, cylindric to sausage-shaped, with a single septum at maturity, smooth, hyaline to yellowish.

FRUITING: scattered, in groups, or in clusters on decaying wood, especially hardwoods; August–November; common.

EDIBILITY: unknown.

Calocera viscosa (Persoon : Fries) Fries

COMMON NAME: Yellow False Coral.

FRUITING BODY: ⅜–1" (1–2.5 cm) wide, 1⅜–4" (3.5–10 cm) high, consisting of erect, pointed, repeatedly forked, antler-like tines; golden yellow to orange-yellow, rubbery-gelatinous, sometimes slimy.

MICROSCOPIC FEATURES: spores 9–13 × 3–5 μm, sausage-shaped, with a single septum at maturity, smooth, hyaline.

FRUITING: scattered or in groups on decaying conifer wood; August–October; infrequent.

EDIBILITY: unknown.

COMMENTS: yellow to orange-yellow coral fungi are brittle and easily broken.

Dacrymyces palmatus (Schweinitz) Bresadola (see photo, p. 14)

COMMON NAME: Orange Jelly, Orange Witches' Butter.

FRUITING BODY: ⅜–2⅜" (1–6 cm) wide, up to 1" (2.5 cm) high; a brain-like to multi-lobed, spreading mass; yellowish orange to orange or reddish orange, whitish near the point of attachment; rubbery-gelatinous at first, becoming soft-gelatinous in age.

MICROSCOPIC FEATURES: spores 17–25 × 6–8 μm, cylindric to sausage-shaped, 7–9 septate, smooth, hyaline.

FRUITING: a dense cluster on decaying conifer branches and logs; year-round; common.

EDIBILITY: edible and rather bland; it may be eaten raw or cooked.

Dacryopinax spathularia (Schweinitz) Martin

FRUITING BODY: ³⁄₁₆–⅜" (5–10 mm) wide, ⅜–1" (1–2.5 cm) high, spatula-shaped to shoehorn-shaped with a wavy margin or fan-shaped with deeply cut lobes, yellow-orange to orange, rubbery-gelatinous; undersurface longitudinally ribbed.

STALK: round at the base, becoming flattened upward, yellow-orange to orange, darkening in age.

MICROSCOPIC FEATURES: spores 8–12 × 3.5–5 μm, sausage-shaped, with a single septum at maturity, smooth, yellowish.

FRUITING: in groups or clusters on decaying wood; July–October; infrequent.

EDIBILITY: unknown.

Exidia nucleata (Schweinitz) Burt

COMMON NAME: Granular Jelly Roll.

FRUITING BODY: ¼–¾" (6–20 mm) wide, up to ⅝" (1.5 cm) high, brain-like to somewhat lobed, fusing together and forming extensive irregular masses up to 4" (10 cm) or more long; rubbery-gelatinous; hyaline to whitish at first, becoming pale yellow to brownish yellow, and finally drying to a thin, reddish brown membrane in age; containing numerous hard, whitish, seed-like granules up to ¹⁄₃₂" (0.5 mm) in diameter.

MICROSCOPIC FEATURES: spores 8–11.5 × 4–4.5 μm, sausage-shaped, smooth, hyaline.

FRUITING: in groups or clusters on decaying hardwood; July–November; fairly common.

EDIBILITY: unknown.

Exidia recisa Fries

COMMON NAME: Amber Jelly Roll.

FRUITING BODY: ⅝–1⅜" (1.5–3.5 cm) wide, up to ¾" (2 cm) high, cushion-shaped to brain-like or irregularly lobed, somewhat erect, with a stalk-like base; forming extensive irregular clusters up to 4" (10 cm) or more long; rubbery-gelatinous, yellowish brown to purplish brown or cinnamon-brown, shiny, smooth to somewhat roughened, and often contorted; margin somewhat thickened and irregular; upper surface typically with scattered, tiny, blackish, wart-like projections (use a hand lens).

MICROSCOPIC FEATURES: spores 11–15 × 3–5.5 μm, sausage-shaped, smooth, hyaline.

FRUITING: in groups or clusters on decaying hardwoods; May–October; fairly common.

EDIBILITY: unknown.

Phleogena faginea (Fries) Link

FRUITING BODY: up to ⁵⁄₁₆" (8 mm) high, consisting of a tiny head and stalk.

HEAD: ¹⁄₁₆–⅛" (1.5–3 mm) wide, nearly round to flattened or somewhat irregular; dry, finely cracked, and roughened, white when young, becoming grayish white to pale brown at maturity.

STALK: up to ¼" (6 mm) long, up to ¹⁄₃₂" (0.5 mm) thick, equal or enlarged in either direction, cylindric to slightly flattened, sometimes with shallow longitudinal furrows.

MICROSCOPIC FEATURES: spores 8–10 μm, nearly round, smooth, thick-walled, brownish.

FRUITING: scattered, in groups, or in clusters on decaying hardwood; August–November; fairly common.

EDIBILITY: inedible.

COMMENTS: because of its small size it is easily overlooked and sometimes mistaken for a slime mold.

Phlogiotis helvelloides (Fries) Martin (see photo, p. 14)

COMMON NAME: Apricot Jelly.

FRUITING BODY: ¾–2¾" (2–7 cm) wide, 1–3⅛" (2.5–8 cm) high, funnel-shaped with a split side or spoon-shaped to tongue-shaped with a wavy margin, rubbery-gelatinous, nearly smooth; pinkish red to apricot, often paler on the margin in age.

MICROSCOPIC FEATURES: spores 10–12 × 4–7 μm, elliptic, smooth, hyaline.

FRUITING: solitary or in groups on the ground or decaying wood in coniferous and mixed woods; May–October; infrequent.

EDIBILITY: edible but rather bland.

COMMENTS: also known as *Tremiscus helvelloides*.

Pseudohydnum gelatinosum (Fries) Karsten

COMMON NAME: Jelly Tooth.

FRUITING BODY: 1–2¾" (2.5–7 cm) wide, ¾–2" (2–5 cm) high, shoehorn to tongue-shaped or hemispheric, with a cap and eccentric stalk; rubbery-gelatinous; upper surface of cap minutely velvety to smooth, gray-translucent, yellowish, or brown; lower cap surface densely coated with tiny, soft, whitish spines; stalk up to 2" (5 cm) long, broad and somewhat flattened at the apex, tapering downward, smooth, colored like the cap, often absent.

MICROSCOPIC FEATURES: spores 5–7 μm, globose, smooth, hyaline.

FRUITING: solitary, scattered, or in groups on decaying wood, especially hemlock; September–November; common.

EDIBILITY: edible but rather bland.

Sebacina incrustans (Fries) Tulasne

> FRUITING BODY: up to 6" (15.5 cm) or more in diameter; a thin, fibrous-tough to leathery, irregular, sheet-like mass, forming small, lateral projections; whitish to pale tan.
>
> MICROSCOPIC FEATURES: spores 11.5–15 × 6–8 μm, elliptic to ovoid, flattened to slightly depressed on one side, smooth, hyaline.
>
> FRUITING: growing on the ground and soon encrusting plant stems, leaves, branches, cones, and debris.
>
> EDIBILITY: inedible.

Syzygospora mycetophila (Peck) Ginns

> COMMON NAME: Collybia Jelly.
>
> FRUITING BODY: ⅛–1" (3–25 mm) wide or larger, consisting of brain-like, cup-shaped, tumor-like, or irregular growths that are often fused together; pale yellow to brownish yellow, rubbery-gelatinous.
>
> MICROSCOPIC FEATURES: spores 6–9 × 1.5–2.5 μm, elliptic to cylindric, smooth, hyaline.
>
> FRUITING: in dense clusters on the cap, gills, or stalk of *Collybia dryophila*.
>
> EDIBILITY: unknown.
>
> COMMENTS: also known as *Christiansenia mycetophila* and previously known as *Tremella mycetophila*.

Tremella concrescens (Fries) Burt

> FRUITING BODY: a soft, whitish to grayish, irregular, membrane-like, rubbery-gelatinous spreading mass up to 5½" (14 cm) or more in diameter.
>
> MICROSCOPIC FEATURES: spores variable, 9–14 × 5–8 μm, elliptic and slightly curved to oval or nearly round, smooth, hyaline.
>
> FRUITING: on the ground and soon enveloping plant stems and leaves; August–October; fairly common.
>
> EDIBILITY: unknown.

Tremella foliacea Persoon : Fries

> COMMON NAME: Jelly Leaf.
>
> FRUITING BODY: 2–10" (5–25.5 cm) wide, 2–4" (5–10 cm) high, a lettuce-like cluster composed of numerous leaf-like, rubbery-gelatinous folds; pale to dark reddish brown or rarely pale brownish yellow.
>
> MICROSCOPIC FEATURES: spores 8–12 × 7–9 μm, oval to nearly round, smooth, hyaline.
>
> FRUITING: solitary or scattered on decaying wood; July–November; infrequent.
>
> EDIBILITY: edible but rather bland.

Tremella lutescens Fries

> COMMON NAME: Witches' Butter.
>
> FRUITING BODY: ⅝–2¾" (1.5–7 cm) wide, up to 1" (2.5 cm) high, a brain-like to multi-lobed spreading mass; rubbery-gelatinous, pale yellow to yellow or orange-yellow.
>
> MICROSCOPIC FEATURES: spores 10–18 × 8–12 μm, broadly oval, smooth, hyaline.
>
> FRUITING: clustered on decaying hardwood branches and logs; year-round; fairly common.
>
> EDIBILITY: edible and rather bland.
>
> COMMENTS: also known as *T. mesenterica*.

434 / Mushrooms of Northeastern North America

Tremellodendron pallidum (Schweinitz) Burt

COMMON NAME: Jellied False Coral.

FRUITING BODY: 2–5½" (5–14 cm) wide, 1–3¾" (2.5–9.5 cm) high, coral-like, densely clustered, composed of numerous broadly flattened to somewhat cylindric branches; branches leathery to fibrous-tough, white to buff, fused together at their bases, with their apices frequently united.

MICROSCOPIC FEATURES: spores 7.5–10 × 4–6.5 μm, ovoid to sausage-shaped, smooth, hyaline.

FRUITING: solitary, scattered, or in groups on the ground in hardwoods; July–November; common.

EDIBILITY: edible.

COMMENTS: also known as *T. schweinitzii*.

Auricularia auricula

Calocera cornea

Calocera viscosa

Dacryopinax spathularia

Exidia nucleata

Exidia recisa

Phleogena faginea

Pseudohydnum gelatinosum

Sebacina incrustans

Syzygospora mycetophila

Tremella concrescens

Tremella foliacea

Tremella lutescens

Tremellodendron pallidum

Crust and Parchment Fungi and Allies

Crust and Parchment Fungi and Allies is a very large and highly variable complex of inedible species that form thin, spreading, crust-like to papery growths, usually on decaying wood. Some species are nearly flat; others have small, projecting, shelf-like caps usually formed by bending backward at the margin. Their fertile surfaces may be rough, warted, wrinkled, cracked, tooth-like, or smooth, but lack true pores and are not finely roughened like sandpaper. Crust-like species with pores on their fertile surfaces are included in the Polypores. We have included some of the more common crust-like polypores and Carbon and Cushion Fungi which are likely to be confused with the crust and parchment fungi.

Several hundred species of crust and parchment fungi have been identified. We have included some of the more common and conspicuous species likely to be encountered.

Key to Species of Crust and Parchment Fungi and Allies

1. Fruiting body dark gray to black, hard, often carbonaceous, with a minutely roughened surface; or cushion-shaped to confluent and spreading like a crust, white, yellow, green, brick-red, reddish brown, purplish brown, or dark purple to black, with a sandpaper-like roughened surface—Carbon and Cushion Fungi (see p. 521).
1. Fruiting body on wood, soft, gray, with a white margin when young, becoming black and brittle with a sandpaper-like roughened surface—*Ustulina deusta* (Fries) Petrak (see Carbon and Cushion Fungi, p. 525).
1. Fruiting body with short spines or flattened teeth, or with small to large, round to angular pores on the fertile surface → 2.
1. Fruiting body fertile surface rough, warted, wrinkled, cracked, or smooth, but lacking teeth, spines, or pores and not sandpaper-like → 5.
   2. Fruiting body fertile surface with short spines or rounded to flattened teeth; a whitish, bright yellow, orange-yellow to yellow-orange or brownish crust, with or without projecting stalkless caps → 3.
   2. Fruiting body fertile surface lacking spines or teeth; with small to large, round to angular pores on their fertile surfaces → 4.
3. Bright yellow spreading crust with crowded waxy spines measuring ³⁄₁₆–³⁄₈" (5–10 mm) long; slowly developing reddish stains when bruised; with an intense, unpleasant odor variously described as sickeningly sweet, fruity, chemical, etc.; on logs and branches of hardwoods, especially apple; spores 5–6 × 3–4 μm—*Sarcodontia setosa* (Persoon) Donk.
3. Bright orange-yellow spreading crust with a whitish margin; with crowded yellow to

orange rounded spines and orange yarn-like runners; on the underside of decaying hardwood logs; spores 3.5–4.5 × 2–2.5 μm—*Phanerochaete chrysorhiza* (Torrey) Gilbert.

3. Olive-brown to reddish brown, leathery to stiff spreading crust with uneven, flattened teeth; on the underside of decaying hardwood branches; spores 4–6.5 × 1.2–1.5 μm—*Hydnochaete olivaceum* (Schweinitz) Banker.

3. Whitish spreading crust with tubes that break into white to creamy white, flattened teeth; on decaying hardwood branches—*Irpex lacteus* (Fries) Fries (see Polypores, p. 378).

3. Yellow-orange, spreading crust; with or without a zoned, hairy, short projecting cap; with a white margin and short yellow-orange spines; on decaying hardwoods; spores 3–4 × 2–2.5 μm—*Steccherinum ochraceum* (Persoon : Fries) S. F. Gray.

    4. Fruiting body a spreading, brown to rusty cinnamon, honeycomb-like crust with a white, downy margin, grayish white strands of runners, with irregular pores that become elongated into tooth-like projections; on conifer logs or structural timbers, bricks, concrete, or stone in poorly ventilated areas, especially cellars—*Serpula lacrimans* (Fries) Schroeter (see p. 442).

    4. Fruiting body a flattened, rusty brown to yellowish brown crust with a distinctly paler margin on mature specimens, not forming a cap; surface uneven, roughened, with 5–6 rounded pores per mm (use a hand lens); growing on downed branches or trunks of hardwoods or rarely conifers—*Phellinus ferruginosus* (Fries) Patouillard.

    4. Fruiting body a woody, spreading, yellow-brown to reddish brown crust with projecting, stalkless, blackish brown caps with tiny hairs and ridges; margin paler, with 2–5 round to angular pores per mm; on trunks of conifers—*Phellinus chrysoloma* (Fries) Donk.

    4. Fruiting body a spreading, creamy white to pinkish white hard crust with tiny hairs on the margin; with 1–4 irregular elongated pores per mm that break apart, forming tooth-like projections; on decaying branches of hardwoods—*Schizopora paradoxa* (Fries) Donk.

5. Fruiting body with projecting, shelf-like caps and a smooth fertile surface → 6.

5. Fruiting body with projecting, shelf-like caps with a rough, wrinkled, or powdery to bristly fertile surface → 9.

5. Fruiting body a spreading crust, lacking projecting, shelf-like caps → 12.

    6. Fruiting body leathery, with a yellowish to reddish brown upper surface and a purplish, waxy, spreading, fertile surface; on hardwoods, especially apple, pear, plum, and poplar; spores 5–6.5 × 2–3 μm—*Chondrostereum purpureum* (Fries) Pouzar.

    6. Fruiting body spongy, flexible, forming a spreading thin crust with shell-like, overlapping caps; medium brown and wrinkled or felt-like on the upper surface; whitish to pale brown on the fertile surface; on decaying hardwood; spores 3.5–6 × 2–3 μm—*Laxitextum bicolor* (Persoon : Fries) Lentz.

    6. Fruiting body woody, composed of many-sided plates resembling broken pieces of dull ceramic tile fused into irregular colonies; outer surface blackish, smooth; fertile surface pinkish tan to pale grayish olive; on hardwood logs, especially oak; spores 3.5–5 × 2.5–3 μm—*Xylobolus frustulatus* (Persoon : Fries) Boidin.

    6. Fruiting body not with the above combinations of characters; shell- to petal-like, often in overlapping clusters, typically attached to the substrate at a single point, not spreading as a flattened thin crust → 7.

7. Fruiting body white to grayish or pinkish buff, growing on the ground under hard-

woods; vase-shaped with a stalk and a torn margin—*Cotylidia diaphana* (Schweinitz) Lentz.

7. Fruiting body silvery to pale gray with tiny, radiating, silky fibers; fertile surface whitish; on decaying branches of American hornbeam; spores 6–8.5 × 2–3.5 μm— *Stereum striatum* (Fries) Fries.

7. Fruiting body not with the above combinations of characters → 8.

    8. Fruiting body ⅛–1⅛" (3–16 mm) wide, thin, overlapping, laterally fused; upper surface radially furrowed, coated with stiff hairs near the base and fine, silky hairs near the margin, zoned, orange-cinnamon to reddish brown, often yellow at the margin; fertile surface orange, fading to creamy yellow; on decaying hardwood or conifer branches; spores 5–6.5 × 2–2.5 μm—*Stereum complicatum* (Fries) Fries.

    8. Fruiting body ⅜–2¾" (1–7 cm) in diameter, thin, overlapping, sometimes laterally fused; upper surface coated with fine, silky hairs, typically concentrically zoned with reddish brown and various other colors, often whitish at the margin; fertile surface reddish brown; on decaying hardwood logs and stumps—*Stereum ostrea* (Blume and Nees) Fries (see p. 443).

    8. Fruiting body ⅜–2⅜" (1–6 cm) in diameter, thin, sometimes overlapping, laterally fused; upper surface densely coated with soft, downy hairs or coarse, stiff hairs, zoned, cinnamon-brown to clay-color, paler at the margin; fertile surface wood-brown, staining to pale blackish brown or cinnamon-brown when injured; on stumps and logs of hardwoods; spores 5–8.5 × 2–3 μm—*Stereum gausapatum* (Fries) Fries.

    8. Fruiting body ¼–¾" (6–20 mm) in diameter, thin, overlapping, laterally fused; upper surface coated with coarse, stiff hairs, zoned, whitish to pale cinnamon-brown or grayish overall; fertile surface yellowish to reddish tan or wood-brown; on decaying conifer wood or hardwood; spores 5–8 × 2–3.5 μm—*Stereum hirsutum* (Willdenow : Fries) S. F. Gray.

9. Fertile surface yellowish orange, pinkish orange, brownish salmon, orange, golden, golden brown to reddish brown; radially wrinkled to folded, often with cross-veins → 10.

9. Fertile surface pinkish brown to dull brown or blackish brown; roughened to bristly and sometimes finely to deeply cracked → 11.

9. Fertile surface whitish to grayish or pale yellow to pale tan, uneven, roughened and finely cracked, bruising pale brown —*Cystostereum murraii* (Berkeley and Curtis) Pouzar (see p. 441).

    10. Upper surface nearly smooth, coral-pink; fertile surface pinkish orange to brownish salmon; typically forming overlapping clusters on hardwood; spores 4–5 × 2–2.5 μm—*Phlebia incarnata* (Schweinitz) Nakasone and Burdsall = *Merulius incarnatus* Schweinitz.

    10. Upper surface hairy to woolly, white to pale yellow; fertile surface yellowish to brownish orange or pinkish orange; often forming overlapping clusters, usually on hardwood but sometimes on decaying conifer logs and stumps; spores 3–4 × 0.5–1.5 μm—*Phlebia tremellosa* (Schrader : Fries) Nakasone and Burdsall = *Merulius tremellosus* Fries.

    10. Upper surface coarsely hairy, reddish brown to grayish, with multicolored concentric zones; fertile surface reddish brown, darkening in age; forming fused clusters on decaying hardwood; spores 6–8 × 3–4 μm—*Punctularia strigoso-zonata* (Schweinitz) Talbot.

    10. Upper surface finely hairy to woolly, orange to golden brown with a paler golden

margin; fertile surface orange to golden brown; forming laterally fused clusters on the underside of barkless conifer logs; spores 3.5–4.5 × 1.2–1.8 μm—*Pseudomerulius aureus* (Fries) Jülich.

11. Fruiting body leathery, somewhat flexible, able to be torn, growing in overlapping clusters, typically fused laterally on branches and logs of hardwoods; upper surface densely matted and woolly, yellowish brown to orange-brown or dark brown, concentrically zoned, becoming orange-yellow toward the margin; fertile surface brown, uneven, smooth to finely cracked or deeply radially cracked on the flattened portion when dry, with numerous dark brown setae—*Hymenochaete tabacina* (Sowerby) Léveille complex (see p. 441).

11. Fruiting body typically leathery and rigid, growing horizontally in overlapping clusters, typically fused laterally on decaying hardwood stumps and logs; upper surface dark reddish brown to blackish brown with an orange-yellow margin; fertile surface orange-brown to reddish brown when young, becoming grayish brown in age, conspicuously bristly under a hand lens, with little rounded, wart-like elevations and numerous dark brown setae; spores 4–6 × 2.5–3 μm—*Hymenochaete rubiginosa* (Dickson : Fries) Léveille.

11. Fruiting body leathery, covered with stiff hairs, growing horizontally in laterally fused clusters on decaying hardwood; upper surface dull reddish tan to grayish; fertile surface pinkish tan to pinkish brown, minutely bristly to slightly cracked, lacking setae; spores 8–16 × 5–10 μm—*Lopharia cinerascens* (Schweinitz) Cunningham.

12. Fruiting body white to pale brownish pink or pale brown; small, ⅟₃₂–¼" (1–6 mm) in diameter, disc-shaped, flattened to slightly elevated at the margin, attached at a single point, fleshy to leathery, sometimes confluent, growing on wood → 13.

12. Fruiting body dark blue to blackish blue, purplish to pale brownish pink or yellow, a confluent, spreading crust, 1⅛–6" (3–15.5 cm) in diameter, sometimes becoming cracked in age; growing on wood → 14.

12. Fruiting body not with the above combinations of characters → 15.

13. Fruiting body chalk-white; pruinose on the margin, darkening underneath; growing on bark of oak, ash, or maple; spores 15–17 × 11–14 μm—*Dendrothele candida* (Schweinitz : Burt) Lemke.

13. Fruiting body whitish on the undersurface of the margin; fertile surface pale brownish pink to pale brown; growing on hardwoods; spores 18–21 × 12–13 μm—*Aleurodiscus oakesii* (Berkeley and Curtis) Hoehnel and Litschauer.

13. Fruiting body whitish on the undersurface of the margin; fertile surface pale brownish pink; growing on balsam fir or spruce; spores 20–27 × 16–21 μm—*Aleurodiscus amorphus* (Persoon) Rabenhorst.

13. Fruiting body deep mouse-gray on the undersurface of the margin; fertile surface pale brownish pink to pale brown; growing on hemlock or balsam fir; spores 13–18 × 9–12 μm—*Aleurodiscus farlowii* Burt.

14. Fruiting body surface reddish purple to grayish purple when young, fading to pale brownish pink, and finally pale brown in age, cracking irregularly, with whitish flesh showing in the cracks; growing on hardwoods—*Laxitextum roseo-carneum* (Schweinitz) Lentz (see p. 442).

14. Fruiting body surface dark blue to blackish blue, paler toward the margin, usually not cracked, velvety; typically growing on the underside of hardwood logs and branches, especially oak—*Pulcherricium caeruleum* (Persoon) Parmasto (see p. 442).

14. Fruiting body surface golden yellow to honey-yellow, cracking irregularly, with

yellow flesh showing in the cracks; growing on conifer logs; spores 4–5 × 2–2.5 μm—*Phanerochaete carnosa* (Burt) Parmasto.

15. Fruiting body red to purplish brown, with a grayish bloom when dry; ¼–⅜" (5–10 mm) wide, waxy, wrinkled to nearly smooth, wart-like; spores 6–9 × 1.5–2.5 μm; gloeo-cystidia abundant; growing on branches of poplar; spores 6–8 × 1.5–2 μm—*Peniophora rufa* (Fries) Boidin.

15. Fruiting body pinkish orange to pinkish red, wrinkled and radially folded, gelatinous to fibrous, with a whitish hairy margin; growing on decaying conifer or hardwood; spores 3–4.5 × 1.5–2 μm—*Phlebia radiata* Fries.

15. Fruiting body white to pale pinkish brown, large, 1¾–12" (4.5–30 cm) in diameter, waxy, parchment-like, with a margin that becomes free and curls in age; growing on conifer wood; spores 4–5 × 2–3 μm—*Peniophora gigantea* (Fries) Massee.

15. Fruiting body dull white to yellowish or orange-yellow, 1¾–10" (4.5–25 cm) in diameter, highly variable but often cushion-shaped or flattened and irregular, powdery to crusty; growing on decaying wood, wood chips, sawdust, compost, leaf litter, or other debris; spores 6–9 μm, globose, minutely spiny—*Fuligo septica* (Linnaeus) Wiggers. Note: this is a very common Myxomycete (slime mold) and is included here because it is often confused with some Crust and Parchment species.

Cystostereum murraii (Berkeley and Curtis) Pouzar (see photo, p. 15)

FRUITING BODY: ¾–4" (2–10 cm) wide, thin, crust-like, becoming confluent and spreading to form patches 8" (20 cm) or more in diameter; fertile surface whitish to grayish or pale yellow to pale tan, uneven, roughened, and finely cracked, dull; margin on the upper side bent backward, forming a dark, brownish black shelf with irregular concentric ridges and radial folds, sometimes covered with moss, projecting ⅛–⅜" (3–10 mm).

SPORE PRINT: white.

MICROSCOPIC FEATURES: spores 4–5 × 2–2.5 μm, oval and flattened on one side to nearly elliptic, smooth, hyaline.

FRUITING: on logs of hardwoods, especially beech; April–October; fairly common.

EDIBILITY: inedible.

COMMENTS: formerly known as *Stereum tuberculosum* and *Stereum murraii*.

Hymenochaete tabacina (Sowerby) Léveille Complex (see photo, p. 15)

FRUITING BODY: a thin spreading crust with small, shelf-like, stalkless caps ⅜–¾" (1–2 cm) wide in overlapping clusters, often fused laterally and extending 8–12" (20–30 cm) or more, projecting ¼–⅜" (6–10 mm), leathery, flexible, able to be torn; upper surface dry, densely matted and woolly, becoming nearly smooth on old specimens, yellowish brown to orange-brown or dark brown, concentrically zoned; margin orange-yellow to bright golden yellow; fertile surface dull brown, uneven, smooth to finely cracked or deeply radially cracked on the flattened portion when dry, with numerous dark brown setae.

FLESH: tough, fibrous, dull brown.

MICROSCOPIC FEATURES: spores 4.5–6 × 1.5–2 μm, sausage-shaped to nearly cylindric, smooth, hyaline; fertile surface bearing setae 60–90 × 6–12 μm.

FRUITING: in overlapping clusters on decaying hardwood species; throughout the year; common.

EDIBILITY: inedible.

COMMENTS: addition of a drop of KOH to the flesh produces a dark brown to black reaction. Some authors recognize *H. badio-ferruginea* as a distinct species with a smooth to minutely cracked fertile surface and describe *H. tabacina* as having a deeply cracked, radial fertile surface. Because considerable variation occurs and much confusion and conflicting information can be found in the literature, we are treating these species as a complex.

Laxitextum roseo-carneum (Schweinitz) Lentz (see photo, p. 15)
FRUITING BODY: ¹⁄₁₆–¼" (1.5–6 mm) wide, thin, crust-like, becoming confluent and spreading to form patches 1⅜–4" (3.5–10 cm) or more in diameter; surface reddish purple to grayish purple when young, fading to pale brownish pink, and finally pale brown in age, somewhat leathery, dull, cracking irregularly, with whitish flesh showing in the cracks; margin uneven, wavy, free, or bent backward for ¹⁄₁₆–⅛" (1.5–3 mm).
SPORE PRINT: white.
MICROSCOPIC FEATURES: spores 7.5–11 × 4.5–6 μm, oval and flattened on one side to nearly elliptic, smooth, hyaline.
FRUITING: on branches of decaying hardwood; April–December; fairly common.
EDIBILITY: inedible.
COMMENTS: previously known as *Stereum roseo-carneum*.

Pulcherricium caeruleum (Persoon) Parmasto
COMMON NAME: Velvet Blue Spread.
FRUITING BODY: ⅛–¾" (3–20 mm) wide, thin, crust-like, rounded to irregular, becoming confluent and spreading to form patches 1½–6" (4–15.5 cm) or more in diameter; surface dry, velvety, dark blue to blackish blue, paler toward the margin.
FLESH: soft, membranous; odor and taste not distinctive.
SPORE PRINT: white.
MICROSCOPIC FEATURES: spores 6–10 × 4–5 μm, elliptic, smooth, hyaline.
FRUITING: usually on the underside of decaying hardwood logs and branches, especially oak; August–December; occasional; found in the middle to southern part of the region.
EDIBILITY: inedible.
COMMENTS: also known as *Corticum caeruleum*.

Serpula lacrimans (Fries) Schroeter (see photo, p. 15)
COMMON NAME: Dry Rot.
FRUITING BODY: a spreading, brown to rusty cinnamon, honeycomb-like crust of indefinite size, but often more than 3 feet in diameter, with a thick, white, downy margin, sometimes forming shelf-like brackets on vertical surfaces, with grayish white strands of mycelial runners up to several yards long that spread the fungus from one area to another; fertile surface resembling an uneven honeycomb, with elongated, angular, tooth-like projections and small pore-like depressions.
SPORE PRINT: brownish.
MICROSCOPIC FEATURES: spores 8–12 × 4.5–6 μm, elliptic, smooth, thick-walled, with a cyanophilous inner wall, pale yellow.
FRUITING: on conifer logs in woods or spreading over structural timber, concrete, brick, stone, and numerous other substances in damp, poorly ventilated areas; throughout the year; common.
EDIBILITY: unknown.
COMMENTS: an unpleasant, musty odor is typically produced by this fungus.

Stereum ostrea (Blume and Nees : Fries) Fries

COMMON NAME: False Turkey-tail.

FRUITING BODY: ⅜–2¾" (1–7 cm) in diameter, shell- to petal-shaped, thin, leathery, overlapping, sometimes laterally fused; upper surface coated with fine, silky hairs, typically concentrically zoned with reddish brown and various other colors, especially gray, yellow, and orange, often whitish at the margin; fertile surface smooth, lacking pores (use a hand lens), reddish brown to reddish buff or buff.

SPORE PRINT: white.

MICROSCOPIC FEATURES: spores 5–7.5 × 2–3 μm, cylindric, smooth, hyaline.

FRUITING: on decaying hardwood branches, logs, and stumps; June–December; common.

EDIBILITY: inedible.

COMMENTS: it is often misidentified as the Turkey-tail, a polypore.

Pulcherricium caeruleum

Stereum ostrea

Puffballs, Earthballs, Earthstars, and Allies

At some stage in their development, all members of this group are round to oval, pear- to turban-shaped, or irregularly rounded in outline. Species included here can be separated into five subgroups based on macroscopic features and habit of growth: earthstars, earthballs, stalked puffballs and allies, hypogeous allies, and puffballs.

The earthstars are nearly round at first, with a short, pointed, beak-like apex; the outer peridium splits open at maturity, forming star-like rays and exposing a thin inner spore case. Earthballs resemble puffballs but have a thick, hard, rind-like spore case. Stalked puffballs have a papery thin spore case and a distinct stalk or a stalk-like base that is often partially or entirely buried. Hypogeous Allies lack a stalk or stalk-like base and are completely or partially buried in soil or duff. Puffballs lack the combined features of the above groups; they have a thin spore case that is often coated with spines and warts and have a white spore mass when young.

Key to Species of Puffballs, Earthballs, Earthstars, and Allies

1. Fruiting body very small, $\frac{1}{16}$–$\frac{1}{8}$" (1.5–3 mm) wide, nearly round, white to yellow-orange, splitting at maturity to form 4–9 star-shaped to tooth-like rays and exposing a single dull yellow or reddish brown to dark brown peridiole; solitary or in groups on decaying wood, sawdust, compost, and dung—*Sphaerobolus stellatus* (Tode) Persoon (see p. 457).
1. Fruiting body at first nearly round, often with a short, pointed, beak-like apex; outer peridium splitting open at maturity to form several star-like rays that often recurve, exposing a thin inner spore case; growing on the ground; Earthstars → 2.
1. Fruiting body round to oval, sometimes flattened or irregular, or resembling a giant earthstar when expanded, with a thick, hard, rind-like spore case; spore mass purple-gray to gray or black in both immature and mature specimens; lacking a stalk, but sometimes with an elongated stalk-like base; on the ground or on decaying wood; Earthballs → 6.
1. Fruiting body with a distinct stalk or an elongated stalk-like base that is sometimes partially or entirely buried, often growing in very sandy soil; spore case not hard and rind-like; spore mass usually white when young; Stalked Puffballs and Allies → 8.
1. Fruiting body lacking a stalk; round to oval, sometimes flattened or irregular, buried or half-buried in soil or duff; Hypogeous Allies → 9.
1. Fruiting body lacking a stalk, round to oval, top-shaped to pear-shaped or irregular and

sometimes lobed; with a thin spore case; spore mass white when young, becoming distinctly colored at maturity; growing on soil, in grass, compost, or sawdust, or on decaying wood; Puffballs → 10.

2.   Spore case ⅜–1¾" (1–4.5 cm) wide, nearly round to somewhat flattened, grayish brown, forming several small pore-like mouths, mostly on the upper surface, elevated above the rays by several short, slender supporting columns; 5–7 rays, not hygroscopic; spores 4–6 μm, round, distinctly warted, brown; solitary or in groups on sandy soil; July–October; rare; inedible—*Myriostoma coliforme* (Persoon) Corda.

2.   Spore case ⅜–1" (1–2.5 cm) wide, nearly round to somewhat flattened; whitish to grayish or grayish brown; roughened; with a single, irregular, pore-like mouth; 6–12 rays, hygroscopic; interior surface typically with numerous fine cracks; spores 7–11 μm, round, distinctly warted, brown; scattered or in groups on sandy soil; August–November; infrequent to locally common —*Astraeus hygrometricus* (Persoon) Morgan (see p. 451).

2.   Spore case ¾–1⅛" (2–3 cm) wide, nearly round to somewhat flattened, sitting in and surrounded by a bowl-like collar; tan to grayish brown or reddish brown, smooth; with a single, elevated, pore-like mouth encircled by a paler zone; 4–8 rays, not hygroscopic, thick, fleshy and brittle, often cracking or breaking; spores 3–4.5 μm, round, distinctly warted, brown; solitary, scattered, or in groups on the ground, usually in hardwoods; August–November; uncommon—*Geastrum triplex* Junghuhn (see p. 453).

2.   Fruiting body not with the above combinations of characters; rays not hygroscopic; spore case stalkless or supported by a distinct stalk → 3.

3.   Spore case stalkless or nearly so, ¼–¾" (6–20 mm) wide, nearly round, smooth, papery thin; whitish to grayish or grayish brown; with a single apical pore that is delimited by a distinct, circular, paler zone; 4–8 rays; spores 3.5–5 μm, round, distinctly warted, brown; solitary or in groups on the ground; July–October; fairly common; inedible—*Geastrum saccatum* Fries.

3.   Fruiting body nearly identical to the above choice but with smaller spores that measure 3–3.5 μm and an apical pore that is fimbriate and not delimited by a distinct, circular, paler area; inedible—*Geastrum sessile* (Sowerby) Pouzar = *Geastrum fimbriatum* Fries.

3.   Spore case supported by a distinct stalk → 4.

4.   Stalk supporting spore case prominent, long and slender, ¼–⅜" (5–10 mm) long; lower side of spore case usually furrowed; pore-like mouth prominent, elevated, distinctly conic, with fine, longitudinal furrows; 5–10 rays; spores 4–6 μm, round, distinctly warted, brown; solitary or scattered on the ground near cedars; July–October; rare; inedible—*Geastrum pectinatum* Persoon.

4.   Stalk supporting spore case prominent, long and slender, up to ¼" (6 mm) long; spore case dark brown, not furrowed on the lower side; pore-like mouth elevated, minutely fringed to distinctly torn, not furrowed; 6–10 rays; spores 3.5–5 μm, round, distinctly warted, brownish; solitary or scattered on sandy soil; July–October; uncommon; inedible—*Geastrum limbatum* Fries.

4.   Stalk supporting spore case not prominent, typically short and thick, usually ⅛" (1–3 mm) long → 5.

5.   Spore case ⅜–¾" (1–2 cm) wide, nearly round to acorn-shaped, grayish brown to brown, usually coated with minute glistening particles with an inflated collar at the base; supported by a short stalk; pore-like mouth elevated, conic, minutely fringed but not furrowed, delimited by a paler circular zone; 4–8 rays, bent downward at maturity;

spores 4–6 μm, round, warted, brown; solitary to scattered on the ground in conifer woods; July–October; uncommon—*Geastrum quadrifidum* Persoon : Persoon = *G. coronatum* (Schaeffer) Schroeter (see p. 453).

5.   Spore case ⅜–1" (1–2.5 cm) wide, nearly round to somewhat flattened, often with an inflated collar at the base; brown to dark brown; velvety; supported by a short stalk; pore-like mouth fairly large and irregularly torn, not delimited by a paler zone; 4–5 rays, arched and elevated, dark brown to blackish, surface often peeling off in strips and irregular patches; spores 3.5–5 μm, round, warted, brown; solitary to scattered or in groups in leaf litter and other organic debris; September–November; rare; inedible—*Geastrum fornicatum* (Hudson) Fries.

5.   Spore case ¾–1⅛" (2–3 cm) wide, nearly round, lacking an inflated collar at the base, pale pinkish brown to reddish brown, slightly velvety to finely granular, supported by a short, thick stalk; pore-like mouth elevated, conic, minutely fringed, not delimited by a paler zone and lacking furrows; 6–9 rays, thick and fleshy, bent downward at maturity, pinkish brown and frequently cracking in age; spores 4–5.5 μm, round, distinctly warted, brownish; solitary or scattered on soil in conifers or hardwoods; July–October; uncommon; inedible—*Geastrum vulgatum* Vittadini = *G. rufescens* Persoon.

   6.   Fruiting body 1½–4½" (4–12 cm) wide, round to oval or irregular when closed, expanding up to 6" (15.5 cm) and resembling a giant earthstar when open; spore case ⅛–⅜" (3–10 mm) thick, hard, rind-like; dingy white to straw-colored or pale yellow-brown; splitting open at maturity into 4–8 star-like rays and exposing the purplish black to dark brown spore mass; solitary or in groups on or partially buried in sandy soil, often along roads or on hillsides; August–November—*Scleroderma polyrhizon* Persoon (see p. 456).

   6.   Fruiting body 1–4" (2.5–10 cm) wide, ¾–2" (2–5 cm) high, nearly round to somewhat flattened; spore case wall 1/16–3/16" (1.5–4 mm) thick, areolate, pale wood-brown to yellow-brown; covered with coarse warts; spore mass marbled, purplish black to black—*Scleroderma citrinum* Persoon (see p. 456).

   6.   Fruiting body smaller than the above choices, spore case wall up to 1/16" (2 mm) thick, smooth or dotted with minute scales → 7.

7.   Fruiting body ¾–3⅛" (2–6 cm) wide, nearly round or irregular, tapered downward and forming a thick, irregular stalk-like base, up to 3½" (9 cm) long, with coarse, blunt projections and trapped sand; spore case conspicuously areolate and warted at maturity, ochraceous tan to bright ochraceous yellow; spore mass gray or brownish gray to blackish gray; spores 12–20 μm, globose, echinulate, reticulate; partially buried in sand; edibility unknown—*S. meridionale* Demoulin and Malençon = *S. macrorrhizon* Wallroth.

7.   Fruiting body ⅝–1¾" (1.5–4.5 cm) wide, nearly round to somewhat flattened, attached to the substrate by a thick, stalk-like mycelium with trapped sand; spore case wall up to 1/16" (2 mm) thick, smooth, becoming finely areolate in age; straw-yellow to pale orange-yellow when young, becoming orange-yellow to reddish brown with olive-gray tints at maturity; spores 9–16 μm, round, echinulate, with a well-developed reticulum; solitary, scattered, or in groups on soil in grassy areas, nutrient poor habitats, or woodlands, especially oak-pine; August–November; frequent; poisonous—*Scleroderma bovista* Fries = *S. lycoperdoides* var. *reticulatum* Coker and Couch.

7.   Fruiting body ¾–3⅛" (2–6 cm) wide, nearly round to distinctly flattened on the upper surface, attached by a thick, stalk-like mass of mycelium with trapped sand; spore case 1/16" (1.5 mm) thick, whitish to straw-colored when young, becoming orange-yellow to yellow-brown at maturity, smooth or coated with tiny scales; staining deep vinaceous when cut and rubbed; spores 7–13 μm, round, echinulate, lacking reticulation, hyaline

to pale yellowish brown; scattered or in groups on sandy soil; August–October; fairly common; poisonous—*Scleroderma cepa* Persoon = *S. flavidum* Ellis and Everhart.

7. Fruiting body ⅝–1¾" (1.5–4.5 cm) wide, nearly round to oval or top-shaped, attached by a thick, stalk-like mass of mycelium; spore case wall thin, up to ⅟₃₂" (0.5–0.8 mm) thick, pale brown with minute, delicate, darker brown scales; spores 11–17 μm, round, strongly echinulate, lacking reticulation, hyaline to pale brown; in groups or clusters on the ground; July–October; fairly common; poisonous—*Scleroderma areolatum* Ehrenberg = *S. lycoperdoides* Schweinitz.

   8. Spore case whitish to grayish or pale reddish brown, ⅜–¾" (1–2 cm) wide and high, nearly round, supported by a thin stalk; with a circular, pore-like mouth elevated by a short, cylindric, tube-like collar—*Tulostoma brumale* Persoon (see p. 457).

   8. Spore case reddish brown and often coated with sand, ⅜–⅝" (1–1.7 cm) wide, nearly round to acorn-shaped, supported by a ⅝–1⅛" (1.6–3 cm) high, rusty brown to gray, scaly stalk with a bulbous base; with a circular, pore-like mouth elevated by a short, cylindric, tube-like collar; spores 4–7 μm, globose to subglobose, with thick coarse warts; edibility unknown—*Tulostoma simulans* Lloyd.

   8. Spore case pale grayish to pale brownish and often coated with sand, ⅜–⅝" (1–1.8 cm) wide, nearly round or somewhat flattened, supported by a ⅝–2¼" (1.7–5.5 cm) high, scaly stalk; pore mouth only slightly elevated; spores 4–7 μm, globose to subglobose, minutely warted; edibility unknown—*Tulostoma campestre* Morgan.

   8. Spore case white, ¾–2¾" (2–7 cm) wide and up to 1½" (4 cm) high, nearly round at first, often flattened and collapsed in age, lacking a pore-like mouth, supported by a thick stalk—*Calvatia elata* (Massee) Morgan (see p. 452).

   8. Spore case dingy yellow to yellow-brown; 1⅜–4" (3.5–10 cm) wide, oval to pear-shaped, tapering downward to form a thick, stalk-like rooting base; spore case splitting at maturity to expose hundreds of tiny, yellowish to brownish peridioles that disintegrate to form a reddish brown to dark brown spore mass; solitary, scattered, or in groups in sandy soil, typically under oak or pine, commonly growing with prickly pear cactus and often partially buried—*Pisolithus tinctorius* (Persoon) Coker and Couch (see p. 455).

   8. Spore case pale reddish brown to yellow-brown, becoming brownish orange as the ground color is exposed in age; entire fruiting body ⅝–1¾" (1.5–4.5 cm) wide, 1⅛–3¾" (3–9.5 cm) high, narrowly to broadly club-shaped; outer surface granular-roughened and coated with matted fibers when young and fresh, becoming nearly smooth at maturity, stalk flesh watery-gelatinous (meat-like and conspicuously marbled or veined when fresh, young specimens are sectioned longitudinally), becoming dry, fibrous, and forming a whitish columella that extends upward and branches through the spore mass; spore mass distinctly chambered with elongated peridioles, gelatinous, pale to dark reddish brown, eventually powdery in age; on decaying organic matter; rare—*Rhopalogaster transversarium* (Bosc) Johnston (see p. 456).

   8. Spore case bright reddish orange with an irregular, slit-like mouth surrounded by bright red, elevated ridges; with a coarsely reticulated and pitted, reddish orange stalk; both the spore case and stalk sometimes covered by a thick, gelatinous layer, especially on young specimens—*Calostoma cinnabarina* Desvaux (see p. 452).

   8. Fruiting body similar to the previous choice except for the following characters: spore case yellow with an irregular, slit-like mouth surrounded by bright red, elevated ridges; stalk elongated, up to 2⅜" (6 cm) long; spores 6–8 μm, round and

pitted, resembling golf balls; August–October; inedible—*Calostoma lutescens* (Schweinitz) Burnap.

9. Fruiting body ¾–1¾" (2–4.5 cm) wide, nearly round to oval, dark yellow-brown to reddish brown or sometimes paler; covered with small, hard warts; spore case wall firm, rind-like, ⅛–³⁄₁₆" (3–5 mm) thick, whitish to grayish brown; interior grayish to purplish and firm when young, becoming dark brown and powdery at maturity; spores 24–48 μm, round or nearly so, with short spines or warts, brown; solitary, scattered, or in groups buried in soil or humus, usually 2–4" (5–10 cm) below the surface, typically under conifers; often parasitized by species of *Cordyceps* (see color illustration on p. 22)—*Elaphomyces granulatus* Fries.

9. Fruiting body not with the above combination of characters; buried or partially buried, round to oval or irregular in outline; stalk typically absent; spore mass usually located in several small chambers—*Rhizopogon* species and many additional allies. Most of these fungi require microscopic examination for accurate identification and are beyond the scope of this work.

   10. Fruiting body very large, usually 8–15" (20–38 cm) or more wide, round to flattened and sometimes indented; white to creamy white—*Langermannia gigantea* (Batsch : Persoon) Rostkovius (see p. 453).

   10. Fruiting body large, typically 3½–7" (9–17.5 cm) wide and 3½–8" (9–20 cm) high, round to oval when young, becoming pear-shaped to top-shaped or irregular in age; spore mass dull purple to purple-brown and powdery at maturity—*Calvatia cyathiformis* (Bosc) Morgan (see p. 452).

   10. Fruiting body large, 3½–8" (9–20 cm) wide, 2⅜–8" (6–20 cm) high, top-shaped to elongated pear-shaped; often deeply wrinkled to grooved on the upper portion; spore mass white when young, becoming yellow-brown and powdery at maturity; spores 2.5–4 μm, round, nearly smooth; growing on the ground in woods, especially under oak; edible when the spore mass is white—*Calvatia craniformis* (Schweinitz) Fries.

   10. Fruiting body small to medium, ⅜–4" (1–10 cm) wide, growing on the ground or on decaying wood, sawdust, mulch, or organic debris → 11.

11. Growing on the ground in grassy areas, in gardens, and in nutrient poor habitats; spore case white with pink tinges, quickly staining bright yellow or orange to reddish orange when cut or bruised; ¾–4" (2–10 cm) wide, ⅝–2" (1.5–5 cm) high, nearly round to somewhat flattened; distinctly furrowed on the lower portion; spore mass olive-yellow to greenish orange and powdery at maturity, with a disagreeable odor; spores 3–4.5 μm, globose with a short pedicel; edible when the spore mass is white—*Calvatia rubroflava* (Cragin) Lloyd.

11. Growing on decaying mossy hardwood logs and stumps; spore case coated with short, purplish brown to reddish brown spines over a pale pinkish brown ground color; ⅜–1⅛" (1–3 cm) wide, nearly round to pear-shaped—*Morganella subincarnata* (Peck) Kreisel and Dring (see p. 455).

11. Growing on decaying wood, sawdust, and organic debris, scattered or in dense clusters; spore case whitish when young, becoming yellow-brown to reddish brown at maturity; ⅝–1¾" (1.5–4.5 cm) wide, ¾–2" (2–5 cm) high, pear-shaped to nearly round—*Lycoperdon pyriforme* Schaeffer : Persoon (see p. 455).

11. Not as above; fruiting body not staining yellow when cut or bruised and usually not growing on wood → 12.

   12. Spore case bright yellow to bright golden orange-yellow when young, becoming dull brown at maturity; covered with minute warts that fall off, becoming nearly

smooth as it ages; ⅝–1½" (1.5–4 cm) wide, round or nearly round, with a small, pinched-off base; spore mass white and firm when young, becoming olive-brown to yellow-brown and powdery at maturity; spores 3–4 μm, globose; solitary or scattered on humus under conifers, especially spruce, or sometimes on decaying wood; August–October; rare—*Lycoperdon coloratum* Peck (see p. 454).

12. Spore case when immature covered by a thin, white outer peridium that tears apart, exposing a papery thin, smooth, purplish brown to grayish brown or bluish gray spore case with a metallic luster; ¾–3⅛" (2–8 cm) wide, nearly round; spore mass white at first, becoming yellowish olive and finally dark brown and powdery at maturity; spores 5–7 × 4–5 μm, oval to nearly round, with a prominent pedicel that measures 8–14 μm long; solitary, scattered, or in groups on the ground in pastures, parks, cemeteries, in old apple orchards, under hawthorns, and in mixed hardwoods; edible when the spore mass is white—*Bovista plumbea* Persoon.

12. Fruiting body nearly identical to the above choice but with a dark brown to bronze spore case at maturity and smaller spores that measure 3.5–4.5 μm with short, broken pedicels; edible when the spore mass is white—*Bovista pila* Berkeley and Curtis (see p. 451).

12. Fruiting body not with the above combinations of characters → 13.

13. Fruiting body ⅜–2" (1–5 cm) wide, nearly round when young, becoming somewhat flattened to pear-shaped, usually broader than tall; spore case white at first, covered with short spines or warts, less than ⅛" (3 mm) long, which break off in irregular sheets, exposing the nearly smooth, pale to dark olive-brown or reddish brown inner surface—*Lycoperdon marginatum* Vittadini (see p. 454).

13. Not as above; fruiting body covered with spines that measure ⅛–¼" (3–6 mm) long → 14.

13. Not as above; fruiting body covered with spines that measure less than ⅛" (3 mm) long → 15.

14. Spines white when young, becoming dark brown and falling off in age, leaving the surface reticulated by scurfy particles; fruiting body ¾–1¾" (2–4.5 cm) wide, nearly round to pear-shaped when young, becoming somewhat flattened in age, pinched off at the base; outer peridium eventually tearing away to expose the smooth, pale brown to dark purple-brown spore case; spore mass white and firm at first, becoming purplish and powdery at maturity; spores 5–6 μm, globose, often with long pedicels; solitary to scattered on leaf litter and soil under hardwoods, especially beech; August–October; fairly common; edible when the spore mass is white—*Lycoperdon echinatum* Persoon.

14. Spines remaining white until they fall off, leaving a smooth, not reticulated, purple-brown spore case; fruiting body ¾–2" (2–5 cm) wide, pear-shaped to nearly round, pinched off at the base; spore mass white and firm at first, becoming dark purple-brown and powdery at maturity; spores 4–5 μm, globose; solitary to scattered on soil and humus in conifer or hardwoods; August–October; uncommon; edible when the spore mass is white—*Lycoperdon pulcherrimum* Berkeley and Curtis.

15. Spines white, short, conic, with smaller spines and granules distributed between them, forming tiny marks where the conic spines break off; fruiting body 1–2½" (2.5–6.5 cm) wide, 1–3⅛" (2.5–8 cm) high, pear-shaped to turban-shaped with a large, stalk-like sterile base; spore case and base white, aging to yellow-brown—*Lycoperdon perlatum* Persoon (see p. 454).

15. Spines purplish lavender at first, becoming pale brown and falling off at maturity, leav-

ing circular spots on the surface; fruiting body ⅝–1½" (1.5–4 cm) wide, 1⅜–2⅜" (2.5–6 cm) high, pear-shaped to turban-shaped with a large, stalk-like sterile base; spore case whitish and nearly smooth on the lower portion, covered on the upper portion with tapering spines that split at their bases into two to four parts and remain united at their tips; spore mass white and firm when young, becoming olive-brown and powdery at maturity; spores 3.5–4.5 μm, globose with a short pedicel; solitary, scattered or in groups on the ground in grassy areas or in woods; August–October; infrequent; edible when the spore mass is white—*Lycoperdon peckii* Morgan.

15. Spines dark brown with smaller warts distributed between them, eventually wearing away and exposing a yellowish or pale yellow-brown, smooth inner surface; fruiting body 1–3" (2.5–7.5 cm) wide and high, nearly round with a flattened upper surface, and a short, thick, stalk-like base; spore mass white and firm at first, becoming yellow-brown to grayish brown and powdery at maturity; spores 3.5–4.5 μm, globose with a short pedicel; solitary, scattered, or in groups on soil or humus under conifers; August–October; infrequent; edible when the spore mass is white—*Lycoperdon umbrinum* Persoon : Persoon.

Astraeus hygrometricus (Persoon) Morgan
COMMON NAME: Barometer Earthstar.
FRUITING BODY: 1½–3½" (4–9 cm) wide, ⅜–1" (1–2.5 cm) high when fully expanded, consisting of a rounded spore case and star-like rays.
SPORE CASE: ⅜–1" (1–2.5 cm) wide, nearly round to somewhat flattened; whitish to grayish or grayish brown, finely roughened; with a single, irregular, pore-like mouth.
SPORE MASS: white when young, becoming brown and powdery at maturity.
RAYS: 6–12, hygroscopic, yellow-brown to reddish brown or grayish to nearly black; interior surface often finely cracked.
MICROSCOPIC FEATURES: spores 7–11 μm, round, distinctly warted, brown.
FRUITING: solitary, scattered, or in groups on sandy soil; year-round; infrequent to locally common.
EDIBILITY: inedible.

Bovista pila Berkeley and Curtis
COMMON NAME: Tumbling Puffball.
FRUITING BODY: 1¼–3½" (3.2–9 cm) wide, nearly round to somewhat flattened, attached to the soil by a single tiny cord.
SPORE CASE: papery thin, smooth, dark brown to bronze, sometimes with a metallic luster; covered when young by a thin, white, nearly smooth outer peridium that becomes pinkish and tears apart in age; at maturity forming one or more pore-like mouths that often tear.
SPORE MASS: white at first, becoming dark brown and powdery at maturity.
MICROSCOPIC FEATURES: spores 3.5–4.5 μm, round, smooth, pale brown, with short, broken pedicels.
FRUITING: solitary, scattered, or in groups on the ground in woods and pastures and near stables.
EDIBILITY: edible when the spore mass is white.
COMMENTS: compare with *B. plumbea* (p. 450), which has larger spores with prominent pedicels.

Calostoma cinnabarina Desvaux

COMMON NAME: Red Slimy-stalked Puffball.

FRUITING BODY: consisting of a spore case supported by a thick, often buried stalk.

SPORE CASE: ⅜–¾" (1–2 cm) wide and high, oval to nearly round, sometimes collapsed; covered by a thick, gelatinous outer peridium with a thin, bright red inner layer when young; splitting into small seed-like pieces and exposing the smooth, thin-walled, bright reddish orange inner peridium, which fades to orange or orange-yellow in age; with an irregular slit-like mouth surrounded by bright red, elevated ridges.

SPORE MASS: white when young, buff and powdery at maturity.

STALK: ⅝–1½" (1.5–4 cm) long, ⅜–¾" (1–2 cm) thick, nearly equal overall, spongy, coarsely reticulate and pitted, reddish orange to pale reddish brown, covered with a thick, gelatinous layer coated with debris.

MICROSCOPIC FEATURES: spores 14–22 × 6–9 μm, oblong-elliptic, pitted, hyaline.

FRUITING: solitary or scattered on the ground in woods, usually in sandy soil, often buried up to the spore case; August–October; occasional and most often collected in the southern portion of the region.

EDIBILITY: inedible.

COMMENTS: *Calostoma ravenelii* (inedible) has a tan to grayish spore case, an elongated stalk, up to 2" (5 cm), and elliptic spores and lacks a gelatinous outer peridium. Also see *C. lutescens* (p. 449).

Calvatia cyathiformis (Bosc) Morgan

COMMON NAME: Purple-spored Puffball.

FRUITING BODY: 2¾–7" (7–18 cm) wide, 3½–8" (9–20 cm) high, round to oval when young, becoming pear-shaped to top-shaped or irregular in age.

SPORE CASE: whitish to pale brown, smooth at first, then becoming areolate on the upper portion, breaking into thin, irregular plates that flake off with age.

SPORE MASS: white and solid when immature, aging to yellowish, and finally dull purple to purple-brown and powdery at maturity.

MICROSCOPIC FEATURES: spores 4–7 μm, round, weakly echinulate, pale lilac.

FRUITING: solitary, scattered, or in fairy rings on the ground in grassy areas and woodland edges; July–November; fairly common.

EDIBILITY: edible when the spore mass is white.

Calvatia elata (Massee) Morgan

FRUITING BODY: consisting of a rounded to somewhat flattened spore case supported by a long stalk.

SPORE CASE: ¾–2¾" (2–7 cm) wide and up to 1½" (4 cm) high, nearly round at first, often flattened and collapsed in age; outer peridium thin, granular to powdery, white, wearing away to expose a thin, nearly smooth, white inner peridium that becomes dingy tan, then brown, and breaks apart at maturity; lower portion of the spore case and upper stalk often pleated in age.

SPORE MASS: white when young, becoming brown at maturity.

STALK: 1½–3½" (4–9 cm) long, up to 1⅜" (3.5 cm) thick, tapered downward to nearly equal or sometimes enlarged downward to an abruptly tapered base, granular to scurfy, white when young, becoming whitish to tan, then brown in age, often shallowly pitted.

MICROSCOPIC FEATURES: spores 3.5–4.5 × 3.5–4.5 μm, globose, minutely warted, with a short pedicel, pale brown.

FRUITING: scattered or in groups on the ground or well-decayed wood remains in nutrient poor habitats, or woods; July–November; occasional.

EDIBLILITY: edible when the spore mass is white.

Geastrum quadrifidum Persoon : Persoon

FRUITING BODY: ⅝–1⅜" (1.5–3.5 cm) wide, 1–1¾" (2.5–4.5 cm) high when fully expanded, consisting of a rounded spore case and star-like rays.

SPORE CASE: ⅜-¾" (1–2 cm) wide, nearly round to acorn-shaped, grayish brown to brown, usually coated with minute glistening particles with an inflated collar at the base, supported by a short stalk; pore-like mouth elevated, conic, minutely fringed but not furrowed, delimited by a paler circular zone.

SPORE MASS: dark brown to purplish brown and powdery when mature.

RAYS: 4–8, bent downward at maturity and often attached to an entangled mass of substrate, fleshy and brittle or fibrous-tough, pointed, scurfy.

MICROSCOPIC FEATURES: spores 4–6 μm, round, warted, brown.

FRUITING: solitary or scattered on the ground in conifer woods; July–October; uncommon.

EDIBILITY: inedible.

COMMENTS: previously known as *G. coronatum*.

Geastrum triplex Junghuhn (see photo, p. 16)

COMMON NAME: Collared Earthstar.

FRUITING BODY: 2¾–4" (7–10 cm) wide, ¾–1¾" (1–4.5 cm) high when fully expanded, consisting of a rounded spore case and star-like rays.

SPORE CASE: ¾–1⅛" (2–3 cm) wide, nearly round to somewhat flattened, sitting in and surrounded by a bowl-like to saucer-shaped collar, not supported by a stalk; tan to grayish brown or reddish brown, smooth; with a single, elevated, pore-like mouth encircled by a paler zone.

SPORE MASS: dark brown and powdery when mature.

RAYS: 4–8, not hygroscopic, thick, fleshy, and brittle, pointed and decurved at their tips; bending downward, cracking into patches and forming a bowl-like collar that surrounds the spore case; pale to dark tan at first, becoming yellowish brown to pinkish brown.

MICROSCOPIC FEATURES: spores 3–4.5 μm, round, distinctly warted, brown.

FRUITING: solitary, scattered, or in groups on the ground, usually in hardwoods; August–November; uncommon.

EDIBILITY: inedible.

Langermannia gigantea (Batsch : Persoon) Rostkovius

COMMON NAME: Giant Puffball.

FRUITING BODY: very large, usually 8–15" (20–38 cm) wide, but sometimes attaining a diameter of 20" (50 cm) or more; nearly round when young, often somewhat flattened and indented at maturity; attached to the ground by a thick, cord-like, basal rhizomorph.

SPORE CASE: white to creamy white, smooth, soft, resembling deerskin; cracking irregularly in age.

SPORE MASS: white, soft, becoming yellowish to yellow-green, and finally greenish brown as the spores mature.

MICROSCOPIC FEATURES: spores 3.5–5 μm, round, weakly echinulate to smooth, pale brown.

FRUITING: solitary, scattered, in groups, or sometimes in fairy rings on the ground in woods, pastures, parks, golf courses, and brushy areas; July–October; fairly common.

EDIBILITY: edible when the spore mass is white.

COMMENTS: also known as *Calvatia gigantea*. At maturity, this puffball is commonly attacked by *Syzygites megalocarpus,* a gray bread mold.

Lycoperdon coloratum Peck

FRUITING BODY: 5/8–1½" (1.5–4 cm) wide, round or nearly round, with a small, pinched-off base, sometimes pear-shaped.

SPORE CASE: bright yellow to bright golden orange-yellow when young, becoming dull brown to dull olive-brown at maturity, covered with minute warts which fall off, becoming nearly smooth and shiny as it ages.

SPORE MASS: firm and white at first, becoming olive-brown to yellow-brown and powdery at maturity.

MICROSCOPIC FEATURES: spores 3.5–4 μm wide, round or nearly so, smooth to minutely roughened, dull vinaceous brown.

FRUITING: solitary, scattered or in groups on humus under conifers, especially spruce, or sometimes on decaying wood or humus along roots; August–October; rare.

EDIBILITY: edible when the spore mass is white.

Lycoperdon marginatum Vittadini

COMMON NAME: Peeling Puffball.

FRUITING BODY: 3/8–2" (1–5 cm) wide, nearly round when young, becoming somewhat flattened to pear-shaped at maturity and usually broader than tall.

SPORE CASE: white at first, covered with short spines or warts that break off in irregular sheets, exposing the nearly smooth, pale to dark olive-brown or reddish brown inner surface; forming a pore-like mouth on the apex at maturity; usually with a tapering, sterile, stalk-like base.

SPORE MASS: firm and white at first, becoming olive-brown to grayish brown and powdery at maturity.

MICROSCOPIC FEATURES: spores 3.5–4.5 μm, round, minutely punctate to smooth, sometimes with a broken pedicel, pale brown.

FRUITING: solitary, scattered, or in groups on the ground, usually on sandy soil, in oak-pine woods, or in nutrient poor habitats; June–October; fairly common.

EDIBILITY: edible when the spore mass is white.

Lycoperdon perlatum Persoon (see photo, p. 16)

COMMON NAME: Gem-studded Puffball, Devil's Snuffbox.

FRUITING BODY: 1–2½" (2.5–6.5 cm) wide, 1–3⅛" (2.5–8 cm) high, pear-shaped to turban-shaped.

SPORE CASE: white, aging to yellow-brown, covered with short, conic, white spines with smaller spines and granules distributed between them; forming tiny marks where the conic spines break off; forming a rounded pore-like mouth on the apex at maturity; with a large, stalk-like sterile base.

SPORE MASS: white and firm at first, becoming yellow to olive, and finally olive-brown and powdery at maturity.

MICROSCOPIC FEATURES: spores 3.5–4.5 μm, round, weakly echinulate, pale brown.

FRUITING: solitary, scattered, or clustered on the ground in conifer or hardwood forests or on mulch or compost piles; July–November; common.

EDIBILITY: edible when the spore mass is white.

COMMENTS: also known as *L. gemmatum.*

Lycoperdon pyriforme Schaeffer : Persoon

COMMON NAME: Pear-shaped Puffball.

FRUITING BODY: ⅝–1¾" (1.5–4.5 cm) wide, ¾–2" (2–5 cm) high, pear-shaped to nearly round.

SPORE CASE: whitish when young, becoming yellow-brown to reddish brown with tiny warts and granules or with spines, eventually forming a pore-like mouth on the apex at maturity; lower portion tapering downward, sometimes forming a sterile, stalk-like base, which is often compressed.

SPORE MASS: white and firm at first, becoming greenish yellow, and finally dark olive-brown and powdery at maturity.

MICROSCOPIC FEATURES: spores 3–4.5 μm, round, smooth, pale brown.

FRUITING: scattered or in dense clusters on decaying wood, sawdust, and organic debris; July–November; very common.

EDIBILITY: edible when the spore mass is white.

Morganella subincarnata (Peck) Kreisel and Dring

FRUITING BODY: ⅜–1⅛" (1–3 cm) wide, nearly round to pear-shaped, often laterally flattened when growing in clusters.

SPORE CASE: coated with short, purplish brown to reddish brown spines over a pale pinkish brown ground color; pitted after the spines have fallen off at maturity; with or without a sterile base; splitting irregularly to release spores.

SPORE MASS: purplish brown and powdery at maturity.

MICROSCOPIC FEATURES: spores 3.5–4 μm, round, weakly echinulate to smooth, brownish.

FRUITING: scattered, in groups, or in clusters on decaying, mossy hardwood logs and stumps; August–October; fairly common.

EDIBILITY: unknown.

Pisolithus tinctorius (Persoon) Coker and Couch

COMMON NAME: Dye-Maker's False Puffball.

FRUITING BODY: 1⅜–4" (3.5–10 cm) wide, oval to pear-shaped or sometimes club-shaped in age, tapering downward to form a thick, stalk-like rooting base.

SPORE CASE: a thin, smooth, shiny peridium, dingy yellow to yellow-brown, splitting irregularly at maturity to expose hundreds of tiny yellowish to brownish peridioles embedded in a black gelatinous matrix.

SPORE MASS: reddish brown to dark brown and powdery at maturity, produced by the disintegrating peridioles.

MICROSCOPIC FEATURES: spores 7–12 μm, round, echinulate, brownish.

FRUITING: solitary, scattered, or in groups in sandy soil, typically under oak and pine, commonly with Prickly Pear cactus, often partially buried; infrequent but most commonly collected along coastal habitats; July–November; occasional to locally common.

EDIBILITY: inedible.

COMMENTS: the common name of this mushroom is a reference to the fact that it can be used to dye wool various shades of brown or black.

Rhopalogaster transversarium (Bosc) Johnston

FRUITING BODY: ⅝–1¾" (1.5–4.5 cm) wide, 1⅛–3¾" (3–9.5 cm) high, narrowly to broadly club-shaped.

SPORE CASE: the enlarged upper portion, granular-roughened to scurfy, with scattered, matted, brownish fibers when young and fresh, becoming nearly smooth in age, pale reddish brown to yellow-brown or olive-brown, becoming pale brownish orange as the ground color is exposed; rupturing irregularly to expose the spore mass.

SPORE MASS: distinctly chambered like a honeycomb, consisting of elongated peridioles formed by the branching columella, gelatinous, pale to dark reddish brown, slowly staining blackish when exposed, becoming olive-brown to dark brown and eventually powdery in age.

STALK: the narrowed lower portion, covered with a dense coat of matted fibers and colored like the spore case when young and fresh, becoming nearly smooth and pale brownish orange, then whitish as the fibers wear away in age; flesh watery-gelatinous, meat-like and conspicuously marbled or veined when fresh, young specimens are sectioned longitudinally, becoming dry, fibrous, and forming a whitish columella that extends upward and branches through the spore mass to the top of the spore case.

MICROSCOPIC FEATURES: spores 5.5–7.5 × 3–4.5 μm, elliptic, smooth, pale brown.

FRUITING: solitary, scattered, or in groups on decaying organic matter, such as wood chips, logs and limbs, leaf litter, and mulch in oak-pine woods; July–September; rare.

EDIBILITY: unknown.

COMMENTS: the specimens shown in the photograph are young and fresh stages. This mushroom is normally encountered from North Carolina to Florida but has been collected in New Jersey. Its exact distribution is unknown. Orson K. Miller, Jr., has recently noted that the addition of a drop of $FeSO_4$ to the surface of the fruiting body produces a pale green staining reaction.

Scleroderma citrinum Persoon (see photo, p. 16)

COMMON NAME: Pigskin Poison Puffball, Common Earthball, Golden Scleroderma.

FRUITING BODY: 1–4" (2.5–10 cm) wide, ¾–2" (2–5 cm) high, nearly round to somewhat flattened.

SPORE CASE: wall 1/16–3/16" (1.5–4 mm) thick, rind-like, areolate, covered with coarse warts, pale wood-brown to yellow-brown or golden brown; opening by an irregular pore on the upper surface; white when sectioned, slowly staining pinkish when rubbed; attached to the substrate by a thick, stalk-like mycelial base.

SPORE MASS: white when very young, soon marbled and dark gray to purplish black to grayish black, solid, firm, becoming powdery and blackish brown in age.

MICROSCOPIC FEATURES: spores 8–12 μm, round, strongly reticulate; dark brown.

FRUITING: solitary, scattered, or in groups on the ground or on decaying wood; July–November; common.

EDIBILITY: poisonous.

COMMENTS: it is the most common species of *Scleroderma* in northeastern North America and is frequently parasitized by *Boletus parasiticus*. It is also known as *S. aurantium*.

Scleroderma polyrhizon (Gmelin) Lévielle

COMMON NAME: Earthstar Scleroderma.

FRUITING BODY: 1½–4½" (4–12 cm) wide, round to oval or irregular when closed, expanding up to 6" (15.5 cm) and resembling a giant earthstar when open.

SPORE CASE: wall ⅛–⅜" (3–10 mm) thick, hard, rind-like, rough, areolate to somewhat scaly, dingy white to straw-colored or pale yellow-brown, splitting open at maturity into 4–8 star-like rays and exposing the spore mass.

SPORE MASS: firm when young, becoming powdery, brown to purplish brown, becoming blackish brown at maturity.

MICROSCOPIC FEATURES: spores 5–10 μm, globose, coated with short spines, sometimes forming a partial reticulum, purple-brown.

FRUITING: solitary or in groups on or partially buried in sandy soil, often along roads or on hillsides; August–November; uncommon.

EDIBILITY: poisonous.

COMMENTS: also known as *S. geaster.*

Sphaerobolus stellatus (Tode) Persoon

COMMON NAME: Cannon Fungus.

FRUITING BODY: very small, $\frac{1}{16}$–$\frac{1}{8}$" (1.5–3 mm) wide, nearly round.

SPORE CASE: white to yellow-orange, splitting at maturity to form 4–9 star-shaped to tooth-like rays and exposing a single, dull yellow or reddish brown to dark brown peridiole, which is forcibly ejected in response to light and moisture.

MICROSCOPIC FEATURES: spores 7–10 × 3.5–5 μm, oblong, smooth, hyaline.

FRUITING: solitary or in groups on decaying wood, sawdust, compost, and dung; May–October; fairly common.

EDIBILITY: unknown, but as David Arora puts it, "several hundred would be needed for a mouthful."

Tulostoma brumale Persoon (see photo, p. 16)

COMMON NAME: Common Stalked Puffball.

FRUITING BODY: consisting of a small spore case supported by a slender stalk.

SPORE CASE: $\frac{3}{8}$–$\frac{3}{4}$" (1–2 cm) wide and high, nearly round to somewhat flattened, sometimes collapsed; outer peridium yellow-brown, often coated with sand, peeling away to expose a brown to whitish inner peridium; with a circular pore-like mouth elevated by a short, whitish, cylindric, tube-like collar.

SPORE MASS: rusty salmon and powdery at maturity.

STALK: $\frac{3}{4}$–2" (2–5 cm) long, $\frac{1}{8}$–$\frac{3}{16}$" (3–5 mm) thick, nearly equal down to a slightly bulbous base; rusty brown to yellow-brown, paler toward the apex, often coated with sand.

MICROSCOPIC FEATURES: spores 3–5 μm, nearly round, minutely warted, some with short pedicels, hyaline to pale yellow.

FRUITING: scattered or in groups on sandy soil, often in nutrient poor habitats; September–December; fairly common.

EDIBILITY: inedible.

Astraeus hygrometricus

Bovista pila

Calostoma cinnabarina

Calvatia cyathiformis

Calvatia elata

Geastrum quadrifidum

Langermannia gigantea

Lycoperdon coloratum

Lycoperdon marginatum

Lycoperdon pyriforme

Morganella subincarnata

Pisolithus tinctorius

Rhopalogaster transversarium

Scleroderma polyrhizon

Sphaerobolus stellatus

Stinkhorns

Stinkhorns are members of the class Gasteromycetes, commonly known as the Stomach Fungi. These mushrooms are unable to forcibly discharge their spores and have developed alternate methods for spore dispersal, including wind, rain, insects, and mammals. The immature stage of stinkhorns is an egg-like structure that resembles a small puffball. As the stinkhorn matures, a hollow stalk with a head or arms arises from the egg. A slimy, foul-smelling spore mass is formed within the greenish, gelatinous slime inside the egg. At maturity, it is released through the stalk and covers the head or inner portion of the arms of the stinkhorn. The foul odor, which attracts many arthropods, is often detected by mushroom hunters before the fungus is seen.

One member of this group, *Phallogaster saccatus* Morgan, does not form a stalk. It retains its pear-shaped to nearly round appearance until it becomes perforated and releases its spores.

Key to Species of Stinkhorns

1. Fruiting body lacking a distinct stalk and head; pear-shaped and narrowed toward the base or nearly round; white to pink or pinkish lilac; forming irregular depressions that eventually perforate to release the spore mass—*Phallogaster saccatus* Morgan (see p. 463).
1. Fruiting body forming a stalk with 3–5 arched, yellow to orange, tapering arms—*Pseudocolus fusiformis* (Fischer) Lloyd (see p. 463).
1. Fruiting body arising from a whitish egg which splits open forming a whitish stalk with a whitish sac-like cup at the base, and a fertile head which splits vertically to form 5–6 pinkish, incurved or slightly spreading arms, exposing an olive-brown spore mass; spores 3–4 × 1–2 μm; edibility unknown—*Lysurus gardneri* Berkeley (sometimes called the Lizard's Claw or *L. borealis*).
1. Fruiting body at first egg-like, giving rise to a distinct stalk, with or without a head → 2.
    2. Fruiting body with a distinct stalk and head; head granular to wrinkled and dishrag-like, not deeply pitted—*Phallus ravenelii* Berkeley and Curtis (see p. 463).
    2. Fruiting body with a distinct stalk and head; head round to somewhat flattened, with an orange to reddish orange, thick, raised net or lattice surrounding an olive-brown spore mass; egg whitish; stalk orange to reddish orange with a whitish sac-like cup at the base; edibility unknown—*Simblum sphaerocephalum* Schlect.
    2. Fruiting body with a distinct stalk and head; head deeply pitted → 3.
    2. Fruiting body with a distinct stalk, lacking a head → 4.
3. Stalk surrounded at the apex by a white, net-like, flaring veil that emerges from beneath the head; inedible—*Dictyophora duplicata* (Bosc) Fischer (see p. 462).
3. Stalk lacking a net-like, flaring veil; egg and volva white; spores 3–4 × 1–2 μm; inedible—*Phallus impudicus* Persoon.

3. Nearly identical to previous choice except egg and volva pink to pinkish purple; inedible—*Phallus hadriani* Venturi : Persoon.

    4. Fruiting body stalk white to pink or pinkish red, tapered at the apex; spore mass covering the upper quarter or less of the stalk—*Mutinus caninus* (Persoon) Fries (see p. 462).

    4. Fruiting body stalk orange to pinkish orange or pinkish red, tapered from the middle in both directions; spore mass covering the upper third or half of the stalk; spores 4–7 × 2–3 μm; inedible—*Mutinus elegans* (Montagne) Fischer.

Dictyophora duplicata (Bosc) Fischer (see photo, p. 17)

COMMON NAME: Netted Stinkhorn.

FRUITING BODY: at first egg-like, resembling a small puffball, nearly round to somewhat flattened, 1¾–2¾" (4.5–7 cm) high and wide, whitish to pale flesh-color or pinkish brown, often grooved on the lower portion; with one or more thick, whitish to pinkish, often branched rhizomorphs; internally gelatinous; giving rise to a distinct head and stalk.

HEAD: 1⅜–2" (3.5–5 cm) wide, 2–2¾" (5–7 cm) high, oval to conic or bell-shaped, pendant, deeply pitted, with a white-rimmed opening at the apex; white, covered with a greenish brown to brownish olive slimy spore mass.

STALK: 3½–7¼" (9–18.5 cm) long, 1⅜–2⅜" (3.5–6 cm) wide, nearly equal, roughened, spongy, hollow, white; surrounded at the apex by a white, net-like, flaring veil that emerges from beneath the head.

VOLVA: whitish to pale flesh-color or pinkish brown, wrinkled to folded, tough.

SPORE MASS: greenish brown to brownish olive, slimy, foul-smelling.

MICROSCOPIC FEATURES: spores 3.5–4.5 × 1–2 μm, elliptic, smooth, hyaline.

FRUITING: solitary or in groups on the ground in hardwoods or mixed woods; June–October; infrequent.

EDIBILITY: inedible.

COMMENTS: this is the largest stinkhorn in northeastern North America.

Mutinus caninus (Persoon) Fries

COMMON NAME: Dog Stinkhorn.

FRUITING BODY: at first egg-like, resembling a small puffball, oval to nearly round or pear-shaped, ⅜–¾" (1–2 cm) high, ⅜–⅝" (1–1.5 cm) wide; white to pink or pinkish red, with white rhizomorphs; internally gelatinous; giving rise to a stalk; head absent.

STALK: 2⅜–4" (6–10 cm) high, ⅜–¾" (1–2 cm) wide, roughened, nearly equal up to the apex, which tapers to a point, hollow, spongy, white with pinkish tints to distinctly pinkish red.

VOLVA: whitish, sometimes with pinkish tints, tough, wrinkled.

SPORE MASS: olive-brown, slimy, foul-smelling, covering the upper quarter or less of the stalk.

MICROSCOPIC FEATURES: spores 3.5–5 × 1.5–2 μm, cylindric, smooth, hyaline.

FRUITING: solitary or in groups on soil, humus, mulch, wood chips, and decaying wood; August–October; fairly common.

EDIBILITY: inedible.

COMMENTS: the amount of pinkish red coloration on the stalk is a highly variable feature. *Mutinus caninus* var. *albus* has an entirely white stalk.

Phallogaster saccatus Morgan (see photo, p. 17)

COMMON NAME: Club-shaped Stinkhorn.

FRUITING BODY: at first egg-like, resembling a small puffball, 1–2" (2.5–5 cm) high, 3/8–1 3/8" (1–3.5 cm) wide, typically pear-shaped and narrowed toward the base or nearly round; white to pink or pinkish lilac on the upper portion, white toward the base; surface smooth, then forming irregular depressions that eventually perforate to release the spore mass; base attached by numerous intertwined, whitish to pinkish rhizomorphs; internally gelatinous.

SPORE MASS: dark green, slimy, foul-smelling.

MICROSCOPIC FEATURES: spores 4–5.5 × 1.5–2 μm, subcylindric, smooth, green-tinted.

FRUITING: solitary or in groups on decaying wood; May–September; fairly common.

EDIBILITY: inedible.

Phallus ravenelii Berkeley and Curtis (see photo, p. 17)

COMMON NAME: Ravenel's Stinkhorn.

FRUITING BODY: at first egg-like, resembling a small puffball, oval to pear-shaped, 1 3/8–2 3/8" (3.5–6 cm) high, 1 1/8–1 3/4" (3–4.5 cm) wide; whitish to pinkish lilac, with pinkish lilac, often branched rhizomorphs; internally gelatinous; giving rise to a distinct head and stalk.

HEAD: 1 1/8–1 3/4" (3–4.5 cm) long, 5/8–1 1/2" (1.5–4 cm) wide, conic, with a granular to wrinkled and dishrag-like surface, lacking distinct pits, with a white-rimmed opening at the apex; covered with a slimy spore mass.

STALK: 4–6 1/4" (10–16 cm) long, 5/8–1 1/8" (1.5–3 cm) wide; nearly equal; roughened, hollow, spongy, whitish.

VOLVA: whitish to pinkish lilac, wrinkled, tough.

SPORE MASS: greenish brown to olive-brown, slimy, foul-smelling.

MICROSCOPIC FEATURES: spores 3–4 × 1–1.5 μm, cylindric, smooth, hyaline.

FRUITING: solitary to scattered or clustered on or near decaying logs and stumps, woody debris, and wood chips; August–October; fairly common.

EDIBILITY: inedible.

Pseudocolus fusiformis (Fischer) Lloyd (see photo, p. 17)

COMMON NAME: Stinky Squid.

FRUITING BODY: egg-like at first, resembling a small puffball, oval to pear-shaped, 1–1 1/2" (2.5–4 cm) high, 3/4–1 3/8" (2–3.5 cm) wide, grayish brown to pale gray or rarely whitish, typically finely areolate on the upper portion, with white rhizomorphs; internally gelatinous; soon splitting open to form a stalk with tapering arms, a volva, and a spore mass.

STALK: 3/4–1 3/4" (2–4.5 cm) long, shorter or equal in length to the arms, 5/8–1 1/8" (1.5–3 cm) thick, roughened, spongy, hollow, giving rise to 3–5 reticulate-pitted, arched arms that taper upward and are often united at their apices, whitish at the base, becoming yellow to orange above; arms 1–2 3/4" (2.5–7 cm) long.

VOLVA: grayish brown to pale gray or rarely whitish; typically finely to coarsely areolate on the upper portion with white showing in the cracks; wrinkled, tough.

SPORE MASS: olive-green to dark green, borne on the inner side of the arms, slimy, foul-smelling, drying nearly black.

MICROSCOPIC FEATURES: spores 4.5–5.5 × 2–2.5 μm, elliptic-ovoid, smooth, hyaline.

FRUITING: scattered or in groups on soil in conifers or mixed woods and in wood chips used in gardens or for landscaping; July–September; infrequent to occasional.

EDIBILITY: inedible.

COMMENTS: also known as *Pseudocolus schellenbergiae* and *Colus schellenbergiae*.

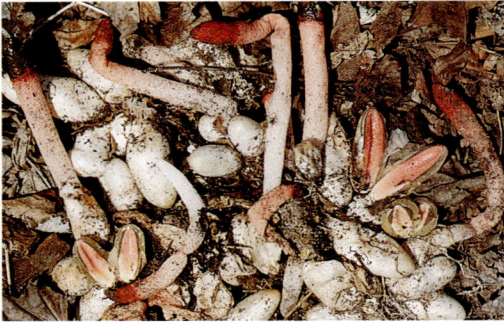

Mutinus caninus

Bird's-Nest Fungi

Bird's-Nest Fungi are very small fungi related to the puffballs; both groups are members of the class Gasteromycetes (Stomach Fungi). They are essentially tiny cups (the shape is generally cylindric to inverted-conic), each filled with numerous peridioles (eggs). Each peridiole is filled with spores and can thus be compared to a miniaturized puffball. When young, the cups are generally protected by a membrane-like lid. Bird's-Nest Fungi, which are decomposers, are especially common in woodchip mulch but are also found on dung, branches, leaves, and other organic debris.

Key to Species of Bird's-Nest Fungi

1. Peridioles white or pallid—*Crucibulum laeve* (Hudson) Kamby (see p. 465).
1. Peridioles gray; inside of cup vertically lined—*Cyathus striatus* Hudson : Persoon (see p. 466).
1. Not as in either of the above choices → 2.
     2. Peridioles dark gray to black, ¹⁄₁₆–⅛" (1–2 mm) in diameter; inside of cup not vertically lined; exterior shaggy-hairy when young; inedible—*Cyathus stercoreus* (Schweinitz) de Toni.
     2. Peridioles grayish brown, about ⅛" (2–3.5 mm) in diameter; inside of cup not vertically lined; exterior smooth to minutely hairy but not shaggy-hairy; inedible—*Cyathus olla* (Batsch) Persoon.

Crucibulum laeve (Hudson) Kamby (see photo, p. 18)
  COMMON NAME: Common Bird's-Nest, White-egg Bird's-Nest.
  CUP: more or less cylindric, tapering somewhat toward the bottom, ¼–⅜" (6–10 mm) wide at the top, ¼–⅜" (6–10 mm) high; when immature, the cup is protected by a white, membrane-like lid that is coated at first with yellowish orange fibers; interior whitish, smooth; exterior hairy to finely velvety, yellowish orange, becoming paler yellow.
  PERIDIOLES: white or pallid, round but flattened, ¹⁄₁₆–³⁄₃₂" (1.5–2 mm) in diameter, each attached beneath by a tiny, coiled cord.
  MICROSCOPIC FEATURES: spores 8–10 × 4–5.5 μm, elliptic, smooth, hyaline.
  FRUITING: scattered to gregarious on wood chips or twigs, also reported on discarded leather and cardboard; July–October; common.
  EDIBILITY: inedible.
  COMMENTS: also known as *C. vulgare*.

Cyathus striatus Hudson : Persoon (see photo, p. 18)

COMMON NAME: Splash Cups.

CUP: inverted-conic, ¼–⁵⁄₁₆" (6–8 mm) wide, ¼–³⁄₈" (6–10 mm) high; when immature, the cup is protected by a white, membrane-like lid and the upper edge of the cap is rolled inward; interior gray to grayish white, shiny, smooth, vertically lined; exterior reddish brown to chocolate-brown or grayish brown, shaggy-hairy to woolly, sometimes faintly to distinctly fluted.

PERIDIOLES: gray, flattened, ¹⁄₁₆–⅛" (1.5–3 mm) in diameter, often vaguely triangular, each attached beneath by a tiny, coiled cord.

MICROSCOPIC FEATURES: spores 15–20 × 8–12 μm, elliptic, smooth, hyaline.

FRUITING: scattered to gregariously grouped on wood chips, twigs, bark, etc.; July–October; frequent to common.

EDIBILITY: inedible.

Blueberry Galls and Azalea Apples

Blueberry Galls and Azalea Apples are a group of parasitic fungi that attack species of the heath family, including blueberry, huckleberry, bilberry, wild azalea, and many others. They invade young shoots, leaves, and flowers, causing the formation of tissue swellings called galls. We have included two of the more common species to represent this genus.

Key to Species of *Exobasidium*

1. Fruiting body a powdery, white to pale gray or reddish mold-like growth on flowers, leaves, and shoots of heath family plants—*Exobasidium vaccinii* (Fuckel) Woronin (see p. 467).
1. Fruiting body a yellow-green to pinkish or whitish gall-like swelling on Rhododendron, especially wild azalea—*Exobasidium rhododendri* (Fuckel) Cramer (see p. 467).

Exobasidium rhododendri (Fuckel) Cramer (see photo, p. 18)
COMMON NAME: Azalea Apples.
FRUITING BODY: ⅜–1¾" (1–4.5 cm) in diameter, an irregular gall-like swelling on leaves; surface yellow-green, smooth, shiny, becoming pinkish, and finally whitish and powdery in age.
FLESH: firm, juicy; odor not distinctive; taste tart to somewhat astringent.
MICROSCOPIC FEATURES: spores 12–16 × 2–4 μm, sausage-shaped to nearly cylindric, with 1 septum at maturity, colorless, inamyloid; aseptate hyaline conidia measuring 5–9 × 1.5–2 μm may also be present.
FRUITING: solitary or in groups on *Rhododendron* species, especially wild azalea (*Rhododendron roseum*) along streams, bogs, and moist woodlands; May–July; fairly common.
EDIBILITY: unknown.
COMMENTS: the gall-like growths produced by this parasitic fungus are fruit-like. Some individuals state that they are edible, but we cannot recommend eating them because of the host plant's documented toxicity. Some authors believe that this species is only one additional form of *E. vaccinii*.

Exobasidium vaccinii (Fuckel) Woronin (see photo, p. 18)
COMMON NAME: Blueberry Galls.
FRUITING BODY: a powdery, white to pale gray or reddish mold-like growth that surrounds and shrouds developing leaves, shoots, and flowers and stimulates the infected parts to form swellings, called galls, in which spores are produced.
MICROSCOPIC FEATURES: spores 11–18 × 2.5–5 μm, nearly cylindric to sausage-shaped,

with 1–6 septa at maturity, colorless, inamyloid; aseptate hyaline conidia measuring 5–9 × 1–2 μm may also be present.

FRUITING: parasitic on many members of the heath family, such as blueberry and huckleberry; May–October; fairly common.

EDIBILITY: unknown.

COMMENTS: many authors have assigned new species names for this fungus based on the host plant attacked; one example is *E. azaleae.*

Rusts and Smuts

Rusts and smuts are two very large groups of parasitic fungi that attack a wide range of plants, expecially grasses and sedges. Rusts get their name from the orange, powdery spore mass that they form at one stage in their complex life cycles. Smuts produce a sooty brown to black spore mass on their host plants. Several hundred species of rusts and smuts have been identified. We have included only a few of the more common and conspicuous species as examples of these fungi.

Key to Species of Rusts and Smuts

1. Fruiting body whitish to silvery gray with black tints, an irregularly shaped, tumor-like gall; parasitic on corn plants—*Ustilago maydis* (De Chambre) Corda (see p. 470).
1. Not as above; fruiting body pale cinnamon-brown to dark brown, orange-yellow, or orange; parasitic on cinquefoils, Jack-in-the Pulpit, or cedar → 2.
   2. Reddish brown to dark brown galls with or without orange, jelly-like horns on cedar—*Gymnosporangium juniperi-virginiana* Schweinitz (see p. 469).
   2. Bright orange, irregular, slightly raised powdery pustules in dense clusters; partially embedded in the leaves and stems of cinquefoils—*Pucciniastrum potentillae* Komarov : Jaczewski, Komarov, and Tranzsche (see p. 470).
   2. Pale yellow, irregular, slightly raised, powdery pustules in dense clusters, which become orange-yellow and finally pale cinnamon-brown; on leaves and stems of Jack-in-the Pulpit—*Uromyces ari-triphylli* (Schweinitz) Seeler (see p. 470).

Gymnosporangium juniperi-virginiana Schweinitz (see photo, p. 19)

COMMON NAME: Cedar-Apple Rust.

FRUITING BODY: a gall measuring 1–2" (2.5–5 cm) in diameter at maturity, at first a small, greenish brown swelling on the upper surface of a cedar needle, enlarging rapidly to form an overwintering stage that turns reddish brown to dark brown and forms small, circular depressions on the surface; during the spring producing conspicuous orange (immature) to orange-brown (mature), jelly-like horns that arise from the small, circular depressions and measure ⅜–¾" (1–2 cm) long.

MICROSCOPIC FEATURES: teliospores 15–21 × 42–65 μm, 2–celled, rhombic-oval to elliptic, thick-walled, brown, on long stalks.

FRUITING: solitary or in groups on cedar needles; May–August; uncommon.

EDIBILITY: the edibility of the horns is unknown, but the galls are too woody.

COMMENTS: this parasitic fungus divides its life cycle between two hosts, cedar and apple, and causes considerable damage to both plants. Galls eventually die but often remain attached to the tree for a year or more. The jelly-like horns are columns of teliospores that germinate to form basidiospores that infect apple leaves.

Pucciniastrum potentillae Komarov : Jaczewski, Komarov, and Tranzsche (see photo, p. 19)
COMMON NAME: Cinquefoil Rust.
FRUITING BODY: at first tiny, rounded, reddish brown outgrowths less than ¹⁄₁₆" (1.5 mm)
wide, partially embedded in plant tissue, expanding and rupturing at maturity, forming
irregular, slightly raised, bright orange, powdery pustules up to ¼" (6 mm) wide.
MICROSCOPIC FEATURES: urediospores 14–20 × 12–15 µm, oval to globose, finely echinulate, hyaline.
FRUITING: in dense clusters; partially embedded in the leaves and stems of cinquefoil
(Potentilla) species; June–September; common.
EDIBILITY: inedible.

Uromyces ari-triphylli (Schweinitz) Seeler (see photo, p. 19)
COMMON NAME: Jack-in-the-Pulpit Rust.
FRUITING BODY: at first very tiny, rounded, pale yellow outgrowths less than ¹⁄₃₂" (0.75
mm) wide, partially embedded in plant tissue, expanding and rupturing at maturity,
forming irregular, slightly raised, orange-yellow, then pale cinnamon-brown powdery
pustules up to ¹⁄₁₆" (1.5 mm) wide.
MICROSCOPIC FEATURES: urediospores 24–35 × 14–24 µm, elliptic, obovoid or nearly
wedge-shaped, sparsely echinulate.
FRUITING: in dense clusters; partially embedded in the leaves and stems of Jack-in-the-
Pulpit *(Arisaema triphyllum)*; June–September; common.
Edibility: inedible.

Ustilago maydis (De Chambre) Corda (see photo, p. 19)
COMMON NAME: Corn Smut.
FRUITING BODY: ¾–3" (2–7.5 cm) long, ⅜–2⅜" (1–6 cm) wide, an irregularly shaped,
tumor-like gall that swells and replaces normal kernels of corn plants; surface smooth,
shiny, delicate, whitish when very young, becoming silvery gray with black tints, and
finally blackish overall, rupturing at maturity to release the spores.
FLESH: at first firm, moist, whitish, then turning black and juicy, becoming dry, powdery,
and dark olive-brown as the galls fill with spores at maturity.
MICROSCOPIC FEATURES: spores 9–11 × 6.5–7.5 µm, round or nearly so, pale olive-brown,
with prominent spines.
FRUITING: solitary or in clusters on corn plants; July–September; common.
EDIBILITY: edible when silvery gray on the surface and black and juicy within.
COMMENTS: gall clusters of this parasitic fungus may reach a diameter of 8" (20 cm). It is
canned and sold commercially in Mexico and is considered a delicacy in many parts of
the world.

Morels, False Morels, and Allies

Morels are hollow-stalked, single-chambered mushrooms with sponge-like, conic to bell-shaped caps with distinct pits and ridges. False morels, also known as lorchels, include mushrooms with brain-like, saddle-shaped, trilobate, mitre-shaped, shield-shaped, or irregularly lobed caps. The stalks of some species are small and terete to compressed or ribbed, while others are massive and multichambered. One species, *Underwoodia columnaris* Peck, lacks a distinct cap and stalk and forms cylindric to spindle-shaped fruiting bodies with longitudinal grooves or wrinkles.

Many species in this group are considered to be choice edibles, but some are poisonous, and one has caused fatalities.

Key to Species of Morels, False Morels, and Allies

1. Fruiting body lacking a distinct cap and stalk; cylindric to spindle-shaped or club-shaped, tapering upward to a rounded tip, with grooves or wrinkles running lengthwise, up to 4" (10 cm) high; cream to tan when young, becoming pale brown in age; spores 25–27 × 12–14 μm, elliptic, coarsely warted, hyaline; solitary or in groups on the ground in hardwoods—*Underwoodia columnaris* Peck (see p. 513).
1. Fruiting body with a cap and stalk; cap nearly round, brain-like to irregularly lobed, composed of cups fused together to form a contorted, compound mass, 2⅜–3⅛" (6–8 cm) wide and tall; pale yellow-brown to grayish brown; stalk 6½–12" (16.5–30.5 cm) long, ¾–1¼" (2–3 cm) wide; brown; spores 32–36 × 12–15 μm; on soil under leaf litter; July–September; rare; edibility unknown—*Wynnea sparassoides* Pfister.
1. Fruiting body with a cap and stalk; cap bell-shaped to oval or cylindric, draping from its point of attachment at the top of the stalk, ⅜–1⅛" (1–3 cm) wide, ¾–1⅛" (2–3 cm) high; yellowish brown, coarsely wrinkled to somewhat pitted; stalk 2–5" (5–12.7 cm) tall, white to brownish yellow, hollow or stuffed with soft, cottony tissue; spores 55–85 × 15–18 μm; on the ground in hardwoods; March–May; edibility questionable and not recommended—*Ptychoverpa bohemica* (Krombholz) Boudier = *Verpa bohemica* (Krombholz) Schroeter.
1. Fruiting body with a cap and stalk; cap oval to conic or cylindric, sponge-like with pits and ridges → 2.
1. Fruiting body not with the above combinations of characters → 3.
   2. Cap broadly conic, attached to the stalk about midway and flaring below; pits elongated and irregular; grayish tan to yellow-brown—*Morchella semilibera* De Chambre : Fries (see p. 477).
   2. Cap oval to conic or somewhat cylindric, attached to and continuous with the

stalk below; whitish, grayish, yellow, or yellow-brown—*Morchella esculenta* Fries (see p. 476).

2. Cap conic to oval, attached to and continuous with the stalk below; dark brown to brownish black—*Morchella elata* Fries complex (see p. 476).

3. Cap thimble- to bell-shaped, broadly rounded to flattened at the center, attached only at the top of the stalk, nearly smooth to finely wrinkled—*Verpa conica* (Müller : Fries) Swartz (see p. 477).

3. Not as above; stalk ½–3½" (1.3–9 cm) thick, granular or smooth, typically ribbed, often chambered or hollow to stuffed; fruiting body medium to robust; cap brain-like to lobed or saddle-shaped, wrinkled to convoluted → 4.

3. Not as above; stalk up to ½" (1.3 cm) thick, terete, compressed or ribbed; fruiting body typically small to medium and not robust; cap saddle-shaped, trilobate, convoluted, shield-shaped, or mitre-shaped → 6.

4. Cap usually saddle-shaped or sometimes trilobate, 1–4" (2.5–10 cm) wide, wrinkled to convoluted, reddish brown to dark brown, lacking distinct violet to lavender tints; margin incurved; stalk ¾–1" (2–2.5 cm) thick, hollow, whitish to pinkish buff; on decaying wood or humus; July–October; spores 18–23 × 7–10 µm, elliptic, smooth—*Gyromitra infula* (Schaeffer : Fries) Quélet (see p. 474).

4. Cap usually saddle-shaped or sometimes trilobate, 1–3⅛" (2.5–8 cm) wide, wrinkled to convoluted, reddish brown with distinct violet to lavender tints; margin incurved; stalk ½–¾" (1.3–2 cm) thick, hollow, whitish with violet to lavender tints; on soil or humus; August–October; spores 22–33 × 7–12 µm, elliptic, smooth; poisonous—*Gyromitra ambigua* (Karsten) Harmaja.

4. Cap typically brain-like to lobed or saddle-shaped, wrinkled to convoluted, robust → 5.

5. Cap yellow-brown to grayish brown; stalk surface purplish red to rosy pink with whitish areas, deeply pitted and ribbed; ribs markedly extending onto the sterile (lower) surface of the cap; spores 9–12 µm, globose, smooth; on decaying logs and stumps in hardwoods or mixed woods or on sawdust piles; May–July; edibility unknown—*Gyromitra sphaerospora* (Peck) Saccardo.

5. Cap red-brown to dark brown or dark yellow-brown; stalk nearly equal, smooth or slightly scurfy, usually not ribbed, white with pinkish tints, hollow or stuffed, with one or two chambers—*Gyromitra esculenta* (Persoon) Fries (see p. 474).

5. Cap yellow-brown to reddish brown, irregularly folded and deeply divided into uplifted wrinkled lobes, sometimes saddle-shaped; stalk massive, 1–3½" (2.5–9 cm) thick, usually enlarged downward, multichambered; chalky white to pale grayish white, typically ribbed—*Gyromitra fastigiata* (Krombholz) Rehm (see p. 474).

5. Cap yellow-brown to pale reddish brown; stalk massive, 1½–3" (4–7.5 cm) thick, usually enlarged downward, mostly concealed by the overhanging cap margin, whitish, with thick rounded ribs extending from the apex to the base, multichambered—*Gyromitra korfii* (Raitvir) Harmaja (see p. 475).

5. Cap reddish brown with several prominent seam-like vertical ridges, usually not saddle-shaped; margin appressed against the stalk; stalk massive, 1–3⅛" (2.5–8 cm) thick, enlarged at the base, distinctly ribbed, multichambered—*Gyromitra caroliniana* (Bosc : Fries) Fries (see p. 473).

6. Stalk terete or somewhat compressed, lacking ribs, smooth except for soft, fine hairs at the apex, cream-colored to pale buff; cap saddle-shaped, with a groove between the ascending lobes, tan to grayish brown; margin curved toward the stalk, entire, sometimes fused with the stalk or itself; sterile (lower) surface white

to tan, paler than the upper surface, smooth; spores 18–24 × 11–14 μm edibility unknown—*Helvella elastica* Fries.

6. Stalk terete to somewhat compressed, lacking ribs, smooth to pubescent, cream-colored, becoming whitish at the apex and darkest at the base; cap saddle-shaped to trilobate, with a groove between ascending lobes, tan to buff or pale orange-buff; margin inrolled when young, becoming curved upward away from the stalk, then flaring at maturity; sterile surface whitish to buff, pubescent when young, pubescent or smooth in age; spores 17–20 × 11–14 μm; edibility unknown—*Helvella stevensii* Peck.

6. Stalk terete to somewhat compressed, lacking ribs, smooth to finely granular, cream-colored, palest at the apex; cap saddle-shaped to trilobate, with a groove between ascending lobes, pale to dark gray-brown; margin inrolled when young, becoming curved upward away from the stalk, then flaring at maturity; sterile (lower) surface cream-colored to whitish, villose when young, nearly smooth in age; spores 20–24 × 12–15 μm; edibility unknown—*Helvella albella* (Quélet) Boudier.

6. Stalk distinctly ribbed, with or without pits → 7.

7. Stalk deeply pitted; ⅜–1⅜" (1–3.5 cm) thick, whitish to pale buff or pinkish buff; cap saddle-shaped to irregularly lobed; fertile (upper) surface pale cream to pale buff—*Helvella crispa* Fries (see p. 475).

7. Stalk typically lacking pits; ¼–1" (5–25 mm) thick, pale to dark gray; cap ⅝–2¾" (1.5–7 cm) wide, saddle-shaped to trilobate or irregularly convoluted, gray to black; sterile (lower) surface pale gray, smooth; spores 15–18 × 10–12.5 μm; on soil, among mosses, or on decaying wood in mixed woods; June–October; edibility unknown—*Helvella sulcata* Fries.

7. Stalk often distinctly pitted; ⅜–1⅜" (1–3 cm) thick, dull white to gray or black, darkest on the ribs and toward the apex; cap mitre-shaped to convex or sometimes saddle-shaped, pale gray to black; on the ground or on decaying wood under conifers or hardwoods—*Helvella lacunosa* Fries (see p. 475).

7. Stalk lacking pits; narrow, ⅛–⅜" (3–9 mm) thick, pale to dark gray, paler near the base; cap shield-shaped to irregularly saddle-shaped; fertile (upper) surface blackish to gray; sterile (lower) surface blackish to gray but paler than the fertile surface; spores 15–21 × 11–13 μm; on mosses, twigs, and needles in wet areas under conifers, especially cedar; edibility unknown—*Helvella palustris* Peck.

Gyromitra caroliniana (Bosc : Fries) Fries

COMMON NAME: Carolina False Morel.

CAP: 2–7" (5–18 cm) wide and tall, convoluted and brain-like, typically with several prominent seam-like vertical ridges, usually not saddle-shaped, moist and lubricous; margin appressed against the stalk, wavy and irregular; fertile (upper) surface reddish brown; sterile (lower) surface white to whitish.

STALK: 1½–6" (4–15.5 cm) long, 1–3⅛" (2.5–8 cm) thick, enlarged at the base, distinctly ribbed, white; interior multichambered.

MICROSCOPIC FEATURES: spores 22–35 × 10–16 μm, elliptic, reticulate, with one or more short projections at each end, hyaline.

FRUITING: solitary or in groups on the ground in hardwoods or mixed woods; April–May; uncommon.

EDIBILITY: questionable and not recommended; although eaten and considered choice by some individuals, this mushroom is suspected to contain toxins.

Gyromitra esculenta (Persoon : Fries) Fries

COMMON NAME: Brain Mushroom.

CAP: 1⅜–4" (3.5–10 cm) wide, 1½–4" (4–10 cm) tall, brain-like to irregularly lobed, deeply wrinkled to convoluted, moist to dry; margin undulating to contorted, often curved toward the stalk; fertile (upper) surface pinkish tan to dark reddish brown or orange-brown, lubricous when fresh; sterile (lower) surface pale pinkish tan to yellowish tan.

STALK: ¾–2¾" (2–7 cm) long, ¾–1⅛" (2–3 cm) thick, enlarging downward or nearly equal, hollow or stuffed with cottony hyphae, sometimes chambered; surface smooth and waxy to slightly granular, dingy white to pinkish tan or tan, often ribbed near the base.

MICROSCOPIC FEATURES: spores 18–28 × 9–13 μm, elliptic, smooth, with 2 oil drops, hyaline.

FRUITING: solitary, scattered, or in groups on the ground under conifers; April–June; common.

EDIBILITY: poisonous; this species contains hydrazines, which can cause serious illness or death.

COMMENTS: the interior of the cap is chambered, and its flesh is very brittle. Many common names have been assigned to this mushroom including Conifer False Morel, Beefsteak Morel, and Lorchel.

Gyromitra fastigiata (Krombholz) Rehm

COMMON NAME: Gabled False Morel, Elephant Ears.

CAP: 1½–4¾" (4–12 cm) wide, 1½–3½" (4–9 cm) tall, irregularly folded and deeply divided into uplifted wrinkled lobes, sometimes saddle-shaped, moist, lubricous; margin wavy and folded; fertile (upper) surface yellow-brown to reddish brown; sterile (lower) surface pale yellow-brown to whitish, slightly granular, often fusing with any part of the cap or stalk.

STALK: 1–4" (2.5–10 cm) long, 1½–3½" (4–9 cm) thick, stout, enlarging downward to nearly equal; surface chalky white to pale grayish white, typically ribbed.

MICROSCOPIC FEATURES: spores 25–30 × 13–15 μm, elliptic, finely reticulate, with several short projections on each end, containing 1–3 oil drops, hyaline; the projections appear only on the spores of mature specimens.

FRUITING: solitary to scattered on the ground in hardwoods; April–May; common in the southern part of the region, rare in northern areas.

EDIBILITY: questionable and not recommended; although eaten by some individuals, this mushroom is known to cause gastric distress and severe headache.

COMMENTS: also known as *G. brunnea* and *Helvella underwoodii*.

Gyromitra infula (Schaeffer : Fries) Quélet

COMMON NAME: Saddle-shaped False Morel.

CAP: 1–4" (2.5–10 cm) wide, ¾–4" (2–10 cm) tall, usually saddle-shaped or sometimes trilobate, margin incurved; fertile surface wrinkled to convoluted or sometimes nearly smooth, moist when fresh, reddish brown to dark brown, lacking distinct violet to lavender tints; interior hollow or chambered; flesh brittle.

STALK: ¾–2⅜" (2–6 cm) long, ¾–1" (2–2.5 cm) thick, dry, hollow, finely granular, whitish to pinkish buff.

MICROSCOPIC FEATURES: spores 18–23 × 7–10 μm, elliptic, smooth, hyaline, with two large oil drops when mounted in water; paraphyses forked and enlarged apically.

FRUITING: solitary, scattered or in groups on decaying wood or humus; July–October; occasional.

EDIBILITY: poisonous.

Gyromitra korfii (Raitvir) Harmaja

COMMON NAME: Korf's Gyromitra.

CAP: 2–5½" (5–14 cm) wide, 1½–3¾" (4–9.5 cm) tall, typically brain-like, wrinkled and deeply folded or sometimes nearly smooth, moist and lubricous; margin undulating to contorted, usually free from the stalk or sometimes fused with it; fertile (upper) surface yellow-brown to reddish brown; sterile (lower) surface somewhat granular, pale yellow-brown to whitish.

STALK: 1½–2¾" (4–7 cm) long, 1½–3" (4–7.5 cm) thick, stout, enlarging downward or nearly equal, chambered and stuffed with cottony white hyphae or hollow; surface whitish, smooth to slightly granular, with thick, rounded ribs extending from the apex to the base.

MICROSCOPIC FEATURES: spores 24–30 × 11–14 μm, spindle-shaped, slightly wrinkled, with blunt projections on each end, containing a large central oil drop and 1–2 smaller ones at each end, hyaline; the blunt projections appear only on the spores of mature specimens.

FRUITING: solitary, scattered, or in groups on the ground in hardwoods or mixed woods; April to early June; common.

EDIBILITY: not established and therefore not recommended.

COMMENTS: this species was named for Richard P. Korf, Professor Emeritus, Cornell University.

Helvella crispa Fries (see photo, p. 20)

COMMON NAME: Fluted White Helvella.

CAP: ¾–2⅜" (2–6 cm) wide, ⅜–1⅝" (1–4 cm) tall, saddle-shaped to irregularly lobed, dry; margin rolled inward when young, then gradually unrolling and expanding at maturity, entire or somewhat lacerated; fertile (upper) surface pale cream to pale buff, smooth to somewhat wrinkled; sterile (lower) surface pale cream to buff, with tiny short hairs.

STALK: 1–3½" (2.5–9 cm) long, ⅜–1⅛" (1–3 cm) thick, typically enlarged in the middle and tapering toward the base and apex, sometimes nearly equal; surface whitish to pale buff or pinkish buff, deeply pitted and ribbed; ribs branching and anastomosing and extending onto the sterile surface of the cap.

MICROSCOPIC FEATURES: spores 14–23 × 10–14 μm, elliptic, smooth, with 1–5 oil drops, hyaline.

FRUITING: solitary, scattered, or in groups on the ground in hardwoods or conifer forests; July–October; fairly common.

EDIBILITY: unknown.

COMMENTS: the pale cream to buff colors of the cap and stalk are diagnostic field characters. It sometimes fruits on moss-covered decaying wood.

Helvella lacunosa Fries

COMMON NAME: Fluted Black Helvella.

CAP: ¾–2" (2–5 cm) wide, ⅜–2¾" (1–7 cm) tall, mitre-shaped to convex or saddle-shaped, dry; margin typically wavy and often incurved, entire or somewhat lacerated; fertile (upper) surface pale gray to black, wrinkled; sterile (lower) surface grayish to black, glabrous, with ribs extending from the stalk apex; ribs often forked and interconnected.

STALK: 2⅜–5½" (6–14 cm) long, ⅜–1⅜" (1–3 cm) thick, nearly equal or enlarged downward, dry, hollow; surface dull white to gray or black, typically darkest on the ribs and toward the apex, deeply pitted and conspicuously ribbed; ribs branching and anastomosing and extending onto the sterile surface of the cap.

MICROSCOPIC FEATURES: spores 16–22 × 10–14 μm, ellipsoid, smooth to slightly wrinkled, with a single oil drop, hyaline.

FRUITING: solitary, scattered or in groups on the ground or on decaying wood under conifers or hardwoods; August–November; occasional.

EDIBILITY: unknown.

COMMENTS: carefully compare with *H. sulcata* (see p. 473).

Morchella elata Fries complex (see photo, p. 20)

COMMON NAME: Black Morel.

CAP: ¾–2⅜" (2–6 cm) wide, 1–3½" (2.5–9 cm) tall, conic to oval, sponge-like, hollow, divided into pits and ridges, continuous with the stalk below; pits elongated and irregular, yellow-brown to gray-brown; ridges anastomosing, dark brown to brownish black.

STALK: 2–4" (5–10 cm) long, ⅜–1½" (1–4 cm) thick, enlarged near the base, hollow; surface white to dingy yellow, granular, sometimes ribbed.

MICROSCOPIC FEATURES: spores 18–25 × 11–15 μm, elliptic, smooth, hyaline.

FRUITING: solitary, scattered, or in groups on soil in a variety of habitats: mixed woods, associated with cherry, poplars, or pines; mixed hardwoods; and burned areas; April–May; fairly common.

EDIBILITY: edible, choice.

COMMENTS: it is usually the first morel to appear in spring. Although choice, it has caused gastrointestinal distress when consumed with alcohol. The Black Morel is a complex of several varieties or species that are nearly indistinguishable. Other commonly applied names include *M. angusticeps* and *M. conica*. Much work remains to be done before the taxonomic problems of this complex will be resolved.

Morchella esculenta Fries

COMMON NAME: Common Morel.

CAP: ¾–2¾" (2–7 cm) wide, ¾–7" (2–17.5 cm) tall, oval to conic or somewhat cylindric, sponge-like, hollow, divided into pits and ridges of variable color, continuous with the stalk below; pits round to elongated and irregular, gray to brown or yellowish; ridges anastomosing, whitish to grayish, yellow or yellow-brown.

STALK: 1–4½" (2.5–11.5 cm) long, ¾–2¾" (2–7 cm) thick, nearly equal or enlarged (sometimes massively) toward the base, hollow; surface whitish, granular, often ribbed.

MICROSCOPIC FEATURES: spores 20–26 × 12–16 μm, elliptic, smooth, hyaline.

FRUITING: solitary, scattered, or in groups on soil in a variety of habitats: near dead elms, in old apple orchards, in burned areas, in mixed hardwoods, and sometimes under conifers; April–June; common.

EDIBILITY: edible, choice; this is one of the most highly prized and sought after mushrooms for the table.

COMMENTS: it has many additional common names, including Yellow Morel, Sponge Mushroom, Land Fish, Pine Cone Mushroom, and Honeycomb. Some authors recognize various color forms and statures as distinct species. As Nancy Weber has stated, more than one hundred names exist in the literature for species in the genus *Morchella*. Examples include the White Morel (*M. deliciosa*), with white ribs and grayish pits, and the Thick-footed Morel (*M. crassipes*), which has a very thick stalk and may attain a height of 12 inches. Extensive work must be done to resolve the taxonomic problems of this genus. Also see the descriptions of *M. semilibera*, the *M. elata* complex, and *Ptychoverpa bohemica* in the comments section of *M. semilibera* (see p. 477).

Morchella semilibera De Chambre : Fries

COMMON NAME: Half-free Morel.

CAP: ⅝–1½" (1.5–4 cm) wide, ⅜–1½" (1–4 cm) tall, broadly conic with a round to blunt apex, sponge-like, hollow, divided into pits and ridges, attached to the stalk about midway and flaring below; pits elongated and irregular, grayish tan to yellow-brown; ridges anastomosing, brownish, darkening to grayish brown or brownish black in age; underside of flaring cap whitish and granular.

STALK: 2–6" (5–15 cm) long, ⅜–1" (1–2.5 cm) thick, enlarging downward or nearly equal overall, hollow; surface whitish, slightly granular, commonly ribbed.

MICROSCOPIC FEATURES: spores 22–30 × 12–17 μm, elliptic, smooth, hyaline.

FRUITING: solitary or scattered on the ground in hardwood forests, especially beech-maple, in old apple orchards, and sometimes associated with a variety of other trees, including poplars and oaks; April–May; fairly common.

EDIBILITY: edible.

COMMENTS: this species typically fruits before the Common Morel and is easily identified because of its half-free cap and hollow stalk. The Wrinkled Thimble-cap, *Ptychoverpa bohemica,* is similar but has a yellow-brown wrinkled cap that lacks true pits and is attached only at the top of the stalk. It can also be differentiated from the Half-free Morel by the smooth undersurface of the flaring skirt-like cap, a stalk stuffed with cottony hyphae, and extremely large, elliptic, smooth spores that measure 55–85 × 15–18 μm. *Ptychoverpa bohemica* is edible for some but poisonous to many others, causing variable reactions, including severe gastrointestinal upset and temporary loss of coordination.

Verpa conica (Müller : Fries) Swartz

COMMON NAME: Smooth Thimble-cap.

CAP: ⅜–1⅛" (1–3 cm) wide, ⅝–1⅛" (1.5–3 cm) tall, thimble-shaped to bell-shaped, broadly rounded to flattened at the center, attached at the top of the stalk, with skirt-like sides that sometimes flare in age, nearly smooth to finely wrinkled; underside of cap whitish to tan and smooth.

STALK: 1½–4½" (4–11.5 cm) long, ¼–⅝" (5–15 mm) thick, nearly equal or tapering toward the apex, hollow or stuffed with white cottony hyphae; surface whitish, smooth to slightly granular, sometimes ribbed.

MICROSCOPIC FEATURES: spores 21–26 × 11–16 μm, elliptic to oval, smooth, hyaline.

FRUITING: solitary to scattered or in groups on the ground in hardwoods, mixed woods, and especially old apple orchards; April–May; fairly common.

EDIBILITY: listed as edible in some guides, but recent reports suggest otherwise, and therefore it is not recommended.

COMMENTS: the flesh of this mushroom is thin and very brittle.

Gyromitra caroliniana

Gyromitra esculenta

Gyromitra fastigiata

Gyromitra infula

Gyromitra korfii

Helvella lacunosa

Morchella esculenta

Morchella semilibera

Verpa conica

Cup and Saucer Fungi

Members of this group produce fruiting bodies that resemble small cups or saucers. These fungi are members of the class Ascomycetes, and their fertile (inner) surfaces, which are often brightly colored, are lined with asci and ascospores. The group is very large, with several hundred species. Classification of the Cup and Saucer Fungi is currently in transition: there is considerable disagreement about the nomenclature and taxonomic position of many species in this group. A discussion of the majority of species included in this group is beyond the scope of this work, and therefore we have included only those species with distinctive features and those most commonly encountered. Many species included in the key can be identified using macroscopic features, but some require microscopic examination for positive identification.

Their flesh is often thin and brittle, and with few exceptions, the edibility of members of this group is unknown or not worth pursuing. In short, they are better viewed than chewed.

Key to Species of Cup and Saucer Fungi

1. Fruiting body 1⅛–5⅞" (3–15 cm) wide, ear-shaped to irregularly cup-shaped, stalkless, rubbery-gelatinous; fertile surface wrinkled, reddish brown; spores 12–15 × 4–6 μm, sausage-shaped, smooth, hyaline, asci and ascospores absent; solitary, in groups, or in fused clusters on decaying wood, especially conifers—*Auricularia auricula* (Hooker) Underwood (see Jelly Fungi, p. 430).

1. Fruiting body ⅜–¾" (1–2 cm) wide, at first cushion-shaped or irregular; fertile surface of young specimens completely covered by the infolded outer surface, eventually splitting open to form an irregular cup or saucer with a deeply indented and incurved margin; fertile surface smooth, yellowish brown to reddish brown; outer surface pale yellowish brown and scurfy; spores 6–12 × 2–4 μm; emerging in clusters on alder and hazel branches; November–June; edibility unknown—*Encoelia furfuracea* (Roth : Persoon) Karsten.

1. Fruiting body 1¾–4¾" (4.5–12 cm) wide, irregularly rounded, dense; outer surface incurved, deeply wrinkled and folded, fibrous-tough, pinkish brown to dingy brown when dry, becoming dark reddish brown to blackish brown when wet; fertile surface located in a depression on the upper surface, smooth to finely cracked, blackish brown; solitary, scattered, or in groups on conifer debris—*Sarcosoma globosa* (Schmidel : Fries) Caspary (see p. 497).

1. Fruiting body ⅛–3⁄16" (3–8 mm) wide, cup- to saucer-shaped; fertile surface blue-green, sometimes tinted yellow; in groups or clusters on decaying hardwood—*Chlorociboria aeruginascens* (Nylander) Kanouse (see p. 491).

1. Fruiting body not with the above combinations of characters; an irregular cup with a split on one side or an elongated half-cup resembling rabbit ears → 2.
1. Fruiting body not with the above combinations of characters → 3.
    2. Fruiting body 1–5½" (2.5–14 cm) wide, 2⅜–5⅛" (6–13 cm) high, composed of several (up to 24) elongated half-cups resembling rabbit ears; fertile surface pinkish orange to pinkish red or orange to brownish orange; outer surface blackish brown to reddish brown; solitary or scattered on soil and arising from a tough underground sclerotium—*Wynnea americana* Thaxter (see p. 497).
    2. Fruiting body ⅜–2" (1–5 cm) wide, irregularly cup-shaped and frequently split on one side; fertile surface bright yellow-orange to orange; outer surface yellow, often stained dark bluish green, especially toward the margin—*Caloscypha fulgens* (Persoon : Fries) Boudier (see p. 491).
    2. Fruiting body ½–1" (1.2–2.5 cm) wide, 1–2" (2.5–5 cm) high, resembling rabbit ears and typically slit down to the base on one side; fertile surface pale yellow to brownish yellow; outer surface yellow to brownish yellow; spores 12–15 × 6–8 μm, elliptic, smooth, hyaline, with 2 oil drops; solitary or in groups on soil and debris in conifer woods; July–September; edibility unknown—*Otidea leporina* (Bataille : Fries) Fuckel.
    2. Fruiting body nearly identical to the above choice except for the following characters: fertile surface with a pinkish tint; outer surface brownish orange to dull orange; spores 10–13 × 5–7 μm; edibility unknown—*Otidea onotica* (Fries) Fuckel.
3. Fertile surface of fruiting body white to creamy yellow, tan, gray, or brownish gray → 4.
3. Fertile surface of fruiting body red → 7.
3. Fruiting body pale pink to pinkish purple, rubbery-gelatinous, and growing on wood, or olive-green with a purplish pink to whitish stalk and growing on sphagnum mosses → 8.
3. Fertile surface of fruiting body black to blackish brown → 9.
3. Fertile surface of fruiting body yellow to orange → 11.
3. Fertile surface of fruiting body yellow-brown, reddish brown, purple-brown, or olive-green to olive-brown → 15.
    4. Outer surface of fruiting body covered with conspicuous hairs, especially on the cup margin; with or without a stalk → 5.
    4. Outer surface of fruiting body lacking conspicuous hairs; with or without a conspicuous stalk → 6.
5. Fruiting body ¾–2" (2–5 cm) wide, ¾–2¾" (2–7 cm) high, distinctly cup-shaped; fertile surface whitish to tan when young, becoming creamy yellow in age, smooth; outer surface pale creamy yellow to pale yellowish brown, covered by a dense layer of soft brown hairs; base of the cup crimped to form a short stalk, ⅜–1" (1–2.5 cm) long, ⅜–¾" (1–2 cm) wide; on soil or decaying wood—*Jafnea semitosta* (Berkeley and Curtis) Korf (see p. 494).
5. Fruiting body ⅜–1⅛" (1–3 cm) wide, ⅜–¾" (1–2 cm) high, cup-shaped; fertile surface whitish to pale gray; outer surface brownish yellow, covered by a dense layer of brownish hairs that project from the margin over part of the fertile surface; scattered or in groups on soil, among mosses, or on decaying wood—*Humaria hemisphaerica* (Wiggers : Fries) Fuckel (see p. 493).
5. Fruiting body 1/16–3/16" (1.5–4 mm) wide, cup-shaped with a distinct stalk; fertile surface white; outer surface white to creamy white, covered by a dense layer of white hairs; scattered, in groups, or in clusters on decaying branches, stems, beechnut burrs, and other debris—*Dasyscyphus virgineus* S. F. Gray (see p. 492).

5. Fruiting body 1⅜–2⅛" (3.5–5.5 cm) high, ¾–1¾" (2–4.5 cm) wide; fertile surface dark gray to brownish gray or gray-brown, smooth; outer surface pale gray to pale brownish gray with grayish hairs; stalk ¾–2" (2–5 cm) long, ⅟₁₆–³⁄₁₆" (1.5–5 mm) thick, enlarging slightly downward, brownish gray to brown, covered with grayish hairs; spores 20–28 × 10–13 μm, elliptic to spindle-shaped, minutely warted, hyaline, with oil drops; solitary, scattered, or in groups on soil or decaying wood; July–October; edibility unknown—*Helvella macropus* (Fries) Karsten.

5. Fruiting body similar to the above choice except for the following characters: typically saucer- to cup-shaped but sometimes saddle-shaped; fertile surface grayish black when young, becoming gray or brownish gray at maturity; outer surface and stalk colored like the fertile surface or paler; spores 16–20 × 10–13 μm, oblong to elliptic, smooth or somewhat warty; on soil under hardwoods; edibility unknown—*Helvella villosa* (Kuntze) Dissing and Nannfeldt.

    6. Fruiting body ⅝–1¾" (1.5–4.5 cm) high, ¾–1¾" (2–4.5 cm) wide, cup- to saucer-shaped, with a distinct or sometimes poorly developed stalk; fertile surface and outer surface pale to dark gray; stalk typically present and distinct, conspicuously ribbed, white to grayish white or pale brownish yellow; ribs repeatedly branched upward and extending over the outer surface, often to the margin; scattered or in groups on soil under hardwoods—*Helvella griseoalba* Weber (see p. 493).

    6. Fruiting body ⅜–2" (1–5 cm) high, ⅜–2⅛" (1–5.5 cm) wide, cup-shaped with a short, often poorly developed stalk; fertile surface gray to dark brownish gray; outer surface gray, usually dark brownish gray near the margin, becoming creamy white below; margin typically toothed and irregularly torn; scattered or in groups on soil under conifers, especially pine—*Helvella leucomelaena* (Persoon) Nannfeldt (see p. 493).

    6. Fruiting body ⅜–1" (1–2.5 cm) high and wide, cup-shaped with a short or conspicuous stalk; fertile surface grayish tan to grayish yellow, smooth; outer surface grayish yellow to pale brownish yellow, distinctly granular; margin toothed and irregularly torn; stalk ⅛–½" (3–12 mm) long, colored like the outer surface and also distinctly granular; spores 18–22 × 10–15 μm, broadly elliptic, smooth, hyaline, with 2 oil drops; on soil and among mosses under hardwoods; June–September; edibility unknown—*Tarzetta cupularis* (Linnaeus : Fries) Lambotte.

    6. Fruiting body ⅜–⅞" (3–22 mm) high, ⅛–½" (3–12 mm) wide, top-shaped when young, becoming cup-shaped, then saucer-shaped to convex with a distinct stalk at maturity; fertile surface whitish or pale orange-yellow, typically with a violet tint; outer surface whitish; stalk ⅛–¾" (3–20 mm) long, tapering downward, whitish near the apex, becoming brownish yellow below and dark brown near the base; spores 10–17 × 3–5 μm, elliptic to spindle-shaped, smooth, hyaline; scattered or in groups on decaying leaves, twigs, branches, and woody debris in wet areas; May–August; edibility unknown—*Cudoniella clavus* (Albertini and Schweinitz : Fries) Dennis.

    6. Fruiting body ½–1⅛" (1.2–3 cm) high, ¾–4" wide, cup- to saucer-shaped with a short stalk; fertile surface whitish when young, becoming yellow-ochre to tan or pale brown at maturity, smooth; outer surface whitish to grayish or yellow-ochre; stalk up to ⅜" (1 cm) when young, often inconspicuous at maturity, whitish to grayish or yellow-ochre to pale yellow-brown; spores 12–15 × 7–10 μm, elliptic, smooth, hyaline, often with small oil drops; solitary, scattered, or in groups in wet or damp areas in houses and apartments, on bath tiles, sheetrock, carpets in cellars, and a variety of other substrates; edibility unknown—*Peziza domiciliana* Cooke.

7. Fruiting body ⅛–⅜" (3–10 mm) wide, ⅜–1¾" (1–4.5 cm) high, cup-shaped with a distinct white stalk; outer surface and margin covered with conspicuous white hairs— *Microstoma floccosa* (Schweinitz) Raitvir (see p. 494).

7. Fruiting body ¼–⅝" (6–16 mm) wide, ½–1¾" (1.2–4.5 cm) high, cup- to saucer-shaped with a distinct white stalk; outer surface and margin lacking conspicuous white hairs— *Sarcoscypha occidentalis* (Sowerby) Saccardo (see p. 496).

7. Fruiting body ¾–3⅛" (2–8 cm) wide, ⅝–1⅜" (2–3.5 cm) high, cup- to saucer-shaped with or without a distinct stalk; outer surface and stalk pale pinkish red to ochre or whitish; margin and outer surface lacking conspicuous white hairs—*Sarcoscypha coccinea* (Jacquin : Fries) Lambotte complex (see p. 496).

7. Fruiting body ⅛–¾" (3–20 mm) wide, ⅜–⅝" (3–15 mm) high, cup-shaped, often distorted, usually stalkless or with a rudimentary stalk; fertile surface bright red to pinkish red, smooth, outer surface pinkish red, lacking conspicuous white hairs; spores 16–20 × 9–11 μm, elliptic, smooth, hyaline, with 2 large oil drops; in groups or clusters on burned ground; June–September; edibility unknown—*Tarzetta rosea* (Rea) Dennis.

  8. Fruiting body ¾–1½" (2–4 cm) wide, ⅜–1" (1–2.5 cm) high, sometimes forming a continuous mass up to 4" (10 cm) or more; rubbery-gelatinous and brain-like with rounded lobes and furrows to saucer-shaped; pale pink to reddish purple; asci tips *not* staining blue in Melzer's; on decaying hardwood branches and trunks— *Ascotremella faginea* (Peck) Seaver (see p. 490).

  8. Fruiting body ⅜–1⅜" (1–3 cm) wide, ⅜–1" (1–2.5 cm) high, forming dense clusters up to 4" (10 cm) in diameter; cushion- to top- or saucer-shaped, often distorted, rubbery-gelatinous, pale pinkish brown or yellowish brown; asci tips staining blue in Melzer's; on decaying hardwoods, especially beech—*Neobulgaria pura* (Fries) Petrak (see p. 494).

  8. Fruiting body ⅛–¾" (3–20 mm) wide, ⅛–⅜" (3–10 mm) high, nearly round when young, soon becoming top- to cushion-shaped, then irregularly saucer-shaped with a lobed or wavy margin; fertile and outer surface pink to reddish purple; flesh rubbery-gelatinous; spores 18–30 × 4–6 μm, elliptic, multiseptate at maturity, smooth, hyaline; in groups or clusters on decaying wood, especially beech—*Ascocoryne cylichnium* (Tulasne) Korf (see p. 490).

  8. Fruiting body nearly identical to the above choice except for the following characters: spores 10–18 × 3–5 μm, elliptic, 0–1 septate, smooth, hyaline; edibility unknown—*Ascocoryne sarcoides* (Jacquin : S. F. Gray) Groves and Wilson.

  8. Fruiting body ⅜–¾" (1–2 cm) wide, ½–1" (1.2–2.5 cm) high, cup- to saucer-shaped with a distinct stalk; fertile surface pale to dark olive-green, smooth; margin wavy to lobed and furrowed; outer surface and stalk purplish pink or whitish, appearing gelatinous; spores 12–16 × 4–6 μm, elliptic, smooth, hyaline; in groups or clusters on sphagnum mosses in bogs; June–October; edibility unknown— *Sarcoleotia turficola* (Boudier) Dennis.

9. Fruiting body ⅛–¾" (3–20 mm) wide, black, irregular to angular or somewhat circular and saucer-shaped, slightly elevated, shiny or dull, typically wrinkled to furrowed and often minutely cracked; on maple leaves—*Rhytisma acerinum* (Persoon) Fries (see p. 496).

9. Fruiting body nearly identical to the above choice but growing on willow leaves; edibility unknown—*Rhytisma salicinum* Fries.

9. Fruiting body distinctly elevated and cup- to saucer- or top-shaped → 10.

  10. Fruiting body ¾–2¾" (2–7 cm) wide, 1¾–4⅛" (4.5–10.5 cm) high, deeply cup-shaped with a distinct stalk; fertile surface blackish brown to black; outer surface black to brownish black, densely matted and woolly, becoming brownish gray in

age; margin toothed and irregularly torn; stalk ¾–1½" (2–4 cm) long, ³⁄₁₆–³⁄₈" (5–
10 mm) thick, tapering downward, black to brownish black; in groups or clusters
on decaying hardwood; March–June—*Urnula craterium* (Schweinitz) Fries (see
p. 497).

10. Fruiting body ¼–⅝" (7–16 mm) wide, saucer-shaped with a short to prominent
stalk, up to ⅜" (1 cm) long, fertile surface smooth, shiny or dull, black, often perforated at the center, usually associated with small, erect, club-shaped fruiting bodies with white to gray heads and black stalks; growing from the cracks in the bark
of fallen hardwood trees, especially basswood—*Holwaya mucida* (Schulzer) Korf
and Abawi (see p. 493).

10. Fruiting body ¼–1⅜" (6–35 mm) wide, ⅛–⅝" (3–16 mm) high, cup- to saucer-
shaped, stalkless; fertile surface blackish brown to black; outer surface coated with
a dense layer of short, matted hairs, black or brownish black; spores 12–14 μm,
round, smooth, hyaline to pale brown, with many small oil drops; tips of the paraphyses straight, not spirally curved; scattered or in groups on moss-covered decaying conifer wood; May–July—*Pseudoplectania nigrella* (Persoon: Fries) Fuckel (see
p. 495).

10. Fruiting body nearly identical to the above choice except for the following characters: fertile surface dark olive-brown to blackish brown; outer surface smooth to
slightly felty; margin remaining incurved over the fertile surface for a long time;
stalk typically present, ⅜–1⅛" (1–3 cm) long, ⅛–¼" (3–6 mm) thick, tapering
downward; tips of some of the paraphyses forked, or forked and spirally curved;
edibility unknown—*Pseudoplectania melaena* (Fries) Saccardo = *P. vogesiaca* (Persoon) Seaver.

10. Fruiting body ⅜–¾" (1–2 cm) wide, ⅜–1" (1–2.5 cm) high, cup-shaped, stalkless
or with a rudimentary stalk, up to ⅜" (1 cm) long and thick; fertile surface blackish
brown to black, smooth; outer surface blackish brown to black, coated with matted hairs that are encrusted with reddish orange granules, especially near the margin; margin short-toothed and sometimes torn, reddish orange; spores 20–28 × 9–
11 μm, spindle-shaped, smooth, hyaline, with or without oil drops; solitary, in
groups, or in clusters on conifer debris; May–June; edibility unknown—*Plectania
melastoma* (Sowerby : Fries) Fuckel.

10. Fruiting body ⅜–1½" (2–4 cm) wide, ½–1⅜" (1.6–3.5 cm) high, nearly round to
top-shaped when young, becoming saucer-shaped in age; fertile surface black,
smooth, shiny; outer surface blackish brown to brown and scurfy; flesh rubbery-
gelatinous; spores 9–16 × 6–7 μm, kidney-shaped to elliptic, smooth, of two color
forms in each ascus: upper four (mature) brown, lower four (immature) hyaline
and typically very poorly developed; in clusters on decaying hardwoods, especially
oak; August–September—*Bulgaria inquinans* Fries (see p. 491).

11. Fruiting body growing in groups or clusters on decaying hardwoods; ⅜–1⅜" (1–3.5 cm)
wide, ½–1⅛" (1.2–3 cm) high, cup-shaped, stalkless or with a short stalk up to ⅜" (1
cm) long; fertile surface pale orange to dull orange, reddish orange, reddish brown, or
tan; outer surface blackish brown, covered with a dense layer of matted woolly hairs—
Galiella rufa (Schweinitz) Nannfeldt and Korf (see p. 492).

11. Fruiting body growing in groups or clusters on moist sandy soil; ⅛–⅝" (3–16 mm)
wide, ¹⁄₁₆–¼" (1.5–6 mm) high, cup- to saucer-shaped, stalkless; fertile surface bright
orange to reddish orange; outer surface orange, covered toward the margin by clumps
of short, reddish brown hairs that resemble small granules—*Melastiza charteri* (W. G.
Smith) Boudier (see p. 494).

11. Fruiting body not with the above combinations of characters; with stiff hairs on the

outer surface or protruding from the margin, sometimes tufted and resembling teeth; fertile surface yellow, golden yellow to yellow-orange, or orange to reddish orange → 12.

11. Fruiting body not with the above combinations of characters; margin lacking stiff hairs; fertile surface bright yellow to orange-yellow, yellowish green, or pale to bright orange, typically not exceeding ¼" (6 mm) wide at maturity; margin and sterile surface sometimes scurfy → 13.

11. Fruiting body not with the above combinations of characters; margin lacking stiff hairs; fertile surface dull orange-yellow to brownish yellow or bright orange, typically greater than ¼" (6 mm) wide at maturity; margin and sterile surface sometimes scurfy → 14.

    12. Fruiting body fertile surface orange-yellow; ⅛–¼" (3–6 mm) wide, saucer-shaped, stalkless; outer surface and margin pale orange-yellow with a sparce coating of long, pale brown, stiff hairs; spores 17–22 × 9–12 μm, elliptic, smooth, hyaline; solitary or in groups on dung; May–October; fairly common; edibility unknown—*Cheilymenia fimicola* (de Notaris and Baglietto) Dennis.

    12. Fruiting body fertile surface pale yellow to yellow; 1/16–⅛" (1.5–3 mm) wide, cup-shaped, stalkless; outer surface and margin pale orange with short, stiff, orange hairs; spores 15–22 × 4–5 μm, spindle-shaped to elliptic, 1–3 septate at maturity, smooth, hyaline; scattered, in groups or clusters on leaves, decaying wood, and soil under hardwoods; May–October; edibility unknown—*Arachnopeziza aurelia* (Persoon) Fuckel.

    12. Fruiting body 1/16–3/16" (1.5–5 mm) wide, saucer-shaped, stalkless; outer surface and margin dull brownish orange with long, stiff, brown hairs; spores 17–22 × 10–14 μm, elliptic, smooth, hyaline; in dense clusters on decaying hardwood—*Scutellinia erinaceus* (Schweinitz) Kuntze (see p. 497).

    12. Fertile surface of fruiting body bright orange to orange-red, ⅛–½" (3–12 mm) wide, saucer-shaped, stalkless; shiny, smooth; outer surface and margin orange, covered with long, stiff, brown hairs; spores 16–21 × 10–14 μm, elliptic, minutely roughened, hyaline, with oil drops when immature; scattered, in groups or clusters, usually on decaying wood but sometimes on soil; June–November; edibility unknown—*Scutellinia scutellata* (Linnaeus : Fries) Lambotte.

13. Fruiting body 1/32–⅛" (1–3 mm) wide, saucer-shaped, stalkless; fertile surface yellow when young, becoming yellowish green in age; scattered, in groups or clusters on decaying wood—*Chlorosplenium chlora* (Schweinitz : Fries) Curtis (see p. 492).

13. Fruiting body 1/32–⅛" (1–3 mm) wide, saucer-shaped, stalkless or with a rudimentary stalk; fertile surface and outer surface bright yellow to golden yellow; asci 100–135 × 7–10 μm; spores 9–14 × 3–5 μm, elliptic, 0–1 septate, smooth, hyaline, with oil drops; in groups or clusters on decaying wood; July–October; edibility unknown—*Bisporella citrina* (Batsch : Fries) Korf and Carpenter (see p. 490).

13. Fruiting body very similar to the previous choice except for the following characters: fertile surface sulfur-yellow; asci 60–90 × 4–5 μm; spores 8–11 × 2 μm, fusiform to elliptic, often slightly curved, 1-septate, smooth, hyaline, with oil drops; edibility unknown—*Bisporella sulfurina* (Quélet) Carpenter.

13. Fruiting body up to 3/16" (5 mm) wide and high, at first cylindric to cushion-shaped or nearly round to top-shaped, pale orange with a whitish scurfy coating overall, soon opening and forming a bright orange to orange-yellow stalkless saucer with an uneven margin; growing in dense clusters covering leaf and needle litter, mosses, twigs, and soil—*Byssonectria terrestris* (Albertini and Schweinitz : Fries) Pfister (see p. 491).

13. Fruiting body 1/16–¼" (1.5–6 mm) wide, saucer-shaped with a distinct stalk; fertile

surface bright yellow to orange-yellow; outer surface pale yellow to whitish, slightly scurfy; stalk ⅛–⅜" (3–10 mm) long, whitish; spores 14–24 × 3–4 μm, spindle-shaped and typically curved, smooth, hyaline, with oil drops; in groups or clusters on partially buried decaying hardwoods, especially beech; August–November; edibility unknown—*Hymenoscyphus calyculus* (Sowerby : Fries) Phillips.

13. Fruiting body nearly identical to the above description, typically with a stalk but sometimes stalkless, on standing or fallen twigs, branches, stems, leaves, fruits, and other plant debris—*Hymenoscyphus, Helotium, Lachnellula* species and allies. (Note: Additional genera and numerous species would also key out here, but further identification is beyond the scope of this work.)

14. Fruiting body ¾–4" (2–10 cm) wide, ¾–1⅜" (2–3.5 cm) high, saucer-shaped with a distinct stalk; fertile surface dull brownish yellow to brownish orange, irregularly wrinkled or folded; outer surface whitish or creamy yellow, minutely felted; margin wavy, sometimes splitting; stalk up to ¾" (2 cm) long and wide, irregularly ribbed, whitish or creamy yellow; spores 33–40 × 12–16 μm, elliptic, reticulate, with a prominent apiculus at each end, hyaline, with one or more oil drops; solitary, scattered, or in groups on soil under hardwoods; April–June; edibility unknown—*Discina leucoxantha* Bresadola.

14. Fruiting body 3/16–⅜" (5–10 mm) wide, saucer-shaped, stalkless; fertile surface bright orange; margin weakly toothed; scattered or in groups on sandy soil among mosses—*Octospora humosa* (Fries) Dennis (see p. 495).

14. Fruiting body ⅜–2" (1–5 cm) wide, irregularly cup-shaped and frequently split on one side; fertile surface bright yellow-orange to orange; outer surface yellow, often stained dark bluish green, especially toward the margin—*Caloscypha fulgens* (Persoon : Fries) Boudier (see p. 491).

14. Fruiting body 3/16–⅝" (5–16 mm) wide, cup- to saucer-shaped, stalkless or with a rudimentary stalk; fertile surface orange to yellow-orange; outer surface yellow, felty, completely covering the fruiting body of young specimens and tearing to form a flap that is discarded to one side, exposing the orange fertile surface, not staining bluish green; in groups or clusters on soil or a mixture of wood chips, hay, and dung—*Acervus epispartius* (Berkeley and Broome) Pfister (see p. 489).

14. Fruiting body ⅜–4" (1–10 cm) wide, cup- to saucer-shaped, stalkless or with a rudimentary stalk; fertile surface bright orange to yellow orange; outer surface pale yellowish orange to yellow, not staining bluish green—*Aleuria aurantia* (Fries) Fuckel (see p. 489).

14. Fruiting body similar to the above choice except for the following characteristics: ⅜–1½" (1–4 cm) wide with a distinct stalk; outer surface pale yellow or whitish; typically clustered with stalks attached at their bases by a dense white mycelium—*Aleuria rhenana* Fuckel (see p. 490).

15. Fruiting body in groups or dense clusters on dung or manured ground, common on horse manure; ¾–3⅛" (2–8 cm) wide, cup-shaped and often distorted, stalkless; fertile surface pale to dark yellowish brown; outer surface pale yellowish brown, scurfy; spores 19–23 × 11–14 μm, elliptic, smooth, lacking oil drops; hyaline—*Peziza vesiculosa* Bulliard (see p. 495).

15. Fruiting body in groups or fused clusters on burned ground, usually under conifers; ¾–4" (2–10 cm) wide, saucer-shaped to flattened or cushion-shaped, stalkless; fertile surface reddish brown to dark brown; margin whitish, wavy; undersurface brownish orange to yellow, with cylindric, forked, whitish, root-like structures attached to the substrate; spores 24–40 × 7–11 μm, spindle-shaped, finely roughened, with a blunt or

somewhat pointed appendage on each end, hyaline, with oil drops; edibility unknown—*Rhizina undulata* Fries.

15. Fruiting body ⅛–¼" (3–6 mm) wide, cup- to saucer-shaped, stalkless; fertile surface greenish at first, soon becoming brown, finally purple-brown and dotted; outer surface yellowish green at first, soon becoming dark brown, coarsely scurfy; spores 20–24 × 12–14 μm, elliptic, warted, purplish brown; on burned ground; April–October—*Ascobolus carbonarius* Karsten.

15. Fruiting body with a distinct central stalk → 16.

15. Fruiting body not with the above combinations of characters → 17.

    16. Fruiting body ¾–2¾" (2–7 cm) wide, cup-shaped; fertile surface pale to dark brown; outer surface pale to dark brown near the margin, becoming whitish near the base; stalk ⅜–2¾" (1–7 cm) long, ⅜–1" (1–2.5 cm) thick, conspicuously ribbed, white; ribs rounded, branched upward, extending over the outer surface; spores 16–19.5 × 11–14 μm, broadly elliptic, smooth, hyaline, with one oil drop; scattered or in groups on the ground under hardwoods; edibility unknown—*Helvella acetabulum* (Fries) Quélet.

    16. Fruiting body ¼–¾" (6–20 mm) wide, cup-shaped; fertile surface and outer surface reddish brown to orange-brown; margin whitish, toothed, sometimes torn; stalk ⅜–1" (1–2.5 cm) long, 1/16–¼" (1.5–6 mm) thick, tapering downward, pale yellow; spores 12–16 × 6–8 μm, elliptic, smooth, hyaline, lacking oil drops; in groups or clusters on burned soil; edibility unknown—*Geopyxis carbonaria* (Albertini and Schweinitz : Fries) Saccardo.

17. Fruiting body ¼–1" (6–25 mm) wide, saucer-shaped with a short stalk; fertile surface olive-green to olive-brown or orange-brown; outer surface dull reddish brown, velvety; stalk ⅛–¼" (3–6 mm) long, tapering downward, dull reddish brown; spores 9–14 × 3–4 μm, elliptic, smooth, hyaline, with 2 oil drops—*Chlorosplenium versiforme* (Persoon : Fries) de Notaris (see p. 492).

17. Fruiting body ¾–4" (2–10 cm) wide, cup- to saucer-shaped, stalkless; fertile surface pale to dark olive-brown, smooth; outer surface reddish brown, scurfy; spores 15–22 × 8–11 μm, elliptic, with a delicate, irregular reticulum, hyaline, often with 2 oil drops—*Peziza badia* Persoon : Mérat (see p. 495).

17. Fruiting body ¾–4" (2–10 cm) wide, cup- to saucer-shaped, stalkless; fertile surface purple-brown to reddish brown or olive-brown; outer surface pale reddish brown to purple-brown, scurfy; spores 16–21 × 8–10 μm, elliptic, finely warted, hyaline, with 2 oil drops—*Peziza phyllogena* Cooke (see p. 495).

17. Fruiting body typically ⅜–2⅜" (1–6 cm) wide, often rubbery-gelatinous, typically growing on wet decaying wood → 18.

17. Fruiting body 1½–8¼" (4–21 cm) wide, typically brittle and growing on soil or wood → 19.

    18. Fruiting body ⅜–2⅜" (1–6 cm) wide, saucer-shaped, stalkless; fertile surface dark reddish brown to olive-brown or yellowish brown, smooth; outer surface watery brown; flesh rubbery-gelatinous to waxy; spores 25–32 × 12–14 μm, elliptic, smooth, hyaline, with 1 or 2 oil drops; scattered or in groups on wet decaying hardwood; edibility unknown—*Pachyella clypeata* (Schweinitz) Le Gal.

    18. Fruiting body nearly identical to the above choice except for the following characters: ⅜–2" (1–5 cm) wide; spores 18–20 × 10–14 μm, elliptic, distinctly warted, hyaline, with 1 or 2 oil drops; edibility unknown—*Pachyella adnata* (Berkeley and Curtis) Pfister.

    18. Fruiting body up to ¾" (2 cm) wide, saucer-shaped, stalkless; fertile surface pale

yellow-brown to reddish brown; outer surface pale grayish yellow; flesh watery and gelatinous; spores 18–23 × 10–15 μm, elliptic, smooth, hyaline, usually with 2 oil drops; scattered or in groups on wet decaying wood, cones, and woody debris; edibility unknown—*Pachyella babingtonii* (Berkeley and Broome) Boudier.

19. Fruiting body 1½–8¼" (4–21 cm) wide, cup- to saucer-shaped with a short stalk; fertile surface reddish brown to yellowish brown with conspicuous vein-like folds and wrinkles; outer surface grayish white to pale brownish yellow; margin wavy to irregular in age; stalk up to ⅝" (1.5 cm) long and thick, typically ribbed; spores 19–26 × 12–15 μm, broadly elliptic, smooth, pale yellow; scattered or in groups on the ground under hardwoods; March–May; edible—*Disciotis venosa* (Persoon : Fries) Boudier.

19. Fruiting body 1½–3¾" (4–9.5 cm) wide, saucer-shaped to ear-like, stalkless or with a rudimentary stalk; fertile surface reddish brown to yellowish brown or tan, deeply folded or wrinkled; outer surface whitish to pinkish brown; margin wavy; spores 24–35 × 10–16 μm, elliptic, minutely warted, with a prominent pointed appendage at each end, pale yellow, often with 3 oil drops; scattered or in groups on decaying wood or humus under conifers; April–July; edible—*Discina perlata* (Fries) Fries.

19. Fruiting body 2–4½" (5–11.5 cm) wide, cup- to saucer-shaped, stalkless or with a rudimentary stalk; fertile surface pale to dark yellow-brown or dark brown, smooth; outer surface whitish; margin wavy; spores 14–16 × 8–10 μm, elliptic, smooth, hyaline, lacking oil drops; solitary, scattered, or in groups on decaying hardwood and adjacent soil; April–October; edibility unknown—*Peziza repanda* Persoon.

19. Fruiting body nearly identical to the above choice except for the following characters: spores 14–19 × 8–10 μm, elliptic, minutely roughened, hyaline, lacking oil drops; growing on the ground under hardwoods, especially beech and maple; April–July; edibility unknown—*Peziza sylvestris* (Boudier) Saccardo = *P. arvernensis* Boudier.

19. Fruiting body ¾–2" (2–5 cm) wide, cup- to saucer-shaped, stalkless; fertile surface grayish brown with an olive tint, smooth; outer surface whitish to grayish or brownish, often yellowish towards the margin, scurfy; margin even to lobed, sometimes wavy; flesh thin, brittle, whitish, exuding a yellow juice when cut; spores 16–22 × 8–12 μm, elliptic, distinctly warted, hyaline, with 2 oil drops; solitary, scattered, or in groups on damp soil in woods; June–September; edibility unknown—*Peziza succosa* Berkeley.

Acervus epispartius (Berkeley and Broome) Pfister

FRUITING BODY: ³⁄₁₆–⅝" (5–16 mm) wide, cup- to saucer-shaped, stalkless or with a rudimentary stalk; fertile surface orange to yellow-orange; outer surface yellow, felty, completely covering the fruiting body of young specimens and tearing to form a flap that is discarded to one side, exposing the orange fertile surface.

MICROSCOPIC FEATURES: spores 6–9 × 3–4 μm, elliptic with blunt ends, smooth, hyaline, usually with 1 oil drop.

FRUITING: in groups or clusters on soil or a mixture of wood chips, hay, and dung; July–October; rare.

EDIBILITY: unknown.

COMMENTS: formerly known as *Acervus aurantiacus*.

Aleuria aurantia (Fries) Fuckel

COMMON NAME: Orange Peel.

FRUITING BODY: ⅜–4" (1–10 cm) wide, cup- to saucer-shaped, stalkless or with a rudimentary stalk; fertile surface bright orange to yellow-orange, smooth; outer surface pale yellowish orange to yellow, slightly scurfy; margin, wavy and often torn at maturity.

MICROSCOPIC FEATURES: spores 17–24 × 9–11 μm, elliptic, covered by a coarse reticulum, usually with one or more projecting spines at each end, hyaline.

FRUITING: solitary, in groups, or in clusters in grassy areas on disturbed soil, in gardens, and along roadsides; June–October; fairly common.

EDIBILITY: edible.

Aleuria rhenana Fuckel

FRUITING BODY: ⅜–1½" (1–4 cm) wide, cup-shaped with a distinct stalk; fertile surface bright orange to yellow-orange, smooth; outer surface pale yellow or whitish, slightly scurfy; margin typically incurved, becoming wavy and torn at maturity; base of the cup compressed and folded; stalk up to ¾" (2 cm) long, ribbed, covered by a dense white mycelium.

MICROSCOPIC FEATURES: spores 20–23 × 11–12 μm, elliptic, covered by a coarse reticulum, hyaline.

FRUITING: typically clustered or in groups on soil in woodlands; July–October; uncommon.

EDIBILITY: unknown.

COMMENTS: probably synonymous with *Aleuria splendens.*

Ascocoryne cylichnium (Tulasne) Korf

FRUITING BODY: ⅛–⅜" (3–10 mm) high, ⅛–¾" (3–20 mm) wide, nearly round when young, soon becoming top- to cushion-shaped, then irregularly saucer-shaped with a lobed or wavy margin; fertile surface and outer surface pink to reddish purple; flesh rubbery-gelatinous.

MICROSCOPIC FEATURES: spores 18–30 × 4–6 μm, elliptic, multiseptate at maturity, smooth, hyaline; sometimes forming one to several tiny secondary spores.

FRUITING: in groups or clusters on decaying wood, especially beech; September–November; fairly common.

EDIBILITY: unknown.

Ascotremella faginea (Peck) Seaver

FRUITING BODY: ⅜–1" (1–2.5 cm) high, ¾–1½" (2–4 cm) wide, rounded to irregularly rounded, sometimes forming a continuous mass up to 4" (10 cm) or more; rubbery-gelatinous, smooth, shiny or dull, brain-like with rounded lobes and deep furrows to saucer-shaped; pale pink when young, becoming pinkish brown to reddish purple in age.

MICROSCOPIC FEATURES: spores 7–8.5 × 4–5 μm, elliptic, often truncate, with 3–4 fine longitudinal striations, hyaline, with 2 oil drops; asci tips not staining blue in Melzer's.

FRUITING: solitary, scattered, or in groups on decaying hardwood branches and trunks, especially beech; June–October; infrequent.

EDIBILITY: unknown.

COMMENTS: may be confused with some jelly fungi, but the presence of asci and ascospores rules them out.

Bisporella citrina (Batsch : Fries) Korf and Carpenter (see photo, p. 20)

COMMON NAME: Yellow Fairy Cups.

FRUITING BODY: ¹⁄₃₂–⅛" (1–3 mm) wide, saucer-shaped, stalkless or with a rudimentary stalk; fertile surface and outer surface smooth, bright lemon-yellow to golden yellow.

MICROSCOPIC FEATURES: asci 100–135 × 7–10 μm; spores 9–14 × 3–5 μm, elliptic, 0–1 septate, smooth, hyaline, with oil drops.

FRUITING: in groups or dense clusters on decaying wood; July–October; common.

EDIBILITY: unknown.

Bulgaria inquinans Fries

COMMON NAME: Black Jelly Drops.

FRUITING BODY: ⅜–1½" (2–4 cm) wide, ½–1⅜" (1.6–3.5 cm) high, nearly round to top-shaped when young, becoming saucer-shaped in age; fertile surface black, smooth, shiny; outer surface blackish brown to brown and scurfy; flesh rubbery-gelatinous.

MICROSCOPIC FEATURES: spores 9–16 × 6–7 μm, kidney-shaped to elliptic, smooth, of two color forms in each ascus: upper four (mature) brown, lower four (immature) hyaline and typically very poorly developed.

FRUITING: in groups or clusters on decaying hardwoods, especially oak: August–September; uncommon.

EDIBILITY: unknown.

Byssonectria terrestris (Albertini and Schweinitz : Fries) Pfister

FRUITING BODY: up to ³⁄₁₆" (5 mm) high and wide, at first cylindric to cushion-shaped or nearly round to top-shaped, pale orange with a whitish scurfy coating overall, soon opening and forming a bright orange to orange-yellow stalkless cup with an uneven and scurfy margin.

MICROSCOPIC FEATURES: spores 18–26 × 8–10 μm, fusiform, smooth, hyaline, with several oil drops.

FRUITING: in dense clusters on a white mycelium covering leaf and needle litter, mosses, twigs, and soil; April–June, found in the northern part of the region; uncommon.

EDIBILITY: unknown.

COMMENTS: because the two stages of development are quite different, you may think you have two separate species. *Byssonectria terrestris* is also known as *Inermisia fusispora*.

Caloscypha fulgens (Persoon : Fries) Boudier

COMMON NAME: Blue-staining Cup.

FRUITING BODY: ⅜–2" (1–5 cm) wide, irregularly cup-shaped and frequently split on one side; fertile surface smooth, bright yellow-orange to orange; outer surface yellow, often stained dark bluish green, especially toward the margin; stalkless.

MICROSCOPIC FEATURES: spores 5–8 μm, round, smooth, hyaline.

FRUITING: scattered or in groups on damp soil or on debris under conifers or in mixed woods; May–July; fairly common.

EDIBILITY: unknown.

Chlorociboria aeruginascens (Nylander) Kanouse

COMMON NAME: Blue Stain, Green Stain.

FRUITING BODY: ⅛–³⁄₁₆" (3–8 mm) wide, cup-shaped to nearly flat and saucer-shaped; fertile surface smooth, blue-green, sometimes tinted yellow; outer surface blue-green, finely roughened; stalk ⅛–¼" (3–6 mm) long, tapering downward, frequently off-center, blue-green.

MICROSCOPIC FEATURES: spores 6–10 × 1.5–2 μm, irregularly spindle-shaped with oil drops, smooth, hyaline.

FRUITING: in groups or clusters on decaying hardwood; June–November; fairly common.

EDIBILITY: unknown.

COMMENTS: also known as *Chlorosplenium aeruginascens*. The mycelium stains the substrate blue-green. *Chlorociboria aeruginosa* (edibility unknown) is nearly identical but has ascospores which measure 8–15 × 2–4 μm and the decaying wood on which it is growing does not stain blue-green.

Chlorosplenium chlora (Schweinitz : Fries) Curtis

FRUITING BODY: 1/32–1/8" (1–3 mm) wide, saucer-shaped, stalkless; fertile surface yellow when young, becoming yellowish green in age, smooth; outer surface bright yellow.

MICROSCOPIC FEATURES: spores 7–9 × 1.5–2 μm, narrowly elliptic to spindle-shaped, smooth, hyaline.

FRUITING: scattered, in groups, or in clusters on decaying wood; August–October; infrequent.

EDIBILITY: unknown.

Chlorosplenium versiforme (Persoon : Fries) de Notaris

FRUITING BODY: 1/4–1" (6–25 mm) wide, saucer-shaped with a short stalk; fertile surface olive-green to olive-brown or orange-brown; outer surface dull reddish brown, velvety; stalk 1/8–1/4" (3–6 mm) long, tapering downward, dull reddish brown.

MICROSCOPIC FEATURES: spores 9–14 × 3–4 μm, elliptic, smooth, hyaline, with 2 oil drops.

FRUITING: scattered or in groups on decaying wood; August–October; uncommon.

EDIBILITY: unknown.

COMMENTS: also known as *Chlorencoelia versiformis* and *Chlorociboria versiformis*.

Dasyscyphus virgineus S. F. Gray

COMMON NAME: Stalked Hairy Fairy Cup.

FRUITING BODY: 1/16–3/16" (1.5–4 mm) wide, up to 1/4" (6 mm) high, cup-shaped with a distinct stalk; fertile surface white, smooth; outer surface white to creamy white, covered by a dense layer of white hairs; stalk 1/16–1/8" (1.5–3 mm) long, covered with white hairs.

MICROSCOPIC FEATURES: spores 6–9 × 1.5–2.5 μm, spindle- to club-shaped, smooth, hyaline.

FRUITING: scattered, in groups, or in clusters on decaying stems, branches, cones, beech-nut burrs, and other debris; May–October; fairly common.

EDIBILITY: unknown.

COMMENTS: other similar species with white cups are not listed in the key and must be differentiated microscopically using some other reference.

Galiella rufa (Schweinitz) Nannfeldt and Korf

COMMON NAME: Hairy Rubber Cup.

FRUITING BODY: 1/2–11/8" (1.2–3 cm) high, 3/8–13/8" (1–3.5 cm) wide, cup-shaped; fertile surface pale orange to dull orange, reddish orange, reddish brown, or tan, smooth; outer surface tough, brown near the margin, blackish brown below, covered with a dense layer of matted woolly hairs; margin finely toothed; flesh rubbery-gelatinous.

MICROSCOPIC FEATURES: spores 18–20 × 8–10 μm, elliptic with narrow ends, finely warted, hyaline.

FRUITING: in groups or clusters on decaying hardwood; July–September; fairly common.

EDIBILITY: unknown.

COMMENTS: also known as *Bulgaria rufa*.

Helvella griseoalba Weber

FRUITING BODY: ¾–1¾" (2–4.5 cm) wide, ⅝–1¾" (1.5–4.5 cm) high, cup- to saucer-shaped, typically with a distinct stalk; fertile surface pale to dark gray, often with yellow tints; stalk ⅜–1⅜" (1–3.5 cm) long, ¼–¾" (6–20 mm) thick, conspicuously ribbed, white to grayish white, pale brownish yellow; stalk sometimes poorly developed; ribs rounded, repeatedly branched upwards and extending over the outer surface, sometimes to the margin.

MICROSCOPIC FEATURES: spores 15–18 × 10–12 μm, oblong to broadly elliptic, smooth, hyaline, with oil drops.

FRUITING: scattered or in groups on soil under hardwoods; June–August; infrequent.

EDIBILITY: unknown.

Helvella leucomelaena (Persoon) Nannfeldt

FRUITING BODY: ⅜–2" (1–5 cm) high, ⅜–2⅛" (1–5.5 cm) wide, cup-shaped with a short stalk; fertile surface gray to dark brownish gray, smooth; outer surface gray to dark brownish gray near the margin, becoming creamy white below and often white at the base; margin typically toothed and irregularly torn; stalk up to ⅝" (1.6 cm) long, conspicuously ribbed to folded or rudimentary.

MICROSCOPIC FEATURES: spores 20–25 × 10–14 μm, oblong, smooth, hyaline, with 1 oil drop.

FRUITING: scattered or in groups on leaf litter or on the ground in conifer or mixed woods; April–June; occasional.

EDIBILITY: unknown.

Holwaya mucida (Schulzer) Korf and Abawi

FRUITING BODY: two distinct stages, an erect club-shaped asexual stage called *Crinula caliciiformis* (Fries) Fries, and a saucer-shaped sexual stage called *Holwaya mucida*.

SEXUAL STAGE: ¼–⅝" (7–16 mm) wide, saucer-shaped with a short to prominent stalk; fertile surface smooth, shiny to dull, black, often perforated at the center; sterile surface dull, smooth to slightly scurfy, black; stalk up to ⅜" (1 cm) long, tapered downward, black.

ASEXUAL STAGE: a head supported by a stalk; head about ⅛" (2–4 mm) wide and high, elliptic to rounded, smooth, shiny to dull, viscid when fresh, white to gray; stalk ¼–¾" (6–20 mm) high, about ⅛" (2–4 mm) thick, shiny to dull, scurfy, black.

MICROSCOPIC FEATURES: spores of the sexual stage (ascospores) highly variable in length, 30–75 × 3–4 μm, needle-like, with 14–20 septations at maturity, smooth, hyaline, often with oil drops; spores of the asexual stage (conidia) approximately 3 × 1 μm, elliptic, smooth, hyaline.

FRUITING: scattered or in groups arising from the cracks in the bark of fallen hardwood trees, especially basswood; September–December; occasional.

EDIBILITY: unknown.

Humaria hemisphaerica (Wiggers : Fries) Fuckel

COMMON NAME: Brown-haired White Cup.

FRUITING BODY: ⅜–1⅛" (1–3 cm) wide, ⅜–¾" (1–2 cm) high, cup-shaped; fertile surface whitish to pale gray, smooth; outer surface brownish yellow, covered by a dense layer of brownish hairs that project from the margin over part of the fertile surface.

MICROSCOPIC FEATURES: spores 25–27 × 12–15 μm, elliptic, with or without tiny warts, hyaline, with 2 oil drops.

FRUITING: scattered or in groups on soil, among mosses, or on decaying wood; July–September; fairly common.

EDIBILITY: unknown.

Jafnea semitosta (Berkeley and Curtis) Korf

FRUITING BODY: ¾–2" (2–5 cm) wide, ¾–2¾" (2–7 cm) high, distinctly cup-shaped; fertile surface whitish to tan when young, becoming creamy yellow in age, smooth; outer surface pale creamy yellow to pale yellowish brown, covered by a dense layer of soft brown hairs; base of the cup crimped to form a short stalk, ⅜–1" (1–2.5 cm) long, ⅜–¾" (1–2 cm) wide, yellowish, covered with soft, brown hairs.

MICROSCOPIC FEATURES: spores 25–35 × 10–12 μm, elliptic to spindle-shaped, warted when mature, hyaline, with 2 oil drops.

FRUITING: solitary, scattered, or in groups on soil or decaying wood in conifer or hardwoods; July–October; infrequent.

EDIBILITY: unknown.

Melastiza charteri (W. G. Smith) Boudier

FRUITING BODY: ¹⁄₁₆–¼" (1.5–6 mm) high, ⅛–⅝" (3–16 mm) wide, cup- to saucer-shaped, often distorted, stalkless; fertile surface bright orange to reddish orange, smooth, shiny; outer surface orange, covered toward the margin by clumps of short, reddish brown hairs that resemble small granules.

MICROSCOPIC FEATURES: spores 17–20 × 9–13 μm, elliptic, ornamented by a coarse reticulum, sometimes with one or more spine-like projections at each end, hyaline, often with oil drops.

FRUITING: in groups or clusters on moist, sandy soil, also reported to occur on sawdust; July–October; infrequent.

EDIBILITY: unknown.

Microstoma floccosa (Schweinitz) Raitvir

COMMON NAME: Shaggy Scarlet Cup.

FRUITING BODY: ⅜–1¾" (1–4.5 cm) high, ⅛–⅜" (3–10 mm) wide, cup-shaped with a distinct stalk; fertile surface bright red, smooth; outer surface and margin pinkish red to bright red, covered by conspicuous white hairs; stalk ¾–1⅜" (2–3.5 cm) long, ¹⁄₁₆–⅛" (1.5–3 mm) thick, nearly equal, smooth to finely roughened, white.

MICROSCOPIC FEATURES: spores 20–35 × 15–17 μm, elliptic to spindle-shaped, smooth, hyaline.

FRUITING: in groups or clusters on decaying hardwood branches; June–August; occasional.

EDIBILITY: unknown.

Neobulgaria pura (Fries) Petrak

FRUITING BODY: ⅜–1" (1–2.5 cm) high, ⅜–1⅜" (1–3 cm) wide, forming dense clusters up to 4" (10 cm) in diameter; cushion- to top-shaped or saucer-shaped, often distorted, rubbery-gelatinous, smooth, shiny; pale pink to pinkish brown or yellowish brown.

MICROSCOPIC FEATURES: spores 7–9 × 3.5–4.5 μm, elliptic, smooth, hyaline, with 2 oil drops; asci tips staining blue in Melzer's.

FRUITING: in groups or clusters on decaying hardwoods, especially beech; July–November; occasional.

EDIBILITY: unknown.

Octospora humosa (Fries) Dennis

FRUITING BODY: ³⁄₁₆–³⁄₈" (5–10 mm) wide, saucer-shaped, stalkless; fertile surface bright orange, smooth, shiny or dull; outer surface whitish to pale brownish yellow; margin wavy, weakly toothed.

MICROSCOPIC FEATURES; spores 19–22 × 11–13 μm, elliptic-cylindric, smooth, hyaline, with one large oil drop and usually several smaller drops.

FRUITING: scattered or in groups on sandy soil among mosses; August–November; infrequent.

EDIBILITY: unknown.

Peziza badia Persoon : Mérat

COMMON NAME: Pig's Ears.

FRUITING BODY: ³⁄₄–4" (2–10 cm) wide, cup- to saucer-shaped, stalkless; fertile surface pale to dark olive-brown, smooth; outer surface reddish brown, scurfy.

MICROSCOPIC FEATURES: spores 15–22 × 8–11 μm, elliptic with a delicate, irregular reticulum, hyaline, often with 2 oil drops.

FRUITING: in dense clusters or groups on sandy soil or sawdust, usually under conifers; June–October; fairly common.

EDIBILITY: unknown.

Peziza phyllogena Cooke

COMMON NAME: Common Brown Cup.

FRUITING BODY: ³⁄₄–4" (2–10 cm) wide, cup- to saucer-shaped, stalkless; fertile surface purple-brown to reddish brown or olive-brown; outer surface pale reddish brown to purple-brown, scurfy.

MICROSCOPIC FEATURES: spores 16–21 × 8–10 μm, elliptic, finely warted, hyaline with 2 oil drops.

FRUITING: scattered, in groups or clusters on soil or decaying wood; May–August; fairly common.

EDIBILITY: edible, according to Gary Lincoff.

COMMENTS: also known as *Peziza badio-confusa*.

Peziza vesiculosa Bulliard

COMMON NAME: Bladder Cup.

FRUITING BODY: ³⁄₄–3¹⁄₈" (2–8 cm) wide, ³⁄₈–2" (1–5 cm) high, cup-shaped and often distorted, margin frequently incurved, stalkless; fertile surface pale to dark yellowish brown, nearly smooth; outer surface pale yellowish brown, scurfy; flesh brittle, brownish, forming tiny blisters in the center of the cup.

MICROSCOPIC FEATURES: spores 19–23 × 11–14 μm, elliptic, smooth, lacking oil drops, hyaline.

FRUITING: scattered or in clusters on manure piles and manured soil; June–October; common.

EDIBILITY: unknown.

Pseudoplectania nigrella (Persoon : Fries) Fuckel

COMMON NAME: Hairy Black Cup.

FRUITING BODY: ¹⁄₈–⁵⁄₈" (3–16 mm) high, ¹⁄₄–1" (6–25 mm) wide, cup- to saucer-shaped, stalkless; fertile surface blackish brown to black, smooth; outer surface coated with a dense layer of short, matted hairs, black to blackish brown.

MICROSCOPIC FEATURES: spores 12–14 μm, round, smooth, hyaline to pale brown, with many small oil drops; paraphyses filiform, multiseptate, tips straight and sometimes forked.

FRUITING: scattered or in groups on moss-covered decaying conifer wood; May–July; fairly common.

EDIBILITY: unknown.

Rhytisma acerinum (Persoon) Fries

COMMON NAME: Tar Spot of Maple.

FRUITING BODY: ⅛–¾" (3–20 mm) wide, black, irregular to angular or somewhat circular and saucer-shaped, slightly raised, shiny or dull, typically wrinkled to furrowed and often minutely cracked.

MICROSCOPIC FEATURES: spores 55–80 × 1.5–2.5 μm, needle-like, smooth, hyaline, with numerous oil drops.

FRUITING: solitary or in groups on maple leaves; September–May; very common.

EDIBILITY: inedible.

Sarcoscypha coccinea (Jacquin : Fries) Lambotte complex (see photo, p. 20)

COMMON NAME: Scarlet Cup.

FRUITING BODY: ⅝–1⅜" (2–3.5 cm) high, ¾–3⅛" (2–8 cm) wide, cup- to saucer-shaped with or without a distinct stalk; fertile surface bright red to orange-red, smooth; outer surface pale pinkish red to ochre or whitish, slightly granular or flaky; stalk typically short and stout, up to ⅜" (1 cm) long and thick, sometimes distinctly longer when the substrate is buried, pale pinkish red to ochre.

MICROSCOPIC FEATURES: spores 20–38 × 9–16 μm, elliptic, with rounded or truncate ends, smooth, hyaline, with oil drops; paraphyses with red contents.

FRUITING: solitary, scattered, or in groups on decaying hardwood branches; March–June; fairly common.

COMMENTS: Frances Harrington (1990) has reported that *S. coccinea* is a species complex and has differentiated taxa using microscopic features. According to her research, *S. coccinea* occurs in the Pacific Northwest and has not been reported from the Northeast. The two species that do occur here are *S. austriaca,* with mostly truncate spores containing many small polar oil drops, and *S. dudleyi,* with mostly rounded spores containing one large polar oil drop and some smaller drops.

Sarcoscypha occidentalis (Sowerby) Saccardo

COMMON NAME: Stalked Scarlet Cup.

FRUITING BODY: ½–1¾" (1.2–4.5 cm) high, ¼–⅝" (6–16 mm) wide, cup- to saucer-shaped, with a distinct stalk; fertile surface bright red, smooth; outer surface pinkish red to pinkish orange, lacking white hairs; stalk ⅜–1⅜" (1–3.5 cm) long, ¹⁄₁₆–⅛" (1.5–3 mm) thick, nearly equal, smooth, white.

MICROSCOPIC FEATURES: spores 20–22 × 10–12 μm, elliptic, smooth, hyaline, with oil drops.

FRUITING: scattered, in groups, or in clusters on decaying hardwood branches; May–September; occasional.

EDIBILITY: unknown.

Sarcosoma globosa (Schmidel : Fries) Caspary

FRUITING BODY: 1¾–4¾" (4.5–12 cm) wide, irregularly rounded, dense; outer surface incurved over the fertile surface, deeply wrinkled and folded, roughened, fibrous-tough and elastic, pinkish brown to dingy brown when dry, becoming dark reddish brown to blackish brown when wet; fertile surface located in a depression on the upper surface, smooth to finely cracked, blackish brown; interior gelatinous with a copious amount of water when fresh.

MICROSCOPIC FEATURES: spores 20–26 × 10–12 μm, elliptic, smooth, hyaline.

FRUITING: solitary, scattered, or in groups on conifer debris; April–June; rare.

EDIBILITY: unknown.

Scutellinia erinaceus (Schweinitz) Kuntze

COMMON NAME: Orange Eyelash Cup.

FRUITING BODY: ¹⁄₁₆–³⁄₁₆" (1.5–5 mm) wide, saucer-shaped, stalkless; fertile surface pale orange-yellow when young, becoming dull orange at maturity, smooth, shiny or dull; outer surface and margin dull brownish orange with long, stiff, brown hairs.

MICROSCOPIC FEATURES: spores 17–22 × 10–14 μm, elliptic, smooth, hyaline.

FRUITING: in groups or clusters on decaying hardwood; July–October; fairly common.

EDIBILITY: unknown.

COMMENTS: formerly known as *Patella setosa*.

Urnula craterium (Schweinitz) Fries

COMMON NAME: Devil's Urn.

FRUITING BODY: 1¾–4⅛" (4.5–10.5 cm) high, ¾–2¾" (2–7 cm) wide, deeply cup-shaped with a distinct stalk; fertile surface blackish brown to black, smooth; outer surface black to brownish black, densely matted and woolly, becoming brownish gray, tough and leathery in age; margin toothed and irregularly torn; stalk ¾–1½" (2–4 cm) long, ³⁄₁₆–³⁄₈" (5–10 mm) thick, tapering downward, black to brownish black.

MICROSCOPIC FEATURES: spores 25–35 × 12–14 μm, broadly elliptic, smooth, hyaline.

FRUITING: in groups or clusters on decaying wood or on the ground attached to buried wood; March–June; occasional.

EDIBILITY: unknown.

Wynnea americana Thaxter

COMMON NAME: Moose Antlers, Rabbit Ears.

FRUITING BODY: 1–5½" (2.5–14 cm) wide, 2⅜–5⅛" (6–13 cm) high, composed of several (up to 24) elongated half-cups resembling rabbit ears; fertile surface smooth, pinkish orange to pinkish red or orange to brownish orange; outer surface covered with small, rounded warts, sometimes wrinkled at maturity, blackish brown to reddish brown; flesh firm, somewhat tough, brown.

MICROSCOPIC FEATURES: spores 32–40 × 15–16 μm, elliptic, extremities apiculate, striately marked by several alternately light and dark bands extending the length of the spore, containing oil drops.

FRUITING: solitary or scattered on soil under hardwoods and arising from a tough, underground, brown sclerotium; July–September; rare.

EDIBILITY: unknown.

Acervus epispartius

Aleuria aurantia

Aleuria rhenana

Ascocoryne cylichnium

Ascotremella faginea

Bulgaria inquinans

Byssonectria terrestris

Caloscypha fulgens

Chlorociboria aeruginascens

Chlorosplenium chlora

Chlorosplenium versiforme

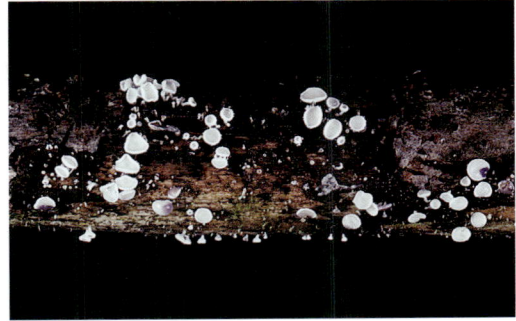

Dasyscyphus virgineus

Galiella rufa

Helvella griseoalba

Helvella leucomelaena

Holwaya mucida and *Crinula calciiformis*

Humaria hemisphaerica

Jafnea semitosta

Melastiza charteri

Microstoma floccosa

Neobulgaria pura

Octospora humosa

Peziza badia

Peziza phyllogena

Peziza vesiculosa

Pseudoplectania nigrella

Rhytisma acerinum

Sarcoscypha occidentalis

Sarcosoma globosa

Scutellinia erinaceus

Urnula craterium

Wynnea americana

Earth Tongues, Earth Clubs, and Allies

Members of this group form erect fruiting bodies that resemble small tongues or clubs. Coral fungi that are typically neither clustered nor branched are also included here. Most species have a distinct cap or a head-like fertile surface and a supporting stalk or stalk-like base. A few members lack a cap or head and consist of a club-shaped to cylindric or irregular stalk. *Cordyceps, Claviceps,* and Allies are similar, but their fertile surfaces are finely roughened like sandpaper.

Many species included in the key can be identified using macroscopic features. Some require microscopic examination for positive identification. Key steps 7–15 require the use of a microscope. Most species in this group are not collected for the table.

Key to Species of Earth Tongues, Earth Clubs, and Allies

1. Fruiting body growing on the ground in hardwoods; up to 4" (10 cm) high and ⅜–1⅛" (1–3 cm) thick, cylindric to spindle-shaped or club-shaped, tapering upward to a rounded tip, with longitudinal grooves or wrinkles, lacking a distinct cap and stalk; cream to tan when young, becoming pale brown in age; spores 25–27 × 12–14 μm, elliptic, coarsely warted, hyaline; solitary or in groups—*Underwoodia columnaris* Peck (see p. 513).

1. Fruiting body growing on partially or completely submerged branches in cold streams; tiny, up to ¾" (2 cm) high, with a distinct cap and stalk, cap 1/16–¼" (1.5–6 mm) wide, hemispheric to convex; yellow, orange, or reddish orange; stalk ¼–⅝" (6–15 mm) long, white to pale translucent gray—*Vibrissea truncorum* Albertini and Schweinitz : Fries (see p. 513).

1. Fruiting body growing in shallow water on decaying needles, twigs, cones, or leaves; head oval, spindle-shaped to elliptic, pear-shaped, irregularly rounded or lobed; smooth or wrinkled, somewhat gelatinous → 2.

1. Fruiting body not with the above combinations of characters → 3.

   2. Fruiting body ¾–2" (2–5 cm) high; head translucent yellow to orange, becoming dull orange in age; stalk 1/16–⅛" (1.5–3 mm) thick, enlarging slightly downward, whitish to pale translucent gray; spores 11–18 × 1.5–3 μm, narrowly elliptic to cylindric or slightly club-shaped, without or with a single septum, smooth, hyaline—*Mitrula elegans* (Berkeley) Fries (see p. 511).

   2. Fruiting body ⅝–1¾" (1.5–4.5 cm) high; head dull orange-yellow to yellow; stalk 1/32–⅛" (1–3 mm) thick, enlarging slightly downward, whitish to pale translucent gray; spores 11–18 × 2.5–5 μm, elliptic to broadly cylindric, without or with a single septum, smooth, hyaline; July–September; edibility unknown—*Mitrula borealis* Redhead.

2. Fruiting body 1–2" (2.5–5 cm) high; head pale watery yellow to pinkish yellow or flesh-pink; stalk ⅟₃₂–⅛" (1–3 mm) thick, enlarging slightly downward, whitish to pale translucent gray, sometimes tinted pink; spores 11–19 × 2–4 μm, crescent- to sausage-shaped or boat-shaped, with or without a septum—*Mitrula lunulatospora* Redhead (see p. 511).

3. Fruiting body ⅟₃₂–⅟₁₆" (1–1.5 mm) wide, ⅛–⅝" (3–15 mm) high, erect, spindle-shaped to cylindric or tine-like; usually simple but sometimes forking, tough, waxy, smooth or longitudinally wrinkled; white to creamy white or pale yellow; stalk poorly defined; growing on algae-covered, wet, decaying wood—*Multiclavula mucida* (Fries) Petersen (see p. 511).

3. Fruiting body similar to the above choice except for the following characters: ⅛–⅜" (3–10 mm) high, typically club-shaped or tine-like; yellow-orange to orange and translucent; scattered or in groups on algae-covered soil—*Multiclavula phycophylla* Leathers (see p. 511).

3. Fruiting body ¼–1" (6–25 mm) high, about ⅛" (2–4 mm) thick, consisting of a tiny head supported by a stalk; head about ⅛" (2–4 mm) thick and high, elliptic to rounded, smooth, shiny to dull, viscid when fresh, white to gray; stalk shiny to dull, scurfy, black; often associated with a small shiny to dull, black, saucer-shaped fruiting body; growing from the cracks in the bark of fallen hardwood trees, especially basswood—*Crinula caliciiformis* (Fries) Fries (see *Holwaya mucida*, p. 493).

3. Fruiting body up to 2" (5 cm) high; head flattened, irregularly wrinkled to folded, fan-shaped, decurrent; yellowish to brownish yellow; stalk ¾–1½" (2–4 cm) long; ¼–⅝" (6–15 mm) thick; tapering downward or nearly equal, solid, reddish brown, velvety, attached to the substrate by a dense, orange, basal mycelium; spores 30–45 × 1.5–2 μm— *Spathularia velutipes* Cooke and Farlow (see p. 512).

3. Fruiting body nearly identical to the above choice except for the following characters: head pale yellow to pale brownish yellow; stalk whitish to pale yellow and smooth, basal mycelium, if present, white to yellowish, not orange; spores 35–50 × 2–2.5 μm; July–October; edibility unknown—*Spathularia flavida* Fries.

3. Fruiting body not with the above combinations of characters → 4.

4. Fruiting body resembling tiny matchsticks, very small, typically less than ⅛" (3 mm) high when mature, black, growing erect on the cap surface of *Trichaptum biformis*, the Violet Toothed Polypore—*Phaeocalicium polyporaeum* (Nylander) Tibell (see p. 512).

4. Fruiting body club-shaped, large, typically 3–8" (7.5–20.5 cm) high when mature, with a rounded to truncate apex, yellowish, orange-yellow to orange or reddish brown, with thick white flesh → 5.

4. Fruiting body consisting of a cap and stalk, 1⅛–4" (3–10 cm) high, fibrous and dry or moist and gelatinous; cap irregularly rounded and flattened, smooth and even to distinctly furrowed or brain-like, with a strongly inrolled margin; pale yellow to olive-brown or green; stalk smooth to finely granular, often fused at the base; scattered or in fused clusters on leaves, litter or soil, usually among mosses → 6.

4. Fruiting body spoon- to lance-shaped or tongue-shaped to cylindric, often compressed to irregularly indented, furrowed or folded; black to brownish black or blackish brown, smooth to minutely spiny → 7.

4. Fruiting body not with the above combinations of characters → 16.

5. Fruiting body 1⅛–2¾" (3–7 cm) high, ¼–¾" (6–20 mm) wide, cylindric to club-shaped or sometimes flattened and tongue-shaped, unbranched, smooth when young, becoming longitudinally wrinkled in age; apex somewhat pointed or blunt; dull yellow to pale orange-yellow or pale salmon to pale brownish orange; flesh thick, firm, white;

odor not distinctive, taste mild to slightly bitter or metallic; stalk poorly defined; spore print white; spores 10–18 × 3–5 μm, narrowly elliptic, smooth, hyaline with yellow oil drops; in groups or clusters conifer needles; July–October; edibility unknown—*Clavariadelphus ligula* (Fries) Donk.

5.   Fruiting body similar to the above choice except for the following characters: 1⅛–6" (3–15.5 cm) high, ¼–1⅜" (6–35 mm) wide, unbranched or branched, apex pale yellow to pale yellow-brown or yellow-green; stalk base covered with abundant white mycelium; spore print brownish orange; spores 15–28 × 4–6 μm, narrowly elliptic and often somewhat curved, smooth, hyaline with yellow oil drops; in groups or clusters on conifer debris; July–October; edibility unknown—*Clavariadelphus sachalinensis* (Imai) Corner.

5.   Fruiting body 2¾–8" (7–20.5 cm) high, ⅜–1¾" (1–4.5 cm) wide, cylindric when young, becoming club-shaped with age, yellowish to orange-yellow, becoming brownish orange to pale reddish brown at maturity; flesh odor not distinctive; taste mild to bitter; scattered or in groups on the ground in woods—*Clavariadelphus pistillaris* (Fries) Donk (see p. 509).

5.   Fruiting body 2–6" (5–15.5 cm) high, 1–2¾" (2.5–7 cm) wide, club-shaped to top-shaped, narrowing downward; apex flattened and often slightly depressed; golden yellow to orange-yellow or pale brownish orange, usually darkest toward the base; flesh white; odor not distinctive; taste sweet or bland; on the ground in conifer woods—*Clavariadelphus truncatus* (Quélet) Donk (see p. 509).

    6.   Cap fleshy-fibrous, not gelatinous, dry, pale yellow to pale brownish orange; stalk ⅝–1¾" (1.5–4.5 cm) long, ⅛–¼" (3–6 mm) thick; spores 40–75 × 2–2.5 μm, needle-like, multiseptate, smooth, hyaline; on leaf litter under hardwoods, especially beech—*Cudonia lutea* (Peck) Saccardo (see p. 509).

    6.   Cap fleshy-fibrous, not gelatinous, dry; fruiting body nearly identical to the above choice except for the following characters: cap creamy yellow to pale brown; stalk pale brown; spores 30–45 × 1.8–2.2 μm, needle-like, multiseptate; on the ground or decaying wood under conifers; edibility unknown—*Cudonia circinans* (Persoon) Fries.

    6.   Cap gelatinous, moist, pale dull yellow to orange-yellow, sometimes with olive tints; stalk pale dull yellow to orange-yellow, nearly smooth; spores 18–25 × 4–6 μm, cylindric-oblong to spindle-shaped, often curved, multiseptate at maturity; in groups or clusters on the ground in conifer or hardwoods; July–October—*Leotia lubrica* Persoon : Fries (see p. 510).

    6.   Cap gelatinous, moist, dark green to olive-green; stalk yellow to orange-yellow, nearly smooth; spores 15–25 × 4–6 μm, spindle-shaped with rounded ends, often curved, multiseptate at maturity; in groups or clusters on the ground or decaying wood; July–October; edibility unknown—*Leotia viscosa* Fries.

    6.   Cap gelatinous, moist, pea-green to bluish green; stalk pea-green to bluish green or pale green to whitish, conspicuously scaly or roughened with tiny granules; spores 18–20 × 5–6 μm, narrowly elliptic to spindle-shaped with rounded ends, straight or curved, multiseptate at maturity; in groups or clusters on soil or among mosses—*Leotia atrovirens* Fries (see p. 510).

7.   Surface of the head appearing minutely spiny when examined with a hand lens; brown setae present among the asci when examined microscopically (continued use of this key requires the use of a microscope) → 8.

7.   Surface of the head appearing smooth or velvety but not minutely spiny when examined with a hand lens; setae absent among the asci when examined microscopically (continued use of this key requires the use of a microscope) → 11.

    8.   Spores needle-like, 8–17 septate or if variable, many with more than 8 septa; on

      soil, decaying wood or sphagnum mosses; usually collected July–October; edibility unknown → 9.

8.   Spores needle-like, 0–7 septate, rarely more; on soil, decaying wood or among mosses; usually collected July–October; edibility unknown → 10.

9.   Asci 4-spored or sometimes fewer; spores 115–145 × 6–7 μm, 0–17 septate but mostly 15-septate—*Trichoglossum tetrasporum* Sinden and Fitzpatrick.

9.   Asci 4-spored; spores 95–150 × 6–7 μm, usually 7–11 septate—*Trichoglossum velutipes* (Peck) Durand.

9.   Asci 8-spored; spores 80–200 × 5–7 μm, mostly 15-septate—*Trichoglossum hirsutum* (Fries) Boudier.

9.   Asci 8-spored; spores 80–150 × 4.5–6 μm, 4–16 septate, most often 10–14 septate—*Trichoglossum variabile* (Durand) Nannfeldt.

    10.   Spores 45–85 × 6–7 μm, 0–6 septate, mostly 3-septate—*Trichoglossum farlowii* (Cooke) Durand (see p. 512).

    10.   Spores 60–125 × 5–6 μm, mostly 7-septate—*Trichoglossum walteri* (Berkeley) Durand.

11.   Fruiting body with both hyaline and brownish needle-like spores; on soil, humus, or decaying wood; edibility unknown → 12.

11.   Fruiting body with brownish needle-like spores only; growing on soil, on decaying wood or among mosses; edibility unknown → 13.

    12.   Spores 50–105 × 5–7 μm; hyaline spores aseptate; brownish spores 0–12 septate—*Geoglossum fallax* Durand.

    12.   Spores 45–75 × 5.5–6 μm; hyaline spores 0–6 septate; brownish spores 7–12 septate—*Geoglossum intermedium* Durand.

    12.   Spores 55–95 × 4–5 μm, spores remaining hyaline for a long time; hyaline spores 0–15 septate; brownish spores 8–15 septate—*Geoglossum alveolatum* (Rehm) Durand.

13.   Spores 3–7 septate → 14.

13.   Spores 10–16 septate → 15.

    14.   Spores 55–102 × 5–6 μm, 0–3 septate when young, typically 7-septate at maturity—*Geoglossum glutinosum* Fries.

    14.   Spores 45–60 × 5–6 μm, mostly 7-septate—*Geoglossum affine* Durand.

    14.   Spores 65–80 × 5–6 μm, mostly 7–8 septate—*Geoglossum umbratile* Saccardo = *G. nigritum* (Fries) Cooke.

    14.   Spores 50–105 × 7–9 μm, mostly 7-septate; upper cells of paraphyses enlarged and rounded to obovoid—*Geoglossum glabrum* Fries.

    14.   Spores 60–102 × 6–9 μm, mostly 7-septate; paraphyses with many barrel-shaped, 2-celled segments—*Geoglossum simile* Peck.

15.   Spores 75–125 × 6–7 μm, mostly 12–15 septate—*Geoglossum difforme* Fries (see p. 509).

15.   Spores 120–175 × 6–7 μm, mostly 12–15 septate—*Geoglossum pygmaeum* Gerard and Durand.

    16.   Fruiting body 1–2⅜" (2.5–6 cm) high, olive-green to dark green, consisting of a head and stalk; head ⅛–½" (3–10 mm) wide, ¾–1⅛" (2–3 cm) high, cylindric to oval or irregularly rounded, dull or shiny, longitudinally furrowed or compressed and irregularly folded; stalk 1⅛–1½" (3–4 cm) long, ⅛–¼" (3–6 mm) thick, scaly-granular; spores 12–22 × 5–6 μm, cylindric-oblong to oblong-elliptic, 0–4 septate, straight or curved, smooth, hyaline; scattered or in groups on the ground in woods; July–September; edibility unknown—*Microglossum viride* Fries.

    16.   Fruiting body similar to the above choice except for the following characters: head

greenish brown to olive-brown or brownish yellow; stalk smooth, shiny, pale brownish yellow to pale brownish olive; spores 12–18 × 4–6 μm, spindle-shaped to oblong, 0–3 septate, straight or curved, smooth, hyaline; solitary, scattered, or in groups on the ground in woods; July–October; edibility unknown—*Microglossum olivaceum* (Persoon) Gillet.

16. Fruiting body similar to the above two choices except for the following characters: head dull yellow-brown to brownish yellow; stalk brownish yellow, smooth and shiny or slightly roughened and dull; spores 20–50 × 5–6 μm, cylindric to needle-like, aseptate when young, becoming 7–15 septate at maturity, curved or straight, smooth, hyaline; July–September; edibility unknown—*Microglossum fumosum* (Peck) Durand.

16. Fruiting body ¾–1⅛" (2–3 cm) high, up to ⅛" (3 mm) wide, consisting of a cylindric to worm-like or spindle-shaped head with a rounded apex and a stalk-like base that tapers downward; head pale yellow when young, becoming reddish blonde in age; growing in groups on soil in plant pots in greenhouses—*Clavaria oronoensis* Petersen and Litten (see p. 508).

16. Not as above; fruiting body yellow or pale brownish yellow to orange → 17.

16. Not as above; fruiting body white to grayish white, yellowish or pale brown → 18.

17. Fruiting body ¾–2¾" (2–7 cm) high, consisting of a head and stalk; head ⅛–⅝" (3–16 mm) wide, ⅜–1⅜" (1–3.5 cm) high, spoon-shaped to tongue-shaped or cylindric, compressed in the center, yellow-orange to orange; stalk yellow, scaly-granular; scattered, in groups or clusters on humus, on decaying wood, or among mosses—*Microglossum rufum* (Schweinitz) Underwood (see p. 510).

17. Fruiting body ¾–2" (2–5 cm) high, ⅛–⅜" (3–10 mm) wide, club-shaped to cylindric, unbranched or sometimes branched, round in cross section when young, becoming flattened in age, with blunt tips, tapering downward as a stalk-like base; pale greenish yellow; scattered or in groups on soil or among mosses under blueberry, sheep laurel, huckleberry, and other heath family plants—*Clavaria argillacea* Persoon (see p. 508).

17. Fruiting body up to 4⅜" (11 cm) high, 1/32–3/32" (0.5–2 mm) thick, cylindric to worm-like; apex typically blunt, smooth, pale brownish yellow, rigid and erect when young, becoming somewhat limp in age; narrowing below to form a stalk-like base; flesh fibrous-tough, odor and taste somewhat sour; spores 6–11 × 3.5–5 μm; scattered or in groups on leaf litter and debris in hardwoods and mixed woods; edibility unknown—*Macrotyphula juncea* (Fries) Berthier.

17. Fruiting body ¾–3⅛" (2–8 cm) high, ⅛–¼" (3–6 mm) wide, consisting of a cylindric to worm-like or spindle-shaped head and a stalk-like base that tapers downward, unbranched, tips pointed or blunt, sometimes flattened, smooth; orange-yellow to orange; spores 4.5–7 × 3.5–5 μm, oval or irregularly rounded with a prominent apiculus, smooth, hyaline to pale yellow; solitary, in groups, or sometimes clustered on soil in woods; July–October; edibility unknown—*Clavulinopsis laeticolor* (Berkeley and Curtis) Petersen = *C. pulchra* (Peck) Corner.

17. Fruiting body nearly identical to the above choice except for the following characters: orange-yellow to yellow with apricot tints, tips pointed, unbranched or sometimes branched; spores 5.5–8 × 4.5–6 μm, broadly elliptic to nearly round, with a prominent apiculus, distinctly warted, hyaline; solitary, in groups, or sometimes clustered on soil or among mosses in woods; July–October; edibility unknown—*Clavulinopsis helveola* (Fries) Corner.

17. Fruiting body ¾–3⅛" (2–8 cm) high, consisting of a broadly club-shaped, spoon-shaped, or lobed to branched head and a stalk-like base that tapers downward; head ¼–

1⅛" (6–30 mm) wide, smooth or longitudinally furrowed to folded; often irregularly contorted or flattened; bright yellow to orange-yellow; spores 6–10 × 3.5–5 μm, elliptic to oval or kidney-shaped, smooth, hyaline; on needles and among mosses in conifer woods—*Neolecta irregularis* (Peck) Korf and Rogers (see p. 512).

17. Fruiting body similar to the above choice except for the following characters: 1⅛–1⅜" (3–3.5 cm) high; head ⅛–⅜" (3–10 mm) wide, narrowed at the apex, pale yellow; spores of two types; ascospores 5.5–9 × 3–4 μm, elliptic to oval or kidney-shaped; conidia 2 μm, round, smooth, hyaline; August–October; edibility unknown—*Neolecta vitellina* (Bresadola) Korf and Rogers.

    18. Fruiting body ⅜–1¾" (1–4.5 cm) high, up to ⅜" (1 cm) wide, consisting of a rounded to compressed, balloon-like head on a tiny stalk; head hollow, smooth, delicate, often partially collapsed and folded or wavy, white, becoming yellowish in age; stalk up to ¾" (2 cm) long, less than 1/32" (0.5 mm) thick; spore print white; spores 4–6 × 2–4 μm; in clusters on decaying wood or sometimes leaves; July–September; edibility unknown—*Physalacria inflata* (Schweinitz) Fries.

    18. Fruiting body up to 5/16" (8 mm) high, consisting of a tiny head and stalk; head 1/16–⅛" (1.5–3 mm) wide, nearly round to flattened or somewhat irregular, finely cracked and roughened, dry, white when young, becoming grayish white to pale brown at maturity; stalk up to ¼" (6 mm) long, less than 1/32" (0.5 mm) thick; in groups or clusters on decaying hardwoods—*Phleogena faginea* (Fries) Link (see Jelly Fungi, p. 432).

Clavaria argillacea Persoon (see photo, p. 21)

COMMON NAME: Moor Club.

FRUITING BODY: ¾–2" (2–5 cm) high, ⅛–⅜" (3–10 mm) wide, cylindric and round in cross section when young, becoming flattened in age, often with shallow, longitudinal furrows at maturity, unbranched or sometimes forked, with blunt tips, tapering downward as a stalk-like base, smooth; pale greenish yellow; flesh pale yellow.

MICROSCOPIC FEATURES: spores 8–12 × 4–6 μm, elliptic to cylindric, smooth, hyaline.

FRUITING: scattered or in groups on soil or among mosses under blueberry, sheep laurel, huckleberry, and other heath family plants; July–November; fairly common.

EDIBILITY: edible.

Clavaria oronoensis Petersen and Litten

FRUITING BODY: ¾–1⅛" (2–3 cm) high, up to ⅛" (3 mm) wide, consisting of a cylindric to worm-like or spindle-shaped head with a rounded apex and a stalk-like base that tapers downward; fertile head portion pale yellow and smooth when young, becoming reddish blonde and longitudinally wrinkled in age, dark reddish brown at the apex; stalk-like base smooth, whitish.

MICROSCOPIC FEATURES: spores 6–8 × 5.5–6 μm, nearly round to broadly elliptic, smooth or distinctly spiny, hyaline.

FRUITING: in groups on soil in plant pots containing ferns, mosses, and lowbush blueberry *(Vaccinium angustifolium* Ait.) in a greenhouse at Utica College of Syracuse University; April–May; September–October; frequency unknown.

EDIBILITY: unknown.

COMMENTS: this species was first observed in plant pots with lowbush blueberry in a greenhouse of the University of Maine at Orono.

Clavariadelphus pistillaris (Fries) Donk

COMMON NAME: Pestle-shaped Coral.

FRUITING BODY: 2¾–8" (7–20.5 cm) high, ⅜–1¾" (1–4.5 cm) wide, cylindric when young, becoming club-shaped with age, unbranched or rarely forked, smooth to longitudinally wrinkled; apex usually rounded and inflated; yellowish to orange-yellow, becoming brownish orange to pale reddish brown at maturity, slowly staining brownish when bruised; flesh thick, firm or spongy, white, staining brownish when cut; odor not distinctive; taste mild to bitter; stalk poorly defined, white.

SPORE PRINT: white to creamy white.

MICROSCOPIC FEATURES: spores 9–16 × 5–9 μm, elliptic, smooth, hyaline with yellow oil drops.

FRUITING: scattered or in groups on the ground, usually in hardwoods; July–October; occasional.

EDIBILITY: edible, but sometimes bitter or unpleasant.

Clavariadelphus truncatus (Quélet) Donk

COMMON NAME: Flat-topped Coral.

FRUITING BODY: 2–6" (5–15.5 cm) high, 1–2¾" (2.5–7 cm) wide, club-shaped to top-shaped, narrowing downward; typically unbranched but sometimes forking, smooth near the base, becoming longitudinally wrinkled upward; apex flattened and often slightly depressed; golden yellow to orange-yellow or pale brownish orange, usually darkest toward the base; flesh thick, firm or spongy, white; odor not distinctive; taste sweet or bland; stalk poorly defined, white, often with a dense, white, basal mycelium embedded in the substrate.

SPORE PRINT: pale brownish yellow.

MICROSCOPIC FEATURES: spores 9–12 × 5–8 μm, broadly elliptic, smooth, hyaline with yellow oil drops.

FRUITING: scattered or in groups on the ground in conifer woods; August–October; infrequent.

EDIBILITY: edible.

Cudonia lutea (Peck) Saccardo

COMMON NAME: Yellow Cudonia.

CAP: ⅜–⅝" (1–1.5 cm) wide, irregularly rounded and flattened, smooth and even to distinctly furrowed or brain-like, with a strongly inrolled margin; fleshy-fibrous, dry, not gelatinous; pale yellow to pale brownish orange.

STALK: ⅝–1¾" (1.5–4.5 cm) long, ⅛–¼" (3–6 mm) thick, finely granular to smooth, pale yellow.

MICROSCOPIC FEATURES: spores 40–75 × 1.8–2.5 μm, needle-like, multiseptate, smooth, hyaline.

FRUITING: in groups or clusters on leaf litter under hardwoods, especially beech; July–September; occasional.

EDIBILITY: unknown.

COMMENTS: compare with *Leotia lubrica,* which is moist and gelatinous (see p. 510).

Geoglossum difforme Fries (see photo, p. 21)

FRUITING BODY: 1⅛–2¾" (3–7 cm) high, black, dull when young, becoming smooth and shiny at maturity, with a head and stalk.

HEAD: ⅛–⅜" (3–10 mm) wide, ¾–1⅜" (2–3.5 cm) high, lance- to spoon-shaped or tongue-shaped to cylindric, compressed at the center or irregularly folded.

STALK: 1–1¾" (2.5–4.5 cm) long, 1⁄16–¼" (1.5–6 mm) thick, nearly equal or enlarging in either direction, slightly roughened or smooth.

MICROSCOPIC FEATURES: spores 75–125 × 6–7 μm, needle-like, mostly 12–15 septate, smooth, brownish; setae absent.

FRUITING: scattered or in groups on soil, decaying wood, and conifer needles or among mosses; August–October; fairly common.

EDIBILITY: unknown.

Leotia atrovirens Fries (see photo, p. 21)

CAP: ¼–½" (6–12 mm) wide, irregularly rounded and flattened, smooth and even to distinctly furrowed or brain-like, with a strongly inrolled margin; moist, gelatinous; pea-green to bluish green.

STALK: ⅜–1¾" (1–4.5 cm) long, ¼–⅜" (6–10 mm) thick, conspicuously scaly or roughened with tiny granules; pea-green to bluish green or pale green to whitish.

MICROSCOPIC FEATURES: spores 18–20 × 5–6 μm, narrowly elliptic to spindle-shaped with rounded ends, straight or curved, multiseptate at maturity, smooth, hyaline.

FRUITING: in groups or clusters on soil or among mosses in woods; July–September; occasional.

EDIBILITY: unknown.

Leotia lubrica Persoon : Fries

COMMON NAME: Ochre Jelly Club, Jelly Babies.

CAP: ¼–1⅛" (6–30 mm) wide, irregularly rounded and flattened, smooth to distinctly furrowed or brain-like, with a strongly inrolled margin; moist, gelatinous, pale dull yellow to orange-yellow, sometimes with olive tints.

STALK: ¾–2" (2–5 cm) long, ¼–⅜" (6–10 mm) thick, enlarged downward, slippery, smooth or nearly so, pale dull yellow to orange-yellow.

MICROSCOPIC FEATURES: spores 18–25 × 4–6 μm, cylindric-oblong to spindle-shaped, often curved, multiseptate at maturity, hyaline.

FRUITING: in groups or clusters on the ground or on decaying wood under conifers and hardwoods; July–October; fairly common.

EDIBILITY: unknown.

Microglossum rufum (Schweinitz) Underwood

COMMON NAME: Orange Earth Tongue.

FRUITING BODY: ¾–2¾" (2–7 cm) high, consisting of a head and stalk.

HEAD: ⅛–⅝" (3–16 mm) wide, ⅜–1⅜" (1–3.5 cm) high, spoon- to tongue-shaped or cylindric, compressed at the center or longitudinally furrowed, smooth, dull or shiny, yellow-orange to orange.

STALK: ⅝–1¾" (1.5–4.5 cm) long, 1⁄16–3⁄16" (1.5–5 mm) thick, nearly equal, yellow to orange-yellow, scaly-granular.

MICROSCOPIC FEATURES: spores 18–38 × 4–6 μm, sausage- to spindle-shaped, aseptate when young, becoming 5–10 septate in age, smooth, hyaline.

FRUITING: scattered, in groups, or in clusters on humus, on decaying wood, or among mosses; July–September; fairly common.

EDIBILITY: unknown.

Mitrula elegans (Berkeley) Fries

FRUITING BODY: ¾–2" (2–5 cm) high, consisting of a head and stalk.

HEAD: ⅛–½" (3–12 mm) wide; ¼–¾" (6–20 mm) high, spindle-shaped to elliptic, pear-shaped or irregularly rounded or lobed; somewhat gelatinous, smooth to slightly wrinkled, shiny; translucent yellow to orange, becoming dull orange in age.

STALK: ¾–1½" (2–4 cm) long, ¹⁄₁₆–⅛" (1.5–3 mm) thick, enlarging slightly downward, smooth, shiny; whitish to pale translucent gray, sometimes tinted pink.

MICROSCOPIC FEATURES: spores 11–18 × 1.5–3 μm, narrowly elliptic to cylindric or slightly club-shaped, single-celled or two-celled with a septum, smooth, hyaline.

FRUITING: scattered or in groups in shallow water on decaying leaves, needles, twigs, and debris in woodlands and bogs; April–August; occasional.

EDIBILITY: unknown.

COMMENTS: this species has been incorrectly identified in some field guides as *Mitrula paludosa*, a species that has broader spores and to date is known only from Europe.

Mitrula lunulatospora Redhead

FRUITING BODY: 1–2" (2.5–5 cm) high, consisting of a head and stalk.

HEAD: ⅛–⅜" (3–10 mm) wide, ¼–¾" (6–20 mm) high, oval to pear-shaped or club-shaped to irregularly rounded; somewhat gelatinous, smooth to slightly wrinkled, shiny; pale watery yellow to pinkish yellow or flesh-pink.

STALK: ¾–1½" (2–4 cm) long, ¹⁄₃₂–⅛" (1–3 mm) thick, enlarging slightly downward, smooth, shiny; whitish to pale translucent gray, sometimes tinted pink.

MICROSCOPIC FEATURES: spores 11–19 × 2–4 μm, crescent- to sausage-shaped or boat-shaped, single-celled or two-celled with a septum, smooth, hyaline.

FRUITING: scattered or in groups in shallow water on decaying scales, leaves, twigs, and cones in woodlands and bogs; May–August; infrequent.

EDIBILITY: unknown.

Multiclavula mucida (Fries) Petersen

COMMON NAME: White Green-algae Coral.

FRUITING BODY: ¹⁄₃₂–¹⁄₁₆" (1–1.5 mm) wide, ⅛–⅝" (3–15 mm) high, erect, spindle-shaped to cylindric or tine-like, usually unbranched, tough, waxy, smooth or longitudinally wrinkled; white to creamy white or pale yellow; stalk poorly defined.

MICROSCOPIC FEATURES: spores 5–7.5 × 2–3 μm, narrowly elliptic, smooth, hyaline.

FRUITING: scattered or in groups on algae-covered, wet, decaying wood; July–September; fairly common.

EDIBILITY: unknown.

COMMENTS: also known as *Clavaria mucida*.

Multiclavula phycophylla Leathers

FRUITING BODY: ¹⁄₃₂–¹⁄₁₆" (1–1.5 mm) wide, ⅛–⅜" (3–10 mm) high, erect, club-shaped to tine-like, usually unbranched, rarely branched, tough, waxy, smooth or longitudinally wrinkled; yellow-orange to orange and translucent; stalk poorly defined.

MICROSCOPIC FEATURES: spores 6.5–10 × 2–3.5 μm, elliptic to sausage-shaped, smooth, hyaline.

FRUITING: scattered or in groups on algae-covered soil; June–September; infrequent.

EDIBILITY: unknown.

COMMENTS: also known as *Clavaria phycophylla*.

Neolecta irregularis (Peck) Korf and Rogers
> COMMON NAME: Irregular Earth Tongue.
> FRUITING BODY: ¾–3⅛" (2–8 cm) high, consisting of a head and stalk-like base; head ¼–1⅛" (6–30 mm) wide, broadly club-shaped, spoon-shaped, or lobed to branched; smooth or longitudinally furrowed to folded, often irregularly contorted or flattened; bright yellow to orange-yellow; stalk-like base whitish to pale yellow, rudimentary or absent, tapering downward.
> MICROSCOPIC FEATURES: spores 6–10 × 3.5–5 μm, elliptic to oval or kidney-shaped, smooth, hyaline.
> FRUITING: solitary, scattered, or in groups on the ground or among mosses in conifer woods; July–October; fairly common.
> EDIBILITY: unknown.
> COMMENTS: also known as *Spragueola irregularis.*

Phaeocalicium polyporaeum (Nylander) Tibell
> FRUITING BODY: very small and often overlooked, resembling tiny black matchsticks, consisting of a tiny head on a very slender stalk, growing erect on the cap surface of the Violet Toothed Polypore, *Trichaptum biforme.*
> HEAD: 1/64–1/32" (0.2–0.6 mm) wide and high, rounded to bluntly oval or cup-like, dull, black.
> STALK: about 1/16" (1–2 mm) long, less than 1/64" (0.3 mm) thick, enlarging downward or nearly equal, dull, black.
> MICROSCOPIC FEATURES: spores 9–18 × 3–4 μm, cylindric to narrowly elliptic with obtuse ends, mostly 1-septate, smooth, pale brown.
> FRUITING: in rows, scattered, or in groups on *T. biforme*; June–October; fairly common.
> EDIBILITY: unknown and most likely very difficult to locate in the frying pan!

Spathularia velutipes Cooke and Farlow (see photo, p. 21)
> COMMON NAME: Velvety Fairy Fan.
> HEAD: ⅜–1⅛" (1–3 cm) wide; ⅜–1½" (1–4 cm) high, flattened, irregularly wrinkled to folded, fan- to spoon-shaped, decurrent; yellowish to brownish yellow.
> STALK: ¾–1½" (2–4 cm) long, ¼–⅝" (6–15 mm) thick, narrowing downward or nearly equal, solid, velvety; reddish brown, with a dense, orange, basal mycelium.
> MICROSCOPIC FEATURES: spores 30–45 × 1.5–2 μm, needle-like, multiseptate, hyaline.
> FRUITING: in groups or clusters on decaying hardwood or on mossy soil; July–September; fairly common.
> EDIBILITY: unknown.
> COMMENTS: also known as *Spathulariopsis velutipes.*

Trichoglossum farlowii (Cooke) Durand
> FRUITING BODY: 1⅛–3⅛" (3–8 cm) high, black to brownish black, with a head and stalk.
> HEAD: ⅛–⅜" (3–10 mm) wide, ½–1⅛" (1.2–3 cm) high, spoon- to lance-shaped or tongue-shaped to cylindric, compressed in the center or irregularly folded, appearing minutely spiny when examined with a hand lens.
> STALK: 1–2⅜" (2.5–6 cm) long, 1/16–3/16" (1.5–5 mm) thick, nearly equal or enlarging in either direction, smooth when young, becoming longitudinally furrowed, velvety in age, appearing minutely spiny when examined with a hand lens.
> MICROSCOPIC FEATURES: spores 45–85 × 6–7 μm, needle-like, 0–6 septate, mostly 3-septate, smooth, grayish or brownish; setae more than 150 μm long, pointed, brown.

FRUITING: scattered or in groups on the ground, on decaying wood, or among mosses; July–October; fairly common.

EDIBILITY: unknown.

Underwoodia columnaris Peck

COMMON NAME: Fluted-stalked Fungus.

FRUITING BODY: 1½–4" (4–10 cm) high, ⅜–1⅛" (1–3 cm) thick, cylindric to spindle-shaped, tapering upward to a rounded tip, with longitudinal grooves or wrinkles, lacking a distinct cap and stalk, cream to tan when young, becoming pale brown in age, interior chambered.

MICROSCOPIC FEATURES: spores 25–27 × 12–14 μm, elliptic, coarsely warted, hyaline.

FRUITING: in clusters with fused bases or sometimes scattered on the ground in hardwoods; June–August; rare.

EDIBILITY: unknown.

Vibrissea truncorum Albertini and Schweinitz : Fries

COMMON NAME: Water Club.

CAP: ¹⁄₁₆–¼" (1.5–6 mm) wide, hemispheric to convex with an inrolled margin, smooth; yellow, orange or reddish orange.

STALK: ¼–⅝" (6–15 mm) long, ¹⁄₃₂–¹⁄₁₆" (0.75–1.5 mm) thick, nearly equal, slightly roughened, white to pale translucent gray.

MICROSCOPIC FEATURES: spores of highly variable length, 120–260 × 1–1.5 μm, needle-like, multiseptate, hyaline.

FRUITING: scattered, in groups, or clustered on partially or completely submerged branches in cold streams; June–August; infrequent.

EDIBILITY: unknown.

Clavaria oronoensis

Clavariadelphus truncatus

Clavariadelphus pistillaris

Cudonia lutea

Leotia lubrica

Microglossum rufum

Mitrula elegans

Mitrula lunulatospora

Multiclavula mucida

Multiclavula phycophylla

Neolecta irregularis

Phaeocalicium polyporaeum

Trichoglossum farlowii

Underwoodia columnaris

Vibrissea truncorum

Cordyceps, Claviceps, and Allies

Cordyceps, Claviceps, and Allies are a group that belong to the class Pyrenomycetes, commonly called the Flask Fungi. Their fertile surfaces are finely roughened like sandpaper due to the protruding necks of numerous flask-shaped reproductive structures called perithecia. *Cordyceps* species are parasitic on other fungi, on the pupae and larvae of moths and butterflies, or on the larvae and adult stages of beetles. *Claviceps purpurea* Fries is parasitic on the inflorescences of many species of grasses. One member, *Podostroma alutaceum* (Persoon : Fries) Atkinson, is not parasitic and grows on the ground or on decaying wood.

Key to Species of *Cordyceps, Claviceps,* and Allies

1. Fruiting body blackish; growing on the inflorescences of grasses—*Claviceps purpurea* Fries (see p. 518).
1. Fruiting body whitish to yellowish, becoming pale yellow-orange to pale brownish orange in age; ⅜–2" (1–5 cm) tall, 3⁄16–⅜" (5–10 mm) wide, cylindric to club-shaped; finely roughened like sandpaper on the upper portion; solitary or in small groups on the ground or on decaying wood—*Podostroma alutaceum* (Persoon : Fries) Atkinson (see p. 519).
1. Fruiting body yellow to reddish orange; cylindric or spindle- to club-shaped, attached to buried insects → 2.
1. Fruiting body with a yellowish brown to reddish brown or olive-black head and a yellow stalk; attached to buried, walnut-shaped, reddish brown false truffles → 3.
    2. Fruiting body pale to bright yellow or golden yellow to pale ochre-yellow; 2⅜–3½" (6–9 cm) tall, ⅜–⅝" (1–1.5 cm) wide, spindle-shaped; head oval to club-shaped, finely roughened like sandpaper; stalk ⅛–⅜" (3–10 mm) wide, smooth; attached to the larval or adult stages of beetles; spores 4–9 × 1–1.5 μm; inedible— *Cordyceps melolonthae* (Tulasne) Saccardo.
    2. Fruiting body reddish orange; ¼–⅜" (6–10 mm) wide; attached to buried pupae or larvae of moths or butterflies—*Cordyceps militaris* (Linnaeus) Link (see p. 518).
3. Head of fruiting body spindle-shaped to oval or club-shaped—*Cordyceps ophioglossoides* (Ehrhart : Fries) Link (see p. 518).
3. Head of fruiting body irregularly rounded; spores 16–28 × 2.5–3.5 μm—*Cordyceps capitata* (Holmskjold : Fries) Link (see p. 518).
3. Head of fruiting body irregularly rounded; spores 30–65 × 3–5 μm; inedible—*Cordyceps longisegmentis* Ginns.

Claviceps purpurea Fries (see photo, p. 22)

> COMMON NAME: Ergot.
>
> FRUITING BODY: ⅝–1⅜" (1.5–3.5 cm) long, cylindric with tapering, rounded ends, typi-
> cally curved, with shallow, longitudinal grooves, wrinkled and finely roughened like
> sandpaper, hard; purplish black to brownish black; white within.
>
> MICROSCOPIC FEATURES: asci elongated-cylindric, thick-walled, tips blueing in Melzer's
> reagent, 8-spored; ascospores 65–95 × 0.5–1 μm, hair-like, smooth, hyaline.
>
> FRUITING: solitary to clustered on the inflorescences of many species of grasses, especially
> rye, wheat, and barley; July–April; fairly common.
>
> EDIBILITY: deadly poisonous, causing ergotism or St. Anthony's Fire, which may be fatal.
>
> COMMENTS: this fungus is believed by some authors to be the cause of the abnormal
> behavior exibited by individuals accused of being "witches" in Salem, Massachusetts,
> during the 1600's.

Cordyceps capitata (Holmskjold : Fries) Link (see photo, p. 22)

> COMMON NAME: Head-like Cordyceps.
>
> HEAD: ¼–¾" (6–20 mm) wide and tall, irregularly rounded, finely roughened like sand-
> paper, dark reddish brown to olive-black.
>
> FLESH: white, thick, firm; odor and taste not distinctive.
>
> STALK: ¾–3⅛" (2–8 cm) long, ¼–⅝" (5–16 mm) thick, nearly equal overall, smooth to
> slightly ridged, fibrous, yellow to dull yellow, becoming olive-brown in age.
>
> MICROSCOPIC FEATURES: spores 16–28 × 2.5–3.5 μm, cylindric to thread-like, smooth, hya-
> line, inamyloid.
>
> FRUITING: solitary or in groups in woods on buried, walnut-shaped, reddish brown fruit-
> ing bodies of *Elaphomyces* species (false truffles); August–November; uncommon.
>
> EDIBILITY: unknown.
>
> COMMENTS: this fungus is parasitic on its host and must be carefully dug up to retrieve
> the entire structure. *Cordyceps longisegmentis* Ginns is macroscopically identical, but is
> easily identified by microscopic examination of its spores, which are 30–65 × 3–5 μm.

Cordyceps militaris (Linnaeus) Link (see photo, p. 22)

> COMMON NAME: Trooping Cordyceps.
>
> FRUITING BODY: ¼–⅜" (6–10 mm) wide, up to 3" (7.5 cm) tall, cylindric to club-shaped
> or spindle-shaped, reddish orange, finely roughened like sandpaper on the upper por-
> tion, smooth and tapering downward on the lower portion.
>
> MICROSCOPIC FEATURES: spores 250–350 × 1–1.5 μm, hair-like, usually breaking into
> barrel-shaped spores, 3.5–6 × 1–1.5 μm, smooth, hyaline, inamyloid.
>
> FRUITING: solitary to grouped on the pupae and larvae of moths and butterflies; August–
> November; fairly common.
>
> EDIBILITY: unknown.
>
> COMMENTS: this parasitic fungus is often collected without its host, which is shallowly
> buried in soil or moss.

Cordyceps ophioglossoides (Ehrhart : Fries) Link (see photo, p. 22)

> COMMON NAME: Goldenthread Cordyceps.
>
> HEAD: ⅜–⅝" (1–1.6 cm) wide, ¾–1" (2–2.5 cm) tall, spindle-shaped to oval or club-
> shaped, finely roughened like sandpaper, yellowish brown to dark reddish brown,
> becoming olive-black in age.
>
> FLESH: whitish, thick, firm.

STALK: 1–3⅛" (2.5–8 cm) long, ⅛–⅜" (3–10 mm) thick, nearly equal overall, smooth; yellow, with golden yellow, cord-like basal threads.

MICROSCOPIC FEATURES: spores 2–5 × 1.5–2 μm, elliptic, smooth, hyaline, inamyloid.

FRUITING: solitary or in groups in woods on buried, walnut-shaped, reddish brown fruiting bodies of *Elaphomyces* species (false truffles); August–November; uncommon.

EDIBILITY: unknown.

COMMENTS: this fungus is parasitic on its host and must be carefully dug up to retrieve the entire structure.

Podostroma alutaceum   (Persoon : Fries) Atkinson

FRUITING BODY: ³⁄₁₆–⅜" (5–10 mm) wide, ⅜–2" (1–5 cm) tall, cylindric to club-shaped; upper portion fertile, finely roughened like sandpaper, whitish to yellowish, becoming pale yellow-orange to pale brownish orange in age; stalk-like base smooth, whitish to pale yellow, sometimes darkening to brownish in age.

MICROSCOPIC FEATURES: spores 3.5–5 × 3–4 μm, variable, often flattened on one end, broadly oval to subglobose, weakly punctate, hyaline.

FRUITING: solitary or in small groups on the ground among leaves and needle litter or on decaying wood; August–October; uncommon.

EDIBILITY: unknown.

COMMENTS: easily confused with *Cordyceps* species, but not attached to insects or hypogeous hosts.

Podostroma alutaceum

Carbon and Cushion Fungi

Most Carbon and Cushion Fungi belong to the class Pyrenomycetes, commonly called the Flask Fungi. They are closely related to the *Cordyceps, Claviceps,* and Allies. Their fertile surfaces are finely roughened like sandpaper, due to the protruding necks of numerous flask-shaped reproductive structures called perithecia. They occur on decaying wood (usually hardwood) and sometimes on the surrounding ground. Three species that grow on leaves and belong to the Cup and Disk Fungi are also keyed out here because they share several characteristics of this group.

Although two species of Carbon and Cushion Fungi may be gelatinous or powdery, most are fibrous-tough to woody, and several are hard, black, and carbonaceous. Most species are inedible. Some resemble Crust Fungi but differ by having a finely roughened, sandpaper-like surface. Most members are cushion-shaped, but a few have distinct stalks. Two species do not form perithecia but are included here because of their cushion shape.

Key to Species of Carbon and Cushion Fungi

1. Fruiting body cushion-shaped or spherical to oval, lacking a stalk; gelatinous to fibrous-tough or hard and woody; green, yellow, orange-yellow, orange, orange-red, brick-red to reddish brown, or black → 2.
1. Fruiting body crust-like, irregular in outline to somewhat circular, spreading over the substrate, lacking a stalk; creamy white to lemon-yellow, brick-red to rusty brown, or black → 5.
1. Fruiting body hard, black, carbonaceous, with or without a stalk, cylindric to club-shaped, irregularly clavate to antler-shaped or with a crown of long, tapering tentacles; growing on or near decaying wood or humus → 7.
    2. Fruiting body ¾–2⅛" (2–5.5 cm) wide, cushion-shaped, oval or irregular, grayish brown to yellowish brown, splitting open to expose a shiny, black, slightly roughened inner surface; flesh reddish brown, not concentrically zoned when cut vertically; odor disagreeable, creosote-like; spores 6–8 × 3.5–4 μm; solitary or in groups on decaying hardwoods, especially elm; fairly common—*Camarops petersii* (Berkeley and Curtis) Nannfeldt = *Peridoxylon petersii* (Berkeley and Curtis) Shear.
    2. Fruiting body ¾–2" (2–5 cm) wide, cushion-shaped to nearly round or irregular, reddish brown to black; flesh concentrically zoned when cut vertically, fibrous to powdery and carbon-like, dark purplish brown alternating with darker or sometimes whitish zones—*Daldinia concentrica* (Bolton : Fries) Cesati and de Notaris (see p. 524).

2.   Fruiting body ⅟₃₂–⅛" (1–3 mm) wide, cushion-shaped to nearly round; pale yellow, developing green dots, and finally green to dark green at maturity; flesh soft, gelatinous—*Creopus gelatinosus* (Tode : Fries) Link = *Hypocrea gelatinosa* (Tode : Fries) Fries (see p. 524).

2.   Fruiting body not as above, orange, pink, yellowish red, orange-red, brownish red to brick-red or reddish brown; not splitting open to expose a shiny, black interior; flesh not concentrically zoned → 3.

3.   Fruiting body ⅛–⅝" (3–16 mm) wide, irregular to elliptic or cushion-shaped to somewhat rounded, usually growing in crowded clusters with individual fruiting bodies separated by furrows; surface wrinkled, reddish brown to brick-red with darker brown dots where the perithecia are embedded; flesh fibrous-tough, white; spores 3.5–6 × 3.5–4.5 μm; August–November; infrequent—*Hypocrea rufa* (Persoon : Fries) Fries.

3.   Fruiting body ⅟₁₆–³⁄₁₆" (1–4 mm) wide, cushion-shaped, nearly round or irregular, waxy, smooth, orange-red to brick-red or pinkish red; fruiting in dense clusters on branches and twigs of hemlock—*Hormomyces aurantiacus* Bonorden (see p. 525).

3.   Fruiting body ¼–⅜" (5–10 mm) wide, cushion-shaped, nearly round or irregular, wart-like, waxy, wrinkled to nearly smooth, red to purplish brown with a grayish bloom when dry; spores 6–9 × 1.5–2.5 μm; gloeocystidia abundant; growing on branches of poplar; inedible—*Peniophora rufa* (Fries) Boidin.

3.   Fruiting body up to ³⁄₁₆" (5 mm) wide and high, at first cylindric to cushion-shaped or nearly round to top-shaped, pale orange with a whitish, scurfy coating overall, soon opening and forming a bright orange to orange-yellow, stalkless saucer with an uneven margin; growing in dense clusters covering leaf and needle litter, mosses, twigs, and soil—*Byssonectria terrestris* (Albertini and Schweinitz : Fries) Pfister (see p. 491).

3.   Fruiting body ⅛–¼" (3–6 mm) wide, a dense, cushion-shaped cluster of individual perithecia attached to a common base; bright yellowish red to dull brownish red; individual perithecia roughened, oval to rounded, with a distinct, nipple-like tip; often intermixed with its asexual stage, *Tubercularia vulgaris* (Tode : Fries) Fries, which forms small, orange to coral-pink fruiting bodies, ⅟₁₆–³⁄₁₆" (1.5–4 mm) wide; on dead branches of many species of hardwoods; year-round—*Nectria cinnabarina* (Tode : Fries) Fries (see p. 525).

3.   Fruiting body not with the above combinations of characters; round to nearly round or angular; hard, woody; brick-red, cinnamon, reddish brown to purplish brown → 4.

4.   Fruiting body ⅛–⅝" (3–16 mm) wide, nearly round, salmon-pink at first, becoming brick-red to cinnamon at maturity and darkening to brown or brownish black in age; with one or two layers of embedded perithecia and a hollow center, as seen when sectioned; spores 11–14 × 5–7 μm, irregularly elliptic, flattened on one side, dark brown, smooth; typically in dense clusters on the bark of dead beech; year-round; common—*Hypoxylon fragiforme* (Persoon : Fries) Kickx.

4.   Fruiting body ⅛–¼" (3–6 mm) wide, irregularly rounded to angular; purplish brown to red-brown; with a single layer of embedded perithecia and sometimes a hollow center, as seen when sectioned; spores 12–15 × 5–7 μm, irregularly elliptic, flattened on one side, dark brown, smooth; typically in dense clusters on the bark of dead hazel, alder, and birch; year-round; common—*Hypoxylon fuscum* (Persoon : Fries) Fries.

4.   Fruiting body ⅟₁₆–³⁄₁₆" (1.5–4.5 mm) wide, nearly round; reddish brown to dark brown; with one or two layers of embedded perithecia, as seen when sectioned; spores 9–10 × 4–5 μm, irregularly elliptic, flattened on one side, dark brown, smooth; typically in dense clusters on the bark of dead hardwoods, especially beech; common—*Hypoxylon cohaerens* (Persoon : Fries) Fries.

5. Fruiting body 3½–8½" (9–21.5 cm) or more wide; creamy white to lemon-yellow with a whitish margin and darker dots marking the embedded perithecia; spores 3–4 μm, round or nearly so, smooth or finely punctate, hyaline; on decaying wood and the surrounding ground; August–October; infrequent—*Hypocrea citrina* (Persoon : Fries) Fries.

5. Fruiting body variable in size and shape but often cushion-shaped; individual fruiting bodies usually fusing and forming crust-like patches up to 4" (10 cm) or more; surface uneven, often furrowed, brick-red to reddish brown or purplish brown; spores 9–12 × 4–5.5 μm, irregularly elliptic, with one side flattened, smooth, dark brown; on many species of barkless hardwoods; year-round—*Hypoxylon rubiginosum* (Persoon : Fries) Fries (see p. 525).

5. Not with the above combinations of characters; fruiting body black, growing on wood or leaves → 6.

    6. Fruiting body ⅛–¾" (3–20 mm) wide; irregular to angular or somewhat circular, slightly raised, shiny or dull, typically wrinkled to furrowed and often minutely cracked; solitary to many on maple leaves; August–May; common—*Rhytisma acerinum* (Persoon) Fries (see p. 496).

    6. Fruiting body ⅛–⅝" (3–15 mm) wide, somewhat circular to irregular or angular, slightly raised, shiny to dull, typically wrinkled to furrowed and often minutely cracked, or sometimes smooth; solitary or scattered on willow leaves; August–May; common—*Rhytisma salicinum* Fries.

    6. Fruiting body wide-spreading, highly variable in shape and size, up to 9" (23 cm) or more long, ¹⁄₃₂–¹⁄₁₆" (1–1.5 mm) thick; purplish brown when young, becoming black, shiny, slightly roughened to nearly smooth, sometimes cracked; spores 6–9 × 1.5–2 μm, sausage-shaped, colorless, smooth; on decaying hardwood branches, especially beech; year-round—*Diatrype stigma* (Hoffmann : Fries) Fries (see p. 524).

    6. Fruiting body 1¾–3½" (4.5–9 cm) wide, ⅛–¼" (3–6 mm) thick, forming extensive sheets up to 15" (38.5 cm) or more, that are irregular in shape; surface at first grayish white, soft, powdery, with multiple lobes, becoming black and brittle at maturity; spores 28–36 × 7–10 μm, irregularly elliptic with one side flattened, smooth, dark brown; on stumps and roots of decaying hardwood trees; July–May—*Ustulina deusta* (Fries) Petrak (see p. 525).

7. Fruiting body unbranched, irregularly clavate or spindle-shaped to flattened, stalk rudimentary, not distinct; typically growing in clusters on stumps of deciduous trees, especially beech, or on the ground nearby—*Xylaria polymorpha* (Persoon : Mérat.) Greville (see p. 526).

7. Fruiting body unbranched, with a fertile, cylindric to clavate head and a long, sterile stalk; spores 13–16 × 5–7.5 μm; typically growing in clusters on decaying branches and stumps of hardwoods or on the ground nearby; fairly common—*Xylaria longipes* (Nitschke) Dennis.

7. Fruiting body with a branched and antler-like, flattened to oval or sometimes rounded stalk, gray to whitish on the upper portion, black toward the base; spores 12–15 × 5.5–6 μm; on decaying wood; common throughout the region—*Xylaria hypoxylon* (Linnaeus) Greville.

7. Fruiting body with a stalk and cylindric head with a crown of long, tapering tentacles, blackish on the stalk, grayish to whitish on the head and tentacles; on organic debris such as decaying wood and humus; fairly common in the most southern part of the region—*Xylaria tentaculata* Berkeley and Broome (see p. 526).

7. Fruiting body spindle-shaped to irregular, tumor-like growths, clasping and enveloping

the branches and twigs of living or dead cherry and plum trees—*Apiosporina morbosa* (Schweinitz) van Arx (see p. 524).

Apiosporina morbosa (Schweinitz) van Arx
COMMON NAME: Black Knot of Cherry.
FRUITING BODY: 1⅜–5½" (3.5–14 cm) long, ⅜–1" (1–2.5 cm) thick, spindle-shaped to clavate or irregular elongated swellings; surface hard, black, carbonaceous, finely roughened, typically furrowed and cracked; stalk absent; flesh white when very young, soon black and brittle; perithecia embedded near the surface in a single layer.
MICROSCOPIC FEATURES: spores 14–22 × 3–6 μm, narrowly elliptic, 1–3 septate, smooth, pale yellowish brown.
FRUITING: solitary or several, clasping and enveloping branches and twigs of cherry and plum trees; year-round; very common.
EDIBILITY: inedible.
COMMENTS: this fungus is a widely distributed and highly destructive pathogen of cherry and plum trees.

Creopus gelatinosus (Tode : Fries) Link (see photo, p. 23)
COMMON NAME: Yellow Cushion Hypocrea.
FRUITING BODY: 1/32–⅛" (1–3 mm) wide, cushion-shaped to nearly round; surface smooth, pale yellow, developing green dots and finally green to dark green at maturity; flesh soft, gelatinous, translucent, surrounding the embedded perithecia which appear as green dots.
MICROSCOPIC FEATURES: spores 4–6 × 3–4 μm, round to elliptic, finely warted, green.
FRUITING: densely clustered on decaying wood; August–October; fairly common.
EDIBILITY: unknown.
COMMENTS: this mushroom is also known as *Hypocrea gelatinosa*.

Daldinia concentrica (Bolton : Fries) Cesati and de Notaris (see photo, p. 23)
COMMON NAME: Carbon Balls or Crampballs.
FRUITING BODY: ¾–2" (2–5 cm) wide, cushion-shaped to nearly round or irregular; surface uneven and furrowed, reddish brown, becoming black, somewhat shiny, finely roughened, often with minute pores; flesh concentriclly zoned when cut vertically, fibrous to powdery and carbon-like, dark purplish brown alternating with darker or sometimes whitish zones; perithecia embedded in a single layer near the surface.
MICROSCOPIC FEATURES: spores 12–17 × 6–9 μm, irregularly elliptic with one side flattened, smooth, dark brown.
FRUITING: solitary or in clusters on decaying hardwood trees; year-round; common.
EDIBILITY: inedible.

Diatrype stigma (Hoffmann : Fries) Fries
FRUITING BODY: a wide-spreading crust, highly variable in shape and size, up to 9" (23 cm) or more long and 1/32–1/16" (1–1.5 mm) thick; surface shiny, finely roughened to nearly smooth, sometimes minutely cracked, purplish to reddish brown when young, soon blackish brown to black; flesh whitish, fibrous-tough; perithecia embedded just beneath the surface in a single layer.
MICROSCOPIC FEATURES: spores 6–9 × 1.5–2 μm, sausage-shaped, smooth, hyaline.
FRUITING: on decaying hardwood branches, especially beech, often surrounding them; year-round; common.

EDIBILITY: inedible.

COMMENTS: the nearly smooth, shiny surface and hyaline spores are characteristic of this species. Other similar black, spreading crusts have more coarsely roughened surfaces or dark brown spores.

Hormomyces aurantiacus Bonorden

FRUITING BODY: $\frac{1}{16}$–$\frac{3}{16}$" (1–4 mm) wide, cushion-shaped, nearly round or irregular, waxy, smooth, orange-red to brick-red or pinkish red.

MICROSCOPIC FEATURES: spores 6–11 × 4–8 μm, obovate to subglobose, smooth, hyaline to pale pinkish.

FRUITING: in dense clusters on branches and twigs of hemlock; August–October; occasional.

EDIBILITY: unknown.

Hypoxylon rubiginosum (Persoon : Fries) Fries

FRUITING BODY: variable in size and shape but often cushion-shaped; individual fruiting bodies usually fusing and forming crust-like patches up to 4" (10 cm) or more; surface uneven, often furrowed, brick-red to reddish brown or purplish brown, in age wearing away and exposing a black interior.

MICROSCOPIC FEATURES: spores 9–12 × 4–5.5 μm, irregularly elliptic, with one side flattened, smooth, dark brown.

FRUITING: in groups or confluent patches on trunks, logs, and stumps of hardwoods; year-round; fairly common.

EDIBILITY: inedible.

Nectria cinnabarina (Tode : Fries) Fries

COMMON NAME: Coral Spot Fungus.

FRUITING BODY: $\frac{1}{8}$–$\frac{1}{4}$" (3–6 mm) wide, a dense, cushion-shaped cluster of individual perithecia attached to a common base; bright yellowish red when young, becoming dull brownish red at maturity; individual perithecia roughened, oval to rounded, with a distinct, nipple-like tip; often intermixed with its asexual stage, *Tubercularia vulgaris* (Tode : Fries) Fries, which forms orange to coral-pink, small, cushion-shaped fruiting bodies, $\frac{1}{16}$–$\frac{3}{16}$" (1.5–4 mm) wide.

MICROSCOPIC FEATURES: spores 12–19 × 5–7 μm, spindle-shaped to elliptic, with a single septum at the center, often slightly constricted, smooth, hyaline.

FRUITING: scattered or in dense clusters on decaying branches of hardwoods; year-round; common.

EDIBILITY: inedible.

COMMENTS: the orange to coral-pink asexual stage forms a cushion-like mass of cells called a stroma, from which the perithecia arise.

Ustulina deusta (Fries) Petrak (see photo, p. 23)

COMMON NAME: Carbon Cushion.

FRUITING BODY: $1\frac{3}{4}$–$3\frac{1}{2}$" (4.5–9 cm) wide, $\frac{1}{8}$–$\frac{1}{4}$ " (3–6 mm) thick, spreading, crust-like, forming extensive sheets up to 15" (38.5 cm) or more, that are irregular in shape; surface at first grayish white, soft, powdery, with multiple lobes (asexual stage), becoming hard, black, finely roughened, and resembling burnt wood at maturity (sexual stage); flesh soft and white at first, becoming black and brittle in age; perithecia embedded just beneath the surface in a single layer.

MICROSCOPIC FEATURES: spores 28–36 × 7–10 µm, irregularly elliptic with one side flat-tened, smooth, dark brown.

FRUITING: on stumps and roots of hardwood trees; year-round; common.

EDIBILITY: inedible.

COMMENTS: mature specimens are easily overlooked. Our photograph shows a young, fresh specimen that represents the commonly observed asexual stage of this fungus.

Xylaria polymorpha (Persoon : Mérat) Greville (see photo, p. 23)

COMMON NAME: Dead Man's Fingers.

FRUITING BODY: ¾–3½" (2–9 cm) high, ⅜–1⅛" (1–3 cm) thick, unbranched, irregularly clavate or spindle-shaped, hard, black, carbonaceous, finely roughened, often wrinkled and minutely cracked; stalk short, indistinct, cylindric; flesh white, fibrous-tough; perithecia embedded just beneath the surface in a single layer.

MICROSCOPIC FEATURES: spores 22–30 × 5–9 µm, spindle-shaped or irregularly elliptic with one side flattened, smooth, dark brown.

FRUITING: densely clustered or sometimes solitary on or near decaying hardwood stumps; June–November; common.

EDIBILITY: inedible.

Xylaria tentaculata Berkeley and Broome

COMMON NAME: Fairy Sparklers.

FRUITING BODY: ¾–1⅜" (1.6–3.5 cm) high, consisting of a stalk and head, with a crown of tentacles; stalk ⅝–1" (1.6–2.5 cm) long, 1⁄16–⅛" (1.5–3 mm) thick, erect, nearly equal to irregular and twisted, frequently curved near the base, surface granular-scurfy, gray to blackish, often with bluish tints, flesh white; head ⅛–⅜" (3–10 mm) long, ⅛–¼" (3–6 mm) wide, cylindric, distinctly scurfy, opening at the apex by a prominent ostiole and crowned by up to 8 or more tentacles, gray to pinkish gray; tentacles ½–1½" (1.3–4 cm) long, up to 1⁄32" (0.75 mm) thick, partially erect, then spreading, tapering toward the tips, sometimes branched, pale gray to pinkish gray, coated with pale gray to whitish powder at maturity.

MICROSCOPIC FEATURES: spores 14–19 × 6–8.5 µm, elliptic, smooth, with 2 large oil drops, hyaline.

FRUITING: solitary, scattered, or in groups, sometimes clustered, on organic debris, such as decaying wood and humus, or among mosses; July–October; fairly common in the most southern part of the region.

EDIBILITY: inedible.

COMMENTS: easily overlooked and very fragile.

Apiosporina morbosa

Diatrype stigma

Hormomyces aurantiacus

Hypoxylon rubiginosum

Nectria cinnabarina

Xylaria tentaculata

Hypomyces and Allies

Hypomyces are a group of fungi that parasitize and disfigure other fungi. These fungi are also known as hyperparasites. About forty species are known to attack various gilled mushrooms, boletes, and polypores, and another half dozen are known to parasitize certain Ascomycetes, including *Leotia lubrica* Persoon : Fries and *Humaria hemisphaerica* (Wiggers : Fries) Fuckel. *Hypomyces* belong to the Pyrenomycetes, a class of Ascomycetes commonly called the Flask Fungi. They produce flask-shaped to rounded sexual structures called perithecia in which asci produce ascospores. The perithecia are often partially embedded in the host tissue, and their protruding necks are responsible for the sandpaper-like texture that covers some hosts.

Many also have an asexual stage that precedes the sexual stage and typically resembles a white mold or a yellow powder. The asexual spores produced are called conidia and aleuriospores. *Hypomyces* are closely related to other Flask Fungi, including the Carbon and Cushion Fungi, and *Cordyceps, Claviceps,* and Allies.

Identification of most species requires microscopic examination. A complete treatment of *Hypomyces* is beyond the scope of this work; therefore, we have included only those species which are commonly encountered or which have conspicuous features. Also included in the key are two species of the genus *Nectriopsis,* which are also parasites of mushrooms, and closely resemble *Hypomyces.*

Key to Species of *Hypomyces* and Allies

1. Fruiting body chalky white, yellowish, or pinkish to pale orange, phallic or club-shaped, roughened like sandpaper, growing on *Amanita* species—*H. hyalinus* (Schweinitz : Fries) Tulasne (see p. 531).
1. Fruiting body orange to reddish orange, roughened like sandpaper, with white flesh inside, growing over the surface of, and deforming, gills, caps and stalks of *Lactarius* and *Russula* species—*H. lactifluorum* (Schweinitz : Fries) Tulasne (see p. 531).
1. Fruiting body yellowish green to dark green, roughened like sandpaper, growing on the gills of *Russula* species—*H. luteovirens* (Fries) Tulasne (see p. 531).
1. Not with the above combinations of characters → 2.
    2. Fruiting body attacking gilled mushrooms → 3.
    2. Not as above → 4.
3. Fruiting body a white to cream moldy coating that covers the caps, gills, and stalks of *Pholiota* species; inedible—*H. succineus* Rogerson and Samuels.

3.   Fruiting body a whitish moldy coating that covers the gills of *Tubaria furfuracea* and causes gross disfiguration of the mushroom—*Nectriopsis tubariicola* W. Gams (see p. 532).

3.   Fruiting body a whitish moldy coating that covers the gills of *Crepidotus* species and causes gross disfiguration of the mushroom; ascospores 7.5–12 × 3–4 μm, elliptical to sole-shaped or somewhat clavate, 2-celled, with fine to coarse warts, hyaline—*Nectriopsis tremellicola* (Ellis and Everhart) W. Gams.

    4.   Fruiting body attacking various polypores, especially *Ganoderma applanatum,* white or brown with white tips, feathery, and radiating from a point of attachment, usually on the pore surface; inedible—*H. chrysostomus* Berkeley and Broome.

    4.   Fruiting body attacking ascocarps of *Leotia lubrica* Persoon : Fries, whitish and moldy at first, then forming partially immersed, flask-shaped perithecia and becoming dark green to black at maturity; inedible—*H. leotiicola* Rogerson and Samuels.

    4.   Fruiting body attacking ascocarps of *Humaria hemisphaerica* (Wiggers : Fries) Fuckel, whitish and moldy at first, then forming pale yellow, flask-shaped perithecia in the mold layer; inedible—*H. stephanomatis* Rogerson and Samuels.

    4.   Fruiting body attacking boletes → 5.

5.   Fruiting body white and moldy at first, partially or completely covering its host, becoming yellow and powdery, finally reddish brown and roughened like sandpaper; inedible—*H. chrysospermus* Tulasne (see p. 530).

5.   Fruiting body white and moldy at first, partially or completely covering its host, becoming pale yellow, then yellow to yellow-brown and roughened like sandpaper; inedible—*H. chlorinigenus* Rogerson and Samuels (see p. 530).

Hypomyces chlorinigenus Rogerson and Samuels
    FRUITING BODY: the asexual stage appears first, is white and moldy, partially or completely covers its host, and produces conidia; it becomes pale yellow, then yellow to yellow-brown and powdery as darker aleuriospores are produced; the sexual stage consists of globose to obovoid, amber perithecia containing asci and ascospores, producing a roughened, sandpaper-like texture.
    MICROSCOPIC FEATURES: conidia 10–11 × 4–5 μm, ovoid to elliptic-cylindric, 1-celled, hyaline; aleuriospores 35–45 × 15–18 μm, cylindric, longitudinally ridged, yellow to yellow-brown; ascospores 7.5–12 × 2.5–5 μm, elliptic-fusiform, 2-celled, smooth to finely warted, hyaline.
    FRUITING: on various boletes; June–November; fairly common.
    EDIBILITY: inedible.
    COMMENTS: the photograph shows the asexual stage attacking specimens of *Boletus pallidus.*

Hypomyces chrysospermus Tulasne (see photo, p. 24)
    COMMON NAME: Bolete Mold.
    FRUITING BODY: the asexual stage appears first, is white and moldy, and partially or completely covers its host, and produces conidia; it becomes yellow to golden yellow and powdery as yellow to yellow-brown aleuriospores are produced; the sexual stage consists of globose to flask-shaped, orange-yellow to red-brown perithecia containing asci and ascospores, producing a roughened, sandpaper-like texture.
    MICROSCOPIC FEATURES: conidia 10–30 × 5–12 μm, elliptic, 1-celled, smooth, hyaline;

aleuriospores 10–25 μm, globose, thick-walled, prominently warted, yellow to golden yellow or yellow-brown; ascospores 15–30 × 4–6 μm, 2-celled, spindle-shaped, hyaline.

FRUITING: on various boletes; June–October; common.

EDIBILITY: inedible.

COMMENTS: the sexual stage is not commonly encountered.

Hypomyces hyalinus (Schweinitz : Fries) Tulasne (see photo, p. 24)

COMMON NAME: Amanita Mold.

FRUITING BODY: 4–12" (10–30 cm) high, up to 2¾" (7 cm) thick, firm, solid, column- to club-shaped, usually quite phallic, chalky white, yellowish or pinkish to pale orange, roughened like sandpaper; host cap, gills, and other structures rarely discernible; odor and taste not distinctive.

MICROSCOPIC FEATURES: ascospores 15–20 × 4.5–6.5 μm, spindle-shaped, 2-celled, prominently warted, hyaline.

FRUITING: on several species of *Amanita,* especially frequent on *A. rubescens;* June–October; frequent and often abundant.

EDIBILITY: reportedly edible if parasitizing an edible species, such as *A. rubescens,* but certainly not recommended.

COMMENTS: the host mushroom's identity can only be presumed; reddish-staining specimens are most likely *A. rubescens.* No asexual stage is known.

Hypomyces lactifluorum (Schweinitz : Fries) Tulasne (see photo, p. 24)

COMMON NAME: Lobster Fungus, Lobster Mushroom.

FRUITING BODY: orange to reddish orange, sometimes with whitish areas, roughened like sandpaper, growing over the surface of, and deforming, caps, stalks, and gills of host mushrooms; parasitized caps are typically dense, often partially buried in conifer debris, measure 2–7⅞" (5–20 cm) wide, and are white and firm within.

MICROSCOPIC FEATURES: ascospores 35–40 × 4.5–7 μm, spindle-shaped, 2-celled, prominently warted, hyaline.

FRUITING: completely covering species of *Lactarius* and *Russula;* July–October; infrequent to fairly common.

EDIBILITY: edible, choice.

COMMENTS: this is a very popular edible mushroom, even though the identity of the host species is usually undetermined. No asexual stage is known.

Hypomyces luteovirens (Fries : Fries) Tulasne (see photo, p. 24)

COMMON NAME: Russula Mold.

FRUITING BODY: yellowish green to dark green, roughened like sandpaper, growing on the gills of host species.

MICROSCOPIC FEATURES: ascospores 28–35 × 4.5–5.5 μm, spindle-shaped, 1-celled, nearly smooth to prominently warted, hyaline.

FRUITING: completely covering the gills of various species of *Russula;* July–November; occasional to frequent.

EDIBILITY: not recommended.

COMMENTS: because this parasite rarely attacks the cap surface of the host mushroom, it can be detected only by picking and inspecting specimens of *Russula.* No asexual stage is known.

Nectriopsis tubariicola W. Gams

FRUITING BODY: a whitish moldy coating that covers the gills of its host and causes gross disfiguration; perithecia almost completely immersed in the white moldy coating, flask-shaped, containing asci and ascospores.

MICROSCOPIC FEATURES: ascospores 12–16 × 3–4 μm, 2-celled, fusiform with rounded to truncate ends, finely warted, hyaline.

FRUITING: on the gills of *Tubaria furfuracea;* September–January; occasional.

EDIBILITY: unknown.

COMMENTS: nonparasitized fruiting bodies of *Tubaria furfuracea* are illustrated on p. 314.

Hypomyces chlorinigenus

Nectriopsis tubariicola

APPENDIXES

GLOSSARY

RECOMMENDED READING

INDEXES

MYCOPHAGY

Collecting for the Table

The same general collecting tips described in the introduction apply to collecting mushrooms for the table. If you are unsure of the identification of your specimens, be certain to collect them carefully in order to preserve structures needed for a positive identification, and keep collections wrapped separately to avoid mixing species. If you are positive of the identification of the mushrooms you are collecting, then you may choose to "field dress" them: cut them close to the ground or other substrate upon which they are growing; cut away dried bits or parts that may house unwanted slugs, insects, or larvae; and brush off dirt, leaves, pine needles, and other debris.

As with any food item, select only mushrooms that are fresh. Be alert for obvious signs of decay: foul odors, soggy stalks, or discoloration. Use common sense to determine the ratio of insect-infested to edible mushroom tissue (the "bug-to-bolete" ratio). Leaving a few "very mature" specimens in the field to help ensure continued spore dispersal is always a good idea. Most importantly, collect gently. Disturb the substrate and the habitat as little as possible. In this way, you minimize damage to the mycelium, help ensure continued fungal growth, and help preserve the environment for others.

Playing It Safe: Guidelines for Eating Wild Mushrooms

The first and most vital rule regarding the consumption of wild mushrooms is this: If in doubt, throw it out! ***Never eat any mushroom unless you are certain of its identification as an edible species!*** It is not worth the risk of illness—or worse, death.

Perhaps nowhere else is superstition and myth as common as it is in regard to mushrooms. Old wives' tales abound, every family has its "secret," and each ethnic group seems to have its ways of telling the "good ones" from the bad. However, almost none of the myths, secrets, or formulas are true. They might even be deadly. Toss your silver spoons and coins away, stop peeling the skin from mushroom caps, and believe us when we tell you that poisonous mushrooms can grow on trees as well as on the ground. The only valid rule of thumb is that there isn't one. If you are interested in identifying wild mushrooms to eat, you must learn them genus by genus, species by species.

Once you are certain that what you have collected is edible, there are some simple considerations of which to be aware. Like any other foods, certain popular edible mushrooms may cause adverse or allergic responses in some individuals—not much different from somebody being sensitive to wheat or peanut butter, for instance. Whenever eating a mushroom species for the first time, eat a small portion sautéed lightly in butter or margarine. Store the rest of your collection for at least 48 hours before eating more of it. This procedure serves two purposes. First, you will discover the taste and texture of that mushroom so you can use it more creatively in your cooking. Second, you will find out if you are sensitive to that particular mushroom. Most adverse reactions consist of mild gastrointestinal upset requiring no medical attention. In the rare instance of a serious reaction indicating possible misidentification of the species, you then have fresh specimens waiting in the refrigerator to bring to the hospital. Mushroom poisoning should not be a concern, however, if proper care is taken. It is also comforting to know that if you should be sensitive to a particular mushroom genus, it does not mean you are sensitive to others. A good friend of ours is allergic to the genus *Agaricus,* including the common grocery store white button mushroom. However, he enjoys eating wild mushrooms from several other genera with impunity and great gusto!

Occasionally, gastrointestinal upsets related to mushroom munching are due to sheer overindulgence. Moderate the amounts of butter, margarine, oil, or cream used in your cooking. Use moderation also in the amount of mushrooms you eat during a meal. Some individuals have adverse reactions if they consume alcohol between 48 hours before and 48 hours after eating certain mushrooms (the Alcohol Inky and Black Morel, for example). Eating infested or spoiled mushrooms can cause unpleasant reactions. Never eat wild mushrooms raw: many contain certain proteins that may cause adverse reactions if not denatured by heat. Always cook your mushrooms thoroughly. Some edible mushrooms are covered with sticky, slimy gluten. This glutinous layer causes gastric distress for some individuals. To be on the safe side, remove the glutinous layer before cooking the mushrooms. Finally, don't use mushrooms from contaminated habitats (chemically treated lawns, the sides of major roadways, railroad right-of-ways, landfills, crop fields, and toxic waste sites) for the table. Mushrooms may concentrate toxic substances.

A multitude of gustatory delights are out there for the picking for those exercising a bit of common sense and caution.

Cleaning, Cooking and Preserving the Harvest

Once you are home with the rewards of a day spent collecting, it is best to remove clumps of dirt and debris from the mushrooms before storing them. Wrap the mushrooms in waxed paper or in brown paper bags and store them in the refrigerator. A final cleaning should be done just before cooking or preserving. Never store mushrooms in plastic bags or plastic wrap, which trap moisture and hasten spoilage.

Mushrooms retain water, so use as little of it as needed when cleaning them. Usually a sharp knife and a soft brush or damp cloth or paper towel are all that is needed. Many culinary shops sell special mushroom brushes for this purpose. If water must be used, allow the mushrooms to drain on paper towels for about ten minutes before cooking. Since a great deal of a mushroom's flavor is located in the outer skin of the mushroom cap, peeling mushrooms is not recommended.

Most mushrooms, if fresh, will keep for up to a week in the refrigerator. There are exceptions, such as the Inky Caps, which become a black, inky liquid within hours, even while refrigerated. You can also preserve mushrooms by drying, freezing, salting, pickling, and canning. There are many fine mushroom cookbooks that give specific information on preserving various mushroom species. Some are listed in the nontechnical section of "Recommended Reading."

Mushrooms lend themselves to a variety of cooking styles. They can be sautéed, baked, grilled, and used in soups, stews, and casseroles. As you become familiar with each mushroom's culinary characteristics, experiment with what you like best. Again, referring to a reputable cookbook is a fine way to begin.

MICROSCOPIC EXAMINATION OF MUSHROOMS

Macroscopic characters, also known as field characters, are features that can be detected without the use of a microscope. Examples include shape, size, color, smell, taste, and spore print. With practice, many mushroom species, such as the Yellow Morel, Chanterelle, and Sulfur Shelf, can easily be identified by using only macroscopic characters. Although most of the keys in this book use only macroscopic features, some include microscopic characters when differentiating between fruiting bodies that are otherwise indistinguishable.

Accurate identification or confirmation of some mushroom species requires the use of a microscope. Black Earth Tongues, species of *Geoglossum* and *Trichoglossum*, are difficult to identify to genus, and nearly impossible to identify to species when using only macroscopic characters. With a microscope, however, it becomes a relatively simple task. Another example of two mushrooms that are best distinguished microscopically is the Red-cracked Bolete, *Boletus chrysenteron,* and its look-alike, *Boletus truncatus.*

If you own a microscope, are thinking of purchasing one, have access to one, or are curious about its use in mycology, consider the following. Proper microscopic examination of mushrooms requires an instrument in good condition with a substage condenser, an artificial light source (preferably built-in) instead of a mirror, and lenses of good quality. The microscope should be equipped with at least three objective lenses: 10×, 40×, and 100× (oil-immersion), each of which produces a sharp image. An ocular micrometer should be inserted into the ocular lens to facilitate the measurement of spores, cystidia, hyphae, and other structures. Before measurements can be calculated, the ocular micrometer must be calibrated using a stage micrometer, a tiny glass ruler with known increments. This is an important step because the magnification of each combination of lenses is different. Most microscope dealers and college and university biology departments have stage micrometers and may assist you with the calibration.

The microscope plays a key role in fungal taxonomy. Numerous microscopic structures are commonly used to differentiate species and genera: spores, cystidia, basidia, clamp connections, hyphae, asci, paraphyses, and others. A thorough discussion of microscopic structures and techniques is beyond the scope of this work. However, for those who have a microscope, we have included spore information in the descriptions and in the keys when it is an important identifying character. Additional microscopic information has been provided when necessary.

Spores from a spore print or from fresh or dried material may be used when determining spore shape, size, and other characters. If a spore print is available, scrape off a minute sample using a razor or knife blade, emulsify it in a small drop of water or in a mounting medium such as 3% KOH, and carefully place a coverslip over the specimen. Using a pencil eraser, gently apply pressure on the coverslip to distribute the spores and remove excess fluid and air bubbles. Place a drop of immersion oil on the coverslip and carefully lower the 100× objective lens into the oil. Adjust the lighting and examine the spores.

If fresh material is used, remove a small section of gill, place it in the mounting fluid, add the coverslip, gently apply pressure, and examine as described above. When using a dried specimen, allow it to soak in the mounting medium for two or three minutes before adding the coverslip. Pieces of dried material may also be placed in wetting agents, such as 70–95% ethyl alcohol, before placing them in the mounting medium. This procedure helps hyphae absorb water and regain their original appearance.

APPENDIX C

CHEMICAL REAGENTS AND MUSHROOM IDENTIFICATION

Chemical reagents have also been used as taxonomic tools in the identification and classification of mushrooms. Two macroscopically similar genera or species may be differentiated according to their reaction to a chemical reagent. For example, *Lentinus* and *Lentinellus,* two wood-inhabiting genera with serrated gills, are differentiated according to the reaction of their spores in Melzer's reagent: the spores of *Lentinellus* stain bluish gray to bluish black, a reaction called **amyloid;** the spores of *Lentinus,* which do not stain bluish gray to bluish black, are **inamyloid.** Two commonly collected boletes, *Boletus spadiceus* and *Boletus subtomentosus,* can be very difficult to differentiate. Adding a drop of ordinary household ammonia to the cap surface provides a rapid distinction between the two species: the cap surface of *Boletus spadiceus* gives a fleeting but vivid blue or blue-green reaction, which then turns reddish brown; *Boletus subtomentosus* immediately stains reddish brown.

The following chemical reagents and solutions and their reactions are mentioned in the text where pertinent and can be very useful in mushroom identification.

NH_4OH = Ammonium hydroxide: 3–14% solution

Household ammonia is perfectly adequate. Ammonium hydroxide, used as a mounting medium for microscopic examination, is ideal for dried material after it has been in 70–95% ethyl alcohol. Ammonium hydroxide is also used to produce macroscopic color reactions when applied to cap, stalk, and flesh tissue.

$FeSO_4$ = Iron sulfate: 10% solution

Iron sulfate produces macroscopic color reactions when applied to cap, stalk, and flesh tissue. For example, the white flesh of the Red-cracked Bolete, *Boletus chrysenteron,* stains lemon-yellow with this solution.

Melzer's Reagent

Melzer's reagent is a special solution of iodine (0.5 g), potassium iodide (1.5 g), chloral hydrate (20 g), and water (20 ml). It is used to separate genera and groups of species that produce hyaline or lightly colored spores. It also serves as a valuable test for identifying some species. Spores, hyphae, paraphyses, and other structures produce specific color reactions when mounted in Melzer's reagent. Spores or other cells that stain bluish gray to bluish black are called **amyloid;** those that stain reddish brown are called **dextrinoid;** and those that stain yellow or remain colorless are called **inamyloid.**

It is possible to determine whether spores are amyloid, dextrinoid, or inamyloid without using a microscope. First, obtain a spore print on glass or another nonporous surface. Place a small drop of Melzer's reagent on a glass slide. Using a razor blade or sharp knife, carefully scrape off some spores and transfer them onto the glass slide next to the reagent. Hold the slide over a piece of white paper in a well-lighted area. Tilt the slide to mix the spores and reagent and observe any color change. (Note: color changes are usually rapid.) If a color change is not evident, check again after a few minutes.

KOH = Potassium hydroxide: 3–5% solution

Potassium hydroxide is used to produce color reactions when applied to cap, stalk, and flesh tissue. For example, the flesh of the Shiny Cinnamon Polypore, *Coltricia cinnamomea,* instantly stains black with the addition of this solution. Potassium hydroxide is also used as a mounting medium for fresh and dried specimens; it stains various microscopic structures, including certain cystidia and hyphae.

APPENDIX D

CLASSIFICATION

Species included in this book are members of the kingdom Fungi and the division Eumycota or "True Fungi." They are classified in three subdivisions: Ascomycotina, the sac fungi; Basidiomycotina, the club fungi; and Zygomycotina, the bread molds and allies.

The subdivision Ascomycotina, the second-largest group represented in this book, includes cup fungi, morels, carbon fungi, *Cordyceps,* and others. They are able to reproduce both sexually and asexually. These fungi produce sexual spores called ascospores, which are formed internally in a closed sac-like container called an ascus, and asexual spores called aleuriospores and conidia, which are produced externally. The asexual stage of some species is often conspicuous and commonly collected. The asexual stage genera are also included on the list for this subdivision.

The subdivision Basidiomycotina, the largest group represented in this work, includes gilled mushrooms, puffballs, polypores, and many others. This group reproduces only sexually and produces basidiospores, which are formed externally on a club-shaped structure called a basidium.

The subdivision Zygomycotina includes mold-like fungi, which often occur on other host fungi. Some are saprobes, and others are parasites. They reproduce both sexually and asexually. They form thick-walled sexual spores called zygospores and asexual spores called sporangiospores. The sporangiospores are produced internally in tiny round to oval containers called sporangia. They are sometimes collected because they occur on host mushrooms (see *Spinellus fusiger* on *Mycena haematopus,* p. 302).

Fungal taxonomy is currently undergoing dramatic changes as new information is uncovered, and therefore extensive information about fungal classification is not included in this work. The following lists contain the genera of Ascomycotina, Basidiomycotina, and Zygomycotina included in this book.

GENERA OF ASCOMYCOTINA

Acervus	*Caloscypha*	*Crinula*
Aleuria	*Camarops*	*Cudonia*
Apiosporina	*Cheilymenia*	*Cudoniella*
Arachnopeziza	*Chlorociboria*	*Daldinia*
Ascocoryne	*Chlorocoelia*	*Dasyscyphus*
Ascotremella	*Chlorosplenium*	*Diatrype*
Bisporella	*Claviceps*	*Discina*
Bulgaria	*Cordyceps*	*Disciotis*
Byssonectria	*Creopus*	*Elaphomyces*

Encoelia
Galiella
Geoglossum
Geopyxis
Gyromitra
Helotium
Helvella
Holwaya
Hormomyces *
Humaria
Hymenoscyphus
Hypocrea
Hypomyces
Hypoxylon
Inermisia
Jafnea
Lachnellula
Leotia
Melastiza

Microglossum
Microstoma
Mitrula
Morchella
Nectria
Nectriopsis
Neobulgaria
Neolecta
Octospora
Otidea
Pachyella
Patella
Peridoxylon
Peziza
Phaeocalicium
Plectania
Podostroma
Pseudoplectania
Ptychoverpa

Rhizina
Rhytisma
Sarcoleotia
Sarcoscypha
Sarcosoma
Scutellinia
Spathularia
Spathulariopsis
Spragueola
Tarzetta
Trichoglossum
Tubercularia
Underwoodia
Urnula
Ustulina
Verpa
Vibrissea
Wynnea
Xylaria

GENERA OF BASIDIOMYCOTINA

Agaricus
Agrocybe
Albatrellus
Aleurodiscus
Amanita
Armillaria
Asterophora
Astraeus
Aurantioporus
Auricularia
Auriscalpium
Austroboletus
Baeospora
Bjerkandera
Bolbitius
Boletellus
Boletinellus
Boletinus
Boletopsis

Boletus
Bondarzewia
Bovista
Callistosporium
Calocera
Calocybe
Caloporus
Calostoma
Calvatia
Cantharellula
Cantharellus
Catathelasma
Cerrena
Chalciporus
Cheimonophyllum
Chlorophyllum
Chondrostereum
Christiansenia
Chromosera

Chroogomphus
Chrysomphalina
Claudopus
Clavaria
Clavariadelphus
Clavicorona
Clavulina
Clavulinopsis
Climacodon
Clitocybe
Clitocybula
Clitopilus
Coltricia
Colus
Conocybe
Coprinus
Coriolopsis
Corticum
Cortinarius

Cotylidia
Craterellus
Crepidotus
Crinipellis
Crucibulum
Cryptoporus
Cyathus
Cyclomyces
Cyptotrama
Cystoderma
Cystostereum
Dacrymyces
Dacryopinax
Daedalea
Daedaleopsis
Dendropolyporus
Dendrothele
Dentinum
Dictyophora
Ductifera
Entoloma
Exidia
Exobasidium
Favolus
Fistulina
Flammulina
Fomes
Fomitopsis
Fuscoboletinus
Galerina
Ganoderma
Geastrum
Globifomes
Gloeophyllum
Gloeoporus
Gomphidius
Gomphus
Grifola
Gymnopilus
Gymnosporangium
Gyrodon
Gyroporus

Hapalopilus
Hebeloma
Hericium
Heterobasidion
Hohenbuehelia
Hormomyces*
Hydnellum
Hydnochaete
Hydnum
Hygrophoropsis
Hygrophorus
Hymenochaete
Hypholoma
Hypsizygus
Inocybe
Inonotus
Irpex
Ischnoderma
Kuehneromyces
Laccaria
Lactarius
Laetiporus
Langermannia
Laxitextum
Leccinum
Lentaria
Lentinellus
Lentinus
Lenzites
Lepiota
Leptonia
Leucoagaricus
Leucocoprinus
Leucopaxillus
Leucopholiota
Limacella
Lopharia
Lycoperdon
Lyophyllum
Lysurus
Macrolepiota
Macrotyphula

Marasmiellus
Marasmius
Megacollybia
Melanoleuca
Melanophyllum
Meripilus
Merulius
Micromphale
Morganella
Multiclavula
Mutinus
Mycena
Mycorrhaphium
Myriostoma
Naematoloma
Nolanea
Oligoporus
Omphalina
Omphalotus
Oxyporus
Panaeolus
Panellus
Panus
Paxillus
Peniophora
Phaeocollybia
Phaeolus
Phaeomarasmius
Phallogaster
Phallus
Phanerochaete
Phellinus
Phellodon
Phlebia
Phleogena
Phlogiotis
Pholiota
Phylloporus
Phyllotopsis
Physalacria
Piptoporus
Pisolithus

Pleurocybella
Pleurotus
Plicaturopsis
Pluteus
Polyozellus
Polyporus
Poronidulus
Porphyrellus
Porpoloma
Pouzarella
Psathyrella
Pseudoclitocybe
Pseudocolus
Pseudocoprinus
Pseudofistulina
Pseudohydnum
Pseudomerulius
Psilocybe
Pucciniastrum
Pulcherricium
Pulveroboletus
Punctularia
Pycnoporellus
Pycnoporus
Ramaria
Ramariopsis

Resinomycena
Resupinatus
Rhizopogon
Rhodocybe
Rhodotus
Rhopalogaster
Rickenella
Ripartites
Rozites
Russula
Sarcodon
Sarcodontia
Schizophyllum
Schizopora
Scleroderma
Sebacina
Serpula
Simblum
Simocybe
Sparassis
Sphaerobolus
Spongipellis
Squamanita
Steccherinum
Stereum
Strobilomyces

Strobilurus
Stropharia
Suillus
Syzygospora
Tectella
Tephrocybe
Thelephora
Trametes
Tremella
Tremellodendron
Trichaptum
Tricholoma
Tricholomopsis
Trogia
Tubaria
Tulostoma
Tylopilus
Tyromyces
Uromyces
Ustilago
Volvariella
Xanthoconium
Xeromphalina
Xerula
Xylobolus

GENERA OF ZYGOMYCOTINA

Spinellus

Syzygites

* The taxonomic status of the genus *Hormomyces* is uncertain at this time because only asexual reproduction has been observed.

Glossary

acrid: producing a burning sensation in the mouth

adnate: attached to the stalk without a notch

adnexed: attached to the stalk and notched

aleuriospore: an asexual pigmented spore supported by a tiny stalk and not enclosed in a container

amatoxins: toxic compounds that cause kidney and liver damage by inhibiting the production of specific proteins; found in some species of *Amanita, Galerina,* and other mushrooms

amyloid: staining grayish to blue-black in Melzer's reagent

anastomosing: fusing to form a network

annular zone: a poorly defined ring

annulus: a ring on the stalk

apex (pl. apices): the uppermost portion of the stalk; the portion of the spore closest to the point of attachment to the basidium

apical: pertaining to the apex

apical callus: a thin-walled apical region

apical pore: a small opening or thin area in the wall at the apex of a spore; also known as a germ pore

apiculus: a short projection at or near the apex of a spore

appendiculate: hung with fragments of the partial veil

appressed: flattened onto the surface

areolate: marked out into small areas by cracks or crevices

ascocarp: the fruiting body of an Ascomycete

Ascomycetes: a major group of fungi that includes those species that produce ascospores in asci

ascospore: a sexual spore formed within an ascus

ascus (pl. asci): a sac-like cell in which ascospores are formed

aseptate: lacking crosswalls in the hyphae

asexual: not resulting from fertilization

attached: joined to the stalk

attenuate: gradually narrowed

azonate: lacking zones

basal: located at the base

base: the lowest portion of the stalk

Basidiomycetes: a major group of fungi that includes those species that produce spores borne on a basidium

basidiospore: a spore formed on a basidium

basidium (pl. basidia): typically a club-shaped cell on which basidiospores are formed

bloom: a dull, thin coating that is typically whitish

bulbous: having a bulb-like base

caespitose: occurring in groups

campanulate: bell-shaped

canescence: a whitish to grayish dust-like bloom

cap: the upper part of a mushroom, which supports gills, tubes, spines, or a smooth surface on its underside

cartilaginous: fibrous-tough, often splitting lengthwise in strands

caulocystidia: cystidia found on the stalk

central: attached to the middle of the cap

cheilocystidia: cystidia that occur on the edge of a gill or the edge of a tube; compare with pleurocystidia

chlamydospores: thick-walled asexual spores that are often nearly round

clamp connection: a semicircular bridge-like structure that connects two adjoining cells in some Basidiomycetes

clavate: club-shaped

collarium: a collar or ring on the stalk apex into which the inner edges of the gills are inserted; the gills do not touch the stalk

collybioid: having the stature of a species of *Collybia*

columella: a persistent sterile column within a sporangium

compressed: flattened longitudinally

confluent: becoming continuous together

conic: shaped more or less like an inverted cone

conidium (pl. conidia): an asexual spore supported by a tiny stalk and not enclosed in a container, hyaline and usually thin-walled

convex: curved or rounded like the exterior of a circle

coprine: an amino acid found in some mushrooms, which when consumed before, with, or after alcohol may cause nausea, vomiting, a flushed feeling, rapid breathing, and other signs and symptoms of coprine poisoning

cortina: a spiderweb-like partial veil

cortinate: appearing spiderweb-like

crenate: finely scalloped

crested: having a showy tuft or projection

crossveined: having tiny veins that connect adjoining gills

cuticle: the outermost tissue layer of the cap; also known as a pileipellis

cyanophilous: staining blue in a solution of cotton-blue; often a reference to a spore wall

cystidia: sterile cells that project between, and usually beyond, the basidia

decurrent: descending or running down the stalk; a form of gill attachment

delimited: marked or bordered by

deliquesce: liquify, as in the gills of the genus *Coprinus*

deliquescent: having the ability to deliquesce

denticulate: having small teeth

depressed: sunken

dextrinoid: staining orange to orange-brown or pinkish red to dark red or reddish brown in Melzer's reagent

disc: the central area of the surface of a mushroom cap

distant: spaced widely apart

doglegged: having an abrupt angle or sharp bend

eccentric: away from the center

echinulate: having small spines

elevated: raised upward above the plane

entire: even; not broken, serrated, or lacerated

equal: having the same thickness over the entire length

eroded: partially worn away and appearing ragged

evanescent: slightly developed and soon disappearing

expanded: enlarged and elongated

faces: the sides of a gill

farinaceous: having an odor of fresh meal or resembling cucumber

fertile surface: the spore-bearing surface

FeSO$_4$: iron sulfate, usually a 10% solution

fetid: having an offensive odor

fiber: a hair-like structure present on the cap or stalk of some mushrooms

fibril: a tiny fiber

fibrillose: composed of fibrils

fibrous: composed of fibers

filiform: thread-like

fimbriate: minutely fringed

flesh: the inner tissue of a fruiting body

floccose: tufted like cotton balls

flocculence: minute floccose decorations

free: not attached to the stalk

fruiting body: the fleshy to hard reproductive structure of a fungus, commonly called a mushroom

fulvous: reddish cinnamon; colored like a red fox

fuscous: dark brownish gray to brownish black

fusiform: spindle-shaped and narrowing at both ends

fusiform-ellipsoid: elliptical but somewhat spindle-shaped

fusoid: somewhat spindle-shaped

Gasteromycetes: a group of Basidio-mycetes, such as puffballs, which produce spores in closed chambers within the fruiting body

genus (pl. genera): taxonomic grouping of closely related species

germ pore: a thin portion of the spore wall through which the hypha passes during germination; also known as an apical pore

gills: thin to thick, knifeblade-like struc-tures on the cap undersurface of some mushrooms

glabrous: bald and smooth

glandular dots: sticky spots on the surface of the stem

glaucous: covered with a thin, whitish bloom that is easily rubbed off

gleba: spore-bearing tissue of stinkhorns, puffballs, and other Gasteromycetes

globose: round

gloeocystidia: cystidia with granular or oily content

gluten: a sticky, glue-like, pectinous material

glutinous: having gluten

granular: resembling tiny grains

granules: tiny grains

gregarious: closely scattered over a small area

H₂SO₄: sulfuric acid; used in concentrated form to determine colorfastness of the spores of *Panaeolus* species

habit: the manner of growth, such as soli-tary, or in fused groups

habitat: the substrate from which the mushroom grows, such as among sphagnum mosses, on wood, or on the ground

hemispheric: shaped like half of a sphere

hoary: covered with very fine, silky down or having a whitish to grayish sheen

homogeneous: composed of uniform cells or tissue

hyaline: transparent; clear and nearly colorless

hydrazine: a colorless, corrosive liquid, used in rocket fuel, that is released from some mushrooms during cooking

hygrophanous: appearing water-soaked when fresh, fading to a paler color as water is lost

hygroscopic: readily absorbing water

hypha (pl. hyphae): thread-like filaments of fungal cells

hypogeous: occurring below the ground surface

ibotenic acid: a toxin found in several *Amanita* species that interferes with the normal metabolism of some amino acids and adversely affects the central nervous system

inamyloid: unchanging or pale yellow in Melzer's reagent; neither amyloid nor dextrinoid

incurved: bent inward toward the stalk

inflorescence: the flowering portion of a stem

inrolled: bent inward toward the stalk and upward

intervenose: having veins on the gill faces which often extend between the gills or from gill to gill

keystone: a stone shaped like a wedge

KOH: potassium hydroxide, usually made up in a 3-5% concentration in water; used to test color reactions

lacerated: appearing torn

lacrymoid: shaped like a teardrop

lamella (pl. lamellae): a gill on the under-side of a mushroom cap

lamellulae: short gills that do not reach the stalk

lateral: attached to the margin of a cap

latex: a watery or milk-like fluid that exudes from some mushrooms when they are cut or bruised

lorchel: a common name for *Gyromitra esculenta* and other false morels

lubricous: smooth and slippery

marcescent: stable and not easily spoiled

margin: the edge of a mushroom cap

marginate: having gill edges that are darker colored than the faces; having a circular ridge on a bulb

marzipan: a confection made with crushed almonds or almond paste

median: reference to a ring located midway down on a stalk

Melzer's reagent: a solution containing iodine used for testing fungal spores for color reactions

multiseptate: having several to many crosswalls

muscimol: a toxin found in several *Amanita* species that interferes with the normal metabolism of some amino acids and adversely affects the central nervous system

mycelium: a mass of hyphae, typically hidden in a substrate

mycophagist: a person who eats mushrooms

mycorrhizal: having a mutually beneficial relationship with a tree or other plant

naked: bald, smooth

NH₄OH: ammonium hydroxide; used to test color reactions on mushroom tissues

nitrous: a reference to nitrogen-containing compounds, which are pungent and unpleasant

obovate: ovate, with the broader end opposite to the point of attachment

obovoid: ovoid, with the broader end opposite to the point of attachment

obtuse: rounded or blunt

ochraceous: pale, brownish orange-yellow

ochre: brownish orange-yellow

ostiole: a small pore-like opening

ovate: shaped like an egg

ovoid: somewhat egg-shaped

paraphysis (pl. paraphyses): a distinctive sterile cell in the spore-producing layer of Ascomycetes that keeps the asci erect

parasite: an organism that obtains its nutrients from a living host

partial veil: a layer of fungal tissue that covers the gills or pores of some immature mushrooms

pedicels: slender stalks

pendant: hanging or draping

peridiole: a tiny, egg-like structure that contains spores

peridium (pl. peridia): a layer of the spore case of puffballs and other Gastero-mycetes; three layers are often present: exoperidium = outermost layer, meso-peridium = middle layer, endoperid-ium = innermost layer

perithecium (pl. perithecia): a minute, flask-shaped structure containing asci

phallotoxins: complex ring compounds, found in some species of *Amanita,* which are not known to cause poisoning

plage: a smooth area near the apiculus on the surface of a spore

plane: flat

pleurocystidia: cystidia that occur on the gill faces or inner surface of the tubes; compare with cheilocystidia

plicate: deeply grooved, sometimes pleated or folded

polar: located at the most distant part

pores: the open ends of the tubes of a bolete or polypore

pore surface: the undersurface of the cap of a bolete or polypore, where the open ends of the tubes are visible

poroid: resembling pores or composed of pores

pruina: powdery particles, flakes, or dots

pruinose: appearing finely powdered

pubescent: having short, soft, downy hairs

punctate: marked with tiny points, dots, scales, or spots

putrescent: soon decaying

radial: pointed away from a common central point, like the spokes of a wheel

radicating: forming a root-like extension in the ground or in wood

recurved: curved backward or downward

reticulate: covered with a net-like pattern of ridges

reticulation: raised, net-like ridges

reticulum: a system of raised, net-like ridges found on the stalk surface or spores of some mushrooms

revolute: rolled up toward the disc, then toward the margin; opposite of inrolled

rhizomorph: a group of thick, rope-like strands of hyphae growing together as a single organized unit

rhombic: having four equal sides and often no right angles

rimose: having distinct cracks or crevices

ring: remnants of a partial veil that remain attached to the stalk after the veil ruptures

saccate: sheath-like or cup-shaped

saprobe: an organism that lives off dead or decaying matter

scabers: small, stiff, granular points on the surface of the stalks of some mushrooms; a distinctive feature of the genus *Leccinum*

scale: an erect, flattened, or recurved projection or torn portion of the cap or stalk surface

sclerotium (pl. **sclerotia**): a small, rounded to irregular body composed of dormant hyphae

scrobiculate: pitted or smeared; having flat, shiny areas

scurfy: roughened by tiny flakes or scales

seceding: attached at first and later separating

sensu: in the sense of, or according to

sensu lato: in the broadest sense

septate: divided by crosswalls

septum: a crosswall

serrate: jagged or toothed like a saw blade

setae: sharply pointed sterile cells that are usually brown or yellow and project on the surface of the stalk or other portion of the fruiting body of some mushrooms

sexual: pertaining to fertilization involving two compatible cells

siderophilous: having granules in basidia that turn blackish purple or violet-black in acetocarmine stain; also known as carminophilous

sinuate: gradually narrowed and becoming concave near the stalk

sordid: dingy or dull

spathulate: spoon-shaped

spines: tapered, downward-pointing projections on the undersurface of some mushroom caps

sporangiospore: an asexual spore formed in a sporangium

sporangium: a sac-like microscopic structure in which asexual spores are produced

spore: a microscopic reproductive cell with the ability to germinate and form hyphae

spore case: a structure containing the spore mass in species of Gasteromycetes

spore mass: a dense layer of spores

spore print: a deposit of spores, on a piece of paper or glass, from a mushroom's gills, tubes, or other spore-producing structures

stalk: the structure that arises from the substrate and supports the cap or spore case of a mushroom

sterile surface: a portion that lacks reproductive structures

sterile tissue: tissue not directly involved with the reproductive process

striate: having small, more or less parallel lines or furrows

strigose: coated with long, coarse, stiff hairs

stroma (pl. **stromata**): dense fungal tissue on or within which fruiting bodies are produced

stuffed: containing a soft tissue which usually disappears in age, leaving a hollow space

subattenuate: abruptly narrowed, but not truncate

subdecurrent: extending slightly down the stalk

subdistant: gill spacing halfway between close and distant

subfusiform: nearly spindle-shaped

subfusoid: somewhat spindle-shaped; tapered slightly at both ends

subglobose: nearly round

substrate: organic matter that serves as a food source for a fungal mycelium

sulcate: grooved; deeper than striate, less than plicate

superior ring: a ring located on the upper stalk surface

teeth: spines that point downward

teliospore: a thick-walled spore formed at the terminal stage of the life cycle of rusts and smuts

terete: rounded like a broom handle

terrestrial: growing on the ground

tiers: sets of equal length

tomentose: coated with a thick, matted covering of hairs

translucent-striate: appearing striate when gill edges are viewed through moist, nearly transparent cap tissue

trilobate: having three lobes

truncate: appearing cut off at the end

tuberculate-striate: having striations that are roughened by small warts or bumps

tubes: narrow, parallel, spore-producing cylinders on the undersurface of the cap of a bolete or polypore

umbo: a pointed or rounded elevation at the center of a mushroom cap

umbonate: having an umbo

uniseriate: arranged in a single row

universal veil: a layer of fungal tissue that completely encloses immature stages of some mushrooms

uplifted: turned upward

urediospore: an asexual spore produced in the life cycle of rusts

veil: a layer of fungal tissue that covers all or part of some immature mushrooms (see *universal veil* and *partial veil*)

ventricose: swollen in the middle and tapering to somewhat of a point

villose: coated with long, tiny, soft hairs

vinaceous: pinkish red to pale purplish red

virgate: streaked with fibrils

viscid: sticky or tacky

volva: a typically cup-like sac that remains around the base of a mushroom stalk when the universal veil ruptures

warts: small patches of tissue that remain on the top of a mushroom cap when the universal veil ruptures

zones: concentric bands of different colors on the surface of the cap or stalk of some mushrooms

zygospore: a thick-walled sexual spore

Recommended Reading

Technical Publications

Bérubé, J. A. and M. Dessureault. 1988. Morphological Characterization of *Armillaria ostoyae* and *Armillaria sinapina* sp. nov. *Can. J. Bot.* 66: 2027-2034.

——. 1989. Morphological Studies of the *Armillaria mellea* Complex: Two new species, *A. gemina* and *A. calvescens. Mycologia* 81: 216–225.

Bessette, A. E., A. R. Bessette, D. P. Lewis, and S. H. Metzler. 1993. A New Substrate for *Strobilurus conigenoides* (Ellis) Singer. *Mycotaxon* 68: 299–300.

Bessette, A. E., and R. L. Homola. 1986. *Lentinus adhaerens,* A Species New to North America. *Mycologia* 78(2): 296–298.

Bigelow, H. E. 1973. The Genus *Clitocybula. Mycologia* 65(5): 1101–1116.

——. 1982. *North American Species of Clitocybe. Part 1.* J. Cramer, Berlin. 280 pp.

——. 1985. *North American Species of Clitocybe. Part 2.* J. Cramer, Berlin. 240 pp.

Bird, C. J., and D. W. Grund. 1979. Nova Scotian Species of *Hygrophorus. Proc. Nova Scotia Inst. of Sci.* 29: 1–131.

Both, E. E. 1993. *The Boletes of North America: A Compendium.* Buffalo Museum of Science, Buffalo. 436 pp.

Coker, W. C., and A. H. Beers. 1943. *The Boletaceae of North Carolina.* Univ. of North Carolina Press, Chapel Hill. 96 pp.

Coker, W. C., and J. N. Couch. 1928. *Gasteromycetes of the Eastern United States and Canada.* Univ. of North Carolina Press, Chapel Hill. 201 pp.

Gilbertson, R. L., and L. Ryvarden. 1986. *North American Polypores Vol. 1.* Fungiflora, Oslo. 433 pp.

——. 1987. *North American Polypores. Vol. 2.* Fungiflora, Oslo. 451 pp.

Gilliam, M. S. 1975. *Marasmius* Section Chordales in the Northeastern United States and Adjacent Canada. *Contr. Univ. Mich. Herb.* 11(2): 25–40.

——. 1976. The Genus *Marasmius* in the Northeastern United States and Adjacent Canada. *Mycotaxon* 4(1): 1–144.

Grund, D. W., and K. A. Harrison. 1976. Nova Scotian Boletes. *Bibliotheca Mycologica.* J. Cramer, Vaduz. Band 47. 283 pp.

Guzmán, G. 1983. The Genus *Psilocybe. Nova hedwigia.* J. Cramer, Vaduz. Heft 74. 439 pp.

Halling, R. E. 1983. The Genus *Collybia* (Agaricales). *Mycologia Memoir No. 8.* New York Botanical Garden. J. Cramer, Vaduz. 148 pp.

Harrington, F. A. 1990. *Sarcoscypha* in North America (*Pezizales, Sarcoscyphaceae*). *Mycotaxon* 38: 417-458.

Harrison, K. A. 1961. *Stipitate Hydnums of Nova Scotia.* Canadian Dept. of Agriculture, Ottawa. 60 pp.

Hawksworth, D. L., P. M. Kirk, B. C. Sutton, and D. N. Pegler. 1995. *Dictionary of the Fungi*. 8th ed. International Mycological Institute, Surrey, England. 616 pp.

Henderson, D. M., P. D. Orton, and R. Watling, eds. 1979. *Coprinaceae Part 1: Coprinus*. Her Majesty's Stationary Office, Edinburgh. 149 pp.

Hesler, L. R. 1965. *North American Species of Crepidotus*. Hafner, New York. 168 pp.

——. 1967. *Entoloma* in Southeastern North America. *Nova Hedwigia*. Verlag Von J. Cramer, Lehre. Heft 23. 196 pp.

——. 1969. *North American Species of Gymnopilus*. Hafner, New York. 117 pp.

Hesler, L. R., and A. H. Smith. 1963. *North American Species of Hygrophorus*. Univ. of Tennessee Press, Knoxville. 416 pp.

——. 1979. *North American Species of Lactarius*. Univ. of Michigan Press, Ann Arbor. 841 pp.

Jenkins, D. T. 1986. *Amanita of North America*. Mad River Press, Eureka, Calif. 198 pp.

Kauffman, C. H. 1971. *The Gilled Mushrooms (Agaricales) of Michigan and the Great Lakes Region*. Vol. 1. Dover, New York. Reprint. 442 pp.

——. 1971. *The Gilled Mushrooms (Agaricales) of Michigan and the Great Lakes Region*. Vol. 2. Dover, New York. Reprint. 481 pp.

Kibby, G., and R. Fatto. 1990. *Keys to the Species of Russula in Northeastern North America*. Kibby-Fatto Enterprises, Somerville, N.J. 61 pp.

Largent, D. L. 1977. The Genus *Leptonia* on the Pacific Coast of the United States. *Bibliotheca Mycologica*. J. Cramer, Vaduz. Band 55. 286 pp.

——. 1986. *How to Identify Mushrooms to Genus I: Macroscopic Features*. Rev. ed. Mad River Press, Eureka, Calif. 166 pp.

Largent, D. L., and T. J. Baroni. 1988. *How to Identify Mushrooms to Genus VI: Modern Genera*. Mad River Press, Eureka, Calif. 277 pp.

Largent, D. L., D. Johnson, and R. Watling. 1977. *How to Identify Mushrooms to Genus III: Microscopic Features*. Mad River Press, Eureka, Calif. 148 pp.

Largent, D. L., and H. D. Thiers. 1977. *How to Identify Mushrooms to Genus II: Field Identification of Genera*. Mad River Press, Eureka, Calif. 32 pp.

Mazzer, S. J. 1976. *A Monographic Study of the Genus Pouzarella*. Bibliotheca Mycologica. J. Cramer, Vaduz. Band 46. 191 pp.

Miller, O. K. 1964. Monograph of *Chroogomphus* (Gomphidiaceae). *Mycologia*. 56: 526–549.

——. 1968. A Revision of the Genus *Xeromphalina*. *Mycologia* 60: 156–188.

——. 1970. The Genus *Panellus* in North America. *Mich. Bot*. 9: 17–30.

——. 1971. The Genus *Gomphidius* Fries, with a Revised Description of the Gomphidiaceae and Keys to the Genera. *Mycologia* 53: 1129–1163.

Miller, O. K., and L. Stewart. 1971. The Genus *Lentinellus*. *Mycologia* 63(2): 333–369.

Miller, O. K., T. J. Volk, and A. E. Bessette. 1995. A New Genus, *Leucopholiota*, in the Tricholomataceae (Agaricales) to Accommodate an Unusual Taxon. *Mycologia* 88(1): 137-139.

Moser, M. 1983. *Keys to Agarics and Boleti*. Gustav Fischer Verlag, Stuttgart. 535 pp.

Motta, J. J., and K. Korhonen. 1986. A Note on *Armillaria mellea* and *Armillaria bulbosa* from the Middle Atlantic States. *Mycologia* 78(3): 471–474.

Mueller, G. M. 1992. Systematics of *Laccaria* (Agaricales) in the Continental United States and Canada, with Discussions on Extralimital Taxa and Descriptions of Extant Types. *Fieldiana*. Pub. 1435 n.s. no. 30. 158 pp.

Overholts, L. O. 1953. *The Polyporaceae of the United States, Alaska and Canada*. Univ. of Michigan Press, Ann Arbor. 466 pp.

Ovrebo, C. L. 1989. *Tricholoma*, Subgenus *Tricholoma*, Section *Albidogrisea:* North American Species Found Principally in the Great Lakes Region. *Can. J. Bot*. 67: 3134–3152.

Pegler, D. N. 1983. *The Genus Lentinus—A World Monograph.* Her Majesty's Stationary Office, London. 281 pp.

Pfister, D. H., and A. E. Bessette. 1985. More Comments on the Genus Acervus. *Mycotaxon* 22: 435–438.

Redhead, S. A. 1986. Mycological observations on *Omphalia* and *Pleurotus. Mycologia* 87(4): 522-528.

Redhead, S. A., J. Ginns, and R. A. Shoemaker. 1987. The *Xerula (Collybia, Oudemansiella) radicata* Complex in Canada. *Mycotaxon* 30: 357–405.

Seaver, F. J. 1928. *The North American Cup-fungi (Operculates).* Lancaster Publishing Co., Lancaster, PA. 284 pp.

———. 1942. *The North American Cup-fungi (Operculates) Supplement.* Lancaster Publishing Co., Lancaster, PA. 90 pp.

———. 1951. *The North American Cup-fungi. (Inoperculates).* Lancaster Publishing Co., Lancaster, PA. 428 pp.

Shaffer, R. L. 1957. *Volvariella* in North America. *Mycologia* 49: 545–579.

Singer, R., and A. H. Smith. 1943. A Monograph on the Genus *Leucopaxillus* Boursier. *Papers of the Mich. Academy of Sciences, Arts and Letters* 28: 85–132.

———. 1947. Additional Notes on the Genus *Leucopaxillus. Mycologia* 39: 725–736.

Smith, A. H. 1947. *The North American Species of Mycena.* Univ. of Michigan Press, Ann Arbor. 521 pp.

———. 1951. The North American Species of *Naematoloma. Mycologia* 43: 467–521.

———. 1972. The North American Species of Psathyrella. *Memoirs of the New York Botanical Garden* 24: 1–633.

Smith, A. H., and L. R. Hesler. 1968. *The North American Species of Pholiota.* Hafner, New York. 402 pp.

Smith, A. H., and R. Singer. 1945. A Monograph of the Genus *Cystoderma. Papers of the Mich. Academy of Science, Arts, and Letters* 30: 71–124.

———. 1964. *A Monograph on the Genus Galerina Earl.* Hafner, New York. 168 pp.

Smith, A. H., and H. D. Thiers. 1964. *A Contribution Toward a Monograph of North American Species of Suillus.* Ann Arbor. 116 pp.

———. 1971. *The Boletes of Michigan.* Univ. of Michigan Press, Ann Arbor. 428 pp.

Snell, W. H., and E. A. Dick. 1957. *A Glossary of Mycology.* Harvard Univ. Press, Cambridge. 181 pp.

———. 1970. *The Boleti of Northeastern North America.* Verlag Von J. Cramer, Lehre. 115 pp.

Stuntz, D. E. 1977. *How to Identify Mushrooms to Genus IV: Keys to Families and Genera.* Mad River Press, Eureka, Calif. 94 pp.

Watling, R. 1977. *How to Identify Mushrooms to Genus V: Cultural and Developmental Features.* Mad River Press, Eureka, Calif. 169 pp.

———. 1982. 3 | Bolbitiaceae: *Agrocybe, Bolbitius* and *Conocybe. British Fungus Flora Agarics and Boleti.* Her Majesty's Stationary Office, Edinburgh. 139 pp.

Weber, N. S. 1972. The Genus *Helvella* in Michigan. *Michigan Botanist* 11: 183–186.

Wells, V. L., and P. E. Kempton. 1968. A Preliminary Study of *Clavariadelphus* in North America. *Michigan Botanist* 7: 35–57.

Zeller, S. M., and A. H. Smith. 1964. The Genus *Calvatia* in North America. *Lloydia* 27: 148–186.

Nontechnical Publications

Ammirati, J. F., J. A. Traquir, and P. A. Horgen. 1985. *Poisonous Mushrooms of the Northern United States and Canada.* Univ. of Minnesota Press, Minneapolis. 396 pp.

Bessette, A. E. 1985. *Guide to Some Edible and Poisonous Mushrooms of New York.* Canterbury Press, Rome, New York. 24 pp.

———. 1988. *Mushrooms of the Adirondacks: A Field Guide.* North Country Books, Utica, New York. 145 pp.

Bessette, A. E., O. K. Miller, A. R. Bessette, and H. H. Miller. 1995. *Mushrooms of North America in Color—A Field Guide Companion to Seldom-Illustrated Fungi.* Syracuse Univ. Press, Syracuse. 188 pp.

Bessette, A. E., and W. J. Sundberg. 1987. *Mushrooms: A Quick Reference Guide to Mushrooms of North America.* Macmillan, New York. 174 pp.

Bessette, A. R., and A. E. Bessette. 1993. *Taming the Wild Mushroom: A Culinary Guide to Market Foraging.* Univ. of Texas Press, Austin. 113 pp.

Fischer, D. W., and A. E. Bessette. 1992. *Edible Wild Mushrooms of North America: A Field-to-Kitchen Guide.* Univ. of Texas Press, Austin. 254 pp.

Groves, J. W. 1979. *Edible and Poisonous Mushrooms of Canada.* 2nd rev. ed. Research Branch, Agriculture Canada, Ottawa. 326 pp.

Katsaros, P. 1990. *Familiar Mushrooms.* Knopf, New York. 192 pp.

Kendrick, B. 1992. *The Fifth Kingdom.* Focus Information Group, Newburyport, Mass. 2nd ed. 406 pp.

Kibby, G. 1992. *Mushrooms and Other Fungi.* Smithmark, N.Y. 192 pp.

Lincoff, G. H. 1981. *The Audubon Society Field Guide to North American Mushrooms.* Knopf, New York. 498 pp.

Lincoff, G. H., and D. H. Mitchell. 1977. *Toxic and Hallucinogenic Mushroom Poisoning.* Van Nostrand Reinhold, New York. 267 pp.

Miller, O. K. 1973. *Mushrooms of North America.* E. P. Dutton, New York. 360 pp.

Miller, O. K., and H. H. Miller. 1980. *Mushrooms in Color.* E. P. Dutton, New York. 286 pp.

Phillips, R. 1991. *Mushrooms of North America.* Little, Brown, Boston. 319 pp.

Smith, A. H. 1949. *Mushrooms in Their Natural Habitats.* Sawyer's, Portland, Ore. 626 pp.

Smith, H. V., and A. H. Smith. 1973. *How to Know the Non-Gilled Fleshy Fungi.* William C. Brown, Dubuque, Iowa. 402 pp.

Smith, A. H., H. V. Smith, and N. S. Weber. 1979. *How to Know the Gilled Mushrooms.* William C. Brown, Dubuque, Iowa. 334 pp.

Smith, A. H., and N. S. Weber. 1980. *The Mushroom Hunter's Field Guide.* Univ. of Michigan Press, Ann Arbor. 316 pp.

Stamets, P. 1993. *Growing Gourmet and Medicinal Mushrooms.* Ten Speed Press, Berkeley, Calif. 552 pp.

Sundberg, W. J., and J. A. Richardson. 1980. *Mushrooms and Other Fungi of the Land between the Lakes.* Tennessee Valley Authority, Knoxville. 60 pp.

Thorn, R. G. 1991. *Mushrooms of Algonquin Provincial Park.* Friends of Algonquin Park, Whitney, Ontario. 32 pp.

Weber, N. S. 1988. *A Morel Hunter's Companion.* Two Peninsula Press, Lansing, Mich. 209 pp.

Weber, N. S., and A. H. Smith. 1985. *A Field Guide to Southern Mushrooms.* Univ. of Michigan Press, Ann Arbor. 280 pp.

Index to Common Names

Page number in *italic* denotes illustration.

Caesar's Mushroom *(Amanita caesarea)*, 62, *272*

Cannon Fungus *(Sphaerobolus stellatus)*, 457, *459*

Carbon Balls *(Daldinia concentrica)*, 23, *524*

Carbon Cushion *(Ustulina deusta)*, 23, *525–26*

Cauliflower Mushroom *(Sparassis herbstii)*, 13, 415–16

Cauliflower Mushroom, Rooting *(Sparassis crispa)*, 13, *415*

Cep *(Boletus edulis)*, 339, *365*

Chanterelle *(Cantharellus cibarius)*, 32, *36*

Chanterelle, Cinnabar-red *(Cantharellus cinnabarinus)*, 32–33, *37*

Chanterelle, Flame-colored *(Cantharellus ignicolor)*, *9*, 33

Chanterelle, Golden *(Cantharellus cibarius)*, 32, *36*

Chanterelle, Scaly Vase *(Gomphus floccosus)*, 36, *37*

Chanterelle, Smooth *(Cantharellus lateritius)*, 34, *37*

Chanterelle, Yellow-footed *(Cantharellus xanthopus)*, 34, *37*

Chicken Mushroom *(Laetiporus sulphureus)*, 386–87, *398*

Chroogomphus, Brownish *(Chroogomphus rutilus)*, 77, *276*

Chroogomphus, Wine-cap *(Chroogomphus vinicolor)*, 78, *276*

Clitocybe, Anise-scented *(Clitocybe odora)*, 85, *277*

Clitocybe, Fat-footed *(Clitocybe clavipes)*, 83, *277*

Clitocybe, Funnel *(Clitocybe gibba)*, 84, *277*

Clitocybe, Wood *(Omphalina ectypoides)*, 215, *303*

Club, Moor *(Clavaria argillacea)*, 21, *508*

Club, Water *(Vibrissea truncorum)*, 513, *515*

Collybia, Clustered *(Collybia acervata)*, 93, *278*

Collybia, Golden Scruffy *(Cyptotrama asprata)*, 114–15, *284*

Collybia, Little Brown *(Collybia alkalivirens)*, 93–94, *278*

Collybia, Oak-loving *(Collybia dryophila)*, 95, *279*

Collybia, Rooted *(Xerula furfuracea)*, 270–71, *314*

Collybia, Spotted *(Collybia maculata)*, 95–96, *279*

Collybia, Tuberous *(Collybia tuberosa)*, 97, *279*

Collybia, Tufted *(Collybia confluens)*, 94, *278*

Coral, Crown-tipped *(Clavicorona pyxidata)*, 420–21, *423*

Coral, Flat-topped *(Clavariadelphus truncatus)*, 509, *514*

Coral, Gray *(Clavulina cinerea)*, 421, *423*

Coral, Magenta *(Clavaria zollingeri)*, *13*, 420

Coral, Pestle-shaped *(Clavariadelphus pistillaris)*, 509, *514*

Coral, Smoky Worm *(Clavaria rubicundula)*, 420, *423*

Coral, Spindle-shaped Yellow *(Clavulinopsis fusiformis)*, *13*, 421

Coral Spot Fungus *(Nectria cinnabarina)*, 525, *527*

Coral, White *(Ramariopsis kunzei)*, 422, *424*

Coral, White Green-algae *(Multiclavula mucida)*, 511, *515*

Cordyceps, Goldenthread *(Cordyceps ophioglossoides)*, 22, *518–19*

Cordyceps, Head-like *(Cordyceps capitata)*, 22, *518*

Cordyceps, Trooping *(Cordyceps militaris)*, 22, *518*

Corn Smut *(Ustilago maydis)*, *19*, 470

Cort, Bracelet *(Cortinarius armillatus)*, 106, *281*

Cort, Pungent *(Cortinarius traganus)*, 112, *283*

Cort, Red-gilled *(Cortinarius semisanguineus)*, 111, *283*

Cort, Saffron-colored *(Cortinarius croceofolius)*, 108, *282*

Cort, Silvery-violet *(Cortinarius alboviolaceus)*, 105–106, *281*

Cort, Spotted *(Cortinarius iodes)*, 109, *282*

Cort, Violet *(Cortinarius violaceus)*, 112–13, *283*

Cort, Viscid Violet *(Cortinarius iodes)*, 109, *282*

Crampballs *(Daldinia concentrica)*, 23, *524*

Crep, Flat *(Crepidotus applanatus var. applanatus)*, 113, *283*

Crimped Gill (Plicaturopsis crispa), *9*, 27

Crinipellis, Zoned (Crinipellis zonata), 114, *284*

Cryptic Globe Fungus *(Cryptoporus volvatus)*, 375, *395*

Cudonia, Yellow *(Cudonia lutea)*, 509, *514*

Cup, Bladder *(Peziza vesiculosa)*, 495, *501*

Cup, Blue-staining *(Caloscypha fulgens)*, 491, *498*

Cup, Brown-haired White *(Humaria hemisphaerica)*, 493–94, *500*, 529, 530

Cup, Common Brown *(Peziza phyllogena)*, 495, *500*

Cup, Hairy Black *(Pseudoplectania nigrella)*, 495–96, *501*

Cup, Hairy Rubber *(Galiella rufa)*, 492, *499*

Cup, Orange Eyelash *(Scutellinia erinaceus)*, 497, *501*

Cup, Scarlet *(Sarcoscypha coccinea)*, 20, *496*

Cup, Shaggy Scarlet *(Microstoma floccosa)*, 494, *500*

Cup, Stalked Scarlet *(Sarcoscypha occidentalis)*, 496, *501*

Dead Man's Fingers *(Xylaria polymorpha)*, 23, *526*

433, *436*
Jelly, Orange *(Dacrymyces palmatus)*, *14*, 431
Jelly Babies *(Leotia lubrica)*, 510, *514*, 529, *530*
Jelly Club, Ochre *(Leotia lubrica)*, 510, *514*, 529, *530*
Jelly Drops, Black *(Bulgaria inquinans)*, 491, *498*
Jelly Leaf *(Tremella foliacea)*, 433, *436*
Jelly Roll, Amber *(Exidia recisa)*, 432, *435*
Jelly Roll, Granular *(Exidia nucleata)*, 431, *435*
Jelly Tooth *(Pseudohydnum gelatinosum)*, 432, *436*

Kurotake *(Boletopsis subsquamosa)*, 382, *395*

Laccaria, Common *(Laccaria laccata)*, 155, *292*
Laccaria, Purple-gilled *(Laccaria ochropurpurea)*, 155–56, *292*
Laccaria, Sandy *(Laccaria trullisata)*, 156, *292*
Land Fish *(Morchella esculenta)*, 476–77, *479*
Lawyer's Wig *(Coprinus comatus)*, 103, *280*
Leccinum, Wrinkled *(Leccinum rugosiceps)*, 352, *369*
Lentinus, Bear *(Lentinellus ursinus)*, 182, *297*
Lepiota, Green-spored *(Chlorophyllum molybdites)*, 76, *276*
Lepiota, Malodorous *(Lepiota cristata)*, 188, *298*
Lepiota, Red-tinged *(Lepiota rubrotincta)*, 189, *298*
Lepiota, Sharp-scaled *(Lepiota acutesquamosa)*, 188, *298*
Leucopax, White *(Leucopaxillus albissimus)*, 192, *299*
Lizard Claw *(Lysurus gardneri)*, 461
Lobster Fungus *(Hypomyces lactifluorum)*, 24, *531*
Lobster Mushroom *(Hypomyces lactifluorum)*, 24, *531*
Lorchel *(Gyromitra esculenta)*, 474, *478*

Magnolia-cone Mushroom *(Strobilurus conigenoides)*, 248, *310*
Marasmiellus, Pointed-stalked *(Marasmiellus praeacutus)*, 196, *300*
Marasmius, Black-footed *(Marasmiellus nigripes)*, 196, *300*
Marasmius, Grass *(Marasmius graminum)*, 200, *300*
Marasmius, Pinwheel *(Marasmius rotula)*, 201–2, *300*
Marasmius, Orange Pinwheel *(Marasmius siccus)*, 202, *301*
Marasmius, Orange-yellow *(Marasmius strictipes)*, 202, *301*
Matsutake *(Tricholoma magnivelare)*, 261–62, *312*

Matsutake, White *(Tricholoma magnivelare)*, 261–62, *312*
Meadow Mushroom *(Agaricus campestris)*, 53, *271*
Mica Cap *(Coprinus micaceus)*, 104, *281*
Milky, Buff Fishy *(Lactarius luteolus)*, 173, *294*
Milky, Burnt-sugar *(Lactarius aquifluus)*, 167, *292*
Milky, Chocolate *(Lactarius lignyotus)*, 172–73, *294*
Milky, Corrugated-cap *(Lactarius corrugis)*, 168–69, *293*
Milky, Deceptive *(Lactarius deceptivus)*, 169, *293*
Milky, Eyespot *(Lactarius oculatus)*, 174, *295*
Milky, Gerard's *(Lactarius gerardii var. gerardii)*, 170, *293*
Milky, Hygrophorus *(Lactarius hygrophoroides var. hygrophoroides)*, 172, *294*
Milky, Northern Bearded *(Lactarius representaneus)*, 176–77, *295*
Milky, Peck's *(Lactarius peckii)*, 175, *295*
Milky, Red-hot *(Lactarius rufus var. rufus)*, 177, *296*
Milky, Silver-blue *(Lactaius paradoxus)*, 174–75, *295*
Milky, Voluminous-latex *(Lactarius volemus var. volemus)*, 180, *297*
Milky, Yellow Latex *(Lactarius vinaceorufescens)*, 179–80, *296*
Moose Antlers *(Wynnea americana)*, 497, *501*
Mop, Decorated *(Tricholomopsis decora)*, 266, *313*
Morel, Black *(Morchella elata)*, *20*, 476
Morel, Common *(Morchella esculenta)*, 476–77, *479*
Morel, Half-free *(Morchella semilibera)*, 477, *479*
Morel, Thick-footed *(Morchella crassipes)*, 477
Morel, White *(Morchella deliciosa)*, 477
Morel, Yellow *(Morchella esculenta)*, 476–77, *479*
Mycena, Bleeding *(Mycena haematopus)*, 211–12, *302*
Mycena, Blue *(Mycena subcaerulea)*, 213, *302*
Mycena, Orange *(Mycena leaiana)*, 212, *302*
Mycena, Pink *(Mycena pura)*, 213, *302*
Mycena, Walnut *(Mycena luteopallens)*, 212–13, *302*

Old Man of the Woods *(Strobilomyces floccopus)*, *10*, 354
Orange Peel *(Aleuria aurantia)*, 489, *498*
Oyster, Black Jelly *(Resupinatus applicatus)*, 242, *308*
Oyster, Elm *(Hypsizygus tessulatus)*, 149, *291*
Oyster, Leaf-like *(Hohenbuehelia petaloides)*, 125–26, *286*

Index to Scientific Names

Species names are in *italic;* genera entries in SMALL CAPS. Page number in *italic* denotes illustration.

abietinum, Trichaptum, 378
abortivum, Entoloma, 117, *284*
abrupta, Amanita, 61
abruptibulbus, Agaricus, 51
abundans, Clitocybula, 89
acadiensis, Psilocybe, 241
acericola, Agrocybe, 56, *272*
acerinum, Rhytisma, 496, *501*
acervata, Collybia, 93, *278*
ACERVUS, 487, 489, *498*
acetabulum, Helvella, 488
acicula, Mycena, 207
acidus var. *intermedius, Suillus,* 356
acre, Tricholoma, 258
acutesquamosa, Lepiota, 188, *298*
acutoconicus, Hygrophorus, 130
adhaerens, Lentinus, 183
adhaerens, Panus. See *adhaerens, Lentinus,* 183
adiposa, Pholiota. See *aurivella, Pholiota,* 227–28
admirabilis, Pluteus, 234, *306*
admirabilis, Polyporus, 388, *399*
adnata, Pachyella, 488
adusta, Bjerkandera, 392
adustum, Mycorrhaphium. See *adustum, Steccherinum,* 402
adustum, Steccherinum, 402
aeruginascens, Chlorociboria, 491–92, *499*
aeruginascens, Chlorosplenium. See *aeruginascens, Chlorociboria,* 491–92
aeruginascens, Fuscoboletinus. See *viscidus, Fuscoboletinus,* 328
aeruginosa, Chlorociboria, 491–92
aeruginosa, Stropharia, 248–49
aestuans, Tricholoma, 258–59, *311*
affine, Geoglossum, 506
affine, Xanthoconium. See *affinis* var. *maculosus, Boletus,* 336–37
affinis, Boletus, 327
affinis var. *affinis, Boletus,* 337
affinis var. *maculosus, Boletus,* 336–37, *364*
affinis var. *affinis, Lactarius,* 166, *292*
affinis var. *viridilactis, Lactarius,* 160

AGARICUS, 51–55, *271*
agathosmus, Hygrophorus, 135, *286*
agglutinatus, Lactarius, 166, *292*
AGROCYBE, 55–57, *272*
alba, Exidia. See *pululahuana, Ductifera,* 429
ALBATRELLUS, 381–82, *395*
albatum, Tricholoma, 253
albella, Helvella, 473
albellum, Leccinum, 331
albiceps, Polyporus, 388
albidum, Hydnum, 404
albissimus, Leucopaxillus, 192, *299*
albisulphureus, Boletus, 325
alboater, Tylopilus, 359, *371*
albocreata, Amanita, 60
albocrenulata, Pholiota, 225
alboflavida, Melanoleuca, 203–4, *301*
alboluteus, Pycnoporellus, 378, *399*
albonitens, Stropharia, 249
alboviolaceus, Cortinarius, 105–6, *281*
alcalina, Mycena, 209
ALEURIA, 487, 489–90, *498*
ALEURODISCUS, 440
alkalivirens, Collybia, 93–94, *278*
allardii, Lactarius, 166–67, *292*
alnicola, Pholiota, 227
alutaceum, Podostroma, 517, 519, *519*
alveolaris, Favolus. See *mori, Polyporus,* 384
alveolaris, Polyporus. See *mori, Polyporus,* 384
alveolatum, Geoglossum, 506
amabilissima, Mycena, 209, *301*
AMANITA, 57–68, *272–74,* 531
ambigua, Gyromitra, 472
americana, Clitocybe, 81
americana, Lepiota, 185
americana, Wynnea, 497, *501*
americanum, Hericium, 402, *408*
americanus, Suillus, 354, *369*
amethystina, Clavaria, 420
amethystina, Laccaria, 154
amianthinum, Cystoderma, 116, *284*
amianthinum var. *rugosoreticulatum, Cystoderma,* 115
amorphus, Aleurodiscus, 440
androrosaceus, Marasmius, 198–99
ANELLARIA. See *solidipes, Panaeolus,* 216–17

563

umbellatus, Polyporus. See *umbellatus, Dendropolyporus,* 386

umbilicatum, Dentinum. See *umbilicatum, Hydnum,* 409–10

umbilicatum, Hydnum, 12, 409–10

umbonata, Amanita, 62

umbonata, Armillaria. See *umbonata, Squamanita,* 248

umbonata, Cantharellula, 74–75, 275

umbonata, Squamanita, 248, 310

umbonatus, Cantharellus. See *umbonata, Cantharellula,* 74–75

umbratile, Geoglossum, 506

umbrinum, Lycoperdon, 451

umbrosum, Porpoloma, 236, 307

umbrosum, Pseudotricholoma. See *umbrosum, Porpoloma,* 236

UNDERWOODIA, 471, 503, 513, *515*

underwoodii, Helvella. See *fastigiata, Gyromitra,* 474

underwoodii, Sarcodon, 411, 413

undulata, Rhizina, 487–88

unguinosus, Hygrophorus, 146, 290

unicolor, Cerrena, 379

URNULA, 484–85, 497, *501*

UROMYCES, *19,* 469, 470

ursinus, Lentinellus, 182, 297

USTILAGO, *19,* 469, 470

USTULINA, *23,* 523, 525–26

uvidus, Lactarius, 179, 296

vaccinii, Exobasidium, 18, 467–68

vaccinum, Tricholoma, 264–65, 313

vaginata, Amanita, 58

validipes, Gymnopilus, 120

variabile, Trichoglossum, 506

variegatus, Coprinus, 105, 281

variicolor, Bolbitius, 73, 275

variipes, Boletus, 326

variipes var. *fagicola, Boletus,* 326

varius, Polyporus, 384

velatum, Hebeloma, 124–25, 286

vellereus, Lactarius, 164

veutina, Lacrymaria. See *velutina, Psathyrella,* 239–40

velutina, Psathyrella, 239–40, 307

velutipes, Collybia. See *velutipes, Flammulina,* 117–18

velutipes, Flammulina, 117–18, 284

velutipes, Spathularia, 21, 512

velutipes, Spathulariopsis. See *velutipes, Spathularia, 512*

velutipes, Trichoglossum, 506

venosa, Disciotis, 489

ventricosa, Catathelasma, 75, 275

ventricosipes, Russula, 247, 309

ventricosum, Catathelasma. See *ventricosa, Catathelasma,* 75

veris, Pholiota, 230, 306

vermicularis, Clavaria, 418

vermiculosoides, Boletus, 347–48, 367

vermiculosus, Boletus, 348, 367

verna, Amanita, 59

verna, Nolanea, 215, 303

vernalis, Kuehneromyces. See *vernalis, Pholiota,* 226

vernalis, Pholiota, 226

VERPA, 472, 477, *479.* See also *bohemica, Ptychoverpa*

versicolor, Trametes, 393–94, 400

versiforme, Chlorosplenium, 492, 499

versiformis, Chlorencoelia. See *versiforme, Chlorosplenium,* 492

versiformis, Chlorociboria. See *versiforme, Chlorosplenium,* 492

vesiculosa, Peziza, 495, 501

vialis, Thelephora, 425

vibratilis, Cortinarius, 109

VIBRISSEA, 503, 513, *515*

vietus, Lactarius, 179, 296

villosa, Helvella, 483

villosavolva, Volvariella, 269

vinaceorufescens, Lactarius, 179–80, 296

vinicolor, Chroogomphus, 78, 276

violaceofulvus, Panellus, 219, 304

violaceus, Cortinarius, 112–13, 283

violascens, Bankera, 407, 412

virgatum, Tricholoma, 258

virgineus, Dasyscyphus, 492, 499

virgineus, Hygrophorus. See *borealis, Hygrophorus,* 128

viride, Microglossum, 506

viridiflavus, Boletus. See *auriporus, Boletus,* 337

virosa, Amanita, 67–68, 274

viscidus, Fuscoboletinus, 328

viscosa, Calocera, 431, 435

viscosa, Leotia, 505

vitellina, Neolecta, 508

vitellinus, Bolbitius, 73

vitellinus var. *olivaceus, Bolbitius.* See *variicolor, Bolbitius,* 73

vitellinus, Hygrophorus, 129

vogesiaca, Pseudoplectania. See *melaena, Pseudoplectania,* 485

volemus var. *flavus, Lactarius,* 180

volemus var. *volemus, Lactarius, 180, 297*

volvacea, Volvariella, 268

VOLVARIELLA, *xii,* 268–69, *314*

volvata, Amanita, 59

volvatus, Cryptoporus, 375, 395

vulgare, Auriscalpium, 12, 406–7

vulgare, Crucibulum. See *laeve, Crucibulum,* 465

PHOTO CREDITS

Except for the following, all photographs were taken by Alan E. Bessette, Arleen R. Bessette, and David W. Fischer. The authors wish to extend their sincere appreciation to the following individuals who provided photographs that contributed immensely to the book.

Arne Benson: *Underwoodia columnaris*

Sheldon Cushing: *Phaeocalicium polyporaeum*

Raymond Fatto: *Boletus fraternus, Russula pulchra, Russula subgraminicolor, Russula tenuiceps*

Emily Johnson: *Agaricus subrufescens, Bovista pila, Calvatia elata, Coprinus plicatilis, Hericium erinaceus, Holwaya mucida/Crinula caliciiformis, Melanophyllum echinatum, Mycena haematopus/Spinellus fusiger, Nectriopsis tubariicola, Tremellodendron pallidum, Tricholomopsis rutilans*

Peter Katsaros: *Cantharellus cinnabarinus, Dacryopinax spathularia, Hypoxylon rubiginosum, Lentinellus omphalodes, Lepiota acutesquamosa, Marasmius pyrrhocephalus, Pholiota veris, Tricholoma myomyces, Volvariella pusilla*

Richard Kay: *Gyromitra caroliniana*

Samuel Ristich: *Byssonectria terrestris, Hormomyces aurantiacus, Ramariopsis lentofragilis*

William Roody: *Amanita brunnescens, Amanita sinicoflava, Auricularia auricula, Baeospora myriadophylla, Boletus frostii, Boletus projectellus, Cantharellus infundibuliformis, Clavariadelphus pistillaris, Coltricia montagnei, Globifomes graveolens, Gymnopilus spectabilis, Helvella griseoalba, Hygrophorus appalachianensis, Laccaria laccata, Lactarius allardii, Leccinum rugosiceps, Lentinus suavissimus, Limacella ilinita, Microglossum rufum, Psilocybe coprophila, Resupinatus applicatus, Scleroderma polyrhizon, Stropharia thrausta, Tricholoma leucophyllum, Tylopilus alboater, Tylopilus intermedius, Vibrissea truncorum*

Walter Sturgeon: *Lycoperdon coloratum*

ALAN BESSETTE is a mycologist and professor of biology at Utica College of Syracuse University. He has published numerous professional papers in the field of mycology and has authored eight books, including *Edible and Poisonous Mushrooms of New York*, *Mushrooms of the Adirondacks*, *Mushrooms: A Quick Reference Guide to Mushrooms of North America* (coauthored with Walter J. Sundberg), and *Edible Wild Mushrooms of North America* (coauthored with David W. Fischer). Alan has presented numerous mycological programs, is the scientific advisor to the Mid-York Mycological Society, and serves as a consultant for the New York State Poison Control Center. He has been the principal mycologist at national and regional forays and was the recipient of the 1987 Northeast Mycological Foray Service Award and the 1992 North American Mycological Association Award for Contributions to Amateur Mycology.

ARLEEN R. BESSETTE is a mycologist and botanical photographer, as well as a psychologist, who has been collecting and studying wild mushrooms for several years. She created and contributed over fifty original recipes for the book *Edible Wild Mushrooms of North America* and is the author of *Taming the Wild Mushroom: A Culinary Guide to Market Foraging*. Arleen has won several national awards for her photography and teaches workshops for the North American Mycological Association and the Northeastern Mycological Federation. Together, Alan and Arleen host regional forays from the Adirondacks to Cape Cod, offering participants the opportunity to collect and study mushrooms, including wild edibles. Their most recent book (coauthored with Orson K. Miller, Jr., and Hope H. Miller) is *Mushrooms of North America in Color: A Field Guide Companion to Seldom-Illustrated Fungi* published by Syracuse University Press.

DAVID W. FISCHER is president of the Central New York Mycological Society and former president of the Northeast Mycological Federation. He serves as a mushroom identification consultant for the Central New York Regional Poison Control Center and for several mycological societies. An award-winning mushroom photographer, David frequently gives slide-illustrated mycology lectures at colleges, museums, and mycological society meetings and conferences. He is coauthor of *Edible Wild Mushrooms of North America* and is a professional editor and graphic designer as well as writer.

DATE DUE

_ DUE DATE SUBJECT TO CHANGE _
IF A RECALL IS REQUESTED